Fine Sediment Dynamics in the Marine Environment

Companion books to this title in the ***Proceedings in Marine Science*** series are:

Volume 1: **Solent Science** – *A Review*
M. Collins and K. Ansell (Eds.)

Volume 2: **Muddy Coast Dynamics and Resource Management**
B.W. Flemming, M.T. Delafontaine and G. Liebezeit (Eds.)

Volume 3: **Coastal and Estuarine Fine Sediment Processes**
W.H. McAnally and A.J. Mehta (Eds.)

Volume 4: **Muddy Coasts of the World: Processes, Deposits and Function**
T. Healy, Y. Wang and J-A. Healy (Eds.)

Proceedings in Marine Science 5

Fine Sediment Dynamics in the Marine Environment

Edited by

Johan C. Winterwerp
WL| Delft Hydraulics, Delft, The Netherlands
also
Delft University of Technology, Delft, The Netherlands
Faculty of Civil Engineering and Geosciences,
Section of Fluid Mechanics

Cees Kranenburg
Delft University of Technology, Delft, The Netherlands
Faculty of Civil Engineering and Geosciences,
Section of Fluid Mechanics

2002
ELSEVIER
Amsterdam – London – New York – Oxford – Paris – Shannon – Tokyo – Boston
San Diego – San Francisco – Singapore – Sydney

ELSEVIER B.V.
Radarweg 29
P.O. Box 211, 1000 AE
Amsterdam, The Netherlands

ELSEVIER Inc.
525 B Street
Suite 1900, San Diego
CA 92101-4495, USA

ELSEVIER Ltd
The Boulevard
Langford Lane, Kidlington,
Oxford OX5 1GB, UK

ELSEVIER Ltd
84 Theobalds Road
London WC1X 8RR
UK

First edition 2002
Reprinted 2005

British Library Cataloguing in Publication Data
A catalogue record is available from the British Library.

Library of Congress Cataloging in Publication Data
A catalog record is available from the Library of Congress.

ISBN: 0-444-51136-9

∞ The paper used in this publication meets the requirements of ANSI/NISO Z39.48-1992 (Permanence of Paper).
Printed in The Netherlands

In memory of RAY B. KRONE

Ray B. Krone, who died on December 7, 2000, was born in 1922 in California, and apart from the years he spent in World War II he lived, studied and worked in California. During those war years he piloted a P-38 aeroplane over Germany in the U.S. Army Air Corps 31st Photo-reconnaissance Squadron. He was fond of recalling his photographing sorties in the war zone, especially one during which he flew over the Eagle's Nest - Adolph Hitler's hideout in the mountains. Piloting small planes remained one of his loves in life, and he never lost his fondness for flying over the state's great Central Valley with its breadbasket farms and orchards.

After returning home from Europe he decided to complete his college education at UC Berkeley. There, in 1950 obtained his B.S. in Soil Science, then M.S. in Sanitary Engineering in 1958, and finally his doctorate in 1962, also in Sanitary Engineering. In 1964 he joined UC Davis as Associate Professor, and the next year founded the environmental engineering programme there. He went on to serve as Chair of the Department of Civil and Environmental Engineering, and as Associate Dean for Research in the College of Engineering. Even while serving in administration, he continued to teach and guide students. Many of his masters and doctoral students went on to become successful engineers and scientists.

Ray retired from UC Davis in 1988, but continued his involvement in teaching and research as professor emeritus, and engineering consulting work through Ray Krone & Associates in Davis, where he lived. David Schoellhamer taught a course on sediment transport once taught by Ray, who continued to run the one-day field trip as part of the course. One stop was a restored wetland. David recalls that on one such trip the group drove their vans up to a locked gate with a large "No Trespassing" sign. With a characteristic sparkle in his eye, Ray quickly climbed the gate and headed off toward the wetland, as the students looked on dumbfounded. So at that point David told them to get over the gate and follow Ray.

For several years, until the end, he served as a most valued consultant on ports and harbors to the Committee of Tidal Hydraulics of the U.S. Army Corps of Engineers, with which he was closely associated since his 1950's work in the San Francisco Bay. He also served on boards of the National Research Council. The American Society of Civil Engineers, of which Ray was a Fellow and in which he served in various professional capacities, presented him the Hans Albert Einstein Award in 1991 for his seminal contributions to sediment transport and sedimentation engineering. This was an apt tribute - Ray worked for almost two decades with Prof. Einstein first as a researcher, then as a student, and finally as a colleague. Ray's scientific work bears a clear mark of the phenomenological and stochastic interpretative approach of Einstein, whom Ray adored both as a scientist and as a person. Among the many memories of their long friendship and partnership, Ray was fond of recounting Einstein's remark to Ray's wife Jane, when Ray decided to accept the professorial position at UC Davis - Einstein said, "Ray will never be rich but he will be happy". Ray also considered Einstein a great mentor – "whenever you wanted to talk to him about research he was all ears", is what Ray would say.

In 1995 Ray achieved another distinction by being elected to the National Academy of Engineering, and in 1996 he was elected a fellow of the American Association for the Advancement of Science for "efforts toward advancing science or fostering applications that are deemed scientifically or socially distinguished."

Jane died in 1999 after 54 years of marriage, the year before Ray's own death, survived by son Ray III, daughter Ann and grandchildren. To those who knew him in his professional life, Ray will be remembered as a founder of the hydraulics of cohesive sediment transport, and as one with the ability to analyse and explain sedimentation related problems with great insight and clarity of communication. He will be remembered even more by the large number of students and colleagues in the U.S. and throughout the world to whom he served as a mentor, and as a source of inspiration through his personal example of human kindness and dedication to profession.

Ray's scientific work, together with Prof. Einstein and later with his students, covered a variety of topics mainly in wastewater treatment and estuarine processes. He contributed to design of ports and harbours to reduce erosion or sedimentation, and hydraulics of marsh restoration. A significant amount of his scientific and engineering work is related to the San Francisco Bay system and tributaries, where he carried out numerous projects on marina design and saltmarsh restoration. He developed an early understanding of the relationship between flow circulation and sedimentation in the bay and its channels, and simple but effective models for simulating the long-term evolution of the peripheral marsh-plains. He served as an expert witness in many cases involving sedimentation and waterline boundary disputes in the bay area.

Sedimentation in the San Francisco Bay has been the focus of interest and research at UC Berkeley from the late 19th century, when the bay and its tributaries, especially the Sacramento-San Joaquin river system, began experiencing excessive sedimentation due to hydraulic mining in the Sierra Nevada range. This sediment raised the bed levels in the rivers and caused flooding of the neighbouring farmland. As a result a legal dispute between the mine owners and farmers, which led the State of California to request several agencies including the U.S. Geological Survey to conduct research on the rivers and the bay to propose appropriate flood control measures. Gustav Karl Gilbert of USGS constructed a flume at the Berkeley campus and conducted his well-known studies on the relationship between flow and sediment (sand) transport. After hydraulic mining was banned, attention shifted to sedimentation in the ports and channels within the bay, which meant looking at the properties of material that was fine-grained and cohesive.

Ray's doctoral work was related to the structure of water, and he revisited that general subject in later years when he pointed out that the classical theory based on molecular (gas) dynamics did not correctly predict the relationship between the viscosity of a liquid and the absolute temperature (Krone, 1983). Using liquid benzene as an example, he carefully reinterpreted molecular dynamics focusing on liquids. Through a momentum exchange model for molecular layers he proposed a better physics-based model for the viscosity-temperature relationship.

Aside for his work on water, the vast majority of Ray's work was related to fine-grained sediments, and of all his contributions in that area, he is most well known for two reports he wrote in the early 1960's (Krone, 1962; 1963). These works included his laboratory studies on the transport of cohesive material from the San Francisco Bay. The need for that work was rooted in a field study carried out in the 1950's in the bay to track the movement of sediment flocs by way of radioisotopic inoculation of the flocs (Einstein and Krone, 1961).

The focus of the 1962 report was the study of deposition of cohesive material from the bay in a flume at Berkeley's Richmond Field Station. The most important observation derived from that work was the demonstration that depositing flocs must not be treated as integral units because due to collisional mechanisms they undergo growth and breakup, or aggregation, which causes their transport properties to change continuously even as they fall out of suspension. Using simple but highly insightful arguments starting from the flocculation theory and experimental observations of Overbeek (1952), Ray developed three equations for the rate of floc deposition, each applicable within a certain range of concentration of the suspended matter. Of these, the equation for the lowest concentration range, with an upper limit of 0.3 kg/m^3, is most commonly used and bears his name. Because of its wide applicability, its utility remains unchallenged in cohesive sediment transport, although in subsequent years it has been extended to enable it to handle fine graded sediment.

The 1963 report, meant to determine the strength and density of flocs in the bay, was a natural extension of the earlier work, and was conducted in a concentric cylinder rheometer. Ray made the observation that when a suspension of bay mud was tested in this apparatus, the derived relationship between the shear rate and the shear stress was piece-wise linear, with the slope of the line, denoting viscosity, decreasing as shear rate increased. This meant, he concluded, that each line corresponded to a class of flocs of constant properties and hence viscosity. Using this observation, and beginning with Albert Einstein's work on the viscosity of a suspension of solid spheres at infinite dilution (Einstein, 1911), Ray developed a method to calculate the density and shear strength of flocs of each class, which he called "order". He further showed that under typical estuarine conditions flocs of different orders can coexist, with the number of orders dependent on the flow condition and sediment composition - the more cohesive the sediment the greater the number of orders. This important method of organising the process of floc aggregation in the natural environment led to further work on the way in which flocs deposit and then consolidate as the deposited flocs are crushed by self-weight and change their order as consolidation proceeds. Subsequent work by others on the fractal representation of flocs is, in sense, a quantification of the order of aggregation concept.

Ray's recognition of the importance of aggregation led to a scientific controversy in the 1960's and 70's based on measurements in some East Coast estuaries where aggregation was not thought to be important. A partial answer to that issue is found in Ray's 1963 report itself, in which he observed that among the several estuarine sediments he examined, material from San Francisco Bay was far more cohesive than most others, which in turn meant that elsewhere aggregation was likely to be less important. Subsequent work has shown that the role of aggregation is indeed site-specific, and where the material is not too cohesive, say due to the presence of a high fraction of silts or organic matter, aggregation tends to be weak, although usually not negligible. It follows that in experiments where the main objective is to examine the effects of cohesion on transport, weakly cohesive materials, even such clays as kaolinite, may not be used.

My first meeting with Ray was in December of 1970, when I had just finished running some fine sediment deposition tests in an annular flume. He came over to look at the data, flying in from the city of Naples in Florida in a small rental plane he piloted. The results using a kaolinite as sediment were somewhat unusual and I expressed my concern. "You should look at the effect of sorting on your results", he said confidently. New to the area of sediments, I did not know what "sorting" meant. And was afraid to ask.

Ashish Mehta
University of Florida

References

Einstein, A., 1911, A new determination of molecular dimensions, *Annals of Physics*, (19), 289-306; (34), 591-592.

Einstein, H. A., and Krone, R. B., 1961, Estuarial sediment transport patterns, *Journal of the Hydraulics Division, ASCE*, (87) 2, 51-59.

Krone, R. B., 1962, Flume studies of the transport of sediment in estuarial shoaling processes, *Final Report*, Hydraulic Engineering Laboratory and Sanitary Engineering Research Laboratory, University of California, Berkeley, CA, 118p.

Krone, R. B., 1963, A study of rheological properties of estuarial sediments, *Technical Bulletin No. 7*, Committee on Tidal Hydraulics, U.S. Army Engineer Waterways Experiment Station, Vicksburg, MS, 105p.

Krone, R. B., 1983, A viscosity-temperature relation for Newtonian liquids, *Chemical Engineering Communications*, (22), 161-180.

Overbeek, J. Th. G., 1952. Kinetics of flocculation. In: *Colloid Science,* Vol. 1, H. R. Kruyt ed., Elsevier, Amsterdam, 278-301.

Preface

Cohesive sediment, or mud, is encountered in most water bodies throughout the world. Often, mud is a valuable resource, synonymous with fertile land, enriching the natural environment and used as an important building material. Yet, mud also hinders navigation. Consequently, dredging operations have been carried out since ancient times to safeguard navigation. Unfortunately, many mud deposits are now contaminated, endangering the eco-system and increasing the costs of dredging operations.

However, the transport and fate of mud in the environment are still poorly understood and the need for basic research remains. This book contains the proceedings of the INTERCOH-2000 conference on recent progress in cohesive sediment research. It was the sixth in a series of conferences initially started by Prof. Ashish Mehta in 1984 as a Workshop on Cohesive Sediment Dynamics with Special Reference to the Processes in Estuaries. Prof. Metha invited most of the experts on cohesive sediments at that time for a relatively small workshop in Tampa, Florida, USA. Since then, conferences have been held in:

- Gainesville, Florida, USA (1987),
- Petersburg, Florida, USA (1991),
- Wallingford, UK, (1994), where the name INTERCOH was first introduced, and
- Seoul, Korea (1998).

During these conferences the character of the first workshop has always been maintained, that is, small scale and dedicated to the physical and engineering aspects of cohesive sediments, without parallel sessions, ample time for discussions during and after the presentations, and a high-quality Book of Proceedings containing thoroughly reviewed papers.

INTERCOH-2000 was the last conference attended by Prof. Ray Krone, who is considered to be one of the founders of modern cohesive sediment research. Regretfully, he passed away a few months after the conference. The obituary on the preceding pages of this book was written by Prof. Mehta.

INTERCOH-2000 was integrated with the final workshop of the COSINUS project. This project was carried out as part of the European MAST-3 programme, and almost all European cohesive sediment workers were involved. An introduction to this project by the project co-ordinator, Prof. Jean Berlamont, can be found in the first chapter of this book. This introduction is followed by five summaries of the tasks around which the project was organised. Further details are given in a number of papers elsewhere in the Proceedings.

INTERCOH-2000 focused on the behaviour and modelling of Concentrated Benthic Suspensions, i.e. high-concentrated near-bed suspensions of cohesive sediment. Special reference was paid to:

- Sediment – turbulence interaction,
- Flocculation and settling velocity,
- High-concentrated mud suspensions,
- Processes in the bed – consolidation,

- Processes on the bed – erosion,
- Field observations on mud dynamics,
- Instrumentation, and
- Numerical modelling.

The various papers of the Proceedings are organised in chapters on these subjects in alphabetical order.

The INTERCOH-2000 conference could not have been organised without the financial support provided by WL | Delft Hydraulics, Rijkswaterstaat / RIKZ, SILT, Rijkswaterstaat / RIZA, the Port of Rotterdam and Delft University of Technology. We also gratefully acknowledge Caroline Sloot and Astrid van Bragt for their skilful organisation of the conference.

Han Winterwerp
Cees Kranenburg

Delft, The Netherlands

Contents

Fine Sediment Dynamics in the Marine Environment
J.C. Winterwerp and C. Kranenburg (Editors)

Prediction of cohesive sediment transport and bed dynamics in estuaries and coastal zones with integrated numerical simulation models (COSINUS)

Jean E. Berlamont[a]

[a]Hydraulics Laboratory, Katholieke Universiteit Leuven, Belgium

The managing authorities of coastal waters and estuaries face a large number of problems related to cohesive sediment transport, sedimentation and erosion, such as:

- How to maintain safe navigable depths (at minimum cost)?
- Where and how to dump dredged material?
- How can the volume of wetlands be maintained or increased?
- What will happen to the location of the turbidity maximum after constructing new harbour basins or deepening the navigation channels ? etc.

To answer these questions, one needs a model capable of simulating the many different and interrelated (cohesive) sediment processes occurring in coastal and estuarine waters, which can predict natural phenomena and the effects of human interference. Unfortunately, the presently used models are unable to simulate accurately the many different and interrelated cohesive sediment processes occurring in coastal and estuarine waters due to too many simplifications. Therefore there is a need for an integrated sediment transport management model in which all relevant physical processes are integrated.

Progress in the understanding and the mathematical description of the different processes and the increasing capacity and speed of modern computers opens new doors toward the operational use of much more detailed models.

Therefore, the goal of the COSINUS project (executed from October 1997 through September 2000) was to contribute to the development of an integrated sediment transport management model. "Integrated" refers to the integration of all relevant physical processes over the entire water column and the sediment bed and their interactions.

COSINUS covers the theoretical, experimental and numerical study of the interaction of the processes which play a crucial role in the flocculation of sediment particles, the interaction between suspended sediment and turbulent flow, the generation and maintenance of concentrated near-bed suspensions (including lutocline formation) and the transition between fluid mud and the sediment bed.

The state-of-the-art knowledge on cohesive sediment transport showed that there was still a lack of experimental data on the role of flocculation and turbulence in the formation and erosion of mud beds and on the formation of CBS (concentrated benthic suspensions, or "fluid mud"). Therefore, an experimental programme has been set up to obtain these data. It consisted of field measurements in the Tamar estuary on floc formation and laboratory experiments on formation and erosion of mud beds and CBS, and the influence of floc structure and turbulence on these processes. All data are available to the public.

Process modules have been developed and implemented into detailed 1D and 2D vertical models which solve the full hydrodynamic, turbulent energy and sediment mass conservation equations. Two different bed models, to be coupled to these hydrodynamic models have been developed as well (1DV POINT MODEL).

The process modules have been parameterised to obtain relatively simple formulations, which can be (and have been) implemented into currently used 3D and 2DH engineering system models. This was the main objective of the research project.

The performance of the improved system models has been tested by application of the models to a schematic estuary, for which a 2DV solution with the detailed research model was used as a reference. Various scenarios have been simulated. The models have also been applied to three real estuaries (Tamar, Loire and Weser). Data to set-up and calibrate the model applications are stored in the database.

From the experience with the large-scale applications feed-back has been produced towards the process module development and their parameterisations.

The following sub-objectives have been formulated:

1. Select the most suitable model for the simulation of cohesive sediment-laden flow. Extend existing turbulence models to allow for the simulation of sediment-laden flow for a wide range of concentrations and turbulence intensities. Establish a formulation for the turbulence damping (buoyancy effect) and turbulence generation (internal wave turbulence production) in concentrated suspensions.
2. Establish a formulation for the floc formation from which the depositional flux and the resulting bed structure can be evaluated. Define the structural floc parameters that govern the rheological properties of the bed and of fluid mud. Develop a floc model which allows the calculation of the settling velocity of flocs in relation to turbulence, concentration, residence time and, if possible, organic content.
3. Establish a formulation for the exchange of mass and momentum at the fluid mud/water interface, unifying the concepts from deposition, entrainment/erosion and interface instabilities (internal waves). Understand why and how benthic suspensions (CBS) can be generated and maintained, and quantify the relevant processes.
4. Establish a formulation for the development of erosion resistance (strength) in mud beds, unifying the concepts from rheology, consolidation, liquefaction and fluidisation.

In order to make the modules applicable for integrated system models, aimed at large scale simulations, further sub-objectives were to:

5. Establish a data base to validate the models.
6. Establish parameterized forms of the above mentioned process models based on numerical and experimental data.
7. Establish guidelines to select and implement the proper model formulation.
8. Establish guidelines to carry out the necessary experiments in the laboratory as well as in the field for calibration of the models.
9. Establish general guidelines for engineers and managing authorities on how to set up the necessary field measurement campaigns to obtain the necessary initial and boundary

conditions and the material parameters in order to validate or apply the integrated system models.

The project has been structured in six sub-tasks:
- Task A - turbulence modelling of sediment laden flow: turbulence damping and turbulence production (internal waves) in concentrated suspensions.
- Task B - flocculation: floc model development.
- Task C - CBS dynamics: generalised entrainment model and generation and properties of CBS.
- Task D - Bed dynamics: bed strength model and erosion / entrainment model.
- Task E - Parameterisation, the implementation of the process models in the schematic estuary and the two test cases.
- Task F - Set-up and management of the data base.

Summaries on Task A through E arepresented in the following sections of these Proceedings and detailed results are presented in a number of papers elsewhere in the Proceedings.

CONCLUSIONS OF THE COSINUS PROJECT

The objective of the research programme was to establish well validated physical and mathematical descriptions of the behaviour and fate of concentrated near-bed suspensions (CBS or "fluid mud") and their interaction with the water and the sediment bed.

An experimental programme has been set up to obtain missing data on floc formation, the formation of mud beds and CBS and the influence of floc structure and turbulence on these processes.

Different processes have been studied in detail: turbulence damping in sediment laden flow; turbulence production due to internal waves in concentrated suspensions; flocculation; generation, properties and entrainment of CBS; bed strength development and erosion of mud beds.

The detailed process models have been parameterised to obtain relatively simple formulations which can be plugged into currently used 3D and 2DH engineering models.

The performance of the improved models has been tested by application of the models to a schematic estuary for which 2DV solutions with the detailed research models were used as a reference. The models have been applied and tested in three real estuaries (Tamar in U.K., Loire in France and Weser in Germany).

All data have been stored in a data base, which is accessible to the public.

It is felt that great progress has been made in the physically based description of cohesive sediment dynamics with respect to amongst others the formulation of turbulence damping functions; the modelling of the rheology of CBS, incl. consolidation; the modelling of flocculation and the modelling of erosion and entrainment of CBS.

Engineering software tools have been improved to enable better predictions of mud dynamics for the benefit of estuarine an coastal managers.

Partners of the COSINUS project:

Katholieke Universiteit Leuven, Leuven, Belgium, co-ordinator
DHI, Lyngby, Denmark
HR Wallingford, U.K.
Laboratoire Nationales d'Hydraulique, Paris, France
LEGI, Grenoble, France
Oxford University, Oxford, U.K.
Technische Universiteit Delft, Delft, the Netherlands
Universität Hannover, Hannover, Germany
University of Plymouth, Plymouth, U.K.
WL | Delft Hydraulics, Delft, the Netherlands

Chapter 1:

COSINUS task A through D summaries

Fine Sediment Dynamics in the Marine Environment
J.C. Winterwerp and C. Kranenburg (Editors)

Interaction of suspended cohesive sediment and turbulence

E.A. Toorman[a], A.W. Bruens[b], C. Kranenburg[b] and J.C. Winterwerp[b,c]

[a] Hydraulics Laboratory, Civil Engineering Department, Katholieke Universiteit Leuven, Kasteelpark Arenberg 40, B-3001 Leuven, Belgium.
[b] Hydromechanics Section, Civil Engineering Department, Delft University of Technology, PO Box 5048, NL-2600 GA Delft, the Netherlands
[c] Delft Hydraulics, PO Box 177, NL-2600 MH Delft, the Netherlands

This paper describes the work done in the COSINUS project, carried out within the framework of the European MAST3 research programme, on the interaction between suspended (cohesive) sediment and turbulence, with particular emphasis on its modelling. Specific attention is given to the modelling of buoyancy damping effects and turbulence production due to internal waves. Finally, some experimental results are presented on the effect of advected turbulence to the entrainment of fluid mud.

KEY WORDS
turbulence modulation, sediment-turbulence interaction, laminarisation, internal waves, entrainment, modelling

1. INTRODUCTION

The presence of suspended particles in turbulent flow alters the eddy viscosity distribution over the water depth as turbulent energy is dissipated by buoyancy destruction. One of the consequences is a significant apparent bottom friction (or drag) reduction. This has important implications for the transport of cohesive sediment by flowing water, in particular for the estimation of advective transport and the entrainment rate. Sediment-turbulence interaction has been studied as part of the MAST3 "COSINUS" project with the help of numerical models using Prandtl mixing-length (PML) and k-ε turbulence closures, as these presently are the turbulence models used for applied modelling of cohesive sediment transport problems. The present paper investigates how the sediment-turbulence interaction can be modelled properly. The basic approach applies to any type of suspended particles, i.e. also non-cohesive sediment. Differences occur at the level of form of the turbulence modulation correction factor, introduced in Section 3.1, which for fine (e.g. cohesive) particles cause damping (see Section 2.1).

First, a brief overview is presented of available experimental evidence and the proposed mechanisms that contribute to the modulation of turbulence. The following three sections deal with modelling aspects and strategies respectively on buoyancy damping, possible subsequent laminarisation of the flow and possible internal waves at the lutocline. In a last section before the conclusions, results are discussed on the entrainment of a dense layer by shear turbulence generated upstream.

2. SEDIMENT-TURBULENCE INTERACTIONS

Literature reviews have been carried out by Winterwerp (1999), focusing on the occurrence and behaviour of concentrated benthic suspensions, and by Toorman (2000b), focusing on the modelling of sediment-turbulence interactions.

2.1. Experimental observations

The fact that suspended particles modify the turbulence characteristics in shear flows is known from experiments for many years. Many velocity and concentration profile data in sediment-laden flows in flume experiments can be found in the literature (e.g. Vanoni, 1946; Einstein and Chien, 1955; Lyn, 1987). With an increase in the ratio $C_m w_s/u_*$ (where C_m is the depth-averaged mean concentration, a measure of the sediment load, w_s the particle settling velocity and u_* the shear velocity) such flows show an increasingly significant deviation from the traditional log-velocity law for clear water. It was surmised quite early that the presence of suspended particles suppresses turbulent fluctuations and that the deviation can be accounted for by reducing the value of the von Karman parameter κ. Subsequently, Coleman (1981) made a different analysis of experimental velocity profiles. He claimed that the deviations can be accounted for by considering a wake component in the velocity profile. Correcting for this wake effect, one can keep the value of κ constant. His analysis shows several weaknesses and has been opposed by various researchers. Further details can be found in literature reviews by Winterwerp (1999) and Toorman (2000b). The discussion on whether or not the von Karman parameter decreases with increasing stratification is still not closed. The results presented below are meant to provide some new insights.

Experimental data on direct measurement of turbulence modulation by suspended particles are scarce. Nearly all experiments are with non-cohesive particles, and the majority is restricted to pipe flows. Size dependence is observed, i.e., (near-wall) turbulence is found to be attenuated by fine particles (i.e., for particle sizes smaller than about 10% of the length scale of the energy containing eddies or the integral length scale), but enhanced by coarse particles (Gore and Crowe, 1989). Relative movement of fluid and particles has been measured by Best *et al.* (1997). Cellino and Graf (1999) recently published the first comprehensive data set for open-channel flow experiments with fine sand in which the fluctuations of all the velocity components and the concentration have been measured. Attempts are currently being undertaken to obtain similar data for cohesive sediments (e.g., Crapper *et al.*, 2000; Crapper and Bruce, 2002).

2.2 Turbulence modulation mechanisms

Various processes are believed to contribute to the modulation of turbulent fluctuations by suspended particles (e.g. Rocabado, 1999). The most important mechanism is the damping by buoyancy forces, i.e., a mechanism in which gravity opposes upward fluctuations of the particles and, in stable stratification ($\partial\rho/\partial z < 0$, with ρ the suspension bulk density and z the vertical distance from the bottom), downward fluctuations are hindered by higher concentrations of particles below. Buoyancy effects can already be significant at very low concentrations.

Furthermore, the presence of particles in a fluid increases the bulk viscosity of the mixture, which in turn enhances the viscous dissipation of turbulent kinetic energy. At high concentrations, turbulence may be dissipated by interaction between the particles, which may manifest as an additional increase in the suspension viscosity. Generally, the suspension viscosity can be semi-empirically expressed as a power law function of the concentration.

3. MODELLING OF BUOYANCY DAMPING

The application of the Prandtl mixing-length (PML) and the k-ε turbulence models in stratified flow conditions has been studied extensively at the Katholieke Universiteit Leuven (Toorman, 1999, 2000). More complex models, such as the Reynolds stress model (e.g. Galland *et al.*, 1997), are not considered as they do not perform any better by lack of proper calibration data in the case of sediment-laden flows (Toorman, 2000b).

3.1. PML turbulence modelling

The Prandtl mixing length model is based on the hypothesis that the mixing length ℓ in simple near-wall shear flow is proportional to the distance from the bottom. Combining this with the stress balance leads to the well-known logarithmic velocity profile. This result has been confirmed by numerous experiments. Considering the equilibrium stress balance over the entire water column in open-channel flow leads to a parabolic eddy viscosity distribution (Toorman, 2000b). However, this is only valid for homogeneous fluids.

The modelling of turbulence damping by buoyancy effects is done by modulating the clear water eddy viscosity ν_0 and eddy diffusivity (or mixing coefficient) K_0 with damping factors. The momentum damping factor can be defined as $F_m = \nu_t/\nu_0$ (with ν_t the actual eddy viscosity) and the mixing damping function as $F_s = K_s/K_0$ (with K_s the actual eddy diffusivity) (e.g. Munk and Anderson, 1948). It is generally assumed that the eddy diffusivity is proportional to the eddy viscosity, i.e., $K_s = \nu_t/\sigma_s$, where σ_s is called the turbulent Schmidt number.

In order to account for the buoyancy effect, the PML has to be corrected with the damping function, i.e., $F_m = \ell/\ell_0$, with ℓ the actual, buoyancy-corrected mixing length and ℓ_0 the mixing length in non-stratified conditions (Toorman, 2000c, 2002). Subsequently, the correct velocity gradient is written as:

$$\frac{\partial U}{\partial z} = \frac{u_*}{F_m \kappa z} \tag{1}$$

and the corresponding eddy viscosity distribution for open-channel flow is given by:

$$\nu_t = \ell^2 \left|\frac{\partial U}{\partial z}\right| = (F_m \ell_0)^2 \left|\frac{\partial U}{\partial z}\right| = F_m \kappa u_* z \left(1 - \frac{z}{h}\right) \tag{2}$$

where κ is the von Karman constant (= 0.41), u_* is the shear velocity, z is the distance from the bottom and h is the water depth. As the basic assumptions are only valid in the vicinity of a wall, real eddy viscosity profiles in steady open-channel flow deviate slightly from the ideal parabolic profile, in particular in the upper half of the water column (Nezu and Nakagawa, 1993).

3.2. *k*-ε turbulence modelling

As the PML model cannot account for the history of turbulence and is only valid for simple shear flows, a more complex turbulence model is preferred in applied sediment transport modelling whenever possible. At present, the k-ε turbulence model seems to be the best compromise between computational cost and complexity, in particular with regard to coastal and estuarine engineering applications.

This model solves the conservation of turbulent kinetic energy k:

$$\frac{\partial k}{\partial t}+U_j\frac{\partial k}{\partial x_j}=\frac{\partial}{\partial x_j}\left((\nu+\frac{\nu_t}{\sigma_k})\frac{\partial k}{\partial x_j}\right)+P+G-\varepsilon \tag{3}$$

and its dissipation rate ε:

$$\frac{\partial \varepsilon}{\partial t}+U_j\frac{\partial \varepsilon}{\partial x_j}=\frac{\partial}{\partial x_j}\left((\nu+\frac{\nu_t}{\sigma_\varepsilon})\frac{\partial \varepsilon}{\partial x_j}\right)+\frac{1}{T_t}(f_1c_1P+c_3G-f_2c_2\varepsilon) \tag{4}$$

where U is the mean velocity, t is the time, x_j are the components of the co-ordinate vector, ν is the kinematic viscosity of the suspension, $\nu_t = f_\mu\, c_\mu\, k^2/\varepsilon$ is the eddy viscosity, $T_t = k/\varepsilon$ is the (high-Reynolds number) turbulence time scale, P is the shear production and G the buoyancy term. The last two are respectively defined as:

$$P=\nu_t\left(\frac{\partial U_i}{\partial x_j}+\frac{\partial U_j}{\partial x_i}\right)\frac{\partial U_i}{\partial x_j} \tag{5}$$

$$G=\frac{g}{\rho}\frac{\nu_t}{\sigma_s}\frac{\partial \rho}{\partial z} \tag{6}$$

with g the gravity constant. The remaining coefficients have been determined semi-empirically (e.g. Rodi, 1980). The value of c_3 in stable stratified shear flows is generally somewhere in the range of 0-0.3 (Rodi, 1980). Here we will adopt $c_3 = 0$. The factors f_μ, f_1 and f_2 are correction functions for the low-Reynolds number formulation (see Section 4); their value is 1 for the standard high-Reynolds number form.

This model assumes isotropic turbulence and is only valid for high-Reynolds number flows. In stable stratification by suspended sediments, i.e., increasing concentration with depth, which usually is the case in natural waters, the buoyancy term is negative, i.e., turbulence is destroyed as gravity works against the turbulent fluctuations. As the turbulent Schmidt number appears in the buoyancy term, empiricism is still required in this model. Finally, the wall boundary conditions of the k-ε model are based on the assumption of equilibrium in the near-wall layer and on the PML model as the bridging function to provide the missing information from the wall boundary layer, which is not resolved by the k-ε model as the corresponding equations are not valid at a solid boundary (e.g., Toorman, 2000c). These boundary conditions introduce further dependence on empirical damping functions, as shown in the following section.

3.3. Determination of the buoyancy damping functions

Traditionally, the damping functions are chosen as simple empirical functions of the gradient Richardson number Ri, which characterises the degree of stratification, of the form $F_m = (1 + A\,Ri)^{-a}$ and $F_s = (1 + B\,Ri)^{-b}$, where A, B, a and b are empirical parameters, such as the well-known damping functions (for free turbulence) proposed by Munk and Anderson (1948).

A distinction needs to be made between damping functions for free turbulence and for wall-turbulence. Those for wall turbulence are often expressed as a function of the Monin-Obukov length-scale $L = z/Rf$ (with Rf the flux Richardson number) and show a stronger decrease with Ri than the free turbulence damping functions. The empirical coefficients are found as $-5 < A < -10$ and $a = -1$ (Rodi, 1980). One of the reasons for the difference in behaviour is the fact that in the neighbourhood of the bottom the development of internal waves is reduced and vertical mixing by internal wave breaking is prevented (Uittenbogaard, 1995a).

As eventually only F_m and σ_s are used in the models, it is proposed to use a similar empirical form for the turbulent Schmidt number, i.e.:

$$\sigma_s = \sigma_0\,(1 + \alpha\,Ri)^{\beta} \tag{7}$$

with σ_0 the neutral Schmidt number, empirically found to have a value of approximately 0.7 (Turner, 1973), and α and β empirical constants. Various experimental data for F_m and σ_s can be found in the literature, the majority of which come from fresh-salt water experiments. When plotted as a function of Ri, the data points show considerable scatter (Fig. 1), suggesting that a dependence on Ri alone is unsatisfactory. Wall-effects and horizontal gradients may partially explain the scatter. A best fit can only be proposed for individual data sets. Furthermore, one can expect that similar data for sediment stratification would show additional dependence on the ratio w_s/u_*.

Kranenburg (1998) has derived theoretical conditions for the empirical constants. After correction for consistent implementation of the damping functions (Toorman, 2000b), the sufficient condition for stability leads to the following condition for the exponents:

$$1 + a - b \geq 0 \tag{8}$$

Equality is obtained in the case of the existence of a critical flux Richardson number Rf_c at which turbulence is completely damped (i.e., total turbulence collapse). The condition $1 + a - b = 0$ is fulfilled for the Munk-Anderson damping functions and yields the same asymptotic behaviour for large Ri as the theoretically derived turbulent Schmidt number relationship by Ellison (1957), given by:

$$\sigma_s = \sigma_0 \frac{(1 - Rf)^2}{1 - Rf / Rf_c} \tag{9}$$

(see also Turner, 1973) and is shown in figure 1. This relation can be converted into a 3rd degree relation between the Schmidt number and Ri. Comparison with the various data sets suggests that the value of Rf_c is case dependent and can vary widely over an order of magnitude (Fig. 1). For $0.5 < Rf_c < 0.9$ this formulation reaches a local minimum of σ_s, as seems to be found in some experiments (i.e. Webster, 1964; Shiono et al., 2000). These observations indicate that there must be other parameters which control the Schmidt number. Notice that the ratio w_s/u_* does not play a role in these data as they are all (except the Schultz ground data) for non-buoyant stratification (i.e., $w_s = 0$, as no particles are involved).

The condition $1 + a - b = 0$ implies that $\beta = 1$ in eq.(7). However, the existence of a critical Rf_c for turbulence collapse causes numerical problems near the free surface in some cases. This is most clearly illustrated with a 1DV case, where the free surface stress-free boundary condition reduces to $\partial U/\partial z = 0$, resulting in a very large Ri. Consequently, the damping at the free surface is over-predicted, compared to reality (wherein the physical free surface boundary conditions are more complicated), in particular for the k-ε model, making it impossible to mix the sediment up to the surface at very high u_*. Therefore it is advised to take $\beta < 1$, e.g. $\beta = 0.8$ as in the proposed curve in fig.1 (Toorman, 2000c). The problem can be overcome by generalising the definition of the Richardson number to include diffusion, following Ivey and Imberger (1991) (Toorman, 2000c, 2002). The latter solution seems to be the better one, as it is physically based, but requires further research.

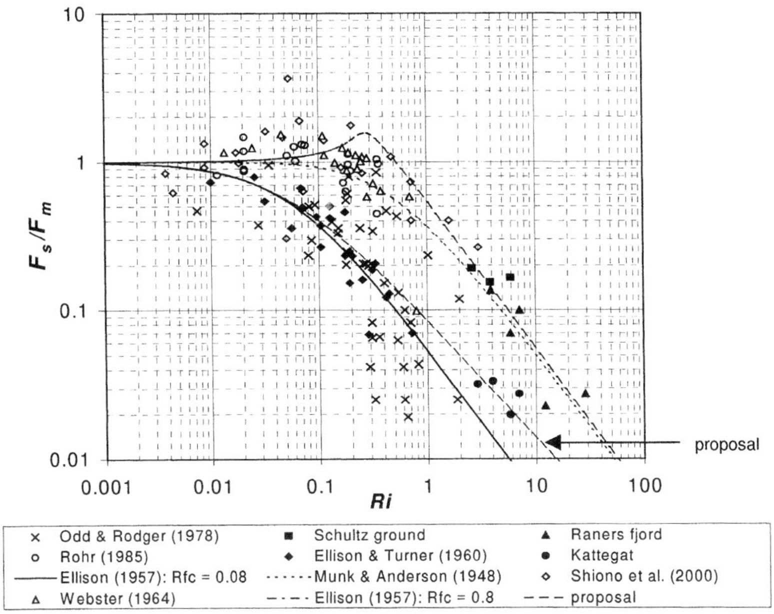

Figure 1. Experimental data for the inverse normalised Schmidt number σ_0/σ_s ($=F_s/F_m$) as a function of the gradient Richardson number and a few proposed closure relationships. Schultz ground and Raners fjord data from (Munk and Anderson, 1948); Kattegat data from (Ellison and Turner, 1960).

The lack of accurate data is one of the major problems which prevents us from proposing a better solution. It is hoped that with the progress in computer capacities, data from numerical experiments with direct numerical simulations (DNS) of sediment-laden flows at realistic scales will become available and will help understand the trends in the experimental data and their possible dependency on other parameters.

In addition to experimental data, the solution of the k-ε model can in principle also be used to determine the damping function F_m by numerical experiments because the buoyancy effect is accounted for by the term G in the k-equation (Toorman, 2000c). Unfortunately, the results (Toorman, 2000c) depend on the choice of the Schmidt number closure, for which no definite solution exists. Nevertheless, they suggest a linear dependence of the form $F_m = 1 - cRi$ for small Ri ($Ri < 0.1$, with c an empirical parameter), similar to the Monin-Obukov relation (Rodi, 1980). Another problem with these experiments is the deviations found near the free surface, where the boundary conditions for the k-ε model are not well established; hence these numerical data are required to be discarded.

Toorman (1999) theoretically found that for the critical flux Richardson number Rf_0 at which the vertical gradient $\partial Rf/\partial z = 0$, the momentum damping function reaches the value $F_m(Rf_0) = \sigma_s(Rf_0)w_s/\kappa u_*$. As this value depends on the Schmidt number for the same condition, the exact value is not known.

The key to progress seems to be in finding the proper closure for the turbulent Schmidt number. The collected pieces of the puzzle presented above are still insufficient to propose a solution which provides the desired accuracy.

3.4. Consistent bottom boundary treatment

Traditionally, numerical models employ so-called wall functions for the determination of the conditions at solid walls, such as the bed. However, they do not account for the effect of turbulence damping. This leads to significant over-estimations of the bottom shear stress or u_*. A simple numerical test is the verification of the shear velocity for open-channel flow driven by a constant pressure gradient, for which theoretical value is $u_* = (\rho^{-1}\ h\ dp/dx)^{1/2}$. The velocity gradient as expressed by eq.(1) is used to calculate u_*. For the k-ε model, a more accurate estimation can be done using the stress balance (Toorman, 2000c, 2002). Hence it is advised that the velocity gradient in the wall node is directly estimated from the computed velocity profile in the grid cell adjacent to the wall, employing simple interpolation functions.

For the bottom boundary conditions of the momentum equations, the velocity in the near-wall node needs to be determined. Integration of eq.(1) yields the velocity profile in the wall layer, introducing an additional integral term to the logarithmic profile due to F_m. However, the damping function is generally not known as a function of z. It is then proposed to write the velocity profile as:

$$U = \frac{u_*}{\kappa z} \ln\left(\frac{z}{\alpha z_0}\right) \tag{10}$$

where z_0 is the roughness height of the bed and α the apparent roughness correction factor, which is related to the damping function by (Toorman, 2000e):

$$\frac{1}{F_m} = 1 - \frac{z}{\alpha}\frac{\partial \alpha}{\partial z} \tag{11}$$

A series of numerical experiments (Toorman, 2000e) suggest that α can be parameterised as:

$$\alpha = \exp\left(-\left(1 + \beta w_s / u_*\right)\left(1 - \exp(-bRi^m)\right)\right) \tag{12}$$

with w_s the settling velocity, *β*, *b* and *m* empirical constants, and *Ri* calculated at the near-wall node. Physically, the apparent change in bottom roughness corresponds to drag reduction, which has been observed both in nature and in the laboratory (Toorman, 2002).

4. LAMINARISATION

When density gradients are large, the damping of turbulence by buoyancy may ultimately become so strong that turbulence cannot be maintained and the flow becomes laminar. Two situations, illustrated by the numerical example shown in figure 2, need to be considered.

Generally, high-density peaks occur at the bottom due to sedimentation and the presence of the viscous sublayer. Turbulence damping occurs along with drag reduction and the subsequent thickening of the viscous sublayer, compared to the clear water case. Toorman (2000e) has shown that the consistent boundary treatment method leads to drag reduction predictions of the same order of magnitude as measured by Li and Gust (2000).

The second possible situation is the generation of a lutocline as the result of the combined effect of hindered settling and buoyancy damping in turbulent shear flows. The occurrence of such two-layer sediment-stratified systems in some estuaries has been observed by Wolanski *et al.* (1988). It should be realised that such two-layer systems cannot be simulated correctly by the PML model, because the lutocline forms a new reference for calculation of the mixing

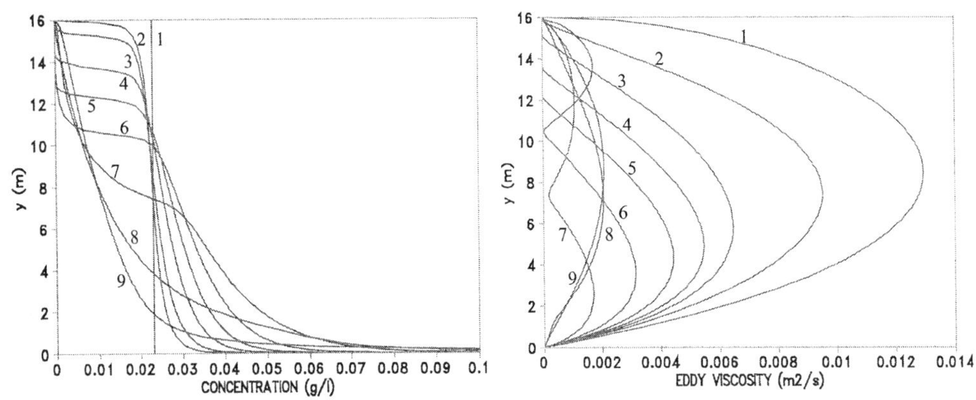

Figure 2. Numerical results (*k*-*ε* model) of the evolution (1⇛9) of concentration and corresponding eddy viscosity profiles of an initially homogeneous suspension (C_0 = 23 mg/l) at a constant flow rate (mean flow velocity = 0.2 m/s), using a constant turbulent Schmidt number ($\sigma_s = \sigma_0 = 0.7$).

length in the upper layer. The second to the fifth eddy viscosity profiles (from left to right) in figure 2 show that turbulence may be completely damped in the upper, sediment-free layer. Whether complete laminarisation really occurs above lutoclines has never been demonstrated experimentally. It is very likely that the density interface becomes unstable, resulting in internal waves, which may increase mixing and turbulence production (see next section). This effect is not included in the model used to obtain the results of fig.2.

Thus far, the modelling of laminarisation has only been successful for relatively simple shear flows where it occurs along the wall, as in the first case. In actuality, because cohesive sediment transport problems are time-dependent, they involve variations of stratification and hence the thickness of the viscous sublayer. In general (in the absence of fluid mud) this thickness will remain much smaller than the vertical grid size, but problems may occur around flow reversal at slack and neap tides, when the sublayer thickness may become very large and fluid mud layers may form. Otherwise, as long as the sublayer thickness remains small, the relevance of modelling near-wall laminarisation may be reduced in commercial models, as they can handle flow reversal due to the fact that a constant horizontal diffusion is used. This is so because the horizontal grid scales are too large compared to the turbulence length scales of the PML or the k-ε model.

In principle, the sublayer requires a different method of solution than the fully turbulent layer, because the assumptions for the above mentioned turbulence models are no longer valid. In the PML model for clear water, investigated at the T.U.Delft (Kranenburg, 1999), a two-layer approach can be applied, where in the near-wall layer another mixing length model, i.e., a modified Van Driest model, is used:

$$\ell(z) = \ell_0(z)\left[1 - \exp\left(-\frac{1}{A}\frac{u_* z}{\nu}\right)\right] \tag{13}$$

where the neutral mixing length distribution for free-surface flow is given by:

$$\ell_0(z) = \frac{\kappa z\sqrt{1 - z/h}}{1 + \Pi(\pi z/h)\sin(\pi z/h)} \tag{14}$$

with Π = Coles' wake strength parameter (Nezu and Rodi, 1986). The interface with the fully turbulent layer is then determined by a new laminarisation criterion:

$$\frac{\nu_{tm}}{\nu} < c\sigma_0 \,\mathrm{Re}_{tc} \tag{15}$$

The selected value of the critical turbulent Reynolds number Re_{tc} is 15 and c = 0.61, as obtained by a numerical calibration procedure. The method has been successfully validated with the simulation of the laminarising duct flow of a homogeneous fluid. The model has also been applied to slowly decelerating, sediment-laden, open-channel flows. The mixing length model (eq.12) then has to be multiplied with the factor F_m to account for turbulence modulation by suspended particles (*cf.* Section 3.1).

In the k-ε model the problem can, in principle, be handled by correcting various constants with near-wall damping functions (f_μ, f_1 and f_2), which is known as a low-Reynolds number turbulence model. This has been investigated at the K.U.Leuven. Toorman (2000d) proposed a new realisable time scale (i.e. a time scale which adapts itself from the laminar flow Kolmogorov scales to the k-ε scales, depending on the turbulent Reynolds number), which defines the damping function f_2. His analysis of direct numerical simulation (DNS) data furthermore indicates that the coefficients σ_k and σ_ε cannot have a constant value within the near-wall layer. Again, however, the damping functions proposed in the literature are only valid for homogeneous fluids and nearly all of them are restricted to smooth walls (Rocabado, 1999). Furthermore, this model requires a very fine grid size near the bottom, which considerably increases the computational cost.

Alternatively, the k-ε model can also be combined with a wall layer model, where a Van-Driest type mixing-length model is used. This is known as the two-layer approach (Toorman, 2000d). A Reynolds number criterion is used to determine the boundary between the two layers. The same problems as for the two-layer PML model apply. Unfortunately, the accuracy of the two-layer approach is generally not very high for coarse grids.

None of the above methods can be applied to laminarisation away from the wall, such as may occur around lutoclines. The so-called "wall-distance free" low-Reynolds number models have recently been proposed in the literature, but they are complicated and are meant for near-wall turbulent shear flows (Toorman, 2000d). Nevertheless, presently used codes seem to be able to handle laminarisation around a lutocline, as figure 2 proves. As the molecular viscosity is too small to stabilise the solution, this is only possible due to numerical diffusion inherent to the numerical schemes, especially in commercially applied codes which require robustness, or to specially designed locally added artificial diffusion in the k-ε turbulence model to stabilise the solution where the actual equations are no longer valid (Toorman, 1999). It is not known how these artificial solutions affect the history of simulated turbulence. Considering furthermore all the uncertainties due to required simplifications following the relatively coarse scales in 3D estuarine modelling, which do not capture sharp lutoclines anyhow, laminarisation away from the wall is not expected occur in applied modelling practice. Furthermore, mixing due to internal waves additionally reduces the chance of laminarisation.

5. INTERNAL WAVES

Internal waves do occur in stratified flows, and are most manifest at the interface of a two-layer flow system. Internal waves contribute to the vertical transport of momentum, but not directly to the transport of mass (sediment in our case). Uittenbogaard (1995a, 1995b) showed that, although internal wave breaking is essential to initiate turbulence at an interface, it is the shearing induced by the internal waves that supplies most of the energy to the turbulence. Hence, enhanced turbulence also affects vertical mass transport.

The effect of internal waves can be incorporated through additional production and dissipation terms in the k-ε turbulence model. In a study by Delft Hydraulics (Winterwerp and Uittenbogaard, 1999), a more parameterised approach is followed in which the effect of internal waves is incorporated in the eddy viscosity (but not in the eddy diffusivity).

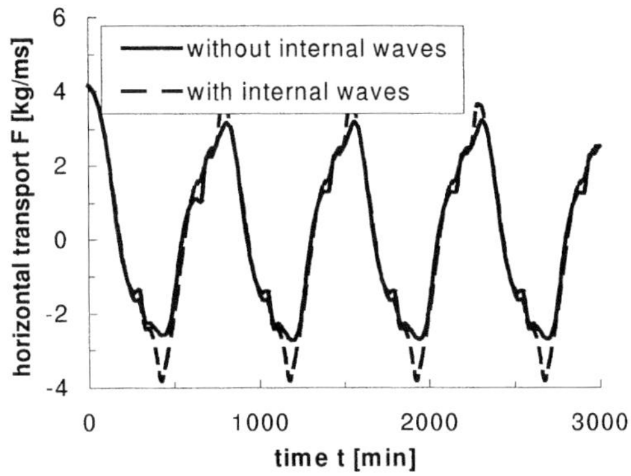

Figure 3. Variation of computed horizontal transport with time.

The effect of turbulence produced by internal waves on the dynamics of concentrated benthic suspensions (CBS) was studied through a one-dimensional approach. The 1DV momentum equation, including the effect of internal waves, reads:

$$\frac{\partial u}{\partial t}+\frac{1}{\rho}\frac{\partial p}{\partial x}=\frac{\partial}{\partial z}\left[\left(\nu+\nu_t+\nu_{IWE}\right)\frac{\partial u}{\partial z}\right] \tag{16}$$

where u is the horizontal flow velocity, p is pressure, x and z are the horizontal and vertical co-ordinate and ν_{IWE} the additional viscosity induced by internal waves.

The additional internal wave dissipation term is parameterised as a function of the buoyancy frequency, the vertical shear rate and a length scale which is related to the Ozmidov (length) scale. This additional viscosity is only relevant in case of large stratification.

This approach was implemented in a 1DV point model (Winterwerp, 1999). The model was run without (as reference condition) and with internal wave effects for a hypothetical open channel flow with a depth of 8 m, a depth-mean velocity of 0.7 m/s and a depth-mean suspended sediment concentration of 0.74 g/l. These conditions are near saturation, hence a strong buoyancy-induced interaction exists between the turbulent flow field and the suspended sediment concentration. As a result the eddy viscosity for the reference case is considerably decreased. The horizontal sediment flux is then particularly affected at larger flow velocities, as the transport is relatively large. The net horizontal transport is defined as $F = \int_0^h uc \, dz$ and the results are presented in fig. 3, with and without internal wave effects. Figure 3 shows differences of about 20 % during maximal velocities, whereas during the rest of the tide the differences are comparatively small. It must be noted that the asymmetric curve at flood and ebb velocity is due to plotting resolution.

From the above simulations it is concluded that the effect of augmented eddy viscosity by internal waves can be considerable for the hydrodynamic conditions examined in this paper. However, it is recommended not to include these effects on a routine basis before turbulent modelling of concentrated benthic suspensions is better understood, as, for instance, low Reynolds number effects may be of greater importance.

For more details the reader is referred to Uittenbogaard (1995a, 1995b) and Winterwerp and Uittenbogaard (1999).

6. DECAY OF NON-LOCALLY PRODUCED TURBULENCE DUE TO DENSITY STRATIFICATION

6.1. Problem definition

In the deeper parts of an estuary, for example in a navigation channel, the deposition rate of cohesive sediment may be high, especially during slack water, forming a CBS, which may behave as a dense fluid for several hours, or longer in the case of stirring by wave action or by passing ships. In the flow upstream of this depression, turbulence is produced mainly as a result of bed friction. In the depression, the production of turbulence is for the larger part suppressed due to the presence of sediment-induced density stratification. In turn, turbulence produced over the rigid bed upstream of the deeper part is advected over the CBS and gradually decays in the downstream direction. While decaying, this non-locally produced turbulence might entrain material from the CBS. This interaction of turbulence and density stratification has been studied at Delft University of Technology. The study consisted of laboratory experiments on the decay of non-locally produced turbulence over a CBS and the subsequent entrainment, as well as of numerical simulations of these processes. The objectives of the experiments were to obtain a relation between hydrodynamic parameters, decay of turbulence and entrainment rates, and secondly, to generate data to validate computer models. This study is summarized next.

6.2. Scaling laws

To scale field conditions based on laboratory experiments, several scaling laws have to be taken into account. The first scaling law is concerned with the turbulence structure. It is required that the Reynolds number ($Re = uh/\nu$, where u is mean velocity, h is depth of the flow and ν is viscosity) in the physical model be sufficiently high. The upstream bed is roughened to obtain hydraulically rough conditions.

The second law is concerned with the scaling for turbulence decay. A lower bound of the length scale L of decay can be derived from a balance of the advection and dissipation terms in the transport equation for turbulent kinetic energy. L scales with the water depth ($L \approx 9h$), and for significant decay to be observable, the length of the depression should be at least several times L.

The third scaling law concerns consolidation time. The consolidation time T_c is proportional to the inverse of the squared depth δ of the CBS layer (i.e., $T_c \propto \delta^{-2}$). This means that in the physical model it is nearly impossible to keep the stationary CBS in a fluidised state. A mud layer, possessing strength (behaving as a non-ideal Bingham fluid when sheared), will be formed and the turbulence production at the water/mud interface then would not be essentially different from that over the upstream rigid bed.

The last scaling law deals with entrainment. The entrainment process is governed by the overall Richardson number

$$Ri_* = \frac{\Delta\rho g h}{\rho u_*^2} \tag{17}$$

where $\Delta\rho$ is the excess density of the CBS, ρ the density of the overlying water, g is the acceleration of gravity, h the height of the water layer and u_* the friction velocity at the upstream rigid bed. Simulation of the entrainment process in a hydraulic model requires scaling at constant Ri_*, which for constant g and ρ yields:

$$\frac{\Delta\rho_m g h_m}{u_{*m}^2} = \frac{\Delta\rho_f g h_f}{u_{*f}^2} \tag{18}$$

where subscript f refers to the field and m to the physical model. Substituting realistic values for the field parameters, it can be shown that extremely low bed shear stresses in the physical model are required. These values are likely to be lower than the yield stress of a concentrated layer in a laboratory flume.

Based on the last two scaling laws, the CBS in the physical model is replaced by salt water. In earlier work it was already shown that in terms of initial entrainment rate, CBS behaves similar to saline water.

6.3. Experimental set-up and results

Two series of experiments were carried out. The first series was concerned with decay of turbulence and the second with entrainment. The tests were conducted in a flume of 30 m length, 1 m width and 0.3 m depth. A longitudinal cross-section of the experimental set-up for the first series is shown in Figure 4. The upstream part of the flume was 0.15 m deep, the deeper part was 0.3 m deep. To prevent flow separation the slope is small (8º). The depth of the saline water in the depression was kept constant (0.1 m) by a continuous inflow (at a small rate) at the upstream slope and a (internal) weir near the end of the depression. After the flow became stationary, turbulent velocities were measured in the upper fresh water layer using laser-Doppler velocimetry. Figure 4 indicates the six positions (with a spacing of 2 m) at which vertical velocity profiles were measured.

The entrainment rate decreased in downstream direction over the depression. To measure the integral entrainment at different distances downstream of the slope, the experimental set-up was slightly changed for the second experimental series by replacing the internal weir by an internal barrier, which could be placed at various positions (for example position 2 to 6 in fig. 4). For each position the rate at which saline water had to be supplied to the flume to keep the height of the saline water layer at 0.1 m was measured.

The experimental program was set up to vary the overall Richardson number (by varying $\Delta\rho_m$ and/or u_{*m}). In figure 5 the non-dimensional turbulent kinetic energy (k/u^2) is plotted against the distance downstream of the ramp for all experiments. The decrease in entrainment rate was more or less proportional to this decay in turbulent kinetic energy. A more detailed analysis of the experimental data can be found in (Wissmann and Bruens, 2000).

6.4. Numerical simulations

A numerical flow model for shallow-water flow (Delft3D) with a standard k-ε turbulence model was used for the numerical simulation of the physical experiments. By comparison with measurements, turbulence decay predicted by the model as well as the predicted entrainment

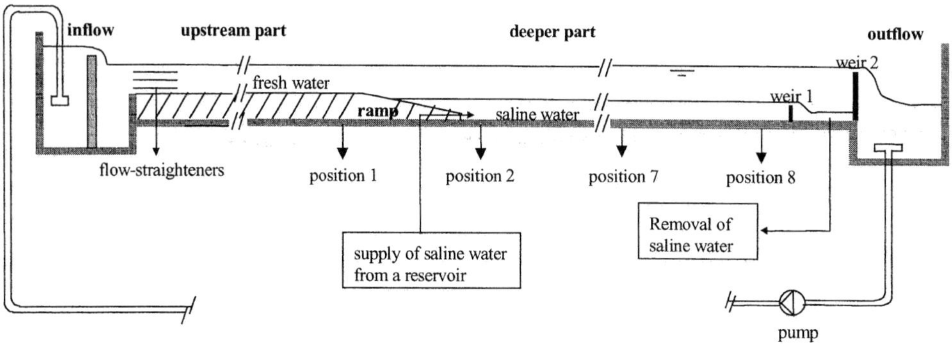

Figure 4. Schematic cross-section of the experimental set-up for the first series of experiments (not to scale).

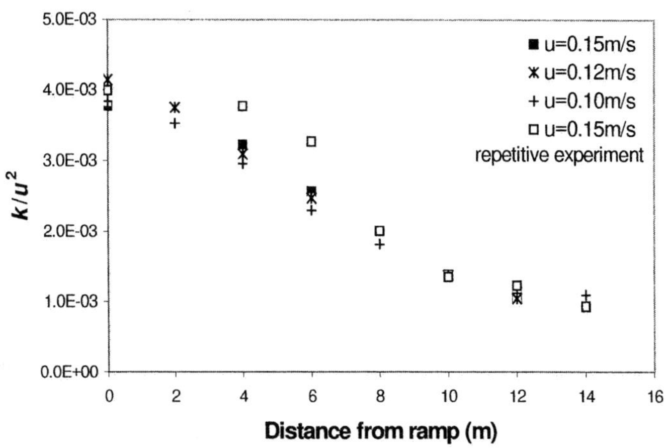

Figure 5. Measured decay of turbulent kinetic energy with distance from the ramp.

of saline water were tested. The model simulated the decay in the upper layer accurately, but the entrainment rate was underpredicted due to the fact that in the model internal waves were not taken into account.

7. CONCLUSIONS

Suspended cohesive sediments cause damping of the turbulent fluctuations in flowing water and alter the apparent bed roughness. Consistent implementation of buoyancy-induced turbulence damping functions allows the modelling of the damping and the bed roughness modification (or drag reduction) and can explain the decrease of the von Karman parameter with increasing stratification. Finding the appropriate turbulent Schmidt number closure seems to be the key to make further progress. A major problem with the further development of a validated modelling strategy for sediment-turbulence interaction in cohesive sediment transport modelling is the lack of data for testing theories, developing and calibrating more accurate damping functions and validating the models. Laboratory experiments at high enough concentrations with (cohesive) sediments require new non-optical (e.g., acoustic) measurement techniques, which are being developed and improved. Another hope is that future direct numerical simulation (DNS) data at realistic scales will become available, once computer power allows it. As long as these validation data are not available, the framework proposed in this paper provides the best possible approximation to be implemented in cohesive sediment transport models.

Turbulence damping by buoyancy can become so strong that the flow laminarises locally, i.e., near the bottom and around a lutocline. At the bottom the viscous sublayer thickens with increasing stratification. When its thickness tends to reach a value on the order of the vertical grid size of the model, it seems advisable that a more comprehensive two-layer approach should be implemented. A more systematic study is required for the numerical implementation of this phenomenon on a realistic coarse grid as used in real applications in order to evaluate the feasibility of this method.

No concern is presently required regarding possible laminarisation around lutoclines, because available models do not capture the gradients accurately enough to lead to the problem. The shortcoming, resulting in an excessive vertical diffusion, is partially compensated by the likely occurrence of internal waves, which increases vertical mixing, which is presently absent in models.

It is recognised that internal waves may become important when the degree of stratification is high. They are an additional source of turbulence production, which is missing in the presently used models. This can explain the underestimation of entrainment in certain simulations. The problem can be handled by the introduction of an additional empirical diffusion coefficient, which is only a rough parameterisation of the complex process. However, in view of the many uncertainties regarding turbulence modelling when sediment is in suspension and the limitations due to the relatively coarse grids for coastal and estuarine applications, it is advised not to implement internal wave corrections at present.

An experimental study on the effect of non-locally produced turbulence on the entrainment of a stationary, localised CBS layer (for example in a depression or a navigation channel) has been carried out. No or only a minor degree of turbulence is generated in the pool as a result of a stable interface. Turbulence produced over the rigid bed upstream of the depression is advected over the depression and decays in the downstream direction. This decay has been measured in a physical model. While decaying it entrains material from the dense layer in the depression. Preliminary results indicate that non-locally produced turbulence can entrain a substantial amount of cohesive sediment. The data obtained have been used to validate a numerical flow model which accurately simulates the decay in the upper layer.

Acknowledgements: This work has been carried out as part of the MAST3 project "COSINUS", partially funded by the European Commission, Directorate General XII for Science, Research and Development under contract no. MAS3-CT97-0082. The first author's post-doctoral position was financed by the Flemish Fund for Scientific Research.

REFERENCES

Best, J., Bennett, S., Bridge, J. and Leeder, M. , 1997, Turbulence modulation and particle velocities over flat sand beds at low transport rates. *ASCE J. Hydr. Eng.*, 123(12), 1118-1129.

Cellino, M. and Graf, W.H. , 1999). Sediment-laden flow in open channels under noncapacity and capacity conditions. *ASCE J. Hydr. Eng.*, 125(5), 456-462.

Coleman, N.L. , 1981, Velocity profiles with suspended sediment, *J. Hydraulic Research*, 19(3), 211-229.

Crapper, M., Bruce, T. and Gouble, C. , 2000, Flow field visualisation of sediment-laden flow using ultrasonic imaging, *Dynamics of Atmospheres and Oceans*, 31, 233-245.

Crapper, M. and Bruce, T. , 2002). Measurement of mud transport processes using ultrasonic imaging, *Proc. INTERCOH-2000*, J.C. Winterwerp and C. Kranenburg eds., Elsevier, this volume.

Einstein, H.A. and Chien, N. , 1955, Effects of heavy sediment concentration near the bed on velocity and sediment distribution, *M.R.D. Sediment Series*, No.8, Missouri River Div., US Army Corps of Engineers.

Ellison, T.H. , 1957, Turbulent transport of heat and momentum from an infinite rough plane, *J. Fluid Mechanics*, 2, 456-466.

Ellison, T.H. and Turner, J.S. , 1960, Mixing of dense fluid in a turbulent pipe flow, *J. Fluid Mechanics*, 8, 514-544.

Galland, J.-C., Laurence, D. and Teisson, C. , 1997, Simulating turbulent vertical exchange of mud with a Reynolds stress model, In: *Cohesive Sediments*, N. Burt, R. Parker and J. Watts, eds., J. Wiley, Chichester, 439-448.

Gore, R.A. and Crowe, C.T. , 1989, Effect of particle size on modulating turbulence intensity, *J. Multiphase Flow*, 15, 279-285.

Ivey, G.N. and Imberger, J. , 1991, On the nature of turbulence in a stratified fluid. Part I: The energetics of mixing, *J. Physical Oceanography*, 21, 650-658.

Kranenburg, C. , 1998, Saturation concentrations of suspended fine sediment. Computations with the Prandtl mixing-length model, Report No.5-98, Faculty of Civil Engineering and Geosciences, Delft University of Technology.

Kranenburg, C. , 1999, Laminarisation in flows of concentrated benthic suspensions. Computations with a low-Reynolds mixing-length model, Report No.1-99, Faculty of Civil Engineering and Geosciences, Delft University of Technology.

Li, M.Z. and Gust, G. , 2000, Boundary layer dynamics and drag reduction in flows of high cohesive sediment suspensions, *Sedimentology*, 47, 71-86.

Lyn, D.A. , 1987, Turbulence and turbulent transport in sediment-laden open-channel flows, *PhD thesis*, California Institute of Technology, Passadena, CA.

Munk, W.H. and Anderson, E.A. , 1948, Notes on a theory of the thermocline, *J. Marine Research*, 3(1), 276-295.

Nezu, I. and Nakagawa, H. , 1993, *Turbulence in open-channel flow*, IAHR Monograph Series, Balkema, Rotterdam.

Nezu, I. and Rodi, W. , 1986, Open-channel flow measurements with a laser Doppler anemometer, *J. Hydr. Eng.*, 112, 335-355.

Odd, N.V.M. and Rodger, J.G. , 1978, Vertical mixing in stratified tidal flows, *ASCE J. Hydraulics Div.*, 104(3), 337-351.

Rocabado, O.I. , 1999, Modelling highly concentrated turbulent flows with non-cohesive sediments, *PhD thesis*, Civil Eng. Dept., Katholieke Universiteit Leuven.

Rodi, W. , 1980, *Turbulence models and their application in hydraulics*, State-of-the-art Paper, IAHR, Delft.
Rohr, J.J. , 1985, An experimental study of evolving turbulence in uniform shear flows with and without stable stratification, PhD thesis, Dept. of Applied Mechanics and Engineering Sciences, University of San Diego, CA.
Shiono, K., Siqueira, R.N. and Feng, T. , 2000, Exchange coefficients for stratified flow in open channel, *Proc. 5th Int. Symp. on Stratified Flows* (G.A Lawrence, R. Pieters and N. Yonemitsu, eds.), 2, 927-932, Dept. of Civil Engineering, University of British Columbia, Vancouver.
Toorman, E.A. , 1999, Numerical simulation of turbulence damping in sediment-laden flow. Part 1. The Siltman testcase and the concept of saturation, *Report HYD/ET99.2*, Hydraulics Laboratory, Katholieke Universiteit Leuven.
Toorman, E.A. , 2000a, Stratification in fine-grained sediment-laden turbulent flow, *Proc. 5th Int. Symp. on Stratified Flows* (G.A Lawrence, R. Pieters and N. Yonemitsu, eds.), 2, 945-950, Dept. of Civil Eng., University of British Columbia, Vancouver.
Toorman, E.A. , 2000b, Sediment-laden turbulent flow: a review, *Report HYD/ET/00.1*, Hydraulics Laboratory, Katholieke Universiteit Leuven.
Toorman, E.A. , 2000c, Parameterisation of turbulence damping in sediment-laden flows, *Report HYD/ET/00/COSINUS3*, Hydraulics Laboratory, Katholieke Universiteit Leuven.
Toorman, E.A. , 2000d, Analysis of near-wall turbulence modelling with the k-ε turbulence model, *Report HYD/ET/00/COSINUS2*, Hydraulics Laboratory, Katholieke Universiteit Leuven.
Toorman, E.A. , 2000e, Drag reduction in sediment-laden turbulent flow, *Report HYD/ET/00/COSINUS5*, Hydraulics Laboratory, Katholieke Universiteit Leuven.
Toorman, E.A. , 2002, Modelling of turbulent flow with suspended cohesive sediment, *Proc. INTERCOH-2000*, J.C. Winterwerp and C. Kranenburg eds., Elsevier, this volume.
Turner, J.S. , 1973, *Buoyancy effects in fluids*, Cambridge University Press.
Uittenbogaard, R.E. , 1995a, The importance of internal waves for mixing in a stratified estuarine tidal flow, PhD thesis, Delft University of Technology, September 1995.
Uittenbogaard, R.E. , 1995b, Observations and analysis of random internal waves and the state of turbulence, *Proc. IUTAM Symp. on Physical Limnology* (Broome, Western Australia, September 1995).
Vanoni, V.A. , 1946, Transportation of suspended sediment by water, *Trans. ASCE*, 111, 67-133.
Webster, C.A.G. , 1964, An experimental study of turbulence in a density stratified shear flow, *J. Fluid Mech.*, 19, 221-245.
Winterwerp, J.C. , 1999, On the dynamics of high-concentrated mud suspensions, *PhD thesis*, Delft University of Technology, Delft.
Winterwerp, J.C. and Uittenbogaard, R.E. , 1999, Effect of internal waves on the saturation of high-concentrated mud suspensions, WL|Delft Hydraulics / Delft University of Technology, Report Z2386.
Wissmann, J. and Bruens, A. W. , 2000, Experiments on the decay of turbulence due to density stratification, Report No.6-00, Faculty of Civil Engineering and Geosciences, Delft University of Technology.
Wolanski, E., Chappell, J., Ridd, P. and Vertessy, R. , 1988, Fluidization of mud in estuaries, *J. Geophysical Research*, 93(C3), 2351-2361.

Fine Sediment Dynamics in the Marine Environment
J.C. Winterwerp and C. Kranenburg (Editors)

Flocculation and settling velocity of fine sediment

J.C. Winterwerp[a], A.J. Bale[b], M.C. Christie[c], K.R. Dyer[c], S. Jones[d], D.G. Lintern[e], A.J. Manning[c], W. Roberts[f]

[a] Delft Hydraulics, PO Box177, 2600MH Delft, Netherlands; also Delft Un. of Techn.
[b] Plymouth Marine Laboratory, Prospect Place, Plymouth PL1 3DH, UK
[c] Institute of Marine Studies, Univ. of Plymouth, Plymouth PL4 1HP, UK
[d] School of Ocean Sciences, Univ. of Wales, Bangor, Gwynedd LL50 5EY, UK
[e] Dept. Engineering Sciences, Univ. of Oxford, Oxford OX1 3PJ, UK
[f] HR Wallingford, Howbery Park, Wallingford OX10 8BA, UK

This paper describes field and laboratory measurements of floc size and settling velocity of cohesive sediment in an estuarine environment and its modelling. The measurements and modelling showed a considerable variation in floc size, hence settling velocity over the tide, and over the spring-neap tide cycle.

The simple flocculation model describes the evolution of floc size at one fraction only. The measurements, however, revealed the existence of a pronounced distribution in floc size, which varies over time. For instance, at the beginning of a flood period, hardly any macroflocs were observed in the field, whereas later in the tide, the fraction of macroflocs increased to about 45 to 80 %.

Also the structure of the flocs appeared to vary over the tide, though this may be caused by advection of sediment from other locations and/or bed erosion.

In the laboratory experiments floc sizes and settling velocities similar to those observed in the field were obtained.

KEY WORDS
flocculation, floc size, floc density, settling velocity

1. INTRODUCTION

The prediction of the transport and fate of cohesive sediment in estuarine and coastal waters is often done with numerical models. Lately, much progress has been made in the development of these models. However accurate the implemented mathematical-physical descriptions may have been the accuracy of the results is still largely dependent on the values of the sediment parameters fed to the model. One of these is the settling velocity of the sediment. This is a difficult parameter, as it is well known for along time that this settling velocity may vary largely over space and in time (e.g. Dyer, 1989). Moreover, the settling velocity is also known to vary with the sediment concentration itself, though very little is known on the evolution of floc size in high-concentrated suspensions.

The COSINUS-project, executed under the framework of the European MAST3 research programme, was aimed at enhancing our understanding of the behaviour of high-concentrated suspensions, and developing mathematical formulations to describe the relevant physical processes, together with their parameterisations. These could then be incorporated in mathematical models for managing authorities and engineering consultants. The present paper describes the work carried out under Task B of the project, focusing on the evolution of floc size and settling velocity in estuarine and coastal environments.

Section 2 describe the results of a field campaign in the Tamar estuary, UK, carried out by Plymouth Marine Laboratory, the Institute of Marine Studies, University of Plymouth, the School of Ocean Sciences, University of Wales, the Department of. Engineering Sciences, University of Oxford, and HR Wallingford. Section 3 describes he results of a laboratory carried out at LEGI in Grenoble and section 4 describes the development of a simple flocculation model carried out at Delft Hydraulics.

2. FIELD WORK

The detailed aims of the fieldwork were to obtain in-situ data on the distribution of floc size, settling velocity and effective density in relation to salinity, suspended solids and turbulence characteristics.

A straight reach of the upper Tamar estuary, within the trajectory of the turbidity maximum was chosen for the experiment, because its characteristics are fairly well documented. In order to separate local and advected effects, two stations, A and B, 970 m apart were occupied with fore-and-aft anchored vessels in the central channel (Figure 1).

The estuary widths and maximum depths at high water were 75 m and 4.5 m respectively at station A, and 50 m and 5.2 m at station B. The tidal range varied from about 3.2 m to about 4.5 m. In this situation it was expected that there would be considerable settling and erosion of mud at times during the tide, and that the suspended sediment dynamics would be dominated by the bed boundary conditions. At these stations the vertical profiles of flow velocity, salinity, temperature, suspended sediment concentration, water level, and a number of floc properties were measured.

The main instrument for the latter was the INSSEV system (Fennessy et al, 1994a) which measured floc size, and settling velocity distributions 0.5 m above the bed at 20 - 30 min intervals, and was located at Station A. The video system had a lower limit of about 20 μm, and could measure floc sizes up to about 1mm. Additionally, a Partech Lasentec P100 laser sizing system (Law and Bale, 1998) was used to obtain vertical profiles of floc size distribution. This instrument senses particles in the size range 2 – 1,000 μm in 38 approximately logarithmically spaced size bands. Profiles were obtained every 30 min. at 0.5 m intervals from the surface to within 0.5 m of the bed, and in between the instrument measured continuously at fixed depths.

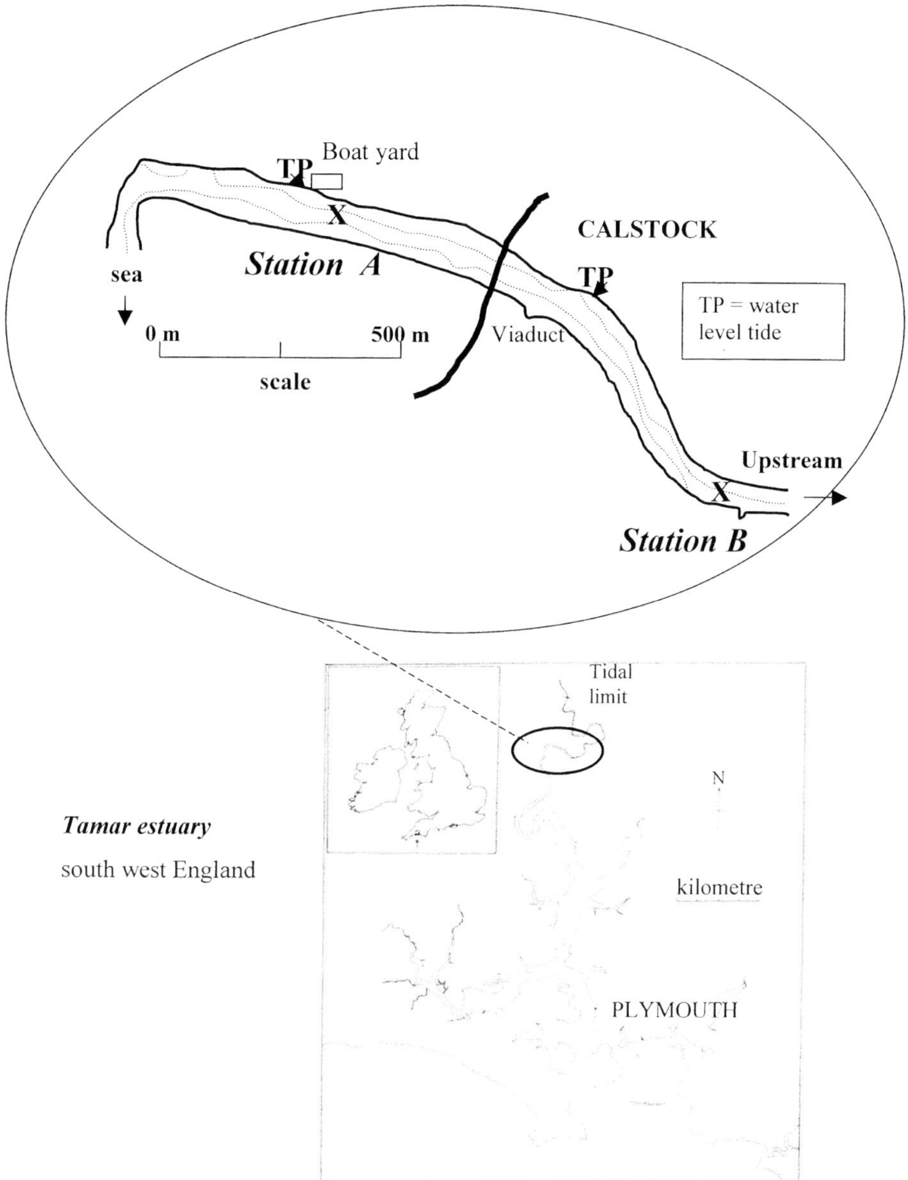

Figure 1. Location diagram of the Tamar estuary showing the position of stations A and B.

At the upper station B, a Sequioa LISST 100 laser diffraction system was used to determine the floc sizes. This instrument is not capable of measuring floc sizes greater than about 250 μm, and becomes saturated in concentrations greater than about 500 mg/l. UWB-QUISSET settling tubes (Jones and Jago, 1996) and the HR Floc Camera (Dearnaley, 1991) were used to quantify floc size and settling velocity distributions from samples taken at about mid depth.

From the measurements it was observed (Dyer et al., 2002a), that during neap tides the turbidity maximum appears to be upstream of the salt intrusion at both stations during the ebb tide. At Station A on the flood tide the turbidity maximum is right at the tip and partly inside the salt intrusion. During the flood tide at Station A, the near bed concentration increased abruptly when the bed shear stress exceeded about 0.2 Pa, and reached a peak of about 0.14 g/l, then decreasing to reach background levels within the saline intrusion.

At Station B, however, the peak of the turbidity maximum appears to be about the same distance upstream of the saline intrusion as on the ebb. Consequently, entrainment appears to be more important than advection of the turbidity on the flood tide. The location of the turbidity maximum appears to be associated with the peak in the currents that occurs upstream of the salt intrusion. The entrainment of sediment from the bed is associated with the velocity peak occurring landwards of the salt intrusion.

On the flood tide the turbidity maximum reached about double the concentration of that on the ebb, and lasted much longer. The concentration started to rise when the dissipation parameter G (for definitions, see van Leussen, 1994, Winterwerp, 1998) exceeded about 3 s^{-1}.

At spring tides the overall concentration in the turbidity maximum is considerably enhanced over that at neaps, and the concentration reached 3 g/l at the peak G. SPM concentration at Station A increased on the ebb tide when the bed shear stress exceeded about 0.25 - 0.3 Pa.
It appeared that erosion of the bed was a slow process, and there was not a bed layer of 'fluid mud' present to be eroded rapidly.

There were several periods during the passage of the turbidity maximum when the turbulent intensities at the upper level were greater than lower in the water column. This is the phenomenon of drag reduction, the shear at the boundary being reduced and the shear being concentrated at a higher level in the flow. This has been considered to result from high concentrations (e.g. Best and Leeder, 1993, Li and Gust, 2000, or Winterwerp, 2001b). The turbulent energy at the upper level can exceed that at the bed by a factor of up to 4. The onset of drag reduction coincides with the decrease in tidal currents, which allows the near bed turbulent flow to become less turbulent and the drag to reduce. During the period of drag reduction the measured gradient Richardson Numbers are less than the critical value of 0.25. This is depicted in Figure 2, showing the difference of TKE over the water column. This indicates that when the concentration gradient exceeds a value of ~4 g/m^4, the drag reduction commences. Then the shear stress does not monotonically decrease from the bed to the water surface, but has a maximum within the flow. This has important implications as it is a feedback mechanism that reduces the potential erosion from the bed, the transporting capacity of the flow, as well as the flocculation process itself.

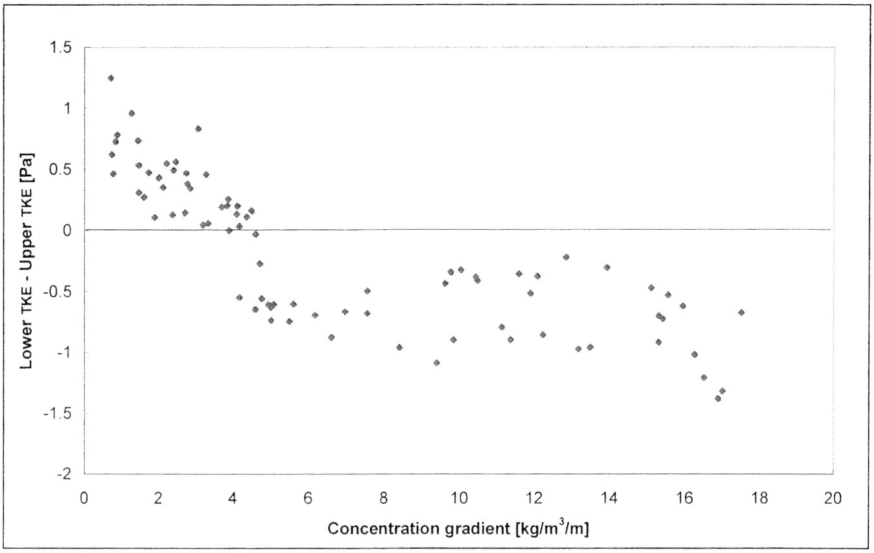

Figure 2. Variation of TKE over the water column as a function of concentration gradient.

Figure 3 shows, as an example, the variation of settling velocity, mean floc size, dissipation parameter G and SPM in the turbidity maximum at Station A during ebb of the neap tide on September 15, 1998. The concentration started to rise when G exceeded a value of about 3 s^{-1}, from a background concentration of about 40 mg/l to reach a maximum of about 280 mg/l within the turbidity maximum. The mean size of the flocs also rose from about 150 to almost 300 μm. The small size flocs in the clearer water before the turbidity maximum had settling velocities of almost 9 mm/s, and these reduced to 1 - 2 mm/s in the turbidity maximum. The increase in size and decrease in settling velocity within the turbidity maximum imply a drastic decrease in effective density (density of floc minus density of water). The effective density fell from about 750 kg/m^3 to less than 100 kg/m^3. This is compatible with a rise in porosity from 70 % to over 90 %.

There was also a change in the percentage contribution of the macro and microflocs to the mass concentration. Macroflocs have been defined from detailed analysis of the results as those flocs >160 μm. Outside the turbidity maximum they were in about equal proportions, but in high concentrations the proportional contributions of the macroflocs rose to over 70 %. This indicates either that a fraction of the microflocs is involved in flocculation to form the macroflocs, or that macroflocs were preferentially eroded from the bed to enhance the concentration

The maximum mean floc size of about 230 μ occurred towards the end of the turbidity maximum. Also, the settling velocity within the turbidity maximum underwent the same enhancement as on the ebb, reaching a peak value of about 2 mm/s. The effective densities were between 50 – 150 kg/m^3, about the same as on the ebb tide.

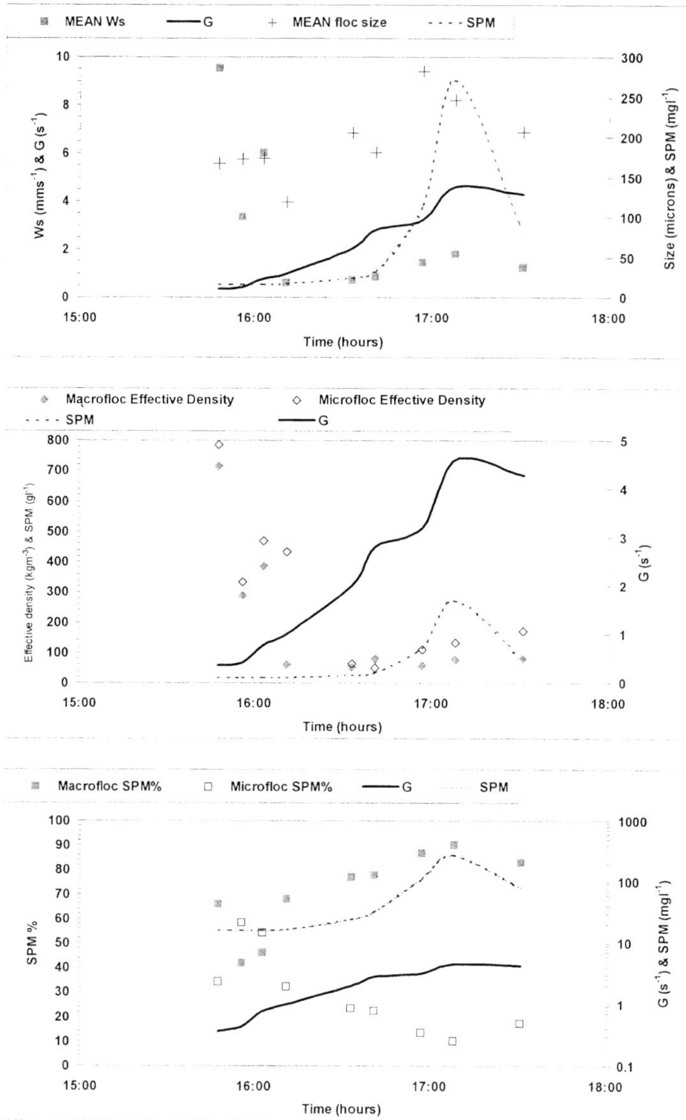

Figure 3. The variation of settling velocity, mean size, dissipation parameter and SPM in the turbidity maximum at Station A during ebb of the neap tide on September 15, 1998.

However, occurrence of the maximum value at the time of maximum concentration and maximum G, and after the maximum of SPM, suggested that additional contributions to the suspended floc population occurred later in the turbidity maximum as a consequence of either advection or resuspension. At the beginning of the flood macroflocs formed only 10 % of the mass, but the proportion increased rapidly with concentration. In the later stages of the turbidity maximum, the proportion of macroflocs varied widely between 80 % and 45 %.

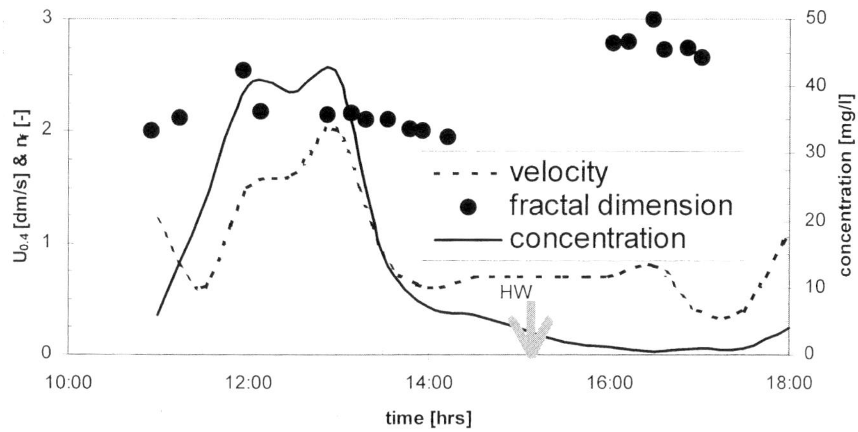

Figure 4. Variation of flow velocity at 0.4 m above the bed, SPM and fractal dimension n_f with time at Station A, September 16, 1998.

The mean floc diameter increased from about 100 μm to 200 μm and the settling velocity increased more or less in line with size to about 2 mm/s. The effective density of the flocs reduced within the turbidity maximum from about 330 to about 150 kg/m^3. There was a large contrast in the distribution of floc sizes, with no macroflocs being sampled before the beginning of the turbidity maximum, but the proportion within the turbidity maximum exceeded 70 %. The contrast in effective densities, size distributions and concentrations imply that not all of the increase in macroflocs could have occurred by flocculation. There must have been flocs created from the sediment eroded from the bed, or by advection.

This variation in floc density or structure is further substantiated by an analysis of the fractal dimension n_f of the flocs, as shown in Figure 4. Before High Water, n_f = 2.0 to 2.2, whereas after High Water n_f increases to about 2.8. This may indeed suggest that the latter flocs stem from bed erosion, whereas the low-n_f-flocs have been formed by flocculation processes in the water column (e.g. Winterwerp, 1999).

On the flood tide the concentration in the turbidity maximum reached in excess of 8 g/l, and the mean floc size was 350 μm. The parameter G was drastically affected by drag reduction (see Dyer et al., 2002a for details), and its value decreased across the turbidity maximum until the concentration fell below about 4 g/l. The settling velocity reached almost 6 mm/s, but together with size, it reduced rapidly towards the end of the turbidity maximum. The effective density stayed almost constant at about 100 - 200 kg/m^3 across the turbidity maximum.

Within the turbidity maximum there is an increase in the mean size of the flocs which, in general, matches the changing concentrations of suspended matter. There was also an increase in settling velocity through the turbidity maximum following the changes in size and concentration. This was generally in the range 0.5 - 2.0 mm/s. The changes in size and settling velocity imply that the effective density must decrease within the turbidity maximum. The results for the early ebb on neap tides showed very high effective density flocs settling at high velocities. These have been previously observed by Fennessy et al. (1994b) and interpreted as the result of the settling of individual tourmaline crystals.

The proportion of flocs with size greater than 160 μm changed significantly across the turbidity maximum. Before the passage of the turbidity maximum the macroflocs contributed about 50 % of the total floc mass on neaps and 0 % on springs. Within the turbidity maximum the macroflocs proportion rose rapidly until they contributed at least 70 % of the floc mass. In general these changes involved greater exchanges of mass between floc size bands than were available within the water. Consequently, many of the macroflocs present in the turbidity maximum must have been brought in by advection, or directly from bed erosion.

With the observations that have been made it is possible to examine the statistical relationships between the variables (e.g. Dyer et al., 2002b). The data set for the main experiment was combined with that for the preliminary experiments in June and August 1998, making a total of 74 simultaneous values. The predictions for settling velocity (W_s in mm/s) of the macroflocs are the ones of most interest for modelling:

$$W_s = -0.243 + 0.567c + 0.98G - 0.093G^2 \qquad (1)$$

which is valid for $0.1 < c < 6$ g/l and $1 < G < 10$ s^{-1}. This has an r^2 of 0.8, which is fairly significant. This relationship has the same form as that proposed by Dyer (1989), with an increase in settling velocity at low shear stresses when aggregation effects are dominant, and floc disruption at higher stresses for the same concentration.

This is further elaborated in Figure 5, comparing the settling velocity data obtained at the lower SPM (e.g. < 0.45 g/l) with an analysis presented by Winterwerp (1998) for mud from the Ems-Dollart in The Netherlands. The two curves in Figure 4 represent the settling velocity as a function of the dissipation parameter G in case of an unlimited residence time, and in case of a limited residence time. The limitation in residence time limits the growth of the flocs at lower values of G.

Manning (2002) has investigated in detail the measured distribution of floc size, settling velocity and effective density. He has also presented statistical relationships of many of those characteristics, such as mass settling flux, macrofloc/microfloc ratio, and settling velocities, against SPM and turbulence. Since most of the measurements were taken in fresh water, salinity is not included in the analysis. Nevertheless, the more continuous Lasentec results show that salinity is also likely to have an effect on the floc properties. Unfortunately, inconsistencies between different instruments did not allow further analysis. This is exemplified by the Lasentec instrument giving mean sizes less than those of INSSEV for concentrations less than about 200 mg/l and greater for higher concentrations. The INSSEV data have been successfully used in modelling (Petersen et al,. 2000).

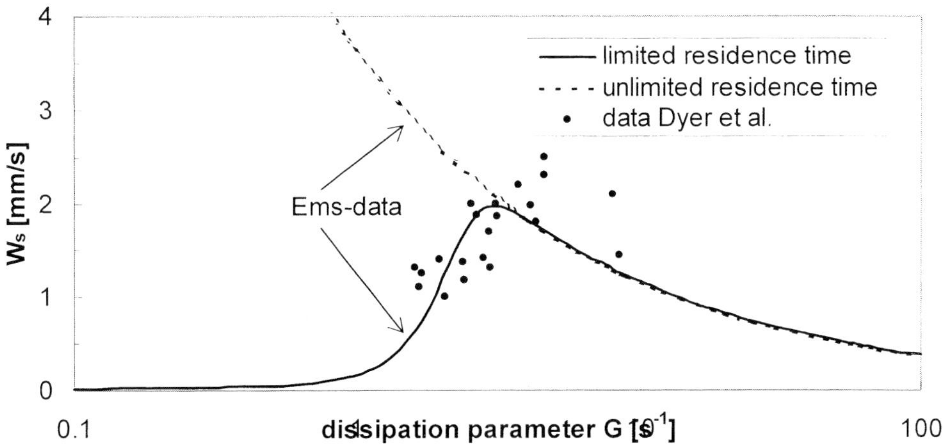

Figure 5. Effect of residence time on floc growth.

The chlorophyll-a was determined by measuring the absorbance of the chlorophyll-a solution at 665 nm and 750 nm, measured in a Philips Pu8720 Series UV/Vis Scanning Spectrophotometer. Total Carbohydrate was measured using the phenol-sulphuric acid assay (Underwood et al, 1995). Carbon/Hydrogen/Nitrogen (C/H/N) analysis was carried out using a Carlo Erba EA1110 CHNS analyser (CE Instruments). Organic analysis was carried out by ignition at 450^0.

Carbohydrate per unit mass of suspended sediment was much higher on neap tides (up to 40 mg/g SPM), which displayed organic contents typically 4 % greater than those measured during the spring tides (up to 7 mg/g SPM). However, the greater abundance of particulate matter in suspension during spring tides produced an overall higher carbohydrate concentration (14 times greater at times). This may have contributed significantly to the higher proportion of large macroflocs.

Hydrogen and nitrogen were greatest during high concentration spring tides, 1.5 % and 0.45 %, respectively, and the dissolved C:N ratios measured fit in well with those reported in other literature.

The organic content at neaps ranged between 13 – 18 %, whereas at spring tides it was significantly lower at 10 - 12.5 %. This is explained by the erosion of sediment with lower organic and chlorophyll-a content from the bed at spring tides, thereby diluting the previously suspended material.

The Owen settling tube samples allowing different settling fractions to be analysed. During neap tides, high carbohydrate levels (17.5 mg/g SPM) acted as an adhesive and assisted in the production of the larger faster settling macroflocs formed during low concentrations. There was then a gradual drop in carbohydrate content to a minimum in the slower settling aggregates. Then there was a significant increase in total carbohydrate (up to 56.7 mg/g SPM) in the very small slow settling microflocs. These are most likely particles that have not been aggregated, due to the combined effects of the lower turbulence, and the reduced collision frequency. It appears that the faster settling macroflocs can selectively scavenge the very small microflocs at a rate faster than that for the medium size flocs.

3. LABORATORY EXPERIMENTS

A joint series of laboratory experiments were conducted in a grid tank at the Laboratoire des Ecoulements Geophysiques et Industriels (LEGI) in Grenoble, France, with the aim of determining the conditions required to generate a Concentrated Benthic Suspension layer as a function of turbulence. During these experiments, measurements of floc sizes, settling velocities and effective densities were made.

The experimental programme used natural mud from both the Tamar (series T) and Gironde estuaries (series Gi), plus mud from the Gironde estuary which had the organic matter removed by chemical pre-treatment (series Gt). Each mud was pre-sieved at 100 μ, and mixed into separate base concentrations of 200 mg/l, 600 mg/l, 1 g/l (1.8 g/l for Gironde estuary natural mud), and 5 g/l (Tamar mud only). These mud slurries were then decanted into the LEGI grid tank and allowed to attain equilibrium with the induced turbulent shear stress of the oscillating grid which was located just above the base of the tank. An acoustic velocity probe measured profiles of the variations in turbulent shear stress during the experiments.

The principle range of turbulent shear used for the grid tank experiments was 5.7 to 16.6 s^{-1}. Van Leussen (1994) stated that an rms of the gradient in turbulent velocity fluctuations (G) of 0.1 to 1.0 s^{-1} was representative of slack water, whilst the region between 1.0 and 10 s^{-1} tended to contribute more to the aggregation growth. G-values beyond 10 s^{-1} were indicative of highly turbulent conditions, which could be expected to occur in the bottom boundary layer during periods of high current velocities.

After a period of 40 min, flocs were withdrawn from the water column by use of a vertically held pipette. They were extracted at three different depths. The flocs were then quickly transferred to a settling column where they could pass from the pipette into a positively charged saline solution. Here, the settling velocities were determined by a miniature underwater video camera.

There was a great deal of similarity in the patterns displayed by the LEGI macroflocs and those observed in-situ. For the Tamar mud, the 200 mg/l base concentration was representative of the advection of the main body of the turbidity maximum through the upper Tamar estuary during neap conditions. At the highest G of 16.6 s^{-1} the largest floc size was 210 μ, which was slightly less than the Kolmogorov eddy size of 251 μm. A fairly even division of floc dry mass between the macroflocs and microflocs was observed. The high velocity particle collisions resulted in both floc groups having settling velocities of 0.6 mm/s. Those experiments with turbulent shear of about 7.7 s^{-1} were a closer reflection of in-situ turbulent shear stress in the upper Tamar, at times of maximum entrainment from the bed. For this situation, the improved coagulation due to the less aggressive inter-particulate impacts raised the macrofloc settling velocity to 1.8 mm/s, and the macroflocs constituting 64 % of the floc mass, and 80 % of the mass settling flux. This is a very similar distribution to that displayed by Tamar estuary neap tide flocs.

At the higher ambient concentrations, equivalent to the Tamar mud CBS layer, the maximum floc diameter was approximately double the corresponding eddy size (400 - 430 mμ). The shear stress of 6.5 s^{-1} combined with an SPM of 8.2 g/l, effectively stimulated aggregate formation to such a high degree, that only 3 % of the floc mass was in the microfloc range. Furthermore, the macroflocs had a settling velocity of 5.75 mm/s. When comparing this to the in-situ Tamar estuary flocs, apart from a 2 – 3 % decrease in the dry floc mass division (with respect to the in-situ macroflocs), the macrofloc settling rates and contribution to the MSF are virtually identical. This demonstrated that the grid tank and settling column experimental set-up simulated in-situ conditions under laboratory conditions with a high level of accuracy and repeatability.

4. SIMPLE FLOCCULATION MODEL

A model description was developed at Delft Hydraulics/Delft Technical University to describe the evolution of floc sizes in a turbulent environment (Winterwerp, 1998). The relation between mass and volumetric concentrations is provided by a fractal description of the mud flocs, implying a power-law behaviour of various mud properties. From this description a new formulation for the settling velocity as a function of floc size is derived, which is consistent with Stokes' law in the case of massive particles with a fractal dimension 3 (e.g. sand), and which agrees well with empirical data from the literature. The hindered settling formula by Richardson and Zaki, derived for fairly large, massive particles, does not account for the floc structure typical for cohesive sediment, and a new formula is proposed. This formula also compares well with empirical data from the literature (Winterwerp, 1999).

The evolution of floc size, hence settling velocity, in a turbulent environment is described through a new flocculation model in a Eulerian framework, that includes the effects of turbulence-induced aggregation and floc breakup. This model predicts that the growth of flocs in open water systems can seriously be limited by a limited residence time of these flocs in the water column, as a result of small water depth and/or of long flocculation times. The model also predicts gelling concentrations in estuaries of the right order of magnitude; gelling values in coastal areas under storm conditions are grossly under-predicted at present. Though this flocculation model compares well with the scarcely available empirical data and yields qualitatively sound results, extensive further validation against comprehensive data sets is required before the model can be deployed with confidence for practical applications.

The various process formulations have been implemented in a one-dimensional vertical numerical model, referred to as the 1DV POINT MODEL. This allows the simulation of the effects of flocculation and gellation, settling, hindered settling and lutocline formation, consolidation, remixing, and sediment-induced buoyancy effects on the turbulence field of high-concentrated mud suspensions in estuarine and coastal environments. The various formulations also provide the relevant scaling parameters of the processes governing the dynamics and appearances of high-concentrated mud suspensions under a wide variety of conditions (Winterwerp, 1999, 2002).

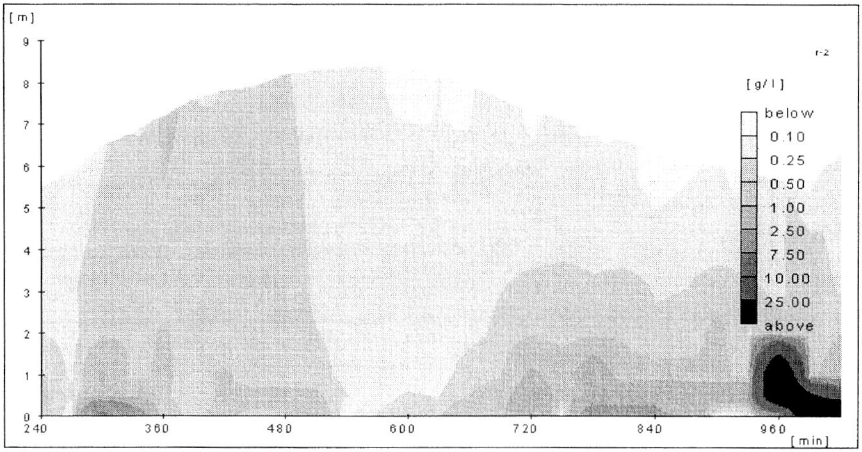

Figure 6. Measured vertical suspended sediment concentrations in the Ems Estuary

Figure 7.Simulation including constant settling velocity and sediment induced buoyancy effects.

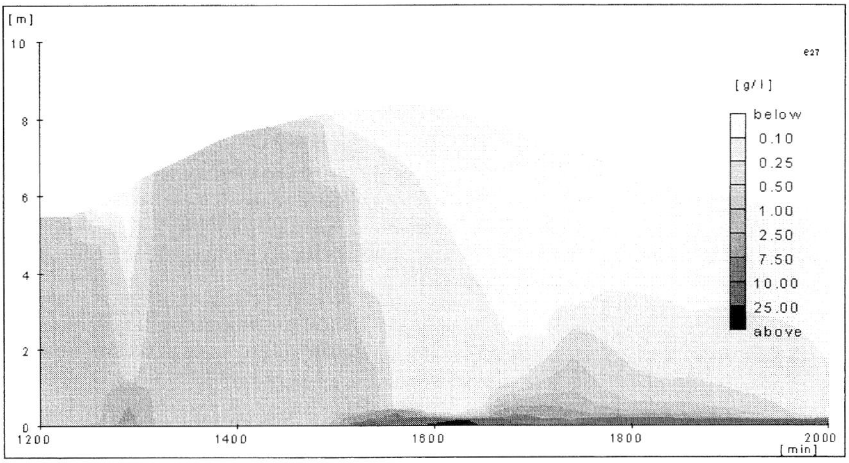

Figure 8. Full flocculation model and sediment induced buoyancy effects.

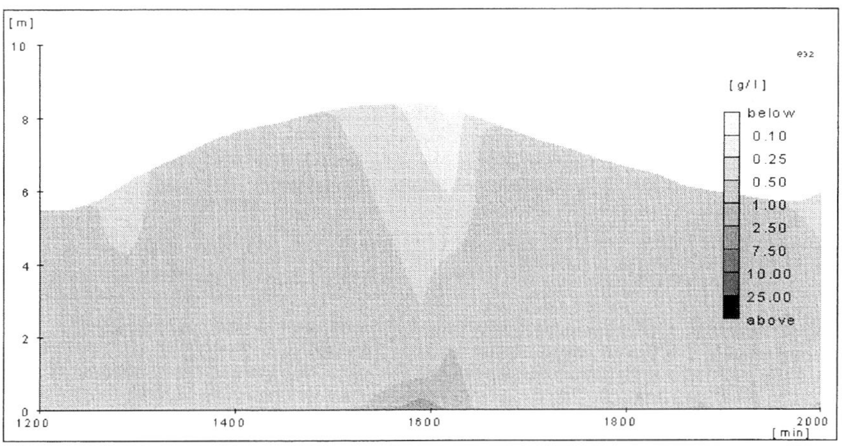

Figure 9. Full flocculation model without sediment induced buoyancy effects.

This model concept was validated through application to well-documented laboratory experiments and field measurements: flume experiments with sand to study the velocity profile in sediment-laden flow, consolidation and entrainment experiments with cohesive sediment in an annular flume, measurements both in and upstream of the turbidity maximum in the Ems Estuary, and measurements during storm conditions in the Maasmond area, the entrance to the Port of Rotterdam. In general, the measurements and model simulations agree well.

The suspended sediment concentration profiles in general and the rapid settling around slack water in particular, as observed in the Ems River, can only be simulated properly if the effects of both flocculation and sediment-induced buoyancy are accounted for in the 1DV POINT MODEL. This is shown in Figures 6, 7, 8 and 9. Figure 6 presents the measured vertical suspendedsediment concentration in the Ems River. Figure 7 - 9 present the results of the model simulations: Figure 7 constant settling velocity and sediment-induced buoyancy effects; Figure 8 full flocculation model and sediment-induced buoyancy effects, and Figure 9 full flocculation model, but without sediment-induced buoyancy effects.

5. DISCUSSION AND CONCLUSIONS

The 1DV-simulations of the sediment dynamics in the turbidity maximum in the Ems-estuary, The Netherlands, showed the importance of a proper formulation of the flocculation behaviour of cohesive sediment. Without such a formulation it was not possible to reproduce the rapid decrease in SPM around slack water, nor the stratified concentration profile during ebbing tide. This conclusion stresses the need for further research into flocculation processes and the collection of proper (field) data.

In the present study, an extensive field campaign has been carried out with exactly this purpose. The data show a considerable variation in floc size and settling velocity over the tidal cycle, and for neap and spring tide.

However, the campaign showed also that it is quite difficult to collect such data in the field. The measurements are biased by advective processes, transporting sediment from other locations to the measuring stations, and by (local) erosion of the riverbed. It is likely that the latter process yields flocs of much larger density, as observed during the surveys. Moreover, a significant variation in the organic content of the flocs was measured, which also may affect floc size and floc structure.

The field measurements also revealed a pronounced variation in floc size distribution during the tidal cycle and during the spring-neap cycle. At present, the effects of such a floc size distribution can not be accounted for in the simple flocculation model, and more research is required for a proper description of size distribution.

The floc sizes obtained in the laboratory experiments were very similar to those observed in the field. Hence, one may conclude that the laboratory is still a suitable environment for cohesive sediment research in general, and for the study on flocculation processes in particular, when the turbulence is adequately controlled.

The field measurements also revealed an important interaction between the suspended sediment concentration and the turbulent flow field, resulting in drag reduction. Further studies with the aforementioned 1DV POINT MODEL revealed that this interaction also affects the flocculation process itself: large positive gradients in settling velocity can result in very stable vertical SPM-profiles (e.g. Winterwerp, 2001).

ACKNOWLEDGEMENTS

This work was partially funded by the European Commission, Directorate General XII for Science, Research & Development through the COSINUS-project within the framework of the MAST-3 programme, contract MASC3-CT97-0082 and by corporate research funds from the various research institutes involved.

REFERENCES

Best, J.L. and Leeder, M.R. 1993. Drag reduction in turbulent muddy seawater flows and some sedimentary consequences. *Sedimentology*,. 40. 1129-1137.

Dearnaley, M.P. 1991. Flocculation and settling of cohesive sediments. HR Wallingford, Report No. SR272.

Dyer, K.R. 1989. Sediment processes in estuaries: future research requirements. *Journal. Geophysical Research*. 94. 14327-14339.

Dyer, K.R., Bale, A.J., Christie, M.C., Feates, N., Jones, S. and Manning, A.J. 2000a, The dynamics of suspended sediment in an estuarine turbidity maximum. Proceedings *INTERCOH-2000, Elsevier, Coastal and Estuarine Fine Sediment Processes*, ed. J.C. Winterwerp and C. Kranenburg, this volume.

Dyer, K.R., Bale, A.J., Christie, M.C., Feates, N., Jones, S. and Manning, A.J. 2000b. The properties of suspended sediment in an estuarine turbidity maximum. Proceedings *INTERCOH-2000, Elsevier, Coastal and Estuarine Fine Sediment Processes*, ed. J.C. Winterwerp and C. Kranenburg, this volume.

Fennessy, M.J., Dyer, K.R. and Huntley, D.A. 1994a. INNSEV: an instrument to measure the size and settling velocities of flocs in –situ. Marine Geology. 117 107-117.

Fennessy, M.J., Dyer, K.R. and Huntley, D.A. 1994b. Size and settling velocity distributions of flocs in the Tamar Estaury during a tidal cycle. Neth. Jour. Aquat. Ecology. 28, 275-282.

Gratiot, N., 2000, Étude expérimentale de la fromation des couches de crème de vase turbulentes, PhD-thesis, L'Université Joseph-Fourier, Grenoble, France (in French).

Jones, S., Jago, C.F., Fox, D. and Bale, A.J. 2000. COSINUS Final report of UWB subcontractor.

Law, D.J. and Bale, A.J. 1998. In-situ characterisation of suspended particles using focused-beam laser reflectance particle sizing. In: Black, K.S., Paterson, D.M. and Cramp, A. (eds) *Sedimentary Processes in the Intertidal Zone*. Geological Society London. Special Publication 139. 57-68.

Li, M.Z. and Gust, G. 2000. Boundary layer dynamics and drag reduction in flows of high cohesive sediment suspensions. Sedimentology. 47. 71-86.

Manning, A.J. 2000. A study of the effect of turbulence on the properties of flocculated mud. PhD Thesis, University of Plymouth. (In prep).

Manning, A.J. and Dyer, K.R. 2000. A comparison of floc properties observed during neap and spring tidal conditions. Proceedings *INTERCOH-2000, Elsevier, Coastal and Estuarine Fine Sediment Processes*, ed. J.C. Winterwerp and C. Kranenburg, this volume.

Peterson, O., Vested, H.J., Manning, A., Christie, M.C. and Dyer, K.R. 2000. Proceedings *INTERCOH-2000, Elsevier, Coastal and Estuarine Fine Sediment Processes*, ed. J.C. Winterwerp and C. Kranenburg, this volume.

Underwood, G.J.C., Paterson, D.M. and Parkes, R.J. 1995. The measurement of microbial carbohydrate exopolymers from intertidal sediments. Limnology and Oceanography. 40. 1243-1253.

Van Leussen, W. 1994. Estuarine macroflocs and their role in fine-grained sediment transport. PhD Thesis, University of Utrecht, The Netherlands.

Winterwerp, J.C., 1998, A simple model for turbulence induced flocculation of cohesive sediment, IAHR, *Journal of Hydraulic Engineering*, Vol 36, No 3, pp 309-326.

Winterwerp, J.C., 1999, On the dynamics of high-concentrated mud suspensions, PhD thesis, Delft University of Technology, The Netherlands, also Delft University of Technology, Faculty of Civil Engineering and Geosciences, Communications on Hydraulics and Geotechnical Engineering, Report 99-3, ISSN 0169-6548.

Winterwerp, J.C., 2001a, Stratification of mud suspensions by buoyancy and flocculation effects, *Proceedings of the XXIX IAHR Congress*, September 2001, Beijing, China, Theme D, pp 235-241.

Winterwerp, 2001b, Stratification effects by cohesive and non-cohesive sediment, *Journal of Geophysical Research*, Vol 106, No C10, pp 22559-22574.

Winterwerp, J.C., 2002, Flocculation and fluid mud formation, *Continental Shelf Research*, in press.

Fine Sediment Dynamics in the Marine Environment
J.C. Winterwerp and C. Kranenburg (Editors)

Dynamics of Concentrated Benthic Suspension Layers

J.C. Winterwerp[a,b], A.W. Bruens[b], N. Gratiot[c], C. Kranenburg[b], M. Mory[d] and E.A. Toorman[e]

[a]WL | delft hydraulics, PO 2600 MH Delft, The Netherlands,
[b]Delft University of Technology, Faculty of Civil Eng. and Geosciences, The Netherlands,
[c]Laboratoire des Ecoulements Géophysiques et Industriels, Grenoble, France,
[d]Ecole Nationale Superieure en Genie des Technologies Industrielles, Pau, France,
[e]Katholieke Universiteit Leuven, Civil Engineering Department, Belgium

This paper describes the dynamics of Concentrated Benthic Suspensions (CBS). CBS is defined as a suspension of cohesive sediment with a notable interaction between the sediment and the turbulent flow field through buoyancy effects, but still displaying near-Newtonian behaviour.

The mechanisms which distinguish CBS from low-concentrated suspensions are described, and the focus is on the (hindered) settling and mixing processes. Experiments were carried out in an oscillating grid tank and in an rotating annular flume, simulating entrainment and mixing associated with the turbulent CBS-layer, as occur in tide-driven flows. It is shown that CBS can be modelled as a viscous fluid, and that the entrainment rates quantitatively match relations described in the literature on salt-fresh water induced stratified systems.

Numerical simulations with one-dimensional vertical models using k-ε and Prandtl mixing length turbulence closures were carried out for hypothetical open channel flows to study the behaviour of CBS through sensitivity analyses. It is shown that high-concentrated mud suspensions may become saturated, generating a CBS-layer prior to the formation of fluid mud.

KEY-WORDS
Concentrated Benthic Suspensions, fluid mud, entrainment, mixing, buyoancy

1. INTRODUCTION

Large siltation rates of navigational channels and harbour basins are often attributed to high-concentrated mud suspensions, and a proper physical description of such suspensions is a necessary condition for a cost-effective maintenance strategy for these fairways. High-concentrated mud suspensions are also encountered in the turbidity maxima of most estuaries, in large stretches of many turbid rivers all over the world and around mud banks in coastal areas.

The COSINUS-project, executed under the framework of the European MAST3 research programme, is aimed at enhancing our understanding of the behaviour of such high-concentrated suspensions, and developing mathematical formulations to describe the relevant physical processes, together with their parameterisations, to be incorporated in mathematical models for managing authorities and engineering consultants. The present paper describes the work carried out under Task C of the project, relating to the dynamics of concentrated near-bed suspensions of cohesive sediment, referred to as Concentrated Benthic Suspensions (CBS). The interaction between sediment and turbulent flow is the characteristic feature of CBS-suspensions, and its modelling is also addressed in a companion paper (Toorman et al., 2001).

Suspensions of cohesive sediment can be classified through their concentration and the flow conditions. Increasing the amount of sediment in suspension, characterised by the overall Reynolds number and the Rouse number, a CBS is obtained upon deposition. As noted, CBS is characterised by a notable interaction between the suspension and the turbulent flow field, and an overall Richardson number is the governing parameter. At higher concentrations, a fluid mud layer occurs, which flows under laminar conditions, and its behaviour is governed by an effective Reynolds number, accounting for plastic yield effects in the mud. At still larger concentrations a stagnant consolidating bed is formed in which the yield strength exceeds the applied stresses. Figure 1 presents a diagram of these four classes, including the relevant exchange processes between the various mud phases. Note that the transfer from one class to another is gradual in general, and that a decrease in flow velocity can result in a sequence similar to the effects of increasing sediment concentration.

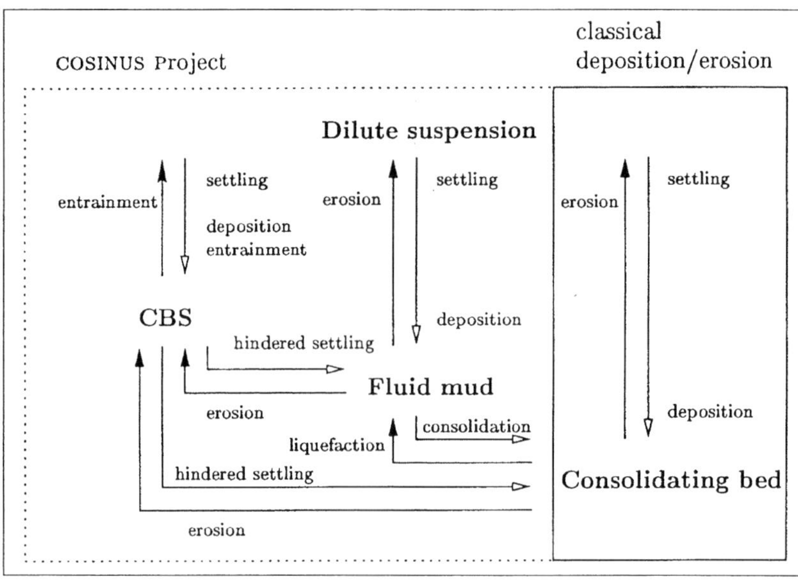

Figure 1. Four classes of suspension and exchange processes.

Whether or not these classes occur depends on the time scales of the physical processes. For instance, the time scales for settling and flocculation determine whether a fluid mud layer is formed during slack water, and the time scale for consolidation determine whether re-entrainment or erosion processes will occur during accelerating flow.

The present paper focuses on the behaviour of CBS-layers, and the processes of entrainment, (hindered) settling and vertical mixing are especially addressed. In Section 2 a phenomenological description of these layers is given, comparing CBS-dynamics with other stratified flow phenomena and with the behaviour of suspensions of non-cohesive sediment. In Section 3 measurements in an oscillating grid tank on equilibrium conditions are presented, together with a series of entrainment experiments carried out in a rotating annular flume. Theoretical and numerical analyses of the behaviour of CBS-layers around and beyond equilibrium are discussed in Section 4. Finally, discussion and an overview of the conclusions is presented in Section 5.

The work described in Section 3.1 was carried out at LEGI, Grenoble, the work presented in Section 4.2 at the Katholieke Universiteit Leuven, and the work described in the Sections 3.2 and 4.1 in Delft.

2. PHENOMENOLOGICAL DESCRIPTION OF CBS-LAYERS

A Concentrated Benthic Suspension can be considered as a turbidity current. Both can be self-sustaining provided turbulence is generated continuously to keep the particles in suspension. It is illustrative to discuss first the behaviour of two miscible fluids of different density, for instance a salt water layer below a fresh water layer. If either of the two, or both layers are sufficiently turbulent, ultimately the two layers will mix completely. The mixing process of such a stratified system is a function of a Richardson number, often characterised as a bulk Richardson number $Ri_*(\equiv B/u_*^2)$, which is the ratio of the total buoyancy $B \equiv \Delta gh$ of the stratified system and the kinetic energy available for mixing, characterised by the friction velocity u_*. Here Δ is the relative density of the sediment particles, g is gravitational acceleration and h is the total water depth.

In the case of a suspension, complete mixing does not necessarily occur, as the particles tend to restore the stratified condition by settling under the effect of gravity. Hence, an equilibrium condition, in which the sediment is not fully mixed over the entire water column, can occur as the result of a balance between the available turbulent mixing energy produced by the bed shear stress and the work to be done to keep the particles in suspension. It can be hypothesised that at equilibrium the effect of the settling velocity of the particles W_s at the interface equals the effect of the entrainment velocity w_e of the interface. The suspended sediment concentration at equilibrium C_e is therefore a function of a reference fluid density ρ, Ri_*, W_s and w_e, or, because it can be shown that $w_e = f(Ri_*)$:

$$C_e/\rho = F_1(Ri_*, \beta) \tag{1}$$

where $\beta(\equiv \sigma_T W_s/\kappa u_*)$ is the Rouse number. Based on a more formal analysis one can show that (Winterwerp, 2001):

$$\frac{C_e}{\rho} = K_s \frac{u_*^3}{\Delta g h W_s} \tag{2}$$

where K_s is a coefficient of order one. C_e can be regarded as a measure for the total load that can be carried by a turbulent shear flow. If by some cause the flow velocity were to decrease, or the water depth were to increase, C_e would decrease, as a result of which part of the suspension would settle. During this process strong buoyancy destruction near the bed occurs, which may cause a thickening of the viscous sublayer. In case of non-cohesive sediment a rigid bed is formed rapidly at which turbulence production remains possible. Thus a new equilibrium is established at a lower load.

It is interesting to note that (2) is very similar to the so-called Knapp-Bagnold criterion (Parker et al., 1986) for the occurrence of submarine turbidity currents:

$$C < \frac{\rho}{\Delta g} \frac{U_t u_*^2}{\delta W_s} \tag{3}$$

where U_t is the mean flow velocity of the turbidity current and δ its thickness. This would be a necessary condition for a self-sustaining turbidity current; it is also known as the auto-suspension criterion.

Next, a suspension of cohesive sediment in equilibrium is considered. If the particles settle because of a decrease in flow velocity or an increase in depth no rigid bed is formed, but a layer of CBS upon which fluid mud occurs, as the large cohesive sediment flocs form a space-filling network at relative low mass concentrations. At the water - fluid mud interface little or no turbulence production is possible. Hence, as the turbulent energy for mixing decreases, C_e decreases further etc. This results in a "snowballing" effect, leading to a complete collapse of the vertical concentration profile and the turbulence field. For cohesive sediments, the equilibrium concentration C_e can therefore be regarded as a saturation concentration, denoted as C_s.

In the next sections the generation of an equilibrium condition for CBS, its mixing characteristics, and its response to changing flow conditions are discussed.

3. EXPERIMENTS ON CBS

Experiments on CBS were conducted in a rectangular tank with an oscillating grid and in an rotating annular flume. It should be noted that the turbulence properties in these two facilities are not identical. In the annular flume, turbulence is generated by shear flow over the entire water column, though the largest production is near the wall. Moreover, the near-wall turbulence is advected by the larger eddies (of the size of the water depth) throughout the water column.

In the tank, the turbulence is generated by an oscillating grid. Near the grid, the large scale eddies have the size of the grid mesh. The turbulent kinetic energy is transported away from the grid higher into the water column by diffusion, increasing more or less linearly in size away from the grid. As a result, the decay of turbulence off the grid must be included in the functional relation (1), which becomes for grid generated turbulence (e.g. Huppert et al., 1995):

$$C_e/\rho = F_2(Ri_*, u_*/W_s, z_0/\delta) \tag{4}$$

,where z_0 is a reference length scale related to the (location of the) oscillating grid, accounting for the decay of turbulence and δ is the thickness of the mixing layer (CBS-layer).

Note that similar several experiments with fresh and salt water and suspensions with non-cohesives have been reported in the literature, e.g. E and Hopfinger (1987) and Huppert et al. (1995). However, literature on experiments with cohesive sediment are rare. Tsai and Lick (1987) used an oscillating grid tank to establish the erosion rate of loosely consolidated mud, and Wolanski et al. (1989) observed steady CBS-layers in mud suspensions in a laboratory experiment in which turbulence was produced by the vertical oscillation along the vertical wall of annular rings having a regular spacing. The CBS thickness and concentration were found to be related to the grid oscillation frequency through a bulk Richardson number. As floc formation and the effects of hindered settling are important in suspensions of cohesive sediments, it was decided to carry out a new series of oscillating grid tank experiments with suspensions of cohesive sediment.

3.1. Experiments in an oscillating grid tank

Experiments were carried out in a square tank (53 cm by 53 cm, 40 cm water depth) in which turbulence was produced by an oscillating grid (see Gratiot et al., 2001, for a detailed description of the apparatus and the procedure). Sediment was led into the tank, while the grid oscillated, with an initial mean concentration C_0. An equilibrium concentration was attained after some time. By increasing C_0 step by step, the formation of a lutocline separating the CBS-layer from the upper layer was observed when C_0 exceeded 2 g/l. Experiments were carried out with organic-rich mud from the Tamar estuary and non-organic mud form the Gironde estuary.

symbol	mud	salinity S [ppt]	grid frequency F [Hz]	initial concentration C_0 [g/l]	concentration in CBS C [g/l]
'□'	Gironde	0	3	85.0	202
'x'	Gironde	0	3	61.0	149
'∇'	Gironde	0	6	30.0	51
'o'	Tamar	16.5	4	3.6	5.1
'.'	Gironde	16.5	4	2.7	3.8

Table 1. Experimental conditions and explanation of symbols.

For equilibrium conditions, vertical profiles of the sediment concentration $C(z)$, the turbulent rms-velocity $u(z)$ and the integral length-scale $\ell(z)$ were measured from the energy spectrum. Figure 2 shows a measured vertical concentration profile. For all conditions, the time-averaged sediment concentration was found to be uniform within the lower 20 cm of the CBS layer. Figure 3 shows vertical profiles of the rms-turbulent velocity (the various symbols are explained in Table 1). The turbulent velocity decreases with increasing distance from the grid, as anticipated from (4). There were no observations suggesting enhanced decay of the turbulent velocity with increasing sediment concentration.

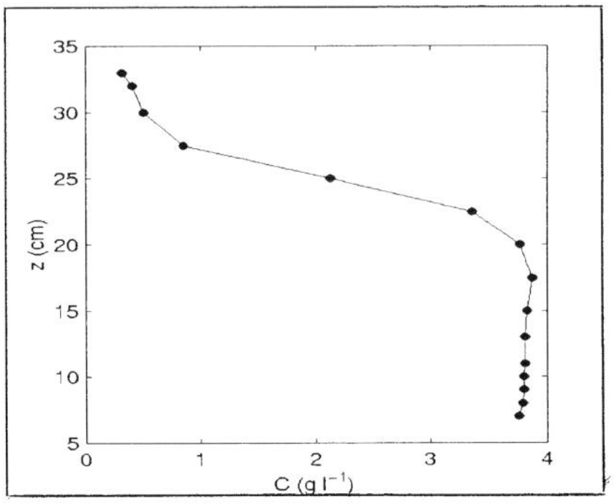

Figure 2. Vertical profile of concentration obtained with Gironde mud for $C_0 = 3.6$ g/l.

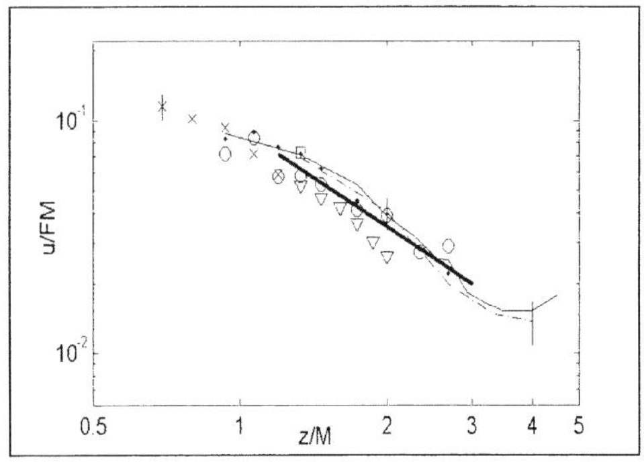

Figure 3. Variation of the turbulent velocity u with distance to the grid z (F is the frequency of grid oscillation, M is the grid mesh size). Full line and dot-dashed line are measurements of turbulent velocity in clear water.

Measurements of the settling velocity W_s were also carried out. Depending on the concentration, the settling velocity was found to vary by two orders of magnitude. The measured values of u, ℓ, C and W_s were used to estimate the flux Richardson number below the lutocline, expressed as (e.g. Gratiot et al., 2001):

$$Ri_f = \frac{\rho_s - \rho_w}{\rho_s} \frac{W_s}{u} \frac{gC\ell}{\rho_w u^2} \Bigg/ \left(A + \frac{\rho_s - \rho_w}{\rho_s} \frac{W_s}{u} \frac{gC\ell}{\rho_w u^2} \right) \tag{5}$$

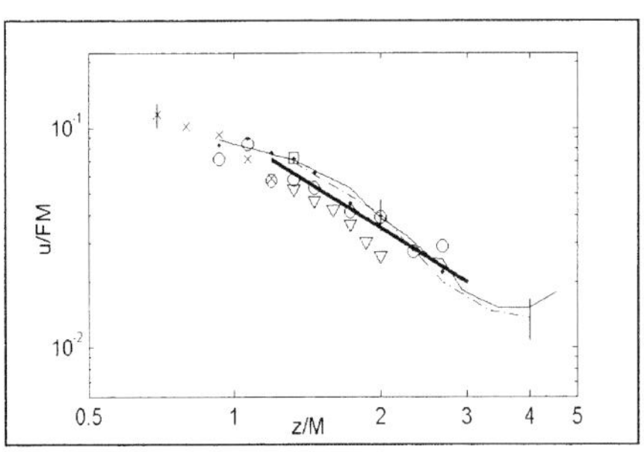

Figure 4. Variation of the flux Richardson number a the lutocline with concentration in the CBS-layer.

For a turbulent flow with no mean velocity component. A is a constant of order one which appears in the rate of dissipation of the kinetic energy $\varepsilon = A\rho_w u^3 / \ell$. Equation (5) gives a simple relationship between the flux Richardson number, the bulk Richardson number and the Rouse number. The value of the flux Richardson number at the lutocline is shown in Figure 4 as a function of the concentration inside the CBS-layer. The flux Richardson number is larger than 0.5 at the lutocline when the concentration in the CBS is less than 50 g/l, and it is of the order of 0.2 when the concentration is above 100 g/l.

3.2. Experiments in a rotating annular flume

The rotating annular flume set-up, with a rotating top lid driving the flow, was appropriately modified to study the mixing/entrainment processes of CBS. For this purpose, the top lid of the flume was replaced by a rotating base plate, supported by streamlined rods. The flume has an outer diameter of 4 m, a width of 0.3 m, and the water height was generally set at 0.25 m. The base plate floated on a bath of mercury to prevent flow around the base plate and the settling of sediment below the base plate.

Base plate and flume could be rotated in opposite directions to minimise the effect of secondary currents. An equation was derived analytically on the basis of the tangential flow momentum equation in the flume to establish the ratio of rotational speeds at which secondary current effects are minimal. This equation was verified with experiments in the flume, visualising the flow and secondary current effects, both for homogeneous and stratified flow conditions. The conditions for homogeneous flow were further substantiated with numerical analyses using the commercial software package PHOENICS. A detailed description of this set-up and the various analyses is given by Bruens et al. (2001).

The experimental procedure was as follows. After the production of a stratified flow in the flume (either salt-fresh water, or turbid-fresh water), flume and base plate were slowly accelerated in the same direction in such a way that the fluid itself was accelerated as a rigid body. Then, when the required rotational speed of the flume was obtained after a period of several tens of minutes, the base plate was accelerated fairly rapidly in the opposite direction until its required rotational speed was obtained. At that point in time the entrainment experiment (measurements) started. In this way, a shear flow in the rotating fluid was generated as long as the denser fluid accelerated through bed friction mainly. Such experiments ended when the stratified fluid was completely mixed over the water depth. Note that, because of the accelerating flow, no equilibrium CBS-height was obtained in this configuration.

During these experiments the following parameters were measured: rotational speeds of flume and base plate, height of the interface, salinity of the upper layer (only insalt-fresh water experiments), (turbulent) flow velocities in longitudinal and vertical directions at two heights and suspended sediment concentration at four heights (only for CBS-experiments).

The friction velocity u_* was not directly measured, but obtained by using the logarithmic law of the wall, a quadratic friction law, and from numerical simulations with PHOENICS. All these approaches yielded similar values.

In the first series of experiments, the mixing of a salt water layer below a fresh water layer in the flume was studied under different flow conditions (in terms of the overall Richardson numbers Ri_*). Next, experiments with China clay in mildly saline water (5 ppt) were conducted. A suspension of 40 g/l concentration was homogeneously mixed over the water column. This suspension was allowed to settle during two hours, resulting in a CBS-thickness of about 0.1 m and a concentration ranging from about 50 g/l at its top to about 200 g/l at its base. Six series of experiments were carried out at five values of Ri_* ranging from 81 to 188 by varying the rotational speeds of flume and base plate (in each case at the optimal ratio).

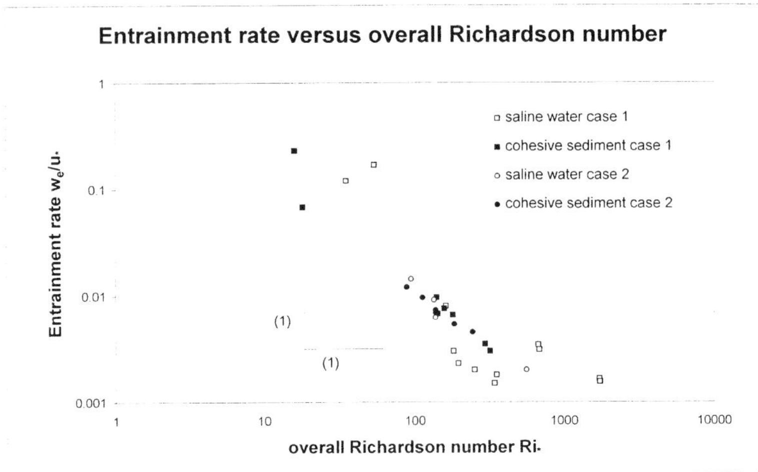

Figure 5. Entrainment diagram from experiments in the rotating annular flume (case 1: upper layer turbulent; case 2: lower layer turbulent).

The results of the entrainment experiments are summarised in Figure 5, where the initial entrainment velocity w_e on the vertical axis is made non-dimensional with the shear velocity u_*. For details the reader is referred to Bruens et al. (2001). This figure also contains data from previous experiments (e.g. Winterwerp and Kranenburg, 1997) carried out in the same rotating annular flume, but with rotating top plate, and it is shown that the results match fairly well. It can be concluded that, as long as the mud has not attained yield strength, a CBS behaves as a viscous fluid.

4. THEORETICAL ANALYSES

The interaction between cohesive sediment and the turbulent flow field may cause a significant reduction in vertical mixing and overall hydraulic resistance. Reduction in vertical mixing was measured in the field by for instance West and Oduyemi (1989) and Van der Ham (1999). A reduction in the overall hydraulic resistance was observed in a number of high-concentrated rivers in China (Dong et al., 1997, Guan et al., 1998 and Wang et al., 1998) and on the Amazon shelf (Beardsley et al., 1995). Theoretical studies on this drag reduction were reported by Zhou and Ni (1995) and Toorman (1999).

Sediment-induced overall drag reduction has a significant effect on the flow field. If the CBS-occurrences are spatially confined, as in navigational channels, at small mud banks, etc., the overall flow rate will remain constant, and the reduction in drag will cause changes in the vertical profiles of flow velocity and mixing.

If the CBS-occurrences are unconfined, as in the large rivers in China and Brazil, the overall flow rate does not necessarily remain constant, and the flow may accelerate, affecting total flow rate, velocity profile and vertical mixing.

In the next two sections distinction is made between these two situations, as this has a large effect on the flow-sediment interaction.

4.1. Confined CBS-occurrences: constant flow rate

Confined occurrence of CBS can be found in navigational channels, in the turbidity maxima in estuaries, or above small mud banks. The flow rate is governed by large scale overall processes, and is not (or only slightly) affected by local CBS-occurrences. In this case a positive feed-back, described in Section 2 of this paper, can be expected, resulting in a total collapse of the vertical concentration profile and turbulence field.

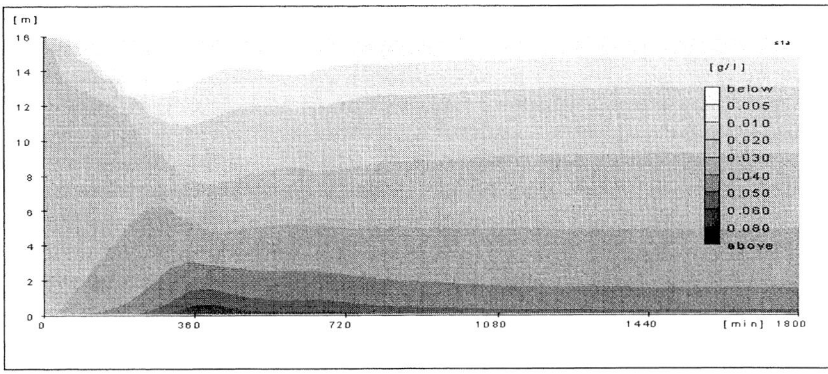

Figure 6. Isolutals for a saturated ($C_0 = 0.023$ g/l) suspension in open-channel flow.

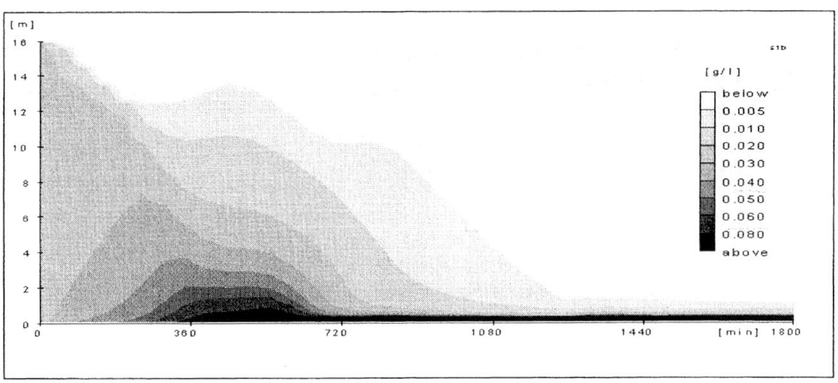

Figure 7. Isolutals for a super-saturated ($C_0 = 0.024$ g/l) suspension in open-channel flow.

Winterwerp (1998, 1999) analysed this feed-back process, starting from an analysis presented by Teisson et al. (1992). The basic idea behind this analysis is that experiments on stratified flow reported in the literature show that a turbulent shear flow collapses when the flux

Richardson number Ri_f exceeds a critical value. The sediment load that is transported just prior to this collapse can be regarded as the transport capacity of the turbulent flow. The related depth-mean suspended sediment concentration is referred to as the saturation concentration C_s in order to differentiate this case from the reversible condition for non-cohesive sediment, where the transport capacity is characterised by an equilibrium concentration C_e. The scaling law for C_s is presented in Section 2, e.g. equ (2).

This saturation behaviour was further studied with the numerical 1DV POINT MODEL, derived from a full three-dimensional water movement and sediment transport model in which all horizontal gradients have been stripped, except for the horizontal pressure gradient. The standard k-ε turbulence model is implemented, including a sediment-induced buoyancy destruction term. The effect of the sediment on the bed boundary condition is implicitly taken into account through the dependency of the friction velocity on the gradients in suspended sediment concentration (fluid density).

Equilibrium and saturation conditions, were studied with the 1DV POINT MODEL for a hypothetical open channel flow having various depths, flow velocities and sediment settling velocity by increasing the total amount of sediment in the flow in small steps, starting from an initially homogeneous concentration profile C_0. The results for two such simulations are shown in Figure 6 and 7 for a water depth h = 16 m, a depth-mean flow velocity U = 0.2 m/s and a settling velocity W_s = 0.5 mm/s. It is observed that for a sediment concentration C_0 = 0.023 g/l a more or less Rousean concentration profile is obtained, whereas a small increase in C_0 to C_0 = 0.024 g/l shows an almost complete collapse of the vertical profile. Above the lutocline, only little turbulence is produced by the local velocity gradients. Apparently, the 0.023 g/l case represents the sediment transport capacity of the flow, whereas the 0.024 g/l case represents super-saturated conditions.

Further simulations showed that the numerical results follow the functional relation (2) properly. These results are not presented herein, and the reader is referred to Winterwerp (1999, 2001).

Kranenburg (1998) studied the equilibrium conditions of CBS-layers with a Prandtl mixing length model, using Munk-Anderson-like damping functions:

$$\overline{u'w'} = -\ell^2 \left|\frac{\partial u}{\partial z}\right| \frac{\partial u}{\partial z} F(Ri) \;, \quad \overline{w'c'} = -\frac{\ell^2}{\sigma_T} \left|\frac{\partial u}{\partial z}\right| \frac{\partial c}{\partial z} G(Ri) \;, \quad \text{with}$$
$$F(Ri) = (1 + ARi)^{-a} \;, \quad G(Ri) = (1 + BRi)^{-b} \tag{6}$$

where $\ell(z)$ is the mixing length. Kranenburg argued that for σ_T = 0.7, A = B = 2.4, and the exponential coefficients should become a = 2 and b = 4, respectively. These damping functions were implemented in the one-dimensional vertical momentum and mass balance equation for suspended sediment, which were solved numerically. The damping function F was also explicitly applied to the shear velocity u_* to include buoyancy effects in the boundary conditions. Simulations were carried out at a constant flow rate and for various amounts of sediment, initially mixed homogeneously over the water column at a concentration C_0. It was shown that at low C_0, a more or less Rousean concentration profile was obtained, whereas for $C_0 > C_{crit}$ the concentration profile collapsed. C_{crit} can be regarded as the saturation concentration C_s, which appeared to be close to the value obtained with a k-ε model in the 1DV POINT MODEL, as noted above.

4.2. Unconfined CBS-occurrences: constant energy slope

Numerical experiments similar to those described in the previous section were carried out with a constant energy slope (equivalent to a constant horizontal pressure gradient and a constant bed shear stress). This allows for steady state conditions, under which the model-predicted shear velocity can be matched against the theoretical one, giving an indication of the error caused by the model. It was shown that the standard bottom boundary conditions, which are in use in the current engineering models, overpredict the bed shear stress at increasing stratification. This problem and its solution are discussed in greater detail in a another paper in this volume (Toorman et al., 2001).

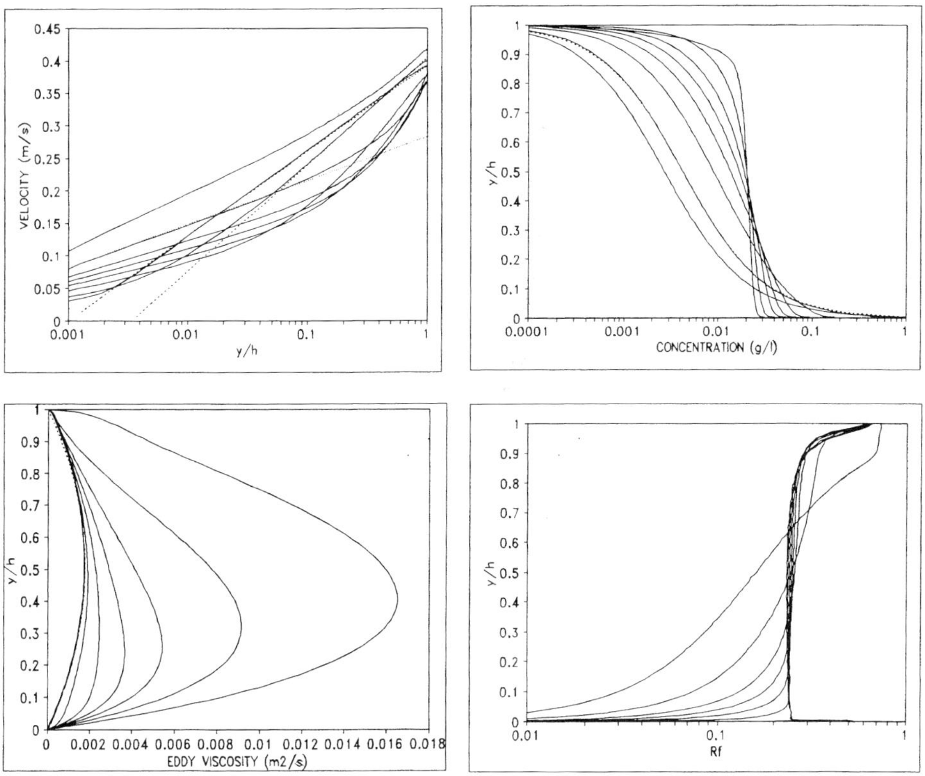

Figure 8. Numerical results for fully-developed sediment-laden turbulent open-channel flow (h = 16 m, W_s = 0.5 mm/s, mean concentration C_m = 37 mg/l) for various shear velocities (u_* = 15.6, 11.8, 9.9, 8.6, 7.9, 6.9, 5.6 and 5 mm/s), using the Munk-Anderson damping functions and the k-ε turbulence model. Profiles of (a) velocity (...: κ = 0.41, --: $\kappa = \sigma_T W_s/u_*$), (b) sediment concentration (--- : Rouse profile with β = 1), (c) eddy viscosity (--- : $\sigma_T W_s\, z(1-z/h)$), and (d) flux Richardson number.

Time-varying isolutals similar to those described in Section 4.1 were computed, yielding qualitatively the same behaviour as described above. The results showed that lutoclines can appear temporarily, forming an interface between nearly clear water at the surface and the CBS-layer. The final steady state stratification depends on the ratio W_s/u_* and the sediment load (Toorman, 1999 & 2001).

In order to better understand the characteristics of CBS-layers, specific attention has been given to steady state calculations in which the influence of input energy (i.e. u_*), settling velocity and sediment load have been studied (Toorman, 1999 & 2001). A typical series of results, where u_* has been varied, is shown in Figure 8. The most noteworthy feature is that, starting from a well-mixed situation at high u_*, the flux Richardson number profile tends to homogenise over the entire water column when decreasing the shear velocity (Figure 8d). A further small decrease of u_* rapidly leads to an apparent total collapse of turbulence at the bottom as soon as $Ri_f = 1$, i.e. turbulence production by shear at the bottom is completely destroyed by buoyancy. What happens physically is a thickening of the viscous sublayer, accompanied by drag reduction (Toorman, 2000).

To describe this process properly, the turbulence models must be extended to include a low-Reynolds module to handle the laminarisation of the flow in the bottom layer, for which a grid refinement and a moving interface are required (Toorman et al., 2001). The value of the flux Richardson number, which corresponds to this condition, lies in the range of 0.2 to 0.25 (when no buoyancy term is used in the dissipation rate equation, and its value is slightly sensitive to the settling velocity). This value corresponds to the empirical critical value for turbulence collapse. Therefore, it seems that the condition $dRi_f/dz = 0$ determines the limiting state at which the minimal energy is provided to keep sediment in suspension. A theoretical proof for this hypothesis, however, is still missing.

Further theoretical analysis of the implementation of the condition $dRi_f/dz = 0$ reveals that the saturated concentration profile is a Rousean profile with Rouse parameter value 1, which is confirmed by the simulations (Figure 8b). These simulations show that any sediment in excess remains in the near-bottom layer where turbulence is no longer fully developed. Hence, Rouse parameter values < 1 correspond to unsaturated and > 1 to supersaturated conditions. The corresponding eddy viscosity profile is parabolic (Figure 8c) and the corresponding velocity profile is logarithmic (Figure 8a), but the value of the von Karman coefficient is reduced to $\sigma_T W_s/u_*$ (where σ_T is the turbulent Schmidt number). Consequently, the exponents of u_* and W_s in eq.(2) would become 4 and 2 respectively (Toorman, 2001).

The sensitivity study on the influence of the shear velocity also shows that the depth-averaged velocity decreases with u_* down to a minimum and then increases again due to drag reduction. This implies that for the same mean flow velocity two steady state solutions exist at a relatively low and high sediment load. This explains why simulations at constant flow rate, as described in section 4.1, performed with two different models for the same input conditions yield a saturated state in one model, but unsaturated in another (Violeau et al., 2001). Further analysis shows that the history of the bed shear stress is an important parameter, which seems to be strongly affected by the numerical scheme used and boundary conditions. Therefore, the interpretation of model results near saturation need to be done with great care.

5. DISCUSSION AND CONCLUSIONS

This paper discusses the dynamical behaviour of Concentrated Benthic Suspensions of cohesive sediment in open channel flow. It is argued that suspensions of cohesive and non-cohesive sediment can both be characterised by capacity conditions, which can be measured by an equilibrium concentration of the suspended sediment. Such capacity conditions for cohesive sediment can only occur if abundant sediment is available. The main difference between the two suspensions at post-capacity conditions is caused by the formation of a layer of fluid mud upon sedimentation of cohesive sediment flocs, whereas the sedimentation of sand results in a rigid bed at which turbulence production remains possible.

It is reasoned that CBS-layers can achieve a state of equilibrium at which the suspended sediment is mixed over only a part of the water depth. This is in contrast to stratified systems of miscible fluids, which will always be mixed completely if at least one of the layers is sufficiently turbulent. At this equilibrium, turbulence production and sediment-induced buoyancy destruction are balanced, which can be achieved only if no positive feed-back occurs between the turbulence production and the suspension (buoyancy destruction). The existence of such an equilibrium was proven experimentally in an oscillating grid tank, showing that the level of the lutocline and the mean (equilibrium) concentration beneath this lutocline are a function of the grid properties (mesh, frequency and amplitude of oscillation). This equilibrium concentration can be regarded as the transport capacity of the flow.

Starting from equilibrium conditions, an increase in turbulence intensity, e.g. by increasing the flow velocity in open channel flow, will result in vertical mixing causing a rise of the lutocline and a lowering of the suspended sediment concentration below the lutocline. Experiments in a rotating annular flume showed that this mixing can be classified as entrainment, and that this entrainment is identical to that which occurs in fresh/salt water stratified systems.

From observations reported in the literature it is known that sediment-induced buoyancy effects can cause a significant reduction in overall hydraulic resistance. This implies that distinction must be made between spatially confined and unconfined CBS-occurrences. Theoretical and numerical analyses show that in the case of confined CBS-occurrences, at which the local flow rate remains constant, post-capacity conditions result in saturation, i.e. a complete and irreversible collapse of the vertical concentration profile and turbulence field. It appears that numerical simulations with a 1DV-model follow the theoretical derived scaling law for saturation properly.

However, the actual value of the saturation concentration, as computed with various numerical models (Violeau et al., 2001), appears to depend on the applied numerical schemes, the bed boundary conditions and their implementation. Furthermore the catastrophic collapse described in Section 4.1 may consist of a narrow concentration range over which the transport capacity of the flow diminishes. In addition, Section 4.2 suggests that the actual value of the saturation concentration and its functional relationship may be more complicated than described in Section 2, i.e. $C_e \propto u_*^n$ (see equ. (2)), where n = 3 or 4. It should be emphasised, however, that at present no direct empirical evidence exists for this "snowballing" effect resulting in the predicted collapse, nor on the actual functional relationship of the saturation value and the hydrodynamic parameters.

A final important observation is that in the case of unconfined CBS-occurrences, drag reduction by sediment-induced buoyancy effects will result in an acceleration of the flow, hence an increase in turbulence production. This implies that significant differences may be expected in the dynamics of confined and unconfined CBS-occurrences.

ACKNOWLEDGEMENTS

This work was partially funded by the European Commission, Directorate General XII for Science, Research & Development through the COSINUS-project within the framework of the MAST-3 programme, contract MASC3-CT97-0082 and by corporate research funds from the various research institutes involved.

REFERENCES

Beardsley, R.C., Candela, J., Limeburner, R., Geyer, W.R., Lentz, S.J., Castro, B.M., Cacchione, D. and Carneiro, N., 1995, The M_2 tide on the Amazon shelf, *Journal of Geophysical Research*, (100), 2283-2319.

Bruens, A.W., Booij, R., Kranenburg, C. and Winterwerp, J.C., 2000, Applicability of the rotating annular flume for entrainment experiments, *Proceedings of the Fifth International Symposium on Stratified Flows*, Vancouver, July, 2000.

Bruens, AW., Kranenburg, C. and Winterwerp, J.C., 2001, Physical and numerical modelling of the entrainment by a turbulent Concentrated Benthic Suspension, *Proceedings INTERCOH-2000, Elsevier, Coastal and Estuarine Fine Sediment Processes*, ed. J.C. Winterwerp and C. Kranenburg, this volume.

Dong, L., Wolanski, E. and Li, Y., 1997, Field and modelling studies of fine sediment dynamics in the extremely turbid Jiaojianng River estuary, China, *Journal of Coastal Research*, (13) 4, 995-1003.

E, X. and Hopfinger E.J., 1987, Stratification by solid particle suspensions, *Proceedings of the third International Symposium on Stratified Flows*, Caltech, Pasadena, 1-8.

Gratiot N., M. Mory M., Manning A.J., Michallet H., 2001, CBS layers in a diffusive turbulence grid oscillation experiment, *Proceedings INTERCOH-2000, Elsevier, Coastal and Estuarine Fine Sediment Processes*, ed. J.C. Winterwerp and C. Kranenburg, this volume.

Guan, W.B., Wolanski, E. and Dong, L.X., 1998, Cohesive sediment transport in the Jiaojiang River Estuary, China, *Estuarine, Coastal and Shelf Science*, (46), 861-871.

Ham, R. van der, 1999, Turbulent exchange of fine sediments in tidal flow, PhD-thesis, *Delft University of Technology, Faculty of Civil Engineering and Geotechnical Sciences*.

Huppert, H.E., Turner, J.S. and Hallworth, M.A., 1995, Sedimentation and entrainment in dense layers of suspended particles stirred by an oscillating grid, *Journal of Fluid Mechanics*, (289), 263-293.

Kranenburg, C., 1998, Saturation Concentrations of Suspended Fine Sediment. Computations with the Prandtl Mixing-Length Model, *Delft University of Technology, Faculty of Civil Engineering and Geosciences*, Report No 5 - 98.

Parker, G., Fukushima, Y. and Pantin, H.M., 1986, Self-accelerating turbidity currents, *Journal of Fluid Mechanics*, (171), 145-181.

Teisson, C., Simonin, O., Galland, J.-C. and Laurence, 1992, Turbulence modelling and mud sedimentation: a Reynolds stress model and a two-phase flow model, *Proceedings of the 23rd International Conference on Coastal Engineering, ICCE*, Venice, (3), 2853-2866.

Toorman, E.A., 1999, Numerical simulation of turbulence damping in sediment-laden flow, *Katholieke Universiteit Leuven, Hydraulics Laboratory*, Report HYD/ET/99.2.

Toorman, E.A., 2000. Modelling of turbulent flow with suspended cohesive sediment, *Proceedings INTERCOH-2000, Elsevier, Coastal and Estuarine Fine Sediment Processes*, ed. J.C. Winterwerp and C. Kranenburg, this volume.

Toorman, E.A., 2001, Suspension capacity of uniform shear flows, Report HYD/ET/00/4, *Hydraulics Laboratory, Katholieke Universiteit Leuven.*

Toorman, E.A., Kranenburg, C., Winterwerp, J.C., Bruens, A.W., 2001, Interaction of suspended cohesive sediment and turbulence, *Proceedings INTERCOH-2000, Elsevier, Coastal and Estuarine Fine Sediment Processes*, ed. J.C. Winterwerp and C. Kranenburg, this volume.

Tsai C.H. and Lick W., 1986, A portable device for measuring sediment resuspension, *Journal of Great Lakes Research*, (12) 4, 314-321.

Violeau, D., Cheviet, C., Markofsky, M., Petersen, O., Roberts, B., Toorman, E. and Weilbeer, H., 2000, Numerical simulation of cohesive sediment transport: intercomparison of several numerical models, *Proceedings INTERCOH-2000, Elsevier, Coastal and Estuarine Fine Sediment Processes*, ed. J.C. Winterwerp and C. Kranenburg, this volume.

West, J.R. and Oduyemi, K.O.K., 1989, Turbulence measurements of suspended solids concentration in estuaries, *ASCE, Journal of Hydraulic Engineering*, (115) 4, 457-474.

Winterwerp, J.C. and Kranenburg, C., 1997, Erosion of fluid mud layers - II: Experiments and model validation, *ASCE, Journal of Hydraulic Engineering*, (123) 6, 512-519.

Winterwerp, J.C., Uittenbogaard, R.E., de Kok, J.M., 2001a, Rapid siltation from saturated mud suspensions, *Proceedings in Marine Science, Coastal and estuarine Fine Sediment Processes*, ed. W.H. McAnally and A.J. Mehta, Proceedings of INTERCOH'98, Elsevier, 125-146.

Winterwerp, J.C., 1999, On the dynamics of high-concentrated mud suspensions, PhD-thesis, Delft University of Technology, also *Delft University of Technology, Faculty of Civil Engineering and Geosciences, Communications on Hydraulic and Geotechnical Engineering*, Report 99-3

Winterwerp, J.C., 2001, Stratification effects by cohesive and non-cohesive sediment, *Journal of Geophysical Research*, (106) C10, 22,559-22,574.

Wolanski E., Asaeda T. and Imberger J., 1989, Mixing across a lutocline, *Limnology and Oceanography*, (34) 5, 931-938.

Zhou, D. and Ni, J.R., 1995, Effects of dynamic interaction on sediment-laden turbulent flows, *Journal of Geophysical Research*, (100) C1, 981-996.

Measurement and modelling of the properties of cohesive sediment deposits

Dearnaley[1], M.P., Roberts[1], W., Jones, S.[2], Leurer, K.C.[3], Lintern, D.G.[4], Merckelbach L.M.[5], Sills[4], G. C., Toorman, E. A.[3] and Winterwerp, J.C.[5,6].

[1] HR Wallingford, Howbery Park, Wallingford, Oxfordshire OX10 8BA, UK.
[2] University of Wales, Bangor, UK.
[3] Hydraulics Laboratory, Katholieke Universiteit Leuven, Belgium.
[4] Dept. of Engineering Science, Oxford University, UK.
[5] Delft University of Technology, the Netherlands.
[6] Delft Hydraulics, the Netherlands.

Research studies undertaken as part of the "Bed Dynamics" Task D of the EC funded COSINUS project are described. The studies undertaken involve the reformulation of sediment exchange equations, in situ field measurements of bed strength, laboratory settling column experiments, bed consolidation modelling, the development of a model of bed dynamics based on generalised Biot theory and the testing of an integrated erosion/entrainment model against laboratory experiments. The results of the various studies are synthesized and overall conclusions drawn.

KEY WORDS
cohesive sediment, erosion, bed strength, consolidation, flocculation, numerical modelling.

1. INTRODUCTION

This paper summarises work undertaken as part of Task D of the European MAST3-COSINUS Project. The aim of Task D on "Bed Dynamics" was to provide a greater understanding of the development of cohesive sediment beds through various laboratory and field based measurements and numerical model development. The work was carried out by a number of organisations – Delft University of Technology, Delft Hydraulics, Katholieke Universiteit Leuven, HR Wallingford, Oxford University and the University of Wales, Bangor. This paper briefly describes the key points of each of the separate studies and attempts to synthesise the results of the various studies by drawing overall conclusions from the results.

2. BACKGROUND

The modelling of cohesive sediment transport in estuaries and coastal areas requires a description of the sediment exchange with the bed through the processes of deposition and erosion. The sediment transport (or sediment mass balance) equation can be written as:

$$\frac{\partial \phi}{\partial t}+U_j\frac{\partial \phi}{\partial x_j}=\frac{\partial}{\partial x_j}\left(\varepsilon_s\frac{\partial \phi}{\partial x_j}\right)-\frac{\partial}{\partial z}(w_s\phi) \tag{1}$$

where: ϕ = concentration by volume of the sediment particles, ε_s = the eddy diffusivity, w_s = the settling velocity, U = flow velocity. At the bottom the boundary condition reads:

$$\varepsilon_s\frac{\partial \phi}{\partial z}-w_s\phi=S_0=S_E-S_D \tag{2}$$

i.e. the sediment exchange flux S_0 consists of two contributions: the upward flux S_E and the deposition flux S_D, which is (proportionally) related to the settling flux, i.e. $S_D = p_D w_s \phi$, where p_D = the fraction which effectively becomes part of the bed. Both fluxes are generally described by an empirical relationship which evaluates the bottom shear stress of the flow (τ_0) against a critical shear stress (e.g. Teisson, 1997).

In practice, the bed exchange module is the weakest part in cohesive sediment transport modelling. There are many reasons for this. One of the fundamental problems is the lack of a general relationship between erosion strength and bed shear strength. But even if the fundamental questions regarding the modelling of sediment exchange between water column and bed were resolved, the fact remains that the bed properties vary in time and space (both horizontally and in depth), for instance due to variations in sediment composition and biological activity.

Although the change of bed level as a result of consolidation is of interest, it is usually less important than the effect of consolidation on increasing the resistance of the bed to erosion. Consolidation, especially over short periods (hours to days), can have an important influence on the behaviour of applied models as this influences whether deposits at slack water or on neap tides can then be re-eroded during periods of faster currents. Effects of consolidation can be taken into account in a number of ways. Many approaches involve a simplified representation of the vertical structure of the bed. At its simplest, this may be only two layers: a surface, unconsolidated layer, with a weak resistance to erosion, and a more consolidated layer below. The time variation in bed properties as a result of consolidation is represented by introducing fresh deposits into the weak top layer, and then gradually transferring that material to the consolidated layer (e.g. Odd & Cooper, 1989; Kusuda & Futawatari, 1992). Different values of the erosion parameters and density are assigned to each layer. A characteristic consolidation time must be specified to determine the rate of transfer of material from one layer to the next.

This general approach can be extended to a multi-layer representation of the vertical bed structure and consolidation effects. This is simple and quick to calculate. A weakness is the discontinuous representation of bed properties and of course the representation of the consolidation process is very crude, in particular with regard to the relationship between density and strength, which are directly related to the bed structure and its history.

Another weakness in these bed models is their inability to account for weakening of the bed by wave induced forces, which under extreme conditions (e.g. storms) may even cause complete loss of erosion strength by liquefaction (shear-induced structural break-up) or fluidisation (pore pressure induced break-up) of part of the bed.

3. FORMULATION OF SEDIMENT EXCHANGE AT THE BED

The manner in which exchange of sediment at the bed is currently characterised in sediment transport models has been investigated by the Katholieke Universiteit Leuven (KUL). These investigations have identified ways in which both the deposition and erosion processes can be more effectively represented. A brief summary of this work is given below but the work is presented in more detail in Toorman (2000).

In practice, the critical shear stress for deposition is a tuning parameter of the model. According to Sanford & Halka (1993), numerical models perform better when *no* threshold is considered for deposition. This makes even more sense if one considers the fraction of the settling flux $w_s\phi$ which does not stick to the bed as a part that is immediately "eroded" (Toorman, 2000). It is then possible to include the fraction that does not stick to the bottom in the erosion flux, i.e.:

$$S_0 = S_E - S_D = \left[S_E + (1 - p_D)w_s\phi\right] - w_s\phi = S'_E - w_s\phi \tag{3}$$

A typical erosion rate equation (for surface erosion) is of the form:

$$S_E = E_0\left[(\tau_0 / \tau_c)^n - 1\right]^m \qquad (\tau_0 > \tau_c) \tag{4}$$

with τ_c the critical stress for erosion. The erosion rate parameter E_0 is expected to be proportional to the bed surface (e.g. volumetric) concentration, as the amount that can be eroded cannot exceed the available amount. The critical stress for erosion is empirically related to a measure of the shear strength of the bed, often the vane shear strength. Toorman (1995) proposed an alternative formulation for erosion strength:

$$\tau_c = a\left(e^{C/C_s - 1} - 1\right) \qquad (C > C_s) \tag{5}$$

which accounts for the fact that there is no structure below the space-filling concentration C_s, which makes this form physically more realistic. If the non-sticking fraction of depositing particles should be included into the erosion law, a contribution without critical erosion stress should be added.

Besides the difficulty in determining the critical stress for erosion, the calculation of the correct bottom shear stress is also crucial, because of buoyancy induced drag reduction. The traditional method in numerical models, which is based on wall functions for homogeneous flow, in the case of a fixed shear velocity overestimates the bed shear stress with increasing sediment load up to a factor of 3 at the saturation condition, whereas a consistent approach, which corrects the near-wall boundary conditions for buoyancy effects, yields the correct value (Toorman, 2002).

4. FIELD MEASUREMENTS OF BED STRENGTH

A field measurement campaign was carried out in September 1998 at Calstock in the Tamar Estuary. The measurements concentrated on suspended sediment properties and hydrodynamics, but HR Wallingford (HRW) and University of Wales, Bangor (UWB) were also involved in measuring properties of the sediment deposits. HRW measured particle size distributions and the resistance to erosion of the sediment deposits exposed on the inter-tidal banks at low water. The sediment was predominantly mud, with a median grain size in the range 10-20 microns and loss on ignition measurements were between 8% and 14%. The water content of the deposits was high and the deposits were accordingly very weak, often fluid. The slope of the inter-tidal banks was steep, at around 10-20°.

Because of the weakness of the sediments and the site conditions the SEDERODE instrument deployed by HRW was unsuccessful at all but 2 of the 16 sites. The critical erosion shear stresses measured were 0.1 Pa and 0.21 Pa for sediments with surface bulk densities of 1230 kg/m^3. The sand content of the surface sediment was about 13% in both cases and the organic content (by loss on ignition) also about 13%.

UWB measured acoustic shear wave velocity in situ at a series of 5 intertidal locations near Sites A and B, on both neaps and springs, using paddle-shaped piezoelectric transducers embedded in the surface sediment (Jones et al., 1993). Acoustic shear wave velocity is a measure of bulk sediment rigidity. No significant difference was found between locations or between springs and neaps. The mean over all sites was 48 m/s. The high degree of variability (31-69 m/s) obtained is indicative of high porosity, low rigidity muds.

UWB also deployed a multi-corer to retrieve five 100 mm diameter cores of up to 210 mm in length from Site A. These were transferred into an instrumented column for measurement of acoustic shear wave velocity and electrical formation factor (Wren, 1996). The electrical formation factor (defined as the ratio of electrical resistivity of the bulk sediment to the electrical resistivity of the pore fluid) is a measure, for a given packing configuration, of the sediment porosity (Jones et al., 1993). In addition, some of the cores were sectioned into 20 mm slices to examine the vertical variation in bulk density.

All five cores exhibited a bioturbated surface layer 60 – 70 mm in thickness overlying a more uniform subsurface layer. Beneath this, bulk density was found to increase down each core and this corresponded with increases in electrical Formation Factor and shear wave velocity. Bulk density and organic matter were similar to those determined at nearby inter-tidal sites although shear wave velocities were significantly higher (107-157 m/s). This may be explained by the fact that transducers of higher resonant frequency were used for the cores.

The main conclusion from the UWB measurements is that physical properties of surficial sediments were not found to vary significantly between Sites A and B or between neap and spring tides, and that depth variation was negligible within the mobile surficial layer. So erosion rate parameters used in the models, which depend on physical properties of the deposits, can be assumed to be constant.

5. SETTLING COLUMN EXPERIMENTS

Laboratory settling column experiments were carried out at both the Delft University of Technology (DUT) and the University of Oxford (UOX).

The DUT experiments concentrated on the consolidation process and examined the variations of density and vane shear strength with time and depth below the sediment surface. Two types of natural mud were used: Caland-Beer mud (from the entrance channel of the Port of Rotterdam) and Dollard mud. The mud beds were allowed to consolidate in short (0.3m) and tall (1.5m) settling columns. The measured parameters for the consolidation process were density and pore water pressure. In order to make accurate strength measurements, segmented settling columns were designed and built.

The segmented columns provided well-defined samples of the bed that were suitable for strength measurement by shear vane testing. Three types of shear vane test were carried out, namely rate controlled, stress controlled and oscillatory rate controlled. Since the shear vane tests are destructive, each series of experiments consisted of multiple settling columns that were set up identically, so that the strength development with time could be monitored.

Data from settling column tests are generally processed in order to obtain empirical relationships of effective stress and of permeability as a function of density. It is now qualitatively understood why these relationships are not unique, but show time-dependence, which is related to the histories of floc and bed structures, depending on the forcing (Sills, 1995; Toorman, 1999). Besides this physical aspect, the accumulation of experimental error in the traditional data processing method contributes to the difficulty in interpretation of the data. A new data processing method has therefore been developed at the Katholieke Universiteit Leuven, based on filtering of errors by using analytical smoothing functions. Simple analytical functions have been derived which give a good approximation of the excess pore water pressure profiles and the constant mass contours in the settling curve plot. The resultant mass gradients are used for the determination of the permeability. This method allows a significant reduction in the error involved in the calculation of the pore pressure gradient and the filtration rate. The method has been applied to experimental data of DUT. Further details can be found in Toorman & Leurer (2000a).

The UOX settling column experiments concentrated on the relationship between the density and strength of the bed and the way in which the deposits are formed. Deposits formed from a slurry were compared with those formed by slow, steady deposition. In the steady deposition experiments a flocculation chamber was used to control the properties of the settling flocs and the effect of floc size and density on the properties of the deposit were investigated.

Table 1
Tamar sediment property tests measured in the UOX tests

Sample	particle density Mg/m^3	organic content (% by mass)	median diameter, D_{50} (μm)	clay content (% conc.)	Silt content (% conc.)	liquid limit	plastic limit
top		24.6	23	5	81		63.3
mid		19.7	20	5	85		65.2
bottom		20	22	5	83		63.7
average	2.570	21	21.6	5	83	88	64.1

Sediment collected from Calstock in the Tamar estuary during the field measurement campaign was used in the UOX experiments. Repeated grain size measurements using a CILAS 920 laser particle sizer indicated that the collected sample had a median grain diameter of 22 μm. This is slightly higher than the 10-20 μm median grain diameter reported in the field. Organic content analyses were conducted according to Head (1992), using a hydrogen peroxide decomposition method, and revealed a mean of 21 % organic content by mass. Again, this is higher than the 8 to 14% reported in the field using a loss on ignition technique. Other parameters measured are reported in Table 1, including particle density, clay and silt content and liquid and plastic limits. The sediment has a British Standards classification as a high plasticity clayey silt.

Details of the instrumentation are given in Lintern (2000). Methods include the use of floc video imaging technology with a laser light source, and indirect density measurement using a non-destructive X-ray technique described in numerous reports (Been, 1980; Been 1981; Sills, 1997; Sills, 1998). Pore water pressures were measured using a technique and apparatus originally developed by Bowden (1988).

Figure 1 shows results of the floc measurements for the experiment COS6. Due to variation in floc shape, the floc size is reported as equivalent spherical diameter (ESD)- the diameter of a sphere that occupies the same volume as the imaged floc. The median floc size (100 μm) is significantly larger than the mean primary particle diameter for Tamar sediment. Floc velocities are calculated from a sequence of images, and using these velocities in a modified form of Stokes' velocity equation the effective floc densities can be calculated.

The experiments demonstrated significant differences in the properties of a bed formed by settlement from a slurry by comparison with one formed by to steady sedimentation. The latter show a much higher degree of aggregation than those sedimented from slurries. Furthermore, the flocculated beds contain larger aggregates than the flocs in the water column which formed them, thus indicating aggregation is continuing during bed development.

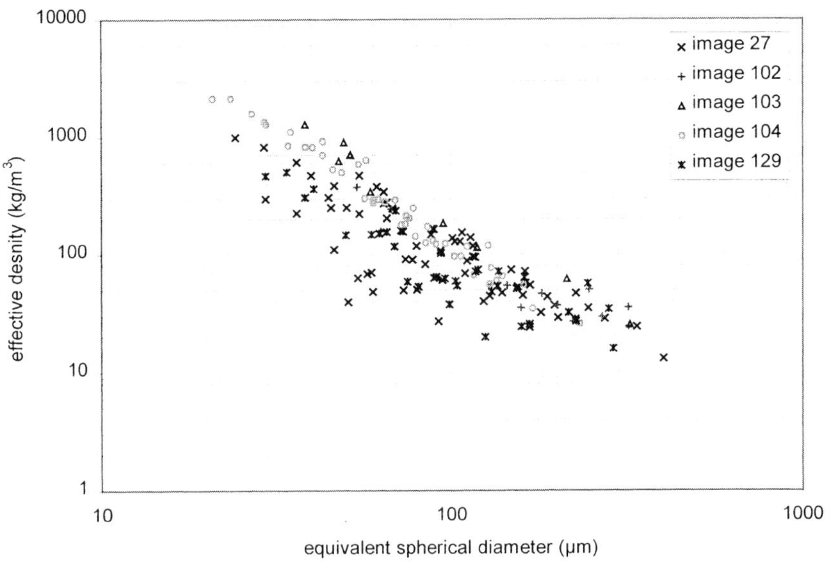

Figure 1. Effective density vs. equivalent spherical diameter for experiment COS6 flocs.

Under self weight consolidation, bed densities range from 1.15 to 1.20 Mg/m^3 at the surface to values above 1.3 Mg/m^3 at depths of 10 cm or more. The flocculation conditions clearly affect the density of the beds. Figure 2 shows density profiles for experiment COS1, in which the sedimentation concentration was varied throughout the experiment. The peaks in density correspond to stages of high sedimentation rates (above 3 g/l), and the troughs are formed during low sedimentation rates (down to 0 g/l). The profiles show that the sedimentation conditions lead to higher variations than either consolidation time, or depth of burial for these self weight experiments. Shear wave velocities ranged from 2 to 30 m/s, increasing with consolidation time. Rigidity moduli calculated from these range between approximately 1 and 5 kPa.

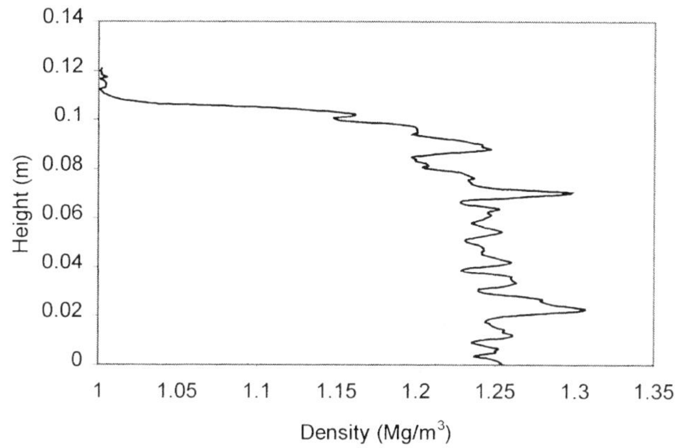

Figure 2. Density profile at the end of consolidation in experiment COS1.

The experiments were designed to simulate mud in its natural state from the Tamar Estuary. The biological components of the mud were not removed for most of the experiments. Within days after the end of sedimentation the bed surfaces became covered with a biological layer. Microscopic analysis showed an abundance of diatoms (many dormant) and other mobile organisms. In one experiment worm burrows became apparent from the outset. Observations show that these worm burrows greatly enhance the settling by providing channels for water to escape upward, and for particles to move downward. The worms then feed on the surface, where they they alter the surface properties by pelletizing the sediment. Other biological factors are also at work in the surface sediment. Density profiles often show a layer of low density 4-15 mm below the surface sediment. Such a layer is thought to arise due to biological activity, and most probably gas production. This layer is found to be rich in diatoms and other algae.

Nine experiments were carried out using the in-situ erosion device ISIS developed by HR Wallingford (Williamson & Ockenden, 1993). The ISIS work and the properties of the biological surface layers are discussed further in a separate paper (Lintern et al., this volume). The ISIS measurements have been made on beds settled from slurries of Tamar mud. The bulk densities of surface of the beds tested were typically about 1.2 Mg/m^3, which is similar to that measured by HRW during the field measurements. Erosion of the bed in the laboratory appears to commence at a similar applied bed shear stress as that in the field, generally around 0.01 Pa.

6. BED DYNAMICS MODELLING

6.1. Consolidation and strength modelling

Delft University of Technology (DUT) modelled the process of consolidation as a one-dimensional process using the Gibson equation (Gibson et al., 1967) written in an Eulerian reference frame and with the particle volume fraction as the dependent variable:

$$\frac{\partial \phi_v}{\partial t} - \frac{\rho_s - \rho_w}{\rho_w} \frac{\partial}{\partial z}\left(k\phi_v^{\,2}\right) - \frac{1}{\rho_w g}\frac{\partial}{\partial z}\left(k\phi_v \frac{\partial \sigma'}{\partial z}\right) = 0 \qquad (6)$$

where ϕ_v is the solids volume fraction, ρ_w the density of water, ρ_s the density of solids, g acceleration due to gravity, k the permeability, σ' the effective stress, z the vertical co-ordinate (positive in upward direction) and t time.

New constitutive equations for effective stress and permeability are derived on the basis of the concept of a scale-invariant bed structure (Merckelbach & Kranenburg, 2000). It is assumed that the volume filling network structure is built by aggregates that consist of clay and silt particles. The structure of the aggregates is assumed scale-invariant. This assumption may be regarded as a generalisation of Krone's concept of orders of aggregation (Krone, 1963). The following relationship between the length scale of an aggregate and the solids volume fraction can be established:

$$R_a^{n_f - 3} \propto \phi_v \qquad (7)$$

where R_a is the length scale of the aggregates and n_f the fractal dimension. During consolidation, excess pore water pressure is transferred to effective stress. The effective stress is assumed to relate to the number of critical bonds within an aggregate. In accordance with the concept of scale invariance, the number of critical bonds per aggregate is independent of the size of the aggregate. Consolidation may be regarded as a condition in which the effective stress is the maximum effective stress that can exist in a network structure. Hence, an increase of effective stress must result in an increase of the number of bonds per unit area. This is achieved by a break-up of aggregates and a reduction in the length scale of the aggregates.

Assuming a linear relationship between the effective stress and the number of bonds as suggested by experimental data presented by Mitchell (1976) and a constant number of bonds per aggregate results in the constitutive equation for effective stress:

$$\sigma' = K_\sigma \phi_v^{\frac{2}{3-n_f}} \qquad (8)$$

where K_σ is an empirical parameter which includes shape effects and the size of clay particles, for example. This relationship, however, does not include time dependency effects, which may play a significant role (Toorman, 1999).

Assuming that the pore water can be modelled as Poiseuille flow and that the size of the virtual tubes is proportionally related to the size of the aggregates, the following relationship for the permeability can be obtained:

$$k = K_k \phi_v^{\frac{-2}{3-n_f}} \tag{9}$$

where K_k is an empirical parameter which includes shape effects and the size of clay particles, and also the viscosity of water.

These new constitutive equations relate the effective stress and permeability to the volume fraction of solids. Effective stress and permeability turn out to be related through the fractal dimension.

Strength may be regarded as resistance against failure. It is assumed that the bed strength is generated by intra- and inter-aggregate particle bonds. The concept of scale invariance implies that the number of intra-aggregate bonds per aggregate is independent of the aggregate size and that the number of inter-aggregate bonds per aggregate is proportional to $R_a^{n_f-1}$. Similar to the procedure followed for the effective stress, the critical shear stress can be expressed in terms of the critical shear stress generated by intra- and inter-aggregate particle bonds, which gives the failure criterion:

$$\tau_c = k_1 \phi_v + k_2 \sigma' \tag{10}$$

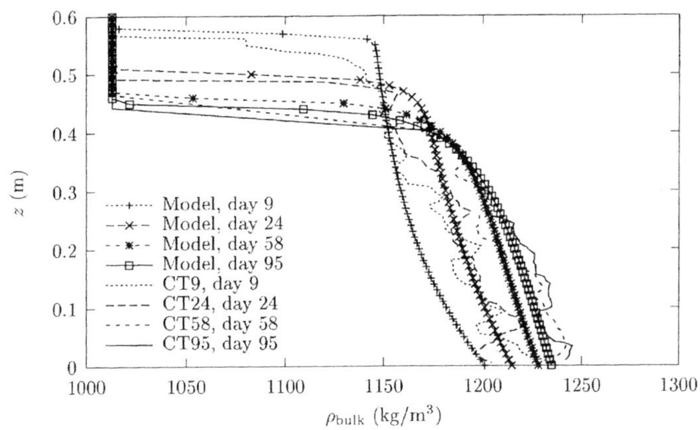

Figure 3. Measured and calculated density profiles for experiment CT.

The empirical coefficients k_1 and k_2 account for inter- and intra-aggregate bond strengths and number of bonds per aggregate. This criterion resembles the Mohr-Coulomb criterion,

$$\tau_c = c' + \tan(\varphi')\sigma' \tag{11}$$

where the true cohesion c' is given by $c' = k_1 \phi_v$ and the angle of internal friction φ' by $\tan(\varphi') = k_2$. The failure criterion can be used in strength modelling. In an accompanying paper (Merckelbach et al. 2000) this is elaborated further by applying the failure criterion to a shear vane test model.

One of the DUT settling column consolidation experiments was simulated using the consolidation equation with the new constitutive equations. The experiment was carried out using Caland-Beer mud. The initial conditions were ρ_l = 1070 kg/m^3 and h_i = 1.53 m. The simulation was carried out using n_f = 2.75, K_σ = 3.2 MPa and K_k = 2.9 10^{-15} m/s. The results are shown in Figure 3 for 9, 24, 58 and 95 days of consolidation. Figure 3 shows a good correspondence between the measured and computed density profiles, which enhances the confidence in the newly derived constitutive equations.

The results of the shear vane test model (Merckelbach et al. 2000) are shown in Figure 4. A good agreement between the measured and modelled strength profiles is observed. The absolute deviation between measured and modelled yield stresses can be as large as 50 Pa for the lower part of the bed on day 9 and day 24, but with respect to the absolute values of the yield stresses, this is an error of maximally 33 %.

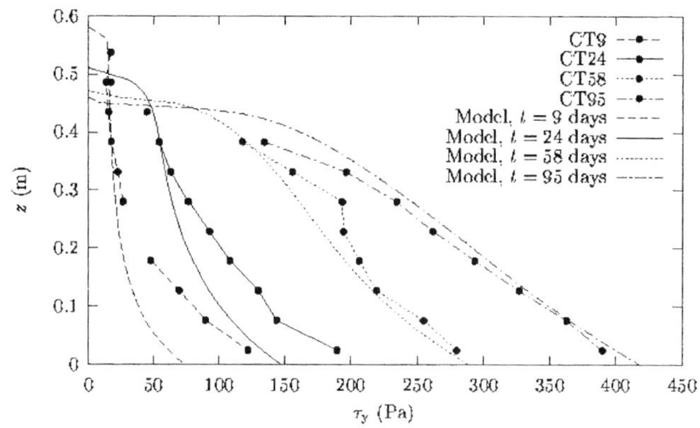

Figure 4. Measured and calculated yield strength profiles for experiment CT.

6.2 Generalised bed dynamics modelling

Traditional bed models, such as the previous one, only describe the strengthening of the sediment bed by self-weight consolidation. Sediment beds in nature are also subjected to shear forces by currents and waves and oscillating pore water pressures due to waves.

At KUL, a model of bed dynamics has been developed based on the generalised Biot theory for saturated porous media, which offers a holistic framework to simulate the combination of all the processes in the bed, i.e. consolidation, liquefaction and fluidisation. It addresses the development in time of density and strength of the bed and can include effects of thixotropy and creep. The major difference with the previous geotechnical model is the replacement of the empirical normal effective stress-void ratio relationship by a rheological model which relates stresses to strains and strain rates. The model can simulate extremely large deformations of fresh mud deposits by implementation of the arbitrary Euler-Lagrange method. Subsequent changes in density and permeability are accounted for.

The model actually combines numerical methods applied in the generalized Biot theory for saturated porous media and in creeping non-Newtonian fluid mechanics, implemented within a mixed Eulerian-Lagrangean co-ordinate system. The model solves the sediment mass balance, stress balance, pore water continuity and a rheological closure relationship. The equations are solved iteratively in three uncoupled groups with unknowns being the bulk density of the bed, solids displacement, pore water pressure and stresses, using the finite element method. In principle, various rheological models can be incorporated, from pure elastic to generalized elasto-plastic and creep, including non-linear material properties. The large deformations and highly non-linear material behaviour pose severe problems regarding numerical stability which are still to be overcome. Thus far, only the idealised case of consolidation of a pure visco-elastic porous material could be simulated without numerical instabilities, already showing that the relationship between effective stress and density is not unique, but time-dependent as expected (Figure 5).

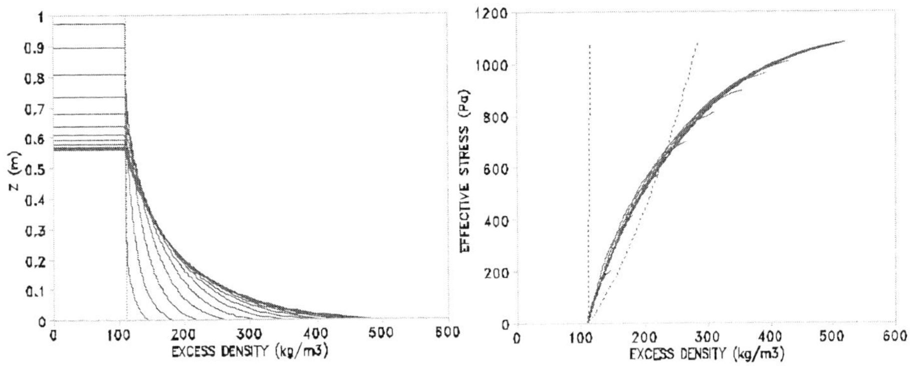

Figure 5. Simulated time evolution of the density (left) for the consolidation of an idealised saturated visco-elastic soil skeleton with initial bulk density Δ_0 = 1110 kg/m^3, shear modulus G = 1000 Pa and viscosity 0 = 100 Pa.s. Right: corresponding effective stress versus excess density ρ -ρ_w (full lines; dotted lines = pure elastic case, 0 = ∞).

The model is not suitable for large-scale 3D applications, but is intended for use as a research tool to better understand the dynamic behaviour of cohesive sediment beds. Further details on this model can be found in a separate report by Toorman *et al.* (2000).

7. INTEGRAL EROSION/ENTRAINMENT MODEL

Modelling work was carried out at Delft Hydraulics, where a 1DV model was used to simulate flume experiments carried out at DUT (Winterwerp and Kranenburg, 1997). These experiments simulated the chain of processes through a tidal cycle consisting of settling, hindered settling, fluid mud formation, consolidation and re-entrainment. Consolidation of the mud layer was modelled with Equation (6) and its strength as a Bingham plastic, its parameters being derived from a fractal description of the mud flocs (as described in the section on consolidation and strength modelling by DUT). It appears that the prediction of the vertical concentration profile in the consolidating mud layer is at present the weak link in simulating this chain of processes. One probable cause is segregation of the fine and coarser fraction during the settling process.

One of the advantages of the aforementioned modelling of the consolidation process in Eulerian co-ordinates using the fractal theory (power law model of material functions) is that the Gibson equation evolves into an advection-diffusion equation for the sediment concentration. This concept was implemented in the 1DV POINT MODEL of Delft Hydraulics to describe consolidation around slack water as part of the settling and mixing processes during successive tidal cycles (Winterwerp, 1999). The resulting mass balance equation accounts for the effects of molecular diffusion (D_s), turbulent mixing (eddy diffusivity Γ_T), hindered settling (f_{hs}-function) and consolidation (f_c- and Γ_c-function) and reads:

$$\frac{\partial c}{\partial t}-\frac{\partial}{\partial z}(X_s c)-\frac{\partial}{\partial z}\left((D_s+\Gamma_T+\Gamma_c)\frac{\partial c}{\partial z}\right)=0 \qquad (12)$$

where: $X_s = f_{hs}+\dfrac{f_c}{1+\eta f_c}$, with : $f_{hs}=w_{s,r}\dfrac{(1-\phi_*)(1-\phi_p)}{1+2.5\phi}$ and $f_c = k\dfrac{\rho_s-\rho_w}{\rho_w}\phi_p$,

$$\Gamma_c \equiv \frac{2K_k K_\sigma}{(3-n_f)g\rho_w},\quad k = K_k \phi_p^{\left(\frac{-2}{3-n_f}\right)}$$

and where c is the sediment concentration by mass, ϕ the volumetric concentration of the flocs ($\phi = c/c_{gel}$), c_{gel} is the gelling concentration at which a space-filling network forms (also referred t as the structural density), ϕ_p the volumetric concentration of the primary mud particles, $\phi_* = \min\{1,\phi\}$, $w_{s,r}$ is a reference settling velocity, and η a parameter ($\eta = 10^5$ s/m).

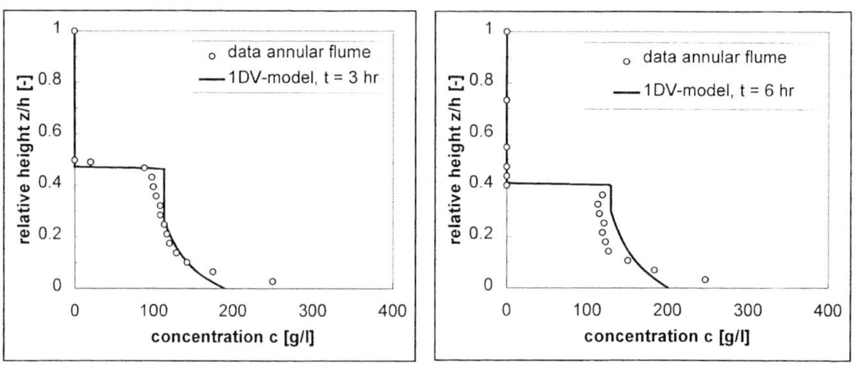

Figure 6. Vertical concentration profiles in flume at $t = 3$ hrs and $t = 6$ hrs.

This model was used to simulate settling, consolidation and remixing measured in a rotating annular flume (Winterwerp and Kranenburg, 1997). After some trial and error, the best results were obtained for $n_f = 2.71$, $c_{gel} = 100$ g/l, $K_k = 1\cdot10^{-14}$ m/s and $K_\sigma = 1\cdot10^9$ Pa, the results of which are presented in Figure 6. This figure shows a reasonable agreement between the measured and computed concentration profiles; however the large concentrations near the base of the profile are not properly predicted, which also affects the concentrations higher in the profile. This deviation is probably caused by the segregation of fine and coarse material that was observed during the experiments.

Further improvement can be obtained by including a second (sand) fraction in Equation (12). This is elaborated in Winterwerp (2002). Note that the consolidating sediment of Figure 6 does not contain a coarse sand fraction, hence no segregation occurs here.

Next, the soft mud layer is subject to erosion by a turbulent flow entraining the sediment. This effect is modelled by an additional stress term in the momentum equation in the 1DV POINT MODEL, an approach very similar to the one deployed by Le Hir (2001):

$$\frac{\partial u}{\partial t}+\frac{1}{\rho}\frac{\partial p}{\partial x}=\frac{\partial}{\partial z}\left[(\nu+\nu_T)\frac{\partial u}{\partial z}+\frac{\tau_{xz}^s}{\rho}\right] \tag{13}$$

where the stress term τ_{xz}^s is described as a Bingham-like plastic model:

$$\tau_{xz}^s=\mu_{mud}\frac{\partial u}{\partial z} \quad \text{with} \quad \mu_{mud}=\frac{a_y\tau_y}{1+a_y|\partial u/\partial z|}+\mu^s \tag{14}$$

in which:

a_y = coefficient (a_y = 0.02 implies $\tau_{xz}^s = 0.95\,\tau_y$ for $\partial u/\partial z = 10^{-3}$ s^{-1}),

μ^s = $K_\mu \varphi_p^n$; n ranges between 2 and 6 for various kinds of mud, and

τ_y = $K_y \varphi_p^{\frac{2}{3-n_f}}$

Note that the computational domain covers the entire water depth including the mud layer; hence entrainment at the interface is not explicitly modelled, and the mud properties change from solid at the flume bottom to liquid at the water surface.

The results of the simulations are presented in Figure 7 and 8 showing the measured and computed increase in suspended sediment concentration in the water column above the soft mud bed. The various parameter settings are given in Table 2.

Table 2

Setting of yield strength parameter in 1DV-simulations (c.o.l.s. is coefficient of lateral stress, i.e. ratio of yield strength and effective stress).

yield strength parameter K_y [Pa]		comments
measured bed density	$1.0\cdot10^8$	after Merckelbach with c.o.l.s. = 0.5
measured bed density	$1.0\cdot10^8$	after calibration
computed bed density	$5.0\cdot10^8$	from consolidation with c.o.l.s. = 0.5
computed bed density	$5.0\cdot10^6$	after calibration

Figure 7 shows that the suspended sediment concentration measured in the water column of the flume can properly be predicted using the measured density profile of the mud layer, using the proper parameters for the bed strength. A proper simulation using the computed density distribution of the bed is only possible for unrealistic strength parameters, e.g. Figure 8. From these results it can be concluded that a proper simulation of the entrainment process requires a proper vertical density profile at a coefficient of lateral stress of 0.5 (Van Kessel, 1997). This sensitivity is of course the result of the high sensitivity of the mud strength τ_y to the mud concentration c: $\tau_y \propto c^7$ for $n_f = 2.71$.

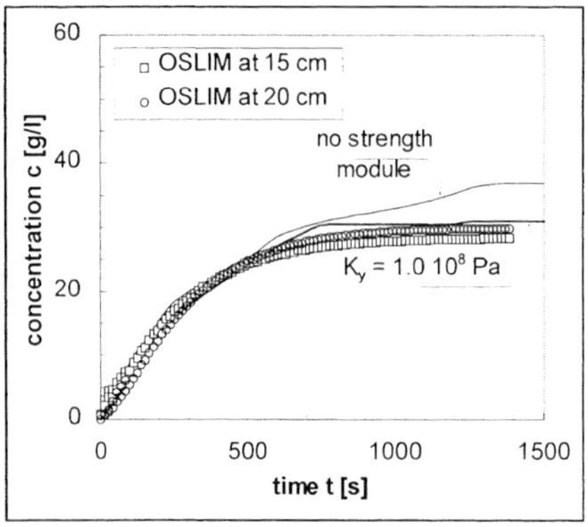

Figure 7: Effect of strength module on entrainment rate; measured initial density.

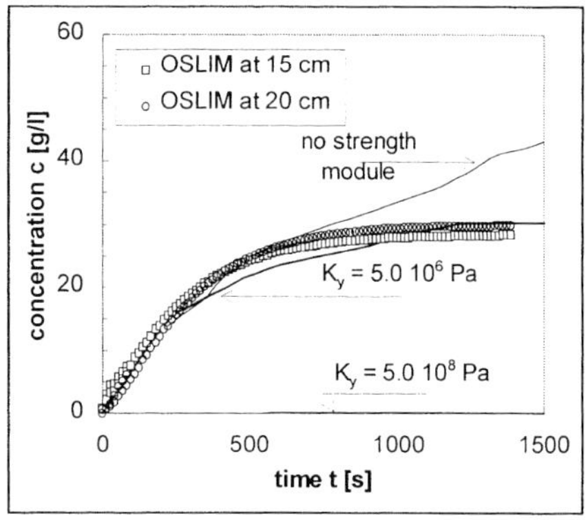

Figure. 8: Effect of strength module on entrainment rate; consolidation model.

It is concluded that the correct computation of the vertical density profile is a necessary, though probably not sufficient, condition for a proper prediction of the strength profile within a soft mud layer.

8. DISCUSSION & CONCLUSIONS

Traditional bed models, if present, coupled to cohesive sediment transport, allow updating the bed surface erosion strength as a function of density and time, by solving a simplified point-consolidation model and assuming a certain empirical relationship between density and shear strength.

In reality, the bed is not only subjected to gravity forces, which result in compaction and subsequent strengthening of the bed structure, but also to weakening shear and oscillating pressure forces due to currents and waves. Particularly in relatively shallow areas, such as estuaries and coasts, these forces may become significant, especially during storms, resulting in liquefaction and fluidisation of the bed, generating fluid mud layers, which can flow, or which are easily entrained. A new general bed dynamics model is developed which can be used to study these processes.

Settling column tests presently provide the only reliable means of examining the development of strength in a consolidating bed. However, such tests themselves may not be sufficiently representative of most natural conditions under which siltation occurs. Certainly biological effects have been demonstrated to significantly affect the bed sediment processes but biological activity in the laboratory may be different from that occurring in nature under perhaps more balanced ecosystem conditions.

The interesting observations of biological activity within the UOX experiments demonstrate that caution is necessary when using laboratory results as the basis for model validation. The models developed during this research project are unable to include the effects of biological activity. Comparison of the model against test results with no biological activity is required for rigour but then application of these models to situations where biological activity occurs will be erroneous.

The settling column work at UOX has provided a link between the field observations of floc behaviour (Dyer et al., 2000) and bed processes. The work has also demonstrated clear differences in the properties of a bed formed by settlement from a slurry by comparison with one formed by to steady sedimentation.

From the modelling undertaken within this project the question of whether flocs can be described as self-similar arises (ie whether is reasonable to assume a constant fractal dimension, as in the Merckelbach/Winterwerp model). The bed structure is not self-similar at every scale, and this assumption may not fully describe the changes in sediment structure that occur with aggregation, but the assumption has been used successfully in modelling work undertaken within this project.

Another important question is what bed surface density is obtained after deposition. Deposited aggregates form a space-filling structure. Assuming a certain averaged floc shape, i.e. spherical, a simple relationship between floc density and bed surface density is found, showing that the bed density will be smaller than the floc density (Toorman, 2000). This also implies that the density of eroded aggregates will be larger than the bed surface density. In addition, observations in the UOX experiments indicate that eroding aggregates are larger than those settling, which suggests that additional inter-aggregate bonds are formed on the bed surface.

It is traditionally assumed that eroded cohesive sediments are all entrained and take part in the suspension transport. However, if bulk erosion of the bed is important, it may be possible that the transport mode might be dominated by true bed load transport of relatively large mud chunks, as has been observed in previous laboratory erosion experiments (e.g. Migniot, 1968; Toorman, 2000). This is a subject which requires further study, i.e. the possible need for the derivation of a bed load function for cohesive sediment should be investigated.

ACKNOWLEDGEMENTS

This work is co-financed by the European Commission Directorate XII for Science, Research & Development, through the COSINUS project within the framework of the MAST-3 programme, contract MASC3-CT97-0082. The post-doctoral position of the co-author Toorman is financed by the Flemish Fund for Scientific Research.

REFERENCES

Been, K., 1980, Stress-strain behaviour of a cohesive soil deposited under water, Ph.D Thesis, Department of Engineering Science, Oxford University.

Been, K., 1981, A non-destructive soil bulk density measurement using X-ray attenuation, *Geotechnical Testing Journal*, December 1981: 69-176.

Bowden, R.K., 1988, Compression behaviour and shear strength characteristics of a natural silty clay sedimented in the laboratory, PhD Thesis, Department of Engineering Science, Oxford University.

Dyer K.R., Christie M.C., Lintern D.G., Manning A.J., Roberts W., Winterwerp J.C., 2002, Measurements and modelling of flocculation and settling, Proceedings *INTERCOH-2000, Elsevier, Coastal and Estuarine Fine Sediment Processes*, ed. J.C. Winterwerp and C. Kranenburg, this volume.

Gibson, R.E., England, G.L., Hussey, M.J.L., 1967, The theory of one-dimensional consolidation of saturated clays, *Géotechnique*, 17:261-273.

Head, K. H., 1992, Manual of Soil Laboratory Testing. London, Pentech Press.

Jones, S.E. and Jago, C.F., 1993, In situ assessment of modification of sediment properties by burrowing invertebrates. *Marine Biology* 115(1):133-142.

Kessel, T. van ,1997, Generation and transport of subaqueous fluid mud layers, PhD Thesis, Delft University of Technology, Department of Civil Engineering

Krone, R.B., 1963, A study of rheological properties of estuarial sediments, Tech. Report 63-8, Hydraulic Engineering Laboratory and Sanitary Engineering Laboratory, University of California, Berkeley.

Kusuda, T. & Futawatari, T., 1992, Simulation of suspended sediment transport in a tidal river, *Water Science Technology*, 26(5): 1421-1430.

Le Hir, P., Bassoulet, P. and Jestin, H., 2001, Application of the continuous modelling concept to simulate high-concentrated suspended sediment in a macrotidal estuary, in *Coastal and Estuarine Fine Sediment Processes*, ed. W.H. McAnally and A.J. Mehta, Elsevier Proceedings in Marine Science, (3), 229-248.

Lintern, D.G., 2000, "Summary of UOX settling column experiments", Unpublished report from COSINUS contract, Department of Engineering Science, Oxford University.

Merckelbach, L.M., & Kranenburg, C., 2000, A constitutive model for soft soils on the basis of scale invariance. Part 1: Consolidation (in preparation)

Merckelbach, L.M., Winterwerp, J.C., & Kranenburg, C., 2002, Strength modelling of mud beds, Proceedings *INTERCOH-2000, Elsevier, Coastal and Estuarine Fine Sediment Processes*, ed. J.C. Winterwerp and C. Kranenburg, this volume.

Migniot, C., 1968, Etude des propriétés physiques de différents sédiments tres fins et de leur comportement sous des actions hydrodynamiques, *La Houille Blanche*, No.7:591-620.

Mitchell, J.K., 1976, *Fundamentals of Soil Behaviour*, John Wiley & Sons.

Odd, N.V.M. & Cooper, A.J., 1989, A two-dimensional model of the movement of fluid mud in a high energy turbid estuary, *J. Coastal Research*, Special Issue No. 5: 185-194.

Sanford, L.P. & J.P. Halka, 1993, Assessing the paradigm of mutually exclusive erosion and deposition of mud, with examples of upper Chesapeake Bay, *Marine Geology*, 114:37-57.

Sills, G.C., 1995, Time dependent processes in soil consolidation, *Compression and Consolidation of Clayey Soils*, ed. Yoshikuni and Kusakabe, A.A. Balkema, Rotterdam, pp. 875-890.

Sills, G.C., 1997, Consolidation of cohesive sediments in settling columns, *Proc. 4th Nearshore and Estuarine Cohesive Sediment Transport Conference INTERCOH 94*, ed. N. Burt, W.R. Parker and J.Watts, J. Wiley & Sons, Chichester.

Sills, G.C., 1998, Development of structure in sedimenting soils, *Phil. Trans. R. Soc. Lond.* A 356: 2515-2534.

Teisson, C., 1997, A review of cohesive sediment transport models, in *Cohesive Sediment,* ed. N. Burt, R. Parker and J. Watts, pp.367-381, J. Wiley & Sons, Chichester.

Toorman, E.A., 1995, A study of erosion and deposition of cohesive sediment with a 1-point transport model, Report HYD147, Hydraulics Laboratory, Civil Eng. Dept., Katholieke Universiteit Leuven.

Toorman, E.A., 1999, Sedimentation and self-weight consolidation: constitutive equations and numerical modelling", *Géotechnique*, 49(6):709-726.

Toorman, E.A., 2000, Some thoughts on the modelling of erosion and deposition of cohesive sediments, Report HYD/ET/00/COSINUS7, Hydraulics Laboratory, Katholieke Universiteit Leuven.

Toorman, E.A., 2002, Modelling of turbulent flow with suspended cohesive sediment, Proceedings *INTERCOH-2000, Elsevier, Coastal and Estuarine Fine Sediment Processes*, ed. J.C. Winterwerp and C. Kranenburg, this volume.

Toorman, E.A. and Leurer, K.C., 2000, An improved data-processing method for consolidation column experiments, Report HYD/ET/00/COSINUS8, Hydraulics Laboratory, Katholieke Universiteit Leuven.

Toorman, E.A., Brenon, I. and Leurer, K.C., 2000, A general model for the study of the dynamic behaviour of cohesive sediment beds with extremely large deformations", Report HYD/ET/00/COSINUS6, Hydraulics Laboratory, Katholieke Universiteit Leuven.

Williamson, H.J. & Ockenden, M.C., 1993, In situ erosion of cohesive sediments, Report ETSU TID 4112, HR Wallingford.

Winterwerp, J.C. & Kranenburg, C., 1997, Erosion of fluid mud layers - II: Experiments and model validation", *ASCE J. Hydraulic Engineering*, 123(6):512-519.

Winterwerp, J.C., 1999, On the dynamics of high-concentrated mud suspensions, PhD Thesis, Faculty of Civil Engineering and Geosciences, Technical University of Delft. Report No. 99-3, ISSN 0169-6548.

Winterwerp, J.C., 2002, Settling, consolidation and re-entrainment of soft mud layers around slack water, (in preparation).

Wren, D.K, 1996, Surficial marine sediment properties: a geophysical investigation of variability and controls. Unpublished PhD thesis, University of Wales, Bangor.

Fine Sediment Dynamics in the Marine Environment
J.C. Winterwerp and C. Kranenburg (Editors)

Numerical simulation of cohesive sediment transport : intercomparison of several numerical models

D. Violeau[1], S. Bourban[2], C. Cheviet[1], M. Markofsky[3], O. Petersen[4], W. Roberts[2], J. Spearman[2], E. Toorman[5], H.J. Vested[4], H. Weilbeer[3]

[1]Laboratoire National d'Hydraulique et Environnement (EDF), 6 quai Watier, 78400 Chatou, France
[2]HR Wallingford, Howbery Park, Wallingford, Oxon OX10 8BA, United Kingdom
[3]Institut für Strömungsmechanik, Applestrasse 9A, D-30167 Hannover, Germany
[4]DHI Water & Environment, Agern Allé 11, DK-2970 Hørsholm, Denmark
[5]Hydraulics Laboratory, Katholieke Universiteit Leuven Kasteelpark, Arenberg 40, B-3001 Leuven, Belgium

Five different numerical models are used to reproduce estuarine cohesive sediment transport and intercomparisons are made of the resulting predictions. Comparison with test cases have shown that the numerical treatment of cohesive sediment is very sensitive to model parameters and formulations, and requires good calibration. Some of the specific processes that have been developed through the MAST3-COSINUS European project are used here to improve the ability of numerical models to reproduce the sediment behaviour in real estuaries. Despite this progress, it is shown that numerical models results still have many limitations, and their results should always be interpreted with great care.

KEY WORDS

Numerical modelling, cohesive sediment processes, schematic cases, estuaries.

1. INTRODUCTION

Within the framework of the MAST3-COSINUS European Project, the aim of Task E (« *Applied Modelling* ») was to compare and improve existing, operational engineering numerical models used for the prediction of cohesive sediment transport in estuaries, utilising the parameterisations gained from the other project tasks. The main goal was to show that the specific processes of cohesive sediment transport are correctly understood, and to provide the numerical models with operational techniques to take into account these phenomena. For this purpose various test cases were defined, in order to compare the results given by several numerical codes. The models have also been tested on field situations. Existing engineering system models have been used, improving the modelling of a number of cohesive sediment processes. However, the models still give diverging and (sometimes) incorrect results. Some explanations are given, as well as recommendations in order to make the numerical predictions more accurate.

Table 1. Description of the numerical codes used in this study.

Institute	Model	Numerical method	Turbulence model	Cases
HR Wallinford	SULIS	3D coupled fully implicit finite differences	Mixing length	Test cases
University of Leuven	FENST-2D (research code)	2DV fully implicit finite elements	k-epsilon	Test cases
DHI Water & Environment	MIKE 3	3D semi-implicit finite differences	Mixing length or k-epsilon	Test cases + Tamar
University of Hannover	TELEMAC-3D	3D semi-implicit finite elements	k-omega	Test cases + Weser
National Hydraulics & Environment Laboratory	TELEMAC-3D	3D semi-implicit finite elements	Mixing length	Test cases + Loire

2. DESCRIPTION OF THE NUMERICAL MODELS

Modelling cohesive sediment transport through a numerical model requires an approach which can adequately represent a number of complex physical processes. The processes that have been considered for this study are turbulence damping by suspended matter, flocculation, and the entrainment of Concentrated Benthic Suspension (CBS). A short description of the method used to represent these processes in the numerical models is presented here, emphasizing the improvements made within COSINUS. For this study, contributions have been received from five laboratories or universities, using different numerical models. Their main characteristics are summarised in table 1.

2.1. Turbulence modelling

Turbulence has been traditionally modelled with a Prandtl Mixing Length model (PML) or a k-ε closure (see below). Suspended particles cause damping of the turbulent fluctuations resulting in a reduction of the eddy viscosity ν_t and the eddy diffusivity K_s. The reduction depends on the Richardson number, the ratio of buoyancy to shear, i.e. of the degree of stratification. This effect is parameterised by implementation of buoyancy damping functions. The eddy viscosity and diffusivity for open-channel flow is then written as (Toorman 2000)

$$\begin{aligned} \nu_t &= F_t \kappa u_* z \left(1 - \frac{z}{h}\right) \\ K_s &= F_s K_0 = \frac{\nu_t}{\sigma_s} \end{aligned} \tag{1}$$

where κ is the von Karman constant, K_0 the neutral sediment diffusion coefficient, u_* the shear velocity, F_t a momentum damping function, F_s a sediment mixing damping function and $\sigma_s = \nu_t / K_s = \sigma_0 F_t / F_s$ is the turbulent Schmidt number, σ_0 being the neutral Schmidt number. Modelling turbulence through a PML model requires the following expressions for eddy viscosity ν_t and diffusivity K_t :

$$\nu_t = \ell_{m0}^2 \cdot F_t(Ri)\left|\frac{\partial u}{\partial z}\right|$$
$$K_s = \frac{1}{\sigma_0}\ell_{m0}^2 \cdot F_s(Ri)\left|\frac{\partial u}{\partial z}\right| \qquad (2)$$

where ℓ_{mo} is the mixing length. Experiments (Taylor 1973) indicate that one can assume $\sigma_0 = 0.7$. As for the damping function, it is usual to consider the genaral form

$$F_t(Ri) = (1 + a \cdot Ri)^{-b}$$
$$F_s(Ri) = (1 + \alpha \cdot Ri)^{-\beta} \qquad (3)$$

A popular choice is that of Munk-Anderson : $a = 10$; $\alpha = 10 / 3$; $b = 0.5$; $\beta = 1.5$, or alternatively, Kranenburg (1998) : $a = \alpha = 2.4$; $b = 2$; $\beta = 4$. However, within COSINUS, it has been shown by Toorman (2000) that these sets of functions are not always consistent. For simplicity it is proposed to use a damping function for σ_s, multiplied by the neutral Schmidt number σ_0, since this approach provides an acceptable fit for the available experimental data. The functions used here are as follows :

$$F_t = (1 + 100 \cdot Ri)^{-1/3}$$
$$\sigma_s = \sigma_0(1 + 21 \cdot Ri)^{0.8} \qquad (4)$$

In the case of the k-ε turbulence model (for a complete description, see Chen et al. 1998), the eddy viscosity is computed according to the definition $\nu_t = C_\mu k^2 / \varepsilon$, and the buoyancy effect is explicitly accounted for by a damping term G in the k-equation. The bottom boundary conditions using the PML theory are based on wall functions and require two damping functions. Furthermore, the non-neutral Schmidt number is used to determine the eddy diffusivity K_s in the diffusion term of the sediment transport equation. A major modelling improvement is obtained through the consistent transfer of the damping functions to the wall functions used to determine the bottom boundary conditions. The theory explains the occurrence of drag reduction in cohesive sediment-laden flows. This new boundary treatment method guarantees the correct bottom shear stress, even when drag reduction at high Richardson number becomes significant, resulting in failure of the traditional method. For a more detailed description of the modelling of sediment-turbulence interaction, see the accompanying paper (Toorman et al. 2000) or Toorman 2000.

In addition to these two turbulence models, a k-ω model was used by the University of Hannover for the purpose of this work. The governing equations of the k-ω model, in a slightly modified form considering the buoyancy term G, are the conservation of turbulent kinetic energy k and the specific energy dissipation rate ω :

$$\frac{\partial k}{\partial t} + u_i \frac{\partial k}{\partial x_j} = \frac{\partial}{\partial x_i}\left[\left(\nu + \frac{\nu_t}{\sigma^*}\right)\frac{\partial k}{\partial x_i}\right] + P + G - \beta^* k\omega$$
$$\frac{\partial \omega}{\partial t} + u_i \frac{\partial \omega}{\partial x_j} = \frac{\partial}{\partial x_i}\left[\left(\nu + \frac{\nu_t}{\sigma_\omega}\right)\frac{\partial \omega}{\partial x_i}\right] + \alpha\frac{\omega}{k}\left(P + C_{\omega 3}G\right) - \beta\omega^2 \qquad (5)$$

where u_i are the velocity components, ν the kinematic viscosity of the fluid, x_i the cartesian co-ordinates, and t the time. The production term P is exactly the same as for the k-ε model (see Chen et al. 1998). Damping of turbulence is considered in the buoyancy term G. In these equations, $\sigma^* = \sigma_\omega = 2$, $\alpha = 5/9$, $\beta = 3/40$, $\beta^* = 9/100$, $C_{\omega 3} = 1$ if $G > 0$ and 0 if $G < 0$. The eddy viscosity is then given by

$$\nu_t = \gamma^* \frac{k}{\omega} \tag{6}$$

where γ^* is a calibration parameter often taken as unity. In addition, k and ω are subject to specific boundary conditions involving (among others) the shear velocity (see Wilcox 1993).

2.2. Flocculation

The flocculation process, which is specific to cohesive sediment, can be modelled through the settling velocity. Various available models exist to evaluate this important parameter, such as a constant formulation with or without hindered settling, or a power law with or without dissipation parameter function (for a complete description of these models, see Spearman & Roberts 2000). In addition, Winterwerp (1999) has developed a framework for the growth and recession of flocs under the influence of changing flow conditions, suggesting that the settling velocity could be expressed through the fractal characteristics of the flocs. All these equations involve calibration parameters. These settling/flocculation models were incorporated into a 1DV sediment-turbulence interaction model, and compared with *in situ* observations from the Calstock field experiment undertaken in the COSINUS project (Spearman & Roberts 2000). Though the general pattern of variation of w_s was reasonably reproduced by the models, none of the models could be said to have reproduced the measured settling velocity well. Even if there is obviously still uncertainty regarding measurement of settling velocities, it is therefore yet to be shown that no particular flocculation model is better than another. However, numerical simulation of the Weser estuary has clearly shown that a flocculation formulation which includes shear floc breakup is preferable to a constant or power law formulation (see Malcherek 1995).Until the quality and quantity of measured settling velocity data allows an informed choice to be made, it is suggested by Spearman & Roberts that the applied modeller use the simplest of flocculation models and only increase the complexity of the model if there is good evidence for doing so.

2.3. Entrainment of CBS

For settled beds and denser suspensions the classical Ariathurai-Partheniades theory depicts a gradual wearing of the sediment interface (Ariathurai 1974, Partheniades 1965) and assumes that the rate of resuspension depends on the excess bed shear stress above a critical value that is characteristic for the bed surface. This theory is described by the following equation:

$$E = E_0\left(\frac{\tau_b}{\tau_c} - 1\right) \quad if\, \tau_b > \tau_c \tag{7}$$

where E is the erosion rate, E_0 is an empirical constant, τ_b is the shear stress acting on the bed and τ_c is the critical shear stress characterising the resistance to erosion of the mud layer.

More recently Kranenburg & Winterwerp (1997) and Mehta (1989) suggested applying an approach analogous with density stratified flows to that of high concentrated mud layers or CBS layers. The suspension is assumed to behave like a denser fluid and the resuspension is seen as an entrainment process, mainly controlled by the flow and the vertical density gradient induced by the mud suspension. Based on experiments, Kranenburg & Winterwerp (1997) demonstrated that the entrainment rate under certain conditions could be approximated using

$$w_e = u_* \left(\frac{c_1}{c_2 + Ri_0} \right)^{0.5} \tag{8}$$

where c_1 and c_2 are constants, w_e is the ascent of the lutocline, u_* is the shear velocity and Ri_0 a Richardson number (itself based on u_*, the density jump across the lutocline and the depth H). By comparison it is seen that the structure of the two relations is similar. However, the last one appears more appropriate for CBS layers, as shown by Petersen & Vested (2000) within the COSINUS project.

3. SCHEMATIC CASES

Two test cases were designed in order to examine the ability of the different numerical models to correctly reproduce cohesive sediment behaviour, and to compare their predictions : a 1DV case and a 2DV schematic estuary case.

3.1. The 1 DV « Siltman » case

This idealised case was defined by Winterwerp (1998) in the Siltman project. It was used here in order to compare the vertical processes within the different models, and in particular the damping of turbulence by suspended sediment (Violeau et al. 2000). The parameters used were as follows : constant water depth (16 m) and mean velocity (0.2 m/s), bed roughness height of 10^{-3} m, constant settling velocity of 0.5 mm/s, and a sediment density of 2650 kg/m^3. Two initial mean concentrations are considered : one concentration below the saturation concentration ($c_0 = 0.010$ g/l), and one around saturation ($c_0 = 0.023$ g/l). The model governing equations are :

$$\begin{aligned} \frac{\partial u}{\partial t} + \frac{1}{\rho}\frac{\partial P}{\partial x} &= \frac{\partial}{\partial z}\left(\left(\nu + \nu_t\right) \frac{\partial u}{\partial z} \right) \\ \frac{\partial c}{\partial t} - \frac{\partial w_s c}{\partial z} &= \frac{\partial}{\partial z}\left(\left(K + K_s\right) \frac{\partial c}{\partial z} \right) \end{aligned} \tag{9}$$

in which u is the horizontal velocity, P the pressure, ν and ν_t the molecular and eddy viscosities, c the suspended sediment concentration, w_s the settling velocity, K and K_s the molecular and eddy diffusivities, x and z the horizontal and vertical upward co-ordinate and t the time. The pressure gradient was adjusted to maintain a constant flow rate :

$$\frac{1}{\rho}\frac{\partial P}{\partial x} = -\frac{u_*^2}{h} + \frac{\bar{u} - u_o}{T_{rel}} \tag{10}$$

in which h is the water depth, u_* the shear velocity, $\bar{u}$ the computed depth-averaged velocity, u_o the prescribed depth-averaged velocity, and T_{rel} a relaxation time. For this test case, there is no flux from the bed and turbulence damping was modelled through Munk-Anderson and Kranenburg functions.

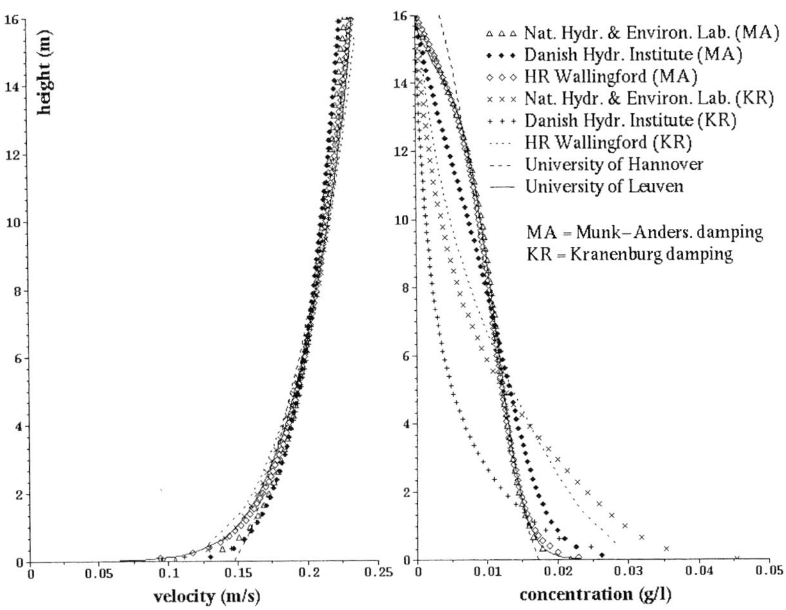

Fig. 1 - 1DV case : velocity and concentration profiles for $c_0 = 0.01$ g/l.

The results are plotted in *figures 1* and *2*. The models show a good agreement regarding velocities, but the concentration profiles are quite different : the models using Munk-Anderson damping functions match best with each other, while Kranenburg damping functions give more discrepancies with a stronger stratification, particularly with mixing length models and when considering an initial concentration of 0.023 g/l. Toorman (2000) produced evidence that the Kranenburg functions are inconsistent and generate two much damping when considering high Richardson numbers. The most likely cause of the differences between numerical results when using the Munk-Anderson functions is that differences between the various numerical schemes generate varying amounts of numerical diffusion. One should keep in mind that the real behaviour of sediment is still subject to a lack on knowledge, even in a simple case. Toorman (1999) has shown that for a certain range of the shear velocities there exist two solutions : a saturated one and a non-saturated one. Which solution a model predicts depends on the implementation of the numerical method (especially with respect to numerical diffusion). In particular, with c_0 = 0.023 g/l LNHE's model leads to a sudden collapse of the sediment profile (not plotted here).

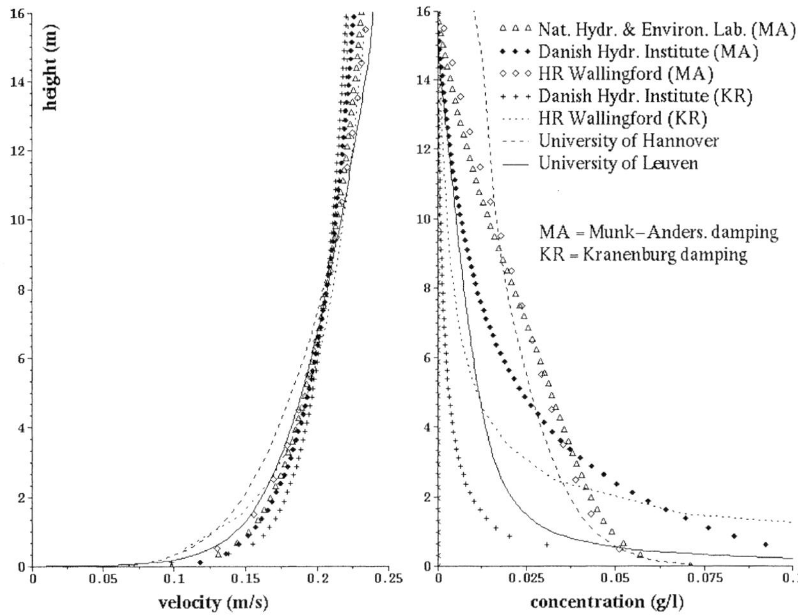

Fig. 2 - 1DV case : velocity and concentration profiles for $c_0 = 0.023$ g/l.

One of the important findings from this exercise was that the shear velocity can be strongly dependent on the scheme, especially when the grid is very coarse. The profiles of eddy viscosity, diffusivity, and flux Richardson number (not plotted here) generally show enormous differences from one code to another. As a conclusion, it seems that the correct prediction of the shear velocity is one of the most important conditions for accurate sediment transport modelling ; this idea was confirmed with the schematic estuary case (see section below). Nevertheless, some reasons for the discrepancies between the model predictions are still unknown, and further extensive testing will be required to identify their root causes.

3.2. The schematic estuary case

A 2DV schematised estuary (Cheviet et al. 2000a) was tested to investigate the horizontal advection processes, using the model designed by Pierre Le Hir (1997) with a representative bathymetry of the Loire river (see *figure 3*). Le Hir's model reproduced numerically the main features of sediment transport in the Loire estuary, which is 106 km long. The width of the estuary was modified here to a constant value of 600 m. Water level was prescribed as a sinusoidal function at the estuary mouth. At the upstream boundary, the river discharge was imposed to be 300 m^3/s. The simulations undertaken covered 12 tidal periods. The initial mass of sediment was imposed as 8 cm of available bed material between x = 10 km and x = 70 km ; there was no initial suspended matter or input of sediment through the open boundaries. In addition, to avoid 3D effects there was no friction on the lateral boundaries. Turbulence damping was modelled through Munk-Anderson damping functions, and an Ariathurai-

Partheniades' model was used for the erosion process. To take into account the effect of hindered settling, a formulation given by Winterwerp (1999) was tested :

$$w_s = w_{sr} \frac{(1-\Phi_*)^m (1-\Phi_p)}{1+2.5\Phi} \tag{11}$$

in which w_{sr} = 1 mm /s is the settling velocity of an individual mud floc in still water, Φ = volumetric concentration of mud flocs = c/c_{gel}, c_{gel} = 80 g / l is the gelling concentration, Φ_* = min $(1,\Phi)$, $\Phi_p = c / \rho_s$, ρ_s = 2650 g/l is the density of primary sediment particles, m = 4 is an exponent accounting for non-linear effects.

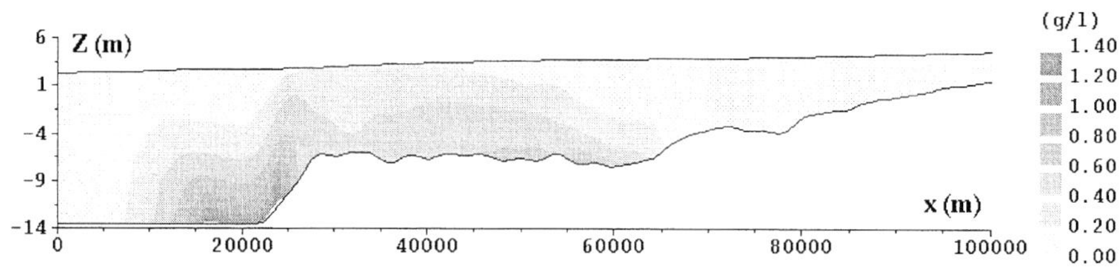

Fig. 3 - Schematic estuary case. Typical sediment distribution one hour before low tide.

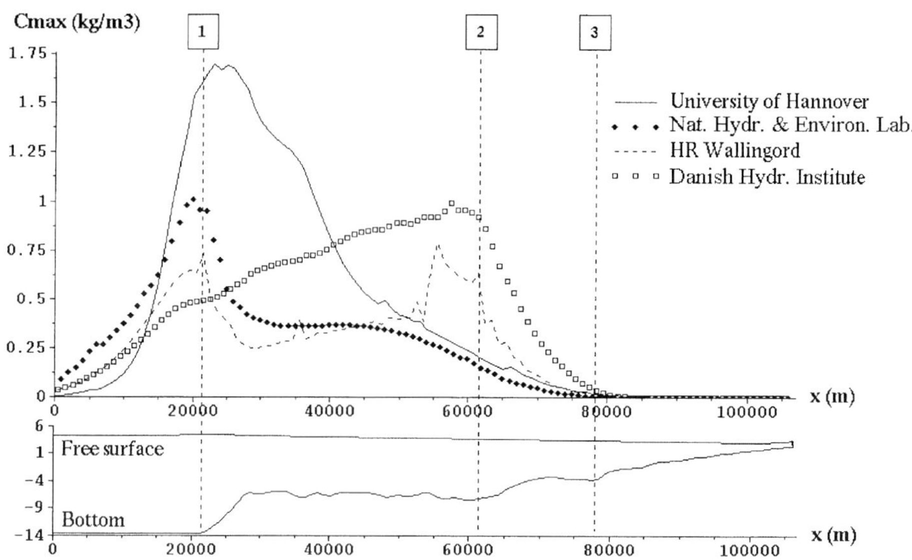

Fig. 4 - Schematic estuary case. Longitudinal plots of maximum depth-averaged concentration during a neap tidal period.

Figure 4 shows the maximum value of depth-averaged concentrations during a neap tidal period, computed with the four models used for this case. All the models (except DHI's) show a peak of concentration around x = 20 km, resulting from the strong bathymetric slope : high turbidities accumulate at this location at low-tide slack, and then are eroded by high currents. HR Wallingford's results reach a second peak around x = 60 km ; Le Hir (1997) had also noticed two separate peaks : he argued that the downstream turbidity patterns are related to trapped sediment generated by the sudden increase of depth, whereas the upstream structures represent the turbidity maximum. The upstream limit of the sediment excursion is similar for all the models, around x = 80 km. One explanation of the different shapes of the curves is high dependence of the models on the horizontal discretization of bathymetry, depending on the numerical methods. As a matter of fact, velocity profiles (not plotted here) show high discrepancies, probably resulting from the bottom boundary shear stress. In order to investigate this hypothesis, a run was performed after a model calibration described as follows : since no measurements were available for this schematic case, LNHE's results were arbitrarily chosen as a reference. DHI and UHA's models were calibrated to match the LNHE surface elevation, by altering the bottom friction coefficient. Sediment transport was then simulated with the new velocities resulting from this calibration. The results are plotted on *figure 5*, showing a much better agreement between the models. However, discrepancies between the results still occur. The reasons are certainly due to the different numerical formulations and to the exact specifications of initial and boundary conditions, as for the 1DV case. Therefore, the precise reasons of the divergences are still unknown : it remains to be determined how much numerical diffusion each model generates, as well as the effects of the different implementations of boundary conditions.

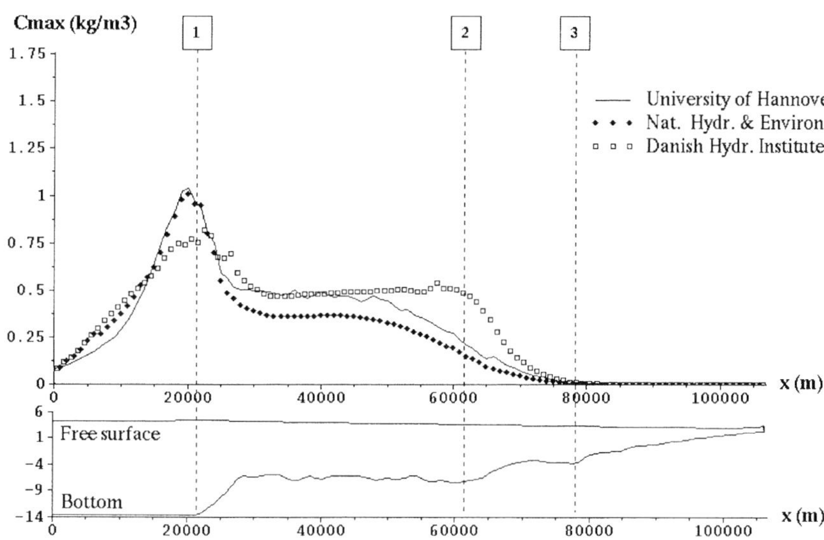

Fig. 5 - Schematic estuary case. Longitudinal plots of maximum depth-averaged concentration during a neap tidal period, after calibration of the models using the bottom friction coefficient.

4. APPLICATION TO REAL ESTUARIES

The models were used in the case of real estuaries, in order to show that the parameterisation models described in section 2 can lead to a correct prediction of the sediment behaviour in field situations. Since two companion papers have been published about the Tamar estuary and the Loire estuary, these are just summarized. The case of the Weser estuary is discussed in more details.

4.1. Two examples : the Tamar and the Loire estuaries

The Tamar Estuary is a macro-tidal estuary on the southern coast of England, with a relatively wide mouth that narrows to an approximately 100 m wide tidal channel in the upper 10 km. The estuary contains considerable amounts of cohesive sediments in the upper 12 km reach, which forms a pronounced turbidity maximum, with SPM concentrations up to 10 g/l. During the COSINUS September 1998 field campaign, extensive observations were made (Dyer et al. 2000). The observed sediment parameters involved *in situ* measurements of particle sizes and fall velocities, continuous profiling of SPM concentrations and measurements of bed sediment properties. It was decided to simulate the flow and sediment processes using a 2-dimensional vertical hydrodynamic model, covering the upper 15 km of the estuary, as this enables a good resolution of the vertical processes and an inclusion of advective processes. The model was set up using DHI's general three-dimensional model system, MIKE 3 (see Petersen & Vested 2000). The results demonstrate that it is possible to reproduce the observed variation of parameters as flow, salinity and suspended sediments ; they also confirmed the qualitative understanding of the dynamics of the estuary. Improvement of the model results has been demonstrated using *i)* a parameterisation describing the dynamics of high concentrated bed layers, developed with reference to laboratory experiments (Bruens 2000 ; Petersen & Vested 2000), *ii)* the effects of turbulence damping (see section 2.1) and *iii)* relations linking fall velocity to flocculation dynamics (see section 2.2).

The Loire estuary is located in the south of French Britain, in the area of Nantes and Saint-Nazaire. Under the combined effects of tide and river discharge, the suspended matter has an oscillating motion, whose excursion depends on the tidal components and the river discharge. For a mean tide and a discharge of about 400 m^3s^{-1}, the excursion of particle motion is about 17.5 km, the effective motion down river being 3.9 km during one tidal period (Le Normant 1995). Concentrations in the upper part of the river vary between 20 and 50 mg/l. The turbidity maximum can reach several g/l. The 3-dimensional numerical model used for this work is TELEMAC-3D, developed by EDF (see Cheviet et al. 2000b & 2000c). The prediction of the centre of gravity of the computed turbidity maximum, its maximum upstream extension, as well as total suspended mass in the estuary and vertical concentration profiles, show good agreement with observations carried out by Migniot (1993) and Gallenne (1974). It is important to note that developments made within the COSINUS project greatly aided the calibration of the numerical model (see Cheviet et al. 2000c).

4.2. The Weser estuary

The Weser estuary is a partially mixed mesotidal estuary located at the southern coast of the North Sea. The tidal range is generally between 2.5 and 3.5 m. The estuary reach of approximately 70-80 km is bounded at the upstream end by the Hemelingen weir, whereas the mouth of the estuary is located near Bremerhaven. A turbidity maximum exists between km 42 and km 60 and was investigated during the MASEX 85 experiment (**M**ud **A**nd **S**uspended sediment **EX**periment) (Riethmüller et al., 1988). A five day period of this experiment was simulated, and *in situ* measurements of velocity, salinity and turbidity (1 m and 3 m above bottom) at different locations were available to be used as boundary condition and for the validation of the model.

The numerical model of the Weser estuary was developed by Lang (1990) and Malcherek (1995, 1996) within previous MAST projects. The mesh extends from the cross-section Q4 at km 59.5 to the Hemelingen weir at km -5. It consists of 1485 nodes and 2276 elements in the horizontal plane and 12 non-equidistant horizontal layers are used. The initial conditions and the open boundary conditions used in the present study are the same as Malcherek used in his model. The most important difference from the former work regards the turbulence modelling. In contrast to the previous used algebraic mixing length model, a k-ω model (see section 2.1) was developed for the present study. First experiences in the application of this turbulence model have shown that the sensitivity to the shear velocity and to the relationship between both turbulent quantities is obviously less than that in the k-ε model and thus the results are more reliable. It has two further improvements when compared to the k-ε model : it is possible to integrate through the viscous sublayer, and the model is known to perform better in flows with adverse pressure gradients (Fredsøe et al. 1999, Patel & Yoon 1995, Wilcox 1993).

At the start of the simulation, a 5 mm thick erodible bed was assumed to be present between km 57 and km 45. This area corresponds to the region of the observed turbidity maximum. As explained in section 2.2, the choice of the settling velocity formulation is of great importance for the simulation of the dynamics of the turbidity zone. The following formulation has been used :

$$w_s = 0.035c\frac{1+aG}{1+bG^2} \qquad (12)$$

where $a = 0.4$ and $b = 0.05$. This formulation incorporates both flocculation and shear induced floc breakup. The absolute velocity gradient G was computed directly from the turbulent quantities k and ω and the kinematic viscosity :

$$G = \sqrt{\frac{\varepsilon}{\nu}} = \sqrt{\frac{\beta^{*}\omega k}{\nu}} \qquad (13)$$

G varies over a tidal cycle in a range from 0 to 60 Hz. During slack tide G becomes small, and thus the settling velocity factor increases. For higher velocities, G becomes greater and the factor changes by an order of magnitude.

The results presented in *figures 6* and *7* are restricted to the application of the k-ω model without considering Richardson damping, i.e σ_s is here treated as a constant (Ri = 0). The agreement between the measured and the calculated suspended sediment concentrations obtained with the new turbulence model is excellent.

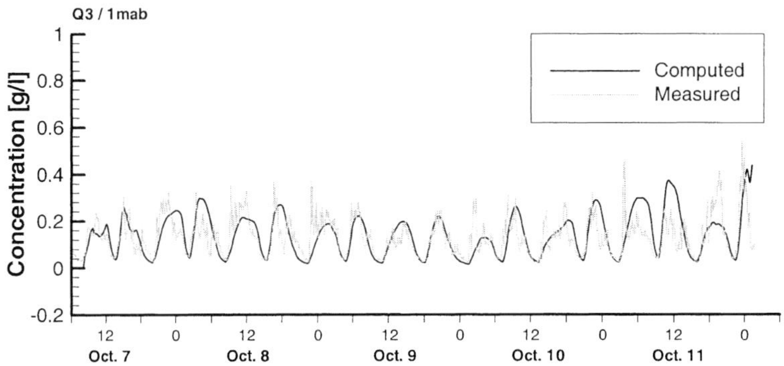

Fig. 6 - Comparison of measured and predicted concentrations at station Q3 (km 52,7) in the Weser Estuary 1 m above bottom.

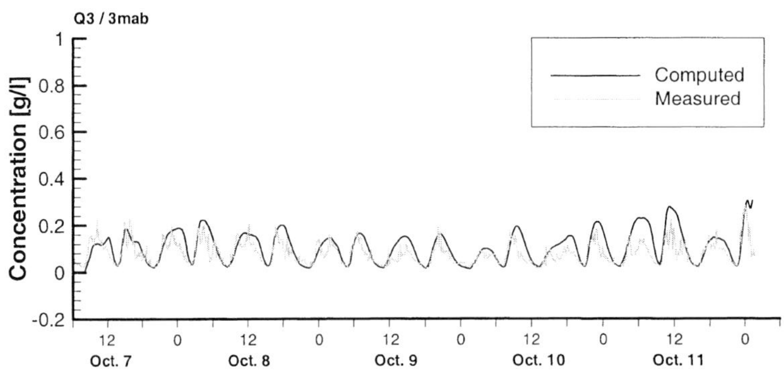

Fig. 7 - Comparison of measured and predicted concentrations at station Q3 (km 52,7) in the Weser Estuary 3 m above bottom.

It should be noted that the *k-ω* model described in this paper is in an early stage of development. First experiences in the application of this model to schematic test cases and to the Weser estuary are promising, and it is the first time in the authors knowledge that a *k-ω* model has been successfully applied to geophysical free surface flows. The advantages of this turbulence model must still be determined (a higher vertical resolution than that used in this study is required in order to allow the generation of a CBS layer in the numerical model). The influence of the vertical discretization as well as the consideration of consistent bottom boundary conditions (Toorman 2000) is also being looked at in ongoing studies.

5. CONCLUSIONS

For improving the numerical prediction of cohesive sediment transport, different parameterisations of specific processes have been implemented into numerical models, which have been tested on various cases, including real estuaries. Despite the latest progress in theory, the numerical models are still inconsistent (and they are inconsistent in a different way and probably to a different degree) for a number of reasons : inaccurate spatial description of the bathymetry, numerical schemes, incomplete and simplified modelling of a number of processes. It is not always possible to know, in any particular instance, which of these problems are relevant. Obviously, all the numerical models still have many limitations, and their results should always be interpreted with great care. Despite this lack of knowledge, this intercomparison between some of the most commonly used models, like the one described above, enables the modeller to better understand the difficulties which can occur when performing the modelling of sediment behaviour in a real estuary. In particular, precise calibration using extensive data seems to be necessary for numerically predicting the sediment behaviour.

ACKNOWLEDGEMENTS

This work was co-financed by the European Commission, Directorate XII for Science, Research & Development, through the COSINUS project within the framework of the MAST-3 programme, contract MASC3-CT97-0082.

REFERENCES

Ariathurai, C.R., 1974, A Finite Element Model for Sediment Transport in Estuaries. *PhD thesis, University of California, Davis, CA.*

Bruens, A.W., 2000, Laboratory Experiments on the Entrainment by a Concentrated Benthic Suspension. *Report No* 3-00, TU Delft.

Chen, C.J. and Jaw, S.Y., 1998, Fundamentals of Turbulence Modelling, *Taylor & Francis, Washington D.C.*

Cheviet, C., Violeau, D., Le Normant, C., 2000, MAST3-COSINUS European Project - Intercomparison of the Results of Several Numerical Models on a Schematic Estuary Case. *Report No* HP-72/2000/026/A, *Electricité de France / LNHE.*

Cheviet, C., Violeau, D., Guesmia, M., 2000, MAST3-COSINUS European Project - 3D-modelling of cohesive sediment transport in the Loire estuary (France). *Report No* HP-72/2000/048/A, *Electricité de France / LNHE.*

Cheviet, C., Violeau, D., Guesmia, M., 2000, Numerical simulation of cohesive sediment transport in the Loire estuary with a three-dimensional model including new parameterisations. *Proceedings* INTERCOH-2000, *Elsevier, Coastal and Estuarine Fine Sediment Processes*, ed. J.C. Winterwerp and C. Kranenburg, this volume.

Dyer, K.R., Bale, A.J., Christie, M.J., Feates, N., Jones, S. and Manning, A.J., 2000, The Properties of Suspended Sediment in an Estuarine Turbidity Maximum. *Proceedings* INTERCOH-2000, *Elsevier, Coastal and Estuarine Fine Sediment Processes*, ed. J.C. Winterwerp and C. Kranenburg, this volume.

Fredsøe, J., Andersen, K.H., Sumer, B.M., 1999, Wave Plus Current over a Ripple-Covered Bed. *Coastal Engineering* **38**, 177-221.

Gallenne, B., 1974, Les accumulations turbides de l'estuaire de la Loire. Etude de la crème de vase. *Doctoral Thesis, University of Nantes.*

Kranenburg, C. and Winterwerp, J.C., 1997, Entrainment of Fluid Mud Layers. I : Entrainment Model. *J. Hydraulic Engineering*, ASCE, **123**(6), 504-511.

Kranenburg, C., 1998, Saturation concentrations of suspended fine sediment. Computations with the Prandtl mixing-length model. *Report No*.5-98, *Faculty of Civil Engineering and Geosciences, Delft University of Technology.*

Lang, G., 1990, Zur schwebstoffdynamik von trübungozonen in astuarien. *Report No* 26, *Institut für Strömungsmechanik und Elektronisches Rechnen im Bauwesen, Hannover University.*

Le Hir, P., 1997, Fluid and sediment « integrated » modelling application to fluid mud flows in estuaries. *Cohesive Sediments, proc.* INTERCOH '94.

Le Normant, C., 1995, Modélisation numérique tridimensionnelle des processus de transport des sédiments cohésifs en environnement estuarien. *Doctoral Thesis for the Institut National Polytechnique de Toulouse, report No* HE-42/95/028/A, *Electricité de France / LNHE.*

Malcherek, A., 1995, Mathematische Modellierung von Strömungen und Stofftransportprozessen in Ästuaren. *Report No.* 44, *Institut für Strömungsmechanik und Elektronisches Rechnen im Bauwesen, Hannover University.*

Malcherek, A., Markofsky, M., Zielke, W., Peltier, E., Le Normant, C., Teisson, C., Cornelisse, J., Molinaro, P., Corti, S., Greco, G., 1996, Three Dimensional Numerical Modelling of Cohesive Sediment Transport in Estuarine Environments. *Final report to the EC contract* MAS2-CT92-0013.

Mehta, A., 1989, On Estuarine Cohesive Sediment Suspension Behaviour. *J. Geophysical Research*, **94**(C10), 14303-14314.

Migniot, C., 1993, Bilan de l'hydrologie et de l'hydrosédimentaire de l'estuaire de la Loire au cours des deux dernières décennies. *Agence pour la Protection de l'Environnement de l'Estuaire de la Loire, Port Autonome de Nantes Saint-Nazaire.*

Partheniades, E., 1965, Erosion and Deposition of Cohesive Soils. *J. Hydraulic Division*, **91**,105-139.

Patel, V.C. and Yoon, J.Y., 1995, Application of Turbulence Model to Separated Flow over Rough Surfaces. *J. Fluids Eng.* **117**, 234-241.

Petersen, O. and Vested, H.J., 2000, An Operational Description of Vertical Exchange Processes in Numerical Mud Transport Modelling. *COSINUS report. DHI 2000, Second Draft version.*

Petersen, O., Vested, H.J., Manning, A.J., Christie, M.J., Dyer, K.R., 2000, Numerical modelling of mud transport in the Tamar estuary. *Proceedings* INTERCOH-2000, *Elsevier, Coastal and Estuarine Fine Sediment Processes*, ed. J.C. Winterwerp and C. Kranenburg, this volume.

Riethmüller, R., Fanger, H.U., Grabemann, I., Krasemann, H.L., Ohm, K., Böning, J., Neumann, L.J.R., Lang, G., Markofsky, M., Schubert, R., 1988, Hydrographic Measurements in the Turbidity Zone of the Weser Estuary. *Physical Processes in Estuaries, edited by J. Dronkers and W. van Leussen, pp. 332-344, Springer Verlag, Berlin Heidelberg.*

Spearman, J.R. and Roberts, W., 2000, Parameterisation of flocculation models for applied sediment transport modelling. *Proceedings* INTERCOH-2000, *Elsevier, Coastal and Estuarine Fine Sediment Processes*, ed. J.C. Winterwerp and C. Kranenburg, this volume.

Taylor, J.S., 1973, Buoyancy Effects in Fluids. *Cambridge University Press.*

Toorman, E.A., 1999, Numerical simulation of turbulence damping in sediment-laden flow. Part 1 : the « Siltman » test case and the concept of saturation. *Report No* HYD/ET/99.2, *Hydraulics Laboratory, Katholieke Universiteit Leuven.*

Toorman, E.A., 2000, Parameterisation of Turbulence Damping in Sediment-Laden Flows. *Report* HYD/ET/00/COSINUS3, *Hydraulics Laboratory, Katholieke Universiteit Leuven.*

Toorman, E.A., Bruens, A.W., Kranenburg, C., Winterwerp, J.C., 2000, Interaction of Suspended Cohesive Sediment and Turbulence. *Proceedings* INTERCOH-2000, *Elsevier, Coastal and Estuarine Fine Sediment Processes*, ed. J.C. Winterwerp and C. Kranenburg, this volume.

Violeau, D., Le Normant, C., Cheviet, C., 2000, MAST3-COSINUS European Project. Siltman 1DV case. Comparison of Several Numerical Models. *Report No* HP-72/2000/042/A, *Electricité de France / LNHE.*

Wilcox, D.C., 1993, Turbulence Modelling for CFD. *DCW Industries, Inc., La Cañada, California.*

Winterwerp, J.C., 1998, *Siltman* - Analysis of Field Measurements. Delft Hydraulics report *No* Z2263.

Winterwerp, J.C., 1999, On the Dynamics of High-Concentrated Mud Suspensions, *Doctoral Thesis for the Technical University of Delft.*

Winterwerp, J.C. and Kranenburg, C., 1997, Entrainment of Fluid Mud Layers. II : Experiments and Model Validation. *J. Hydraulic Engineering*, ASCE, **123**(6), 512-519.

Chapter 2:

High-concentrated mud suspensions

Fine Sediment Dynamics in the Marine Environment
J.C. Winterwerp and C. Kranenburg (Editors)

Tidal asymmetry and variability of bed shear stress and sediment bed flux at a site in San Francisco Bay, USA

Matthew L. Brennan[a], David H. Schoellhamer[b], Jon R. Burau[b] and Stephen G. Monismith[a]

[a]Environmental Fluid Mechanics Laboratory, Dept. Civil & Environmental Engineering, Stanford University, Stanford, CA, 94305-4020 USA.

[b]U. S. Geological Survey, Placer Hall, 6000 J St., Sacramento, CA 95819 USA

The relationship between sediment bed flux and bed shear stress during a pair of field experiments in a partially stratified estuary is examined in this paper. Time series of flow velocity, vertical density profiles, and suspended sediment concentration were measured continuously throughout the water column and intensely within 1 meter of the bed. These time series were analyzed to determine bed shear stress, vertical turbulent sediment flux, and mass of sediment suspended in the water column. Resuspension, as inferred from near-bed measurements of vertical turbulent sediment flux, was flood dominant, in accordance with the flood-dominant bed shear stress. Bathymetry-induced residual flow, gravitational circulation, and ebb tide salinity stratification contributed to the flood dominance. In addition to this flow-induced asymmetry, the erodibility of the sediment appears to increase during the first 2 hours of flood tide. Tidal asymmetry in bed shear stress and erodibility help explain an estuarine turbidity maximum that is present during flood tide but absent during ebb tide. Because horizontal advection was insignificant during most of the observation periods, the change in bed mass can be estimated from changes in the total suspended sediment mass. The square wave shape of the bed mass time series indicates that suspended sediment rapidly deposited in an unconsolidated or concentrated benthic suspension layer at slack tides and instantly resuspended when the shear stress became sufficiently large during a subsequent tide. The variability of bed mass associated with the spring/neap cycle (about 60 mg/cm^2) is similar to that associated with the semidiurnal tidal cycle.

KEY WORDS
estuaries, San Francisco Bay, suspended sediment, bed shear stress, stratification

1. INTRODUCTION

The flux of sediment between the bed and the water column, in response to variations in bed shear stress that occur in a partially stratified estuary, is examined in this paper. The net bed flux changes from erosion to deposition at tidal time scales, thereby controlling the amount of suspended sediment available for transport by the flow. When flow direction alternates with the tides, the tidally averaged (net) transport determines the ultimate fate of sediments. This net transport is created by asymmetry of flow and sediment response between flood and ebb tide. For example, Dronkers (1986) hypothesized that the peak suspended sediment concentration at the landward limit of salt intrusion in an estuary may be created by an asymmetry in bed shear stress that creates landward flow to counteract the seaward river discharge.

Erosion from the bed to the water column is initiated by the shear stress exerted on the bed by the flow. As sinusoidal tidal currents accelerate, sediment is resuspended from the bed. Subsequently, the flow decelerates to the point that the net bed flux becomes dominated by deposition. However, the timing and magnitude of bed shear stress differs with each tidal phase, leading to a corresponding variability in bed flux. In the field, measurements of the Reynolds stress in the near-bed region provide a reasonable estimate of bed shear stress (Trowbridge et al., 1999).

The sources of variability that are examined in this paper are flood and ebb tides, spring and neap tides, and salt stratification and destratification. When flood and ebb react differently to channel bathymetry, a residual flow is created (Fischer et al., 1979). Spring-neap variations in barotropic forcing are caused by phasing differences between solar and lunar tidal components.

Vertical density gradients contribute to the bed shear stress asymmetry in a complex interaction with the barotropic forcing. The horizontal density gradient produces gravitational circulation that is directed landward at the bed and seaward at the surface, thereby strengthening bed shear stress on flood tide and weakening it on ebb tides. In addition to horizontal density gradients, vertical density gradients intermittently damp turbulent eddies, thereby reducing bed shear stress. Ebb tides tend to create stratification because vertical velocity shear advects fresher surface water over denser, saltier water, a process termed "strain induced periodic stratification" (SIPS) by Simpson et al. (1990). However, tidal flow also generates mixing, which breaks down stratification. Therefore, stratification is most prevalent during slack tides or less energetic neap tides (Stacey et al., 1999).

The balance between turbulence generation by fluid shear and turbulence suppression by density stratification is given by the gradient Richardson number, R_i

$$R_i=-(g\partial\rho/\partial z) / (\rho_0 (\partial u/\partial z)^2)$$

where g=gravitational constant, ρ=fluid density as a function of only salinity, ρ_0=mean fluid density, and z is the coordinate directed positive upwards from the bed. Turbulence tends to mix the water column when R_i is near zero. When $R_i>0.25$, turbulence is suppressed by density stratification (Itsweire et al., 1993), greatly diminishing mixing and allowing suspended particles to settle.

The relationship between bed flux and bed shear stress during a pair of field experiments conducted at a site in San Francisco Bay, California, is examined in this paper. Flow velocity, vertical density profiles, and suspended sediment concentration (SSC) were measured continuously throughout the water column over a 24-hour period in July 1997 and over a 9-day period in October 1999. These time series were analyzed to determine bed shear stress, vertical turbulent sediment flux, and mass of sediment suspended in the water column.

2. METHOD

2.1 Site bathymetry, hydrodynamics, and sediment characteristics

All the field data presented in this paper were collected in Suisun Cutoff, a tidal channel in northern San Francisco Bay, California (Figure 1). This channel is 2 km long and 500 m wide and has a nearly rectangular cross-section. While most of the channel is 10 m deep, a sill 2 m deep bounds the channel to the west and the east end deepens to 20 m. Approximately 65 km of water separates the site from the ocean.

A preliminary study was conducted on July 29-30, 1997, when the river discharge was approximately 300 m^3/s and spring tides prevailed. The second, more extensive, study from October 15-27, 1999, occurred just 1 month before the winter storm season, so river discharge at 100 m^3/s was near the year's minimum. In contrast, the annual peak flow averages approximately 3,000 m^3/s, and typically occurs between January and March. This second study began during a neap tide with a tidal range of 1.2 m and continued to a spring tide with a range of 1.5 m. Both experiments were situated in the center of Suisun Cutoff (approximately the top vertex of the triangle in Figure 1).

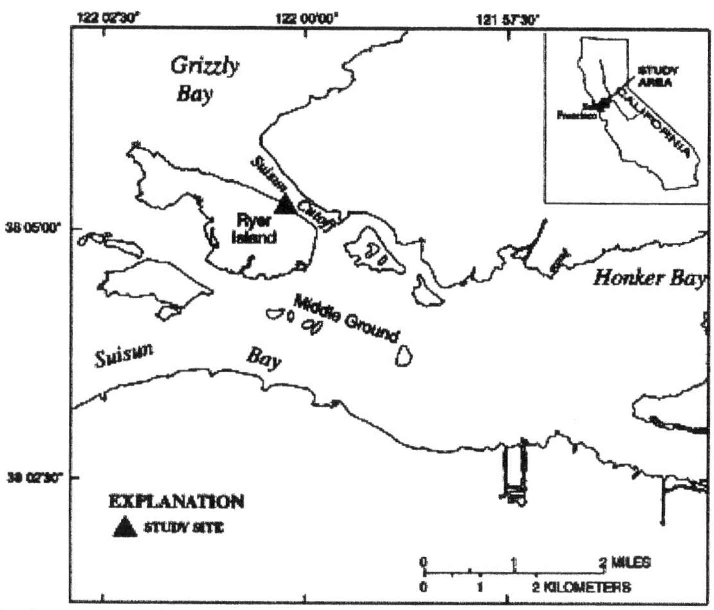

Figure 1. Study area - northern San Francisco Bay, California.

During the 1997 experiment, the balance created between the river discharge and the intruding salt resulted in salinity values of 4 to 9 on the practical salinity scale. With less river discharge during the 1999 study, the salinity increased, spanning a range from 8 to 16. Vertical gradients in salinity varied between well-mixed and stratified conditions, as is typical of a partially stratified estuary. Because maximum SSC observed during this study was approximately 200 mg/l and the vertical salinity gradient varied from 0 to 1 m^{-1}, salinity and temperature were the only constituents assumed to affect water density.

Although this work did not include sediment analysis, work by other researchers at nearby sites provides an estimate of sediment characteristics in Suisun Cutoff. Disaggregated size class analysis by Kineke and Sternberg (1989) indicates that suspended sediment is approximately 40% clay, 55% silt, and 5% sand, while bed sediment is 32% clay, 52% silt, and 16% sand. Sediment settling velocity has been estimated to be 10^{-3}-10^{-2} m/s (Sternberg et al, 1986, Kranck and Milligan, 1992). Kranck and Milligan (1992) also hypothesize that all but the largest flocs (>370 μm) are "relatively stable particles which settled and were resuspended without much floc breakup."

2.2 Instrumentation

In 1997, we collected vertical profiles of velocity, density, and SSC at an anchor station for nearly 24 hours. A boat-mounted acoustic Doppler current profiler (ADCP) returned velocity vectors from every 0.25 m of the water column at a frequency of approximately 1 Hz. Concurrently, we sampled the vertical structure of salinity and SSC from the water surface to 0.25 m above the bed every 10 minutes with a manually operated instrument package. This instrument package contained probes for conductivity and temperature to estimate salinity, an optical backscatter sensor (OBS) to estimate SSC, as well as a pressure transducer to measure depth.

The 1999 experiment's instrumentation consisted of a pair of bed-mounted instrument frames and an autonomous water column profiler. These three instrument packages were located within 75 m of each other. One instrument frame carried an acoustic Doppler velocimeter (ADV) and an OBS that sampled the same measurement volume at 0.97 m above the bed. The ADV measured the three components of velocity (u, v, w) and acoustic backscatter intensity at 25 Hz. The OBS collected data at approximately 6 Hz. The frame also carried a conductivity sensor and a temperature sensor for determining salinity. The second frame carried an ADCP mounted 0.4 m above the bed. This instrument sampled velocity vectors in 25-cm intervals from 1.25 m above the bed to just below the water surface, collecting a complete profile at approximately 2 Hz. Communication cables connected these instruments to computers on board a boat anchored at the site.

The autonomous profiler consisted of a mechanical winch assembly mounted on the moored boat that lowered and raised an instrument package through the water column every 15 minutes. The instrument package measured vertical profiles of salinity and SSC, as in the 1997 experiment. However, the autonomous profiler used in 1999 only traversed water column from 1 m to 7 m below the surface, leaving the bottom 3 m unsampled. The 1997 data set provides the most complete picture of the near-bed region.

2.3 Data processing

All of the flow and sediment statistics were averaged into 10-minute blocks. This time interval reflects a balance between reducing uncertainty by increasing the number of samples and maintaining stationarity (Gross and Nowell, 1983). Whereas the ADCP has an internal compass and tilt sensor to establish its reference frame, the ADV data were rotated such that the mean vertical and cross-stream velocities in each 10-minute block were zero. In instances when the data from the autonomous profiler and ADCP were used simultaneously, the velocity was interpolated to coincide with the 15-minute interval of the autonomous profiler.

Backscatter intensity was converted to SSC through comparison with SSC estimates from gravimetric analysis of water samples collected in situ. For an OBS, the linear relationship between SSC and output voltage was determined using a repeated median fit (Buchanan and Ruhl, 2000) to decrease the influence of outliers on the regression. For the ADV, the intensity of returned acoustic energy is proportional to the logarithm of SSC (Thorne et al., 1993; Kawanisi and Yokosi, 1997). Before comparing ADV backscatter to water samples, the raw ADV backscatter data were treated to remove outliers. First, when measurements collected simultaneously from each of the ADV's three receivers differed by more than 5% of the mean intensity, the disparate values were discarded. Then, after averaging the three measurements, points further than five standard deviations from the 10-minute mean also were discarded. Between these two filters, no more than 2% of the data was eliminated.

The ADV estimates (SSC_{ADV}) of mean SSC are compared to the OBS estimates (SSC_{OBS}) in Figure 2. SSC_{ADV} only diverges significantly from SSC_{OBS} for approximately 1% of the data pairs. Values of SSC_{ADV} tend to be less than SSC_{OBS} at lower SSC values and greater than SSC_{OBS} at higher SSC values. The variance between the two measurements is a function of SSC, as expected from laboratory tests that indicate the direct relationship between OBS variance and SSC (Downing and Beach, 1989). When the absolute value of the difference between the two estimates is normalized by SSC_{OBS}, the mean percent error is 16%, which is comparable to the agreement of 10% between optical and acoustic backscatter sensors for sand, reported by Osborne et al. (1994).

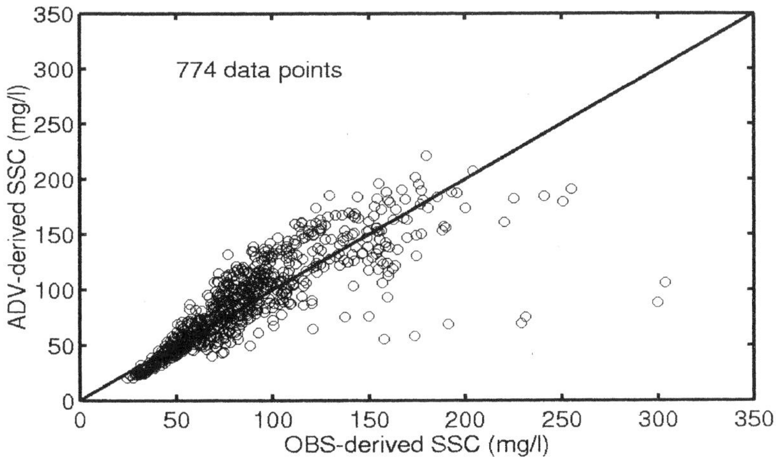

Figure 2. Comparison of ADV-derived SSC and OBS-derived SSC.

Concentration measurements derived from the ADV backscatter intensity were combined with the coincident velocity measurements from the ADV to estimate the turbulent vertical sediment flux, <w'c'>, where < > indicate a time average and primes indicate fluctuations from the mean.

Suspended mass was calculated from the vertical SSC profiles. For the 1997 data, suspended mass was calculated by summing SSC and multiplying by the uniform vertical

SSC sampling interval of 0.25 m, and the lowest sample point was 0.25 m above the bed. The 1999 data had nonuniform vertical spacing, suspended mass was calculated with the trapezoidal rule, and the lowest sample point was 0.97 m above the bed. Suspended mass will be underestimated because SSC measurements did not extend down to the bed, especially for the 1999 data.

Vertical profiles of density gradient were calculated from the CTD casts and vertical profiles of velocity gradient were calculated from the ADCP velocity profile data. The density and velocity gradient profiles then were used to calculate vertical profiles of the gradient Richardson number (R_i), which vary with time.

3. TIDAL ASYMMETRY OF BED SHEAR STRESS AND ERODIBILITY

3.1 Neap tide

The bed shear stress opposes the mean streamwise currents. These mean currents (U) measured at 0.97 m above the bed during the 1999 experiment, were flood dominated, as indicated in Figure 3A. This flood dominance runs counter to the estuary's overall seaward discharge of river water. This flood asymmetry probably results from residual flow created by the region's complex bathymetry (Stacey, 1996) and gravitational circulation, which contributes to landward flow in the near-bed region (Hansen and Rattray, 1966).

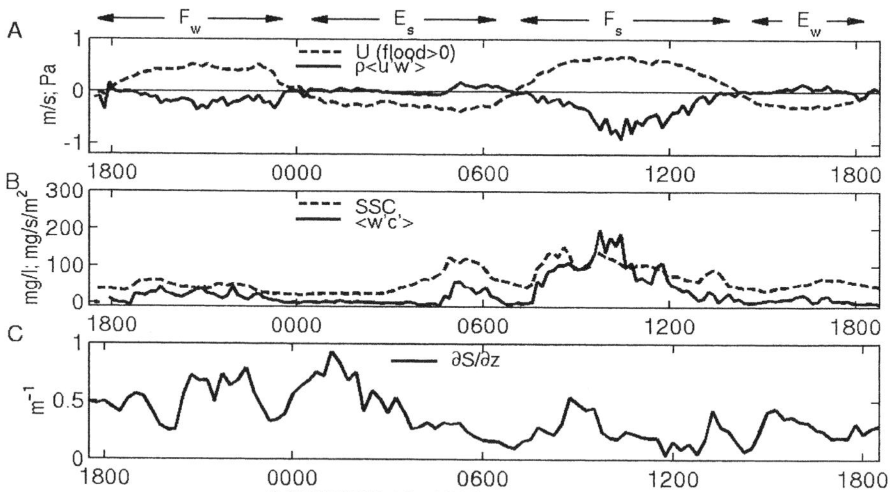

Figure 3. Measurements from October 1999 at 0.97 m above the bed: A) Time series of mean velocity (U) and Reynolds stress (<u'w'>) during neap tides. B) Time series of mean suspended sediment concentration (SSC), turbulent sediment flux (<w'c'>). C) Top-bottom vertical salinity gradient (∂S/∂z).

As expected in light of the mean currents, the bed shear stress, as measured by the Reynolds stress, ρ<u'w'>, also favored flood tides. In fact, the diurnal inequality created a cycle of four tides with different bed shear stress characteristics. The stronger flood (F_s) attained peak bed shear stress of 0.9 Pa and the weaker flood (F_w) attained a peak just over

0.4 Pa (Figure 3A). In contrast, the bed shear stress during the stronger ebb (E_s) and weaker ebb (E_w) reached a peak of just 0.1 Pa. As indicated in Figure 3C, strong salinity stratification existed between 1800 hours and 0430 hours. This period of strong stratification reduced bed shear stress during both the weaker flood and stronger ebb. This salinity stratification was particularly strong on the ebb tide, from 0000 to 0430 hours, when bed shear stress vanished even though the mean velocity reached speeds of 30 cm/s. This complete suppression of turbulence and bed shear stress were precipitated by SIPS, the ebb tide advection of fresh water on top of saltier water (Stacey et al., 2000).

ADV-derived time series of the turbulent sediment flux, <w'c'>, and the mean SSC are shown in Figure 3B. Over the course of a neap tidal day, the strong flood dominated the bed flux asymmetry. Turbulent sediment fluxes during the strong flood exceeded 150 mg/s/m^2, a magnitude at least three times the peak flux during the other tidal phases.

In the period 1800 to 0430 hours depicted in Figure 3B, changes in SSC correlated more strongly with ρ<u'w'> and <w'c'> than the mean flow, because the flow was stratified. The turbulent sediment flux, <w'c'>, was considerably less during the weaker flood (1800-0000 hours) than <w'c'> during unstratified spring tides of only slightly larger magnitude (e.g. Figure 4B, 0000-0600 hours). At the beginning of the strong ebb in Figure 3B (0000-0430 hours), the mean flow accelerated to a magnitude that readily induced erosion during other tidal periods, but <w'c'> vanished, indicating that erosion may have been eliminated completely. After 0430 hours, when stratification had diminished, <w'c'> and SSC both increased rapidly.

Even though ρ<u'w'> and <w'c'> are smaller and last for a shorter duration during the strong ebb, as compared to the weak flood, the peak SSC attained at the end of the strong ebb (0530 hours) is 30-50 mg/l larger than the peak attained during the weak flood. This suggests that horizontal advection contributed to SSC during the ebb. The rising edge of SSC that occurred during 0300-0430 hours, before the sharp increase in ρ<u'w'> and <w'c'>, is another signal of advection-induced SSC.

3.2 Spring tide

The more energetic spring tides later in the experiment produced larger bed shear stresses in all phases of the tidal day (Figure 4A). The flood still dominated, with the peak bed shear stresses of 1.1 Pa and 0.8 Pa during the stronger and weaker floods, respectively. Bed shear stress on the stronger ebb was comparable to that of the weaker flood, while the weaker ebb yielded peaks of only 0.2 Pa. Turbulent mixing was greater than during neap tide, such that periods of vertical stratification were weaker and lasted only 1-2 hours around slack tides (Figure 4C), minimizing stratification's influence on sediment bed flux.

In contrast to only one large resuspension event during a neap tidal day (Figure 3B), significant resuspension events occurred on three of four tides comprising the spring tidal day (Figure 4B). The turbulent sediment flux peaked at 120-180 mg/s/m^2 during the weaker flood, the stronger flood, and the stronger ebb, as compared to the peak of only 20 mg/s/m^2 during the weaker ebb. Hence, the relative absence of resuspension during the weaker ebb leads to a net flood asymmetry in bed flux.

Around the time of peak flow velocity, decreasing SSC coincided more closely with the decline in the turbulent quantities, ρ<u'w'> and <w'c'>, than the deceleration of the mean flow. For example, SSC, <w'c'>, and ρ<u'w'> began to decline at 1400 hours on October 25, but the mean velocity stayed nearly constant for another hour (Figure 4B). This behavior parallels oobservations by Sanford and Halka (1993). These authors propose that this response may be modeled best by continuous deposition, rather than the widely-used model (Krone, 1962), in which deposition occurs only when the bed shear stress falls below a certain value. Sanford and Halka (1993) suggest two possible explanations for the disjoint between these deposition models; differences between laboratory and field flow scaling or the effect of a distribution of sediment sizes.

While <w'c'> was proportional to bed shear, the relation was not temporally uniform. Turbulent sediment flux was most responsive to ρ<u'w'> at the beginning of flood tides, as indicated by the steeper slope of data points from the first 2 hours of flood (Figure 5). The slope's steepness was less during all other times of the tidal cycle, which may indicate the sediment's degree of erodibility.

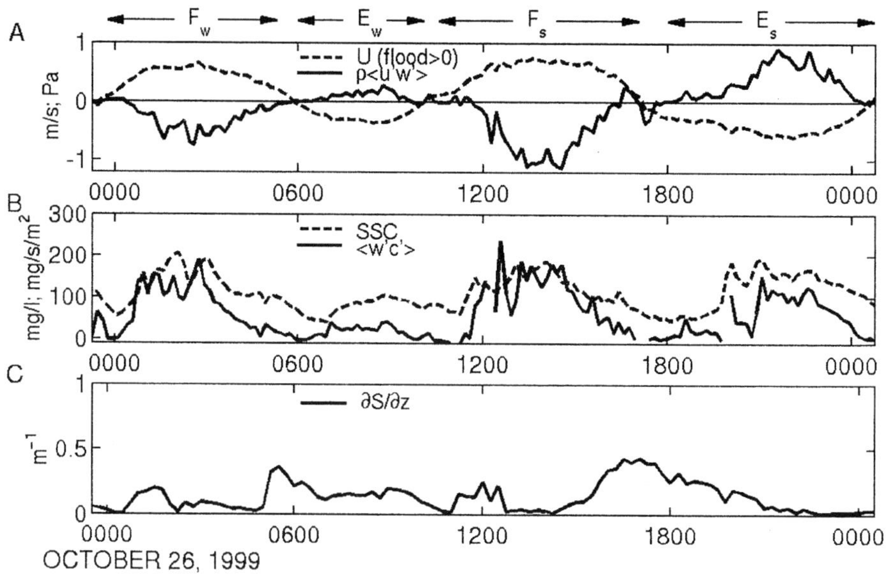

Figure 4. Measurements from October 1999 at 0.97 m above the bed: A) Time series of mean velocity (U) and Reynolds stress (r<u'w'>) during spring tides. B) Time series of mean suspended sediment concentration (SSC), turbulent sediment flux (<w'c'>). C) Top-bottom vertical salinity gradient (∂S/∂z).

Salinity enhancement of bottom sediment shear strength or bioturbation may explain this asymmetry in erodibility. Sediment deposited at the end-of-ebb salinity of 8 had less shear strength than sediment deposited at the greater end-of-flood salinity of 14. Sediment shear strength has increased to salinities of 10 in laboratory experiments (Parchure and Mehta, 1985). Or perhaps clams (*Potamocorbula amurensis*), which inhabit the bed in Suisun Cutoff at densities of several thousand per square meter, are a source of bioturbation during the beginning of flood tides (J. Thompson, U.S. Geological Survey, oral communication, 2000). They typically feed on the bed surface during ebb tides, which carry a larger nutrient load from river discharge. When the tide switches to flood, the clams actively bury themselves into the mud. Finally, the asymmetry may not be related to the sediment erodibility at all; it may be caused by horizontal advection. Because the magnitude of concentration fluctuations scale with the mean SSC, the increase in <w'c'> at the start of flood may result from advection of a more turbid water mass into the ADV control volume. Most of the time, however, SSC increases correspond to increased bed shear stress, so advection does not seem to be a likely explanation.

Whatever the mechanism, the asymmetry in erodibility combined with the asymmetry of bed shear to create a mean landward flux of sediment of 14 $g/s/m^2$ over the 8 days of ADV data from this experiment.

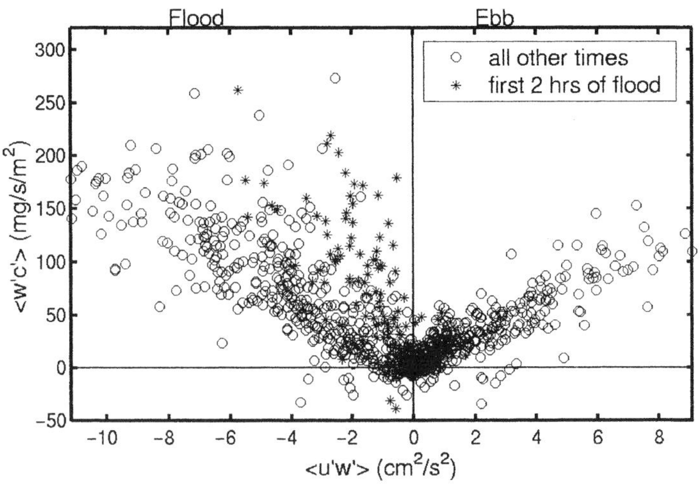

Figure 5. Turbulent sediment flux (<w'c'>) response to Reynolds stress (ρ<u'w'>) during the first two hours of flood (negative ρ<u'w'>) and all other times during the October 1999 experiment. Measurement volume is 0.97 m above the bed.

3.3 Effect on the estuarine turbidity maximum

Tidal asymmetry of bed shear stress and erodibility help explain an estuarine turbidity maximum in Suisun Bay at salinities of 1 to 6 during flood tides that is absent during ebb tides (Schoellhamer, 2001). During neap tides, tidal asymmetry of bed shear stress is most pronounced, and during (neap) flood tides, an estuarine turbidity maximum is regularly observed in the eastern side of Suisun Bay at salinities less than 5.

4. TIDAL VARIABILITY IN SUSPENDED AND BED MASS

Variability of SSC on the tidal time scale during the 1997 experiment was controlled by density stratification of the water column, vertical mixing, resuspension, and deposition. SSC varied inversely with the gradient Richardson number, R_i. As R_i decreased, turbulence and vertical mixing increased, which increased SSC. In 1997, SSC decreased during a slack after flood tide at 1500 hours on July 29 as turbulence in the water column diminished (Figure 6B). A weak ebb tide followed until 1800 hours that failed to disrupt stratification and resuspend the deposited bottom sediment. At 2000 hours, a flood tide mixed the water column and increased SSC. Turbulence vanished during a slack tide at 0000 hours on July 30 and SSC decreased as suspended sediment deposited. A stronger ebb tide mixed the water column and resuspended bottom sediment at 0300 hours.

Changes in SSC were controlled by local stratification, turbulent mixing, and settling — not horizontal advection. At slack tides, turbulence vanished and suspended sediment deposited on the bed. The time scale of settling (depth divided by settling velocity) is on the order of 1 hour, which is about the duration of suppressed turbulence at slack tide. If the subsequent tide was strong enough to break down stratification, then bottom sediment resuspended into the water column. The 1997 data were collected during a spring tide when increased tidal energy mixed the water column during most tides. Horizontal advection of suspended sediment did not appear to be significant, as indicated by rapid decreases in SSC at slack tide, when advection would be small and rapid increases in SSC were concurrent with generation of turbulence in the water column.

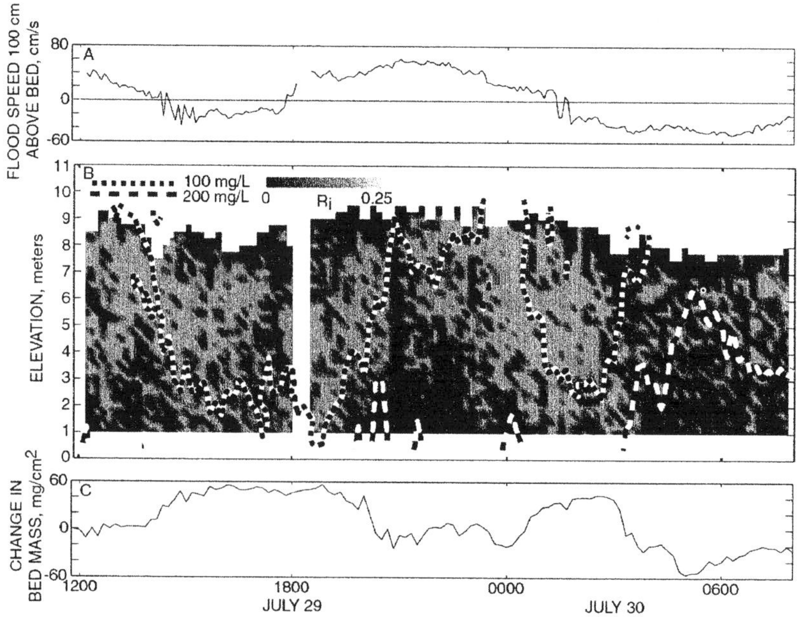

Figure 6. A) Flood current speed 100 cm above the bed. B) Vertical profiles of Richardson number and contours of suspended-sediment concentration. C) Change in mass of sediment on the bed, Suisun Cutoff, July 1997.

Because local deposition and resuspension controlled SSC, the change in bed mass can be assumed to be opposite of the change in suspended mass. Mass of suspended sediment was calculated by integrating the vertical profiles of SSC as described previously.

The time series of bed mass appears similar to a square wave (Figure 6C) due to bed flux. A greater quantity of sediment was on the bed near slack tides and during the weak ebb tide on July 29. When the velocity was great enough to resuspend sediment, about 60 mg/cm^2 were resuspended almost instantly. After this initial resuspension (sometimes called bulk erosion), there was no significant change in bed mass until the following slack tide because the remaining bottom sediment had sufficient shear strength to resist (mass) erosion. When the current and turbulence decreased, the same quantity of sediment (60 mg/cm^2) rapidly deposited. Assuming that the dry density of the deposit was on the order of 100 mg/cm^3, the depth of the deposit was on the order of several millimeters. After this initial deposition, the bed mass remained roughly constant because turbulence in the water column was able to keep the remaining suspended sediment in suspension. Thus, about 60 mg/cm^2 appeared to rapidly deposit in an unconsolidated layer with little shear strength. This layer was instantly resuspended when the shear stress became sufficiently large. Otherwise, the rates of erosion and deposition were relatively small.

The observed weak transient deposit may be better defined as a concentrated benthic suspension (CBS), which Winterwerp (2002) predicted will form for similar water depth (8 m), tidal velocity (50 cm/s), and settling velocity (0.5 mm/s), but greater mean SSC (290 mg/L), and no salinity stratification. Mean SSC at our study site was smaller, which would lessen the chance that a CBS would form. Turbulence damping by salinity stratification at our study site, however, may increase the chance that a CBS would form. Mehta (1989) presented idealized vertical profiles of SSC that included fluid mud, similar to CBS, at the interface of the water column and cohesive bed.

5. SPRING/NEAP VARIABILITY OF SUSPENDED AND BED MASS

Suspended and bed mass also vary with the spring/neap cycle. The October 1999 instrument deployment began during a neap tide and finished during a spring tide 9 days later. Stratification and turbulent mixing are described in detail by Stacey et al. (2000).

During the neap tides from October 16-20 the water column usually was stratified, turbulence usually was suppressed, and SSC and bed flux were small (Figure 7B). Salinity-induced periodic stratification (SIPS) was created during ebb tides and vanished during strong flood tides, which was when turbulence was generated and SSC and suspended mass increased. SSC and suspended mass also increased during ebb tides, though there was little turbulence; this was probably due to advection of more turbid water from shoals at Middle Ground and the shallow subembayment, Honker Bay. During neap tides, resuspension was smaller and, therefore, did not mask advection, unlike spring tides. Measurements of $\langle w'c' \rangle$ (Figure 3B) also help to differentiate increases in suspended mass due to resuspension and advection. The change in suspended mass during neap tides probably is equal to the change in bed mass during slack tide deposition and flood tide resuspension when local resuspension or deposition dominate horizontal advective transport. The change in bed mass during these phases of the tidal cycle ranges from about 0-20 mg/cm^2.

The more energetic spring tides around October 23-27 periodically generated turbulence, mixed the water column, resuspended bottom sediment, and increased SSC (Figure 7B). Most of the changes in suspended mass were associated with increased turbulence, resuspension, and mixing or turbulence suppression, slack tide, and deposition. As in 1997, suspended mass, and, thus, usually bed mass, varied similar to a square wave with a height of about 60 mg/cm^2 during spring tide (Figure 7C).

Assuming that changes in suspended mass generally are opposite to changes in bed mass, the tidal variability and total quantity of bed mass change from neap to spring tide. Tidal variability of bed mass increases from 0-20 mg/cm^2 during neap tide to about 60 mg/cm^2 during spring tide because the more energetic spring tides increase resuspension, SSC, and deposition flux. These quantities represent the size of the unconsolidated layer that deposits during slack tide and erodes easily during the subsequent tide. During slack tides, when suspended mass is smallest and bed mass is greatest, bed mass is about 20 mg/cm^2 less during spring tides than during neap tides. During maximum tides, when bed mass is smallest, bed mass is about 80 mg/cm^2 less during spring tides than during neap tides. Assuming that the dry density of the deposit was on the order of 100 mg/cm^3, the elevation of the bed varied by several millimeters between neap and spring tide. Therefore, bed flux associated with the spring/neap cycle is similar to that associated with the semidiurnal tidal cycle.

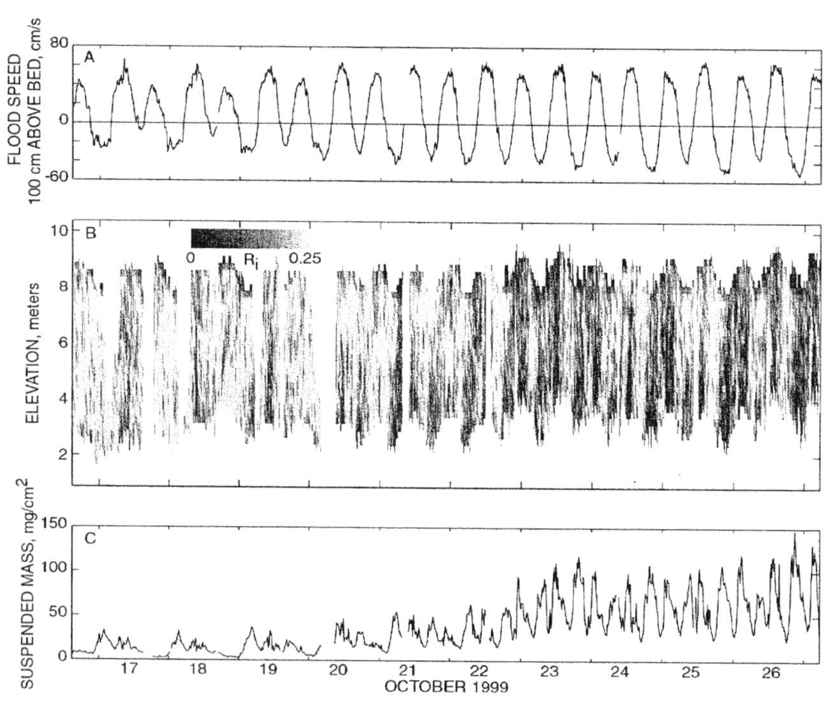

Figure 7. A) Flood current speed 100 cm above the bed. B) Vertical profiles of Richardson number. C) Mass of suspended sediment, Suisun Cutoff, October 1999.

The variations in total suspended mass agree qualitatively with the ADV suspended sediment measurements. For example, the suspended mass data (Figure 7C) and the ADV data (Figure 3B and Figure 4B) both show increases by a factor of two to three between neap tide and spring tide. Additionally, the second peak in suspended mass on October 26 (Figure 7C) is considerably less than the three other peaks on that day, which corresponds to the much lower <w'c'> shown in Figure 4B during the weaker ebb (0600-1000 hours). Finally, the <w'c'> time series concurs with the square wave behavior of the total suspended mass. For instance, the rapid increase in <w'c'> at the beginning of each of the stronger tides in

Figure 4B (0100, 1200, and 2100 hours), corresponds to the period of rapid resupension which generates the steep rises of the square wave in Figure 7C. Also, the constant peak in the suspended mass which appears as the square wave plateaus (e.g. Figure 6C, 1500-2000 hours) are matched by similar peak plateaus in <w'c'> (e.g. Figure 4B, 1200-1500 hours). These plateaus of constant suspended mass are produced by the equilibrium condition of constant, upwards <w'c'> balancing downwards settling flux (product of particle settling velocity and mean concentration). Quantitative comparison between the ADV data and the total suspended mass data are not viable, because the ADV's small sampling volume (~1 cm^3) does not capture significant vertical variations over the 7 meters of water column integrated to obtain total suspended mass.

6. CONCLUSIONS

The bed shear stress, as inferred from the Reynolds stress, ρ<u'w'>, is flood-dominated at this study's field site. Tidal pumping around the site's complex bathymetry and gravitational circulation induced by horizontal density gradients contribute to the mean flow flood asymmetry. Stratification contributes to this asymmetry by preferentially suppressing turbulent mixing, particularly on ebbing neap tides.

These three factors produce a corresponding asymmetry in bed flux, as estimated from the vertical turbulent sediment flux, <w'c'>. During neap tides, bed flux was approximately three times larger during the stronger flood, as compared to the three other tidal periods of the tidal day. This flood dominance is created, in part, by periods of strong stratification on ebbing neap tides. The stratification completely suppresses turbulence and resuspension for two-thirds of the stronger ebbs, though mean velocity exceeds values that induce erosion during other phases of the tide. During spring tides, the net flood bed flux asymmetry is created, in large part, by the lack of resuspension during the weaker ebb. In addition, the flood asymmetry is augmented by changes in sediment erodibility. During the first two hours of flood tide, sediment erodibility apparently increases, as compared to other times in the tidal cycle. Tidal asymmetry in bed shear stress and erodibility help explain an estuarine turbidity maximum that is present during flood tide but absent during ebb tide.

During periods of decelerating flow (e.g. 1400 hours in Figure 4), SSC declines in concert with ρ<u'w'> and <w'c'>, while mean velocity maintains its peak value for about another hour. This behavior replicates the sediment response reported by Sanford and Halka (1993). These authors modeled this SSC behavior with continuous deposition, as opposed to the standard model of bed-shear-stress dependent deposition.

During spring tides and some phases of neap tides, changes in SSC were controlled by local stratification, turbulent mixing, and settling--not horizontal advection. When local bed flux controlled SSC, the change in suspended mass can be assumed opposite of the change in bed mass. At slack tide, suspended sediment rapidly deposited in an unconsolidated layer with little shear strength. This layer, which may be a concentrated benthic suspension, was instantly resuspended when the shear stress became sufficiently large during a subsequent tide. Otherwise, the rates of erosion and deposition were relatively small. The resulting time series of bed mass appear similar to a square wave. This tidal variability of bed mass increased from 0-20 mg/cm^2 during neap tide to about 60 mg/cm^2 during spring tide because the more energetic spring tides increase resuspension. This would produce changes of several millimeters in bed elevation. The variability of bed mass associated with the spring/neap cycle is similar to that associated with the semidiurnal tidal cycle.

ACKNOWLEDGMENTS

The authors thank Jon Yokomizo, Jim George, Jay Cuetara, Jessica Lacy, Cary Troy, Catherine Ruhl, Rob Sheipline, Mark Stacey, Jim DeRose, and Brad Sullivan for helping collect these data. Mark Stacey and Ray Krone reviewed previous versions of this paper. This study was supported by the U.S. Geological Survey Place-Based Program, the U.S. Office of Naval Research, the Interagency Ecological Program for the Sacramento-San Joaquin Estuary, and the Burt and Deedee McMurtry Stanford Graduate Fellowship.

REFERENCES

Buchanan, P.A. and Ruhl, C.A., 2000, Summary of suspended-solids concentration data, San Francisco Bay, CA, water year 1998, *U.S. Geological Survey Open File Report* 00-88, 41 p.

Downing, J.P. and Beach, R.A., 1989, Laboratory apparatus for calibrating optical suspended solids sensors, *Marine Geology*, (86) 243-249.

Dronkers, J., 1986, Tide-induced residual transport of fine sediment, *Physics of Shallow Estuaries and Bays*, ed. J. Van de Kreeke, Springer-Verlag, 228-244.

Fischer, H.B., List, E.J., Imberger, J. and Brooks, N.H., 1979, *Mixing in inland and coastal waters*, Academic Press Inc., 483 p.

Gross, T.F. and Nowell, A.R.M., 1983, Mean flow and turbulence scaling in a tidal boundary layer, *Continental Shelf Research*, (2) 2/3, 109-126.

Hansen, D.V. and Rattray, M., 1966, Gravitational circulation in straits and estuaries, *Journal of Marine Research*, (23) 2, 104-122.

Itsweire, E.C., Koseff, J.R., Briggs, D.A., and Ferziger, J.H., 1993, Turbulence in stratified shear flows: implications for interpreting shear-induced mixing in the ocean, *Journal of Physical Oceanography*, (23), 1508-1522.

Kawanisi, K. and Yokosi, S., 1997, Characteristics of suspended sediment and turbulence in a tidal boundary layer, *Continental Shelf Research*, (17) 8, 859-875.

Kineke, G.C. and Sternberg, R.W., 1989, The effect of particle settling velocity on computed suspended sediment concentration profiles, *Marine Geology*, (90) 3, 159-174.

Kranck, K. and Milligan, T.G., 1992, Characteristics of suspended particles at an 11-hour anchor station in San Francisco Bay, California, *Journal of Geophysical Research*, (97) C7, 11,373-11,382.

Krone, R. B., 1962, Flume studies of the transport of sediment in estuarial shoaling processes, *Final Report*, University of California, Berkeley, Hydraulic and Sanitary Engineering Research Lab. 110 p.

Mehta, A.J., 1989, On estuarine cohesive sediment suspension behavior, *Journal of Geophysical Research*, (94) C10, 14,303-14,314.

Osborne, P.D., Vincent, C.E. and Greenwood, B., 1994, Measurement of suspended sand concentrations in the nearshore: field intercomparison of optical and acoustic backscatter sensors, *Continental Shelf Research*, (14) 2-3, 159-174.

Parchure, T.M., and Mehta, A.J., 1985, Erosion of soft cohesive sediment deposits, *Journal of Hydraulic Engineering*, (111) 10, 1308-1326.

Sanford, L.P. And Halka, J.P., 1993, Assessing the paradigm of mutually exclusive erosion and deposition of mud, with examples from upper Chesapeake Bay, *Marine Geology*, (114) 1-2, 37-57.
Schoellhamer, D.H., 2001, Influence of salinity, bottom topography, and tides on locations of estuarine turbidity maxima in northern San Francisco Bay, *Coastal and Estuarine Fine Sediment Transport Processes*, ed. W.H. McAnally, and A.J. Mehta, 343-357.
Simpson, J.H., Brown, J., and Matthews, J. and Allen, G., 1990, Tidal straining, density currents, and stirring in the control of estuarine stratification, *Estuaries*, (13) 125-132.
Stacey, M.T., 1996, Turbulent mixing and residual circulation in a partially stratified estuary, Ph.D. Dissertation, Stanford University.
Stacey, M.T., Monismith, S.G., and Burau, J.R., 1999, Observations of turbulence in a partially stratified estuary, *Journal of Physical Oceanography*, (29) 1950-1970.
Stacey, M.T., Burau, J.R., Brennan, M.L., Lacy, J., Tobin, C. C., and Monismith, S.G., 2000, Spring-neap variations in stratification and turbulent mixing in a partially stratified estuary, Proceedings 5th *International Symposium on Stratified Flows*, ed. G.A. Lawrence, R. Pieters, and N. Tomemitsu, 939-944.
Sternberg, R.W., Cacchione, D.A., Drake, D.W. and Kranck, K., 1986, Suspended sediment transport in an estuarine tidal channel within San Francisco Bay, California, *Marine Geology*, (71) 3-4, 237-258.
Thorne, P.D., Hardcastle, P.J., and Soulsby, R.L., 1993, Analysis of acoustic measurements of suspended sediments, *Journal of Geophysical Research*, (98) C1, 899-910.
Trowbridge, J.H., Geyer, W.R., Bown, M.M. and Williams, A.J., 1999, Near-bottom turbulence measurements in a partially mixed estuary: turbulent energy balance, velocity structure and along channel momentum balance, *Journal of Physical Oceanography*, (29) 12, 3056-3072.
Winterwerp, J.C., 2002, Scaling parameters for high concentration mud suspensions under tidal conditions. Proceedings *INTERCOH-2000, Elsevier, Coastal and Estuarine Fine Sediment Processes*, ed. J.C. Winterwerp and C. Kranenburg, this volume.

Fine Sediment Dynamics in the Marine Environment
J.C. Winterwerp and C. Kranenburg (Editors)

Physical modelling of entrainment by a Concentrated Benthic Suspension

A.W. Bruens[a], C. Kranenburg[a] and J. C. Winterwerp[a,b]

[a]Fluid Mechanics Section, Faculty of Civil Engineering and Geosciences, Delft University of Technology, P. O. Box 5048, 2600 GA Delft, The Netherlands.

[b] WL|delft hydraulics, P.O. Box 117, 2600 MH Delft, The Netherlands.

The process of entrainment of an upper clear water layer by a highly turbulent lower Concentrated Benthic Suspension (CBS) is studied in this paper. A rotating annular flume was adapted to carry out experiments on this type of entrainment. Velocities have been measured at two heights. The results indicate that the lower layer is more turbulent than the upper layer. As a result of entrainment, the density of the lower layer decreases and the height of the interface increases with time. The results of the experiments show that a freshly deposited mud layer behaves in a similar manner as a viscous fluid. The results also indicate that the non-dimensional entrainment rate is proportional to the inverse of an overall Richardson number.

KEYWORDS
cohesive sediments, concentrated benthic suspension, entrainment, turbulence, density stratification, siltation.

1. INTRODUCTION

1.1. General

Cohesive sediment is transported from fluvial and marine sources to depositional environments. The deposition of cohesive sediments can lead to a range of managerial problems. Examples of problem fields are maintenance of navigation channels, pollution and effects on construction works. To deal with these problems, it is essential to understand the behavior of cohesive sediment and its interaction with the flow field.

In this paper the classification of sediment-water mixtures as proposed in Bruens, 1999, is applied. This classification is based on a distinction of four physical states in which a water-sediment mixture can exist. For very low concentrations the characteristics of the mixture do not substantially differ from that of clear water and the mutual interference between sediment particles and the turbulent flow field is negligible. Only minor vertical gradients in sediment concentration are present and in most cases buoyancy effects can be neglected. Such a mixture is classified as a dilute suspension. At higher sediment concentrations areas of large vertical gradients in sediment concentration develop, leading to density stratification. Buoyancy becomes important and turbulence is partially damped in these regions of high concentration gradients. This type of water-sediment mixture is classified as a concentrated benthic suspension, henceforth abbreviated as CBS. For higher sediment concentrations the number of interfloc contacts increases and the flocs become space-filling. If the concentration reaches the gelling point, the flocs start to form a network, called a gel, and the volume concentration of floc in the water column becomes equal to one. The water-sediment mixture is then classified as fluid mud. The motion of fluid mud due to (hydrodynamic) forcing is laminar by definition. Further compaction of the material is due to self-weight consolidation; the excess pore water pressure will gradually decrease, resulting in an increase in effective strength. If the sediment is deposited for a longer period, the material has enough time to build-up strength large enough to withstand the driving forces. The material of this so called consolidating bed is not flowing, the only possible motion is creep, which contributes to the consolidation process.

1.2. Entrainment processes

Under tidal flow conditions, cohesive sediment is deposited around slack water and a concentrated near-bed layer may be formed, herein called a Concentrated Benthic Suspension (CBS). Due to the tidal cycle, streamwise pressure gradients increase after some time period. This time period is often too short for consolidation to occur in the CBS. As no or only little strength builds up in this layer, it cannot resist the increase in pressure gradient and a flow is generated in the CBS. Such a flow has been observed in the field, for example in the Loire Estuary (Le Hir, 1997). The left-hand side of Figure 1 represents a measured concentration profile during a spring tide, showing a CBS with a concentration of nearly 30 g/l and a thickness of approximately 50 cm. The right-hand side of this figure represents a velocity profile, indicating velocities of about 0.75 m/s in the CBS. Turbulence is generated at the base of and within this CBS. Due to buoyancy effects the turbulence is damped at the interface between CBS and water. If there are no other sources of turbulence such as wind action, the

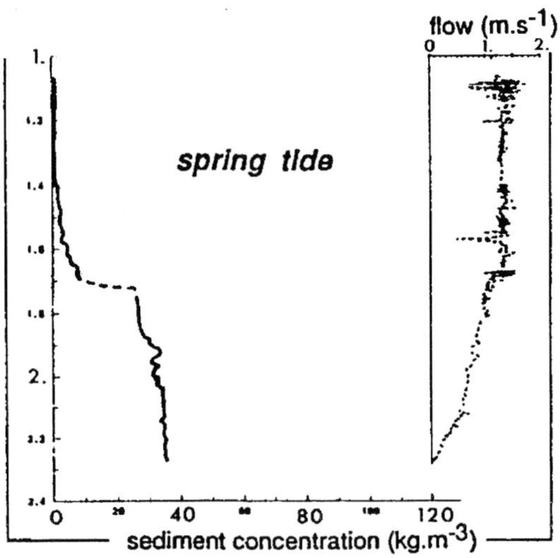

Figure 1: Measurements of concentration and velocity profiles in the Loire Estuary (Le Hir, 1997).

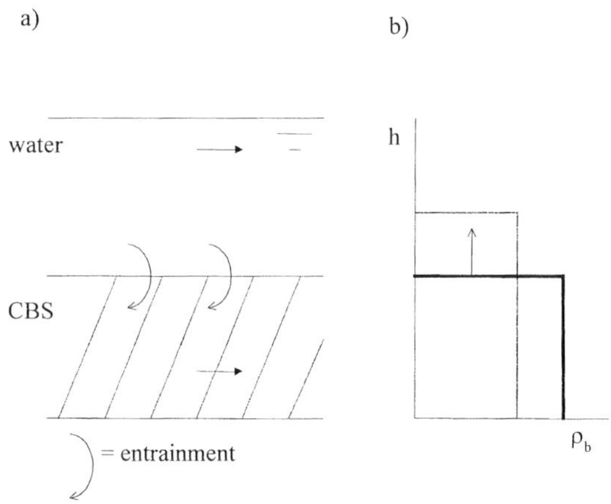

Figure 2: Case 1: Processes acting in case of a turbulent CBS, (a) a schematic cross section indicating the entrainment of water, (b) the density profile showing an increase in thickness and a decrease in density of the CBS.

flow in the upper water layer will be less turbulent than in the lower layer. The more turbulent CBS will now start to entrain water from the upper layer, resulting in an increase in thickness and a decrease in density of the CBS. These processes are described in Bruens (1999), and schematically represented in Figure 2 (case 1).

If, on the other hand, cohesive sediment is deposited for longer time periods, for example during neap tide, strength can develop in the concentrated near-bed layer and a fluid mud or consolidating bed is formed. This layer will be strong enough to resist the tidal pressure gradient, and flow is only generated in the upper water layer. Turbulence is now generated in this layer, and it will entrain (or erode) material from the fluid mud or consolidating bed. The result is an increase in density of the upper layer and a decrease in thickness of the fluid mud. These processes are schematically represented in Figure 3 (case 2). The processes in this figure are also representative of the situation where turbulence is generated at the water surface due to wind action (Kranenburg, 1994), or the situation where confined amounts of cohesive sediment are trapped in a deeper part.

1.3. Present study

In the past, experiments on the entrainment of a concentrated near-bed layer by a turbulent water layer (case 2) were carried out (Mehta & Partheniades, 1993; Winterwerp & Kranenburg, 1997). The study described in this paper was concerned with the entrainment by a turbulent CBS (case 1). The objective of the study was to improve the understanding of the behavior of cohesive sediments and to generate data to validate computer models of CBS dynamics. As far as is known to the authors, no experiments on the entrainment by a CBS have been described in the literature.

The present study consisted of the following parts: 1) designing a device to carry out experiments on the entrainment by a CBS, 2) optimizing the experimental facilities, 3) carrying out experiments on the entrainment by a lower, saline water layer, 4) carrying out experiments on the entrainment by a CBS, 5) analyzing the experimental results by using numerical simulation models. In the present paper, the results of the experiments with cohesive sediment are presented. The results of the numerical simulations will be published in a future paper.

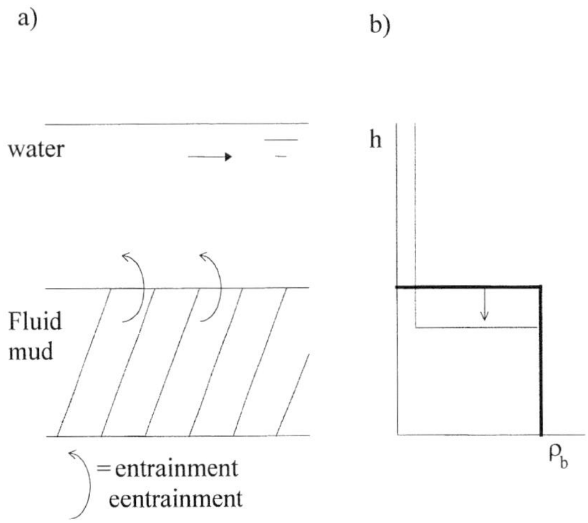

Figure 3: Case 2: Processes acting in case of a turbulent water layer, (a) a schematic cross section indicating the entrainment of sediment, (b) the density profile showing a decrease in thickness of the fluid mud and an increase in density of the upper layer.

2. EXPERIMENTAL FACILITIES

2.1. Rotating annular flume with top lid

A device often used to carry out erosion experiments on mud in the laboratory is the annular flume. The main advantages of annular flumes are the absence of inflow and outflow conditions, as present in straight flumes, and the absence of pumps, which break down the sediment flocs. A rotating top lid usually drives the flow in annular flumes. The disadvantage of annular flumes is their curvature; fluid with higher tangential velocity near the lid is driven away from the centerline of the flume and a secondary flow is generated in the flume. In order to minimize these secondary currents, a so-called *rotating* annular flume can be deployed. In a rotating annular flume the flume itself can rotate in a direction opposite to that of the top lid. Thus secondary currents (important for deposition studies (Mehta & Partheniades, 1973a; Mehta & Partheniades, 1973b)) can be minimized and uniform distributions of tangential velocity and near-bottom shear stress across the flume width (important for erosion studies (Mehta & Partheniades, 1979; Parchure & Mehta, 1985)) can be obtained.

A cross section of the rotating annular flume at Delft University of Technology is shown in Figure 4. The flume has a width of 0.3 m, a height of 0.4 m and a mean diameter of 3.7 m. The flow in the flume is driven by the top lid. In case the flume is filled with a mud layer and a water layer, turbulence in the upper water layer is mainly generated at the top lid and material from the mud layer is entrained (case 2, as sketched in Figure 3).

In the past an extensive study on the optimum ratio (U_l/U_f) of rotational speeds of top lid (U_l) and flume (U_f) was carried out in the rotating annular flume at Delft University of Technology. This study consisted of velocity measurements with a backscatter laser-Doppler velocimeter system (Booij, 1994) and computations with the flow simulation system PHOENICS of CHAM Ltd (Booij & Uijttewaal, 1999). The results of the flow simulations were in good agreement with the velocity measurements. From this study an optimum ratio for entrainment studies (i.e. minimum secondary currents in the upper, turbulent layer) was derived. This ratio was applied during previous entrainment experiments in the flume with top lid. The results of these experiments were in good agreement with an entrainment model derived by integrating the turbulent kinetic energy equation across the mixed upper layer (Winterwerp & Kranenburg, 1997).

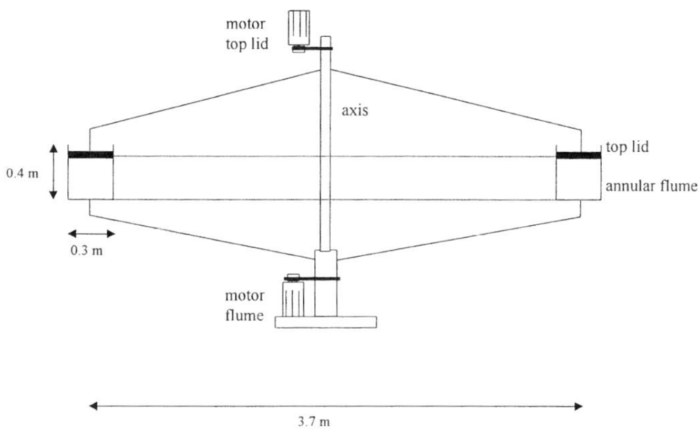

Figure 4: Cross section of the rotating annular flume with top lid at Delft University of Technology

2.2. Rotating annular flume with base plate

For the present experiments, the top lid of the rotating annular flume was replaced by a rotatable base plate (Figure 5). This base plate is fixed to the supporting gear, used to drive the top lid, by eight vertical, streamlined rods with a chord of 0.1 m and a maximum thickness of 5 mm. The slit between base plate and flume is filled with mercury in order to prevent deposition as well as flow of cohesive sediments under the plate.

For the new configuration of the annular flume, another optimum ratio of the rotational speeds of base plate and flume had to be assessed. The objective was to find a ratio, for which secondary currents in the lower layer are sufficiently weak and shear stresses near the interface are uniformly distributed across the flume. The integral momentum equation was used to derive an equation for such a ratio. For this derivation it was assumed that due to the rotation of base plate and flume, a core exists in which hardly any velocity gradients exist. This core takes up approximately 90 % of the flow and is surrounded by boundary layers where velocity gradients are high. For annular flumes with a top lid, secondary currents where shown to be minimal for a zero velocity of this core (U_c=0) with reference to the non-rotating reference frame (Booij, 1994; Spork, 1997). A zero velocity was therefore applied for the present derivation. To validate this theoretically derived equation, computations with the flow simulation system PHOENICS as well as experiments with dye were carried out. The results of the experiments agreed (qualitatively) with the numerical simulations. Both demonstrated that, by using the derived equation, the rotational speed of flume (U_f) and base plate (U_b) can be set such that secondary currents are sufficiently weak for carrying out entrainment experiments. This set-up was further tested by entrainment experiments with salt and fresh water. For details the reader is referred to Bruens et al., 2000.

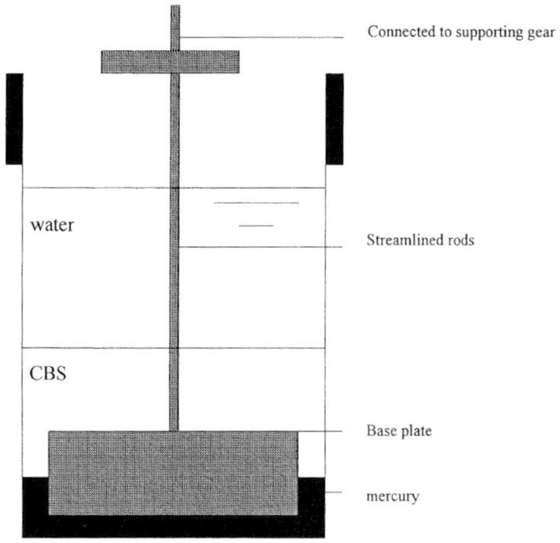

Figure 5: Cross section of the rotating annular flume with base plate.

3. EXPERIMENTS WITH COHESIVE SEDIMENT

3.1. Introduction

Natural cohesive sediment usually contains a certain amount of organic material. This organic content influences the properties and behavior of natural mud. Therefore experiments with natural cohesive sediment are often not fully reproducible. For this reason it was decided to use an artificial clay in saline water during the present series of entrainment experiments. From previous entrainment experiments in the rotating annular flume with top lid, it was concluded that the artificial China Clay behaved qualitatively similar to samples of natural mud from the Maasmond area in the Netherlands. Based on these experiences, China Clay was selected for the present experiments. China Clay and saline water (salinity of 5 ppt) were premixed during two weeks. As several processes, e.g. formation of diffuse double layers, are initialized when China Clay and saline water are mixed, this pre-mixing was necessary in order to obtain an equilibrium condition. The properties of this China Clay were extensively determined by De Wit (1995); the volumetric density of the solids is 2622 ± 1 kg/m^3, the specific surface area is 23.9 ± 0.1 m^2/g and CEC is 3.3 ± 0.1 meq per 100 g. Table 1 summarizes the viscosity and Bingham strength for various concentrations (De Wit, 1995).

For CBS layers, which possess hardly any strength, the entrainment processes as shown in Figure 2 take place (case 1). With an initial, uniform sediment concentration of about 43 g/l (i.e. a density of about 1028 kg/m^3) a CBS was formed after approximately two hours with a thickness of 0.1 m and a concentration ranging from 50 g/l at its top to 200 g/l at its base (yielding a Bingham strength less than 0.6 Pa (Table 1)).

Concentration (g/l)	58	113	172	237	275
Viscosity (mPas)	1	2	3	4	5
Bingham strength (Pa)	-	0.1	0.4	1.0	1.5

Table 1: Viscosity and Bingham strength of China Clay used (De Wit, 1995).

3.2. Experimental procedure and program

For the present experiments, the flume was filled with a well-mixed suspension of China Clay (43 g/l). Flume and base plate were initially set to rotate at the same speed in the same direction resulting in a rigid body rotation of the suspension. From this suspension, a CBS was formed by the settling of the flocs. Changing the rotational direction of the base plate then started an entrainment experiment. Due to inertia, the CBS maintained its initial speed and as a result of the friction between base plate and CBS, turbulence was generated at the base plate. Because of buoyancy effects the turbulence was damped near the interface between CBS and water and the turbulent CBS started to entrain water from the non-turbulent upper water layer. Initial concentration profiles were measured prior to the entrainment phase. The following parameters were measured during the entrainment phase:

- Height of the interface.
- Sediment concentrations at four different heights, using turbidity sensors (at the three lowest measurement locations available i.e. 0.033, 0.083 and 0.133 m above the base plate) and a portable density meter (at the upper measurement point at 0.25 m).
- Flow velocities in both tangential and vertical directions at two heights, using electromagnetic flow meters; mean velocities as well as turbulent fluctuations were measured.

The experimental program is summarized in Table 2. This program was set up such that the overall Richardson number, which is a function of the hydrodynamic parameters, was varied. The overall Richardson number (Ri_*) is defined as $Ri_* = (\Delta\rho g h)/(\rho\, u_*^2)$, where h is the thickness of the turbulent mixing layer, $\Delta\rho$ is the density difference between the two layers, u_* is the friction velocity where the turbulence is mainly produced (in the present case at the base plate), ρ is the density of the turbulent layer and g is the gravitational acceleration. The initial density of the upper layer, with a sediment concentration of 0 g/l and a temperature of 22 °C, was approximately 998 kg/m^3. The overall Richardson number can be regarded as the ratio of potential energy required to establish mixing and kinetic energy available for mixing. Ri_* was varied by varying u_*, obtained by variations in rotational speeds of flume and base plate (Δu). The friction velocity was not directly measured, but obtained by using the logarithmic law of the wall, the quadratic friction law and numerical simulations with Phoenics, all leading to the same values as given in Table 2 (Bruens, 2000). Δu was set to vary between 0.7 and 1.1 m/s.

Exp.	ρ **g/m^3)**	***H*** **(m)**	**Δu (m/s)**	**u_* (m/s)**	**Ri_***
C1	1065 ± 3	0.1 ± 0.01	0.7 ± 0.01	0.016±0.01	234
C2	1065 ± 3	0.1 ± 0.01	1.1 ± 0.01	0.026±0.01	86
C3	1065 ± 3	0.1 ± 0.01	0.9 ± 0.01	0.021±0.01	136
C4	1065 ± 3	0.1 ± 0.01	0.9 ± 0.01	0.021±0.01	136
C5	1065 ± 3	0.1 ± 0.01	0.8 ± 0.01	0.018±0.01	185
C6	1065 ± 3	0.1 ± 0.01	1.0 ± 0.01	0.023±0.01	113

Table 2: Summary of the experimental program for China Clay.

A velocity difference of 0.7 m/s appeared to be the lower limit at which the CBS layer became turbulent, as this velocity difference resulted in a bottom shear stress that was only slightly larger than the bingham strength at the base of the sediment layer (bingham strength as given in Table 1). A velocity difference of 1.1 m/s was the maximum velocity at which the entrainment process could be observed and recorded accurately. Experiment C4 was planned as a repetitive experiment to investigate the reproducibility of the experiments. In the second column of Table 2 the mean bulk density of the CBS at the start of an entrainment experiment is given.

3.3. Settling phase

Although the rotational speed during the rigid body rotation differed for the six experiments, the settling curve is equal for all six experiments (Figure 6), as expected. It is shown that no equilibrium is yet obtained after 2.5 hours (9000 s).

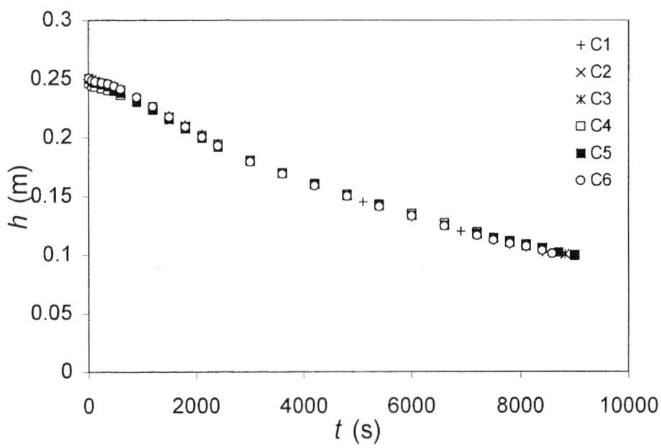

Figure 6: Settling curves for all experiments with China Clay.

3.4. Initial concentration profile

After a settling period of approximately 9000 s the CBS had a thickness of 0.1 m (Figure 6). At that time the concentration profile was measured with an electric conductivity probe (ECP). These profiles are shown in Figure 7. The initial conditions for all experiments were similar. The concentration ranged from approximately 200 g/l at the base of the CBS (z = 0 m) to approximately 50 g/l in the upper part of the CBS (z = 0.1 m). Due to an inaccurate calibration of the ECP for experiment C6, data for this experiment is not available.

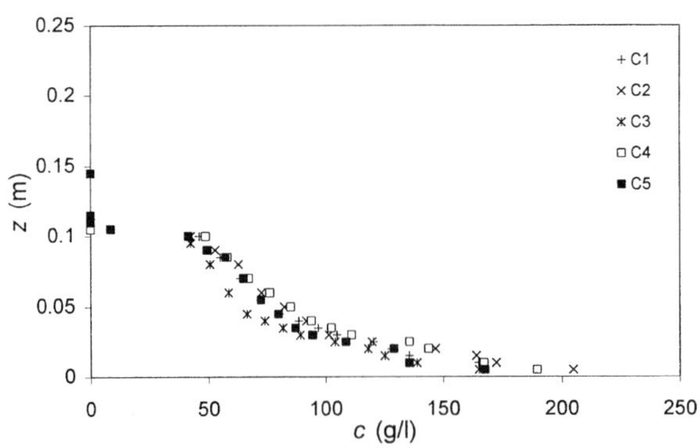

Figure 7: concentration profiles for all experiments with China Clay.

3.5. Entrainment phase

All entrainment experiments were started after 9000 s, i.e. from the density profiles given in Figure 7. The interface between CBS and water was observed with the naked eye. A picture of the interface (Figure 8) shows the entrapping of water by Kelvin Helmotz instabilities.

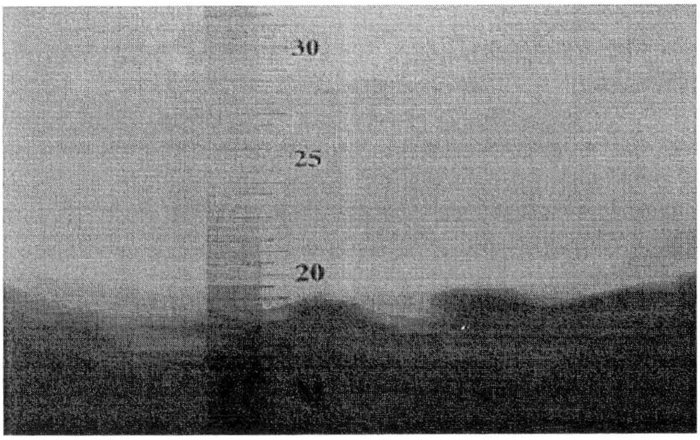

Figure 8: Picture of the interface showing the entrapment of water by Kelvin-Helmotz instabilities (the flow direction in the lower layer is from left to right).

The heights of the interface as functions of time are plotted in Figure 9 for all six experiments. During all experiments, the height of the interface could be observed up to a level of approximately 0.17 m. The slopes of the graphs in Figure 9 are more or less constant with time, as is expected from entrainment theory. Mean entrainment rates can be calculated from the mean slopes ($W_e = \Delta h/\Delta t$); values are given in Table 3. The graphs of experiments C3 and C4 are identical, proving the reproducibility of the experiments.

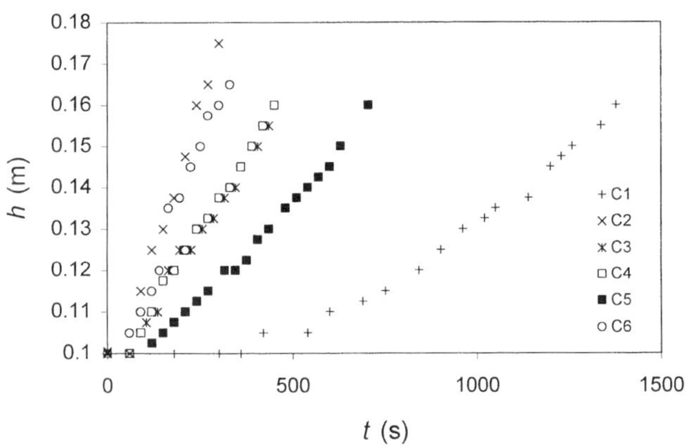

Figure 9: Heights of the interface during the entrainment phase for all experiments with China Clay.

Sediment concentrations at four different heights during experiment C5 are plotted in Figure 10. The trend in concentration increase/decrease is equal for all six experiments; in this paper only the results for experiment C5 are presented and discussed in detail. The lowest turbidity meter (z = 0.033 m) was initially situated within the CBS. Initial sediment concentrations at this height exceeded the maximum recordable concentration of 80 g/l. As long as sediment concentrations at this position remained higher than 80 g/l, the sensor recorded this maximum concentration ($t < 300$ s). Because of entrainment, the concentrations became lower than 80 g/l at $t \approx 300$ s, and the sensor started to record the decrease in concentration of the CBS ($t > 300$ s). The concentration decreased until well-mixed conditions prevailed: for $t > 1000$ s sediment concentrations were constant at approximately 43 g/l.

The turbidity meter at $z = 0.083$ m was initially located in the upper part of the CBS layer. The meters were switched on approximately 30 to 60 seconds before the base plate was set to rotate in a direction opposite to that of the flume (i.e. before the entrainment experiment started). During this time the sensor at 0.083 m registered a decrease in sediment concentration ($t < 100$ s). This decrease was a consequence of settling during the rigid body rotation. After the base plate was set to rotate in opposite direction, turbulence was generated and the CBS was mixed. As a result concentrations at $z = 0.083$ m increased ($100 \text{ s} < t < 300$ s). After $t \approx 300$ s, the sensor recorded the decrease in concentration of the CBS due to the entrainment of overlying, less dense suspension until the concentration became approximately 43 g/l ($t \approx 1000$ s).

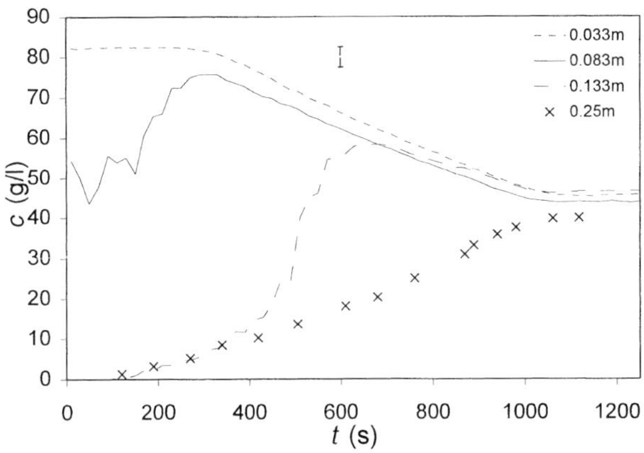

Figure 10: Concentrations at various heights during experiment C5.

The turbidity meter at $z = 0.133$ m was initially located in the upper, clear water layer. Therefore, the initial concentration was zero ($t < 100$ s). The interface passed the height of $z = 0.133$ m at $t \approx 500$ s. Between the start of the experiment and the passing of the interface ($100 \text{ s} < t < 500$ s), the concentration at $z = 0.133$ m (i.e. in the upper layer) slowly increased. This increase was caused by the additional mixing induced by the streamlined rods. These rods rotated in a direction opposite to that of the flume and their drag generated flow in the upper layer. As the flow velocity in the upper layer was no longer equal to the rotational speed of the flume, turbulence was generated at the sidewalls. A small amount of turbulence was also generated at the rods itself. As there exists a mean shear across the interface, internal waves may be produced, resulting in the generation of turbulence (Uittenbogaard, 1988). This turbulence mixed material from the lower layer into the upper layer, resulting in the increase in concentration between $t \approx 100$ and $t \approx 500$ s. After the interface had passed $z = 0.133$ m ($t > 600$ s), the turbidity meter recorded the decrease in sediment concentration of the CBS until it became approximately 43 g/l ($t \approx 1000$ s). For $t > 600$ s, the concentration at $z = 0.133$ m appears to exceed concentrations at $z = 0.083$ m and $z = 0.033$ m, but this is caused by the inaccuracy of the measurements (see also error bar in Figure 10).

Additional concentration measurements with the portable density meter were carried out just beneath the free surface ($z \approx 0.25$ m). These measurements show the increase in concentration in the upper layer, as the effect of the streamlined rods led to mixing, until a concentration of 43 g/l was reached. Until $t \approx 400$ s, concentrations at $z = 0.133$ m and $z = 0.25$ m are equal, indicating that the upper layer is a well mixed turbulent layer. Sediment concentrations of approximately 43 g/l for $t \geq 1000$ s at all four positions indicate that all the sediment was well mixed through the entire fluid at that time.

The data in Figures 9 and 10 indicate that both mixing and entrainment took place (at the same time). Taking both processes into account, the mass balances for the upper and lower layers become:

$$\frac{\mathrm{d}(H-h)C_2}{\mathrm{d}t} = -w_e C_2 - w_m\left(C_2 - C_1\right) - w_s C_2 \tag{1}$$

$$\frac{\mathrm{d}hC_1}{\mathrm{d}t} = w_e C_2 + w_m\left(C_2 - C_1\right) + w_s C_2 \tag{2}$$

where H is the total depth of both layers (0.25 m), h is the height of the lower layer, C_1 is sediment concentration in the lower layer, C_2 is concentration in the upper layer, w_e is the entrainment velocity, w_m is a mixing velocity and w_s is a fall velocity. The fall velocity is obtained from Figure 6 (w_s $2.5\cdot10^{-5}$ m/s). Mixing implies that equal amounts of material are exchanged between the two layers (resulting in an increase in concentration of the upper layer and a decrease in the lower layer), whereas entrainment implies that material is only transported in one direction. As a result of mixing, the interface between the two layers becomes less distinct, but mixing does not result in a change in height of the interface. The increase in height of the interface in Figure 9 results from entrainment of material from the upper layer in the, more turbulent, lower layer. The fact that the interface is no longer visible above a height of approximately 0.17 m is a result of the mixing process. Using the data presented in Figures 9 and 10, w_e and w_m can be solved from Equations 1 and 2. Their values are given in Table 3, as well as the entrainment velocities obtained from Figure 9 directly (W_e). w_e and w_m are of the same order of magnitude. Except for experiment C3, w_e is slightly larger than w_m. W_e obtained from visual observations is slightly larger than the calculated w_e. Using equation 1 and 2 it is assumed that the two layers are well-mixed and can both be represented by one homogenous concentration (C1 and C2). The height of the interface is then defined as the height at which the concentration increases abruptly from C1 to C2. In reality though, mixing results in the formation of a thin interfacial layer in which the concentration gradually increases from C2 to C1. It is not unequivocal what 'interface' is observed with the

Exp.	w_m m/s) from Eq. 1 & 2	w_e(m/s) from Eq. 1 & 2	W_e (m/s) from Figure 9
C1	$4.1\cdot10^{-5}$	$4.8\cdot10^{-5}$	$5.0\pm0.3\cdot10^{-5}$
C2	Not available	Not available	$2.9\pm0.2\cdot10^{-4}$
C3	$9.7\cdot10^{-5}$	$7.7\cdot10^{-5}$	$1.4\pm0.1\cdot10^{-4}$
C4	$1.1\cdot10^{-4}$	$1.1\cdot10^{-4}$	$1.5\pm0.1\cdot10^{-4}$
C5	$6.4\cdot10^{-5}$	$6.8\cdot10^{-5}$	$9.3\pm0.1\cdot10^{-5}$
C6	$1.3\cdot10^{-4}$	$2.0\cdot10^{-4}$	$2.3\pm0.4\cdot10^{-4}$

Table 3: Entrainment velocities and mixing velocities obtained from Equations 1 and 2 and entrainment velocities directly obtained from Figure 9.

naked eye; the highest position at which the concentration starts to increase, the height at which the density increase is maximum, or any other position in the density profile? In case of a changing interfacial layer (growing as a result of mixing) this may result in the difference in entrainment rates obtained from visual observations and from equation 1 and 2.

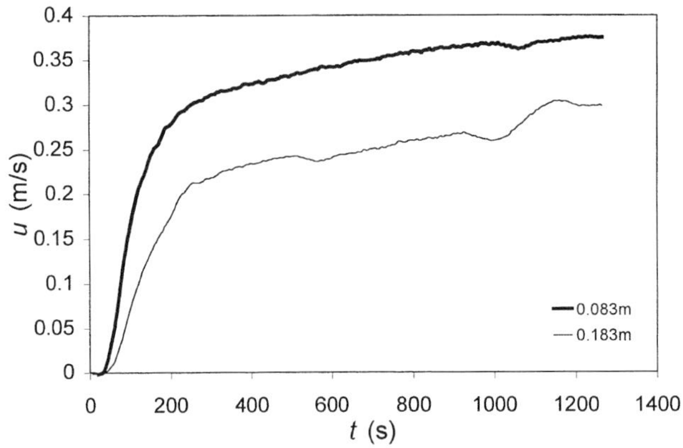

Figure 11: Horizontal velocities (running mean), with reference to the flume, at two heights during experiment C5.

During the entrainment phase, velocities were measured by electromagnetic flow meters at 0.083 m and 0.183 m above the base plate. In Figure 11 tangential velocities at these heights during experiment C5 are presented. The flow meters were connected to the flume; therefore velocities are relative to the flume. The rotating speed of the flume for experiment C5 was -0.35 m/s, the speed of the base plate was 0.45 m/s.

Because of the rigid body rotation preceding the entrainment phase, the initial velocities relative to the flow meters were zero. Once the base plate started to rotate at a different speed, flow velocities increased, first at z = 0.083 m and somewhat later also at z = 0.183 m. The velocity in the upper layer (z =0.183 m) is not equal to zero, indicating that the velocity is not equal to the rotational speed of the flume. This is caused by the drag of the streamlined rods, as mentioned before.

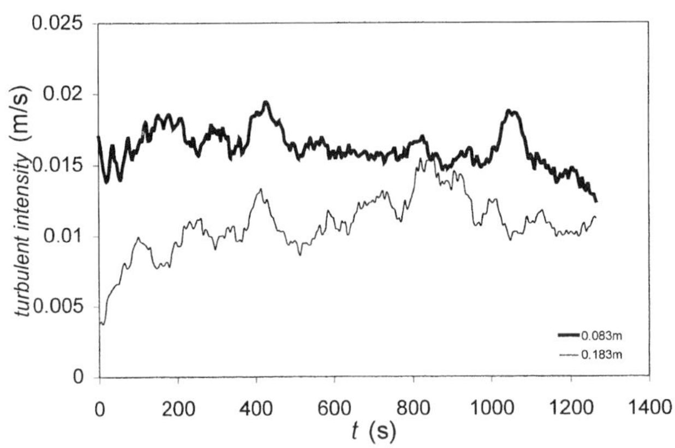

Figure 12: Horizontal turbulent intensities at two heights during experiment C5.

For experiments C5 and C6, high frequency velocity measurements (20 Hz) were also carried out. By subtracting the mean flow velocity from the instantaneous velocity, velocity fluctuations are obtained (u') from which turbulent intensities $\langle u'^2 \rangle^{1/2}$ averaged over 50 s can be calculated. These intensities are plotted in Figure 12. This figure shows that as turbulence is produced at the rods and the walls of the flume, the upper layer became turbulent as well, leading to mixing of the two layers (w_m in Equations 1 and 2). The intensity in the upper layer (thin line) determines the mixing rate. The difference between the intensities of the two layers (the lower layer, thick line, being higher) determines the entrainment rate.

4. DISCUSSION

Since long it has been recognized that the rate at which a well-mixed layer entrains material from an ambient layer is a function of friction velocity, depth of the layer and buoyancy across the interface, this is characterized by a bulk Richardson-number. Usually, experiments were carried out in which one layer was turbulent, and the entrainment velocity was related to the Richardson number of this turbulent layer. In previous studies on the entrainment of a concentrated layer (case 2), a relation between the non-dimensional entrainment rate E ($=w_e/u_*$) and Ri_* in the form $E \propto Ri_*^{-n}$ was found (Kantha et al., 1977, Kato & Phillips, 1969, Srinivas & Mehta, 1989, Mehta & Partheniades, 1993, Winterwerp & Kranenburg, 1997). In case side-wall friction did not play a role $n \approx 0.5$. If side-wall friction was more important, $n >$ 0.5, and often the entrainment rate became proportional to Ri_*^{-1}.

During the present experiments, both layers were turbulent and the measured entrainment- and mixing rates were functions of the Richardson number of both layers. In Figure 13 the non-
dimensional entrainment rate E is plotted against the overall Richardson number (Ri_*) of the more turbulent, lower layer for the experimental series on case 1 entrainment with cohesive sediments and for the experimental series with saline water (for more details on the experiments with saline water the reader is referred to Bruens, 2000). This figure shows an identical relation for both series, indicating that for the present experiments mud properties such as fall velocity and viscosity did not influence the entrainment processes. This result agrees with the findings of Mehta & Srinivas (1993) and Winterwerp & Kranenburg (1997), who concluded that a freshly deposited mud layer behaves as a viscous fluid, as long as no strength develops. For the present experiments also a relation in the form $E \propto Ri_*^{-1}$ was found.

In Figure13 also the results are plotted from experiments in the annular flume with top lid at Delft University of Technology on case 2 (Winterwerp & Kranenburg, 1997). In these experiments only the upper layer was turbulent. All available turbulence in this layer could be used for entrainment, whereas in the case of two turbulent layers part of the turbulence would result in mixing and part of it (i.e. the difference in turbulence between the two layers) would lead to entrainment. This means that entrainment rates will be lower in case of two turbulent layers. In a future paper the effect of two turbulent layers on the entrainment rate will be quantified by using a numerical simulation model.

The entrainment rate is a function of the aspect ratio (Kantha et al., 1977, Deardorff & Willis, 1982, Kranenburg & Winterwerp, 1997 amongst others); higher aspect ratio's lead to lower entrainment rates. The aspect ratio for the present experiments is not constant in time, as the height of the turbulent layer increases from 0.1 m to 0.2 m while the width of the flume is constant (0.3 m). The mean aspect ratio for the experiments on case 2 was larger than for the experiments on case 1 (0.225/0.3 and 0.15/0.3 respectively). This is expected to result in slightly higher entrainment rates for case 1 (a factor of 1.2 higher according to Kranenburg & Winterwerp, 1997), counteracting the effect of two turbulent layers. Figure 13 shows that data for case 1 and case 2 coincide as a result of the aforementioned counteracting effects.

The experiments differ from the conceptual model presented in Figure 2 by the fact that the upper layer is also turbulent and subsequent mixing results in the increase of density of the upper layer. Under field conditions, it is likely that some degree of turbulence is present in the upper layer (due to for example wind forcing or any other shear invoked in this layer) and that both mixing and entrainment takes place (simultaneously).

Interpreting the experimental results in terms of field conditions is not straightforward. Obviously, the aspect ratio in the field goes to zero as the width approaches infinity. The driving force in the field (mainly tidal pressure gradients) results in higher flow velocities in the upper layer (Figure 1). In the experiments the flow is driven at the base plate resulting in higher velocities in the lower layer with reference to a non-rotating reference frame. With reference to the rotating base plate though, velocities in the upper layer are higher, comparable to the field situation. In a future paper, the physical experiments will be simulated using a numerical model. Using the experimental data to calibrate and validate the numerical model, the model can be applied to simulate field conditions.

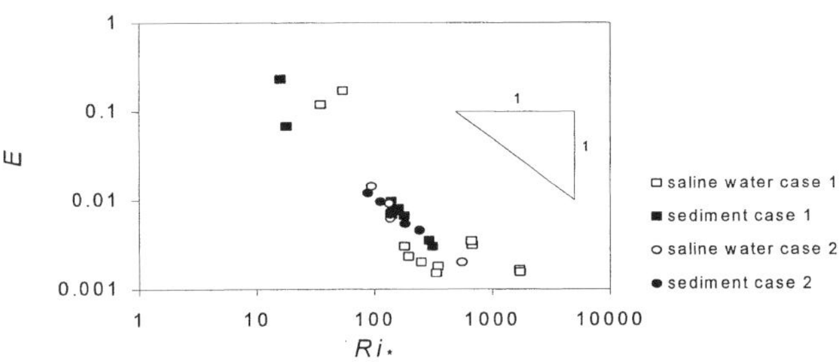

Figure 13: Entrainment rate versus Ri_* for experiments with saline water and cohesive sediments for case 1 and case 2.

5. SUMMARY AND CONCLUSIONS

The process of entrainment of an upper clear water layer by a highly turbulent lower Concentrated Benthic Suspension (CBS) is studied in this paper. A rotating annular flume was adapted to carry out experiments on this type of entrainment. To simulate mixing in tidal flow after slack water, the flow in the flume is now driven by a rotating base plate. Due to friction between base plate and CBS, turbulence is generated at the base plate and the CBS becomes turbulent and starts to entrain material from the upper layer. During the entrainment phase, velocities are measured at two heights. The results indicate that the lower layer is more turbulent than the upper layer. Concentration measurements are carried out at four heights. As a result of entrainment, the density of the lower layer decreases and the height of the interface increases with time. Due to the generation of a small amount of turbulence in the upper layer, mixing between the two layers takes place as well. By using the mass balances for the two layers, entrainment rates as well as mixing rates could be calculated. The derived entrainment rates agree with the measured entrainment rates (obtained by visual observations). The results of the experiments show that a freshly deposited mud layer behaves in a similar manner as a viscous fluid. The results also indicate that the non-dimensional entrainment rate is proportional to the inverse of an overall Richardson number.

One of the objectives of the experiments described in this paper was to enhance the understanding of the behavior of cohesive sediments and the interaction with the flow field. From the experimental results presented in this report, it is concluded, that a turbulent lower, dense layer, either a saline layer or a CBS layer, can entrain water from an upper, less dense layer. Processes indicated in Figure 2 take place: The density of the lower layer decreases as the thickness of this layer increases. When dealing with managerial problems, for example maintenance of navigation channel, this type of entrainment and transport of cohesive sediments have to be considered.

The second objective of the experiments was to obtain data to validate computer models of CBS dynamics. The data presented in this paper accomplished this objective and the results of numerical simulations will be presented in a future paper.

ACKNOWLEDGMENT

This work was partially funded by the European Commission, Directorate General XII for Science, Research & Development through the COSINUS-project within the framework of the MAST-3 program, contract MASC3-CT97-0082. The authors owe many thanks to Rob Booij for his contribution on the subject of secondary flow patterns and Phoenics simulations. Sincere thanks are due to Prof. Dr J. A. Battjes for his critical and helpful comments and suggestions. Stefan Vollmer, of the technical University of Aachen, is gratefully acknowledged for his contribution to the experiments and data processing. The technical staff of the Laboratory of Fluid Mechanics at Delft University of Technology is acknowledged for the support in building the experimental set-up.

REFERENCES:

Booij, R. 1994. Measurements of the flow field in a rotating annular flume. *Communication on hydraulic and getechnical engineering,* nr 94-2

Booij, R. & Uijttewaal, W. S. J. 1999. Modelling of the flow in rotating annular flumes. Pages 339-348 of: Rodi, W. & Laurance, D. (eds), *Proc. of the 4th Int. Symp. on Engineering Turbulence Modelling and Experiments.*

Bruens, A. W. 1999, Transport of cohesive sediments. Classification and requirements for turbulence modelling. *Tech. Rept. 2-99. Delft University.*

Bruens, A. W., Booij, R., Kranenburg, C. & Winterwerp, J. C. 2000. Applicability of the rotating annular flume for entrainment experiments. Pages 1173-1178 of: G. A. Lawrence, Pieters, R. & Yonemitsu, N. (eds), *Proc. of the fifth international symposium on stratified Flows.*

De Wit, P. J. 1995. Liquefaction of Cohesive Sediments caused by Waves. *Ph.D. thesis, Delft University of Technology, the Netherlands.*

Kantha, L. H., Phillips, O. M., and Azad, R. S., 1977. On turbulent entrainment at a stable Density interface. *Journal of Fluid Mechanics,* 79(4), 753-768.

Kato, H. and Phillips, O. M., 1969. On the penetration of a turbulent layer into stratified fluid *Journal of Fluid Mechanics*, 37, 643-655.

Kranenburg, C. & Winterwerp, J. C. 1997. Ersoion of Fluid Mud layers. I: Entrainment Model. *Journal of Hydraulic Engineering*, ASCE, 123, 504-511.

Le Hir, P. 1997. Fluid and sediment 'integrated' modelling application to fluid mud flows in estuaries. In: *Cohesive Sediments. 4th Nearshore and Estuarine Cohesive Sediment Tranport Conference INTERCOH '94.*

Mehta, A. J. & Partheniades, E. 1973a. Depositional behavior of cohesive sediments. *Tech. Rep. nr 16, Coastal and Oceanographic Engineering Department, University of Florida, USA.*

Mehta, A. J. & Partheniades, E. 1973b. Effect of physico-chemical properties of fime Suspended sediment on the degree of deposition. Pages 465-476 of: *Proc. Int. Symp. River Mechanics, IAHR, Bangkok, Thailand.*

Mehta, A. J. & Partheniades, E. 1979. Kaolinite resuspension properties. *Journal of Hydraulic Division, ASCE,* 105, 411-416.

Mehta, A. J. & Partheniades, E. 1993. Observations on the entrainment of fluid mud by shear flow. *Coastal and Estuarine studies*, 42, 224-246.

Mehta, A. J. & Srinivas, R., 1993. Observations on the entrainment of fluid mud by shear flow. *Coastal and estuarine sturdies*, Mehta, A. J., ed., Am. Geohys. Union, Washington, D.C., 42, 224-246.

Parchure, T. & Mehta, J. A. 1985. Erosion of soft cohesive sediment deposits. *Journal of Hydraulic Engineering*, ASCE, 114-120.

Schlichting, H. 1968. Boundary Layer Theory. McGraw-Hill Book Company.

Spork, V. 1997. *Untersuchung der Transporteigenschaften von Flusscheb in einem Kreisgerinne,* Tech. Rep. Theinisch Westfalische Technische Hochschule Aachen.

Srinivas, R., and Mehta, A. J., 1989. Observations on estuarine fluid mud entrainment. *International Journal Sediment Research,* 5, 15-22.

Uittenbogaard, R. E., 1988. Measurement of turbulence fluxes in a steady, stratified, mixing layer. *Proc of In. symp. on Refined Flow Modelling and Turbulence Measurements, IAHR, Tokyo*, pp. 725-732.

Uittenbogaard, R. E., Winterwerp, J. C., van Kester, J. A. T. M., & Leepel, H. 1996. 3D Cohesive sediment transport; a preparatory study about implementation in DELFT3D. *Tech. Rept. Z1022. Delft Hydraulics.*

Winterwerp, J. C. & Kranenburg, C. 1997. Ersoion of Fluid Mud layers. II: Experiments and model validation. *Journal of Hydraulic Engineering, ASCE,* 123, 512-519.

Winterwerp, J. C. & Uittenbogaard, R. E. 1997. Sediment transport and fluid mud flow; Physical mud properties and parameterisation of vertical transport processes SILTMAN-set-up of a POINT-MUD MODEL. *Tech. Rep. nr z2005, Delft Hydraulics.*

Winterwerp, J. C. 1999. On the dynamcis of high-concentrated mud suspension. *Ph.D. thesis, Delft University of Technology.*

Fine Sediment Dynamics in the Marine Environment
J.C. Winterwerp and C. Kranenburg (Editors)

Interfacial instabilities at the lutocline in the Jiaojiang estuary, China

J. Jiang[a] and A. J. Mehta[b]

[a]ASL Environmental Sciences Inc.,
1986 Mills Road, Sidney, BC V8L 5Y3, Canada

[b]Department of Civil and Coastal Engineering, University of Florida,
345 Weil Hall, Gainesville, FL 32611-6580, USA

Interfacial instabilities at the lutocline in the Jiaojiang estuary, China, were examined on a preliminary basis. Internal waves at two distinct frequencies were detected from acoustic records. The 0.38 m high low-frequency (0.09 rad/s) wave, close to the shallow water wave regime, was near the buoyancy frequency. The high-frequency (1.33 rad/s) wave of 0.21 m height was in deep water with sharp crests and flat troughs possibly due to interfacial shear. With increasing Richardson number the height and frequency of both wave types decreased, whereas the celerity and wave length increased. A straightforward phenomenological framework is used to examine these trends.

KEY WORDS
Buoyancy, cohesive sediment, erosion, sediment-water interface, sediment transport

1. INTRODUCTION

Interfacial instabilities at the lutocline are critically important in governing the vertical mixing of suspended sediment in high-concentration estuaries (Kirby, 1986, Wolanski et al., 1989, Mehta and Srinivas, 1993 and Winterwerp and Kranenburg, 1997). Such instabilities exhibit two distinct internal wave modes riding on the suspended sediment interface, i.e., the lutocline (Jiang and Wolanski, 1998 and Shi, 1998). Here we present preliminary observations on such waves in the turbid Jiaojiang estuary in China.

Jiaojiang is located on the east coast of China, about 200 km south of the Yangtze River (Figure 1). It is close to 200 km in length with an estuarine segment of about 35 km. The mean water depth below mean sea level is 4-7 m, and the mean width is about 1.2 km with a maximum of 1.8 km at the mouth. Seaward of the mouth the width increases rapidly, forming Taizhou Bay in shallow coastal waters. Semi-diurnal macro-tides prevail there, with a mean range of about 4 m and a spring range of 6.3 m at the mouth of the Jiaojiang. In the estuary the depth-mean peak tidal current can be as high as 2 m/s, and the tidal wave is strongly distorted within a short distance from the mouth. Partly as a result the duration of ebb exceeds flood by 1-2 h in the estuary, and at Haimen (Figure 1) the measured peak flood and ebb currents are 2.1 and 1.8 m/s, respectively (Zhou, 1986, Dong et al., 1997 and Guan et al., 1998). The suspended sediment is a cohesive silt-clay with a median (dispersed) particle size of about 5 μm (Li et al., 1993).

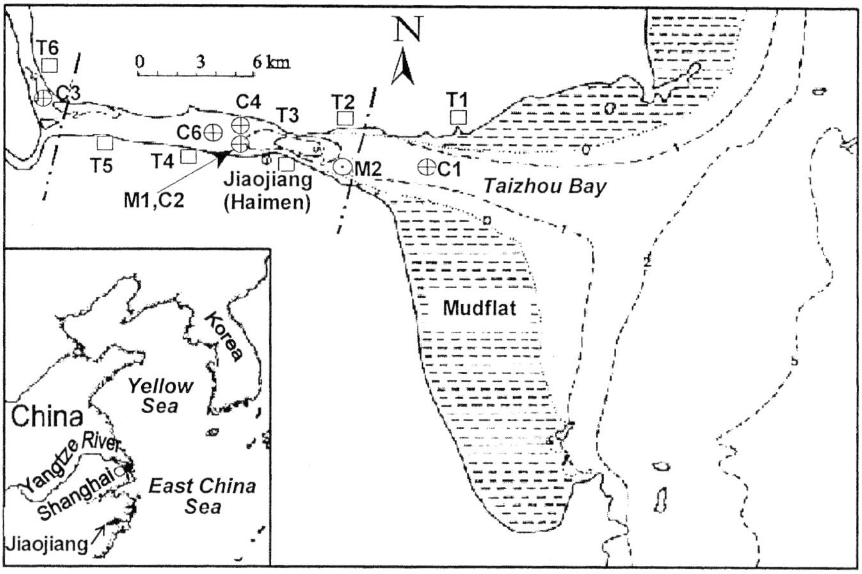

Figure 1. Jiaojiang estuary, China. Depths are in meters below lowest astronomical tide. M1 and M2 are mooring sites; C1-C4 are velocity measurement and SSC profile sampling stations; C6 is the site of ASSM and T1-T6 are tide stations. The region between the double dotted lines is the modeled domain.

2. METHOD

A ship-borne Acoustic Suspended Sediment Monitor (ASSM) (made by the Shanghai Acoustics Laboratory, Academia Sinica) was used to observe the lutocline at site C6 (Figure 1) during 0600-1600 h on November 15, 1995 (a neap tide). The ASSM consisted of a 0.5 MHz acoustic transducer/receiver. The acoustic probe was deployed 1-2 m below the water surface, and the entire system was under control of a PC for synchronization of sampling, preliminary data reduction and storage. The device had a pulse duration of about 40 s, and measured the vertical profiles of sound scattered from suspended sediment particles in the range bins at 0.6 s interval with a vertical resolution of 5 cm. Data were sampled at a rate of approximately 75 kHz in 9 min bursts. Each burst consisted of 900 profiles of backscattered acoustic energy from particles between the bottom and the probe.

In addition to acoustic records, interpretation of data in terms of the physical properties of the internal waves relied on measurements of velocity and suspended sediment profiles collected at and in the proximity (Figure 1). The examination of these profiles was facilitated by the use of three-dimensional numerical codes COHYD-UF and COSED-UF (Jiang, 1999) within the domain shown in Figure 1, which made it possible to correlate the acoustic records with suspended sediment concentrations on a synchronous basis. Interpretation of interfacial instabilities was carried out through a straightforward phenomenological examination.

3. EXPERIMENTAL RESULTS

An example of the acoustic record at site C6 is shown in Figure 2. As observed the ASSM signal showed a distinct step-structure representing the actual lutocline position detected independently by a turbidity meter. Also observed are lutocline undulations, suggesting noteworthy internal wave generation and likely sediment entrainment activity. The suspended sediment concentration (SSC), c, was calculated from the ASSM reading, F, using

$$c = k_a F^2 (h + h')^\alpha \exp(\beta h) \tag{1}$$

where h is the depth below the acoustic probe and k_a, h', α and β are sediment-dependent constants (Thorne et al., 1994). From the time-series of lutocline elevation $\zeta(t)$, a spectral analysis of internal waves was carried out, and related to local hydrodynamics.

Figure 3 shows three typical time-series Examples (a), (b) and (c) of $\zeta(t)$ traced from the 9 min ASSM records. It is evident that there are two distinct wave modes riding on the lutocline, namely, low-frequency waves with a period on the order of 1 min and high-frequency ones with a period of about 5 s.

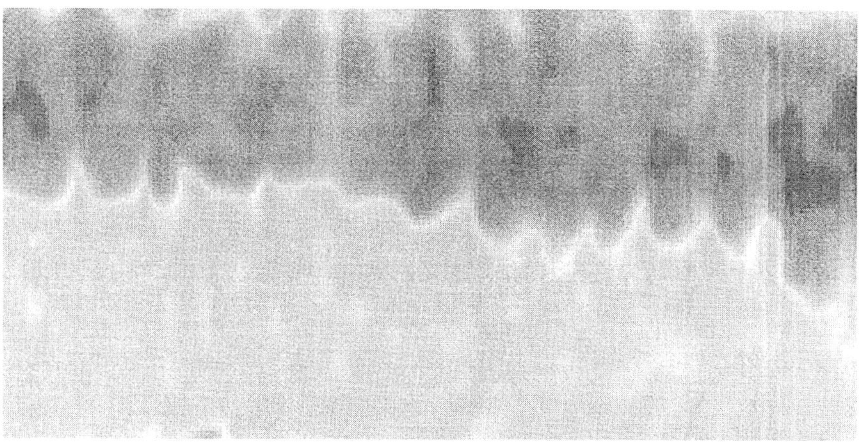

Figure 2. Raw, 1 min long and approximately 1 m high acoustic record showing a sediment-induced density interface (lutocline) during a neap ebb tide on November 15, 1995. Sampling interval was 0.6 s.

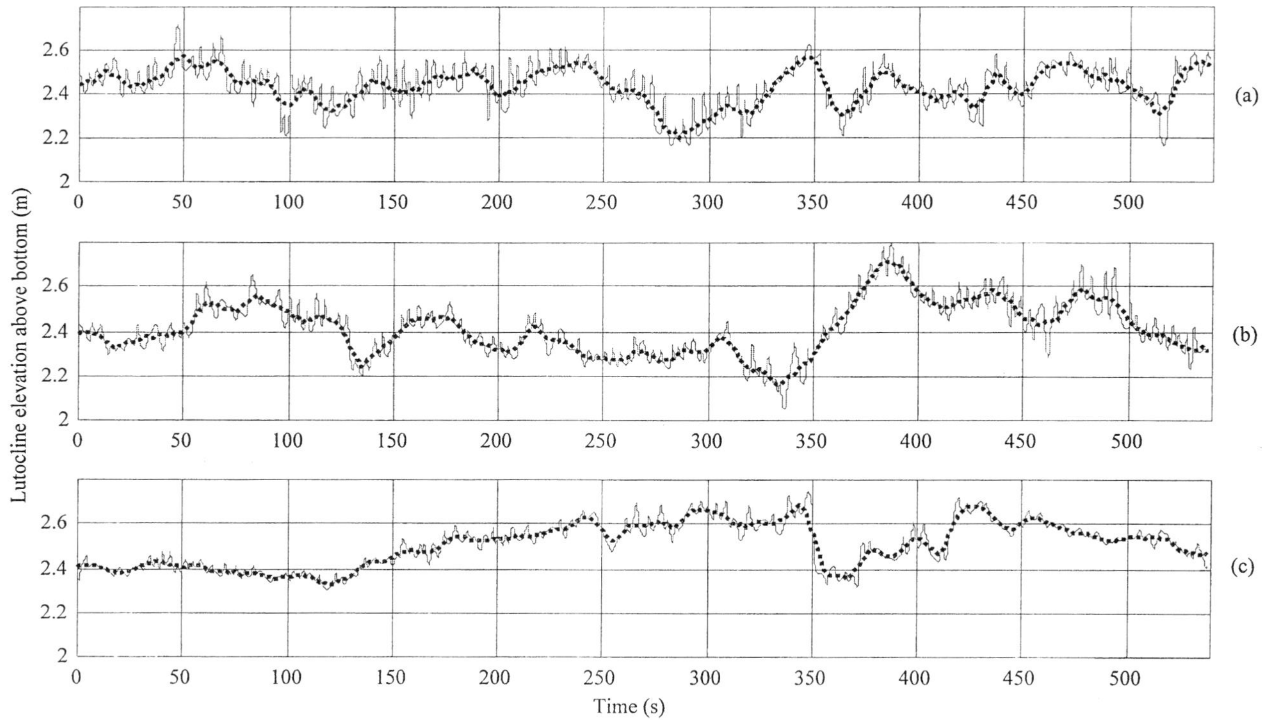

Figure 3. Time-series of lutocline elevation. (a) and (b) were sampled during flood at a Richardson number Ri_0 of about 1, and (c) during ebb with Ri_0 of about 150. Solid lines are instantaneous elevation and dashed lines are mean trends.

For the spectral analysis of the high-frequency waves, $\zeta(t)$ was processed through an approach used by Costa 1989) for filtering out the low-frequency signature. Figure 4 shows an enlarged portion of the time-series accentuating the undulations after trend removal. It is seen that the waves often had sharp crests and flat troughs similar to those resulting from Holmboe (HO) type instability (Broward and Wang, 1972). In the laboratory experiments of Srinivas (1989), both HO and Kelvin-Helmholtz (K-H) instabilities appeared. However, in the present case it was difficult to isolate K-H from HO.

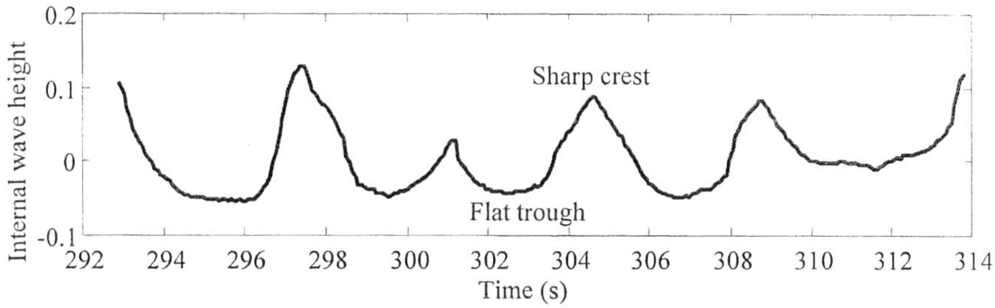

Figure 4. Typical profiles of internal waves exhibiting sharp crests and flat troughs. Data were taken from example (a) in Figure 3. Wave heights range from 0.07 m to 0.23 m.

A net upward-asymmetric mass mixing can be expected to be associated with these instabilities, and mixing can be further enhanced if and when the waves begin to break. For an analysis of the time-series, a global Richardson number, Ri_0, was calculated using measured current velocity and SSC profiles (Jiang, 1999)

$$Ri_0 = \frac{g(\rho_s - \rho_0)(C_2 - C_1)h_{mix}}{\rho_0^2(U_2 - U_1)^2} \tag{2}$$

where g is the gravitational acceleration, ρ_s is the sediment granular density, ρ_0 is water density, C_1 is the depth-mean sediment concentration (dry mass per unit volume) in the upper, mixed layer of height h_{mix}, C_2 is the corresponding value for lower layer, and U_1 and U_2 are the respective depth-mean flow velocities.

Spectral densities as functions of wave angular frequency ω corresponding to the data shown in Figure 3 are plotted in Figure 5. The maximum density is located around the modal angular frequency ω_m of 1.3 rad/s during flood, with Ri_0 of about 1 [Example (a) and (b)], and around 1.1 rad/s during ebb, with Ri_0 of about 150 [Example (c)]. It is apparent that ω_m and the spectral density decrease with increasing Ri_0, replicating the observations of Jiang and Wolanski (1998). In order to obtain an insight into this behavior, the root-mean-square (rms) wave height, H_{rms}, as well as ω_m were plotted against Ri_0. H_{rms} of the high-frequency waves and the corresponding ω_m were obtained for 1-min segments of the ASSM output. One segment was selected from each 9-min ASSM burst (totally 21 bursts) for this purpose. H_{rms} and ω_m of the low-frequency wave were derived for each 9-min segment of ASSM burst. The Ri_0 value was averaged over 1 min for the high-frequency waves and 9 min for low-frequencies.

Results for the high-frequency waves are shown in Figures 6 and 7, and summarized in Table 1 for both high- and low-frequency waves. Notwithstanding the paucity of data as well as scatter, one may conclude that for both wave modes H_{rms} and ω_m show identifiable mean trends with respect to Ri_0. In general, the lower the Ri_0, the higher the H_{rms} and ω_m values. On that basis, the significance of these trends is briefly examined next.

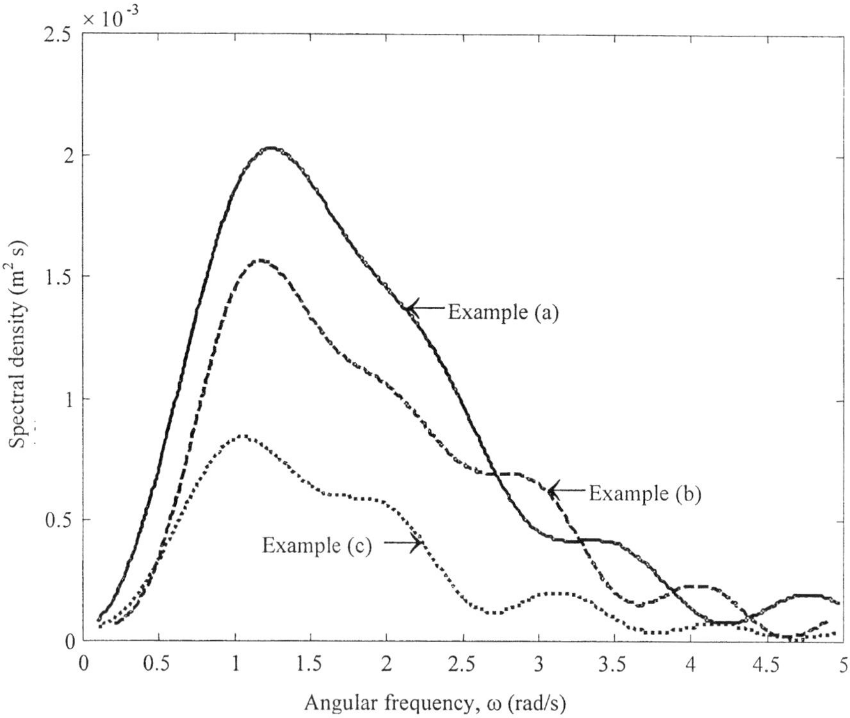

Figure 5. Spectra of high-frequency internal waves corresponding to Examples (a), (b) and (c) in Figure 3.

4. WAVES PROPERTIES

4.1. Buoyancy frequency

Following Wright et al. (1988), a possible relationship between the internal wave frequency and the Brunt-Väisälä or buoyancy frequency is examined here. This frequency is defined as $\omega_v = \sqrt{-g/(\rho\, \partial\rho/\partial z)}$, where ρ is suspension density and z is the vertical coordinate. In consonance with (2), one can approximate ω_v as

$$\omega_v = \sqrt{\frac{g(\rho_s - \rho_0)(C_2 - C_1)}{\rho_0^2 h_{mix}}} = \sqrt{\frac{g'}{h_{mix}}} \tag{3}$$

where g' is reduced gravity. To estimate ω_v using (3), SSC profiles measured at the same time as the ASSM data were analyzed (Jiang, 1999). The results indicated that ω_v was in the range of 0.12-0.26 rad/s. Note that the low-frequency waves ranged from 0.06 to 0.12 rad/s with a mean value of about 0.09 rad/s. Thus, the low-frequency waves had modal frequencies near the local buoyancy frequency, as found at haloclines and thermoclines (Neumann and Pierson, 1966 and Phillips, 1977). Hence it is believed that the low-frequency wave exhibited a characteristic free oscillation about the lutocline.

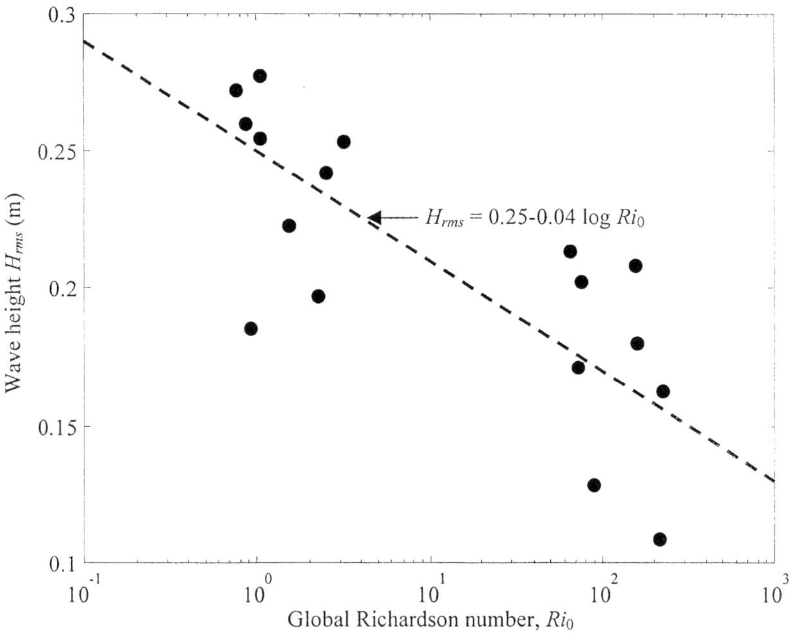

Figure 6. rms of high-frequency internal waves height as a function of global Richardson number.

4.2. Effect of Ri_0 on wave height

Since high Ri_0 implies high (negative) buoyancy-induced stabilization of the lutocline, increasing Ri_0 should correlate with decreasing H_{rms}. Such a correlation has been observed in laboratory experiments on lutoclines (Mehta and Srinivas, 1993) and other pycnoclines (Chou, 1975). By assuming that the interfacial undulations are due to the energy-containing mixed-layer eddies impinging on the density interface, Narimousa and Fernando (1987) established an empirical relationship between H_{rms} and Ri_0 of the form

$$\frac{H_{rms}}{h_{mix}} \propto Ri_0^{-1/2} \tag{4}$$

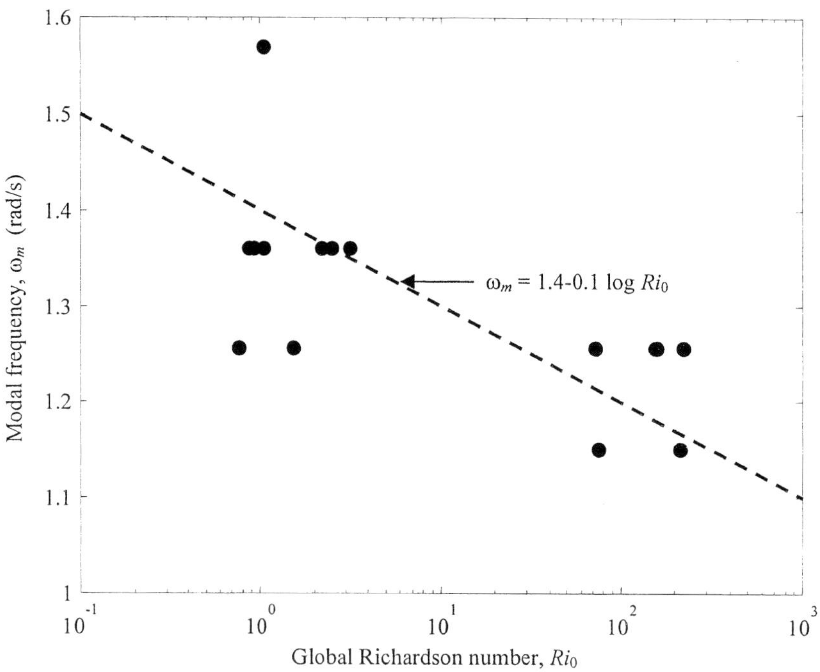

Figure 7. Modal frequency of high-frequency internal waves as a function of global Richardson number.

Accordingly, H_{rms} decreases with increasing Ri_0 with a slope of 0.5 on a log-log plot. In the Jiaojiang, slopes much smaller than 0.5 were found (Figure 6 and Table 1). This difference is believed to be mainly due to different physical scales and associated hydrodynamic effects including the degree of turbulence and eddy lengths in the prototype in relation to the laboratory.

Table 1. Mean trends in rms height and frequency as functions of Richardson number

	Mean trends	
Wave mode	H_{rms} (m)	ω_m (rad/s)
High-frequency	$H_{rms} = 0.25 - 0.041 \log (Ri_0)$	$\omega_m = 1.4 - 0.11 \log (Ri_0)$
Low-frequency	$H_{rms} = 0.438 - 0.053 \log (Ri_0)$	$\omega_m = 0.103 - 0.013 \log (Ri_0)$

4.3. Effect of Ri_0 on wave frequency

In order to examine the influence of Ri_0 on ω_m one may introduce the work of Lamb (1945), who analytically examined waves at the interface of two inviscid fluids of densities ρ_1 and ρ_2, one beneath the other and moving parallel to the x-axis with velocities U_1 and U_2, respectively (Figure 8). By assuming both fluids to be of unlimited depth and taking the profile of the instantaneous wave elevation $\eta(x,t)$ as

$$\eta(x,t) = 0.5H_0\,[i(\omega t - kx)] \tag{5}$$

where H_0 is a representative wave height and k is the wave number, the following wave dispersion relationship is obtained (Lamb, 1945):

$$\frac{\omega}{k} = \frac{\rho_1 U_1 + \rho_2 U_2}{\rho_1 + \rho_2} \pm \sqrt{\frac{g(\rho_2 - \rho_1)}{k(\rho_2 + \rho_1)} - \frac{\rho_2 \rho_1}{(\rho_2 + \rho_1)^2}(U_2 - U_1)^2} \tag{6}$$

The first term on the right-hand side of (6) is the average velocity U of the two layers. It is seen that the values of wave frequency ω given by (6) are imaginary if

$$\frac{g(\rho_2 - \rho_1)}{k\rho_1 (U_2 - U_1)^2} < \frac{\rho_2}{\rho_2 + \rho_1} \approx \frac{1}{2} \tag{7}$$

and also, for two fluids of nearly equal densities such as water and turbid suspension, $\rho_2/(\rho_2 + \rho_1) \approx 0.5$.

Under the condition imposed by (7), two possible cases can arise with respect to the sign of the second term on the right hand side of (6). Considering (5) and (6), and taking the plus sign in (6), H will dissipate with time, which inherently implies that the interface will be stable. On the other hand, if the minus sign is taken H will grow with time. In other words, the interface will be unstable. For the present analysis, only the unstable mode is of interest, i.e., with the minus sign relative to the second term in (6). Thus, as soon as (7) is satisfied the interface will become unstable, resulting in K-H instability (Delisi and Corcos, 1973). Assuming the applicability of this theory to HO instability as well, if one considers waves of all likely lengths in the estuarine setting such as the Jiaojiang, it can be concluded that sufficiently short interfacial waves will be present to cause instability. Therefore, this two-layered estuarine (shear) flow will be characteristically unstable.

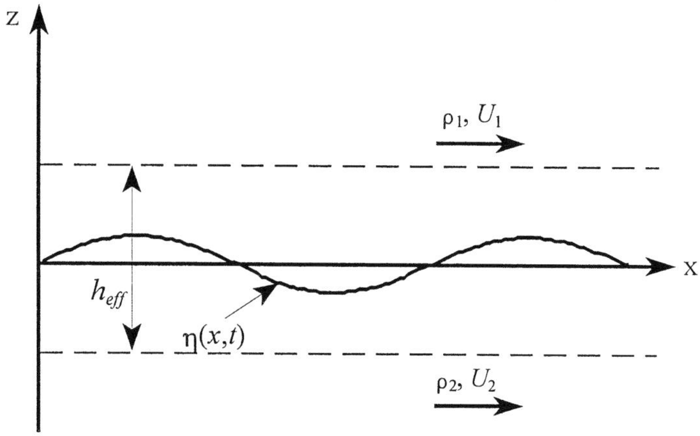

Figure 8. Definition sketch of a two-layered flow system.

Based on the above, holding ρ_1, ρ_2, k and U constant and introducing an effective (and notional) water depth h_{eff} (Figure 8) scaled by the internal wave height, one can express (6) in terms of a stream Richardson number, Ri_s, as

$$\frac{\omega}{k} = U - \left[A - \frac{Bh_{eff}}{Ri_s}\right]^{\frac{1}{2}} \tag{8}$$

where, $A = g(\rho_2 - \rho_1)/k(\rho_2 + \rho_1)$, $B = g\rho_2(\rho_2 - \rho_1)/(\rho_2 - \rho_1)^2$ and

$$Ri_s = \frac{g(\rho_2 - \rho_1)h_{eff}}{\rho_1(U_2 - U_1)^2} \tag{9}$$

where Ri_s is conceptually analogous to Ri_0. From (9) it is seen that ω decreases as Ri_s increases, an observation that is consistent with measurements (Table 1).

4.4. Wave regime and interfacial instability

The wave regime in reference to interfacial instability is now examined based on calculation of wave celerity and length. As before the flow is treated as two-layered system. Knowing reduced gravity $g' = g(\rho_2 - \rho_1)/[0.5(\rho_2 + \rho_1)]$ and the mean lutocline elevation ζ, the wave celerity $C = \omega/k$, for which the wave length $\lambda = 2\pi/k$ can be very approximately calculated from

$$\lambda = \frac{2\pi g'}{\omega^2} \tan h(k\zeta) \tag{10}$$

given that (10) is based on the assumption that the influence of current on λ is minor and can be ignored. From the measured SSC profiles, lutocline elevation (Figure 3) and the modal frequency, calculations for Examples (a), (b) and (c) in Figure 3 are summarized in Table 2 (Jiang, 1999). For these calculations, ζ was taken for each ASSM segment, and an iterative method was used to solve (10).

Table 2. Low- and high-frequency internal wave-related parameters

					Low-frequency wave			High-frequency wave			
No	ζ (m)	h_{mix} (m)	g' (m/s^2)	Ri_0	ω_V (rad/s)	ω (rad/s)	C (m/s)	λ (m)	ω (rad/s)	C (m/s)	λ (m)	λ_c (m)
1	2.40	1.38	0.096	1.6	0.26	0.105	0.460	27.6	1.24	0.078	0.39	2.71
2	2.43	1.50	0.097	2.5	0.25	0.093	0.470	31.5	1.18	0.082	0.44	1.88
3	2.51	2.86	0.119	150	0.20	0.070	0.536	48.2	1.06	0.112	0.65	0.06

From (7) one can calculate the critical wave length λ_c below which the interface will become unstable in the K-H or the HO mode due to interfacial shear. By equating the two sides of (7) one obtains

$$\lambda_c = \frac{2\pi\rho_1\rho_2(U_2 - U_1)^2}{g(\rho_2^2 - \rho_1^2)} \approx \frac{\pi h_{mix}}{Ri_0} \tag{11}$$

Values of λ_c calculated from (11) are given in Table 2. Also given is the buoyancy frequency calculated from (3). It is seen that the high-frequency waves were characteristically in deep water, with the ratio ζ/λ on the order of 5 (»0.5), whereas the low-frequency waves were close to the shallow water regime, with ζ/λ on the order of 0.07 ($\approx$0.05). It is also observed that the celerity and wave length increased with increasing Ri_0 for both wave modes. Observe further that λ_c decreased with increasing Ri_0 (ranging from 2.71 m to 0.06 m). Thus, the wave lengths of high-frequency internal waves (ranging from 0.39 m to 0.65 m) were between the maximum and minimum critical wave lengths for stability, which in turn suggests that the high-frequency waves were generated by interfacial shear.

5. CONCLUSIONS

Low- and high-frequency internal waves were detected at the lutocline in the Jiaojiang. The shallow water low-frequency wave had a representative rms height of 0.38 m and a modal frequency of 0.09 rad/s, which was near the local buoyancy or Brunt-Väisälä frequency. The deep water high-frequency wave was characterized by sharp crests and flat troughs, with an rms height of 0.21 m and a modal frequency of 1.33 rad/s. This latter wave mode was possibly induced by interfacial shear at the lutocline.

The height and the angular frequency of both high- and low-frequency waves decreased with increasing Richardson number, implying the role of increasing SSC on wave damping (Jiang and Mehta, 2000). It was further found that the height and angular frequency versus log Ri_0 plots exhibited linear trends. The corresponding the celerity and wave length increased with increasing Ri_0.

REFERENCES

Broward, F. K. and Wang, Y. H., An experiment on the growth of small disturbances at the interface between two streams of different densities and velocities, Proc. Int. Symp. Strat. Flows, Novosibirsk, USSR (1972) 491.

Chou, I. B., An experimental investigation of interfacial waves generated by low frequency internal waves, M. S. thesis, University of Florida, Gainesville, 1975.

Costa, R. G., Flow-fine sediment hysteresis in sediment-stratified coastal waters, M. S. thesis, University of Florida, Gainesville, 1989.

Delisi. D. and Corcos, G. M., A study of internal waves in a wind tunnel, Bound. Layer Meterol., 5 (1973) 121.

Dong, L. X., Wolanski, E. and Li, Y., Field and modelling studies of fine sediment dynamics in the extremely turbid Jiaojiannng River estuary, China, J. Coast. Res., 13(1997) 995.

Guan, W. B., Wolanski, E. and Dong, L. X., Cohesive sediment transport in the Jiaojiang River estuary, China, Estuar., Coast. and Shelf Sci., 46 (1998) 861.

Jiang, J. H., An examination of estuarine lutocline dynamics, Ph.D. thesis, University of Florida, Gainesville, 1999.

Jiang, J. H. and Wolanski, E., Vertical mixing by internal wave breaking at the lutocline, Jiaojiang River estuary, China, J. Coast. Res., 14 (1998) 1426.

Jiang, J. and Mehta, A. J., Lutocline behavior in high-concentration estuary, J. Watwy, Port, Coast., and Oc. Eng., 126 (2000) 324.

Kirby, R., Suspended fine cohesive sediment in Severn estuary and Inner Bristol channel, U.K., Rept. ETSU-STP-4042, UK Atomic Energy Authority, Harwell, 1986.

Lamb, H., Hydrodynamics, Sixth ed., Dover Publications, New York, 1945.

Li, Y., Wolanski E. and Xie Q. C., Coagulation and settling of suspended sediment in the Jiaojiang River estuary, China, J. Coast. Res., 9 (1993) 390.

Mehta, A. J. and Srinivas, R., Observations on the entrainment of fluid mud in shear flow, Nearshore Estuarine Cohesive Sediment Transport, A. J. Mehta (ed.), American Geophysical Union, Washington, DC (1993) 224.

Narimousa, S. and Fernando, H. J. S., On the sheared interface of an entraining stratified fluid, J. Fluid Mech., 174 (1987) 1.

Neumann, G. and Pierson, W. J., Principles of physical oceanography, Prentice-Hall, Englewood Cliffs, New Jersey, 1966.

Phillips, O. M., The dynamics of the upper ocean, 2nd ed., Cambridge University Press, London, 1977.

Shi Z., Acoustic observations of fluid mud and interfacial waves, Hangzhou Bay, China, J. Coast. Res., 14 (1998) 1348.

Srinivas R., Response of fine sediment-water interface to shear flow, M. S. thesis, University of Florida, Gainesville, 1989.

Thorne, P.D., Hardcastle, P.J., Flatt, D. and Humphrey, J. D., On the use of acoustics for measuring shallow water suspended sediment processes, IEEE, J. Oceanic Eng., 9 (1994) 48.

Winterwerp, J. C. and Kranenburg, C., Erosion of fluid mud layers. II: experiment and model validation, J. Hydr. Eng., 123 (1997) 512.

Wolanski, E., Asaeda, T. and Imberger, J., Mixing across a lutocline, Limnol. and Oceanogr., 34 (1989) 931.

Wright, L. D., Wiseman, W. J., Bornhold, B. D., Prior, D. B., Suhayda, J. N., Keller, G. H., Yang, Z. S. and Fan, Y. B., Marine dispersal and deposition of Yellow River silts by gravity-driven underflows, Nature, 332 (1988) 629.

Zhou, Y. K., Some characteristics of stream-like macro-tidal estuary (Jiaojiang), Geographical Study, 5 (1986) (in Chinese).

Fine Sediment Dynamics in the Marine Environment
J.C. Winterwerp and C. Kranenburg (Editors)

CBS layers in a diffusive turbulence grid oscillation experiment

M. Mory[a], N. Gratiot[b,*], A. J. Manning[c] and H. Michallet[b]

[a]Ecole Nationale Supérieure en Génie des Technologies Industrielles,
rue Jules Ferry, 64000 Pau, France

[b]Laboratoire des Ecoulements Géophysiques et Industriels,
BP 53, 38041 Grenoble Cédex 9, France

[c]Institute of Marine Studies, University of Plymouth, Drake Circus,
Plymouth, Devon PL4 8AA, United Kingdom

The generation of a Concentrated Benthic Suspension (CBS) of a fluid mud mixture is investigated in the laboratory in a grid stirred experiment. A sediment concentration in the CBS layer in the range 3 g/l to 200 g/l was considered. The sediment concentration, the turbulence properties, and the settling velocity have been measured. The decay of turbulence with increasing distance from the grid is not found to vary with the concentration inside the CBS layer except in the vicinity of the lutocline. The settling velocity decreases rapidely with increasing sediment concentration and appears to be the most sensitive parameter in the experiment. A clear difference in the value of the flux Richardson number at the lutocline is observed depending on the sediment concentration. Ri_f is about 0.1 for the highest concentrations and increases above 0.5 for the low concentration cases.

KEYWORDS
Fluid mud, turbulence, floculation, cohesive sediment suspensions

1. INTRODUCTION

Special attention has recently focused on the existence of Concentrated Benthic Suspensions (CBS) of cohesive sediments in numerous natural sites. The concentration is large in these suspensions which are subjected to a high level of turbulence. The concept of a saturation concentration for cohesive sediment was introduced in 1996 by Winterwerp (see Winterwerp, 1999). For a turbulent shear flow, the latter obtained a relationship between the thickness of a CBS layer, the concentration, and the hydrodynamic properties of the flow. The

---

* present address: IRD Guyane, BP 165, Rte de Montabo, 97323 Cayenne, France

concept of saturation is based on the principle (Turner, 1973) that a turbulent shear flow collapses when the flux Richardson number Ri_f reaches a value of about 0.15.

We investigate in this paper the concept of saturation concentration in an oscillating grid experiment. Natural mud sediment was introduced in a tank where turbulence was produced by an oscillating grid. Grid tank experiments were used before to study the resuspension of cohesive sediments by turbulence (Tsai and Lick, 1986; Wolanski *et al.*, 1989). In a grid tank experiment similar to ours, E and Hopfinger (1987) observed that the layer containing non-cohesive particles resuspended by turbulence reached an equilibrium thickness, an observation which is in a qualitative agreement with the concept of saturation. Although oscillating grid experiments are far from producing flow conditions achieved in the field, they allow to study turbulence conditions with no mean flow. Turbulence is transported by turbulent diffusion. The isolated contribution of turbulence to mixing can be studied in oscillating grid devices.

A specific feature of the present study is that experiments were performed in a grid tank facility using fluid mud mixtures, implying that the settling of sediments is modified by the flocculation of particles. Because the sediment concentration is sufficiently high, it is assumed in this study that the effect of particles on turbulence is solely related to density-stratification. The focus of the paper is on the interaction between the turbulence and density-stratification. By performing settling velocity measurements, the effect of turbulence on the settling velocity is however accounted for. Turbulence properties, sediment concentration and settling velocities were all measured in the laboratory. The flux Richardson number below the interface was determined from these measurements in order to investigate a relationship between the thickness of the CBS layer and the sediment and turbulence properties. The paper is divided as follows. A review of the effect of stratification on turbulence is presented in the next section. The flux Richardson number is defined and its relation to the Rouse number and the Richardson number as usually defined in oscillating grid turbulence is given. The different measurements are presented in sections 3 to 5. The results are used in section 6 to determine the flux Richardson number at the interface.

2. THE EFFECT OF STRATIFICATION ON GRID TURBULENCE IN SEDIMENT LADEN FLOW

The evolution of the turbulent kinetic energy in a sediment laden flow is subjected to four mechanisms: the production of turbulent kinetic energy by energy exchange with the mean flow, the diffusion, the dissipation, and gravity effects. The balance equation for the turbulent kinetic energy k is often written in a symbolic form (see Winterwerp, 1999),

$$\frac{dk}{dt} = P + D - \varepsilon - B \quad , \tag{1}$$

in order to distinguish the four processes. P and D denote the production of turbulent kinetic energy by the mean flow and the diffusion of turbulent kinetic energy, respectively. They are simply expressed in the framework of the k-ε model of turbulence. ε is the dissipation rate of the turbulent kinetic energy. The buoyancy term is

$$B = \frac{\gamma}{\rho_w} g\overline{w'c'} \tag{2}$$

(Sheng and Villaret, 1989), with $\gamma = (\rho_s - \rho_w)/\rho_s$. ρ_s and ρ_w are the densities of solid particles and water, respectively. g is the acceleration of gravity. B is the rate of increase of potential energy required to maintain the sediment in suspension. Particles in a sediment laden fluid have a non-zero settling velocity W_s. Steady conditions are obtained when the turbulent flux of sediment is equal to the settling flux of sediment as deduced from the mass balance equation , i.e.

$$\overline{w'c'} = W_s C \quad . \tag{3}$$

The buoyancy term B has therefore a simple dependence with the mean concentration C:

$$B = \frac{\gamma}{\rho_w} g W_s C \, . \tag{4}$$

Following Ivey and Imberger (1991), the flux Richardson number

$$Ri_f = \frac{B}{P + D} \tag{5}$$

quantifies the relative importance of gravity in the balance equation of the turbulent kinetic energy in steady conditions. For turbulent shear flows the diffusion term is often omitted (Villaret and Trowbridge, 1991; Winterwerp, 1999) because the production of turbulence by the mean flow gradient is usually the dominant term, except in the vicinity of the free surface. In the absence of mean flow, such as in grid turbulence, the production term vanishes and the diffusion of turbulence is the only process by which the turbulent kinetic energy is transported. For a steady turbulence with no mean shear flow, the flux Richardson is simply rewritten as

$$Ri_f = \frac{B}{B + \varepsilon} \quad . \tag{6}$$

Equation (6) is more convenient than (5) since the dissipation term is simply expressed in terms of the turbulent rms velocity u and turbulent lengthscale ℓ

$$\varepsilon \approx A \frac{u^3}{\ell} \quad , \tag{7}$$

where A is a constant of order O(1). Using equation (4), we then deduce

$$Ri_f = \frac{\dfrac{W_s}{u}\dfrac{\gamma g C\ell}{\rho_w u^2}}{A + \dfrac{W_s}{u}\dfrac{\gamma g C\ell}{\rho_w u^2}} \quad . \tag{8}$$

The flux Richardson number is a local quantity which varies over the vertical as a function of the turbulent rms velocity $u(z)$, the turbulent lengthscale $\ell(z)$ and the mean concentration $C(z)$. For the conditions considered in this paper, our experimental observations will show that the concentration in the CBS layer is uniform (see section 4). We therefore write $C(z) = \Delta C$ in the CBS layer, ΔC being the sediment concentration jump across the interface. The flux Richardson number $Ri_f(z)$ is therefore expressed as a function of the Richardson number $Ri(z)$, as usually defined for grid turbulence experiments (Hopfinger & Toly, 1976),

$$Ri(z) = \frac{\gamma g \Delta C \ell(z)}{\rho_w u^2(z)} \quad , \tag{9}$$

and the Rouse number

$$Rou(z) = \frac{W_s}{u(z)} \quad . \tag{10}$$

Considering steady conditions, the integration of equation (1) over a layer having its lower boundary h_0 below the lutocline and its upper boundary h_1 above the lutocline gives

$$F(z = h_0) = w_c \gamma g \Delta C (h_1 - h_0) + \int_{h_0}^{h_1} \varepsilon(z).dz \quad . \tag{11}$$

$F(z)$ denotes the flux of turbulent kinetic energy at the position z. The flux of turbulent kinetic energy is written

$$F(z) = A' \ u^3(z) \ , \tag{12}$$

where A' is a constant of order O(1). Because turbulence has vanished above the lutocline, the flux of turbulent kinetic energy at h_0 balances in eq. (11) the rate of increase of potential energy and the total dissipation within the layer. Equation (11) is of little use for determining the position of the lutocline unless the vertical variation of the dissipation rate is known. However it explains the existence of a maximum position of the lutocline for given conditions since the buoyancy term in eq. (11) increases linearly with the layer thickness.

By comparing in eq. (11) the buoyancy term to the flux of turbulent kinetic energy, we observe that the parameter driving the lutocline occurrence is the product of the Richardson number by the Rouse number $Ri(z)\ Rou(z)$ as already anticipated in the expression of the flux Richardson number (eq. 8). When $Ri(z).Rou(z) << 1$, gravity effects are negligible and the evolution of u and ℓ with z is not modified by the presence of sediments. Gravity effects will be active when $Ri(z).Rou(z)$ reaches a value of order O(1).

Different authors have considered before the significance of the flux Richardson number. Turner (1973) and Ivey and Imberger (1991) have shown that turbulence collapses when the flux Richardson number Ri_f reaches a value of the order of 0.15 to 0.25. This experimental evidence was taken by Winterwerp (1999) as a basis for establishing the concept of a saturation concentration for cohesive sediments. The critical value of Ri_f in the range 0.15 to 0.25 was obtained by Ivey and Imberger (1999) in view of experimental data for salt-stratified or temperature-stratified shear flows. The applicability of this result to fluid mud mixtures in a grid tank turbulence is questionable. On the one hand, the settling velocity of particles tends asymptotically to zero for salt stratification. In oscillating grid tank experiments, no steady state is obtained and the interface is continuously displaced upward for this type of stratification. On the other hand, diffusing turbulence and shear produced turbulence are different. To our knowledge, estimates of Ri_f in purely diffusive flows are still lacking.

Previous experiments (E & Hopfinger, 1987; Huppert *et al.*, 1995) show the existence of a saturation concentration for sediments maintained in suspension by a diffusing turbulence. The above considerations provide a physical understanding of the process, but the quantitative determination of the position of the lutocline is not available*. In our laboratory experiments, measurements of the vertical variations of the turbulent rms velocity, the integral lengthscale, the mean concentration and the settling velocity have allowed to estimate the value of Ri_f at the lutocline for various CBS experimental conditions.

3. THE EXPERIMENTAL SET-UP, PROCEDURES AND OBSERVATIONS

The experiment was performed in a square tank, 53 cm wide. For all experiments, the water depth was 40 cm. The horizontal grid, made of 1.5 cm square bars with a mesh $M = 7.5cm$, oscillated vertically. The mean position of the grid was 5 cm above the bottom of the tank. The grid oscillation conditions were determined by the stroke $S = 4.5cm$ (twice the amplitude) and the frequency F of oscillation. Three different values $F = 3Hz$, $F = 4Hz$ and $F = 6Hz$ were considered in the study. The tank and the grid geometry were similar to these used previously by Hopfinger and Toly (1976). The latter authors have shown that these experimental conditions ensure that there is no mean flow at distances from the grid larger than $2M$. The turbulence is transported away from the grid by a diffusive process. The turbulence properties are 1DV, i.e. they vary only with the distance z from the grid. The position $z=0$ is here the mean position of the grid and the Oz axis is oriented upward. Figure 1 is a sketch of the experimental set-up.

* E & Hopfinger (1987) and Huppert *et al.* (1995) have determined from an equation similar to (11) a relationship between the layer thickness, the sediment concentration and the turbulence properties, but they assume that turbulence is not modified by gravity effects.

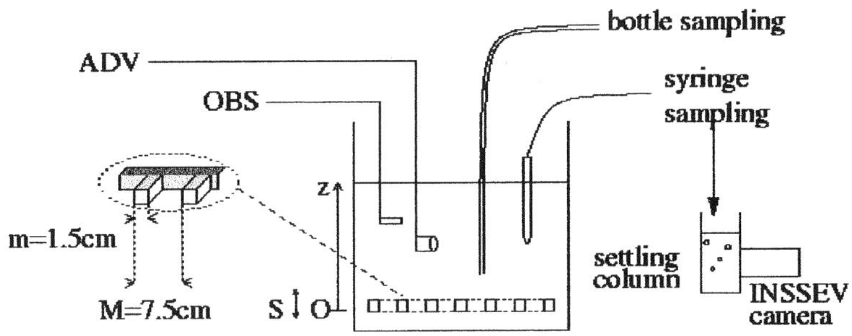

Figure 1. Experimental set-up.

An homogenous premixed fluid mud mixture was introduced in the tank while the grid oscillated. The vertical distribution of sediments was redistributed during a transient evolution in which settling of sediments and turbulent transport of sediments both occurred. Settling occurred in the upper part of the tank where the turbulence level was not sufficient to maintain the sediment in suspension. A lutocline appeared that moved downward until it became stable. Oscillations of the lutocline were observed but the mean level of the lutocline did not appear to change anymore. When the amount of sediment introduced inside the tank exceeded a certain level, the turbulence produced in the CBS layer was also not sufficient to maintain the whole amount of sediment in suspension. Sediment settled on the bottom to form a deposited bed. The equilibrium state consisted of three superposed layers: the deposited bed, the CBS layer and the clear water layer. This is called the saturated regime. When only a limited quantity of sediments was introduced in the tank a quasi-uniform suspension with no lutocline could be observed, and eventually no deposited bed was seen. This is the non-saturated regime.

Experiments were also carried out by starting the grid oscillation with the sediment initially deposited on the bottom and left to consolidate. For the same grid oscillation conditions, a smaller amount of sediment was maintained in suspension by turbulence when the sediment was initially on the bottom of the tank than when the sediment was introduced in suspension while the grid oscillated. We attribute this observation to consolidation effects in the deposited bed.

The tank was made of transparent Perpex walls which enabled the visualisation of the different sediment layers. Figure 2 shows a typical picture of the three layers stratification inside the tank taken in saturated steady conditions. Internal wave oscillations are visible at the lutocline.

Figure 3 illustrates the process that leads from non-saturated to saturated conditions. Vertical profiles of concentration, in steady conditions, are shown for four different initial concentrations $\overline{C}$. $\overline{C}$ is defined as the mass of dry sediment divided by the total volume of fluid inside the tank. When $\overline{C}$ is less than $1g/l$ the concentration is clearly uniform over the vertical. A small gradient of concentration is noticed in the upper part for $\overline{C} = 1.3g/l$ and the lutocline is clearly identified for $\overline{C} = 2.7g/l$. A further increase of $\overline{C}$ enhances the gradient across the lutocline which is simultaneously displaced downward.

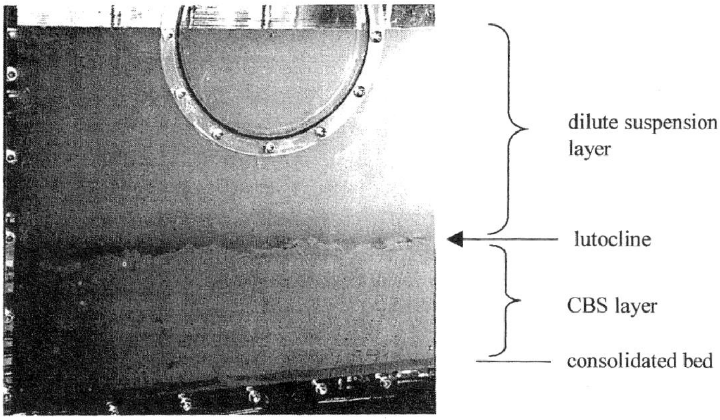

Figure 2. Photograph of the three layers observed for steady saturated conditions. The concentration in the CBS layer is $C_{CBS} \approx 80 g/l$.

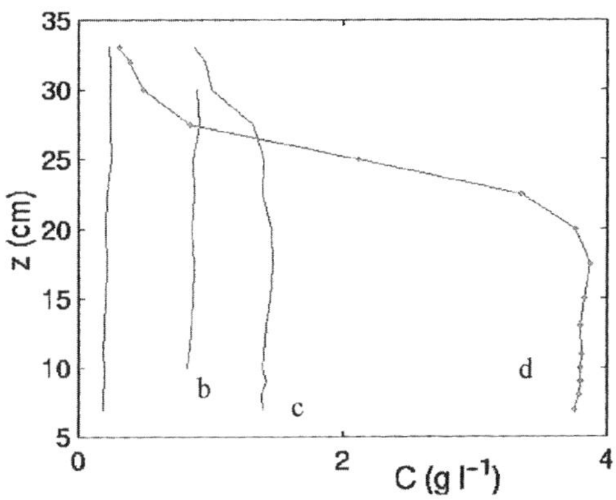

Figure 3. Vertical profiles of concentration measured for increasing values of the averaged concentration of sediments introduced in the tank. $F = 4Hz$. (a) $\overline{C} = 0.2 g/l$; (b) $\overline{C} = 0.9 g/l$; (c) $\overline{C} = 1.3 g/l$; (d) $\overline{C} = 2.7 g/l$. Inorganic mud from Gironde.

Three different natural mud have been used from the Gironde estuary (organic and inorganic) and from the Tamar estuary (organic). The inorganic mud from Gironde was chemically treated with potassium permanganate and passed through a 100 μm sieve. Sediment analysis performed by de Croutte et al. (1996) revealed that the sediment's median diameter was about D_{50}=12 μm and the sand content (63-100 μm) was below 3%. The clay fraction was comprised of illite (35%), kaolinite (27%), smectite (24%) and chlorite (14%).

4. CONCENTRATION AND TURBULENCE MEASUREMENTS FOR SATURATED CONDITIONS

Vertical profiles of the mean sediment concentration $C(z)$ and of the rms turbulent velocity $u(z)$ were measured for different saturated conditions. The integral lengthscale $\ell(z)$ was also determined from velocity measurements for selected cases.

Saturated conditions were obtained when a sufficient amount of sediment $\overline{C}$ was initially introduced in the tank. Table 1 summarises the different experimental conditions. Only the conditions that vary are given. Table 1 also contains the mean concentration C_{CBS} measured in the CBS layer in steady conditions and the distance z_l of the lutocline from the mean position of the grid. A wide range of concentration C_{CBS} was investigated.

Table 1
Experimental conditions of saturated concentration experiments.

mud	Sa (ppt)	F (Hz)	$\overline{C}$ (g/l)	C_{CBS} (g/l)	z_l (cm)	symbol
Gironde I	0	3	85	202	10.5	□
Gironde I	0	3	61	149	9.8	x
Gironde I	0	6	30	51	17.0	▽
Gironde I	0	4	2.8	4.9	18.0	▷
Gironde I	0	4	7.1	11.6	17.0	◁
Tamar O	16.5	4	5.0	8.4	19.0	+
Gironde I	16.5	4	3.1	4.7	21.5	◇
Tamar O	16.5	4	5.1	8.1	20.0	△
Tamar O	16.5	4	3.6	5.1	23.5	○
Gironde I	16.4	4	2.7	3.8	23.5	•

Sa: salinity, F : frequency of grid oscillation, $\overline{C}$: averaged sediment concentration in the tank, C_{CBS} : mean sediment concentration in the CBS layer, z_l : distance of the lutocline from the grid. The symbols are those used in Figures 4, 5 , 6 and 9.

Vertical profiles of the mean sediment concentration were measured from samples analysis and using an Optical Backscatter System (OBS) developed at the Institute of Marine Studies (Christie and Dyer, 1997). The calibration curve was determined for each mud. The voltage output of the OBS increases monotonically with increasing concentration when the concentration is below 40 g/l while it decreases with increasing concentration above 45 g/l. The OBS was therefore not used for concentrations in the range 35 – 50 g/l. Sample analysis were made to determine the appropriate part of the calibration curve.

Time series measurements of the velocity were made using an Acoustic Doppler Velocimeter (ADV) developed at Laboratoire des Ecoulements Géophysiques et Industriels. This system determines the velocity from the changes in the time shift of pulse-to-pulse back-scattered signals. The principle of operation is not new in itself, but a specific study (Gratiot *et al.*, 2000) was conducted to check its ability to measure the turbulent velocity in concentrated fluid mud mixtures. The adjustment of parameters of the ADV enabled to measure instantaneous velocities with a data rate of the order of 70 data/s for a concentration of 50 g/l. The data rate decreases to about 40 data/s for a concentration of 100 g/l. The rms turbulent velocity u was computed from time series of about 4000 measurements. The ADV system could measure simultaneously the velocity component parallel to the beam axis at 16 positions. The distance between two successive measurement volumes was 3 mm. The integral lengthscale ℓ was deduced from the computation of the correlation function of the velocity measurements made simultaneously at the different locations.

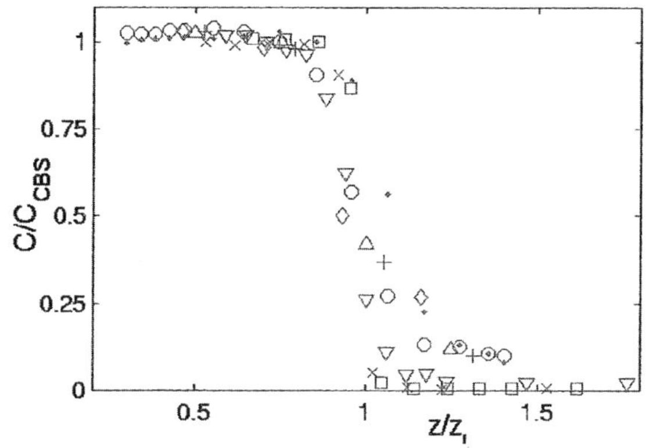

Figure 4. Vertical profiles of concentration measured in saturated steady conditions. Experimental conditions corresponding to the symbols are given in Table 1.

Figure 4 shows different vertical profiles of the mean concentration. The distance z from the grid is made dimensionless using the distance z_l of the lutocline from the grid, so that all profiles are centred around the lutocline position. We observe that the concentration is homogenous inside the CBS layer. The homogeneity of the concentration is apparently even better for the highest concentration ($C_{CBS} = 202 g/l$); the interface thickness is smaller. This is partly because internal wave oscillations of the lutocline have a larger amplitude for smaller values of C_{CBS}, contributing to make the interface look thicker in Figure 4 when C_{CBS} is small.

The decay of the rms turbulent velocity u with increasing distance from the grid is shown in Figure 5 for different saturated experimental conditions. The concentration in the CBS layer considered for the plots in Figure 5 ranges between 3.8 g/l and 202 g/l. Also superimposed are two vertical profiles of the turbulent rms velocity made in clear water. It is observed that the decay of u is similar for all concentrations in the CBS layer considered. The uncertainty in measured data is rather large, as indicated by error bars, but no general trend indicating that u decreases for increasing values of C_{CBS} is observed. Actually the measurements made for the largest values of C_{CBS} collapse on the decay plots obtained in clear water, and the highest difference is between measurements in clear water and for $C_{CBS} = 51 g/l$. For analysis of the data (section 6) the decay laws were interpolated by the power law function

$$\frac{u(z)}{FM} = 0.092\left(\frac{z}{M}\right)^{-1.4} \tag{13}$$

in the range $1.2 \le z/M \le 3.2$ (M is the grid mesh) and for $z<z_l$. Although the exponent of the power is of the order of those obtained by different authors for oscillating grid turbulence, no particular attention should be paid to the value of the exponent, considering the narrow range of values of z/M. Because of the lutocline, turbulent velocity measurement could not be made at large distances from the grid.

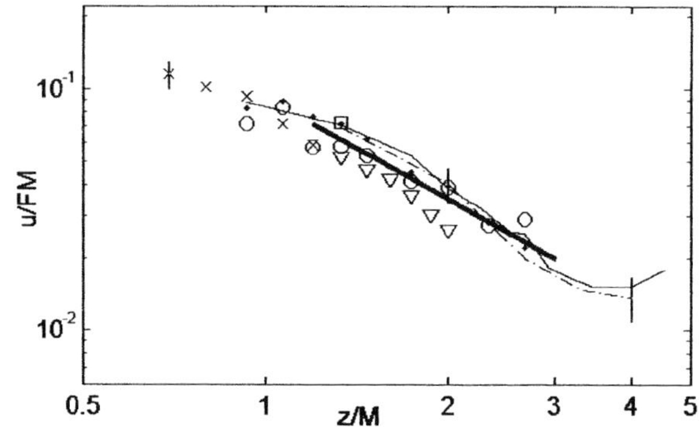

Figure 5. Vertical variations of the rms turbulent velocity in the CBS layer for saturated steady conditions. Experimental conditions corresponding to the symbols are given in Table 1. The solid light line and dashed dotted line are measurements in clear water for frequencies of grid oscillation $F = 6Hz$ and $F = 3Hz$, respectively. The heavy solid line is eq. (13).

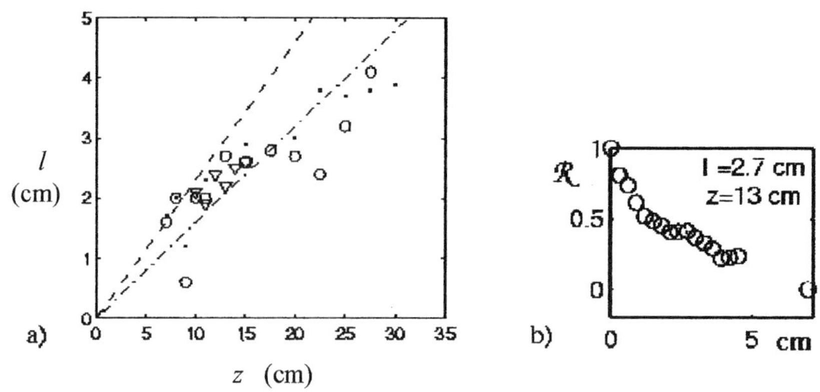

Figure 6. 6a. Integral lengthscale variations versus the distance from the grid in saturated steady conditions. Experimental conditions corresponding to the symbols are given in Table 1. The dashed line and the dashed dotted line are the results obtained by Hopfinger and Toly (1976) for $S = 4cm$ while $M = 10cm$ and $M = 5cm$, respectively.
6b. Correlation function measured at $z = 13cm$ for $C_{CBS} = 5.1g/l$.

Measurements of the integral lengthscale evolution with the distance from the grid are less common in the literature than measurements of the rms turbulent velocity, but there is a general agreement that the integral lengthscale varies linearly with the distance from the grid. This experimental result is supported by k-ε modelling (Sonin, 1983) in clear water, when the decay of the turbulent kinetic energy follows a power law, anyhow the value of the exponent is. Measurements of the integral lengthscale were made for selected cases. The results are shown in Figure 6a together with the results obtained by Hopfinger and Toly (1976). A correlation function R is also shown in Figure 6b. A number of correlation functions that were computed did not display such a nice and smooth behaviour as the case shown in Figure 6b. This explains the large scatter in Figure 6a. Nevertheless, the data indicate a clear increase of the integral lengthscale with increasing distance from the grid. Our measurements are not inconsistent with the results of Hopfinger and Toly. In view of this, the evolution of the integral lengthscale with the distance from the grid was taken for subsequent use of the data as

$$\ell(z) = 0.2z \quad . \tag{14}$$

5. SETTLING VELOCITY MEASUREMENTS

The settling velocity of cohesive sediments varies very much with the concentration. When the concentration in the CBS layer was below 15 g/l, the settling velocity was determined using the INSSEV system video camera (Fennessy *et al.*, 1994). This apparatus was developed at the Institute of Marine Studies for in situ measurements of the settling velocity of individual flocs. The size and the settling velocity of flocs are measured from the analysis of images of settling flocs in a column recorded by a video camera (Manning and Fennessy, 1997). Whereas the system is immersed in the water column for in situ measurements, samples were taken from the grid tank using a syringe of length 450 mm and diameter 8 mm, and introduced in the settling column for image recording. By comparing the floc properties measured in situ and measured after sampling, Manning (2001) concluded that the procedure of sampling did not modify the flocs. Figure 7 shows a typical diagram relating the floc diameter D_f, to the settling velocity W_s. The diameter D_f of the flocs measured by INSSEV is the square root of the length normal to settling direction times that measured in the settling direction. An essential feature is the dispersion in floc diameter and in settling velocity, indicating that the floc density is as well subjected to a significant variability. For further analysis, the settling velocity of each fluid mud suspension was determined by computing the mass flux of the suspension using the settling velocities measured for individual flocs. Error bars in the estimates were determined from the scatter in measurements.

For concentrations in the CBS layer larger than 15 g/l, the sedimentation of the suspension inside a column was recorded and the settling velocity was determined as a function of the concentration using the Kynch method (1952). Toorman (1999) recently reported the relevance of this method applied to highly concentrated suspensions of cohesive particles.

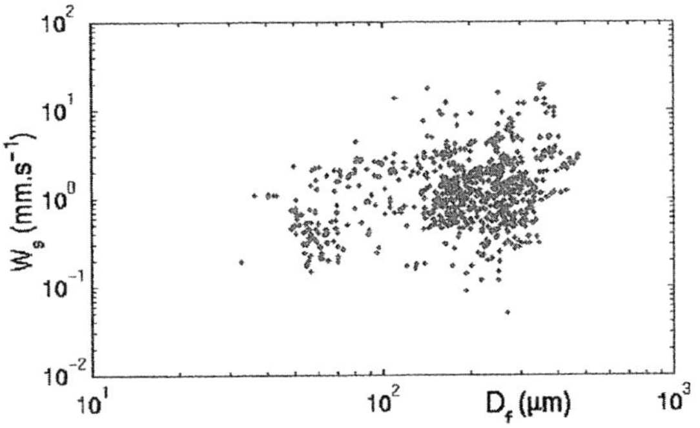

Figure 7. Variation of the settling velocity with the size of floc measured using INSSEV in a fluid mud suspension: Tamar mud, $\overline{C}$ = 1 g/l.

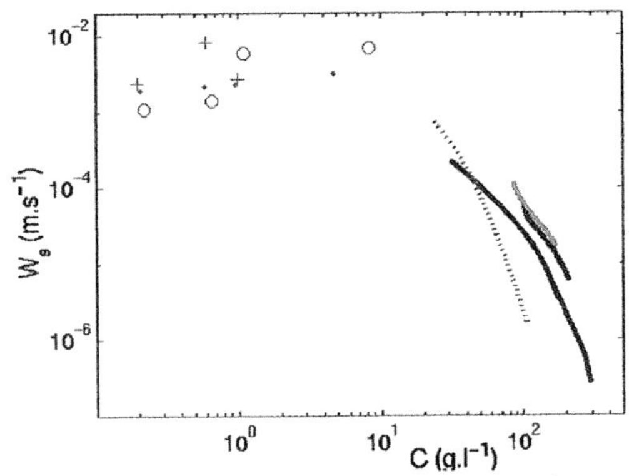

Figure 8. Variation of the settling velocity in fluid mud suspensions as a function of concentration. — Inorganic mud from Gironde ; + organic mud from Gironde; O and ··· organic mud from Tamar.

Figure 8 shows the variation of the settling velocity with the concentration determined for the different mud considered in the study. Results obtained using INSSEV appear using symbols, whereas results obtained from sedimentation experiments are displayed as curves. Because INSSEV was available only during a limited period of the experimental study, the settling velocity has not been measured for all fluid mud suspensions concentration below 15 g/l. The variation of the settling velocity with the concentration in the CBS layer was interpolated from Figure 8. In spite of the large scatter in settling velocity estimates, Figure 8 clearly shows that the settling velocity decreases by two orders of magnitude when the concentration increases from 4 g/l to 200 g/l. The techniques used to measure the settling velocity in low concentrated suspensions (INSSEV measures the settling velocity of individual flocs) and in high concentrated suspensions (analysis of sedimentation experiments) are very different in their principles. It is however expected that the settling velocity decreases significantly when the sediment concentration is larger than 15 g/l, because of hindered settling effects.

6. THE FLUX RICHARDSON NUMBER AT THE INTERFACE

For all experiments listed in Table 1, the rms turbulent velocity u and the integral lengthscale ℓ were determined at the lutocline from eqs. (13) and (14), respectively. Their values, as well as the mud concentration C_{CBS} and the settling velocity W_s, are given in Table 2. These data were used to estimate the flux Richardson number at the lutocline (eq. 8). The variations of Ri_f with the concentration in the CBS layer are shown in Figure 9. The error bars result mainly from the uncertainties in the measurements of the settling velocity and of the rms turbulent velocity u (the dissipation rate varies as u^3). The value of $A = 1$ for estimating the turbulent kinetic energy dissipation rate (eq. 7) was used for the plot in Figure 9. Similar curves were obtained for $A = 0.5$ and $A = 2$. They are not shown here, as the conclusions of the paper do not depend on this value.

Table 2
Turbulence properties at the lutocline and settling velocity for the CBS conditions given in Table 1

C_{CBS} g/l	202	149	51	4.9	11.6	8.4	4.7	8.1	5.1	3.8
z_l cm	10.5	9.8	17.0	18.0	17.0	19.0	21.5	20.0	23.5	23.5
u mm/s	12.9	14.2	13.1	8.1	8.8	7.5	6.3	7.0	5.6	5.6
ℓ cm	2.1	2.0	3.4	3.6	3.4	3.8	4.3	4.0	4.7	4.7
W_s mm/s	0.007	0.016	0.18	3.3	1.5	6.8	3.2	7.4	9.8	3.3
Rou	0.0005	0.0011	0.014	0.41	0.17	0.91	0.51	1.1	1.8	0.59
Ri	159	90.1	62.6	16.8	32.0	35.4	31.7	41.5	48.2	35.9

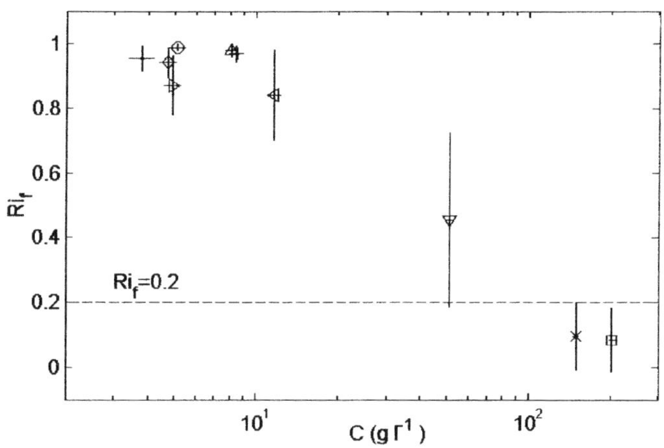

Figure 9. Variation of the flux Richardson number at the lutocline with the concentration in the CBS layer for steady conditions. Experimental conditions corresponding to the symbols are given in Table 1.

The major result of Figure 9 is that the flux Richardson number decreases from about 1 at the interface when the concentration in the CBS layer is low (a few g/l) to 0.1 when the concentration in the CBS layer increases to 200 g/l. This behaviour is due to the strong decrease in the settling velocity when the concentration in the CBS layer is larger than 50. The estimates of the Rouse and Richardson numbers at the lutocline are included in Table 2. The Rouse number is indeed very small for the large concentration cases, which correspond simultaneously to the highest values of the Richardson number (i.e. the strongest stratification).

In spite of the limited number of experiments carried out for large concentration conditions, their interpretation is less subjected to experimental uncertainty. For such conditions, the Richardson number is high (see Table 2). Because the amplitude of oscillations at the lutocline was small, velocity measurements could be performed close to the interface as shown in Figure 5. The rms turbulent velocity in the CBS layer displays no significant decrease as compared to measurements in clear water. Table 2 indicates that the turbulent Reynolds number based on u and ℓ is of the order of 200 for all conditions. The estimates of turbulence properties at the lutocline are based on measurements made close to the interface, enabling a confident estimate of the flux Richardson number at the interface. On the other hand, turbulence measurements could not be made close to the interface for the low concentration cases (less than 12 g/l) because of the oscillations of the lutocline. The estimates of u and ℓ at the lutocline given in Table 2 for the low concentration cases were extrapolated from eqs. (13) and (14). Since the product of the Rouse and the Richardson number is high, the flux Richardson number is about one at the interface. We cannot exclude that a smaller value would be obtained using measurements closer to the lutocline. Nevertheless, using measured values of turbulence, we estimate that $Rou.Ri = 1$ at a distance from the grid ranging between 11.5 cm and 13.8 cm for the low concentration range $(3.8g/l \le C_{CBS} \le 11.6g/l)$ and at a distance of 14.5 cm for $C_{CBS} = 51g/l$. The flux Richardson number is therefore larger than 0.5 at the lutocline for the low concentration cases, a value which is clearly different from that obtained for the large concentration cases.

7. CONCLUSION

Concentrated Benthic Suspensions have been generated in a grid tank experiment using fluid mud mixtures of concentration in the range 4 g/l to 200 g/l. The concentration inside the CBS layer was nearly homogeneous. An important result of the study was that the decay of turbulence inside the CBS layer does not vary with the concentration, excepted in the vicinity of the lutocline. The decay of turbulence are quantitatively similar in a CBS layer of concentration as high as 200 g/l and in clear water.

The observation of CBS layers of thickness independent of time is not new. We have considered if the saturation effect that fixes the lutocline position can be characterised in terms of the criterion involving the flux Richardson number. The flux Richardson number at the interface (Figure 9) has been computed from measurements of turbulence and concentration inside the CBS layer and estimations of the settling velocity. Although our experiments have not been successful for determining a criterion for the lutocline position, a clear difference was found between the high concentration range $(C_{CBS} > 100g/l)$ for which $Ri_f \approx 0.1$ and the low concentration cases $(C_{CBS} < 50g/l)$ for which $Ri_f > 0.5$. A numerical study is currently under investigation using a 1DV k-ε model in order to compute the vertical variation of Ri_f below the lutocline for the different cases considered in this study.

A significant part of the study was the measurement of the settling velocity. A very strong decrease of the settling velocity with increasing concentration was measured in our fluid mud mixtures. The theoretical analysis of the effect of stratification in a diffusing turbulence (section 2) shows that the driving parameter is the product of the Rouse number by the Richardson number, and therefore the settling velocity is indeed the most sensitive quantity when large concentration are considered.

8. ACKNOWLEDGEMENT

This work was carried as part of the COSINUS Program, which is funded by the European Commission (contract MAS3-CT97-0082). We express our gratitude to E. Hopfinger, who made the grid tank facility available, and J.M. Barnoud for his continuous technical assistance. A close collaboration was established with the Institute of Marine Studies (University of Plymouth) regarding measurements techniques. K. Dyer is thanked for his support.

REFERENCES

Christie, M.C. and Dyer, K.R., 1997, Optical backscatter turbidity sensor, Institute of Marine Studies, University of Plymouth, internal report.

De Croutte, E., Gallissaires, J.M. and Hamm, L., 1996, Flume measurements of mud processes over a flat bottom under steady and unsteady currents, Rapport R3, Sogréah Ingénierie.

Fennessy, M.J., Dyer, K.R. and Huntley, D.A., 1994, INSSEV: an instrument to measure the size and settling velocity of flocs in situ, *Marine Geology*, (117) 107-117.

Gratiot, N., Mory, M. and Auchère D., 2000, An acoustic Doppler velocitymeter (ADV) for the characterisation of turbulence in concentrated fluid mud, *Continental Shelf Research*, (20) 1551-1567.

Hopfinger, E.J. and Toly, J.A., 1976, Spatially decaying turbulence and its relation to mixing across density interfaces, *J. Fluid Mech* ., (78) 155-175.

Huppert, H.E., Turner, J.S. and Hallworth, M.A., 1995, Sedimentation and entrainment in dense layers of suspended particles stirred by an oscillating grid., *J. Fluid Mech.*, (289) 263-293.

Ivey, G. and Imberger, J., 1991, On the nature of turbulence in stratified fluids ; Part I : the energetics of mixing, *J. Physical Oceanography*, (21) 5, 650-658.

Kynch, G.J., 1952, A theory of sedimentation, *Tram Faraday Soc.*, (48) 166-176.

Manning, A.J., 2001, *A study of the effect of turbulence on the properties of flocculated mud*, PhD Thesis, University of Plymouth.

Manning, A.J. and Fennessy, M.J., 1997. INSSEV (In Situ Settling Velocity instrument) - 1.3: Operator Manual. Institute of Marine Studies, University of Plymouth, 25pp.

Sheng, Y.P. and Villaret, C., 1989, Modelling the effect of suspended sediment stratification on bottom exchange processes, *J. of Geophysical Research*, (94)C10, 14,429-14,444.
Sonin, A.A., 1983, Calibration of the k-ε turbulent model for the diffusion of turbulence, *Physics Fluids*, (26)10, 2769.
Toorman, E., 1999, Sedimentation of self weight consolidation: constitutive equations and numerical modelling, *Géotechnique*, (49) 709-726.
Tsai, C.H. and Lick, W., 1986, A portable device for measuring sediment resuspension, *J. Great Lakes Res.*, (12)4, 314-321.
Tuner J.S., 1973, *Buoyancy effects in fluids*, Cambridge University Press.
Villaret, C. and Trowbridge, J.H., 1991, Effect of stratification by suspended sediments on turbulent shear flows, *J of Geophysical Research*, (96)C6, 10,659-10,680.
Winterwerp J.C., 1999, *On the dynamics of high-concentrated mud suspensions*, PhD Thesis, Technical University Delft.
Wolanski, E., Asaeda, T. and Imberger, J., 1989, Mixing across a lutocline, *Limnol. Oceanogr.*, (34)5, 931-938.
E., X. and Hopfinger, E.J., 1987, Stratification by solid particle suspensions, Proceedings *3rd Int. Symp. On Stratified Flows*, Caltech, Pasadena, 1-8.

Fine Sediment Dynamics in the Marine Environment
J.C. Winterwerp and C. Kranenburg (Editors)

Modelling of turbulent flow with suspended cohesive sediment

Erik A. Toorman

Hydraulics Laboratory, Civil Engineering Department, Katholieke Universiteit Leuven
Kasteelpark Arenberg 40, B-3001 Leuven, Belgium

Traditional (cohesive) sediment transport models contain several simplifications which are no longer justifiable when sediment concentrations or stratification effects become significant. This paper gives a rather technical overview of various modifications to a sediment transport model with k-ε turbulence closure, used as research tool, in order to improve the physics described by the model. The attention is focussed on the modelling of sediment-turbulence interactions.

KEY WORDS
sediment-laden flow, turbulence modulation, buoyancy damping, saturation, drag reduction, modelling

1. INTRODUCTION

The numerical modelling of suspended sediment transport in turbulent flows has been studied since the late seventies. However, the simplifications made in the early models, which are still commonly used, are only acceptable for very low concentrations and low levels of stratification. Several consequences of the sediment-turbulent interactions have never been taken into account. Within the framework of the MAST3 COSINUS project this subject has been studied in greater depth (Toorman *et al.*, 2002). This paper focuses on the contribution of the Hydraulics Laboratory of the Katholieke Universiteit Leuven, where a generalized 2-dimensional numerical model based on the Reynolds-averaged Navier-Stokes (RANS) equations with eddy viscosity turbulence closure, the code FENST, has been developed over the past years. The primary purpose of the model is to use it as a virtual laboratory to study sediment transport processes in a vertical longitudinal cross-section of an open-channel (e.g. an estuary). The model generated data are more detailed than can be obtained experimentally and can help understand phenomena which are observed in the field. In this context it is very important to realise that the interpretation of the model results has to be done with great care in view of the limitations and shortcomings of the model (e.g. the description of the physics therein and the numerical discretisation).

The model restricts itself to the implementation of the mixing-length and the two-equation k-ε turbulence models, which are presently used in engineering models applied to coastal and estuarine sediment transport. This will allow a relatively easy transfer of new implementation methods and improvements into these models. Higher-order turbulence models, such as

Reynolds stress models (e.g. Galland *et al.*, 1997), do not yet perform any better and at a much higher computational cost.

Details of the numerical model FENST are presented, showing the major differences with the traditional models. First the model equations are described in greater detail. Following, the modelling of buoyancy effects and the treatment of boundary conditions is discussed.

Model results of steady open-channel flow are presented and discussed. Certain results, which reveal some remarkable features, are studied analytically in further detail. Their implications are shown to help understand the concepts of drag reduction and saturation in sediment-laden turbulent flows.

2. MODEL DESCRIPTION

The present study has been carried out with the research code "FENST-2D" (Finite Elements for Navier-Stokes, Sediment Transport and Turbulence – 2-Dimensional), developed by the author. It solves the full hydrodynamic equations for the velocity components and the pressure, and the sediment transport equation (or sediment mass balance) for the sediment concentration. Various turbulence closures are implemented: the Prandtl mixing-length (PML) model, the standard k-ε model for high turbulent Reynolds number shear flows, various low-Reynolds number (low-Re_t) k-ε models and a hybrid two-layer approach, where a one-equation turbulence model coupled to a mixing-length model is applied in a layer near solid boundaries. The model is second-order accurate in space (using 9-noded Lagrangean elements) and uses first order implicit time stepping. Various numerical stabilisation techniques have been implemented (streamline-upwinding for advection-dominated flows, self-eliminating artificial diffusion, pseudo-time stepping and relaxation techniques for the k-ε model). Nevertheless, the intrinsic numerical diffusion of the various schemes is very low, in particular compared to commercial CFD codes where robustness is the primary concern. Moving boundaries (e.g. tidal effects) can be dealt with using the Arbitrary Lagrange-Euler method (Huerta and Liu, 1988; Toorman, 1993).

2.1. Equations

The hydrodynamics of the sediment-laden flowing water is modelled using the mixture theory. As both the fluid and solid phases are assumed incompressible, the suspension mass conservation reduces to the continuity equation, despite the fact that the suspension density ρ may vary (Toorman, 1996). The suspension continuity reads:

$$\frac{\partial U_j}{\partial x_j} = 0 \tag{1}$$

The momentum conservation of the suspension is described by the Reynolds-averaged Navier-Stokes equations:

$$\rho\left(\frac{\partial U_i}{\partial t} + U_j \frac{\partial U_i}{\partial x_j}\right) = \frac{\partial}{\partial x_j}\left(\rho(\nu + \nu_t)\left(\frac{\partial U_i}{\partial x_j} + \frac{\partial U_j}{\partial x_i}\right)\right) - \frac{\partial p}{\partial x_i} - \rho g \delta_{iz} \tag{2}$$

Conservation of sediment is expressed by the sediment transport equation:

$$\frac{\partial C}{\partial t}+U_j\frac{\partial C}{\partial x_j}=\frac{\partial}{\partial x_j}\left((\nu+\frac{\nu_t}{\sigma_s})\frac{\partial C}{\partial x_j}+w_s C\delta_{jz}\right) \tag{3}$$

In these equations the following notations are used: U = the mean velocity, p = pressure, t = time, x_j = the components of the co-ordinate vector, ν = the kinematic viscosity of the suspension, ν_t = the eddy viscosity, σ_s = the turbulent Schmidt number (the ratio of ν_t to the eddy diffusivity of the sediment particles), C = concentration by mass, w_s = the representative mean settling velocity, δ_{ij} = the Kronecker delta. For cohesive sediments, the settling velocity preferably is obtained from a flocculation model.

This set of equations requires a turbulence closure. In this paper the focus is on the application of the k-ε turbulence closure, where the eddy viscosity is calculated as:

$$\nu_t = f_\mu c_\mu k^2/\varepsilon \tag{4}$$

The k-ε model solves the conservation of turbulent kinetic energy k:

$$\frac{\partial k}{\partial t}+U_j\frac{\partial k}{\partial x_j}=\frac{\partial}{\partial x_j}\left((\nu+\frac{\nu_t}{\sigma_k})\frac{\partial k}{\partial x_j}\right)+P+G-\varepsilon \tag{5}$$

and its dissipation rate ε:

$$\frac{\partial \varepsilon}{\partial t}+U_j\frac{\partial \varepsilon}{\partial x_j}=\frac{\partial}{\partial x_j}\left((\nu+\frac{\nu_t}{\sigma_\varepsilon})\frac{\partial \varepsilon}{\partial x_j}\right)+\frac{1}{T_t}\left(f_1c_1P+c_3G-f_2c_2\varepsilon\right) \tag{6}$$

where: $T_t = k/\varepsilon$ is the (high-Reynolds) turbulence time scale, P is the shear production and G the buoyancy term, respectively defined by:

$$P=\nu_t\left(\frac{\partial U_i}{\partial x_j}+\frac{\partial U_j}{\partial x_i}\right)\frac{\partial U_i}{\partial x_j} \tag{7}$$

$$G=\frac{g}{\rho}\frac{\nu_t}{\sigma_s}\frac{\partial \rho}{\partial z} \tag{8}$$

with g the gravity constant and z the vertical co-ordinate. The remaining coefficients have been determined semi-empirically and are taken as $c_\mu = 0.09$, $c_1 = 1.44$, $c_2 = 1.92$, $\sigma_k = 1.0$ and $\sigma_\varepsilon = 1.3$, which are commonly used values for turbulent shear flows (Rodi, 1980; Chen and Jaw, 1998). The value of c_3 in stable stratified shear flows is generally somewhere in the range 0-0.3 (Rodi, 1980). Uittenbogaard *et al.* (1992) argue that the Richardson number effect is negligible for the scales where ε is important. Hence, they suggest $c_3 = 0$, which is chosen here also. It is computationally advantageous and seems to perform satisfactorily (e.g. it yields the expected saturation value of the Richardson number; see below). The factors f_μ, f_1 and f_2 are correction functions for the low-Reynolds formulation (see further); their value is 1 for the standard high-Reynolds form.

The above conservation equations neglect as usual certain contributions with regard to density variation effects. For the applications under consideration (i.e. sediment-laden open-channel flow without density fronts) these additional terms are not important.

2.2. Buoyancy effects

Despite the presence of an explicit buoyancy term in the TKE equation, empirical damping functions, i.e. F_m for momentum and F_s for mixing, such as required for the PML model, are still needed for the determination of the turbulent Schmidt number $\sigma_s = \sigma_0 F_m/F_s$, (where σ_0 = the neutral turbulent Schmidt number) which occurs in the sediment transport equation and the buoyancy term, and for the ε near-wall boundary condition, eq.(12) (see further). Traditionally, they are determined as a function of a Richardson number, usually the gradient Richardson number Ri. Details on the determination of the damping functions can be found in (Toorman *et al.*, 2002).

However, this Richardson number may not be the most suitable one. The general flux Richardson number is defined as the ratio of buoyancy to all production mechanisms (Ivey and Imberger, 1991), i.e.:

$$Rf = \frac{1}{1 - \varepsilon / G} \tag{9}$$

and is even valid under non-equilibrium conditions. Traditionally, inertia and diffusion are neglected, resulting in the definition $Rf_P = -G/P$. This is acceptable for slowly varying shear flows, except in open-channel flows near the free surface where diffusion dominates over shear production, because there is no shear at the surface (in the absence of wind stresses). Consequently, Richardson numbers based on P only produce too high values near the surface, resulting in excessive turbulence damping which does not allow mixing up to the surface. This can be avoided by defining a generalised gradient Richardson number, in analogy with the generalised Schmidt number, as:

$$Ri = \frac{-\sigma_s G}{D + P} = \sigma_s Rf \tag{10}$$

where D is the diffusion term in the TKE equation. Damping functions based on the generalised flux Richardson number seem to be better founded (e.g. Ellison, 1957) and more general, but need to be converted into a form where the Schmidt number can be computed explicitly. This is still under investigation (Toorman, 2000b).

2.3. Free surface boundary treatment

The free surface boundary conditions for the k-ε model are not well established. From measurements it is known that both k and ε are non-zero at the surface (Nezu and Nakagawa, 1993). The semi-empirical method for the determination of k and ε at the surface, proposed by Rodi (1980), has become popular. Nevertheless, setting k and ε zero gives comparable results and is preferred because of its simplicity. It should also be emphasised that in 2DV calculations, the simulated flow is not completely realistic, because turbulence is always 3D and channel widths are finite, generating secondary currents, which are not taken into account by the present model. This will be studied in the future with a quasi-3D model.

2.4. Bottom boundary treatment

As the standard k-ε model is only valid for high turbulent Reynolds numbers (i.e. $Re_t = k^2/\nu\varepsilon > 100$), numerical problems occur where this condition is not fulfilled. This is the case near a solid boundary, where a viscous boundary layer is formed as the velocity at the boundary itself is zero. Therefore, the wall layer is skipped by the k-ε turbulence model and boundary conditions are imposed far enough from the wall where the turbulence is assumed to be fully developed. From shear flow experiments it is known that this distance can be estimated as $\delta = 60\ \nu/u_*$ (with u_* the shear velocity $\equiv (\tau_0/\rho)^{1/2}$ and τ_0 the bottom shear stress). The boundary conditions are obtained from assuming local equilibrium between production, buoyancy destruction and dissipation of TKE. Toorman (1999) shows that diffusion remains negligibly small until saturation (see section 3.2 for its characterisation). Hence, the boundary condition for ε then is:

$$\varepsilon = P + G = P(1 - Rf) = \nu_t \left(\frac{\partial U}{\partial z} \right)^2 (1 - Rf) \tag{11}$$

The eddy viscosity and velocity gradient are estimated using the consistent PML approximation (Toorman *et al.*, 2002). Hence, the boundary condition becomes:

$$\varepsilon = \frac{u_*^3}{F_m \kappa z_b} (1 - Rf) \tag{12}$$

where F_m = the momentum damping function, κ = the von Karman constant (= 0.41) and z_b = the level of the near-bottom boundary node. The boundary condition for k is obtained from combining the previous expression with the definition of the eddy viscosity, eq.(4):

$$k = \frac{u_*^2}{\sqrt{c_\mu}} \sqrt{1 - Rf} \tag{13}$$

These boundary conditions require the knowledge of the shear velocity. Traditionally, u_* is obtained from the wall shear stress, approximated by the PML model, where the velocity gradient is obtained by elimination of u_* between the log-velocity profile and its derivative. This method works adequately as long as stratification effects are negligible. When buoyancy damping at the near-wall node is significant, unacceptable errors are made and the shear velocity is overestimated (figure 1).

Usually, the layer between the fully-developed turbulent layer and the wall, which contains the viscous sublayer and the transition layer, is not solved. Because the highest sediment concentration usually is found at the bed, this wall layer may contain a non-negligible amount of the total sediment load in suspension. Therefore, this layer should be included into the computational domain. In the present model a single-element layer is assigned to the wall layer.

In this layer the eddy viscosity is computed with the PML model. Boundary conditions for the velocity field are then imposed both at the bottom (non-slip condition, i.e. zero velocities) and at the interface between the wall layer and the turbulent layer. This is a cheap alternative for using low-Reynolds near-wall turbulence models which require too much grid refinement and which are not adapted for sediment-laden flow (see section 4.2).

The boundary condition for the velocity (tangential to the bottom slope) at the interface is obtained using the consistent law-of-the-wall profile:

$$U = \frac{u_*}{\kappa} \ln\left(\frac{z}{\alpha z_0}\right) \tag{14}$$

where z_0 = the characteristic roughness height of the bottom and α = the friction correction factor, which accounts for the buoyancy effects and can be related to the momentum damping function. α can be expressed by an empirical relationship, derived from numerical experiments, which is a function of the Richardson number and the ratio w_s/u_* (Toorman, 2000; Toorman *et al.*, 2002). Equation (14) remains applicable up to saturation.

Another source of error is the mesh size near the wall. A sensitivity analysis shows that with increasing thickness of the computational wall layer, compared to the actual thickness of the real sublayer, the error in the estimation of u_* increases considerably. Besides the fact that the estimation of the velocity gradient is less accurate, another source of error is the traditional neglection of the pressure gradient contribution. The estimation of u_* can be improved considerably by solving the local stress balance, i.e.:

$$u_* = \sqrt{\left(\nu + c_\mu \frac{k^2}{\varepsilon}\right)\frac{\partial U}{\partial z} - \frac{z}{\rho}\frac{\partial p}{\partial x}} \tag{15}$$

The corresponding error on u_* remains < 2% for z_+ up to 1000 (for unstratified conditions; for stratified conditions < 10% in the extreme case of saturation).

Sediment exchange with the bed is possible by equalling the bottom flux to the net deposition/erosion flux (a Neumann type boundary condition). The deposition flux is assumed to equal the settling flux w_sC at the bottom, while the erosion flux is modelled with a traditional empirical erosion law extended with a contribution of the deposited sediment which remains mobile.

3. FULLY-DEVELOPED OPEN-CHANNEL FLOW

Steady-state calculations of sediment-laden turbulent open-channel flow have been performed. The water depth h = 16 m. A sensitivity analysis has been carried out for the parameters shear velocity (u_*), settling velocity (w_s) and sediment load (or depth-averaged concentration). Various observations have been made.

3.1. Drag reduction

The effect of the implementation of the consistent bottom boundary treatment is significant, as is illustrated by figure 1. The inconsistent method overpredicts the shear velocity increasingly with increasing stratification in the near-wall node up to 70% at the saturation state. This implies that the bed shear stress in the latter case is three times overestimated, which would have enormous implications for the correct prediction of erosion and deposition. Notice furthermore the change (i.e. steepening) of the slope of the profiles, which is equivalent to the decrease of the von Karman coefficient from κ to $F_m\kappa$ (Toorman, 1999).

The reduction of the shear velocity for the same energy input is known as drag reduction. Various conditions are now known which can cause this phenomenon. Drag reduction in shear flows of clay suspensions has been reported in the literature (e.g. Gust, 1976; Wang *et al.*, 1998; Li and Gust, 2000). Its practical importance is illustrated by the fact that the friction coefficient for some sediment-laden rivers in China had to be modified to those for bottoms smoother than glass (Wang and Larsen, 1994).

The exact mechanism for drag reduction is still not understood. Gust (1976) tried to explain drag reduction in terms of energy dissipation within the flocs, in analogy with polymers. This, however, seems unlikely considering their weak strength. Best and Leader (1993) proposed that the increased spacing between low-speed streaks and decreasing bursting rate in the inner wall layer, observed in polymer flow studies, could also occur in clay suspensions. Li and Gust (2000) conclude that there is no experimental evidence to support any of the hypothesized mechanisms for drag reduction in clay suspension flow.

Model results with the inconsistent treatment produce drag reduction one order of magnitude smaller than observed. Therefore, it was believed up till now that buoyancy could not account for the drag reduction. However, the present results yield the right order of magnitude and even correspond quantitatively very good with the recent data of Li and Gust (2000), who also found a maximal drag reduction of 70%.

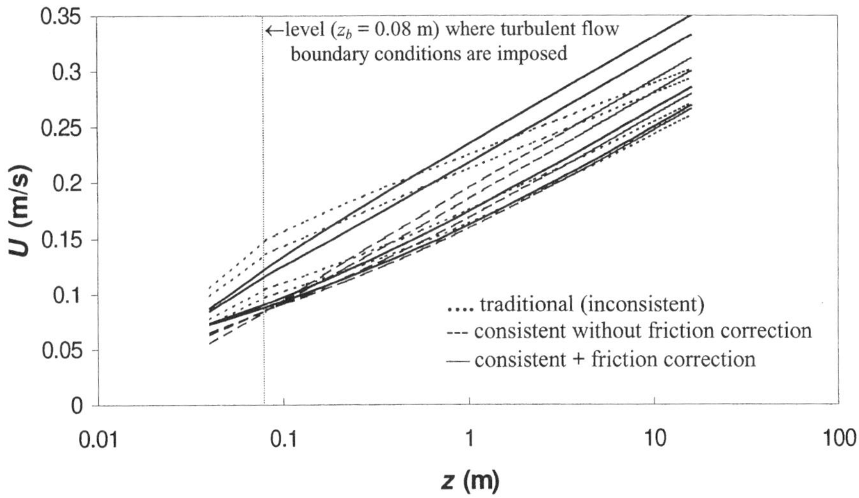

Figure 1. Computed velocity profiles (on semi-log scale) of fully-developed open-channel flow driven by a constant horizontal pressure gradient (dp/dx = 0.04 Pa/m, h = 16 m) for four different sediment loads (mean concentration from bottom to top = 19, 26, 68 and 118 mg/l).

A systematic study of steady state calculations with stepwise decreasing shear velocities clearly shows how the depth averaged velocity decreases down to a minimum and then increases again due to the dominance of the drag reduction.

Another reason for the thickening of the sublayer is the increase of the suspension viscosity due to the high concentrations near the bottom, which is calculated according to the Krieger-Dougherty model (Krieger, 1972):

$$\nu = \nu_w (1 - \phi / \phi_{max})^{-2.5\phi_{max}} \tag{16}$$

where ν_w = the kinematic viscosity of water and ϕ = the solids volume fraction. The maximum volume fraction ϕ_{max} is assumed to be 0.74, corresponding to the maximum packing fraction of spheres. For cohesive sediments this relationship needs to be modified, because the suspended particles are flocs, containing immobilised pore water. Hence, ϕ needs to be replaced by the volume fraction of the flocs. ϕ_{max} then corresponds to the gelling volume concentration (again of the flocs), but may need to multiplied by a factor >1 to account for the fact that the space-filling structure itself is not rigid.

3.2. Saturation

By reducing u_* or by increasing w_s or the depth-averaged concentration $\overline{C}$, while keeping the other parameters constant, qualitatively similar results are obtained, i.e. an increase of the stratification with subsequent increasing buoyancy damping of the turbulent kinetic energy. This cannot be continued endlessly. The total amount of available TKE can only keep a limited amount of sediment in suspension. This limit is known as the saturation or capacity limit; any amount in excess will deposit (Cellino and Graf, 1999).

This is illustrated by figure 2, which shows the computed steady state concentration profiles at the same energy input (constant u_*) for various sediment loads. For small loads the concentration is relatively homogeneous. With increasing load, the profile is more and more approaching a limiting profile, starting near the surface.

A remarkable feature is the fact that the corresponding Richardson number tends to homogenise over the entire water column, except near the free surface, when approaching the saturation limit (figure 3). This has been investigated analytically by looking at the implications of the condition $dRf/dz = 0$, which can be rewritten as $dG/dz = -Rf\, dP/dz$. After substitution of (7) and (8) and making a few simplifying assumptions (i.e. w_s and κ are constant, $\rho/\rho_w \cong 1$ and $du/dz = u_*/\kappa z$), this leads to a parabolic eddy viscosity distribution, given by (Toorman, 1999):

$$\nu_t = \sigma_s w_s z (1 - z/h) \tag{17}$$

This is like the PML eddy viscosity distribution in which the von Karman constant κ has been replaced by $\sigma_s w_s / u_*$. The corresponding concentration profile can be computed by integration of the equilibrium sediment flux, where (17) is used as eddy viscosity distribution. It is found to be a Rouse profile with Rouse parameter $Z = \sigma_s w_s / \kappa u_* = 1$, which is the fully saturated limiting profile (figure 2). Attempts to increase the sediment load beyond fail because the turbulence damping at the bottom becomes too strong due to the increasing sediment load which cannot be entrained into the turbulent flow.

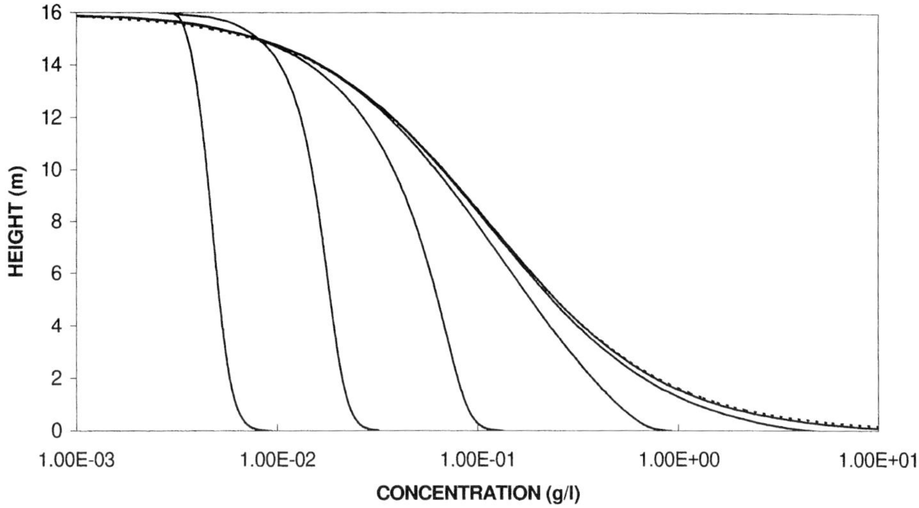

Figure 2. Computed concentration profiles for various sediment loads corresponding to a depth-averaged concentration (from left to right) of 4.88, 16.3, 48.8, 163, 326 and 488 mg/l in fully-developed turbulent sediment-laden open-channel flow with u_* = 0.01 m/s and w_s = 0.5 mm/s, using Munk-Anderson damping functions. Dashed line = limiting Rouse profile (Z = 1).

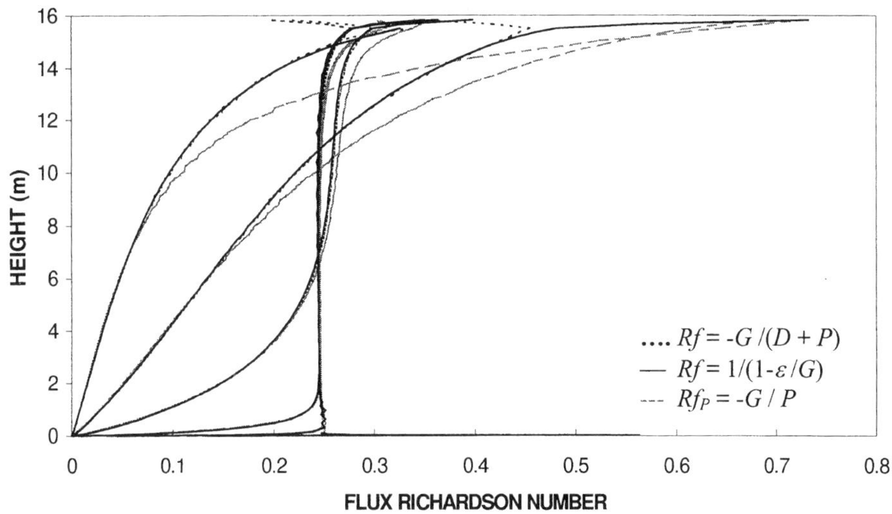

Figure 3. Generalised flux Richardson number Rf and corresponding Rf_P for the same conditions as in figure 2. Increasing sediment load from left to right.

The physical meaning of the value of Rf for which this limit occurs is still unclear. The value found for the example is $Rf_s = 0.245$, which is close to the magical value of 0.25 for which total collapse of turbulent mixing is claimed by some. There seems to be some dependence on certain model parameters (e.g. for $c_3 = 1$, one obtains $Rf_s \cong 0.5$). The model results indicate that stable turbulent flow can exist near the surface where diffusion dominates for $Rf > Rf_s$ and > 0.25. Therefore, Rf_s cannot be the critical value for total turbulence collapse. That critical value seems to be simply $Rf_c = 1$. It is therefore suggested that the results can only be understood in terms of the vertical gradient of Rf: a stable situation seems only possible when $\partial Rf/\partial z \geq 0$. The condition $\partial Rf/\partial z = 0$ should be related to the minimal energy required to keep a certain amount of particles in suspension. The range $0.25 < Rf < 1$ probably corresponds to the "Richardson number hysteresis" phenomenon, described by Woods (1969). It is hypothesised that Rf_s is only an apparent critical number, because the model results indicate the existence of a thin near-wall layer over which Rf rapidly reaches the physically expected critical value 1. A further discussion can be found in (Toorman, 2000a).

The slope of the corresponding velocity profile on a logarithmic z-scale has also increased from $1/\kappa$ to $u_*/\sigma_S w_S$. This is new theoretical evidence that the von Karman coefficient indeed decreases with increasing sediment load. All these theoretical findings are reproduced by the model (Toorman, 1999).

From the sediment transport equation one can determine the local equilibrium between settling and turbulent mixing. Teisson *et al.* (1992) call the corresponding concentration for the critical value of Rf where "turbulence collapse occurs" the (local) saturation concentration. As argued above, the interpretation of the critical value of Rf should be relaxed, because turbulence may persist for higher values of Rf. Therefore, saturation is physically related to optimal energy conditions instead of turbulence collapse. Since Rf is constant at the saturation limit, one can integrate this relationship over the water depth, giving the total maximum suspended sediment load per unit area:

$$\overline{C_{sat}}h = \frac{\rho_w u_*^2 \overline{U}}{w_s g \Delta\rho_s / \rho_s} Rf_{sat} \tag{18}$$

where $\overline{U}$ is the depth-averaged velocity. With exception of the factor Rf_{sat}, this is identical to the auto-suspension criterion for density currents, as given by Parker *et al.* (1986).

It should be realised that the described numerical experiment is not very representative for most field conditions. Steady flow conditions do not occur in coastal and estuarine areas, where the flow is subjected to tidal variations, i.e. periodic acceleration and deceleration. Therefore, concentrations in the field may be higher or lower than the steady state values depending on the flow conditions and the time scales for entrainment and settling relative to the tidal period. Even when saturation would have occurred at steady state, higher concentrations than the corresponding saturation value may be found in the field.

4. LOW-REYNOLDS EFFECTS

4.1. Modelling the wall layer

In the previous model description the wall layer was treated on a single-element layer grid. The parabolic interpolation function over this layer cannot follow the exact profiles of any of the variables. As long as the thickness of the sublayer is much smaller than the wall layer grid size, the error remains acceptably small. Problems occur when the sublayer thickness grows due to the reduction of the shear velocity and its dimension becomes of the same order of magnitude as the grid size. In physical practice, the problem of the thickening of the sublayer may occur during flow reversal around slack water. In numerical practice, problems may be encountered when approaching saturation due to the large gradients near the bottom. The only way to improve the model's performance is to solve the wall layer more correctly on a refined grid in order to approximate the profiles better.

A first approach, which has been investigated, is the low-Reynolds k-ε model, where the damping functions f_μ, f_1 and f_2 are introduced to correct the original k-ε equations, as shown in equations (4) and (6). Various forms have been proposed in the literature, most of which contain correction factors which are a function of the distance from the wall. These have been evaluated against Direct Numerical Simulation (DNS) data (Toorman, 2000c), which resulted in finding a relationship between f_2 and a new formulation of the realisable time scale, which allows transition from the fully-developed turbulent time scale k/ε to the Kolmogorov time scale $(\nu/\varepsilon)^{1/2}$:

$$T = f_T \frac{k}{\varepsilon} = \sqrt{\frac{\nu_t / f_\mu + \nu}{c_\mu \varepsilon}} \tag{19}$$

where $f_T = f_2^{-1} = (1+1/c_\mu Re_t)^{1/2}$ the realisable time scale factor, with $Re_t = k^2/\nu\varepsilon$ the turbulent Reynolds number. The analysis also revealed that the model parameters σ_k and σ_ε cannot be constant, because the assumption of isotropy of turbulence is no longer correct in the wall layer. It would not be surprising that these model constants need to be modified as well, i.e. made a function of Rf, in the turbulent layer when highly stratified. Because of the high degree of refinement in the wall layer, this method is not suitable for large-scale applications.

Another method which has been investigated is the two-layer approach, in which the wall layer is solved with a modified mixing-length model in replacement of the ε equation (e.g. Chen and Patel, 1988; Rodi, 1991). The eddy viscosity in the wall layer is computed as:

$$\nu_t = c_\mu \sqrt{k} \ell_\mu \tag{20}$$

and the dissipation rate as:

$$\varepsilon = k^{3/2} / \ell_\varepsilon \tag{21}$$

The models in the literature use a length scale of the form:

$$\ell_\mu = \kappa c_\mu^{-3/4} z \left(1 - \exp(-R_z / A_\mu)\right) \tag{22}$$

where $R_z = k^{1/2} z/\nu$ = a turbulent Reynolds number. For the length scale ℓ_ε two different forms have been proposed, either like the one for ℓ_μ, or:

$$\ell_\varepsilon = \frac{\kappa c_\mu^{-3/4} z}{1 + A_\varepsilon / R_z} \qquad (23)$$

The values of the model parameters A_μ and A_ε chosen by various researchers differ. The wall model and the full k-ε model are matched at a distance where $R_z > 200$ ($z_+ \approx 135$) or where $\nu_t/\nu > 30$. The model still requires about 10 grid points in the wall layer. Further details and references can be found in Section 4.4 of (Chen and Jaw, 1998). The model has been tested successfully for fully-developed turbulent shear flow (Toorman, 1998).

The major problem with both methods is the limitation of their validity to homogeneous fluids for which the damping functions have been calibrated. Effects of buoyancy are not included and require suitable data, in particular within the wall layer, which, as far as known, are not available. Hopefully, future DNS simulations of particle-laden flows at realistic Reynolds numbers will provide useful data. Based on these data damping functions can be proposed in analogy with those for the Prandtl mixing length turbulence closure.

The other concern is the required grid refinement. More work needs to be done in order to find ways to minimise the number of grid points in the wall layer without loss of accuracy.

4.2. Modelling lutoclines

Model results for unsteady open-channel flow simulations indicate that laminarisation may also occur around lutoclines when the density gradient is too large. Results obtained with the present model are shown in (Toorman *et al.*, 2002). Low-Re_t formulations suitable for these conditions are not yet available. Therefore, a simple, heuristic method has been implemented consisting of the introduction of local artificial diffusion which is inverse proportional to the local value of ε. This added diffusion only plays a role where turbulence is excessively damped and stabilises the solution of the k-ε model. The occurrence of the numerical problem and its cure also depends on the grid size at the location of the lutocline. Internal wave corrections may be necessary and helps reducing the problem. This is discussed in more detail in (Toorman *et al.*, 2002).

5. CONCLUSIONS

The consistent modelling of turbulent flows with suspended sediment particles, using the k-ε turbulence model, requires several modifications to the standard implementation. The bed boundary conditions are adapted and the computation of the shear velocity is corrected following the consistent implementation of buoyancy damping functions. Previously, it was thought that several features, such as the decrease of the von Karman coefficient and drag reduction, could not be explained by buoyancy effects alone. This conclusion was based on application of the traditional, inconsistent modelling approach. Numerical experiments with the consistent implementation shows that this is not true, i.e. buoyancy effects are very important, even at very low concentrations. Other turbulence modulation mechanisms, which occur also in homogeneous suspensions, are less important in the present context because of the low concentrations. From the analysis of experimental data of Cellino (1998), it is concluded that these effects are no longer negligible in the oversaturated bottom layer (Toorman, 2000a).

Another correction, based on the local stress balance, is carried out allowing a more accurate estimation of the shear velocity on a coarse grid. The wall layer is explicitly solved using the

Prandtl mixing length concepts in order to include the sediment in this layer. The Richardson number has been generalised in order to account for the dominance of diffusion of turbulent kinetic energy at the free surface. All these modifications can easily be implemented into presently used engineering models which use the k-ε turbulence closure. Many of them can even be implemented for models using the PML model (Toorman *et al.*, 2002). Coastal and estuarine applications generally need to be solve in three dimensions and require a large number of grid points, which results in a relatively coarse vertical discretisation. Its implications in view of the presently proposed improvements are discussed in (Toorman *et al.*, 2002).

Numerical simulations of steady-state conditions reveal that saturation can be obtained and seems to be characterised by the condition of a zero vertical gradient of the Richardson number. The theoretical analysis of this condition shows that the von Karman parameter indeed decreases with increasing sediment load. The data also suggest that turbulent flow can be maintained for flux Richardson numbers between the saturation value and the maximum value of 1 where buoyancy destroys shear production, as long as the vertical gradient of Rf remains positive.

Other steady-state computations confirm that drag reduction takes place as a result of the buoyancy damping and its magnitude is found to be of the same order as observed in laboratory and field conditions. This contradicts the previous hypothesis that buoyancy damping cannot account for the high degree of drag reduction measured.

The model starts having numerical problems when saturation is approached or when the shear velocity becomes small, because the wall-layer is no longer solved properly. This topic needs further study.

Acknowledgements: This work has been carried out within the framework of the MAST3 project "COSINUS", partly funded by the European Commission, Directorate General XII for Science, Research and Development under contract no. MAS3-CT97-0082. The author's post-doctoral position is financed by the Flemish Fund for Scientific Research.

REFERENCES

Best, J.L. and Leeder, M.R. , 1993, Drag reduction in turbulent muddy seawater flows and some sedimentary consequences, *Sedimentology*, 40, 1129-1137.

Chen, C.J. and Jaw, S.Y. , 1998, *Fundamentals of Turbulence Modelling*, Taylor and Francis, Washington D.C.

Chen, H.C. and Patel, V.C. , 1988, Near-wall turbulence models for complex flows including separation, *AIAA J.*, 26(6), 641-648.

Cellino, M. , 1998, Experimental study of suspension flow in open-channels, PhD thesis, Dépt. de Génie Civil, Ecole Polytechnique Fédérale de Lausanne.

Cellino, M. and Graf, W.H. , 1999, Sediment-laden flow in open channels under noncapacity and capacity conditions, *ASCE J. Hydr. Eng.*, 125(5), 456-462.

Ellison, T.H. , 1957, Turbulent transport of heat and momentum from an infinite rough plane, *J. Fluid Mechanics*, 2, 456-466.

Galland, J.C., Laurence, D. and Teisson, C. , 1997, Simulating turbulent vertical exchange of mud with a Reynolds stress model, In: *Cohesive Sediments* , N. Burt, R. Parker and J. Wats, eds., J. Wiley and Sons, Chichester, 439-448.

Gust, G. , 1976, Observations on turbulent drag reduction in a dilute suspension of clay particles, *J. Fluid Mech.*, 75(1), 29-47.

Huerta, A. and Liu, W.K. , 1988, Viscous flow with large free surface motion, *Comp. Meth. Appl. Mech. and Engrg.*, 69, 277-324.

Ivey, G.N. and Imberger, J. , 1991, On the nature of turbulence in a stratified fluid. Part I: The energetics of mixing, *J. Physical Oceanography*, 21, 650-658

Krieger, I.M. , 1972, Rheology of monodisperse latices, *Advanc. Colloid Interface Sci.*, 3, 111-136.

Li, M.Z. and Gust, G. , 2000, Boundary layer dynamics and drag reduction in flows of high cohesive sediment suspensions, *Sedimentology*, 47, 71-86.

Nezu, I. and Nakagawa, H. , 1993, *Turbulence in Open-Channel Flows*, IAHR Monograph Series, A.A. Balkema, Rotterdam.

Parker, G., Fukushima, Y. and Pantin, H.M. , 1986, Self-accelerating turbidity currents, *J. Fluid Mechanics*, 171, 145-181.

Rodi, W. , 1980, *Turbulence models and their application in hydraulics*, State-of-the-art Paper, IAHR, Delft.

Rodi, W. , 1991, Experience with two-layer models combining the k-ε model with a one-equation model near the wall, *AIAA 29th Aerospace Sciences Meeting*, Paper AIAA-91-0216.

Teisson, C., Simonin, O., Galland, J-C. and Laurence, D. , 1992, Turbulence modelling and mud sedimentation: a Reynolds stress model and a two-phase flow model, *Proc. 23rd Int. Conf. Coastal Eng.*, ASCE, New York, 3, 2853-2866,.

Toorman, E.A. , 1993, Simulation of free surface flow of mud, *Proc. 8th Int. Conf. on Finite Element Methods in Flow Problems* , Morgan *et al.* eds., CIMNE/Pineridge Press, Barcelona/Swansea, 1211-1220.

Toorman, E.A. , 1996, Sedimentation and self-weight consolidation: general unifying theory, *Géotechnique* 46(1), 103-113.

Toorman, E.A. , 1998, A study of modelling (near-wall) turbulence damping, *COSINUS Annual General Meeting Grenoble 1998*, Book of Abstracts, pp.13-16.

Toorman, E.A. , 2000a, Suspension capacity of uniform shear flows, Report HYD/ET/COSINUS4, Hydraulics Laboratory, Katholieke Universiteit Leuven.

Toorman, E.A. , 2000b, Parametrisation of turbulence damping by suspended sediment, Report HYD/ET/COSINUS3, Hydraulics Laboratory, Katholieke Universiteit Leuven.

Toorman, E.A. , 2000c, Analysis of near-wall turbulence modelling with k-ε models, Report HYD/ET/COSINUS2, Hydraulics Laboratory, Katholieke Universiteit Leuven.

Toorman, E.A., Bruens, A.W., Kranenburg, C. and Winterwerp, J.C. , 2002, Interaction of suspended cohesive sediment and turbulence, *Proc. INTERCOH-2000*, J.C. Winterwerp and C. Kranenburg eds., Elsevier, this volume.

Uittenbogaard, R.E., van Kester, J.A.Th. and Stelling, G. , 1992, Implementation of three turbulence models in TRISULA for rectangular horizontal grids, Report Z162, Delft Hydraulics.

Wang, Z.Y. and Larsen, P. , 1994, Turbulent structure of water and clay suspensions with bed load, *J. Hydraulic Eng.*, 120(5), 577-600, ASCE.

Wang, Z.Y., Larsen, P., Nestmann, F. and Dittrich, A. , 1998, Resistance and drag reduction of flows of clay suspensions, *J. Hydr. Eng.*, 124(1), 41-49.

Woods, J.D. , 1969, On Richardson's number as a criterion for laminar-turbulent-laminar transition in the ocean and atmosphere, *Radio Science*, 4(12), 1289-1298.

Fine Sediment Dynamics in the Marine Environment
J.C. Winterwerp and C. Kranenburg (Editors)

Scaling parameters for High-Concentrated Mud Suspensions in tidal flow

J.C. Winterwerp[a]

[a]Delft University of Technology, Dep. of Civil Engineering and Geosciences, PO Box 5048, 2600 GA Delft, The Netherlands, also: Delft Hydraulics, The Netherlands

This paper summarises the relevant scaling laws for the behaviour of High-Concentrated Mud Suspensions (HCMS) under tidal flow conditions. A HCMS is defined as a suspension where the suspended sediment and the turbulent flow field interact strongly through sediment-induced buoyancy effects. It is shown that these HCMS's can become super-saturated as a result of a positive feed-back between the sediment, the turbulent flow field and the formation of a fluid mud layer. The results of 1DV numerical simulations agree well with a theoretical analysis, based on an integral entrainment formulation.

KEY WORDS
high-concentrated mud suspensions, buoyancy effects, scaling laws

1. INTRODUCTION

Many harbour basins and navigational channels all over the world suffer from large siltation rates. This is often attributed to the presence of high-concentrated mud suspensions. In the present study, High-Concentrated Mud Suspensions (HCMS) are defined as suspensions of cohesive sediment with a near-Newtonian behaviour and a distinct interaction with the turbulent flow field. Winterwerp and Kranenburg (1997) have shown that HCMS and recently deposited fluid mud layers behave as viscous fluids, with a somewhat augmented viscosity. Typical concentrations of HCMS range from a few hundred to a few thousand mg/l, depending on the prevailing flow conditions. In line with this definition, fluid mud herein is defined as a suspension of cohesive sediment at a concentration beyond the gelling point, i.e. of the order of several 10 to 100 g/l. This suspension exhibits profound non-Newtonian behaviour, and it is either stationary or mobile. In the latter case the fluid mud flow may be laminar or turbulent and its dynamics will be fairly independent of the flow in the water column above.

The author has shown in earlier studies (Winterwerp, 1998, 2001b) that under steady flow conditions, HCMS can become super-saturated, i.e. the sediment load can become so large, or the turbulent mixing capacity of the flow can become so small, that the flow is no longer able to carry the sediment in suspension. As a result both the sediment concentration profile and the turbulence field collapse, upon which a two-layer fluid system is formed, the lower layer existing of fluid mud. This saturation condition is characterised by the depth-averaged saturation concentration C_s, which was shown to scale as:

$$C_s \equiv \frac{1}{h}\int_0^h c_s \mathrm{d}z = K_s \frac{\rho}{\Delta g}\frac{u_*^3}{hW_s} \tag{1}$$

where h is the water depth, c_s a local equilibrium concentration, ρ the water density, Δ the relative sediment density, g the gravitational acceleration, u_* the shear velocity, W_s a characteristic settling velocity, and K_s is a proportionality constant of order unity. If c_s is multiplied with the flow velocity u, and integrated over the depth h, a transport formula for HCMS is derived, which represents the maximal load of cohesive sediment that can be transported by turbulent flow through its turbulent energy: $F_s = \int_h c_s u \mathrm{d}z \propto \tau_b u_* U / \Delta g W_s$. Remarkably, this formula is almost identical to the transport formula by Bagnold (1966) for suspended sand transport:

$$F_{Bag} = \frac{e_s(1-e_b)\tau_b U^2}{\Delta g W_s} \tag{2}$$

in which e_b and e_s are efficiency parameters.

The latter equation was obtained from energy considerations for uniform, steady flow in open channels with an unlimited amount of sediment on the bed. The corresponding concentration is known as the equilibrium concentration for suspended (non-cohesive) sediment.

The major difference between the transport of cohesive and non-cohesive sediment is that upon deposition of non-cohesives a rigid bed is formed directly. At the water-bed interface turbulence production is still possible, whereas in the case of cohesives at super-saturated conditions a fluid mud layer is formed, destroying turbulence production. As long as the fluid mud layer does not consolidate, the collapse of the turbulence field is irreversible.

Saturation can also occur under tidal conditions, as will be shown in this paper. Towards and around slack water a near-saturated suspension settles, as the decelerating flow contains too little energy to mix all sediment over the entire water depth. In this period layers of fluid mud are formed, that consolidate slightly in the period prior to re-acceleration of the tide, when the sediment is remixed over the water column. If this pattern is repeated every tidal cycle, i.e. if a dynamic equilibrium between settling and mixing occurs, the suspension is sub-saturated.

When the sediment load becomes too large, the flow will no longer be able to carry this load, and the layer of fluid mud formed around slack water cannot be re-entrained anymore during the subsequent accelerating flow and will consolidate further to build up strength. This condition is referred to as saturation and it may occur towards neap tide, in diverging channels and/or in harbour basins.

The present paper focuses on the behaviour of HCMS near saturation conditions. Section 2 describes the scaling parameters relevant for the settling and remixing of HCMS during a tidal cycle. In Section 3 the conditions and relevant scaling laws for saturation of HCMS under tidal conditions are derived, which are tested with a numerical model (the 1DV POINT MODEL), the results of which are presented in Section 4.

2. SETTLING AND REMIXING OF HCMS

The settling and remixing of HCMS is discussed for four phases of the tide: around maximal flow velocity, towards slack water (decelerating flow), around slack water and during accelerating flow.

The time scale of the primary driving force is the tidal period T, or more accurately, the period of the decelerating and accelerating phase of the tide. It can be shown that the local adaptation time of the turbulence parameters, like the eddy viscosity and eddy diffusivity, to changes in hydrodynamic conditions is small compared to other time scales, so that the turbulence time scale $T_T \approx 0.06h/u_*$ can be neglected in the analyses. Note that non-locally generated turbulence can affect the behaviour of the suspension, especially in the vicinity of confined patches of fluid mud (e.g. Bruens, 2002).

2.1. Around maximum flow velocity

It is assumed that around maximum flow velocity, no immediately mobilisable fluid mud is found on the river/sea bed, and full turbulence production is possible at the water-bed interface. In coastal and estuarine environments, the size of mud flocs of the suspension (hence the settling velocity W_s) is probably more or less in equilibrium with the local hydrodynamic conditions over the entire water depth (Winterwerp, 2001c), and the vertical concentration profile has a more or less Rousean shape, governed by the Rouse number $\beta \equiv w_s/u_*$. Its precise form though is also a function of the sediment-turbulence interaction, affecting the vertical eddy viscosity profile; hence the bulk Richardson number $Ri_* \equiv \Delta gh/u_*^2$ becomes a relevant parameter as well.

2.2. Towards slack water (decelerating flow)

During decelerating flow the flocs grow, hence the settling velocity increases. At the same time the mixing capacity of the flow decreases, as a result of which rapid sedimentation takes place, giving rise to buoyancy effects. If the sedimentation rate exceeds the consolidation rate, layers of fluid mud are formed.

The flocculation process determines the size of the mud flocs D_f, hence the local settling velocity w_s of the suspension. The equilibrium size of the mud flocs scales as $D_e \propto c(h/U^3)^{1/4}$ (Winterwerp, 1998, 2001c), implying a decrease with increasing flow velocity (shear level). Vice versa, an increase in floc size, hence settling velocity is expected with decreasing flow velocity, i.e. towards slack water. However, because the flocculation time T_f, which scales as $T_f \propto c^{-m}(U^3/h)^{-n}$, where $1 \leq m \leq 2$ and $1/4 \leq n \leq 1/2$, is large around slack water, the equilibrium floc size, c.q. settling velocity, will probably not be attained (Winterwerp, 2001c); T_f/T is therefore an important scaling parameter.

As the settling velocity increases and the mixing capacity of the flow decreases (the latter being a function of the sediment-induced buoyancy effects, i.e. Ri_*), the sediment starts to settle, forming a fluid mud layer if the siltation rate exceeds the consolidation rate, i.e. $T_s < T_c$, where the relative sedimentation time is defined as $T_s = h/W_sT$, and the relative consolidation time T_c is defined in a next paragraph. In the case that $T_s > T_c$, a rigid bed is formed upon deposition.

The near bed floc size also affects the gelling concentration of the fluid mud layer (i.e. the concentration at which a space-filling network is formed, e.g. Winterwerp, 2001c), hence the thickness of the fluid mud layer $\delta = h\bar{c}/c_{gel}$, where $\bar{c}$ is the suspended sediment concentration averaged over the sedimentation depth (see below), and c_{gel} has a value between a few 10 g/l and 100 to 150 g/l, depending on the sediment properties, the local hydrodynamic conditions and the history of floc formation.

2.3. Around slack water

The rapid settling, started during decelerating flow, continues as the mixing capacity of the flow has decimated around slack water. Layers of fluid mud form when the sedimentation rate exceeds the consolidation rate.

It is convenient to define the sedimentation depth h_s as the vertical distance a particle can fall during decelerating tide: $h_s \equiv W_sT$. All sediment in a water column of thickness h_s will settle during decelerating tide (also during the first phase of the accelerating tide some sediment may settle) to form a layer of fluid mud. During the first phase when the consolidation process is governed by the permeability of the skeleton, the consolidation time of such a fluid mud layer scales as $T_c \propto \delta\rho_b^m$, with $4 \leq m \leq 8$, whereas in a later phase, when the effect of the effective stresses becomes dominant $T_c \propto \delta^2$ (Winterwerp, 1999).

During this consolidation process, a strength within the mud layer builds up (Merckelbach and Kranenburg, 2001). This process is of particular importance during the period towards neap of a spring-neap tidal cycle. However, the processes during such a cycle are not further elaborated upon, and the reader is referred to e.g. Villaret and Latteux (1992).

2.4. Accelerating tide

During accelerating tide, the fluid mud layer may be eroded again, either through floc erosion or entrainment processes, provided the exerted stress exceeds the strength of the mud layer. When the exerted stress exceeds the critical stress for (floc) erosion, floc erosion will be the eroding agent. If the exerted stress increases further beyond the yield strength of the bed, entrainment becomes important (see Section 3), and the relevant scaling parameters are a bulk Richardson number Ri_*, the Rouse number β and the relative stress level $\theta_y \equiv \rho u_*^2/\tau_y$ (e.g. Kranenburg and Winterwerp, 1997).

Whether the fluid mud layer becomes turbulent, and entrains the upper water column, depends on whether the effective Reynolds number of the lower layer, defined as $1/Re_e = 1/Re + 1/Re_y$, where $Re = 4U_m\delta_m/\nu_m$ is the common Reynolds number, in which ν_m is the effective viscosity, augmented by non-Newtonian effects and $Re_y = 8\rho_m U_m^2/\tau_B$ the yield Reynolds number, with τ_B the Bingham strength of the mud layer, exceeds a critical value of about 2,000 to 3,000 (Liu and Mei, 1989).

3. SATURATION UNDER TIDAL CONDITIONS - THEORY

Under tidal conditions, the saturation concentration C_s is conveniently defined as the amount of sediment, initially distributed homogeneously over the water depth, that can be carried by the turbulent flow in the form of a Rousean-like vertical profile during just one instant within the tidal cycle. From a formal dimensional analysis it can be shown that in this case the magnitude of C_s is governed by a series of non-dimensional parameters, i.e. by a bulk Richardson number $Ri_* \equiv \Delta gh/u_*^2$, the Rouse number $\beta \equiv W_s/u_*$, the relative settling time $T_s' \equiv T_s/T = h/W_sT$ and the relative mixing time $T_m' \equiv T_m/T = h^2/\Gamma_T T \propto h/Tu_*$, where Γ_T is the vertical eddy diffusivity.

The variation of the sediment concentration profile with time during a tidal cycle in a water column of large depth ($h_s < h$) is sketched in Figure 1. The analysis is started at maximal flow velocity from a homogeneous suspension of sediment of concentration C_0 equal to the saturation concentration for tidal conditions C_s. The sediment will settle during decelerating tide. However, all sediment initially above the level $z > h_S$ will remain in the water column. The sediment below this level will form a layer of fluid mud on the bed at $z = 0$ (see Figure 1). The maximal thickness of this layer δ_m around slack water amounts to $\delta_m = h_s c_{gel}/C_0 = h_s c_{gel}/C_s$. The gelling concentration c_{gel} is defined as the mass concentration at which a space filling network of cohesive sediment flocs is formed, i.e. the concentration of the fluid mud layer. It is a function of the sediment properties, the local hydrodynamic conditions and the history of the floc forming process (Winterwerp, 2001c). During accelerating tide this fluid mud will be remixed over the water column by entrainment processes.

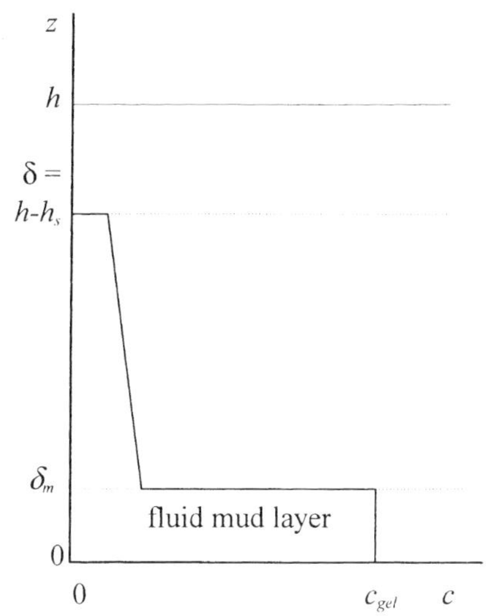

Figure 1. Schematic concentration profile around slack water ($h_S < h$).

In the present analysis, consolidation or thixotropic effects in the fluid mud layer are not accounted for, hence the fluid mud layer is treated as a Newtonian fluid, possibly with an augmented viscosity. When the fluid accelerates, it is therefore the fluid mud layer that becomes the more turbulent layer because of friction between the fluid mud layer and the (rigid) bed. This is illustrated in the series of graphs in Figure 2, showing the evolution of the turbulent kinetic energy and eddy diffusivity profiles with time (t = 0, 60, 90, 120, 180, 300, 450, 600, 750 and 900 minutes) in a flow with constant acceleration ($U = 1.67\cdot10^{-5}\times t$ m/s), together with the related velocity and concentration profiles, as computed with the 1DV POINT MODEL. This model is based on Delft Hydraulics' full three-dimensional hydrostatic code DELFT3D, in which all horizontal gradients have been stripped, except for the longitudinal pressure gradient. It contains the momentum equation, the advection-diffusion equation for cohesive sediment including the effects of hindered settling, and a description to account for the effects of surface waves on the bed shear stress and the vertical mixing. Uittenbogaard (1995) has shown that the k-ε turbulence closure model is applicable to fairly stratified conditions. This model is therefore used with a sediment-induced buoyancy term. It is implicitly assumed that the suspension may be treated as a single-phase fluid, which is allowed for HCMS, as discussed in Winterwerp (1999, 2001a, 2001b). For further details the reader is referred to these references as well. The simulations are started with zero velocity over the entire depth of 10 m and a fluid mud layer of 20 % of the total water depth at an initial concentration of 5 g/l; the settling velocity of the sediment is set at 0.1 mm/s.

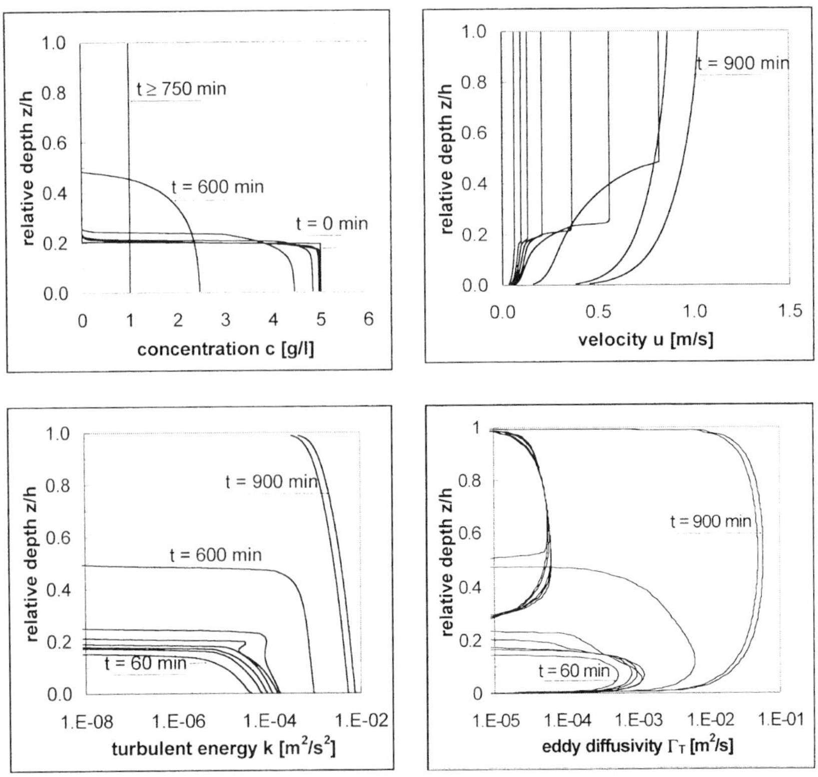

Figure 2: Entrainment of fluid mud layer during accelerating flow; results of simulations with 1DV POINT MODEL

These results clearly show that turbulence is generated at the bed, and that the water column is entrained by the fluid mud layer, which is thickened and diluted. A similar behaviour in the field was observed in the Severn estuary (Crickmore, 1982, Parker, 1987) and in the Loire estuary (Le Hir, 1997). The eddy diffusivity higher in the water column is not exactly zero, but one to two orders of magnitude smaller than within the fluid mud layer: the vertical sediment transport is effectively damped by the large density gradients near the water - fluid mud interface. Only at $t \geq 900$ min., when the sediment is fully mixed over the water column, a parabolic eddy diffusivity profile is obtained (note that the abscissa of Figure 2d has a logarithmic scale). This process has been validated experimentally by Bruens et al. (2002).

The entrainment process described is elaborated by Kranenburg (1994, 1997), and his analysis is used to establish the scaling laws for C_s under tidal conditions. Following Kranenburg, the balance equation for the turbulent kinetic energy k, neglecting horizontal advection reads:

$$\frac{\partial k}{\partial t} = D + P - B - \varepsilon \tag{3}$$

where D is a diffusion term, P the turbulence production, B represents buoyancy destruction, i.e. the work required to keep the sediment in suspension, and ε the energy dissipation rate per unit mass. When this balance equation is integrated over the turbulent layer, i.e. over the mud layer, using various approximations, an integral model describing the entrainment of water into the fluid mud layer is obtained. This is elaborated by Kranenburg (1994, 1997) to yield:

$$\begin{aligned} c_q \frac{\mathrm{d}}{\mathrm{d}t}\left(\delta u_*^2\right) &= c_q \delta \frac{\mathrm{d}u_*^2}{\mathrm{d}t} + c_q u_*^2 \frac{\mathrm{d}\delta}{\mathrm{d}t} = \\ &= c_s \left(U_w - U_m\right)^2 \frac{\mathrm{d}\delta}{\mathrm{d}t} + c_s' \left|U_w - U_m\right| u_*^2 + c_\sigma u_*^3 - 2\int_0^\delta B \mathrm{d}z \end{aligned} \tag{4}$$

where $\delta(t)$ is the thickness of the mixing layer with concentration c - see also Figure 1. The various empirical coefficients have the following values: $c_q = 5.6$, $c_s = c_s' = 0.25$, and $c_\sigma = 0.42$. U_w is the mean velocity in the water column above the fluid mud layer and U_m is the mean velocity within the mixing (fluid mud) layer. These two velocities have to be established from the momentum equations for the water and fluid mud layers, which should be solved simultaneously with equ. (4). In the case where the upper (water) layer is turbulent, a similar integral model is obtained, which has been validated extensively against laboratory experiments (Winterwerp and Kranenburg, 1997).

In the following paragraphs the last term of equ. (4) is elaborated. Two regimes have to be distinguished: REGIME I: $h \leq h_s$ and REGIME II: $h > h_s$. In REGIME I all sediment in the water column can settle during decelerating tide to form a fluid mud layer and no sediment remaining above the fluid mud layer will have to be dealt with. In this case the initial thickness of the mixing layer δ in equ. (4) equals the thickness of the fluid mud layer δ_m.

The definition of C_s implies that during accelerating tide, all fluid mud has to be remixed over the entire water depth. This means that the entrainment velocity w_e has to be sufficiently large to mix the entire fluid mud layer during accelerating tide over the water depth. Thus:

$$\int_0^T w_e \mathrm{d}t = h, \quad \text{hence } w_e \approx h/T \tag{5}$$

The elaboration is presented in Appendix A, together with the resulting scaling law for REGIME I, yielding $C_s/\rho_w = F(Ri_*, \beta, T'_m)$.

Under REGIME II not all sediment settles during decelerating tide. This implies that the fluid mud layer obtains a thickness $\delta_{m,0} = h_s c_{gel}/C_0$, and that work has to be done to remix the sediment also beyond $z = h_s$ over the rest of the water column. Two phases are distinguished. In Phase 1, the fluid mud layer becomes turbulent and is remixed over the sedimentation depth; the related entrainment velocity is referred to as w_{e1}. The fact is ignored that during this process the sediment beyond $z = \delta_m$ may continue to settle because of the small eddy diffusivity above the water - fluid mud interface. In Phase 2, all the sediment is remixed over the entire water column; the related entrainment velocity is referred to as w_{e2}. The definition of C_s implies that during accelerating tide, both the fluid mud and the suspension above have to be remixed over the entire water depth. This means that:

$$T \approx \frac{h_s}{w_{e1}} + \frac{h - h_s}{w_{e2}} \tag{6}$$

In Phase 1, the concentration difference Δc_m over the water - fluid mud interface for a saturated suspension follows form $\Delta c_m \delta_m \approx (c_m - C_0)\delta_m = C_s W_s T - C_0 \delta_m \approx C_s W_s T$, which is correct for the major part of the accelerating tide. The storage term in equ. (4) may be neglected. In Phase 2 of the entrainment process, the concentration difference Δc over the water - fluid mud interface for a saturated suspension follows from $\Delta c \delta \approx c\delta = C_0 h = C_s h$. In Phase 2, the Richardson number is not too large in general, as a consequence of which the storage term in equ. (4) may not be neglected a priori. Again, the elaboration is presented in Appendix A, together with the resulting scaling law for REGIME II, yielding $C_s/\rho_w = G(Ri_*, \beta, T', T'_s)$.

It is obvious that the formulae in Appendix A are not very practical; even the explicit form of equ. (A.5) for REGIME I is too complicated to get a clear picture of the functional relation between C_s and the various physical parameters like U, u_*, h and W_s. However, some workable relations can be established for extreme conditions. For this purpose, REGIME II is divided in a subregime with $h_s/w_{e1} >> (h-h_s)/w_{e2}$, which implies $h \approx h_s$, but $h > h_s$, and a subregime with $h_s/w_{e1} << (h-h_s)/w_{e2}$, which implies $h >> h_s$. The case $\beta >> 1$ will not be elaborated upon, as this represents the suspension of (coarse) sand, which, of course, does not form fluid mud layers upon deposition. This analysis yields the scaling relations summarised in Table 1.

Table 1 shows that the saturation conditions are governed by four sets of non-dimensional parameters, i.e. a bulk Richardson number Ri_*, the relative sedimentation time T'_s (or relative sedimentation depth h_s/h), the Rouse number β and the relative mixing time T'_m. In the next section it is studied whether these scaling relations are supported by simulations with the IDV POINT MODEL.

	$\beta << 1$		
	$T'_m << 1$ $T'_m << \beta$	$T'_m << 1$ $T'_m >> \beta$	$T'_m >> 1$ $T'_m >> \beta$
REGIME I: $h \le h_s$	$C_s \propto \frac{1}{\beta Ri_*}$ $\propto \frac{\rho_w u_*^3}{\Delta g h W_s}$	$C_s \propto \frac{1}{T'_m Ri_*}$ $\propto \frac{\rho_w T u_*^3}{\Delta g h^2}$	$C_s \propto \frac{1}{Ri_*}$ $\propto \frac{\rho_w u_*^2}{\Delta g h}$
REGIME II; phase 1: $h > h_s$	$C_s \propto \frac{T'_s}{\beta Ri_*} \propto \frac{\rho_w u_*^3}{\Delta g T W_s^2}$		
REGIME II; phase 2: $h >> h_s$	$C_s \propto \frac{1}{\beta Ri_*}$ $\propto \frac{\rho_w u_*^3}{\Delta g h W_s}$	$C_s \propto \frac{1}{T'_m Ri_*}$ $\propto \frac{\rho_w T u_*^3}{\Delta g h^2}$	$C_s \propto \frac{1}{Ri_*}$ $\propto \frac{\rho_w u_*^2}{\Delta g h}$

Table 1. Scaling relations for C_s under tidal conditions.

4. SATURATION IN TIDAL FLOW - NUMERICAL SIMULATIONS

Similar to the simulations described in Winterwerp (2001b), the evolution of the sediment concentration profile is studied as a function of time for various initially homogeneous concentration profiles, increasing C_0 in small steps until the flow is no longer able to remix the sediment for just one instant over the entire water column. The value of C_0 just prior to this collapse is defined as the saturation concentration for tidal flow conditions.

These simulations are carried out for a constant water depth of 8 m, a 12.5 hrs period semi-diurnal, sinusoidal tide with a velocity amplitude of 0.5 m/s, and a settling velocity of 0.5 mm/s including hindered settling with c_{gel} = 80 g/l; the other parameters are listed in Table 2. Figure 3 presents the evolution of the suspended sediment concentration in the form of isolutals for initial concentrations of C_0 = 0.28 and 0.29 g/l, showing a dynamic equilibrium for the 0.28 g/l case with alternating periods of fluid mud formation and periods of complete mixing over the water column, and a complete collapse for the 0.29 g/l case. The 0.28 g/l case obviously represents the saturation conditions defined in Section 3.

It is concluded that, similar to the case of steady state flow situations, also for tidal flow conditions a saturation concentration C_s can be defined which represents the maximal sediment load that can be carried in suspension by a turbulent flow. At a sediment load beyond this saturation condition, the flow is not able to mix the entire fluid mud layer over the whole water depth. As a result, the vertical concentration (density) gradient grows with time, eventually resulting in a complete collapse of the turbulence. This process is irreversible, as long as the fluid mud has not gained sufficient strength to re-enable turbulence production at the water - fluid mud interface.

parameter		value	remarks
water depth	h	variable	time-independent
tidal flow amplitude	U_m	variable	
bed roughness	z_0	1 mm	hydraulically rough
water density	ρ_w	1020 kg/m^3	
sediment density	ρ_s	2650 kg/m^3	
initial sediment concentration	C_0	variable	homogeneous profile
settling velocity	W_s	0.1 & 0.5 mm/s	constant
hindered settling		yes	c_{gel} = 80 g/l
water-bed exchange		no	
Prandtl-Schmidt-number	σ_T	0.7	
number of layers		109	logarithmic/equidistant
time step	Δt	1 min	

Table 2. Reference parameter settings in numerical simulations.

Prior to establishing the variation of C_s as a function of the physical parameters like flow velocity depth, etc., the effect of numerical parameters was studied by varying the time step and/or the vertical discretisation (i.e. number of vertical grid points and their spacing). It was shown that the numerical results were unaffected by these parameters as long as the relative grid size at $\Delta z/h < 0.01$; away from the bed a logarithmic increase of the grid size could be applied. The reader is referred to Winterwerp (1999) for further details.

Next, the results of a series of simulations are presented to show the variation of C_s as a function of flow parameters. The first grid size near the bed is set at $\Delta z/h$ = 0.02 %. The results are presented in Figure 4 and 5 - the parameter settings are summarised in Table 2.

The variation of C_s with h for a semi-diurnal tide with a velocity amplitude of U_m = 0.5 m/s shows a fairly irregular trend - see Figure 4 - especially for W_s = 0.1 mm/s. However, this trend is consistent with the classification of Table 1. By increasing the water depth, as in Figure 4, the path represented by the dotted arrow in Table 1 is followed. It is observed that at small depth, C_s scales with $1/h$. In REGIME II, Phase 1, the h-dependency disappears, but C_s becomes proportional to W_s^{-2}: indeed Figure 4 shows that the influence of variations in settling velocity becomes more and that of variations in water depth less important. Finally, at large h, C_s scales with $1/h^2$, and becomes independent of W_s, which is also shown in Figure 4.

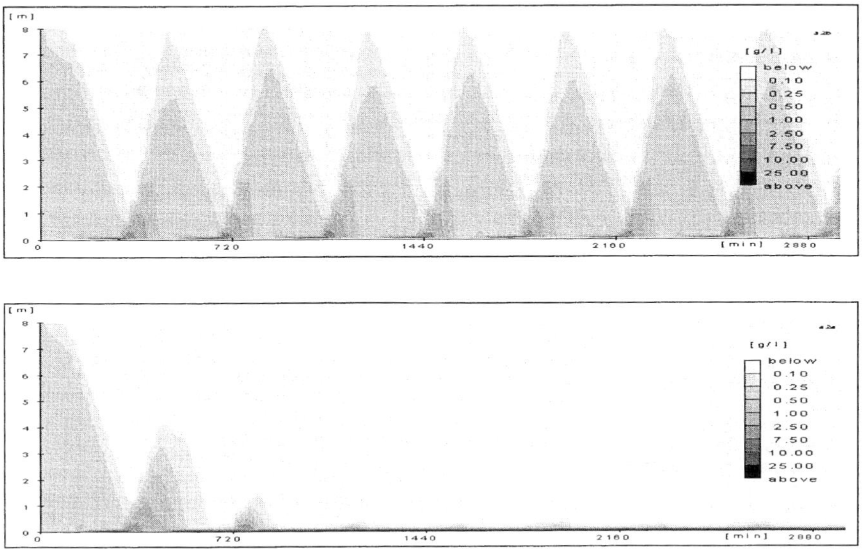

Figure 3. Time evolution of sediment concentration profile for saturated conditions (upper panel: $C_0 = 0.28$ g/l) and for super-saturated conditions (lower panel: $C_0 = 0.29$ g/l).

Figure 5 represents conditions for which $T'_m \approx T'_s \approx 1$. The variation of C_s with U_m can therefore not be read from Table 1. Instead, the full formulation given by equ. (A.5) has to be used, from which it follows that C_s scales with $U_m{}^n$, where $2 < n < 3$ (but n is almost 3).

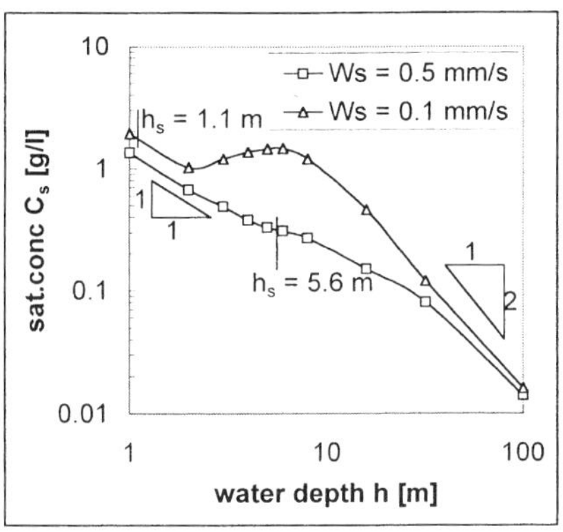

Figure 4. Variation of C_s with h for various settling velocities.

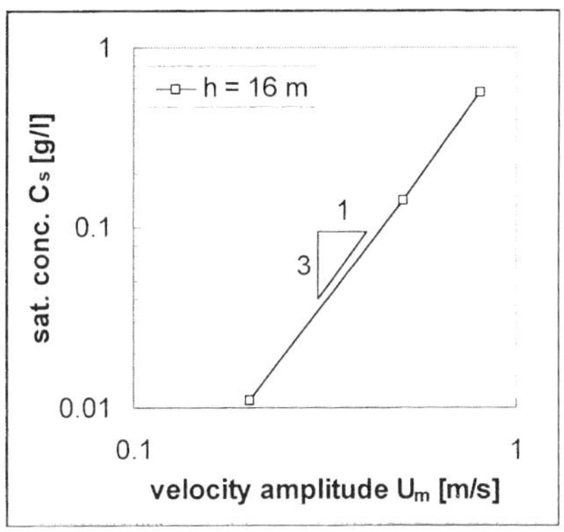

Figure 5. Variation of C_s with U_m for $W_s = 0.5$ mm/s.

5. DISCUSSION AND CONCLUSIONS

It is concluded that the behaviour of High-Concentrated Mud Suspensions is governed by a series of dimensionless parameters which determines whether or not layers of fluid mud are formed and saturation occurs. It appears that also for tidal flow conditions a saturation concentration C_s can be defined, and that the numerical analyses with the 1DV POINT MODEL support the theoretical analyses of Section 3. For extreme conditions simple scaling relations for C_s can be derived, but in general this is not the case and C_s has to be established separately for each site, and probably for varying flow conditions at that site as well.

The relevant dimensionless scaling parameters, discussed and derived in this paper, are summarised in Table 3.

scaling parameter		comments / effects on:
tidal period	T	major driving agent
turbulence generation	T_T/T	negligible for tidal flow
effective Reynolds number	Re_e	turbulence level fluid mud layer
Rouse number	β	concentration profile; saturation parameter
particle Richardson number	Ri_*	overall buoyancy effects; saturation parameter
relative sedimentation time	T_s/T	relaxation time; saturation parameter
relative sedimentation depth	h_s/h	sediment available for fluid mud formation
relative mixing time	T_m/T	relaxation time; saturation parameter
relative flocculation time	T_f/T	relevance of variations in settling velocity; initial fluid mud concentration
relative consolidation time	T_c/T	strength in fluid mud & turbulence production
relative erosion strength	θ_e	floc erosion of consolidating bed
relative bed irregularities	δ_m/k_s	saturation; damping of turbulence production

Table 3. Scaling parameters for HCMS-dynamics.

If the hydrodynamic conditions are cyclical and the amount of sediment is constant, the behaviour of the near-saturated HCMS is cyclical as well, in the sense that a period of settling and fluid mud formation during decelerating tide is followed by a period of re-entrainment of the fluid mud and the subsequent mixing over the water column during accelerating tide. If, however, by some mechanism, either the tidal flow velocity or the wave effects decrease, or the amount or properties of the sediment change, the HCMS may become super-saturated, resulting in a total collapse of the turbulence field and the vertical concentration profile. Such effects may occur in the following cases, for instance:

1) an increase in water depth in the direction of the flow, for instance a cross current over a navigation channel,
2) a decrease in flow velocity, for instance in a harbour basin, or towards neap tide,
3) a decrease in wave height, for instance after a storm, resulting in a decrease in vertical mixing capacity,
4) an increase in wave height, for instance during a storm, eroding consolidated mud deposits,

5) dredging works, increasing the volume of sediment in the environment,
6) algae bloom, increasing the concentration of poly-saccharides, hence the floc size,
7) a decrease in bed friction, for instance above mud deposits, decreasing the vertical mixing capacity, etc.

This collapse will result in fluid mud formation; the turbulent mixing capacity of the flow above the fluid mud will be very small until the fluid mud has gained sufficient strength to re-enable turbulence production at the water-mud interface. The time scale at which this fluid mud is formed is the sedimentation time. Prior to fluid mud formation, the HCMS may behave as a density current, entering harbour basins for instance.

The results of the theoretical and numerical analyses presented in this paper are mutually consistent and agree qualitatively with field observations. The occurrence of a saturation condition, beyond which the vertical profiles of suspended sediment concentration and eddy viscosity collapse, is predicted as a result of a positive feed back between the suspended sediment, the turbulent flow field and the formation of a layer of fluid mud. However, direct observations of such a collapse in nature have not (yet) been reported in the literature, though this feed back itself has been observed (van der Ham et al., 2001 and West and Oduyemi, 1989). Hence, experimental evidence, either through laboratory experiments or through field measurements is required to obtain more certainty about the existence of saturation conditions and the governing scaling laws.

ACKNOWLEDGEMENTS

This work was partially funded by the European Commission, Directorate General XII for Science, Research & Development through the COSINUS-project within the framework of the MAST-3 programme, contract MASC3-CT97-0082 and by corporate research funds from Delft Hydraulics. The valuable advice from and fruitful discussions with Dr. C. Kranenburg of Delft University of Technology is gratefully acknowledged. Mr. J.M. Cornelisse helped with the implementation of the various equations in the 1DV POINT MODEL and Mr. R. Bruinsma with the various graphs; both from Delft Hydraulics. I like to thank Prof. J.A. Battjes of Delft University of Technology for his comments and encouragement.

REFERENCES

Bagnold, R.A., 1966, An approach to the sediment transport problem from general physics, *Geological Survey Professional Paper* 422-I, Physiographic and hydr. studies of rivers.

Bruens, A.W., Kranenburg, C. and Winterwerp, J.C., 2002, Physical and numerical modelling of the entrainment by a turbulent Concentrated Benthic Suspension, *Proceedings INTERCOH-2000, Elsevier, Coastal and Estuarine Fine Sediment Processes*, ed. J.C. Winterwerp and C. Kranenburg, this volume.

Crickmore, M.J., 1982, Data collection - tides, tidal currents and suspended sediment, *The Dock & Harbour Authority*, (LXIII) 742, 183-186.

Ham, R. van der, Fontijn, H.L., Kranenburg C. and Winterwerp, J.C., 2001, Turbulent exchange of fine sediments in a tidal channel in the Ems/Dollard estuary - Part I: Turbulence measurement, *Continental Shelf Research*, (21), 1605-1628.

Hir, P. le, 1997, Fluid and sediment "integrated" modelling application to fluid mud flows in estuaries, in: *Proceedings of the 4th Nearshore and Estuarine Cohesive Sediment Transport Conference*, INTERCOH'94, Wallingford, UK, July 1994, ed. by N. Burt, R. Parker and J. Watts, John Wiley & Sons, 417-428.

Kranenburg, C., 1994, An entrainment model for fluid mud, *Delft University of Technology, Faculty of Civil Engineering, Communications on Hydraulic and Geotechnical Engineering*, Report 93-10.

Kranenburg, C. and Winterwerp, J.C., 1997, Erosion of fluid mud layers - I: Entrainment model, *ASCE, Journal of Hydraulic Engineering*, (123) 6, 504-511.

Liu, K. and Mei, C.C., 1989, Effects of wave-induced friction on a muddy seabed as a Bingham plastic fluid, *Journal of Coastal Research*, (5) 4, 777-789.

Merckelbach, L.M. and Kranenburg, C., 2001, Constitutive equations for effective stress and permeability for mud-sand mixtures on the basis of a scale-invariant bed structure, submitted to *GeoScience*.

Uittenbogaard, R.E., 1995, Observations and analysis of random internal waves and the state of turbulence, *Proceedings of IUTAM Symposium on Physical Lymnology*, Broome, Australia, September 1995.

Villaret, C. and Latteux, B., 1998, Long-term simulation of cohesive sediment bed erosion and deposition by tidal currents, in: *International. Conference Computer Modelling for Seas and Coastal Regions*, ed. P.W. Partridge, 363-378.

West, J.R. and Oduyemi, K.O.K., 1989, Turbulence measurements of suspended solids concentration in estuaries, *ASCE, Journal of Hydraulic Engineering*, (115) 4, 457-474.

Winterwerp, J.C. and Kranenburg, C., 1997, Erosion of fluid mud layers - II: Experiments and model validation, *ASCE, Journal of Hydraulic Engineering*, (123) 6, 512-519.

Winterwerp, J.C., 1999, On the dynamics of high-concentrated mud suspensions, PhD-thesis, Delft University of Technology, also *Delft University of Technology, Faculty of Civil Engineering and Geosciences, Communications on Hydraulic and Geotechnical Engineering*, Report 99-3

Winterwerp, J.C., Uittenbogaard, R.E., de Kok, J.M., 2001a, Rapid siltation from saturated mud suspensions, *Proceedings in Marine Science, Coastal and estuarine Fine Sediment Processes*, ed. W.H. McAnally and A.J. Mehta, Proceedings of INTERCOH'98, Elsevier, 125-146.

Winterwerp, J.C., 2001b, Stratification effects by cohesive and non-cohesive sediment, *Journal of Geophysical Research*, (106) C10, 22,559-22,574.

Winterwerp, J.C., 2001c, On the flocculation and settling velocity of estuarine mud, *Continental Shelf Research* (in press).

APPENDIX A: INTEGRATION OF THE ENTRAINMENT MODEL

In this appendix the entrainment model equ. (4) is elaborated. It is assumed that $U_m = e_m U_w$, where e_m is a (time dependent) proportionality parameter, and $u_*^2 = \lambda U_m^2$. Hence $(U_w - U_m) = (1-e_m)/e_m \sqrt{\lambda} u_* = \sqrt{\lambda'} u_*$. As $e_m \approx 0.2$ to 0.5 (see Figure 3) and $\lambda \approx 0.0018$, and $\lambda' \approx 0.002$ to 0.03.

In Section 3 distinction was made between two regimes: REGIME I: $h \leq h_s$ and REGIME II: $h > h_s$. In REGIME I all sediment in the water column can settle during decelerating tide to form a fluid mud layer and no sediment remaining above the fluid mud layer will have to be dealt with. In this case the thickness of the mixing layer δ in equ. (4) equals the thickness of the fluid mud layer δ_m and the buoyancy term is described by:

$$\int_0^{\delta} B\mathrm{d}z = \int_0^{\delta_m} \left[\frac{\Delta g}{\rho_w} \overline{w'c'} \right] \mathrm{d}z \approx \frac{\Delta g}{\rho_w} \left[\delta_m W_s c_m + \frac{1}{2} \delta_m \Delta c_m \frac{\mathrm{d}\delta_m}{\mathrm{d}t} \right] \tag{A.1}$$

where $c_m(t)$ is the mean concentration in the fluid mud layer and $\Delta c_m(t) = c_m(t) - c_{z>\delta} = c_m(t)$; $c_{z>\delta}$ is the sediment concentration above the fluid mud layer. In deriving equ. (A.1), we have used the mass balance equation - see Kranenburg (1994) or Kranenburg and Winterwerp (1997) for details. Substituting equ. (A.1) into equ. (4) and using conservation of mass ($c_m\delta_m = c_{m,0}\delta_{m,0} = hC_0$), where the subscript $_{,0}$ refers to initial conditions (i.e. prior to entrainment, hence $c_{m,0} = c_{gel}$), a relation for the entrainment velocity $w_e = \mathrm{d}\delta_m/\mathrm{d}t$ is found:

$$\begin{aligned} &\left[c_q u_*^2 + \Delta g \delta_m \frac{c_m}{\rho_w} - c_s (U_w - U_m)^2 \right] \frac{\mathrm{d}\delta_m}{\mathrm{d}t} = \\ &\quad = c_s' |U_w - U_m| u_*^2 + c_\sigma u_*^3 - 2\Delta g \frac{c_m}{\rho_w} W_s \delta_m - c_q \delta \frac{\mathrm{d}u_*^2}{\mathrm{d}t} \end{aligned} \tag{A.2}$$

From equ. (A.2) Kranenburg (1998) concludes that entrainment can occur if:

$$c_s'(U_w - U_m) u_*^2 + c_\sigma u_*^3 > 2\Delta g \frac{c_{gel}}{\rho_w} W_s \delta_{m,0} + c_q \delta_{m,0} \frac{\mathrm{d}u_*^2}{\mathrm{d}t} \tag{A.3}$$

i.e. some time after slack water. This is also shown in Figure 2. The entrainment rate w_e would become very large if:

$$c_s (U_w - U_m)^2 \uparrow c_q u_*^2 + \Delta g \delta_m c_m / \rho_w \tag{A.4}$$

However, for non-stratified conditions, w_e cannot exceed $0.28u_*$ (e.g. Tennekes and Lumley, 1994). An order of magnitude estimate shows that in stratified flow, the first and last terms of (A.2) are small in comparison to the other terms.

For saturated conditions in REGIME I $C_s = C_0$, and upon substitution of equ. (5) into equ. (A.2) the following relation for the saturation concentration is found:

$$\frac{C_s}{\rho_w} \approx \frac{\dfrac{c_s}{\lambda'}\dfrac{h}{u_* T} + \dfrac{c'_s}{\sqrt{\lambda'}} + c_\sigma}{\dfrac{\Delta gh}{u_*^2}\dfrac{h}{u_* T} + \dfrac{2\Delta gh}{u_*^2}\dfrac{W_s}{u_*}} = \frac{\dfrac{c_s}{\lambda'}T'_m + \dfrac{c'_s}{\sqrt{\lambda'}} + c_\sigma}{\left(T'_m + 2\beta\right)Ri_*} \tag{A.5}$$

with the Rouse number $\beta \equiv W_s/u_*$ and a bulk Richardson number $Ri_* \equiv \Delta gh/u_*^2$.

Under REGIME II not all sediment settles during decelerating tide. This implies that the fluid mud layer obtains a thickness $\delta_{m,0} = h_s c_{gel}/C_0$, and that work has to be done to remix the sediment beyond $z = h_s$ over the rest of the water column. Two phases are distinguished. In Phase 1, the fluid mud layer becomes turbulent and is remixed over the sedimentation depth; the related entrainment velocity is referred to as w_{e1}. The fact is ignored that during this process the sediment beyond $z = \delta_m$ may continue to settle because of the small eddy diffusivity above the water - fluid mud interface. In Phase 2, all the sediment is remixed over the entire water column; the related entrainment velocity is referred to as w_{e2}. The definition of C_s implies that during accelerating tide, both the fluid mud and the suspension above have to be remixed over the whole water depth.

In Phase 1, the concentration difference Δc_m over the water - fluid mud interface for a saturated suspension follows form $\Delta c_m \delta_m \approx \left(c_m - C_0\right)\delta_m = C_s W_s T - C_0 \delta_m \approx C_s W_s T$, which is correct for the major part of the accelerating tide. The storage term in equ. (4) may be neglected in this case. Hence, upon integration of the buoyancy term of equ. (4) over the thickness of the fluid mud layer δ_m, as in equ. (A.1), the entrainment rate w_{e1} becomes:

$$\frac{w_{e1}}{u_*} \approx \frac{\dfrac{c'_s}{\sqrt{\lambda'}} + c_\sigma - \dfrac{2\Delta ghC_s}{\rho_w u_*^2}\dfrac{W_s}{u_*}\dfrac{W_s T}{h}}{\dfrac{\Delta ghC_s}{\rho_w u_*^2}\dfrac{W_s T}{h} - \dfrac{c_s}{\lambda'}} = \frac{\left(\dfrac{c'_s}{\sqrt{\lambda'}} + c_\sigma\right)T'_s - 2Ri_*\beta\dfrac{C_s}{\rho_w}}{Ri_*\dfrac{C_s}{\rho_w} - \dfrac{c_s}{\lambda'}T'_s} \tag{A.6}$$

In Phase 2 of the entrainment process, the concentration difference Δc over the water - fluid mud interface for a saturated suspension follows from $\Delta c\delta \approx c\delta = C_0 h = C_s h$. In Phase 2, the Richardson number is not too large in general, as a consequence of which the storage term in equ. (4) may not be neglected a priori. Hence, the entrainment velocity w_{e2} becomes:

$$\frac{w_{e2}}{u_*} \approx \frac{\dfrac{c'_s}{\sqrt{\lambda'}} + c_\sigma - \dfrac{2\Delta ghC_s}{\rho_w u_*^2}\dfrac{W_s}{u_*} - \dfrac{2c_q}{u_*}\dfrac{\delta}{u_*}\dfrac{\mathrm{d}u_*}{\mathrm{d}t}}{c_q + \dfrac{\Delta ghC_s}{\rho_w u_*^2} - \dfrac{c_s}{\lambda'}} = \frac{\dfrac{c'_s}{\sqrt{\lambda'}} + c_\sigma - Ri_*\beta\dfrac{C_s}{\rho_w} - \dfrac{2c_q}{u_*}\dfrac{\delta}{u_*}\dfrac{\mathrm{d}u_*}{\mathrm{d}t}}{c_q + Ri_*\dfrac{C_s}{\rho_w} - \dfrac{c_s}{\lambda'}} \tag{A.7}$$

Substitution of equ. (A.6) and equ. (A.7) into equ. (6) yields a scaling equation for the saturation concentration under REGIME II.

Chapter 3:

Flocculation and settling velocity

Fine Sediment Dynamics in the Marine Environment
J.C. Winterwerp and C. Kranenburg (Editors)

Direct observation of the formation and break-up of aggregates in an annular flume using laser reflectance particle sizing.

A.J. Bale, R.J. Uncles, J. Widdows, M.D. Brinsley and C.D. Barrett

Plymouth Marine Laboratory, Prospect Place, West Hoe, Plymouth, PL1 3DH, UK.

Experiments have been carried out to examine the aggregation of natural estuarine suspended sediment under controlled conditions in an annular flume programmed to simulate oscillating tidal currents in an estuary. The size distribution of the suspended particles was measured in-situ using a Lasentec P-100 laser-reflectance particle sizer with the sensing probe inserted directly through the wall of the flume. Parallel measurements of the solids concentration were made using a calibrated OBS sensor. The flume was filled with river water collected from above the influence of salt water. Various quantities of natural estuary sediment were added to the flume to provide solids concentrations of nominally 100, 800 and 4000 mg l^{-1}. A series of experiments was performed in which the flume was run through consecutive, four hour cycles where the mean current velocity in the flume changed sinusoidally from 5 to 45 cm sec^{-1}. For each sediment concentration the experiment was repeated with some of the fresh water replaced by particle-free seawater to give salinities of 0, 0.2, 2.0 and 10. Over a typical velocity cycle, suspended sediment concentrations decreased with decreasing current velocity, initially slowly, and then more rapidly. The concentration and size of material in suspension minimised over the low velocity period as particles settled. After a certain lag, sediment erosion occurred with increasing velocity and suspended solids concentration increased to a point where all the sediment was in suspension. The particle size data showed that during declining velocity conditions the median size of the particles initially increased as velocity decreased and then decreased as settling of the larger particles from suspension outweighed the aggregation process. During the erosion phase the median diameters increased initially but then decreased and levelled off as current velocity increased further. This was interpreted as mobilisation of aggregated particles followed by breakage as velocity increased. In general the degree of aggregation, and thus deposition, increased with salinity and with solids concentration. Solids concentration had by far the greatest effect on aggregation and deposition rate.

Key Words: suspended particles, aggregation, estuaries, particle sizing, annular flume

1. INTRODUCTION

The gravitational circulation and hydrodynamics of estuaries leads to trapping of fine, cohesive sediment particles in the low salinity region (Officer, 1981; Dyer, 1997). These particles have a large impact on water quality because of their propensity to sorb contaminants (Morris, 1986; Ackroyd et al., 1986; Stewart and Thomson, 1997) and because they exert a pronounced oxygen demand (Uncles, 1998) and there is consequently pressure to model their behaviour. However, the behaviour of suspended particles in estuaries, where tide and wave action leads to periodic erosion and resuspension, is complex. Suspended sediment behaviour is further complicated by the tendency of particles to undergo cycles of aggregation and break-up with consequences for settling and sedimentation which are poorly understood (Eisma, 1986; 1991; Law et al., 1997).

The complexity of particle behaviour in estuaries is compounded by the diversity of natural particles (mineral grains, biogenic debris, diatom frustules and faecal pellets; Eisma et al., 1991; Fennessy et al., 1993). Additionally, particles become coated with organic macro-molecules derived from terrestrial plant degradation and exopolymer exudates from bacteria and diatoms which make the surfaces complex (Eisma et al., 1983). These coatings are known to affect the efficiency of aggregation once collisions have occurred but are likely to vary with seasonal influences on supply and possibly with microbial activity. Furthermore, natural aggregates are known to be extremely fragile and virtually impossible to sample or size without physical disruption (Gibbs, 1981; Gibbs and Konwar, 1982; 1983). For this reason considerable effort has been given to direct, non-intrusive, *in situ* observations in recent years (Bale and Morris, 1987; Fennessy et al., 1994; Bale, 1996; Eisma et al., 1990). Using a novel, *in-situ* laser reflectance apparatus, Law et al. (1997) showed that aggregation processes, combined with resuspension and settling, resulted in large changes in the size distribution of suspended particles over a tidal cycle in the turbid Humber Estuary. Relatively large particles were observed to form in the water column after slack water, when suspended particulate material (spm) concentrations were high, which led to rapid sedimentation.

Estuaries, however, are extremely dynamic systems; the vagaries of climate on wind conditions (waves) and rainfall (river flow) combined with cyclical tides mean that the estuary mixing system in general, and the turbidity maximum in particular, are continually re-equilibrating to changing physical conditions, (Morris et al., 1982; Bale et al., 1985; Uncles et al., 1985; Uncles and Stephens, 1993). This natural variability makes it difficult to study the factors that influence the behaviour of particle aggregation in a systematic way.

Laboratory flumes offer an alternative means with which to study natural suspended particle characteristics and behaviour under controlled physical conditions (Ockenden, 1993; Manning and Dyer, 1999). This paper describes a series of experiments where a non-intrusive laser particle sizer was used with an annular flume to study the effect of particle concentration and salinity on the formation and break up of aggregates under various current velocity conditions. Although similar experiments have been performed previously (Manning and Dyer, 1999), this is the first time that non-intrusive optical particle characterisation methods have been employed at particle concentrations which are typical of turbid estuaries ($> g\ l^{-1}$).

2. METHODS AND MATERIALS

2.1. Annular flume

This work was carried out using the annular flume described by Widdows et al. (1998a; 1998b). The flume has an outer diameter of 660 mm, a channel width of 100 mm and a typical water depth of 280 mm which gave a sample volume of 50 l. For these experiments the motor driving the stirring plate was programmed to adjust the rotation speed through successive, sinusoidal cycles. Current velocities generated within the flume ranged from 5 cm sec^{-1} to 45 cm sec^{-1} (equivalent to bed shear stresses of 0.02 to 1.6 Pa) over a period of 2 hours, i.e., a full cycle in 4 hours. This was shorter than the 6 hour, quarter-diurnal tidal cycle but approximated to the flood tide period of about 4 hours in the low salinity region of the Tamar Estuary (Uncles, 1985). An optical backscatter sensor (OBS) (Downing, 1983) was mounted in the flume to monitor spm concentrations. A data logger recorded the stirring plate velocity and the OBS values at 10 second intervals for each experiment.

2.2. Particle size measurements

The size characteristics of the suspended particles within the flume were monitored using a Partec 100 laser reflectance instrument (Lasentec, Redmond WA) with the sensor probe inserted through a gland in the wall of the flume to allow direct, in-situ observation of the particles (Fig. 1). The sensor window was set flush with the wall of the flume to minimise localised turbulence. This instrument has been previously evaluated for use with estuarine particles and compared with a number of other sizing methods (Law et al., 1997). Size spectra were collected at 16 second intervals and logged on a PC.

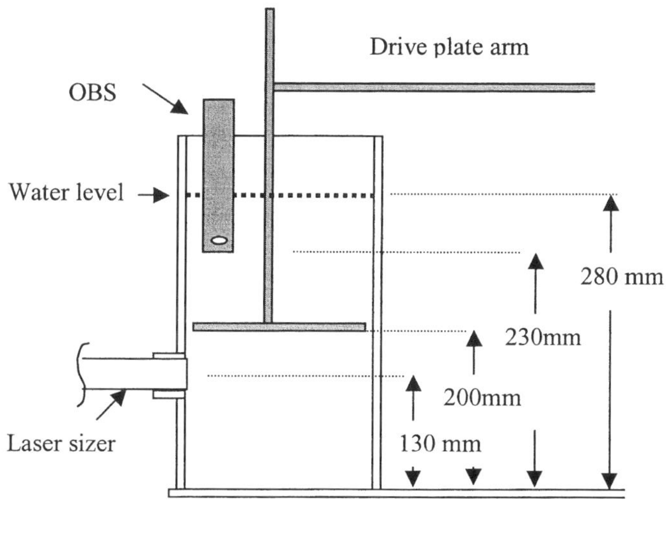

Figure 1. Shows a section of the annular flume channel with the locations of the sensors and water surface given in mm above the base of the flume. OBS is optical backscatter sensor; this drawing is not to scale.

2.3. Materials

A volume of river water (spm <2 mg l^{-1}) was collected from above a weir on the Tamar that limits the intrusion of saline water; this and a smaller quantity of offshore seawater (salinity 35; spm <1 mg l^{-1}) were stored at 4°C over the duration of the experiments. A stock of fine, superficial sediment (20 l) taken from an intertidal mud bank in the turbidity maximum zone of the Tamar at low water was also stored at 4°C. The water content of the stock sediment was 90 % and 82 % of the particles were less than 63 μm on sieving.

2.4. Experimental procedure

All the experiments were carried out at 15° $^{+}/_{-}$ 1°C in a constant temperature room. Each experiment was started with the flume filled with river water and with a volume of stock sediment added to provide a particular concentration of spm when evenly mixed. The maximum current velocity used in the flume completely mobilised all the sediment. Experiments were carried out at nominal solids concentrations of 100, 800 and 4000 mg l^{-1}. Samples of suspension were taken from the flume at various current velocities and analysed gravimetrically to calibrate the OBS response. Different salinities were obtained by allowing the particles to settle completely and replacing a volume of the overlying fresh water with seawater to give a higher salinity when mixed. Salinity values were checked using a precision bench salinometer. For a given concentration of solids, experiments were carried out at salinities of fresh, 0.2, 2 and 10. Individual experiments were run overnight and four complete cycles of oscillating velocity were achieved in this period of approximately 16 hours. Each experiment was carried out under identical physical conditions where the sediment had been subject to identical consolidation of less than one day between experimental runs. Control experiments were run at each particle concentration for four consecutive days with no alterations to salinity to check whether repeated stirring, or degradation of the particles within the flume, led to changes in the size and settling characteristics.

3. RESULTS

For each experiment, the particle size data were calibrated using an algorithm described by Law et al. (1997) and the median value (D_{50}) of each size spectra was calculated. The logged results for suspended solids (from OBS calibrated by gravimetry) and flume rotation speed (converted to current velocity) were merged with the particle size data using a spread sheet routine to bring the data to a common sampling frequency.

3.1 General response of particle size to changes in current velocity

At least four complete current velocity cycles were obtained in each experiment. The results of a typical experiment (Fig. 2) show that the flume speed control was extremely precise over the 16-18 hour period. The behaviour of the particles was also very reproducible from cycle to cycle although a small downward drift in the maximum D_{50}, and upward drift in the minimum D_{50}, were recorded over the experimental period. The general response of both spm concentration and median size to velocity changes was consistent from experiment to experiment changing only in the magnitude of the signal. Typically, sediment erosion lagged behind velocity until a critical velocity had been reached at which point the concentration of solids increased quickly until all the settled material had been eroded and mixed through the

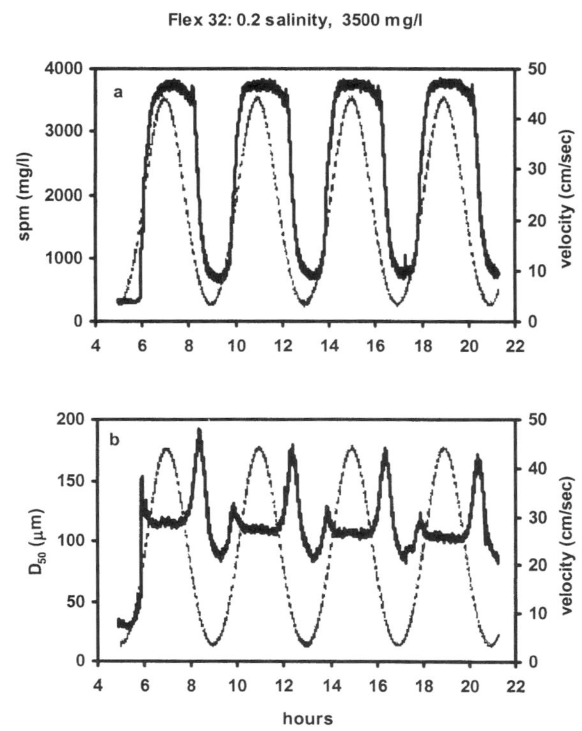

Figure 2. (a) Concentration of spm and (b) the D_{50} of the particles measured over 4 cycles of oscillating current velocity (broken line) during a typical, 16 hour flume experiment.

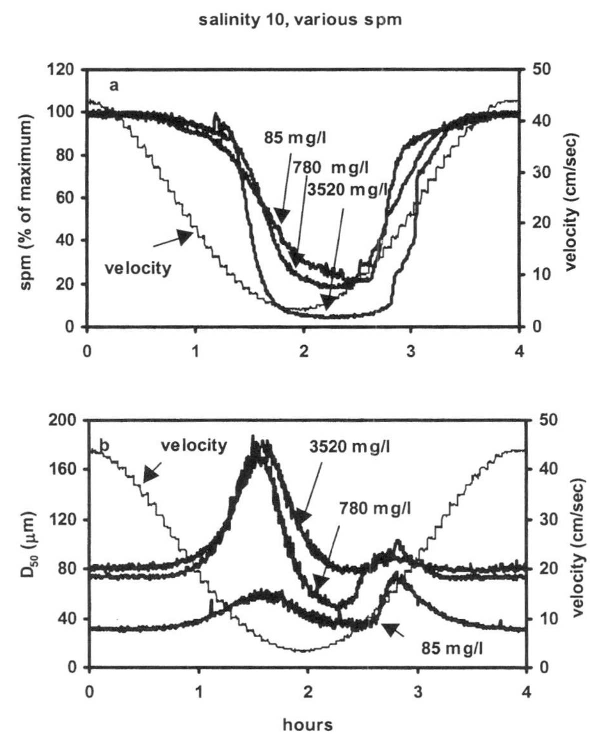

Figure 3. (a) Comparisons of the changes in the concentration of spm and (b) the D_{50} of the particles over one current velocity cycle at three sediment concentrations but with all other factors constant. The salinity was 10 and the change in spm concentration has been shown as a percentage of the maximum concentration in suspension. Current velocity is shown by the light line.

water column. The D_{50} of particles in suspension was smallest at minimum velocity but increased initially as particles were eroded from the bed but then decreased as velocity increased further. Over the period of maximum velocity, when concentration levelled off, the D_{50} remained constant (120 decreasing to 100 μm dia., Fig. 2). As velocity reduced the D_{50} started to increase and the concentration of spm decreased slightly. This process then became more pronounced and the D_{50} of the particles in this case increased from around 100 μm to about 180 μm. Concomitantly, the concentration of spm fell quickly, presumably because of enhanced settling of the larger aggregates. As a result of the enhanced settling, a corresponding reduction in D_{50} was observed.

3.2. Effect of solids concentration on particle behaviour

To enable comparisons of experiments, only the last complete current velocity cycle of the respective experiments were examined. The stock sediment additions used in these experiments gave final spm concentrations of about 90, 770 and 3700 mg l^{-1} respectively, when all the material was suspended but, for comparison, the concentrations are shown as percentages of the maximum value measured. The behaviour of particles in the flume at three particle concentrations, but at the same salinity (fresh), are broadly similar for all the concentrations (Fig. 3); the solids concentration remained constant over the maximum velocity period and only started to fall when velocity declined to 28 cm sec^{-1}. The rate of removal from suspension was greatest for the highest concentration and the transition from slow removal to fast removal occurred at about the same velocity (15 cm sec^{-1}). At 85 mg l^{-1}, 80 % by weight of the particles settled over the minimum velocity period; at 780 mg l^{-1}, slightly more (82%) of the material settled and, at 3500 mg l^{-1}, 95% of the material had settled. The rate of settling also increased with particle concentration. The changes in the D_{50} all showed the same general pattern but the effect of concentration was less regular in that there was a stepwise increase in maximum D_{50} from 85 to 780 mg l^{-1} but little difference then between 780 and 3500 mg l^{-1}.

3.3. Effect of salinity on particle behaviour

The effect of salinity on particle behaviour was studied at all three particle concentrations; for the 780 mg l^{-1} suspension (Fig. 4a) the D_{50} of the suspension was 80 μm at maximum current velocity at all four salinities. However, the D_{50} of the aggregates that formed as velocity declined grew larger with increasing salinity (from 140 to 179 μm dia.). The percentage of total particle material settling over slack water therefore increased from 60 % at 0 salinity to 80 % at a salinity of 10 (Fig 4a).

3.4. Control experiments

Experimental controls were run at spm concentrations of 200, 780 and 3500 mg l^{-1} and were each monitored on four consecutive days; there was no added salt and the experimental suspensions remained essentially constant. The control results at 780 mg l^{-1} (Fig. 5) show a slight change of particle character with each successive experiment. In general, the maximum D_{50} of the aggregates that formed when velocity was decreasing grew slightly bigger over four days (from 130 to 150 μm dia.) and correspondingly more material settled at minimum current periods. This result was slightly accentuated in the 200 mg l^{-1} controls but reversed at 3500 mg l^{-1} where progressively smaller aggregates were formed, and the degree of settling reduced, each day. However, the changes in properties and behaviour observed in the controls over four consecutive days were small compared with the variation in size and settling behaviour measured over a velocity cycle.

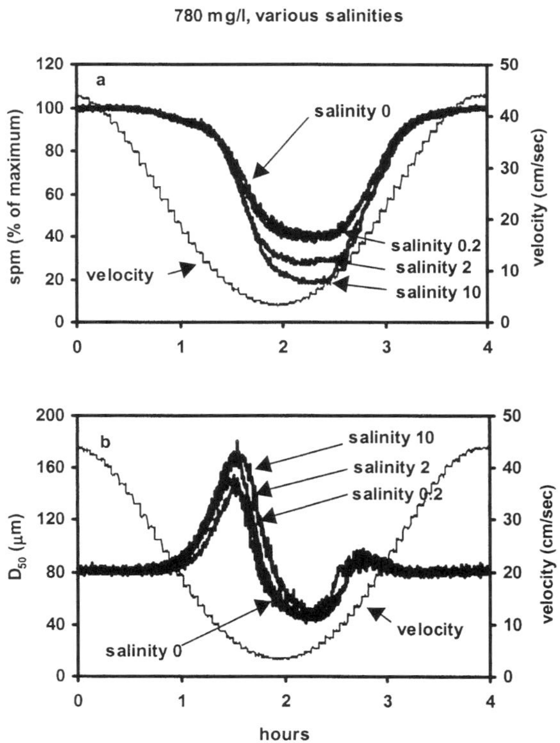

Figure 4. (a) Comparisons of the changes in the concentration of spm and (b) the D_{50} of the particles over one current velocity cycle at four salinities but with all other factors constant. The concentration of spm was 780 +/- 10 mg l^{-1} in each experiment and the change in concentration has been shown as a percentage of the maximum concentration in suspension. Current velocity is shown by the light line.

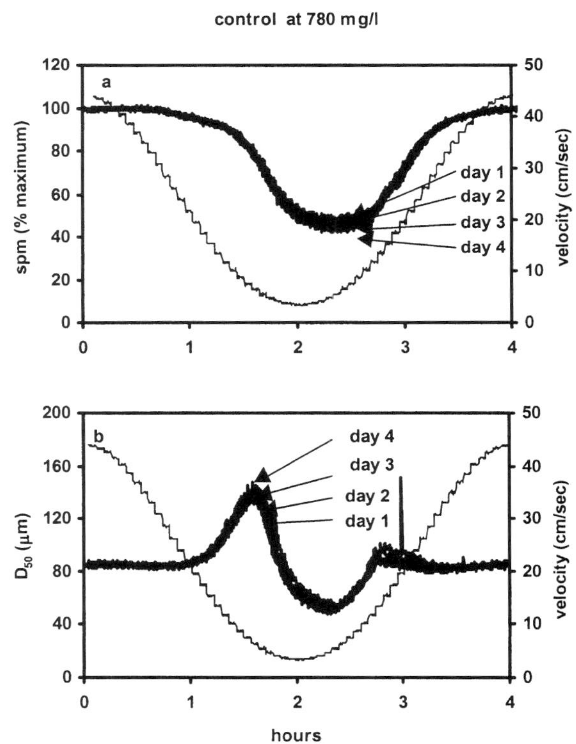

Figure 5. Comparison of four control experiments at a spm concentration of 780 +/- 10 mg l^{-1} and 0 salinity with all physical factors, other than velocity, maintained constant. The flume was stopped for 8 hours between each experiment. Current velocity is shown by the light line.

Table 1. Deposition rates calculated from the exponential decline in spm concentration at various salinity and initial spm concentrations. * Data from the control experiments.

salinity	max'm spm concentration mg.l^{-1}	deposition rate g.$m^{-2}min^{-1}$	max'm aggregate size μm
0	98	0.12	30
0.2	92	0.23	38
2.0	88	0.32	45
10	85	0.41	57
0	810	3.00	150
0.2	830	3.31	150
2.0	790	4.51	165
10	780	5.41	173
0*	200	0.39	45
0*	196	0.54	50
0*	774	2.73	140
0	3670	10.51	140
0.2	3780	17.12?	170
2	3730	17.19	160
10	3520	26.66	179

3.5. Sedimentation rates

The effect of salinity and spm concentration on the aggregation that was observed over minimum current periods in the flume have been analysed in terms of the deposition of spm. For each experiment the observed reduction in spm concentration with time was fitted to an exponential relationship that allowed the deposition rate to be calculated. The results are given in Table 1 and plotted against spm concentration in Fig. 6. Deposition rates increase linearly with spm from 0.12 g $m^{-2}min^{-1}$ at 100 mg l^{-1} to 10.5 g $m^{-2}min^{-1}$ at 3500 mg l^{-1} in fresh water (5 values) and the slope of the line increased with salinity although the deposition rate increased non-linearly with salinity.

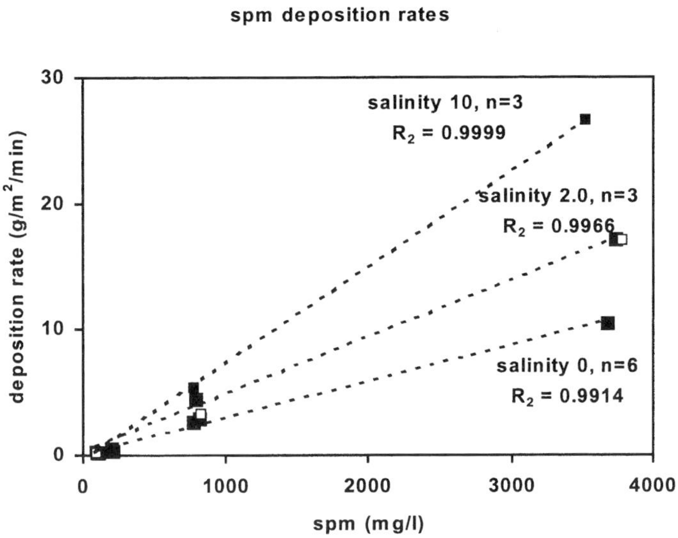

Figure 6. The relationship between deposition rate derived from the decrease in spm concentration and the maximum concentration of spm in each experiment. Note: the regression for 0.2 salinity (open squares) falls very close to the results obtained at a salinity of 2.0 due to one anomalous deposition rate value at 3500 mg l^{-1}.

4. DISCUSSION

This series of laboratory measurements has shown that aggregation of estuarine particles occurs under conditions of high particle concentration and low current velocity (turbulent shear). It has removed the possibility that similar observations in estuaries may be due solely to advection or to localised differences in particle properties. The measurements presented here are consistent with some previous results from experimental work on particle sizes. Lick et al. (1993) derived an empirical expression between floc diameter, D, suspended particle concentration, P, and turbulent shear (i.e. root mean square of the gradient in turbulent velocity fluctuations), G, (s^{-1}) of the form:

$$D = c(PG)^{-d} \tag{1}$$

In which c and d are positive regression constants (the regression constant, c, depends on the units employed) and where the turbulent shear, G, is related to the turbulent energy dissipation rate per unit mass, ε, and the molecular kinematic viscosity, ν, by (Dyer and Manning, 1999):

$$G = (\varepsilon/\nu)^{1/2} \tag{2}$$

Values of G were derived using the method described by Manning and Dyer (1999) and, as such, provide no spatial detail but rather represent a bulk, or global measure for the flow as a whole. Manning and Dyer (1999) derived a similar regression relationship to Equation (1), with $d = 0.37$. Similar results were also observed during the early-to-late decelerating phase of our highest concentration flume experiments, where concentrations were nominally 3700 mg l^{-1} and 800 mg l^{-1}. Particle sizes could be represented by $d = 0.26$ (88% of variance explained) and $d = 0.19$ (86% of variance explained), respectively. At this phase of the experiment, all of the sediment had started off in suspension at the highest current velocities and current shears, and then subsequently aggregated and settled towards the bed as the shears decreased with time. In the case of the highest concentration experiment, the effective fall velocity (speed of sediment deposition) at the time of maximum particle size was 0.23 mm sec^{-1}, whereas the estimated fall velocity (McCave, 1984) of the observed flocs was 0.3 mm sec^{-1} at that time.

Improved regressions between particle sizes and the physical variables could be achieved using more complex relations that are consistent with the conceptual model described by Dyer (1989) in which low shears result in increased aggregation whereas strong shears cause floc disruption. For the highest concentration, 3700 mg l^{-1} experiment, on the early to late phase of decelerating currents:

$$D_{50} = 0.08\left[(1+29G)/\left(1+125G^{4.6}\right)\right]P^{-1.1} + 0.86 \qquad (3)$$

Where all variables in this equation have been divided by experimentally determined mean values in order to normalise them and simplify the units. For the 800 mg l^{-1} experiment the corresponding relationship is:

$$D_{50} = 0.06\left[(1+33G)/\left(1+138G^{3.8}\right)\right]P^{-3.5} + 0.88 \qquad (4)$$

Both equations explain more than 99.8% of the variance in particle size during this stage of the experiment. Therefore, at low shears (G small) the sizes increase with increasing shear, whereas at high shears the sizes decrease. Increasing particle concentrations tend to reduce particle size, in accord with the result by Lick et al. (1993) and Manning and Dyer (1999). At the highest experimental shears there appears to be an asymptotic aggregate size that reflects the existence of very strong microflocs.

There are indications that the particles used in this work became ‘conditioned’ within the flume during individual 16 hour experiments and during the control experiments, where the maximum size of aggregates decreased slightly with time. However these effects were small compared with the principal aggregation and break up signal observed over the current velocity cycle. It is also possible that organic material associated with particle surfaces may have been degraded during storage with the prospect that the surfaces of the stock particles were less ‘sticky’and therefore their aggregation efficiency reduced. Both of these factors mean that, in the natural system, the degree of aggregation is likely to be greater than observed in these experiments. Indeed, a comparison of aggregate sizes measured in the field compared with those measured during this work, when examined as a function of particle concentration, suggests that, at concentrations over 780 mg l^{-1}, D_{50} values in the flume are possibly a factor of 2 to 3 times smaller than observed in field data. However, this difference may be due in part to the fact that much of the field data was measured near bed in typically 4m of water where settling aggregates would have had the opportunity to grow larger through differential settling. Conversely, in the flume, all the size measurements were made at one depth (150 mm below surface) determined by the position of the port in the flume wall. There was a poor relationship between the maximum D_{50} observed and the deposition rate of spm (Table 1), with observed aggregate sizes appearing to reach a maximum D_{50} of 180μm. Therefore the aggregate sizes observed in the flume are possibly the result of a dynamic balance between aggregation and settling in which larger aggregates are settling below the level of the sensor too rapidly to observe. Using settling rates determined by Fennessy et al. (1994) in the turbidity maximum of the Tamar Estuary, the time taken for aggregates of 300 μm to settle through a 4m water column decreases to 0.6 hours from over 100 hours for 63 μm particles. This interval is consistent with the observed rapid settling of suspended solids from the turbidity maxima in the Tamar (Uncles and Stephens, 1993) and Humber-Ouse estuaries (Uncles et al., 1998) over slack water at high tide. Although this example relates to larger aggregates than we observed in the flume, field data (Law et al., 1997) indicates that the formation of aggregates with a D_{50} of 300 μm or larger are not unrealistic at low shear and high particle concentrations.

5. CONCLUSIONS

Direct observations of estuarine particle aggregation and aggregate breakage have been made using laser reflectance particle sizing in an annular flume during simulated cycles of tidal currents and at realistic particle concentrations.

Marked particle aggregation occurred at a specific phase in the simulated tidal cycle when certain criteria were met: a) velocity, and thus shear, was decreasing, b) the suspended solids concentration was high. These conditions occurred as velocity decreased on each cycle. This observation is consistent with the rapid reduction in suspended solids concentration observed over slack water in the turbidity maximum of turbid estuaries.

Under constant physical conditions, the degree of aggregation, and rate of sedimentation over minimum velocity periods increased with particle concentration and, to a lesser extent, with salinity.

As the current flow field within the flume was not rigorously defined in relation to secondary currents, these observations have to be viewed as qualitative, albeit intuitively 'sensible' and consistent with the pattern of field observations.

The degree of aggregation observed in the field was always greater than simulated in the flume. This was almost certainly an artefact of the experimental conditions within the flume and may reflect the lack of biological polymers (through microbial degradation).

There is a need for further, systematic work with this combination of apparatus in order to investigate the effect of variables such as biological polymer (EPS) and settling depth within the flume channel.

6. ACKNOWLEDGEMENT

This work formed part of the NERC CCMS Restricted Exchange Environments Core Research Programme. The authors are grateful to James Strong who carried out a large number of the laboratory flume experiments.

REFERENCES

Ackroyd, D.R., Bale, A.J., Howland, R.J.M., Knox, S., Millward G.E. and Morris, A.W., 1986. Distributions and behaviour of Cu, Zn and Mn in the Tamar Estuary. *Estuarine, Coastal and Shelf Science* **23**: 621-640.

Bale, A.J., Morris, A.W. and Howland, R.J.M., 1985. Seasonal sediment movement in the Tamar Estuary. *Oceanologica Acta* **8**: 1-6.

Bale, A.J. and Morris, A.W., 1987. In situ measurement of particle size in estuarine waters. *Estuarine, Coastal and Shelf Science* **24**: 253-263.

Bale, A.J., 1996. In situ laser optical particle sizing. *Journal of Sea Research* **36**: 31-36.

Downing, J.P., 1983. An optical instrument for monitoring suspended particles in ocean and laboratory. Proceedings: *Oceans '83*, (San Francisco). IEEE & MTS. pp 199-202.

Dyer, K.R. and Manning, A.J., 1999. Observation of the size, settling velocity and effective density of flocs, and their fractal dimensions. *Journal of Sea Research* **41**: 87-95.

Dyer, K.R., 1997. Estuaries a physical introduction 2nd Edition. John Wiley, Chichester. pp 157-160.

Dyer, K.R., 1989. Sediment processes in estuaries: future research requirements. *Journal of Geophysical Research* **94** (C10): 14327-14339.

Eisma, D., Boon, J., Groenewegen, R., Ittekkot, V., Kalf, J. and Mook, W.G., 1983. Observations on macro-aggregates, particle size and organic composition of suspended matter in the Ems Estuary. *Mitt. Geol.-Palaont. Inst. Univ. Hamburg*. SCOPE/UNEP Sonderband, **55**: 295-314.

Eisma, D., 1986. Flocculation and de-flocculation of suspended matter in estuaries. *Netherlands Journal of Sea Research* **20**: 183-199.

Eisma, D., 1991. Particle size of suspended matter in estuaries. *Geo-marine Letters,* **11**: 147-153.

Eisma, D., Bale, A.J., Dearnaley, M.P., Fennessy, M.J., van Leussen, W., Maldiney, M-A., Pfeiffer, A. and Wells, J.T., 1996. Intercomparison of in situ suspended matter (floc) size measurements. *Journal of Sea Research* **36**: 3-14.

Eisma, D., Schuhmacher, T., Boekel, H., van Heerwaarden, J., Franken, H., Laan, M., Vaars, A., Eijgenraam, F. and Kalf, J., 1990. A camera and image analysis system for *in situ* observation of flocs in natural waters. *Netherlands Journal of Sea Research* **27**: 43-56.

Fennessy, M.J., Dyer, K.R. and Huntley, D.A., 1994. INSSEV: An instrument to measure the size and settling velocity of flocs in situ. *Marine Geology* **117**: 107-117.

Gibbs, R.J., 1981. Floc breakage by pumps. *Journal of Sediment Petrology* **51**: 670-672.

Gibbs, R.J. and Konwar, L.N., 1982. Effect of pipetting on mineral flocs. *Environmental Science and Technology* **16**: 119-121.

Gibbs, R.J. and Konwar, L.N., 1983. Sampling of mineral flocs using Niskin bottles. *Environmental Science and Technology* **17**: 374-375.

Law, D.J., Bale, A.J. and Jones, S.E., 1997. Adaption of focused beam reflectance measurement to in-situ particle sizing in estuaries and coastal waters. *Marine Geology* **140**: 47-59.

Lick, W., Huang, H. and Jepson, R., 1993. Floculation of fine-grained sediments due to differential settling. *Journal of Geophysical Research* **98** (C6): 10279-10288.

Manning, A.J. and Dyer, K.R., 1999. A laboratory examination of floc characteristics with regard to turbulent shearing. *Marine Geology***160**: 147-170.

McCave, I.N., 1984. Size spectra of aggregates of suspended particles in the deep ocean. *Deep-Sea Research* **31**: 329-352.

Morris, A.W., Bale, A.J. and Howland, R.J.M., 1982. Chemical variability in the Tamar Estuary, south-west England. *Estuarine, Coastal and Shelf Science* **14**: 649-661.

Morris, A.W., Howland, R.J.M. and Bale, A.J., 1986. Dissolved aluminium in the Tamar Estuary, southwest England. *Geochimica et Cosmochimica Acta* **50**: 189-197.

Morris, A.W., 1986. Removal of trace metals in the very low salinity region of the Tamar Estuary, England. *Science of the Total Environment* **49**: 297-304.

Ockenden, M.C., 1993. A model for the settling of non-uniform cohesive sediment in a laboratory flume and an estuarine field setting. *Journal of Coastal Research* **9**: 1094-1105.

Officer, C.B., 1981. Physical dynamics of estuarine suspended sediment. *Marine Geology* **40**: 1-14.

Stewart, C. and Thompson, J.A.J., 1997. Vertical distribution of butyltin residues in sediments of British Columbia harbours. *Environmental Technology* **18**: 1195-1202.

Uncles, R.J., Elliott, R.C.A. and Weston, S.A., 1985. Observed fluxes of water, salt and suspended sediment in a partially mixed estuary. *Estuarine, Coastal and Shelf Science* **20**: 147-167.

Uncles, R.J., Joint, I. and Stephens, J.A., 1998. Transport and retention of suspended particulate matter and bacteria in the Humber-Ouse estuary, UK, and their relationship to hypoxia and anoxia. *Estuaries* **21**: 597-612.

Uncles, R.J., and Stephens, J.A., 1993. Nature of the turbidity maximum in the Tamar Estuary, UK. *Estuarine, Coastal and Shelf Science* **36**: 413-431.

Widdows, J., Brinsley, M.D., Bowley, N. and Barrett, C., 1998a. A benthic annular flume for *in situ* measurement of suspension feeding/biodeposition rates and erosion potential of intertidal cohesive sediments. *Estuarine, Coastal and Shelf Science* **46**: 27-38.

Widdows, J., Brinsley, M.D. and Elliott, M., 1998b. Use of in situ flume to quantify particle flux (biodeposition rates and sediment erosion) for an intertidal mudflat in relation to changes in current velocity and benthic macrofauna. *In*: Black, K.S., Paterson, D.M. and Cramp, A., (Eds.) *Sedimentary Processes in the Intertidal Zone.* Geological Society, London, Special Publications **139**, 85-98.

The Turbidity Maximum in a Mesotidal Estuary, the Tamar Estuary, UK: I. Dynamics of Suspended Sediment.

K.R.Dyer[a], A.J.Bale[b], M.C.Christie[a*], N.Feates[c], S.Jones[d], A.J.Manning[a]

[a] Institute Marine Studies, University of Plymouth, Plymouth PL4 8AA, UK.
[b] Plymouth Marine Laboratory, Plymouth PL1 3DH, UK.
[c] HRWallingford, Howbery Park, Wallingford, OX10 8BA, UK.
[d] School Ocean Sciences, University of Wales Bangor, Menai Bridge, Gwyedd LL59 5EY, UK.

A series of detailed simultaneous measurements of current velocity, salinity, suspended sediment concentration and turbulence, at two stations in the upper Tamar Estuary, has revealed that the turbidity maximum is centred about 1km upstream of the salt intrusion, and is created mostly from entrainment from the bed. Suspended sediment concentrations at about 0.5m above the bed reach about $6gl^{-1}$ on spring tides, but only about $0.3gl^{-1}$ on neaps. Based on measurements of turbulent energy, the threshold of sediment erosion is estimated as $0.17\text{-}0.2Nm^{-2}$. Drag reduction was experienced when the suspended sediment concentration gradients exceeded $4kgm^{-4}$. Analysis of surface water slopes in conjunction with the currents suggests that there is a minimum in velocity between the salt intrusion and the turbidity maximum which coincides with a reduction in the water level, together forming a 'bow-wave' effect, upstream of the salt intrusion. These results form the background for detailed analysis of the distribution of floc properties, particularly settling velocity, within the turbidity maximum (discussed in Part II).

Keywords: Turbidity maximum, suspended sediment concentration, turbulence, flocculation, salinity intrusion, shear stress, water slopes.

1. INTRODUCTION

It has only been comparatively recently that field studies have been possible on the characteristics of suspended fine sediment when in transport in the estuarine environment. The basic problem relates to the variable degree of flocculation that makes settling velocity, size and density non-constant. Flocculation occurs because of chemical ionic bonding, as well as by biological 'glues'. The strength of the bonds, and the size of the flocs has been shown to change markedly with concentration of particles, and hydrodynamic forces creates disruption because of turbulent shearing. Initial laboratory studies suggested a

---

• Now at: South Devon NHS Trust, Torquay. UK.

strong effect of the fluid salinity in enhancing flocculation, but in the field this appears to be less important. Comparative laboratory and field studies showed that almost any attempt

to sample the flocs led to their disruption. Consequently, in-situ measurement of their properties is essential for proper parameterisation of size, settling velocity and effective density. Also this raises questions regarding the importance of turbulent shearing, occurring naturally in the water, in breaking up the flocs forming at times of lower shear. The use of settling tubes of various designs has established relationships between concentration and settling velocity, which were generally of an exponential form for concentrations below the hindered settling limit $\sim 10 gl^{-1}$. However, settling velocity measurement in-situ still involves problems associated with disruption (Dyer et al, 1996). To some extent these have been overcome by the development of the INSSEV system (Fennessy et al, 1994a), and use of this instrument in conjunction with precise turbulence measurements makes it possible to derive empirical relationships and algorithms.

Within an estuarine turbidity maximum there are changes of concentration and shear, as well as regular erosion of material from the bed and settling back to the bed. Prediction of the transport of fine suspended sediment depends on a good specification of the settling velocity, and of the mass settling flux, as well as of the erosion properties. This can only be done by using field measurements to develop empirical algorithms that can be implemented in transport models.

The MAST III funded project COSINUS (Prediction of cohesive sediment transport and bed dynamics in estuaries and coastal zones with integrated numerical simulation models) aimed to study, by laboratory and field experiments, the flow of high concentration bed layers. The flocculated state of the layers, the size and settling velocity of the flocs, and their relationships to the driving variables are important facets of this study. As part of the COSINUS project a two week long experiment was carried out within the turbidity maximum (TM) of the Tamar Estuary, UK. The aims were to measure the effect of velocity shear and turbulence, salinity, and concentration on the size, settling velocity and effective density of flocs, and the relationship to the concentration and velocity profiles. The relationships were to be tested against theory and the results of laboratory experiments, and in order to put the results in the context of modelling fine sediment transport, also implemented within a model of the Tamar Estuary (Peterson et al, 2002).

This paper reports the overall objectives of the experiment, the scope of the data and what it revealed of the dynamics of the turbidity maximum, the erosion, movement and deposition of fine sediment during passage of the TM through the section on flood and ebb, neap and spring tides. The results of analysis of floc properties obtained with the INSSEV system is reported in Part II (Dyer et al, 2002). More detail of the floc properties are reported in Manning (2001), and of modelling in Peterson et al (2002). Further results of the turbulence measurements will be presented in Christie et al (2002). The data is accessible as part of the COSINUS database at http://www.bwk.kuleuven.ac.be/bwk/cosinus/cosinus.html.

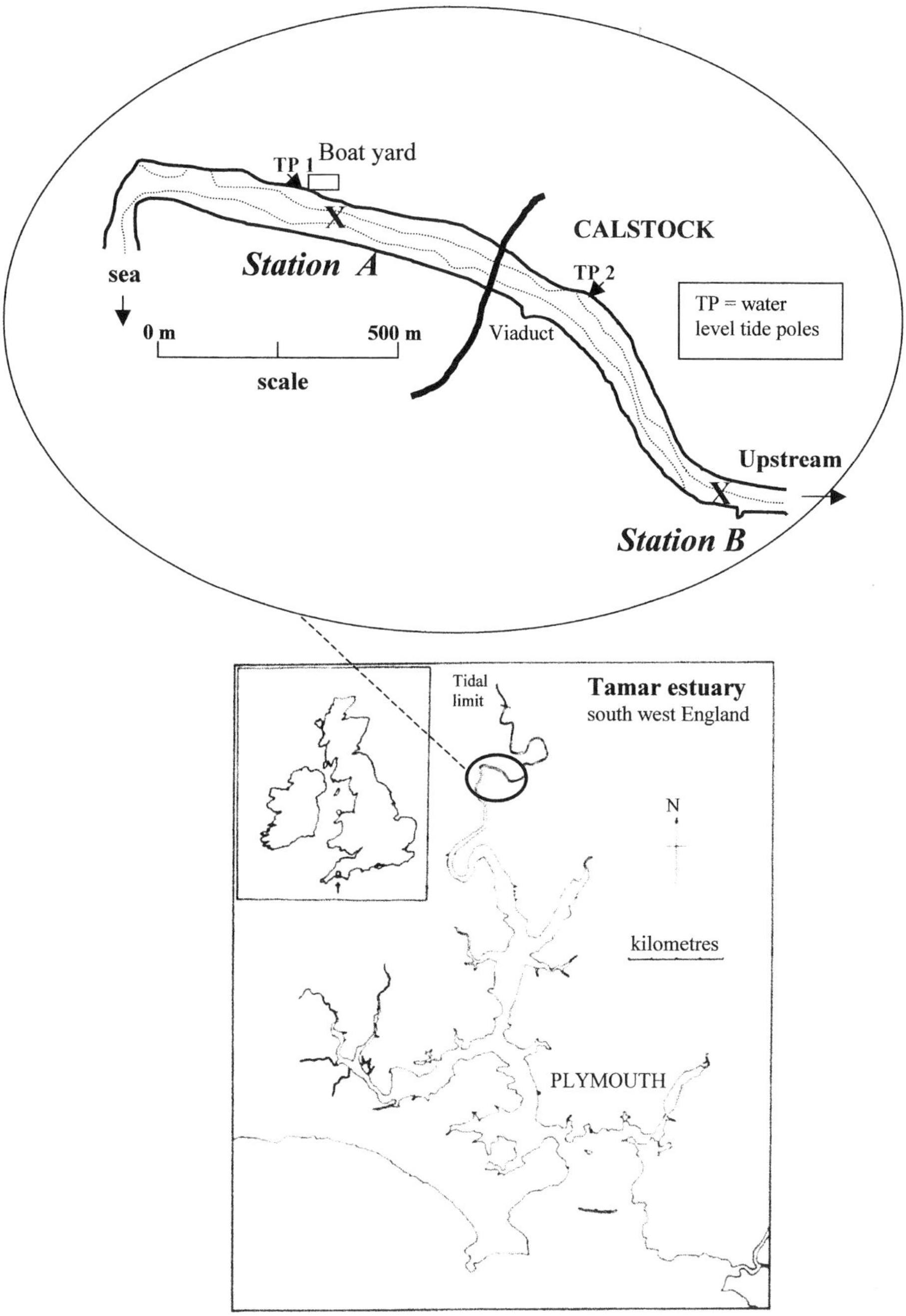

Figure 1. Location diagram of the Tamar estuary showing the positions of Station A and B.

2. METHODOLOGY

A straight reach of the upper Tamar estuary, within the tidal trajectory of the turbidity maximum was chosen for the experiment, because its overall characteristics are fairly well documented (Uncles and Stephens, 1993). In order to separate local and advected effects, two stations 970m apart were occupied with fore-and-aft anchored vessels in the central channel (Figure 1). Station A was approximately 400m below the Calstock Railway Viaduct, and Station B about 500m upstream of it. The estuary widths and maximum depths at high water were 75m and 4.5m respectively at the lower section and 50m and 5.2m at the upper. The tidal range varied from about 3.2m to about 4.5m. In this situation it was expected that there would be considerable settling and erosion of mud from the bed at times during the tide, and that the suspended sediment dynamics would be dominated by the bed boundary conditions.

It was not the intention to obtain complete tidal cycles of measurements, but to investigate in detail processes on a sub-tidal scale. Consequently, measurements were obtained for sections of the tide when conditions were reasonably steady, or gradually changing. Eight sampling runs, of approximately three hours duration, were carried out simultaneously at both stations during the two-week experimental period 14-24th September 1998, which covered spring and neap tides. River discharge was around the annual mean of about 20 m^3s^{-1}.

At each station, profiles of salinity, temperature, and suspended solids concentration were taken at least every 15min by hand profiling Seabird systems CTD and Downing optical backscatter gauges (OBS). Hourly water samples were obtained for calibration of the OBS's. Current velocities were measured by direct reading current meters suspended at a constant depth below the vessels; near surface at Station A and mid-depth at Station B. Also intercalibration was attempted by use of all instruments simultaneously at station A on one day of neap tides and another of springs.

At Station A, the POST system (Christie et al, 1997) was mounted on the same bedframe as the INNSEV system (Fennessy et al, 1994a). The former system comprises an array of 4 EM flowmeters and 8 OBS's. The EM flowmeters were arranged in pairs to obtain the three orthogonal components of the flows at 0.25 and 0.75m above the bed. The OBS's were arranged approximately logarithmically through the bottom 1m of the flow. The data from the POST system was logged at 18Hz and filtered at 5Hz. From these measurements the velocity shear, the turbulence characteristics and the mean and fluctuating concentration could be interpolated at the INSSEV sampling height (0.5m). Water surface slopes were measured during some experimental runs from visual water levels at two positions within the experimental reach. These were accurately surveyed into benchmarks with closure errors of less than 0.01m.

3. RESULTS

3.1 Location of the turbidity maximum.

Neap tides.

The turbidity maximum was located within the fresh water landwards of the salt intrusion on both neap and spring tides. This agrees with the studies of Uncles and Stephens (1993).

Figure 2 shows the salinity variation during the ebb tide at Station A. The salinity reached 10 on the bottom at 14:00hr, with a halocline at about 2m height and almost fresh water on the surface. The halocline elevation reduced steadily until about 15:00hr, but a salinity of 5 persisted until about 16:20hr, followed by a further steady decrease until 17:00hr. The convoluted form of the halocline may be the result of the contouring system, but could suggest that considerable interfacial activity occurred during retreat of the salt intrusion. Comparison of the timing of disappearance of salt at the two stations showed a delay of 30-45min, which is in agreement with the time taken for advection over the distance.

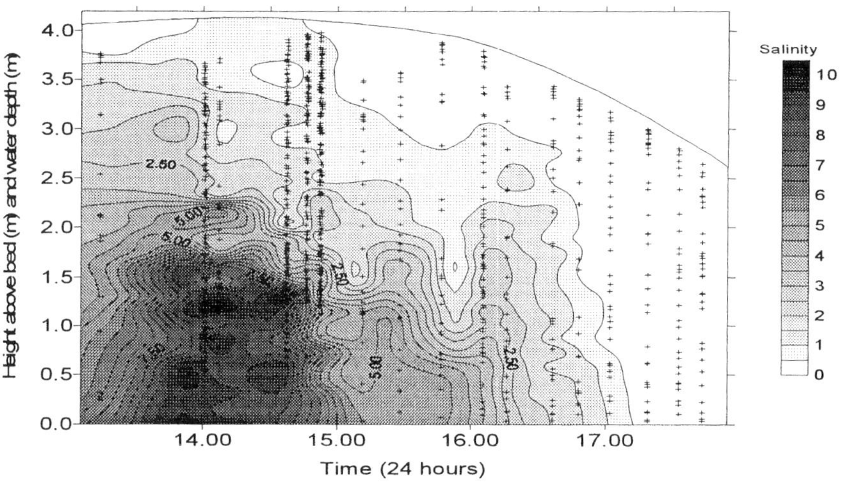

Figure 2. Contour plot of the salinity variation during the early ebb tide at Station A on a neap tide on 15th September 1998. Crosses show times and heights of measurements.

Figure 3a shows the neap tide variation of ebb velocity at about mid-depth at both stations, together with the near-bed measurements at Station A. The ebbing surface current accelerated almost linearly with the falling elevation until the first peak at about 16:30hr. The velocity at Station A was slightly greater than at the upstream station, though this may be caused by the former measurement being taken slightly higher in the flow. The near bed current (Fig 3b), however, showed a two-layer flow was present, with low but slightly decreasing currents near the bed until 16:10hr, after which the flow also accelerated to a peak at 16:30hr. The period of low currents coincided with the presence and gradual thinning of the saline layer near the bed. This phenomenon is a regular occurrence at this location (McCabe et al 1993, Fennessy et al 1994b).

Following the first peak in the currents there was a broad second peak about an hour later which was of the same magnitude on the surface as the first. Near the bed, however, the second peak involved a velocity increase from 0.32 to 0.55ms^{-1}, with a concomitant increase in current shear near the bed. The reason for the dip in currents between the first and second peaks will be considered later.

The suspended sediment concentration SPM (Figure 4) very rapidly increased after 16:50hr. This was after the first peak in the surface current, and was about the same time that the last traces of salt disappeared from near the bed. The SPM reached a maximum at 24cm above the bed of 0.65 gl^{-1} at 17:15hr. This was about 30min before maximum current and it is conjectured that this could represent the erosion of a layer of surface material and exposure of more resistant sediment beneath. On the other hand, the current at 25cm above

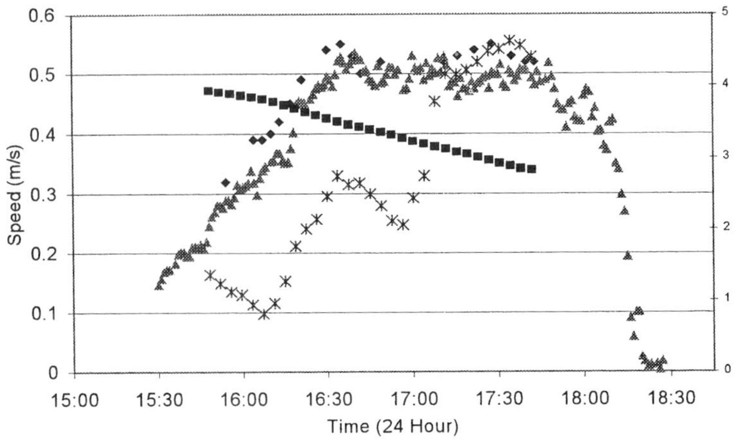

Figure 3a. Current speed measurements at Station A and Station B during the ebb tide on 15th September. The water elevation is from Station A.

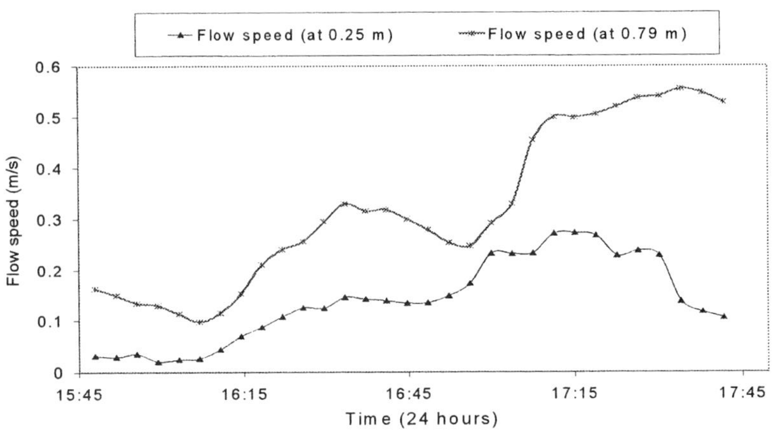

Figure 3b. Current speed measurements at the two heights on the POST rig 15th September 1998. The speed at 0.79m is the same as that plotted on Figure 3a.

the bed decreased after 17:15hr even though the current higher in the flow continued increasing. Thus it is possible that the suspended sediment load may have produced a distortion of the velocity profile, and a degree of reduction in shear which may have limited further erosion.

At Station B there was no sign of a distinct turbidity maximum with concentrations greater than 65mgl^{-1} before low water slack, raising the probability that the TM observed at A was almost entirely produced in the reach between the two stations.

The mean shear stress calculated from the turbulent energy at the two EM levels, using the proportionality arguments of Soulsby (1998), suggest that the threshold of erosion of the mud involved in the TM was 0.17Nm^{-2}.

The time delay between the disappearance of the salinity and the peak of the turbidity maximum was of the order of 30min, again about equivalent to an advection

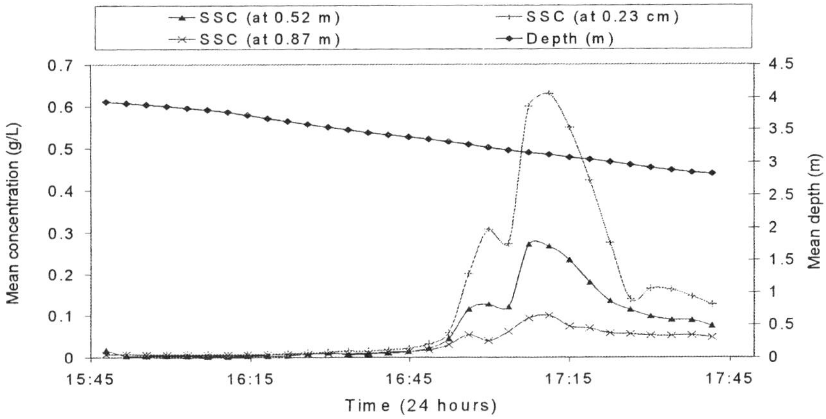

Figure 4. Suspended sediment concentration measured at three heights on the POST rig during the ebb tide on 15th September 1998.

distance of 1km at 0.55ms^{-1}. The TM was entirely within the fresh water, and it is suggested that erosion of the bed sediment to form the TM was not possible until the salt had retreated from the near bed water, allowing the current to penetrate to the bed.

During the flood tide the saline intrusion advected up the estuary as an almost vertically well-mixed 'wall'. The first salinity appeared at about 12:25hr on 16th September at Station A (Figure 5a) and rose to about 10 at high water. At Station B the water was fresh until after 14:05hr. The currents rose almost linearly to a first peak at about 12:00hr (Figure 6), though it appeared to be slightly earlier at Station B. There was then a drop of about 5cms^{-1} followed by a rise to a further peak at about 12:40hr. The maximum currents in this second peak were about 10 cms^{-1} faster at Station B than at A. The rapid reduction in the near bed current at Station A after about 12:20hr coincided with the appearance of the saline intrusion, and the surface current reduced about 20 min later. At Station A the near bed concentration increased abruptly at about 11:20hr (Figure 5b) when the bed shear stress exceeded about 0.2 Nm^{-2}, and reached a peak at about 12:15hr, then declined to reach background levels within the saline intrusion soon after 1320hr. At Station B the increase in

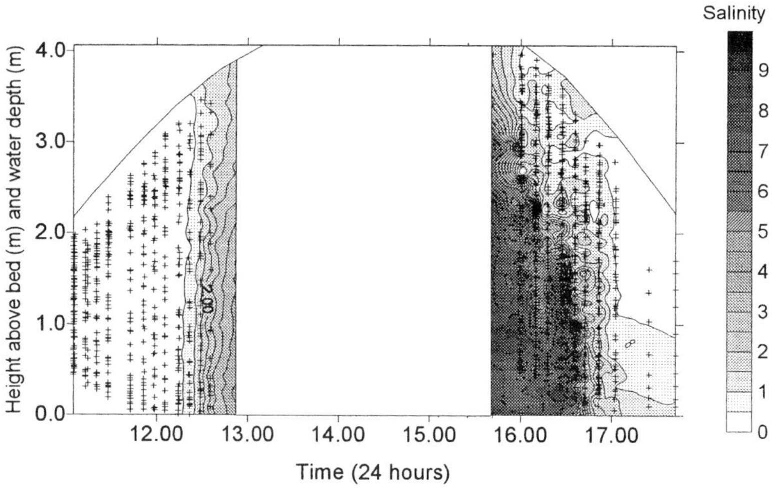

Figure 5a. Contours of salinity measured on the neap tide on the 15th September 1998 at Station A. The crosses show the times and heights of the measurements.

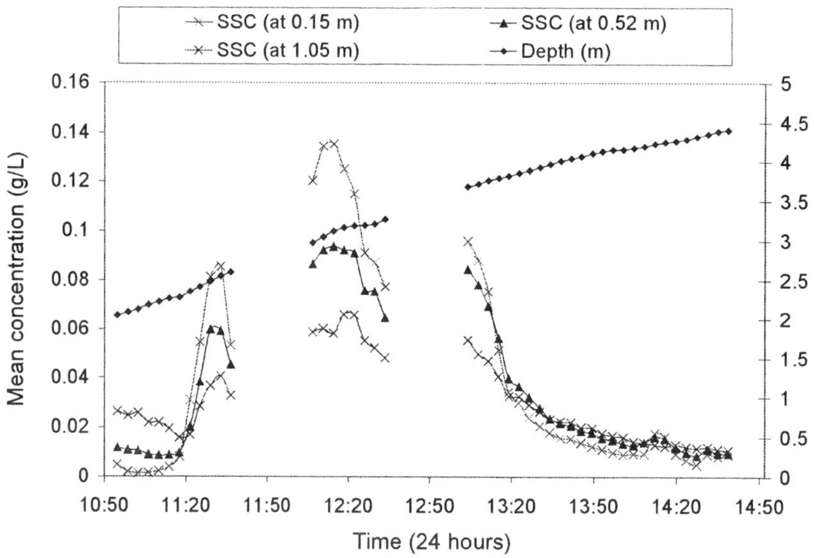

Figure 5b. Suspended sediment concentrations measured during the flood tide on 16th September at Station A.

concentration also started about 11:20hr, and the peak concentration was reached at about 12:45hr, more or less coincident with the peak surface current and well before the appearance of salt water.

Thus, during neap tides the turbidity maximum appears to be upstream of the salt intrusion at both stations during the ebb tide. At Station A on the flood tide the TM is right at the tip and partly inside the tip of the salt intrusion. At Station B, however, the peak of the turbidity maximum appears to be about the same distance upstream of the saline

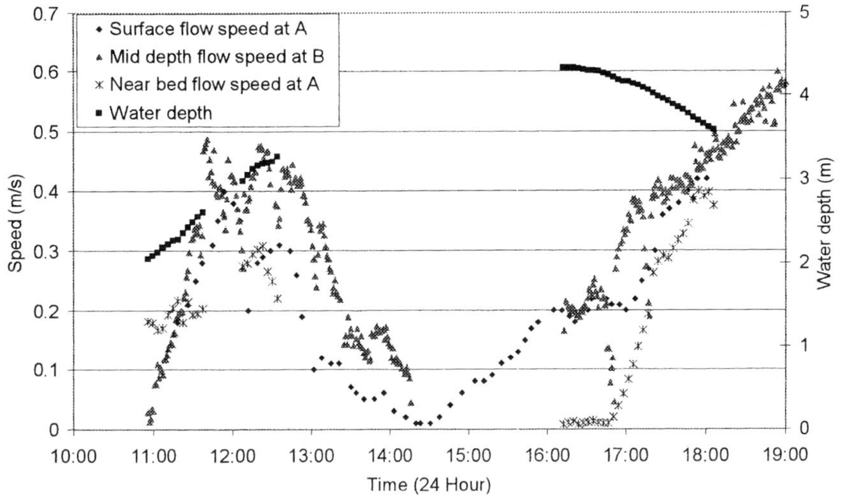

Figure 6. The current speeds measured at Station A and Station B on 16th September 1998.

intrusion on the flood tide as on the ebb. Consequently, entrainment appears to be more important than advection of the turbidity on the flood tide. The location of the turbidity maximum appears to be associated with the peak in the currents that occurs upstream of the salt intrusion. Between there and the tip of the salt intrusion there is a reduction in current speed, but it increases again as the intrusion arrives. The entrainment of sediment from the bed is associated with the velocity peak occurring landwards of the salt intrusion. The presence of even small amounts of salt appear to limit the amount of sediment erosion, and enhances the reduction in concentration.

Spring Tides

The ebb and flood salinity variations for a spring tide on the 22nd September at Station A is shown in Figure 7. The last traces of salinity disappeared on the ebb soon after 10:00hr, and the main saline intrusion reappeared on the flood at about 18:00 hr. At Station B the water was fresh throughout the ebb (09:30-12:50hr) and salinity only rose above a value of 0.5 after 17:45hr (about 30min after the same conditions at Station A). The second velocity peak during the morning ebb at Station A (Figure 8) reached 0.9 ms^{-1} at about

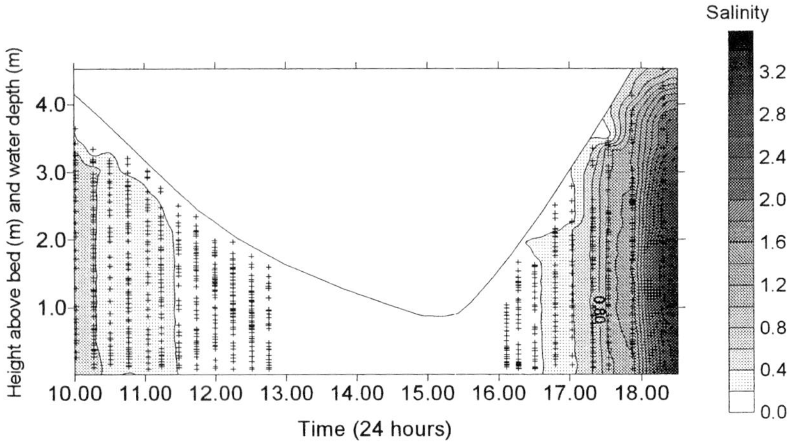

Figure 7. Contours of salinity measured at the end of the ebb and the early flood tide during a spring tide on 22nd September at Station A. The crosses show the times and heights of the measurements.

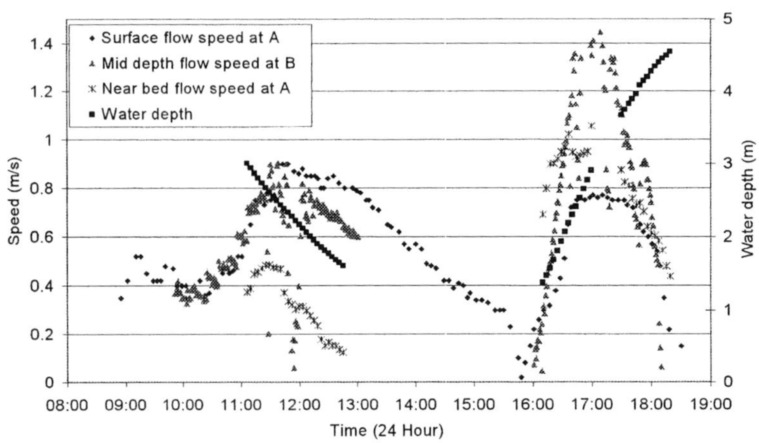

Figure 8. Current speeds measured at Station A and Station B on the 22nd September 1998.

11:45hr, and the duration of the velocity decrease to slack water was longer than on neap tides. The first peak at 09:20hr was considerably less strong than at neaps. Again the start of the second peak coincided with the final withdrawal of the saline water.

The flood phase of spring tides was relatively compressed, lasting only about 3hr, but with higher velocities than the ebb. The currents at Station B were almost double those at A, due to the changing width, or shadowing of the near surface current meter by the platform. There was little sign of the double velocity peak on the flood, except briefly near the bed at about 16:40hr.

Concentration at Station A increased on the ebb tide between 10:35-11:05hr (Fig 9), when the bed shear stress exceeded about 0.25-0.3Nm^{-2}. Again this was associated with the velocity peak when salt was not present. The first peak in concentration at 11:30hr coincided with the peak in near bed currents. Thereafter there was a reduction at about 11:40hr by ~1.5gl^{-1}, followed by a rise to over 4gl^{-1} that lasted for over an hour, despite the currents falling steadily over that period. This implies that the continued high concentrations are the result of significant advection.

On the flood tide, the concentrations rose simultaneously at both stations at 16:16hr and decreased at about 17:15hr when the current fell below about 0.7ms^{-1} (Fig 8). In this instance erosion determined the onset of the turbidity maximum, followed by advection, but once the current decreased, settling occurred rapidly.

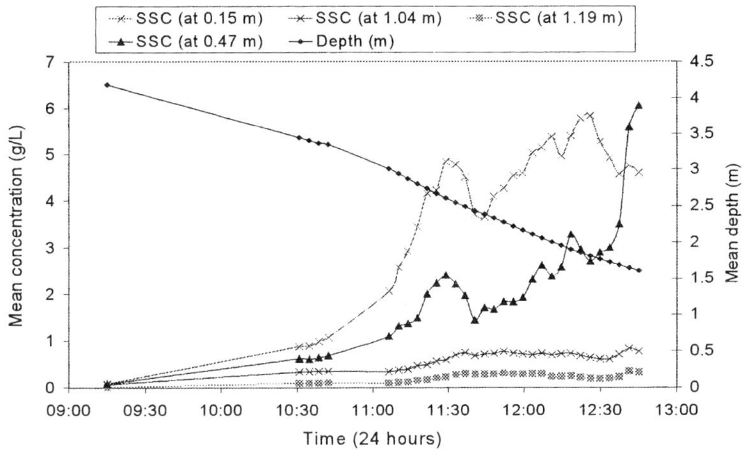

Figure 9. Suspended sediment concentration measured at four heights on the POST rig during the ebb on 22nd September 1998, together with the water depth.

3.2 Water surface slopes

The surface water slopes were measured at spring and neap tides at intervals during each of the runs, in order to examine the balance between the various terms in the equation of motion. It is apparent that water slope and friction did not always balance, and at times there were significant contributions from inertia and the density field. It is estimated that the error in slopes is +/- 0.5 x 10^{-5}. A positive slope occurs when the level at Station B exceeds that at Station A, and the surface slope is down-estuary.

On the flood tide during the neaps there was a steady rise in water elevation of about 2mhr^{-1}. Initially the water slope was landwards, driving the incoming flood tide.

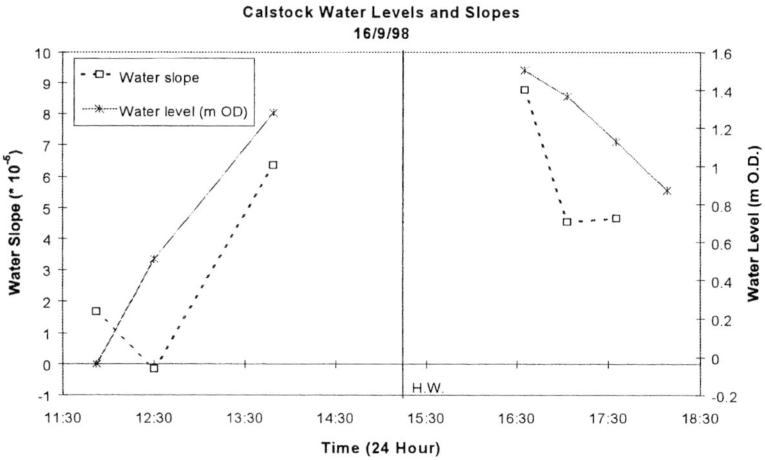

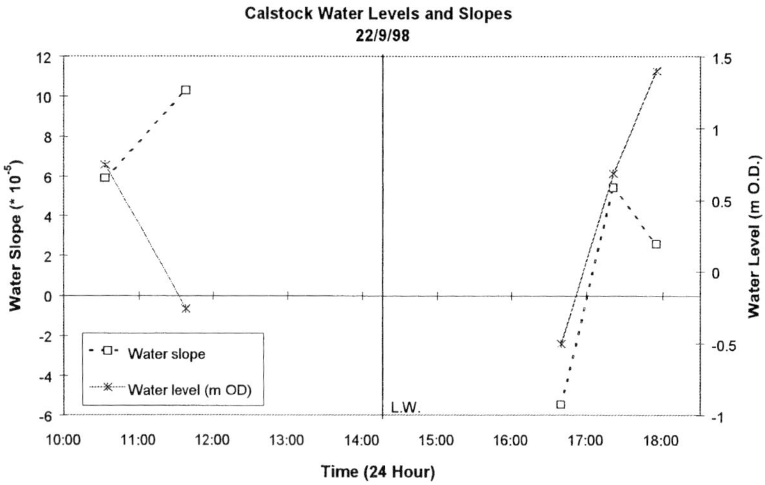

Figure 10. Water slope between the lower and upper measuring positions on a neap tide (16th September) and a spring tide (22nd September, and the water level at the lower position. Water levels are shown relative to Ordnance Datum. Slopes are positive when the water slopes downwards towards the sea.

However, after about 2hr of the flood, the slopes reversed, with the elevation at Station B exceeding that at Station A. This super-elevation of the water surface must have been caused by the 26% reduction in cross sectional area between A and B. The effect, of course, does not occur on the ebb tide, and the slope was consequently positive throughout this phase of the tide.

During the incoming flood tide on the 16th Sept the water surface sloped downwards towards the sea after about 11:30 (Fig 10). There was an abrupt decrease in slope so that at about 12:30hr the water was virtually horizontal. This was quickly followed by a return to a strong down estuary slope. This wave-like reversal occurred between the transit of the TM and the salt intrusion, and coincided with the reduced velocities between the two peaks in the current. Similarly on the ebb tide, a change in the gradient occurred at about 17:00hr, again coincident with the interval between withdrawal of the salt, and the appearance of the TM.

On the spring tides, on the ebb, the slope changed from negative to positive at about 17:00hr on the 22nd Sept, went through a maximum soon after 17:00hr and reduced at 18:00hr. The changes again appeared to be connected to the separation zone between the turbidity maximum and the salt intrusion.

The overall slopes agree with those found by Oduyemi (1986), though his stations were 8km apart. He found that the slopes were towards the sea for much of the tide, accept at the very beginning of the flooding tide. However, the closer station spacing here reveals that penetration of the salt intrusion up the estuary on the flood tide is preceded by a bulge in the water surface, that temporarily reduces the slope. The presence of the bulge produces a decrease in the current speeds. The reverse effect occurs on the ebb tide with recession of the salt intrusion. The analogy is thus like a ‘bow-wave’ occurring up-estuary of the salt intrusion. The wavelength of the feature must be of the order of 1km, and its presence would not be seen in water levels or slopes at widely spaced stations. The wave appears to control the occurrence of two peaks in the current, with the minimum between them separating the salt intrusion from the turbidity maximum. Since the timing of the appearance of the saline intrusion varies with position along the estuary, and with river flow, the reduction in velocity could appear with very variable phase relative to the maximum tidal currents. Thus the ‘bow wave’ need not always control the generation of the turbidity maximum. However, it should have appreciable influence on the timing and intensity of the suspended sediment concentrations involved.

3.3 Drag reduction

Measurements of the turbulent velocity fluctuations at about 25 and 75cm above the bed allowed calculation of the turbulence properties of the flow. Because the Reynold’s stress is very sensitive to alignment errors in the EM heads (Heathershaw, 1979), the turbulent shear stresses were calculated from the total turbulent kinetic energy at each level, that can be considered as proportional to the shear stress (Soulsby, 1998). Under normal circumstances within the constant stress near-bed layer, the intensities would be equal. If the constant stress layer was thinner than 75cm, it would be expected that the intensity would decrease with height. However, there were several periods during the passage of the turbidity maximum when the reverse occurred, and the intensities at the upper level were greater. This is the phenomenon of drag reduction, the shear at the boundary being reduced and the shear being concentrated at a higher level in the flow. This has been considered to result from the presence of high concentrations (Best and Leeder, 1993, Li and Gust, 2000).

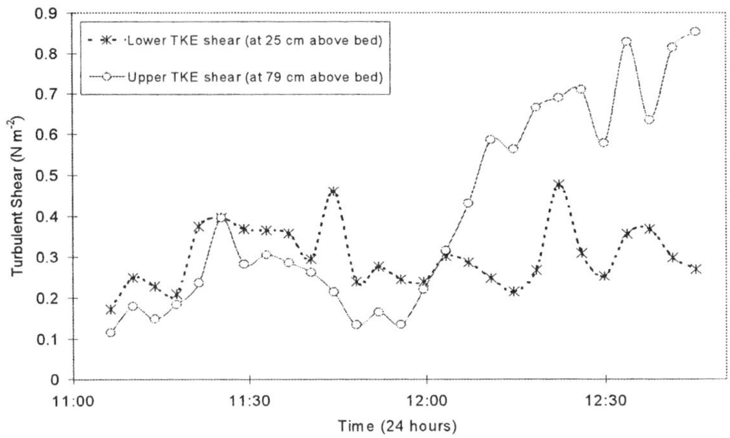

Figure 11. Turbulent shear stresses at two heights on the POST rig on the ebb tide on the 22nd September showing the shear stresses at the upper level exceeding that at the lower level.

An example of the effect is shown by comparison between Figures 9 and 11. In Fig 9 the reduction in concentration at 11:40hr appeared at all levels. It is almost as if the concentration increase was offset and lowered by about $1.5 gl^{-1}$ for about an hour. This reduction coincided with the period when the turbulent energy at the upper level increased whilst that at the lower level remained roughly constant. As a consequence the energy at the upper level exceeded that near the bed by a factor of up to 4. There appears to be a slight delay in the effect, as the decrease in concentration occurred when the intensity at the upper sensor started to increase relative to the lower level, rather than exceeding it (Fig. 11). During the period of drag reduction the gradient Richardson Numbers were less than the critical value of 0.25. However, they were greater than the value of 0.03 that limits neutral flow stability. The Ri values tend to follow the varying concentrations denoting that the stability is affected by the density variations, but intensity of the drag reduction suggests that it must be mainly due to an inherent dissipation of energy within the suspension by collisions, floc break-up, and changes in floc rotation.

A plot of the difference in turbulent energies at the two levels against concentration did not show clear relationships, but one against the vertical concentration gradient (Fig 12) is significant. This implies that when the concentration gradient exceeds a value of ~ $4 kgm^{-4}$, drag reduction commences. The shear stress does not then monotonically decrease from the bed to the water surface, but has a maximum within the flow. This has important implications as it is a feed-back mechanism that reduces the amount of potential erosion from the bed, as well as the transporting capacity of the flow.

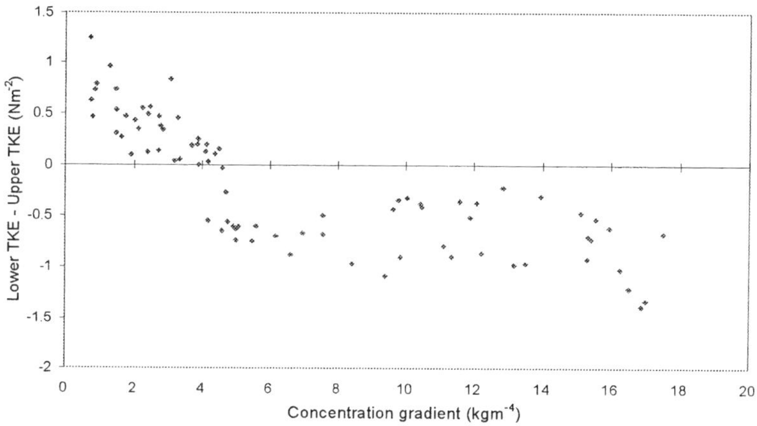

Figure 12. Plot of the turbulent kinetic energy at two levels on the POST rig for all periods of measurement, neap and spring tides, against the concentration gradient between the two levels.

4. CONCLUSIONS

The experiment showed that the turbidity maximum for average river flows and for one distinctive reach of the Tamar estuary had a length of about 1km and was created largely by suspension from the bed during parts of the tide when the bed shear stress exceeded about 0.17-0.2 Nm^{-2}. The current on both flood and ebb tides had two pulses separated by a slight reduction in velocity which came between the passage of the turbidity maximum past the measurement points and the appearance of the saline intrusion. This current minimum coincided with a reversal in the surface water slopes caused by a 'bow-wave' effect upstream of the intrusion. At spring tides the turbidity maximum lengthened and tended to overlap with the salinity intrusion on the flood tide. The turbulence measurements showed significant drag reduction at suspended sediment concentration gradients above about $4gm^{-4}$. The current, salinity and concentration profiles, together with the water slopes and the quantification of the turbulent stresses should allow a comprehensive model of the system to be validated.

Acknowledgements

The authors would like to thank all of the many people who took part in the organisation of the field experiment and in carrying the measurements, particularly the boat operators and the instrumentation support team. The work was funded by the EC MAST programme as part of contract MAS3-CT97-0082 COSINUS.

REFERENCES

Best, J.L. and Leeder, M.R., 1993, Drag reduction in turbulent muddy seawater flows and some sedimentary consequences. *Sedimentology*. (40), 1129-1137.

Christie, M.C., Dyer, K.R., Turner, P. and Manning, A.J. 2002, The effects of density gradients upon water column turbulence within an estuarine turbidity maximum. (in preparation)

Christie, M.C., Quartley, C.P. and Dyer, K.R., 1997, The development of the POST system for in-situ intertidal measurements. *In 7th Conf. Electronic Engineering in Oceanography. IEE Conf. Publ* 439. (734), 39-45

Dyer, K.R., Cornelisse, J., Dearnaley, M., Fennessy, M.J., McCave, I.N., Puls, W., van Leussen, W. & Wolfstein, K., 1996, A comparison of in-situ techniques for estuarine floc settling velocity measurement. *Jour. Sea Research* (36), 15-29.

Dyer, K.R., Bale, A.J., Christie, M.C., Feates, N., Jones, S. and Manning, A.J. 2002. The turbidity maximum in a mesotidal estuary, the Tamar Estuary, UK: II The floc properties. *INTERCOH-2000, Elsevier, Coastal and Estuarine Fine Sediment Processes,* ed. J.C. Winterwerp and C. Kranenburg, this volume.

Fennessy, M.J., Dyer, K.R. and Huntley, D.A., (1994a), INNSEV: an instrument to measure the size and settling velocities of flocs in–situ. *Marine Geology,* (117), 107-117.

Fennessy, M.J., Dyer, K.R. and Huntley, D.A., (1994b), Size and settling velocity distributions of flocs in the Tamar Estaury during a tidal cycle. *Neth. Jour. Aquat. Ecology,* (28*),* 275-282.

Heathershaw, A.D., (1979), The turbulent structure of the bottom boundary layer in a tidal current. *Geophys. Jour. R. Astronomical Soc.* (58), 395-430.

Li, M.Z. and Gust, G., (2000), Boundary layer dynamics and drag reduction in flows of high cohesive sediment suspensions. *Sedimentology*. (47), 71-86.

Manning, A.J., (2001),.A study of the effect of turbulence on the properties of flocculated mud. *PhD Thesis, University of Plymouth.*282pp.

McCabe, J.C., Dyer, K.R., Huntley, D.A. and Bale, A.J., (1993), The variation of floc sizes within a turbidity maximum at spring and neap tides. *Coastal Engineering 1992. Proc. 23rd Int. Coastal Engineering Conf. Venice,* 3178-3188.

Oduyemi, K.O.K., (1986), Turbulent transport of sediments in estuaries. *PhD Thesis. University of Birmingham.* 124pp.

Peterson, O., Vested, H.J., Manning, A., Christie, M.C. and Dyer, K.R., (2002). Numerical modelling of mud transport in the Tamar Estuary. *INTERCOH-2000, Elsevier, Coastal and Estuarine Fine Sediment Processes,* ed. J.C. Winterwerp and C. Kranenburg, this volume.

Soulsby, R.L., (1998). Dynamics of Marine Sands: a manual for practical applications. *Thomas Telford, London.* 250pp.

Uncles, R.J. and Stephens, J.A., (1993). Nature of the turbidity maximum in the Tamar Estuary, UK. *Estuarine Coastal Shelf Science.* (36), 413-431.

Fine Sediment Dynamics in the Marine Environment
J.C. Winterwerp and C. Kranenburg (Editors)

The Turbidity Maximum in a Mesotidal Estuary, the Tamar Estuary, UK: II. The Floc Properties.

K.R.Dyer[a], A.J.Bale[b] , M.C.Christie[a*], N.Feates[c], S.Jones[d], A.J.Manning[a]

[a] Institute Marine Studies, University of Plymouth, Plymouth PL4 8AA, UK.
[b] Plymouth Marine Laboratory, Plymouth PL1 3DH, UK.
[c] HRWallingford, Howbery Park, Wallingford, OX10 8BA, UK.
[d] School Ocean Sciences, University of Wales Bangor, Menai Bridge, Gwynedd LL59 5EY, UK.

Measurements have been carried out in the turbidity maximum of the Tamar Estuary at neap and spring tides of floc size, settling velocity and effective density. These measurements were complemented by suspended sediment concentration and turbulence profiles. The overall dynamics of the turbidity maximum have been presented in the companion paper (Part I), and form a context for the results presented here. The floc size increased from about 50 to about 350 microns across the turbidity maximum, and settling velocity changed from 0.0-6.0mms^{-1}. The equivalent change in floc effective density was 50-750kgm^{-3}. Within the turbidity maximum the proportion of macroflocs (>160microns) increased. The changes in floc properties were correlated with changes in concentration and the turbulence parameter G, the root mean square of the gradient in turbulent velocity fluctuations. An empirical equation representing the settling velocity Ws (mms^{-1}) in terms of SPM (mgl^{-1}) and G (s^{-1}) is:

$$Ws = -0.243 + 0.000567SPM + 0.981G - 0.0934G^2$$

Keywords: Flocculation, settling velocity, floc size, turbulence, suspended sediment concentration, turbidity maximum, Tamar Estuary.

1. INTRODUCTION

Within the turbidity maximum (TM) of an estuary there is a considerable increase in the suspended sediment concentration above background levels, and this may reach values of several gl^{-1}. The size of the flocs rises with concentration, and it can therefore be anticipated that the sizes of the flocs will increase. Even though the larger flocs are more porous, and have a lower effective density, these macroflocs nevertheless have a greatly enhanced settling velocity, therefore significantly contributing to the mass settling flux. This flux is a crucial factor in sediment accumulation and morphological change, and predicting it is the aim of many

---

* Now at: South Devon NHS Trust, Torquay, UK.

mathematical models. The three floc properties; size, settling velocity and effective density, are all variables, and it is seldom that more than one is measured at any one time. Size and settling velocity can be related through Stokes' Law, providing an assumption is made of density. There is a great deal of laboratory and field work that has been devoted to producing algorithms for the relationship between size and settling velocity as a function of concentration. However, macroflocs may also be disrupted by turbulent shearing, breaking up into microflocs, especially within the lower layers of the flow where the shearing is greatest. There are many laboratory-based studies on the effects of shearing on floc size, but few produce a quantification of the combined effects of shearing and concentration. Dyer (1989) proposed a conceptual model of the relationship, which included flocculation at low shear and low concentration, and disruption at higher shears. Lick et al (1993) found that the mean size varied as a function of a power of the product of the concentration and the shear. Additionally there are a number of useful theoretical approaches (Van Leussen, 1994; Winterwerp, 1996; Hill, 1996), but they would benefit from simultaneous in situ measurements of size and settling velocity, from which estimates could be made of the variation in effective density of the flocs. As flocs are multi-component, being composed of grains of various minerals as well as organic material, the effective density is not a simple function of size.

The distribution of floc size therefore changes with concentration and with shear, with the large macroflocs being formed from smaller microflocs, which are normally considered sufficiently dense and strong to withstand the ambient disruptive shearing. The boundary between the two types of flocs is at a size of about 125 microns (Eisma, 1986), or 160 microns (Manning 2001).

Additionally, in the turbidity maximum there may be contributions to the suspension of macro or microflocs from erosion of bed material. Also, because the timescales involved in the floc aggregation process and in break-up may become similar to those of change in the driving forces, a non-steady floc distribution may develop. Consequently there is need to quantify the variation of floc size and settling velocity in terms of the concentration and turbulence changes throughout the turbidity maximum.

The MAST III funded project COSINUS (Prediction of cohesive sediment transport and bed dynamics in estuaries and coastal zones with integrated numerical simulation models) aimed to study, by laboratory and field experiments, the characteristics and flow of high concentration bed layers. As part of the project a two week long experiment was carried out within the turbidity maximum (TM) of the Tamar Estuary, UK. The aims were to measure the effect of velocity shear and turbulence, salinity, and concentration on the size, settling velocity and effective density of flocs, and the relationship to the concentration and velocity profiles. Of particular interest were the floc characteristics in high concentration near bed layers. Within COSINUS the relationships were to be tested against theory and the results of laboratory experiment, and in order to put the results in the context of modelling fine sediment transport, also within a model of the Tamar Estuary (Peterson et al, 2002).

The overall concepts of the experiment, the scope of the data and what it revealed of the dynamics of the turbidity maximum, the erosion, movement and deposition of fine sediment, are reported in Part I of this paper (Dyer et al 2002). The main conclusion from that analysis was that the turbidity maximum was present in fresh water about 1km upstream of the salt water intrusion within which salinities did not exceed 10 at high water, and were either zero, or considerably lower than 10 when the main measurements were obtained. Consequently, the relationship of floc properties with salinity cannot be fully explored. The TM was separated from the saline intrusion by a zone in which there was a slight reduction in the currents. Maximum concentrations reached about $8gml^{-1}$, so that a wide range of concentrations were sampled. The turbidity maximum appeared to be formed of sediment that was largely eroded and re-deposited on the bed during each tide, though there was a contribution from advection particularly at spring tides. Additionally, drag reduction occurred when concentration gradients exceeded about $5kgm^{-4}$.

This paper reports the mean size and mean settling velocities derived from measurements of distributions of individual flocs within the TM on both neap and spring tides, together with salient details of the floc effective densities and size distributions. In particular, empirical relationships are investigated between settling velocity of flocs with suspended sediment concentration and turbulent shear. Unfortunately, because the TM was almost entirely in fresh water, the influence of salinity awaits additional measurements. Further detail of the results of the turbulence measurements, and floc properties are reported in Christie et al (2002), and Manning (2001), and of modelling in Peterson et al (2002). The data is accessible as part of the COSINUS database at http://www.bwk.kuleuven.ac.be/bwk/cosinus/cosinus.html.

2. METHODOLOGY

A straight reach of the upper Tamar estuary, within the tidal trajectory of the turbidity maximum was chosen for the experiment. The details of the location are shown in Dyer et al (2002). In order to assist in understanding the relative importance of local and advected effects, two stations 970m apart were occupied with fore-and-aft anchored vessels in the central channel. Station A was approximately 400m below the Calstock Railway Bridge, and station B about 500m upstream of it. The estuary widths and maximum depths at high water were 75m and 4.5m respectively at the lower section and 50m and 5.2m at the upper. The tidal range varied from about 3.2m to about 4.5m. In this situation it was expected that the suspended sediment dynamics would be dominated by the bed boundary conditions.

Measurements were obtained for sections of the tide covering the transit of the TM. Eight sampling runs, of approximately three hours duration were completed during the two-week experimental period 14-24th September 1998, which covered spring and neap tides. River discharge was around the annual mean of about 20 m^3s^{-1}.

At each station profiles of salinity, temperature, and suspended solids concentration were taken at least every 15min by hand profiling Seabird systems CTD and Downing optical backscatter gauges. Hourly water samples were obtained for calibration. Current velocities were measured by direct reading current meters suspended at a constant depth below the vessels; near surface at Station A and mid-depth at Station B. Also intercalibration was attempted by use of all instruments simultaneously at station A on one day of neap tides and another of springs.

The main instrument for measurement of floc characteristics was the INSSEV system (Fennessy et al, 1994a), which comprises a sample deceleration chamber from which flocs were allowed to settle into a settling column. The flocs settling in still water were observed and recorded by a miniature video system. This allowed measurement of floc size, and settling velocity distributions at 20-30 min intervals. The instrument sampled at a height of 0.5m above the bed, and was located at Station A. The video system had a lower limit of about 20 microns, and could measure floc sizes up to about 1mm. From the measured diameter and settling velocity the effective density could be estimated from Stokes' Law. Additionally, a Partech Lasentec P100 laser sizing system (Law and Bale, 1998) was used to obtain vertical profiles of floc size distribution. This instrument senses particles in the size range 2-1000 microns in 38 approximately logarithmically spaced size bands. Profiles were obtained every 30min at 0.5m intervals from the surface to within 0.5m of the bed, and between times the instrument measured continuously at a variable fixed depth.

At the upper station B, a Sequioa Instruments LISST 100 laser diffraction system was used to determine the floc sizes (Law and Bale, 1998). This is not capable of measuring floc sizes greater than about 250 microns, and saturated in concentrations greater than about 500 mgl^{-1}. UWB-QUISSET settling tubes (Jones and Jago, 1996) and the HR Floc Camera (Dearnaley, 1991) were used to quantify floc size and settling velocity distributions from samples taken at about mid depth. The former is a 1m long settling tube, operated vertically, with samples withdrawn from the base at intervals up to 180 minutes. This was capable resolving settling velocities of less than 0.001 mms^{-1}. The Floc camera system measures the size and settling velocity of floc within a settling tube with a resolution of between 7-17 microns. Typically 5-10 pairs of images were observed during the first 3 min of settling.

Estimates of the turbulence parameters were obtained from the POST system which was mounted on the same bedframe as the INSSEV system (Christie et al, 1997). This comprised an array of 4 EM flowmeters and 8 optical backscatter sensors (OBS). The EM flowmeters were arranged in pairs to obtain the three orthogonal components of the flows at 0.25 and 0.75m above the bed, with the INSSEV sampling at 0.50m. The OBS's were arranged approximately logarithmically through the bottom 1m of the flow. The data from the POST system was logged at 18Hz and filtered at 5Hz with files of 3 min 47 seconds length. From these measurements the velocity shear, the turbulence characteristics and the mean and fluctuating concentration could be interpolated at the INSSEV sampling height, and integrated with the timing of the INSSEV sampling.

The close proximity of the INSSEV system and the turbulence and concentration measurements from POST allows direct comparison between the results. The values of concentration and of turbulent shear are interpolated linearly at the INSSEV measuring height. The results are the subject of this paper. The profiles of floc measurements, and the spatial variations in properties between stations will be reported elsewhere (Jones et al, in prep).

There are many ways in which the turbulence can be characterised. The value of G is considered to be a major parameter in models of floc size (eg. Van Leussen, 1994, Lick et al, 1993, Malcherek, 1995), where G is the root mean square of the gradient in turbulent velocity fluctuations. This can be defined as:

$$G = (U_*^3/(\kappa \nu z))^{0.5} \qquad 1.$$

where U_* is the friction velocity, ν is the kinematic viscosity, κ the von Karman constant, and z is the height above the bed.

3. RESULTS

3.1 Neap tides: Ebb

Figure 1a shows the distribution of mean mass weighted floc size, mean mass weighted settling velocity, concentration, and G. The mass of each floc was calculated from the size and the effective density was calculated from Stokes' Law. The effective density is the density of the floc minus the density of water. The mean mass weighted size was determined as the mean size of the integral of the mass times the size of the individual flocs. Similarly, the mean mass weighted settling velocity was the mean settling velocity of the integral of the settling velocity times the mass of the individual flocs.

The concentration started to rise when G exceeded about 3 s^{-1} at about 16:45hr. From a background concentration of about 40mgl^{-1} it rose to reach a maximum of about 280mgl^{-1} within the turbidity maximum. Over the same time the mean size of the flocs also rose from about 150 to almost 300 microns. However, in the clearer water before the onset of the TM, the small size flocs had high settling velocities of almost 9mms^{-1}, but these reduced to 1-2mms^{-1} in the TM. However, the settling velocity was lowest at the beginning of the TM when shear and concentration were relatively low, but increased as shear and concentration increased.

There is also a contrast in the percentage contribution of the macro and microflocs to the mass concentration. The critical size between macro and microflocs was determined from the complete set of results to average 160 microns (Manning, 2001). Figure 1b shows the effective density of all flocs within the macrofloc and microfloc size bands. The first sample shows a higher percentage of mass in the microfloc sizes. However, subsequent samples show increasing mass in the macroflocs and decreasing mass in the microflocs with increasing G values. The proportions of both fractions approach 50%, and then cross over at a G value of about 3s^{-1}. At peak concentrations and G, the proportional contribution of the macroflocs to the mass concentration rose to over 70%. This indicates either that a fraction of the microflocs are involved in flocculation to form the macroflocs, or that macroflocs were preferentially eroded from the bed to enhance the concentration.

The combined increase in size and decrease in settling velocity across the TM implies a drastic decrease in effective density. Figure 1c shows that the microflocs are consistently denser than the macroflocs. Effective densities fell rapidly from about

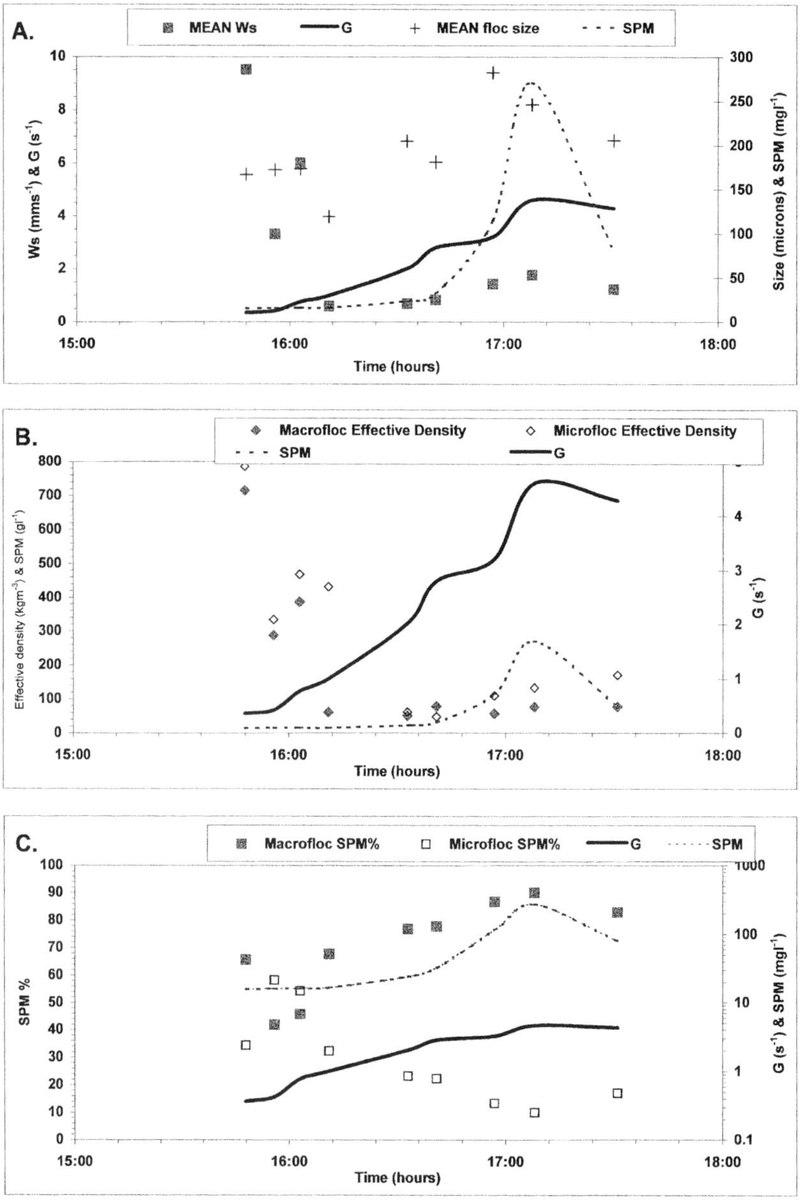

Figure 1. Time series showing the variations of the turbulence parameter G, and suspended particulate matter concentration in the turbidity maximum during the ebb on a neap tide on the 15th September 1998 at Station A. ***A***) includes mean settling velocity and mean floc size data; ***B***) includes macrofloc and microfloc effective density data; and ***C***) includes the division of macrofloc and microfloc suspended particulate matter concentration.

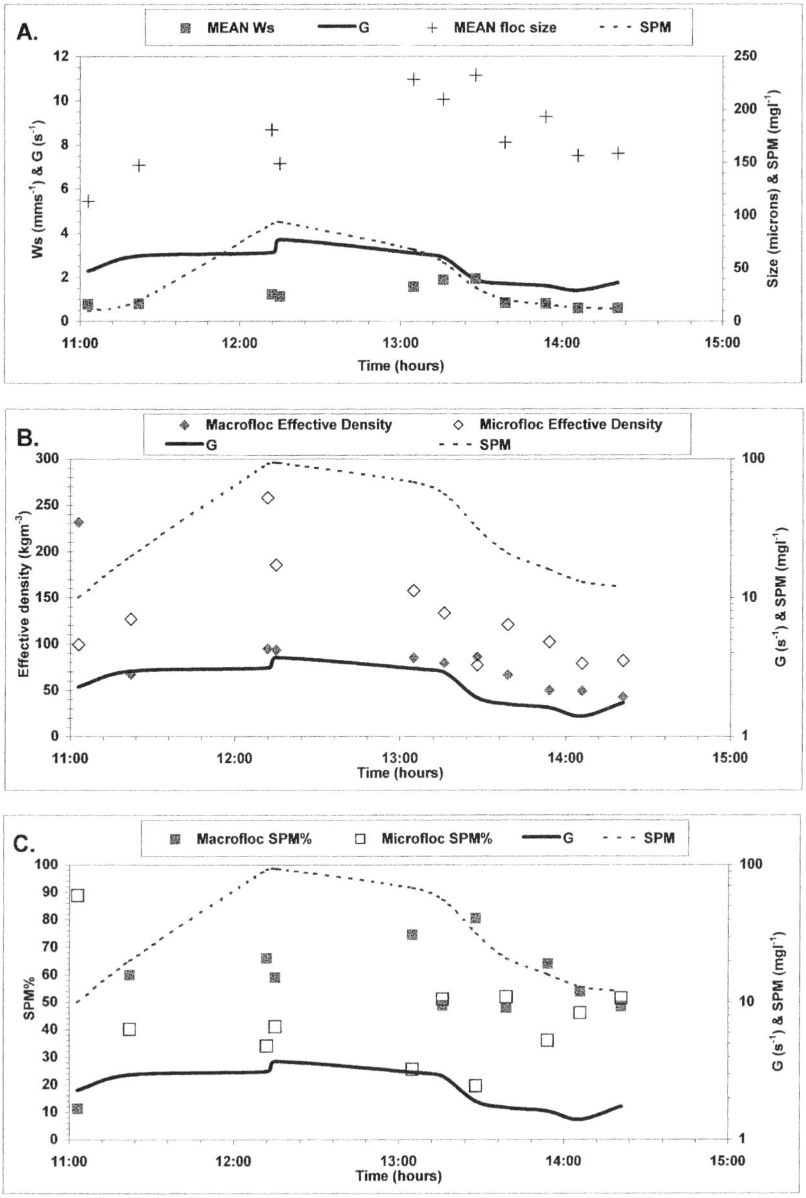

Figure 2. Time series showing the variations of the turbulence parameter G, and suspended particulate matter concentration in the turbidity maximum during the flood on a neap tide on the 16th September 1998 at Station A. ***A***) includes mean settling velocity and mean floc size data; ***B***) includes macrofloc and microfloc effective density data; and ***C***) includes the division of macrofloc and microfloc suspended particulate matter concentration.

750kgm^{-3} with increasing G, but became almost constant at around 100kgm^{-3} within the TM. This is compatible with a rise in porosity of the flocs from 70% to over 90%.

3.2 Neap tides: flood

On the flood tide the TM reached less than half the concentration of that on the ebb, reaching less than 100mgl^{-1}, but lasted much longer (Figure 2a). Again the concentration started to rise when G reached about 3s^{-1} at about 11:10hr. The maximum mean floc size of about 230 microns occurred towards the end of the TM. Also, the settling velocity within the TM underwent the same enhancement after the suspended sediment concentration rise as on the ebb, with low values at low shear and concentrations, but reaching a peak value of about 2mms^{-1} towards the end of the TM. The effective densities were between 50-150kgm^{-3} within the TM, about the same as on the ebb tide. However, the maximum value of effective density occurred at the time of the maximum concentration and maximum G, before the maximum in size and settling velocity. This suggests that additional contributions to the suspended floc population may have occurred later in the TM as a consequence of either advection or resuspension.

The effective density of the microflocs was again higher than that of the macroflocs, reaching 250kgm^{-3} (Figure 2b).The exception again was the first sample which had higher density macroflocs. Additionally, the macroflocs had a greater spread of densities than the microflocs.

At the beginning of the record macroflocs formed only 10% of the mass, and microflocs dominated the distribution (Figure 2c). However, the proportions crossed over at the beginning of the TM, as they did on the ebb. The macrofloc proportion exceeded 60% for most of the TM, but towards the end the proportions were about equal for some of the samples, though macroflocs exceeded 70% for the rest. This contrasts significantly with the situation during the ebb TM when the largest proportion of macroflocs coincided with the highest concentration and shear.

3.3 Spring tide: ebb

At spring tides the overall concentration in the TM is considerably enhanced (Figure 3a). The value of G was above about 5s^{-1} until about 11.75hr when it rose steadily until about 12.25hr. The concentration also rose steadily and reached 3gl^{-1} at the peak G. This suggests that erosion of the bed was a slow process, and there was not rapid erosion of a bed layer of 'fluid mud'. The mean floc diameter increased from about 100 microns to 200 microns and the settling velocity increased more or less in line with size to about 2mms^{-1}. The effective density reduced within the TM from 330 to about 150kgm^{-3}.

There was a large contrast in the distribution of floc sizes, with no macroflocs being sampled before the beginning of the TM, but the proportion within the TM exceeded 70%. The contrast in effective densities, size distributions and concentrations imply that not all of the increase in macroflocs could have occurred by flocculation from microflocs, since there was insufficient in those sizes available.

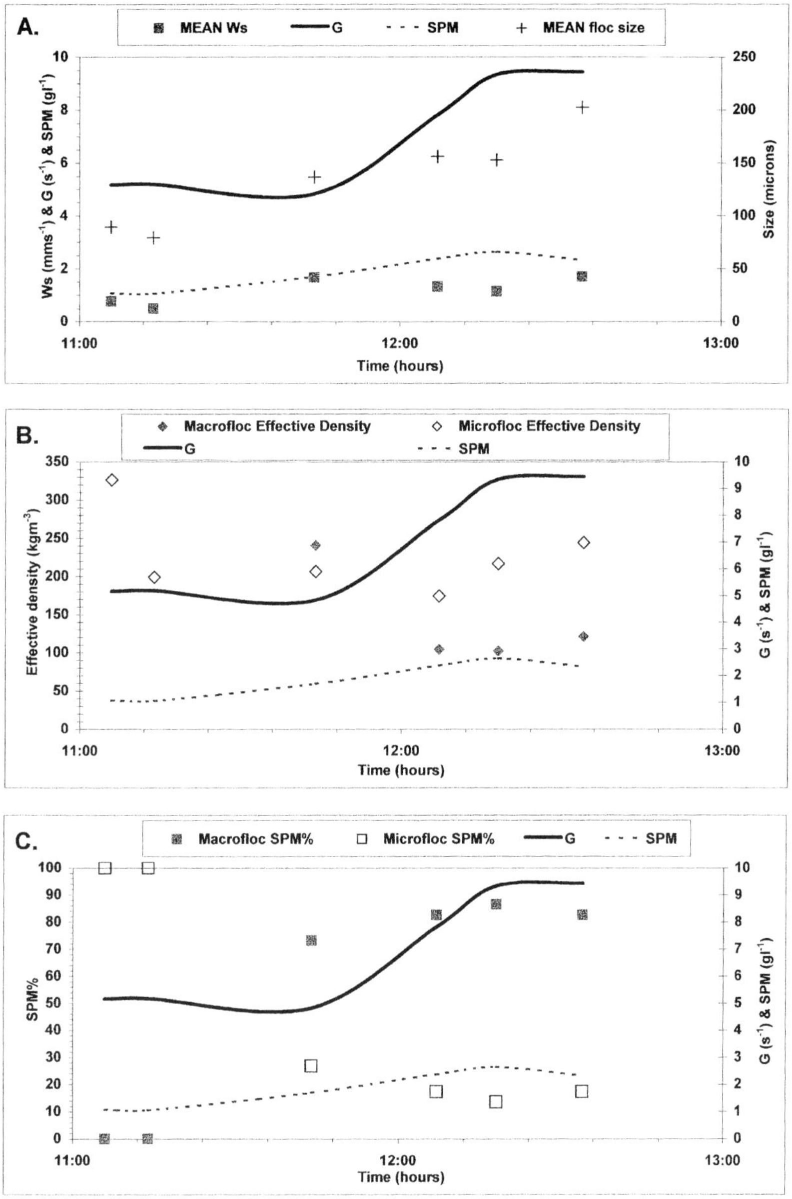

Figure 3. Time series showing the variations of the turbulence parameter G, and suspended particulate matter concentration in the turbidity maximum during the ebb on a spring tide on the 22nd September 1998 at Station A. ***A***) includes mean settling velocity and mean floc size data; ***B***) includes macrofloc and microfloc effective density data; and ***C***) includes the division of macrofloc and microfloc suspended particulate matter concentration.

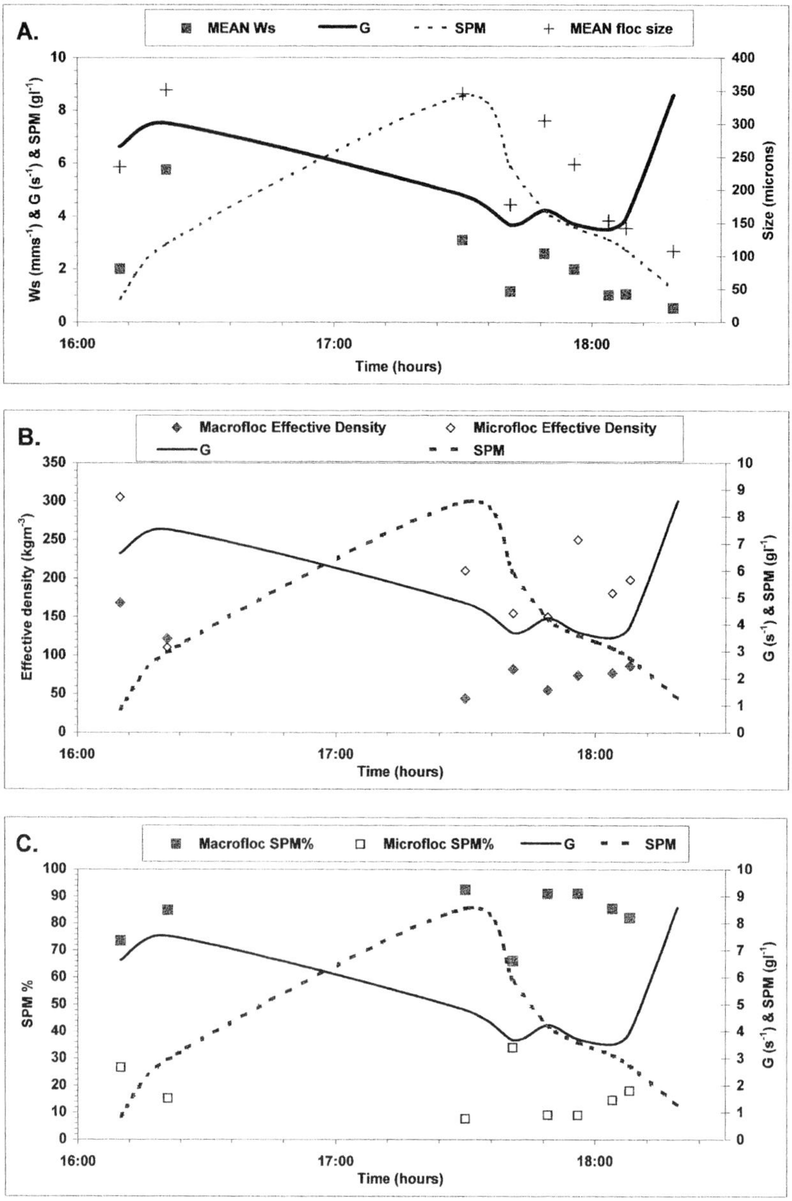

Figure 4. Time series showing the variations of the turbulence parameter G, and suspended particulate matter concentration in the turbidity maximum during the flood on a spring tide on the 22nd September 1998 at Station A. ***A***) includes mean settling velocity and mean floc size data; ***B***) includes macrofloc and microfloc effective density data; and ***C***) includes the division of macrofloc and microfloc suspended particulate matter concentration.

There must have been additional flocs created from the sediment eroded from the bed, or by advection.

The effective densities (Figure 3b) of the microflocs were up to 320kgm^{-3} for the first two samples, and there were no macroflocs observed (Figure 3c). For subsequent samples there appeared to be a cross-over of densities at about 12:00hr, with the macroflocs then becoming denser than the microflocs, and exceeding 80% as a proportion of the total SPM mass.

3.4 Spring tide: flood

The concentrations on the flood tide were significantly greater than on the ebb tide, and reached in excess of 8gl^{-1}, and the corresponding mean floc size was in excess of 300microns (Figure 4a). The most distinctive effect was that of the high concentrations on the turbulence (see Dyer et al, 2002 for more details). The turbulence parameter G was drastically affected by drag reduction once concentrations exceeded about 3gml^{-1} and the G values decreased across the TM until the concentration fell below about 4gl^{-1}.at which time the G value increased rapidly as turbulence became more active. At the tail end of the TM, once concentrations started reducing, the mean floc size decreased to about 150microns, with a further reduction to about 100 microns when G increased drastically. The settling velocity reached almost 6mms^{-1}, but reduced rapidly to about 1mms^{-1} towards the end of the TM.

The proportion of macroflocs was 73-85% at the beginning of the TM (Figure 4b), and exceeded 90% for most of the rest of the TM, though there was indication that the proportion might have been decreasing towards the end of the high concentration period.

The effective densities (Figure 4c) showed that the macroflocs were significantly less dense than the microflocs (70 as against 200kgm^{-3}). This indicates that the macroflocs were loose in structure. However, the densities were very similar to those at neap tides in much lower concentrations.

4. DISCUSSION

Since the TM was within the fresh water landwards of the salt intrusion, analysis is restricted to consideration of flocculation in the absence of salt. Other measurements using the Lasentec and the LISST 100 system over the high water period, when the INSSEV system was not operating suggest that there is also an effect of salinity on floc size (Jones et al in prep). However, there is insufficient information taken with INSSEV during times when salt was present to include it in the analysis.

Within the TM there was an increase in the mean size of the flocs which, in general, matched the increasing concentrations of suspended matter. There was also an increase in settling velocity throughout the TM that followed the increases in size and concentration. The change in the settling velocities was generally in the range 0.5-6.0mms^{-1}, rather wide in comparison with the value of 1mms^{-1} often assumed in modelling simulations.

The proportion of flocs with size greater than 160 microns changed significantly across the TM. There appears to be a fairly consistent pattern of changing floc proportions at the beginning of the TM. Just before erosion of sediment from the bed the proportion of microflocs exceeds that of macroflocs. Close to the threshold the proportions change and macroflocs contribute the higher mass to the total SPM. Within the TM the macroflocs proportion rises rapidly until they contribute at least 70% of the floc mass. On spring tides the macrofloc proportion reached in excess of 90% during the TM. At the end of the TM the macrofloc proportion decreases, largely due to the preferential settling of the denser, faster settling macroflocs. At spring tides the increasing concentrations and turbulent shearing enhances the production of macroflocs of low density. This contrasts with the higher proportion of denser microflocs at neap tides when the concentration of suspended sediment was insufficient to cause floc collisions and aggregation.

The changes in mean size and mean settling velocity imply that the mean floc effective density must decrease within the TM. The changes in effective density were in the range 50-300kgm^{-3}, with the densities at neap tides being generally similar to those at spring tides. The changing proportion of macroflocs to microflocs is therefore required to explain the changes in mean size and mean settling velocity between the contrasting tidal states. At spring tides there is a high proportion of macroflocs to microflocs, and they are of lower density. On neap tides the proportion of macroflocs is lower (70% as against >90%). The mean effective density consequently is higher on neap tides than on spring tides.

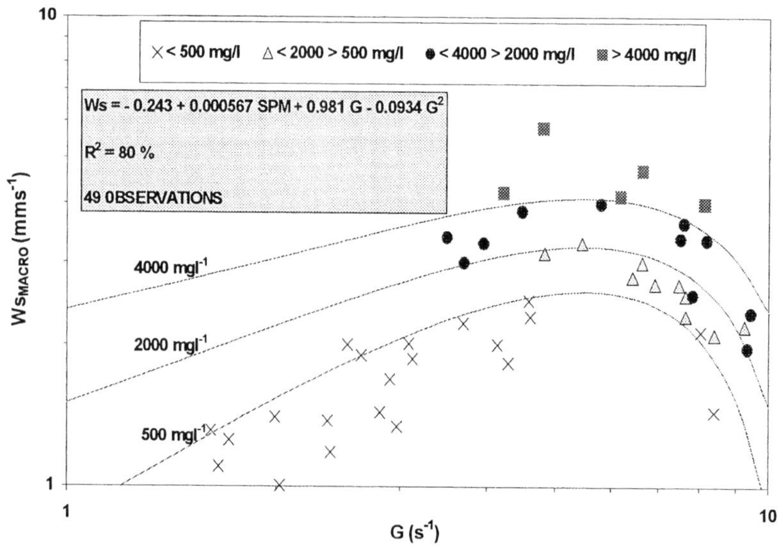

Figure 5. A plot of the mean mass weighted settling velocity for flocs over 160 microns in diameter against the turbulence parameter G for different suspended particulate matter concentrations. The best fit curves are of equation 2 for specific concentration values.

At neap tides just after high water, and before the TM appears, both macro and microflocs had very high settling velocities and effective densities, as shown by the first few samples on Figure 1. For the first sample the proportion of the SPM contributed by the macroflocs was also greater than that of the microflocs. This situation has also been reported by Fennessy et al (1994b) who observed needle-like mineral grains settling end on at high velocities. These were interpreted as being of tourmaline, and originating from break-up of macroflocs in the high shear at the salinity interface immediately above the instruments.

With the observations that have been made it is possible to examine the statistical relationships between the variables, and to produce empirical algorithms. Those for settling velocity (Ws) of the macroflocs are of most interest for modelling, since they represent the velocities of the fraction dominating the suspended sediment fluxes to the bed. For the 47 simultaneous observations of settling velocity, concentration and turbulent shear, the best fit relationship shown in Figure 5 is:

$$Ws = -0.243 + 0.000567SPM + 0.981G - 0.0934G^2 \qquad 2.$$

Where SPM is the suspended sediment concentration (mgl^{-1}) and G is the turbulence parameter (s^{-1}). This has an r^2 of 0.80, which is highly significant. The relationship has the same form as that proposed by Dyer (1989), with an increase in settling velocity at low shear stresses due to flocculation enhanced by shear, or by limited residence time of the flocs, and floc disruption at higher stresses for the same concentration. The maximum occurs at about $G = 4s^{-1}$.

Acknowledgements

The authors would like to thank all of the many people who took part in the field experiment for their invaluable help. The work was funded by the EC MAST programme as part of contract MAS3-CT97-0082 COSINUS.

REFERENCES

Christie, M.C., Quartley, C.P. and Dyer, K.R., (1997), The development of the POST system for in-situ intertidal measurements. *In 7th Conf. Electronic Engineering in Oceanography. IEE Conf. Publ 439.* (734), 39-45.

Christie, M.C., Dyer, K.R. and Turner, P., (2002), The effects of density gradients upon water column turbulence within a turbidity maximum. (in preparation).

Dearnaley, M.P., (1991), Flocculation and settling of cohesive sediments. *HRWallingford,* Report No. SR272.

Dyer, K.R., (1989), Sediment processes in estuaries: future research requirements. *J. Geophysical Research,* (94), 14327-14339.

Dyer, K.R., Bale A.J., Christie, M.C., Feates, N., Jones, S. and Manning, A,J., 2002. The turbidity maximum in a mesotidal estuary, the Tamar Estuary, UK: I Dynamics of suspended sediment. *INTERCOH-2000, Elsevier, Coastal and Estuarine Fine Sediment Processes,* ed, J.C. Winterwerp and C. Kranenburg, this volume.

Eisma, D., (1986), Flocculation and de-flocculation of suspended matter in estuaries. *Neth. J. Sea Research,* (20), 183-199.

Fennessy, M.J., Dyer, K.R. and Huntley, D.A., (1994a), INSSEV: an instrument to measure the size and settling velocity of flocs in-situ. *Marine Geology,* (117), 107--117.

Fennessy, M.J., Dyer, K.R. and Huntley, D.A., (1994b), Size and settling velocity distributions of flocs in the Tamar Estuary during a tidal cycle. *Neth. Jour. Aquatic Ecology*, (28), 275-282.

Hill, P.S., (1996). Sectional and discrete representations of floc breakage in agitated suspensions. *Deep-Sea Research*, (43), 679-702.

Jones, S.E. and Jago, C.F., (1996), Determination of settling velocity in the Elbe Estuary using UWB-QUISSET tubes. *Journal Sea Research,* (36), 63-67.

Law, D.J. and Bale, A.J., (1998), In-situ characterisation of suspended particles using focused-beam laser reflectance particle sizing. In: Black, K.S., Paterson, D.M. and Cramp, A. (eds) *Sedimentary Processes in the Intertidal Zone.* Geological Society London. Special Publication (139). 57-68.

Lick, W., Huang, H. and Jepsen, R., (1993), Flocculation of fine-grained sediments due to differential settling. *Journal Geophysical Research,* (98, C6), 10279-10288.

Malcherek, A., (1995), Mathematische Modellierung von Stromungen und Stofftranportprozessen in Astuaren. *PhD Thesis. University of Hannover.*

Manning, A.J., (2001), A study of the effect of turbulence on the properties of flocculated mud. *PhD Thesis. University of Plymouth.* 282pp.

Peterson, O., Vested, H.J., Manning, A.J., Christie, M.C. and Dyer, K.R., (2002), Numerical modelling of mud transport in the Tamar Estuary. *INTERCOH -2000, Elsevier, Coastal and Estuarine Fine Sediment Processes,* ed, J.C. Winterwerp and C.Kranenburg, this volume.

Winterwerp, J.C., (1996), A simple model for turbulence induced flocculation of cohesive sediment. *IAHR. J. Hydraulic Eng.* (36), 309-326.

Van Leussen, W., (1994), Estuarine macroflocs and their role in fine-grained sediment transport. *PhD thesis. University of Utrecht*, 488p.

Fine Sediment Dynamics in the Marine Environment
J.C. Winterwerp and C. Kranenburg (Editors)

A Comparison Of Floc Properties Observed During Neap and Spring Tidal Conditions.

A.J.Manning and K.R.Dyer

Institute of Marine Studies, University of Plymouth, Drake Circus, Plymouth PL4 8AA, United Kingdom

It is recognised that in order to properly understand how suspended particulate matter behaves during different tidal conditions within an estuary, high quality in-situ data is of a prime requirement. This paper initially presents floc data sets collected in the upper reaches of the Tamar estuary in south-western England. All floc samples were obtained using the in-situ sampling device INSSEV. The floc data was supplemented by simultaneous time series of near-bed profiles (using the high frequency POST system) of: turbulent shear stress (TSS), suspended particulate matter (SPM) and current velocity. To enable a comparison of typical spring and neap tidal conditions, respective data sets were collected (on a sub-tidal duration) on 24th June 1998 and 5th August 1998. The spring tides experienced nearly twice the annual mean river flow (~ 40 m^3s^{-1}), and salinity did not exceed 0.5 at anytime during sampling. The afternoon flood saw surface currents approaching 1.1 ms^{-1}, and a maximum TSS of 0.7 Nm^{-2} (at 25 cm).Throughout this period a concentrated benthic suspension layer developed, which displayed a peak particle concentration of 6 gl^{-1} (50 cm above the bed) and a lutocline ~ 40-60 cm above the bed. For the 5th August the annual mean river flow allowed the near-bed salinity at Station A to reach 8 during the afternoon flood. Surface currents did not exceed 0.55 ms^{-1} and the SPM remained under 190 mgl^{-1}, with the exception of the turbidity maximum (TM) formation at sampling Station A 1.5 hours into the flood, where the near-bed SPM rose to 1.15 gl^{-1}. The maximum flood TSS 25 cm above the bed was 0.74 Nm^{-2} and occurred just prior to the TM formation. An abundance of fast settling macroflocs (> 160 microns) from spring tides, accounted for a time series average of 89% of the mass settling flux (MSF). Whereas during neap tides, the macroflocs contributed 16% less to the MSF rate. This was partly due to a time series averaged macrofloc settling velocity of 4.6 mms^{-1} from the spring tidal data; 2.8 mms^{-1} higher than for neap tide conditions. During the TM passage at spring tides, macroflocs reached 1.5 mm in diameter; these flocs had settling velocities of up to 16.6 mms^{-1}, but effective densities were less than 50 kgm^{-3}, which means they would be prone to break-up when settling to a region of high shear. At the opposite end of the scale, low SPM and quiescent conditions severely restricted floc production. A multiple parametric analysis identified both the TSS and SPM concentration as significant controllers of the settling velocity of the macroflocs, and these parameters must be included within any quantitative empirical algorithms.

KEYWORDS
Cohesive sediment, flocculation, turbulent shear stress, floc size, settling velocity, effective density, suspended particulate matter, INSSEV instrument, Tamar estuary, turbidity maximum.

1. INTRODUCTION

The implementation of over-simplified settling velocity parameterisations has a major effect on the accuracy of predictive estuarine sedimentation models. This is primarily due to the individual particles flocculating into larger aggregates which are significantly less dense than their component primary particles. An individual floc may constitute up to 10^6 individual particles, and flocculation is a dynamically active process which is directly affected by its environmental conditions; this results in a continual process of aggregation and disaggregation, and hence a continual change in floc properties. Smaller microflocs join to form larger more fragile aggregates referred to as macroflocs. Turbulent shear within the water column has been identified as a prime contributor to particle collisions, but very little work has quantified its influence on *in-situ* floc formation (Manning and Dyer, 1999). Concentration and current velocities vary considerably during neap and spring tides within and outside the turbidity maximum. As a consequence the floc properties also vary. This paper utilises data collected from field experiments conducted during June and August 1998 in the Tamar estuary, with the aim of examining the differences exhibited by the floc populations occurring during spring and neap tidal conditions, respectively. Specific points in the tidal cycle are identified where there are significant changes in the spectra of the various floc properties, such as floc size, effective density, settling velocity, fractal dimension, floc shape, porosity, and the respective particulate mass distributions.

2. METHODOLOGY

Located on the south-western peninsula of England, the Tamar estuary provides a natural division between the counties of Devon and Cornwall. As a drowned Quaternary drainage channel, the Tamar is topographically dendritic in shape; it has numerous meanders and wide mud flats exposed at low water, and so can be classified as a drowned river valley. The main contributor is the River Tamar which has a 470 km^2 catchment zone, and from source to estuary mouth it is approximately 75 km in length. However, the tidal influence only extends 31 km inland. The Tamar estuary experiences semi-diurnal tides with mean neap and spring ranges of 2.2 m and 4.7 m respectively. This classifies the Tamar as mesotidal (Davies, 1964). The annual mean river discharge is about 20 m^3s^{-1}. Sampling was conducted in a straight section in the upper Tamar estuary, near Calstock (Figure 1), within the tidal reaches of the turbidity maximum (TM). The water depth varied between 1-4 m at this site, and the water column was predominantly fresh. This would mean that peak suspended concentration levels together with maximum current velocities would both be experienced. When there was an absence of stratification, processes in this region would be dominated more by vertical fluxes driven by boundary friction, rather than internal shear. With approximately 80% of the turbulent energy generated by the water flow nominally occurring within the lower 15% of

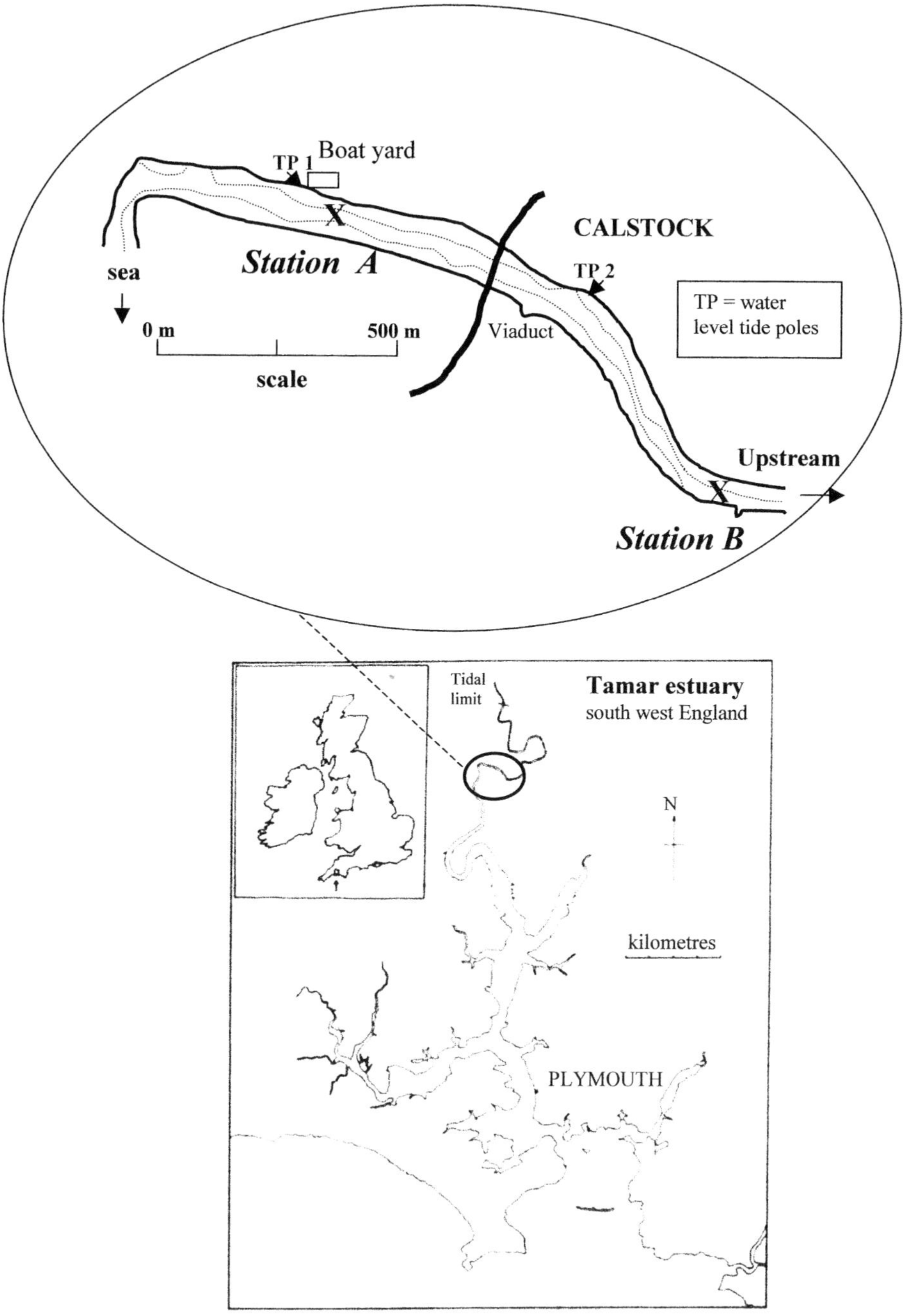

Figure 1. Location diagram of the Tamar estuary showing the positions of Station A and B.

the water column, Eulerian measurements were made in this near-bed region at times during the tidal cycle when flow conditions were either steadily changing or uniform.

Floc settling velocity and size was measured simultaneously by the video floc sampler INSSEV: IN-Situ Settling Velocity (Fennessy et al., 1994) instrument. Once the instrument was aligned with the ambient current flow, flocs were sampled at 0.5 m above the channel bed every 10-30 minutes, where complementary gravimetric water samples were also collected. Simultaneous measurements of current velocity and suspended particulate matter (SPM) were obtained using the POST: Profile Of Sediment Transport system (Christie et al., 1997). This consisted of four two-channel miniaturised electro-magnetic current meters (EMCM), eight optical backscatter (OBS) sensors, and a pressure transducer to measure the water depth, which all recorded continuously at 18 Hz and were filtered at 5 Hz. The sensors were attached to a vertical pole laterally off-set from the INSSEV sampling unit. EMCM pairs measured the three orthogonal components of the flow at both 25 cm and 75 cm above the bed. Vertical profiles of salinity, temperature and turbidity were also taken. Bio-chemical tests were conducted on selected gravimetric samples.

Floc sizes and settling velocities were both measured directly from the video monitor display (manually), and these were then converted into the actual dimensions determined by the initial video image calibration. The flocs were measured for their dimensions both along the axis in the direction of settling (D_Y), and the axis normal to it (D_X) from which a height:width ratio could be determined. This was then translated into a spherical equivalent floc diameter, D:

$$D = (Dx \cdot Dy)^{0.5} \tag{1}$$

A re-arranged Stokes' Law relationship (equation 2) was applied to the floc images sampled by INSSEV, which enabled the calculation of individual floc effective density (ρ_e) values

$$\rho_e = (\rho_f - \rho_w) = \frac{Ws\ 18\mu}{D^2\ g} \tag{2}$$

where W_S is the settling velocity, g is gravity, and μ the dynamic molecular viscosity. The effective density is the difference between the floc bulk density (ρ_f), and the water density (ρ_w). The water density was calculated from the measured salinity and water temperature using the International Equation of State of Sea Water, 1980 (Millero and Poisson, 1981). These values were also used to determine dynamic molecular viscosity.

3. RESULTS

The three components of flow measured by the EMCMs enabled values of turbulent shear stress (TSS), τ, to be calculated at each sensor group height using the Turbulent Kinetic Energy (Soulsby, 1983) method shown by equation 3:

$$\tau = 0.19\ (0.5\ \rho_W\ (u'^2 + v'^2 + w'^2)) \tag{3}$$

where ρ_w is the ambient water density and u', v', w' are the three turbulent Reynolds components computed from each 3 minute 47 second duration file. The turbulence could also be described in the format of G, which is the root mean square of the gradient in the turbulent velocity fluctuations with the unit s^{-1}

$$G = \left[\frac{U_*^3}{(\kappa \cdot \nu \cdot z)} \right]^{0.5} \quad (4)$$

where κ is the Von Karmon constant, z is the distance above the estuary bed, ν is the kinematic viscosity (molecular viscosity divided by the density of water), and U_* is the friction velocity.

For the floc spectra, a series of algorithms were used (Fennessy et al., 1997) to reference the floc populations with the ambient suspended particulate matter (SPM) concentration. The floc samples were then segregated in to bandings of size (SB) spectra (see Table 1). An assessment of each INSSEV sample, for their level of representative quality, was undertaken using equation 5:

$$Q_c = \frac{C_{INSSEV}}{C_{filtered}} \quad (5)$$

where C_{INSSEV} is the SPM computed from the floc population measured by INSSEV and $C_{filtered}$ is the SPM obtained from a filtered gravimetric sample collected simultaneously at the time of the floc sample acquisition. Thus Q_c is a comparison of the ratio of the SPM determined from the INSSEV samples to that of the filtered SPM, and it showed that the majority of the INSSEV samples gave a Q_c value of around unity. This indicated that those samples were a near-ideal representation of their respective populations. A small number of samples, mainly from high ambient SPM conditions, had Q_c ratios that were between 0.7-0.8. These underestimates were primarily on samples that contained a large number of flocs (typically over 800 flocs) and the potential for image over-lapping increases. However, these were still deemed representative samples, due to the actual number of flocs measured. Fractal dimensions can provide a different perspective on floc formation (Dyer and Manning, 1998) and they were computed by the formula advocated by Winterwerp (1997), with an assumed primary particle size of 4 microns. The ratio of the macrofloc to the microfloc sub-populations present within an INSSEV sample can be calculated using equation 6, whilst equation 7 is implemented to determine the ratio of the macrofloc to microfloc SPM distribution of a sample.

$$\text{Ratio of floc numbers} = \frac{\text{number of macroflocs}}{\text{number of microflocs}} \quad (6)$$

$$\text{Ratio of SPM division} = \frac{\text{SPM of macroflocs}}{\text{SPM of microflocs}} \quad (7)$$

Where appropriate, all calculated mean floc parameters presented in this paper were dry mass weighted means.

Table 1
INSSEV floc size bandings

Size Band	1	2	3	4	5	6	7	8	9	10	11	12
Floc size (μm)	20-40	40-80	80-120	120-160	160-200	200-240	240-320	320-400	400-480	480-560	560-640	>640

3.1. Spring Tide Conditions

The spring tidal data was comprised of two separate runs conducted on the 24th June 1998. The period of the tidal cycle covered was the end of the ebb, through the low water period and up to the middle of the flood. The runoff into the upper estuary was above the seasonal average (~ 40 m^3s^{-1}), and consequently salinity values remained below 0.5 for the entire sampling period, with no significant stratification present (Figure 2A). The first detection of the saline water reaching Station A was at 16:00hr when the near-bed salinity was 0.25.

Figure 2B shows the variations in current velocity for the experimental duration. The surface current was ebbing at a velocity of 0.8 ms^{-1} at 12:15hr. The asymmetric distortion in the tidal curve, primarily due to the estuary topography, prolonged the ebb beyond the predicted time of low water which was 13:15hr. The surface flow was still ebbing at a speed of 0.45 ms^{-1} at 13:55hr. The rapid flood which is a well documented characteristic of the Tamar estuary was demonstrated by a peak surface velocity of 1.1 ms^{-1} being attained at 15:38hr; this was only 70 minutes after the initial transition to the flood. This peak was only sustained for 15 minutes whereafter a gradual deceleration in the surface flow occurred, resulting in a speed of 0.75 ms^{-1} at 16:55hr. Closer to the bed, a rapid decrease in the current speed was observed from the record of the EMCMs located 25 cm above the bed. Between 15:27hr to 15:38hr the flow speed at 25 cm slowed from 0.31 ms^{-1} down to 0.08 ms^{-1}.

The variations in SPM concentration are shown in Figure 2C, and a maximum concentration of 1.27 gl^{-1} was observed 22 cm above the bed at the start of the run (12:15hr), when the current velocities were at their strongest. The gradually decelerating flow velocity corresponded well with a reduction in suspended concentration throughout the water column. An SPM of 400 mgl^{-1} was measured just below the water surface at 13:55hr indicating the presence of a high background turbidity composed of very small slow settling (possibly of near-neutral buoyancy) particles during these spring tidal conditions. On the flood tide, the slowing of the bottom current (at 15:27hr) coincided with the rapidly growing near-bed concentrated benthic suspension (CBS) layer, which was ~ 5 gl^{-1} at this stage and was approximately 40 cm thick. By 16:00hr the near-bed current had slowed even more to a minimum rate of 0.03 ms^{-1}. At this point in the tide the lutocline had risen to approximately the height of the INSSEV decelerator chamber roof above the bed (~ 0.6 m), with a flood peak CBS layer concentration of 6 gl^{-1}.

The formation of the CBS layer had the net effect of reducing the turbulent shear stress (TSS) 25 cm above the bed (Figure 2D) from a rate of 0.7 Nm^{-2}, at 15:27hr, down to 0.14 Nm^{-2} by 16:00hr. The damping of the turbulent energy in the CBS layer was probably partly due to the high abundance of particulates causing a high increase in the effective water density. This then significantly increased the particle collision frequency, which in turn used

A.

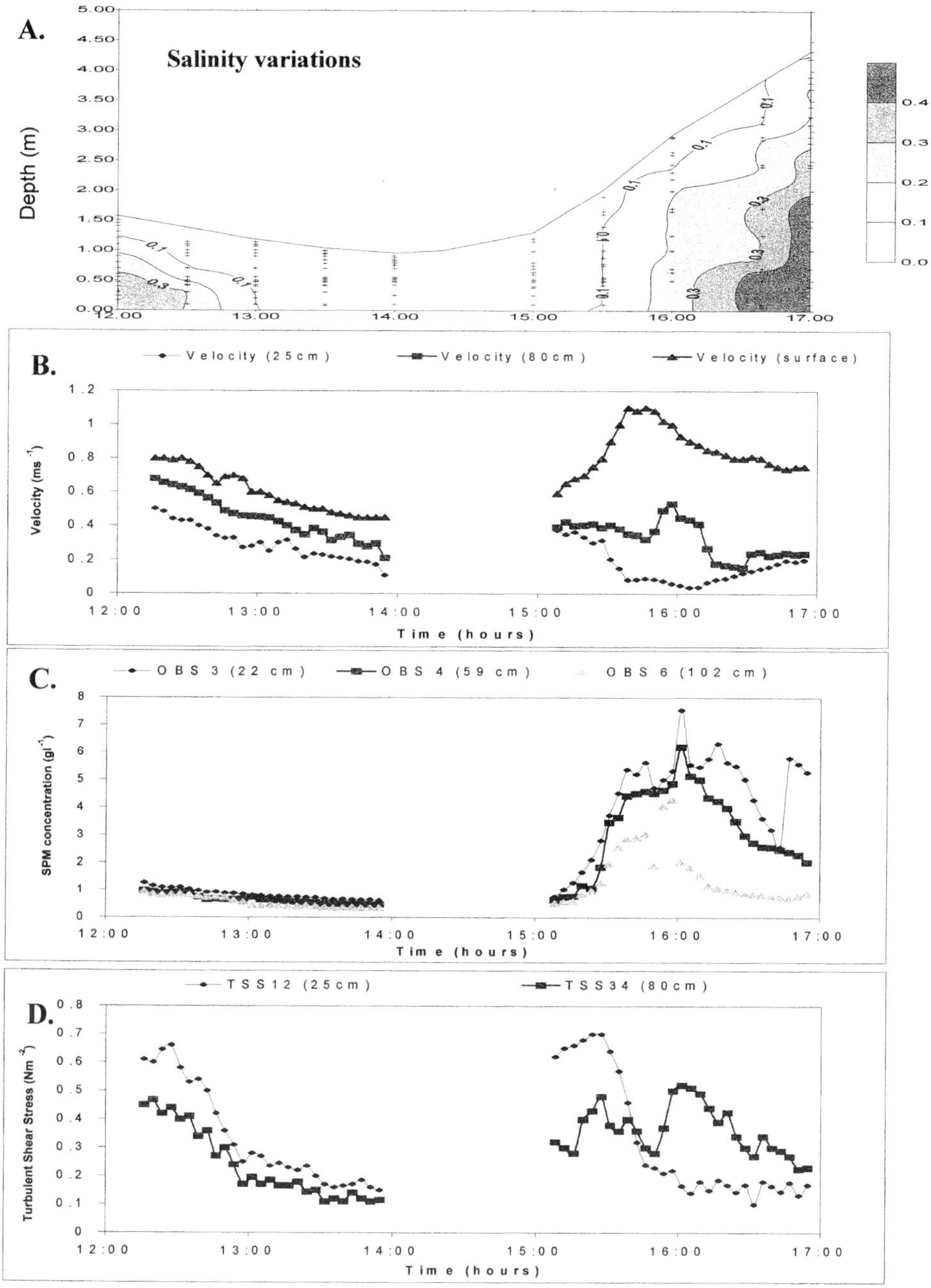

Figure 2. Time series of variations in: *A.* salinity, *B.* velocity, *C.* suspended particulate matter concentration, and *D.* turbulent shear stress for the spring tide on the 24th June 1998.

more of the hydrodynamic potential energy to either enhance floc formation or aid de-flocculation. This reduction in shear at the boundary, and the consequent transfer of an increased turbulent shear intensity to a higher level in the water column, is the phenomenon of drag reduction. This was also observed by Dyer et al. (2002a) during other experiments conducted in the turbidity maximum in the upper Tamar estuary, and a comprehensive review of this aspect of natural flow dynamics is given by Christie et al. (2001). It is interesting to observe that the concentration continued to increase 22 minutes after the maximum surface current velocity was achieved, highlighting that a large portion of the particulate matter observed at Station A was a result of advection.

3.2. Spring Tide Flocs - Ebb

The gradual reduction in TSS coupled with the decrease in ambient SPM, in the run period leading to low water, resulted in a general progressive reduction in maximum floc size. This was illustrated by the comparison of two INSSEV samples. The first floc sample, 24-2 (Figure 3), was obtained at 12:35hr where the ambient SPM and TSS were 885 mgl^{-1} and 0.47 Nm^{-2} (G value of 6.9 s^{-1}), respectively. The scatter plot shows the presence of a single floc 650 microns in diameter with a settling velocity of 7 mms^{-1}. At the smaller size fraction, the flocs became more numerous and the spread in settling velocities increased. This suggests that the level of hydrodynamical stirring present, was not permitting the particulate matter to achieve a fully optimum state of flocculation. As the flocs grew, the number of constructive collisions would be progressively reduced. This was well demonstrated by the 160-200 micron division (SB5) containing the largest portion (25 %) of particulate mass.

Moving closer to low water (Figure 4), the decrease in concentration to 487 mgl^{-1} at 13:35hr saw a reduction in the maximum floc size of sample 24-5. The flocs did not exceed 350 microns in diameter, and this was a product of the much lower turbulent shear stress (0.14 Nm^{-2}; G value of 2.8 s^{-1}) which was not physically capable of bringing the lower level of particulate matter together at a sufficient rate to promote flocculation. There was a more regular distribution in settling velocity with decreasing size, however there was a much greater number of SB1-4 flocs (68% of the total population), and these aggregates had rates of fall mainly between 0.1-0.9 mms^{-1}. Even so, SB5 (160-200 microns) again contained the highest portion of dry floc mass (29%). This is a recurring observation, especially during spring tide conditions, which suggests that the 160 micron size flocs creates the principle dividing line between the mean characteristics displayed by the larger and smaller floc fractions (Manning, 2001). The lower amount of flocculation occurring within sample 24-5 meant that the mean settling velocity for the flocs exceeding 160 microns in diameter was only 1.6 mms^{-1}; a fall rate decrease of 2 mms^{-1} from sample 24-2, which would hold significant ramifications in the rates of mass settling flux (MSF).

3.3. Spring Tide Flocs - flood

The most evident changes in floc properties throughout the flood occurred during the advection of the turbidity maximum through Station A. The combination of varying levels of concentration together with the associated effects of drag reduction at times of high concentration, had a significant effect on the aggregates formed. At 15:33hr (Figure 5) the turbidity level had risen to 2.6 gl^{-1} at the INSSEV sampling height of 0.5 m, and signified the head of the TM. At this point in the tidal cycle, the corresponding TSS of 0.6 Nm^{-2} (G value of 8.2 s^{-1}) was entraining a large amount of solids into suspension, and floc sample 24-7 was representative of the floc community at that time. The initial impression gained from

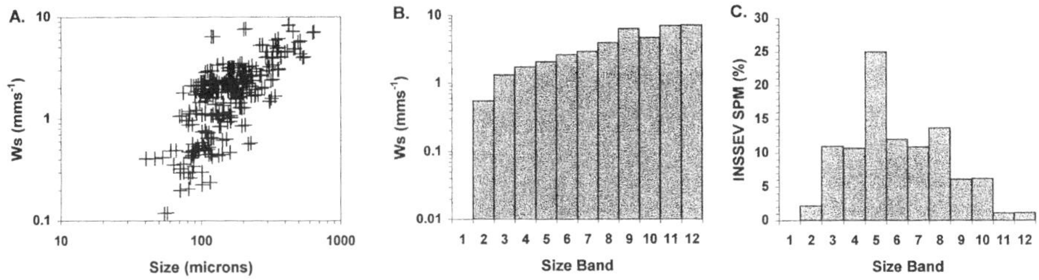

Figure 3. Floc characteristics for INSSEV sample 24-2 collected at 12:35hr.

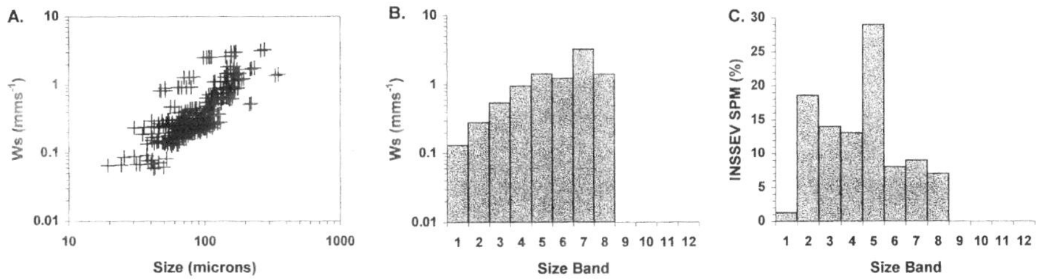

Figure 4. Floc characteristics for INSSEV sample 24-5 collected at 13:35hr.

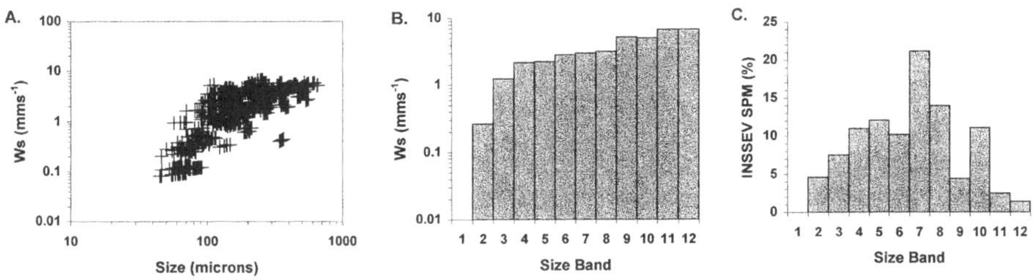

Figure 5. Floc characteristics for INSSEV sample 24-7 collected at 15:33hr.

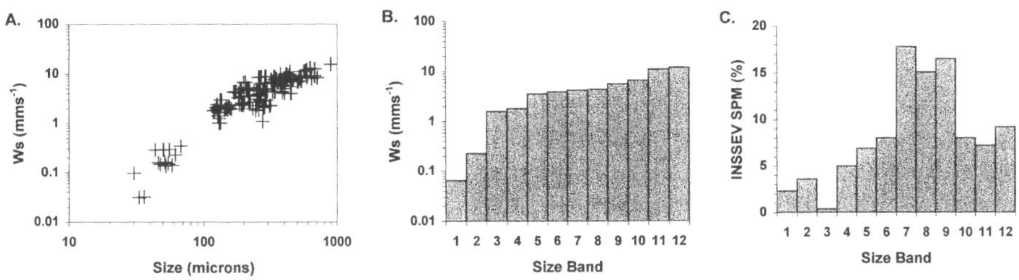

Figure 6. Floc characteristics for INSSEV sample 24-9 collected at 16:00hr.

examining the scatterplot (Figure 5A) showing the floc size against the settling velocity of the individual flocs, was that the number of flocs decreased rapidly with increasing size. Although flocs up to 650 microns in diameter had been constructed within the ambient conditions, the very turbulent water column had also induced a high proportion of destructive three-particle contacts causing de-flocculation. This was evident with nearly 47% of the 942 flocs being situated in the sub-160 micron size range. The primary mode of the SPM distribution occurred at SB7 (240-320 microns), however the fraction > 240 microns in diameter, only constituted 54% of the particulate mass. This resulted in a mean settling velocity for the flocs over 160 microns in size of 2.9 mms^{-1}.

In contrast, the main body of the TM produced an ambient concentration of 5.6 gl^{-1} at 16:00hr (Figure 6), and consequently restricted the turbulent energy to a TSS of 0.36 Nm^{-2} (G value of 5.7 s^{-1}) at 0.5 m above the bed. This transformed the flocs into a weakly bi-modal population, as the record of sample 24-9 shows. Although the scatter plot showed a sub-group of slow settling (settling velocity < 0.3 mms^{-1}) flocs in the 30-70 micron size range (Figure 6A), they only constituted 9.5% of the total population. The majority of the particulate mass, 49%, was contained within the larger sized aggregates of SB7-9 (240-480 microns), with a further 25% of the dry floc mass contained in the three largest size bands (SB10-12). This cluster of large flocs had settling velocities between 4-8 mms^{-1}, which produced a resultant mean settling velocity for the fraction > 160 microns of 5.5 mms^{-1}, which was an increase of over 2.5 mms^{-1} when compared to the earlier spring flood sample. This highlighted a significant optimisation in the flocculation process during the latter sample ambient conditions. The high percentage of floc mass contained by the macroflocs situated within the TM was also observed by Dyer et al. (2002b). As with the aggregates observed from within the CBS layer during the September 1998 spring conditions, these larger flocs also exhibited a height:width ratio of 0.9-0.8. This may have been a product of partial compaction due to the closeness of over-lying flocs in the thick high concentration CBS layer. Also the floc fractal dimensions were predominantly 2.4-2.5.

3.4 Neap Tide Conditions

For the August neap tidal run, the last part of the ebb was covered continuously through to the latter stages of the flood. The river flow was around the annual mean of 20 m^3s^{-1} which was much lower than observed during the June experiment. Figure 7A shows the salinity variations during the neap tidal conditions of the 5th August 1998. The water column for the latter part of the ebb at Station A, was completely fresh. The first saline (salinity ~ 1) water started had arrived at Calstock by 14:30hr. The salt intrusion continued to move through Station A as a well mixed saline wall, and a near-bed salinity of 8.45 was measured at 16:30hr.

The current velocity record (Figure 7B) shows the surface ebb velocity was 0.5 ms^{-1} at 10:40hr. As the ebb flow slowed with the reducing water level, a dip in the near-linear velocity record was observed at 11:00hr. This led to a minor peak in the surface current 10 minutes later, and then a progression towards slack water at 11:55hr, although no slack period occurred in the lower part of the water column. The near-bed flow 80 cm above the bed tended to be approximately 0.1 ms^{-1} greater than the surface flow between 11:20hr-11:45hr. At 12:00hr the surface flow started to steadily accelerate to a flood velocity peak of 0.54 ms^{-1} which was reached at 13:25hr. This was then followed by a dip in current speed down to 0.45 ms^{-1} at 13:47hr. A second peak of 0.51 ms^{-1} then occurred at 14:10hr, which was then followed by a progressive decrease in surface current speed to slack conditions by 15:45hr.

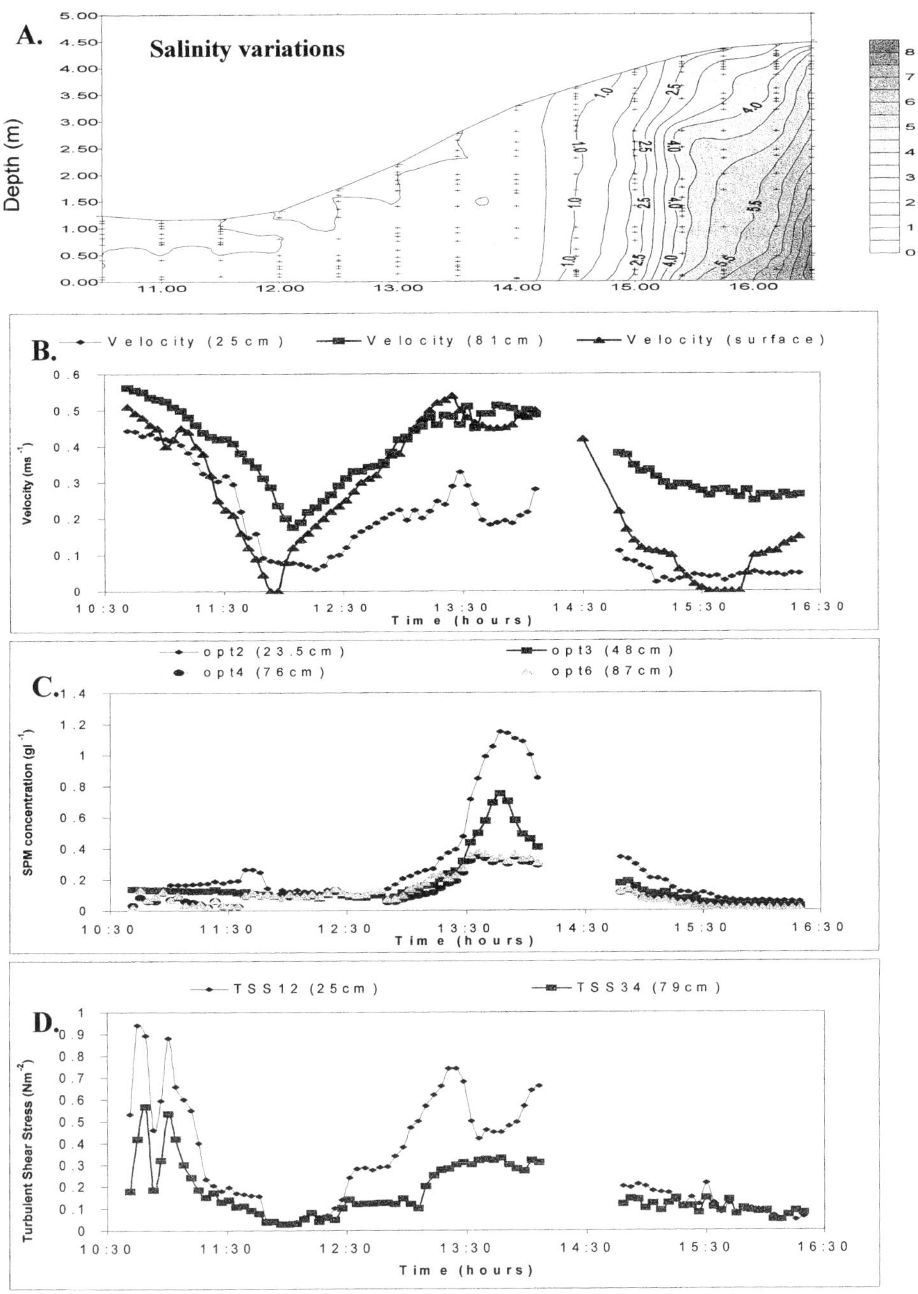

Figure 7. Time series of variations in: *A.* salinity, *B.* velocity, *C.* suspended particulate matter concentration, and ***D.*** turbulent shear stress for the neap tide on the 5th August 1998.

The SPM concentrations (Figure 7C) throughout the ebb were typically under 190 mgl^{-1} at 23.5 cm above the bed. On the flood, the occurrence of the velocity peaks coincided with the entrainment of bed sediment into the water column. At 13:28hr the near-bed concentration gradient was 0.32 kgm^{-4}, with an SPM of 475 mgl^{-1} measured 23.5 cm above the bed. The minimum velocity dip in both the surface and near-bed current coincided with the peak turbidity maximum formation at Station A. At this point in the flood tide (13:47hr) the SPM at the 23.5 cm height had risen to 1.15 gl^{-1}, resulting in the concentration gradient increasing to 1.74 kgm^{-4}. As the concentration reduced, so the current velocity increased to the second peak at 14:10hr. This suggests that there was minimal particulate advection present at the Calstock site during the passing of the TM. The appearance of these two peaks in velocity are the "bow-wave" phenomena reported by Dyer et al. (2002a), where it was suggested that the presence of this feature controls the separation of the salt intrusion from the turbidity maximum. The concentration at 23.5 cm had fallen to ~ 400 mgl^{-1} by 14:30hr, where the first saline water was detected. Once the salinity had risen to 5 at 15:40hr, the near-bed concentration (at 23.5 cm) further decreased to 77 mgl^{-1}.

The corresponding variations in turbulent shear stress are illustrated in Figure 7D. At 10:45hr, a high shear stress was present in the bottom layer of the water column (0.53 Nm^{-2} at 25 cm above the bed). Although the velocity deceleration through the run was relatively linear, the TSS at the 25 cm height demonstrated two peaks during the early part of the record. The first peak was at 10:45hr with a TSS of 0.95 Nm^{-2}, and the second peak with a TSS rate of 0.9 Nm^{-2}, occurred 15 minutes later. After the second peak, the reduction in flow velocity causes a marked reduction in the TSS in the near-bed region to 0.04 Nm^{-2}. During the flood the turbulent shear stress at the 25 cm height was no more than 0.1 Nm^{-2} for the initial 25 minutes. The TSS increased with the accelerating surface current to an initial peak rate of 0.74 Nm^{-2} by 13:24hr. The development of the high near-bed concentration gradient had the effect of reducing the turbulent energy dissipation, and consequently by 13:35hr the TSS reduced to 0.42 Nm^{-2}. As the turbidity started to fall, the increasing current resulted in a second peak TSS of 0.66 Nm^{-2} at 14:06hr. After this point in the tidal cycle, the TSS continued to steadily decrease, in response to the slowing surface flood current.

3.5. Neap Tide Flocs - Ebb

As with the spring ebb tide, the steady decrease in turbulent shear stress that occurred during the latter part of the ebb on the 5th August, had a direct influence on the resultant floc populations. Figure 8 shows the floc variations for sample 05-2 which was at 10:53hr. The SPM concentration was 131 mgl^{-1} and the fast flowing current produced an ambient TSS of 0.32 Nm^{-2} (G value of 5.2 s^{-1}). The level of turbulent mixing was of a sufficient strength to encourage floc growth up to a maximum size of 300 microns. These flocs were low in effective density (~ 45 kgm^{-3}), predominantly "comet" shaped and exhibited settling velocities in the region of 3 mms^{-1}. The flocs became more numerous with decreasing size band, and demonstrated a negative skewing of the dry floc mass distribution, with 55% in the sub-160 micron fraction. As the ebb approached low water at 12:20hr, the reduction in turbulent energy dissipation gave a resultant TSS of 0.07 Nm^{-2} (G value of 1.7 s^{-1}). The concentration was generally unaltered at ~ 100 mgl^{-1}. As the record for sample 05-7 (Figure 9) shows, this had the net effect of severely restricting floc growth, with most aggregates under 80 microns in diameter. This aspect will be considered in the discussion.

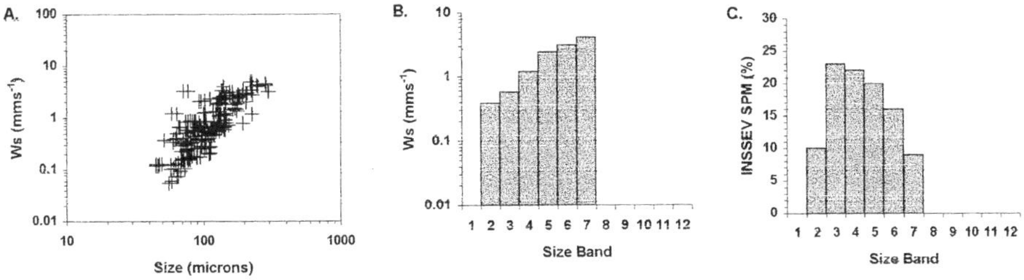

Figure 8. Floc characteristics for INSSEV sample 05-2 collected at 10:53hr.

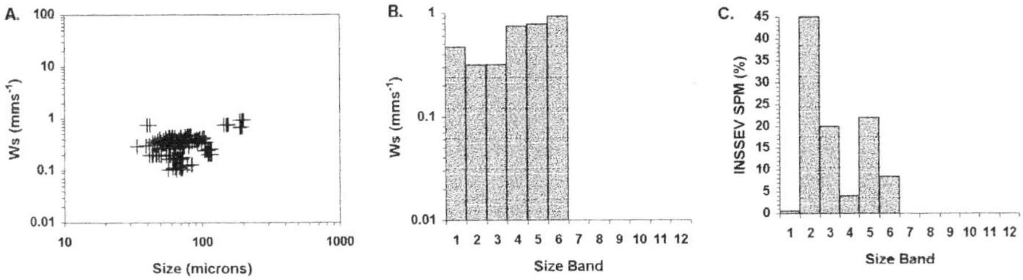

Figure 9. Floc characteristics for INSSEV sample 05-7 collected at 12:20hr.

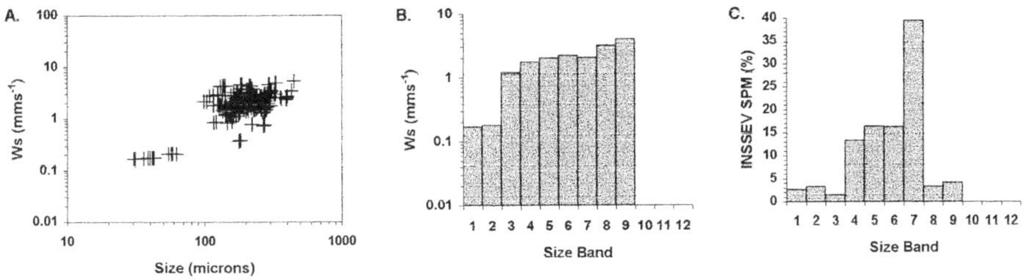

Figure 10. Floc characteristics for INSSEV sample 05-9 collected at 13:25hr.

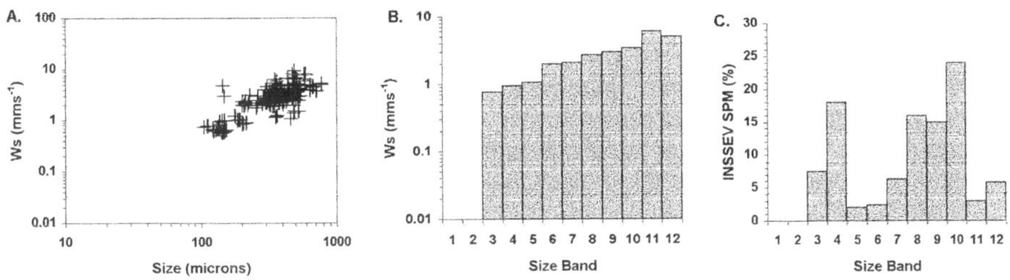

Figure 11. Floc characteristics for INSSEV sample 05-10 collected at 13:46hr.

3.6. Neap Tide Flocs - Flood

The simultaneous concentration and TSS increases of the early part of the flood, brought about a progressive increase in the rate of orthokinetically induced floc growth. This was demonstrated by sample 05-9 (Figure 10) which was obtained at 13:25hr when the TSS had attained a peak of 0.55 Nm^{-2} (G value of 7.5 s^{-1}) at the INSSEV height of 0.5 m, and the SPM was 262 mgl^{-1}. The high collision frequency created by the ambient conditions had encouraged a cluster of flocs to form in the 120-320 micron size range. These flocs accounted for 85% of the dry floc mass, with 40% of the total mass in SB7 (240-320 microns). With only 3.8% of the total population attaining a floc size of up to 430 microns, it was possible that the high turbulence level had imposed a limit on the maximum floc size.

A concurrent drop in the turbulent shear stress to 0.4 Nm^{-2} (G value of 6.1 s^{-1}) occurred as the main body of the TM advected through Station A at 13:46hr, raising the SPM to 750 mgl^{-1}. Figure 11 shows the floc variations for sample 05-10 which was collected at that point in the flood tide. The combination of the ambient conditions had further stimulated the flocculation. Now the main cluster of floc mass had moved to the 320-560 micron flocs, which contained 55% of the particulate matter. These flocs, together with aggregates which had grown to a maximum size of 778 microns, all exhibited settling velocities of ~ 3-5 mms^{-1}. This translated into a mean settling velocity of 3.1 mms^{-1} for the flocs > 160 microns in diameter, which was a fall rate increase of 0.8 mms^{-1} over the larger fraction of the preceding INSSEV sample. As the TM passed through, so the slowing current caused the SPM to reduce back to ~ 200 mgl^{-1}, and the TSS to drop to 0.16 Nm^{-2} (G value of 3.1 s^{-1}) by 14:48hr. This permitted the macroflocs with large settling velocities to settle to the bed. The net effect at the INSSEV sampling height was that the reduced concentration produced less numerous constructive collisions. The maximum floc size was only 280 microns, and there was no real exaggerated growth towards the higher size banded flocs, but a generally even distribution throughout the size range. The settling velocity of the larger fraction was 1.8 mms^{-1}. The majority of the flocs observed throughout this neap flood had fractal dimensions ranging between 2.25-2.35.

4. DISCUSSION

Primarily, the characteristics of the largest flocs from each sample will be focused upon, as they have the potential to contain a high proportion of the particulate mass. In order to examine this facet, mean floc parameters were calculated from the six largest flocs of each INSSEV sample, referred to as MAX6 parameters. For the spring conditions the MAX6 floc size ranged between 304-1598 microns. The largest of these aggregates were formed at the time the main body of the TM advected through the Calstock sampling location, and these flocs of very low density (effective densities < 50 kg m^{-3}) had settling velocities of up to 16.6 mms^{-1}. In comparison, the MAX6 neap flocs were both 50% smaller in size and in settling velocity (at the occurrence of peak concentration). It was also noticeable that the MAX6 fractal dimensions were generally 0.1 lower during neap conditions, giving values of ~ 2.2.

To obtain the best impression of how each floc population varies, a divisional boundary as defined by Manning (2001) was applied at the 160 micron mark, which discriminated the macroflocs from the microfloc sub-grouping. Averaging through both time series showed that 73% of the particulate mass was held in the macrofloc portion during the spring tide. This translated into mean macrofloc:microfloc ratios of floc numbers and SPM division of 1.7 and

3.4, respectively. Comparably, the same ratios computed for the neap conditions were 1.6 and 0.8, meaning that only 56% of the ambient SPM was contain within the macrofloc sub-population throughout the neap tidal conditions. The result of this indicated that the fast settling macroflocs from the spring tide accounted for a time series average of 89% of the MSF. Whereas during neap tides, the macroflocs contributed 15.7% less to the MSF rate. This was partly due to a time series averaged macrofloc settling velocity of 4.6 mms^{-1} from the spring tides. This was an average settling rate 2.8 mms^{-1} faster than that computed for neap conditions. The contrasting absolute rates of MSF for the macrofloc sub-population, at the time the main body of the TM advected through the sampling location, were 37800 $mg.m^{-2}s^{-1}$ and 2000 $mg.m^{-2}s^{-1}$ for spring and neap tides, respectively. The high contribution of the macroflocs towards the MSF agrees with the conclusions of a recent study conducted within the Gironde estuary, in France by Manning et al. (2001).

During the neap tide conditions, near-bed SPM ranged between 50-500 mgl^{-1}, with an organic content of between 14.5-28%. These relatively low SPM conditions coincided with higher levels of mean chlorophyll-a (up to 0.98 μgl^{-1}), when compared to the spring tidal conditions (chlorophyll-a up to 0.24 μgl^{-1}) which produced concentration levels up to a order of magnitude greater than those of the neap tides, with significantly lower average organic levels (10-14%). This is explained by the erosion of the more compact lower organic and lower chlorophyll-a content sediment (in various stages of decomposition) from the bed during the stronger spring tidal flow conditions, diluting the previously suspended more easily erodible material. Similarly the mean carbohydrate per mass of sediment was higher on the neap tides (up to 34 mg/g of SPM) than the spring tides (up to 5.9 mg/g of SPM). However, with a significantly higher abundance of suspended particulate matter experienced on the spring tides, the overall carbohydrate concentrations were seen to be up to seven times greater during the spring tidal conditions (i.e. maximum values of carbohydrate concentration of 2.5 mgl^{-1} at neap tides, compared to 17.5 mgl^{-1} during spring tides).

The basic mechanism controlling the rate of flocculation is the number of positive inter-particle collisions that occur during hydrodynamically induced stirring within the water column. However, the combination of a low particulate abundance and a quiescent water column, results in an extremely low collision frequency. A good example of this effect was demonstrated by the slack lower water period during neap tides, when the SPM was ~100 mgl^{-1}, but the TSS was only 0.07 Nm^{-2} (sample 05-7). Three quarters of the flocs were less than 80 microns in size and these had effective density values ranging between 200-1000 kgm^{-3}. In early work on estuarine flocculation, both Krone (1963, 1986) and Partheniades (1965) suggested that flocs were constructed in a progressive order. Primary particles glued together form zero order flocs, these in turn combine to form 1st order flocs etc; this was known as the order of aggregation theory. This indicated that the vast majority of particulate matter did not progress beyond the zero or first order flocculation stage i.e. the smallest floc size the flow could break up. Those which had evolved into aggregates of up to a maximum size of 202 microns, were in the minority. Comparing this sample (sample 05-7) to 05-02 (where the SPM was 131 mgl^{-1}, but the TSS was 0.32 Nm^{-2}), the mean Ws of the > 160 micron fraction for sample 05-2 was 2.8 mms^{-1}, whereas the latter sample was only 0.8 mms^{-1}.

The increased particulate collision frequencies which occurred with the high turbulent mixing activity and SPM levels of the spring tidal conditions, was seen to optimise the production of a high percentage of fast settling macroflocs by the added presence of a high total carbohydrate concentration, hence causing greater inter-particle adhesion. This was particularly evident with the observation of a large number of stringer type flocs during the spring conditions. These are smaller aggregates connected together by a network of fine organic strands. Qualitatively, this suggests that the total amount of sugars present seems to be a more important factor with regards to flocculation than that of the mean organic content. To fully quantify the significance of carbohydrates (particularly extra-cellular polymetric substances) on flocculation and floc settling rates, would require a further series of experiments with this as the main aim.

A two week long experiment was also conducted at the same Calstock location in September 1998, and was part of a European Commission MAST III funded project COSINUS (Prediction of cohesive sediment transport and bed dynamics in estuaries and coastal zones with integrated numerical simulation models). Data from this study together with results obtained from the COSINUS experiment (Dyer et al., 2002b) were used to produce empirically derived flocculation algorithms. The statistical variations in settling velocity (with the units mms^{-1}) of the macroflocs (Ws_{MACRO}) is shown in equation 8:

$$Ws_{MACRO} = -0.243 + 0.000567SPM + 0.981G - 0.0934G^2 \qquad (8)$$

where the suspended particulate matter (SPM) and the turbulence parameter G (see equation 4), had the units of mgl^{-1} and s^{-1}, respectively. This formula was derived from a multiple linear regression analysis of 74 simultaneous observations collected over the three experimental periods (Manning, 2001), and has an R^2 of 0.80 which is highly significant. Curves for this algorithm are shown in Figure 12. The data points for the 24th June and 5th August experiments have been plotted in Figure 12 for comparison. A number of these algorithms have been tested and implemented in a recent numerical model of the Tamar estuary (Peterson et al., 2002). It must be noted that equation 8 does not include the effects of sugars or salinity (as described earlier in this paper) on the settling velocity.

5. CONCLUSIONS

This study successfully measured the variations in floc properties throughout spring and neap tidal conditions, together with the hydrodynamics within the turbidity maximum zone of the Tamar estuary. The high inter-particulate collision frequencies which occurred during the simultaneously high turbulent mixing activity and SPM levels of the spring tidal conditions, was seen to optimise macrofloc production. The flocculation was further enhanced by the high total carbohydrate concentration present during spring conditions, which would have added more cohesion to the constituent particulates.

The fast settling macroflocs from the spring tide accounted for a time series average of 89% of the MSF. Whereas during neap tides, the macroflocs contributed 16% less to the MSF rate. This was partly due to a time series averaged macrofloc settling velocity of 4.6 mms^{-1} from the spring tidal data. This was an average settling rate 2.8 mms^{-1} higher than for neap tidal conditions. The largest flocs were formed at the time the main body of the TM advected through the Calstock sampling location. During spring tides the MAX6 flocs reached 1.5 mm in diameter. These flocs had settling velocities of up to 16.6 mms^{-1}, but their effective densities were less than 50 kgm^{-3}, which means they would be prone to break-up when settling through a region of high shear. At the opposite end of the scale, low SPM and more quiescent conditions severely restricted floc production. A multiple parametric analysis identified both the TSS and SPM concentration as significant controllers of the settling velocity of the macroflocs, and these parameters must be included within any quantitative empirical algorithms.

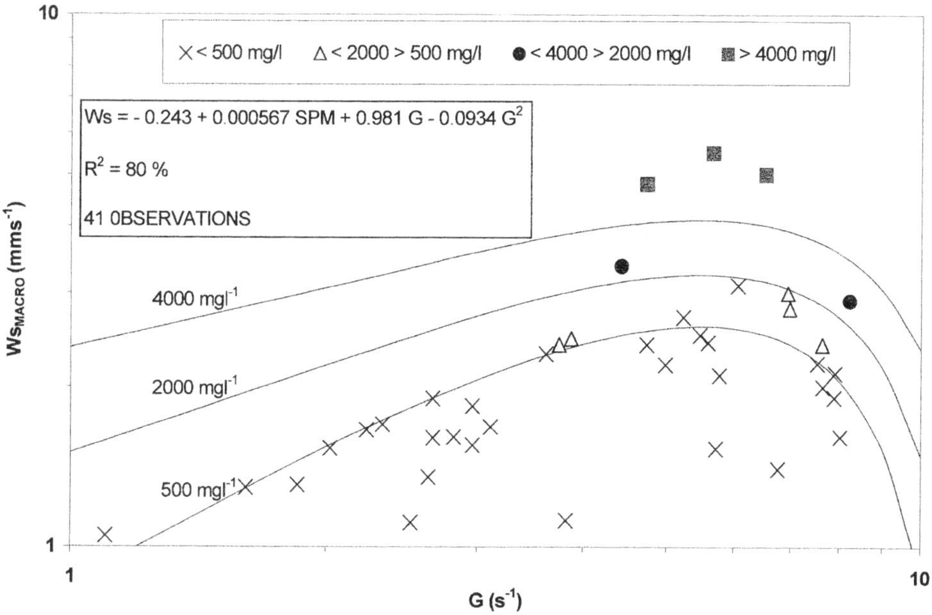

Figure 12. A statistical representation of macrofloc settling velocity (Ws_{MACRO}), based on equation 8, at various levels of SPM concentration plotted against the turbulent shear parameter G. Ws_{MACRO} values from the 24th June and 5th August 1998 experimental data sets have been sorted according to SPM concentration sub-grouping and added to the plot.

ACKNOWLEDGEMENTS

The authors wish to acknowledge the participation of the numerous personnel who assisted with the collection of the data sets examined in this paper. The work was partly funded by the EC MAST III contract MAS3-CT97-0082 COSINUS.

REFERENCES

Christie, M.C., Quartley, C.P. and Dyer, K.R., 1997. The development of the POST system for in-situ intertidal measurements. *The 7th Int. Conf. On Elec. Eng. In Oceanography, 23-25 June 1997, p.39-45 of Conference Publication No. 439, by Inst. Elec. Eng., London.*

Christie, M.C., Dyer, K.R., Turner, P. and Manning, A.J., 2001. The effects of density gradients upon water column turbulence within a turbidity maximum (*In prep*).

Davies, J.H., 1964. A morphogenetic approach to world shorelines. *Z. Geomorphol.*, 8: 127-142.

Dyer, K.R. and Manning, A.J., 1998. Observation of the size, settling velocity and effective density of flocs, and their fractal dimensions. *Journal of Sea Research*, 41: 87-95.

Dyer, K.R., Bale, A.J., Christie, M.C., Feates, N., Jones, S. and Manning, A.J., 2002a. The turbidity maximum in a mesotidal estuary, the Tamar estuary, UK. Part I: Dynamics of suspended sediment. Proceedings *INTERCOH-2000, Elsevier, Coastal and Estuarine Fine Sediment Processes*, ed. J.C. Winterwerp and C. Kranenburg, this volume.

Dyer, K.R., Bale, A.J., Christie, M.C., Feates, N., Jones, S. and Manning, A.J., 2002b. The turbidity maximum in a mesotidal estuary, the Tamar estuary, UK. Part II: The floc properties. Proceedings *INTERCOH-2000, Elsevier, Coastal and Estuarine Fine Sediment Processes*, ed. J.C. Winterwerp and C. Kranenburg, this volume.

Fennessy, M.J., Dyer, K.R. and Huntley, D.A., 1994. INSSEV: an instrument to measure the size and settling velocity of flocs in-situ. *Marine Geology*, 117: 107-117.

Fennessy, M.J., Dyer, K.R., Huntley, D.A. and Bale, A.J., 1997. Estimation of settling flux spectra in estuaries using INSSEV. *Proc. INTERCOH'94, Wallingford, England. Lohn Wiley & Son, Chichester*, 87-104.

Krone, R.B., 1963. A study of rheological properties of estuarial sediments. *Hyd. Eng. Lab. and Sanitary Eng. Lab., University of California, Berkeley, Report No*. 63-68.

Krone, R.B., 1986. The significance of aggregate properties to transport processes. In *Estuarine Cohesive Sediment Dynamics* (Mehta, A. J., ed.), *Springer-Verlag, Berlin*, pp. 66-84.

Manning, A.J., 2001. A study of the effect of turbulence on the properties of flocculated mud. *Ph.D. Thesis, University of Plymouth*, pp. 282.

Manning, A.J. and Dyer, K.R., 1999. A laboratory examination of floc characteristics with regard to turbulent shearing. *Marine Geology*, 160, 147-170.

Manning, A.J., Dyer, K.R. and Christie, M.C., 2001. Properties of macroflocs in the lower reaches of the Gironde estuary. Coordinateurs: Elbee, J. (d') and Prouzet, P. *Oceanographie Du Golfe De Gascogne. VIIe Colloque International, Biarritz, France, 4-6 Avril 2000.* Ed. Ifremer, France, *Actes de Colloques No.* 31: 230-235.

Millero, F.J. and Poisson, A., 1981. International one-atmosphere equation of state seawater. *Deep-sea Research*, 28 (A): 625-629.

Partheniades, E., 1965. Erosion and deposition of cohesive soils. *J. Hydr. Div., Proc. Am. Soc. Civ. Engrs.*, 98: 79-99.

Peterson, O., Vested, H.J., Manning, A.J., Christie, M.C. and Dyer, K.R., 2002. Numerical modelling of mud transport in the Tamar Estuary. Proceedings *INTERCOH-2000, Elsevier, Coastal and Estuarine Fine Sediment Processes*, ed. J.C. Winterwerp and C. Kranenburg, this volume.

Soulsby, R.L., 1983. The bottom boundary layer of shelf seas. In: B. Johns (Editor), *Physical oceanography of coastal and shelf seas. Elsevier, New York, N.Y.* 189-266.

Winterwerp, J.C., 1997. A simple model for turbulence induced flocculation of cohesive sediment. *IAHR., J. Hydraulic Eng.,* Vol. 36, No. 3, 309-326.

Particle Size Distribution in an Estuarine Turbidity Maximum Region

S. B. Mitchell[a] and J. R. West[b]

[a] School of the Environment, University of Brighton, Lewes Road, Brighton, BN2 4GJ, U.K.
[b] School of Civil Engineering, University of Birmingham, Edgbaston, Birmingham, B15 2TT, U.K.

Observations are presented of primary, deflocculated particle size and suspended solids concentration in the turbidity maximum region at Burringham, on the Trent estuary, UK, during two tidal cycles, in July 1996 and July 1997. Pumped samples were obtained at intervals throughout both tidal cycles, for subsequent analysis for suspended solids concentration and particle size distribution. Both deployments took place for spring tide conditions, but the antecedent fresh water flow conditions were much higher in July 1997 than in July 1996. As a result the turbidity maximum region was located further downstream in July 1997 than in July 1996. A generally higher flood-tide median particle size (D_{50} = 50 μm) in July 1997 compared with July 1996 (D_{50} = 10 μm) points to the possibility of sediment size sorting as a mechanism for the maintenance of larger particles at the landward end of the turbidity maximum. The difference in particle size regime also helps to explain the higher flood-tide suspended solids concentrations (8-10 g/l as opposed to 5–7 g/l), together with higher particle-induced density gradients during the ebb tide, in July 1997, for a similar hydraulic regime. One important consequence of the presence of larger particles is the nature of the relationship between suspended solids concentrations obtained by gravimetric analysis of pumped samples (g/l) and turbidity, measured by optical transmissometer. At high slack water, preferential settling by these larger particles below the level of the lens of the turbidity sensor may lead to anomalously high readings caused by the presence of high concentrations and flocs of smaller particles. Since optical turbidity sensors are often used to estimate trends in fine-sediment transport in estuaries, this preferential settling has important consequences for calibration of these sensors in highly turbid estuarine environments. For the data presented, a simple straight-line relationship can be demonstrated for all times during the tidal cycle except for slack water periods, when a different calibration should be applied.

KEYWORDS

Particle Size; Turbidity Maximum; Suspended Solids Concentration; Trent estuary

1. INTRODUCTION

The behaviour of high concentrations of fine sediment in estuarine turbidity maximum regions has important implications for problems associated with siltation, navigation, recreation, water quality and the ecosystem. The fine balance of processes which contribute to the formation and maintenance of the turbidity maximum varies from estuary to estuary. For each estuary, this balance must be well understood in order that management decisions relating to the estuary may be taken of an informed nature based on an appreciation of the widest possible understanding of the interactions between the estuarine hydraulics, solute and sediment transport. In particular, understanding of the temporal and spatial distribution of particle size is required in order that predictions can be made relating to flocculation processes in estuary systems. Field data obtained of this nature is essential in providing calibration data for mathematical models, which can be used as management tools.

Previous work relating to estuarine sediments within turbidity maxima has focused mainly on the depth and temporal variation of sediment transport under various conditions of tidal range and fresh water flow (e.g. Uncles et al. 1998a; Grabeman and Krause, 1989). By taking detailed measurements of the variations in concentrations of suspended solids over the depth during individual tidal cycles, estimates may be made of the net sediment flux over a tidal cycle. Additionally, these surveys help to provide evidence for a mechanistic interpretation of the behaviour of the turbidity maximum (Guezennec et al., 1999; Mitchell et al., 1998). For macrotidal systems, such studies have noted the importance of vertical density gradients in promoting tidal pumping of the turbidity maximum in an upstream direction under low fresh water flows, and of downstream flushing of sediment under high fresh water flows. These processes are summarised more fully in Dyer (1997).

The importance of flocculation (Dyer, 1986; Puls et al., 1988) of sediment particles also has a bearing on the net flux of sediment in an estuary, as this affects sediment settling. Inspection of floc sizes using video techniques (Fennessy et al., 1994; ten Brinke, 1997) and an in situ laser device (Law et al., 1997) has demonstrated the complex nature of the patterns of floc formation and destruction under different hydrodynamic conditions. This in turn has been related to the nature of the turbulent eddies in the flow (van Leussen, 1997). Research carried out in two different estuaries has demonstrated that the nature and cohesiveness of flocs also depends on the nature of the primary particles which constitute those flocs (Wolanski and Gibbs, 1995; Li et al., 1999). Thus, sediment trapping in these estuaries may be a function not only of the hydrodynamics of the estuary, but also of the flocculation characteristics of the particles themselves.

Much consideration has also been given to the effect of particle characteristics on the relationship between optically measured turbidity and gravimetrically determined suspended solids concentration (SSC). This may be a non-linear relationship, depending on particle size, shape, colour and concentration (Vanous et al., 1982; Lawler, 1995), and any calibrations made between the two parameters should always be site specific, and treated with caution. Consideration of the influence of primary particle size characteristics, in addition to the in-situ floc size, may help to provide some generic characteristics of the shape of the calibration curve which could be relevant for other systems.

The UK Natural Environment Research Council (NERC) Land-Ocean Interaction Study (LOIS) was concerned in part with trying to understand and predict estuarine sediment transport processes in and through the Humber estuary system, and this research formed a part of that effort. It is intended that the data presented in this paper will show that the distribution of primary particle size in the turbidity maximum of the Trent estuary depends on the location of the turbidity maximum relative to the observation point. This in turn is related to the recent history of antecedent fresh water flow. Through this appreciation of the temporal variation of primary particle size within the tidal cycle, it will be demonstrated that for this system, the relationship between optically-measured turbidity and grams-per-litre suspended solids concentration is modified at slack water periods.

2. STUDY SITE AND METHODS

Observations have been made of the distributions of SSC and primary particle size in the region of the turbidity maximum in the tidal section of the river Trent, UK at Burringham, Northern England (see Figure 1). At this point, the river is subject to a high degree of tidal asymmetry, with high-velocity flood tides of a shorter duration than lower velocity ebb tides (Mitchell et al, 1998). During spring tides a tidal bore is often observed. The channel has a maximum depth of approximately 3.5 m at low tide, low fresh water flow conditions, and is approximately 60 m wide, with steep banks. Due to land drainage activity in the coastal part of the Trent catchment, and shipping activity, the Trent is canalised, having only gradual reduction in cross section size between the Trent-Ouse confluence at Trent Falls and the tidal limit at Cromwell Weir. The study site at Burringham is located approximately 80 km from the estuary mouth at Spurn Head, at grid reference 0° 45′ W, 53° 35′ N.

The Trent is a macrotidal estuary with mean tidal range (predicted for Immingham) of between 3.2 m (neap tides) and 6.4 m (spring tides) and approximate mean fresh water inputs of between 30 m^3/s (summer) and 400 m^3/s (winter). Fresh water flow in the Trent is supplemented with effluent from several high-volume waste water treatment plants located within the catchment. The characteristics of the turbidity maximum in the tidal Trent, and the mechanisms behind its formation and maintenance are described in Mitchell et al. (1998).

Experiments were carried out during two spring tidal cycles on 29/30 July 1996 and 21 July 1997. The tidal range, predicted at Immingham, was 7.1 m for the July 1996 tidal cycle and 6.5 m for the July 1997 tidal cycle. Mean 5-day antecedent fresh water flow, obtained from the UK Environment Agency and measured at Cromwell, at the tidal limit of the Trent, was 30.5 m^3/s on 31 July 1996 and 48.4 m^3/s on 21 July 1997. The sampling period in July 1996 occurred at the end of a prolonged dry weather period, with correspondingly persistent low fresh water flows entering the Trent catchment. Furthermore, immediately prior to the sampling period of July 1997, a prolonged period of wet weather was observed during much of June and the early part of July.

Samples were obtained using a small submersible pump attached to a 10 mm diameter flexible plastic tube, which fed individual 250 ml plastic sample bottles. In July 1996 the pumps were deployed from a UK Environment Agency research vessel, the *Sea Vigil*,

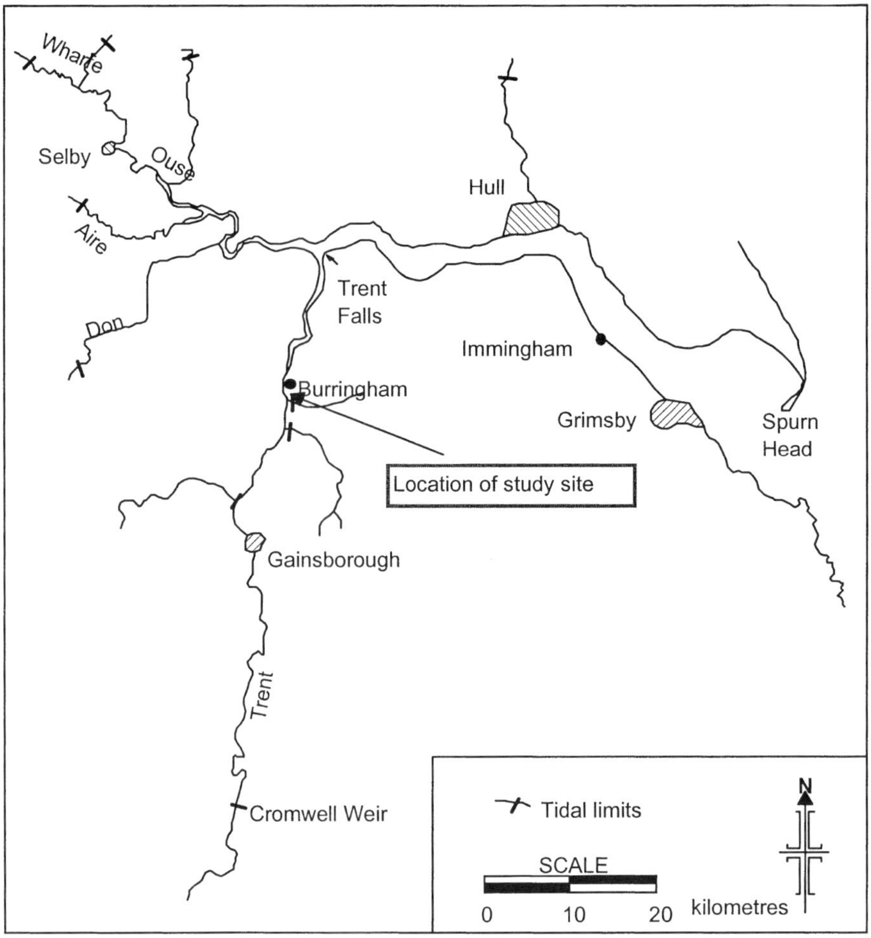

Figure 1 The Humber Estuary System Showing Location of Study Site

attached to some mooring near the west bank of the river. In July 1997, samples were collected from a culvert platform on the east bank opposite the mooring dolphins which protruded into the main channel. Thus in both cases samples and readings were taken in the main flow channel. Data are presented for a fixed sampling depth, at approximately mid-depth at high water, which is a near-surface value at low water. The SSC was obtained by gravimetric analysis of pumped samples by filtering the sample through glass fibre filter paper and weighing the dried residue. All laboratory analyses were carried out at the School of Civil Engineering at the University of Birmingham. Samples were stored in the dark at 4° C prior to analysis and the filter papers dried for 24 hours at 105°C before and after filtration.

Particle size analysis was carried out on all the collected samples, using a Malvern Mastersizer μ to obtain size distributions for samples of suspended sediment, which had previously been deflocculated with 2 g/l sodium hexametaphosphate and subjected to overnight agitation. This sizing involved use of a laser diffraction technique. Rapid stirring was carried out immediately prior to and during the sizing analysis to help to ensure that the particles were deflocculated, and to try to ensure homogeneity of the sample.

Salinity was measured *in situ* by a Valeport CTD probe, and in the July 1997 deployment a Partech IR15C transmissometer was also used to give readings of turbidity at the same depth, and at the same time as the pumped samples were removed for gravimetric analysis. Water level data was obtained from a pressure transducer connected to a data logger adjacent to the culvert platform at Burringham. During the July 1996 deployment, measurements were made of the depth mean water velocity by first obtaining velocity profiles over the depth at the sampling point in the main channel. This was done using a 125 mm diameter 8011 series Braystoke impeller connected to an automatic counter set to record the number of rotations in 50 seconds. A factory calibration was used to convert these readings into metres per second. Further details of all experimental procedures used may be found in Mitchell (1998).

3. RESULTS

The variation in SSC, measured at a fixed point in the water column, together with the primary particle size distribution measured for a range of samples, is shown for 29/30 July 1996 (Figure 2) and 21 July 1997 (Figure 3). The upper graphs in each figure (Figures 2a and 3a) show the particle size distribution for a few representative samples (3 and 6 respectively), with the particle size in microns along the bottom axis, on a log scale for clarity. The lower graphs in each figure (Figures 2b and 3b) show the temporal variation in water level and SSC on each deployment. For the July 1997 deployment, the flood tide median particle size was larger (D_{50} = 50 μm) than for the July 1996 deployment (D_{50} = 10 μm). In each figure the lower graph is a time series of water level, in metres above ordnance datum, and SSC in g/l. Figure 2 shows a complete tidal cycle, while in Figure 3 data were only available from the flood tide to the early part of the ebb tide. Also shown in Figure 3b are measured transmissometer readings in millivolts (mV), which may be related to turbidity.

The salinity readings obtained using the Valeport CTD probe reached a peak of 8.0 in the July 1996 deployment, and little variation was observed in salinity over the depth (Figure 4). For approximately the last 4 hours of the ebb tide, observed salinities were less than 1.0. The maximum observed salinity during the July 1997 deployment was 3.0. For the July 1996 deployment, the observed maximum depth mean velocity during the flood tide was approximately 1.8 m/s, while during the ebb tide the maximum depth mean velocity was observed to be approximately 1.0 m/s (Figure 5).

The faster flood tide velocities and SSC's shown in Figure 5 are estimates based on a limited amount of data collected in only the top 2-3 m of the water column, due to the difficulties of deploying the instrument array at high current speeds. The highly turbulent conditions were thought to lead to nearly well-mixed conditions during the high velocity flood tide, thus the

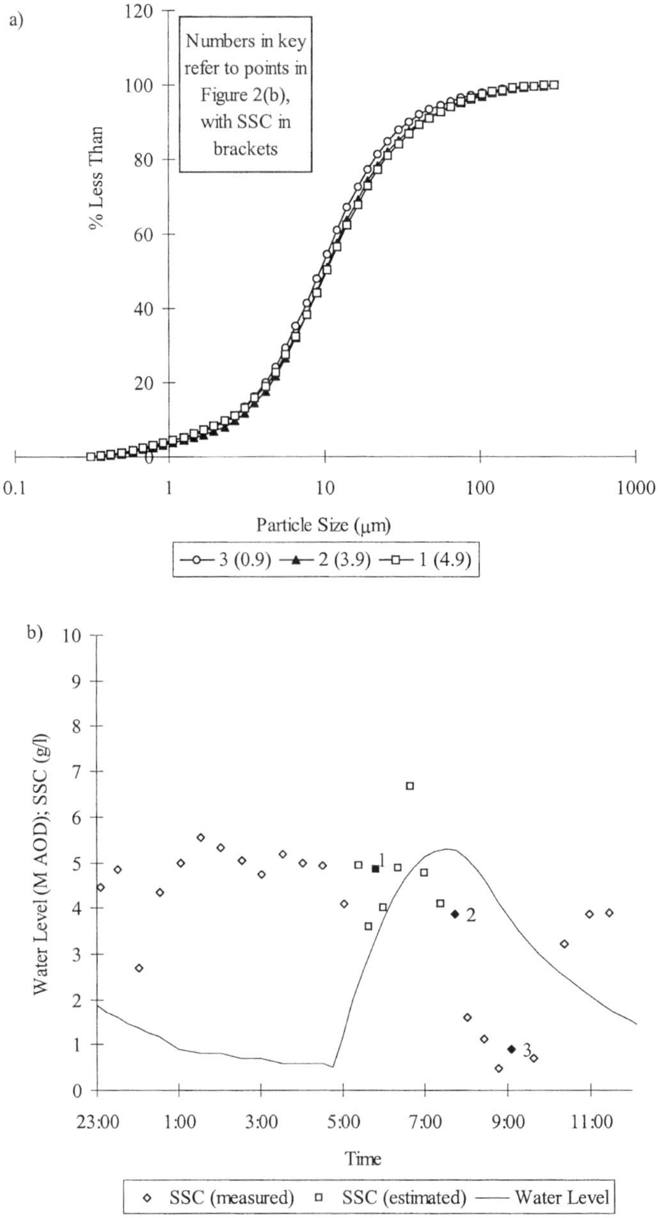

Figure 2 (a) Particle size distributions with sample concentrations (in g/l) in brackets and (b) water level and suspended solids conc. (SSC), Burringham, 29-30 July 1996

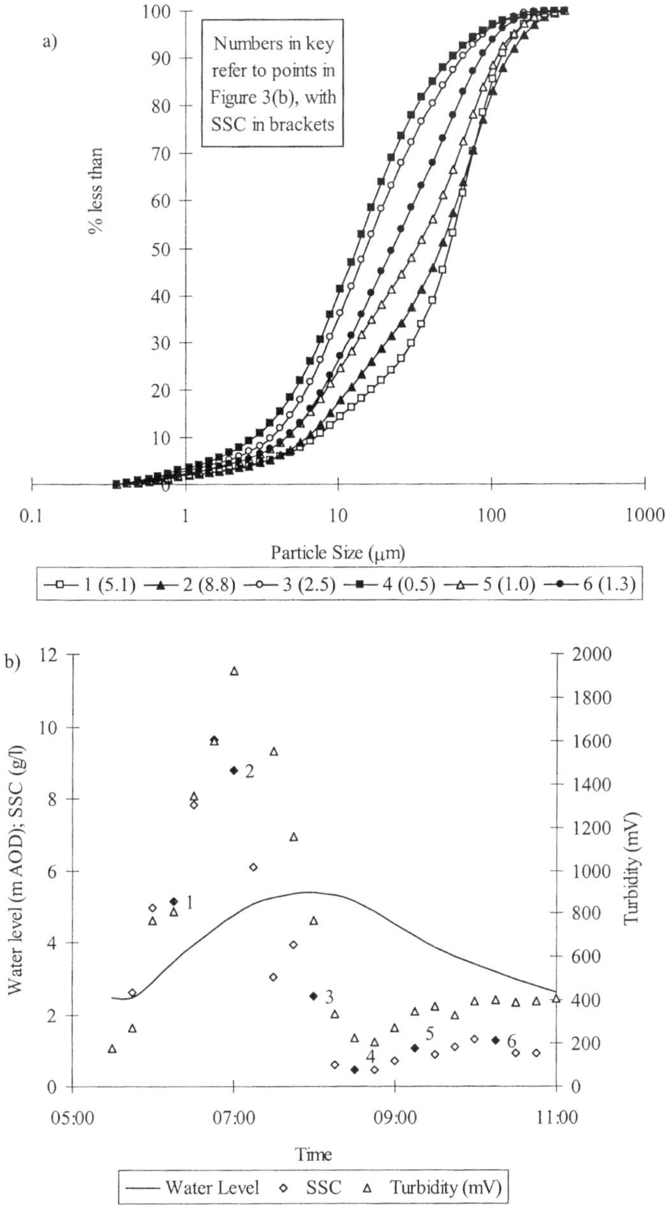

Figure 3 (a) Particle size distribution of samples with sample concentrations (in g/l) in brackets and (b) water level, turbidity and suspended solids concentrations (SSC) at Burringham, 21 July 1997

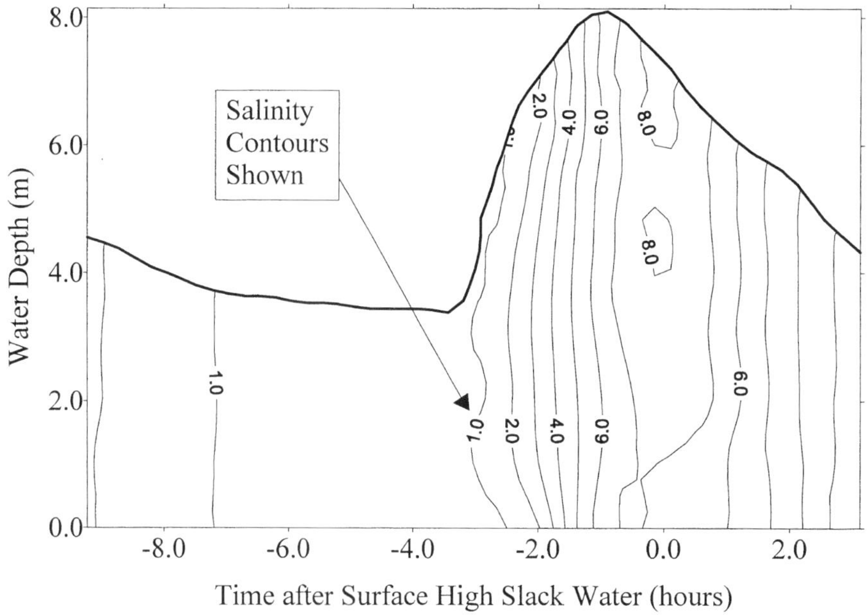

Figure 4 Vertical Salinity Profile against time 29-30 July 96. Surface High Slack Water is at 08:10

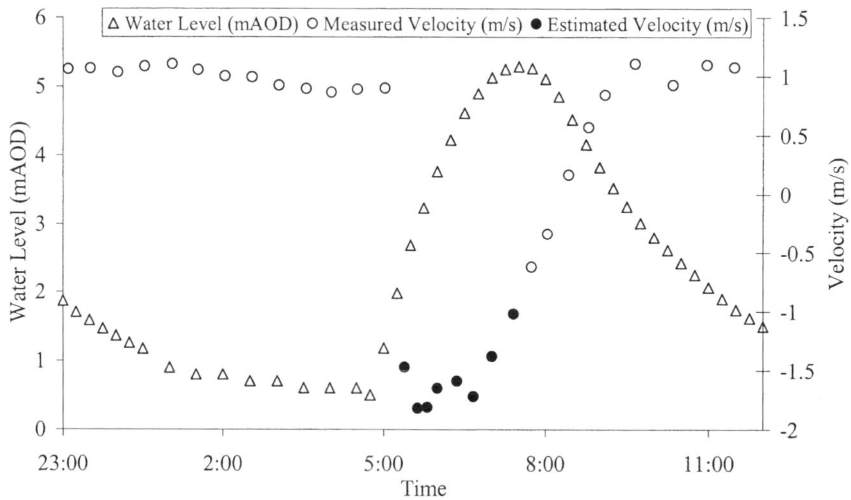

Figure 5 Variation in Depth Mean Velocity and Water Level (mAOD), Burringham, 29-30 July, 1996. Positive Velocities in Direction of Ebb Tide

near-surface readings obtained in this way were considered to be close to the depth mean values.

In both deployments the SSC varied considerably over the tidal cycle. Detailed analysis of the time-varying particle-induced vertical density gradient at this location (Mitchell et al., 1998) showed that during spring flood tide conditions, the water is generally well mixed from the start (often accompanied by a tidal bore) until after the time of maximum velocity. After this, vertical gradients are formed during the approximately 2-hour depositional phase associated with high slack water. During the subsequent ebb tide, substantial particle-induced density gradients can exist.

4. DISCUSSION

Consideration of the SSC measured at a fixed point showed that there was a higher flood tide concentration during July 1997 compared with July 1996. Consideration of the whole water column (Mitchell et al., 1998) showed that there was a higher degree of ebb-tide stratification during July 1997 than July 1996. The similarity between observed values of SSC between flood and late ebb tide (Figure 2b) demonstrates the generally well-mixed nature of the flow in July 1996. Furthermore, the lower peak tidal salinity in July 1997 reveals that high fresh water flows have led to flushing of the turbidity maximum further downstream than for July 1996. The greater variation in observed SSC in July 1997 could therefore partly be due to the difference in location of the turbidity maximum relative to the observation point. This has been brought about by the flushing of sediment downstream under the higher antecedent fresh water flow conditions of July 1997. Thus, although the *hydrodynamic* conditions are similar (due to the strong influence of the tide, compared with the river flow, on water level at this location), different *antecedent flows* contribute to different locations of the turbidity maximum, thus differences in the sediment characteristics between the two deployments.

Analysis of the distribution of primary particle size (Figures 2a and 3a) shows that the particle size exhibited more variation with time in July 1997 than in July 1996. In particular, sediment obtained during the flood tide in July 1997 has a much higher median particle size (D_{50} = 50 μm) than that obtained during the flood tide in July 1996 (D_{50} = 10 μm). During, and immediately after, slack water in July 1997, median particle size decreases markedly (points 3 and 4 on Figure 3b), providing evidence for preferential settling of larger, heavier particles, below the level of the sampling point. The absence of these larger particles in July 1996 could also help explain why the water is better mixed for a longer time during the ebb tide, compared with the July 1997 deployment, even though the hydraulic conditions are similar. This is suggested because smaller particles are more easily re-suspended under these conditions than the larger ones. Previous research carried out in the nearby lower tidal Ouse suggests that a high proportion of very fine sand particles (D_{50} = 63-125 μm) may be present on the bed towards the tidal limit under low fresh water flow conditions (Uncles et al., 1998b). The results obtained in July 1996 show that little or no transport of these very fine sand particles occurs under these conditions at this location, since the observed D_{50} value is much lower than 63 μm. This suggests that most of the 50 μm material observed in July 1997 was of a transitory nature, being re-suspended and transported along with the turbidity maximum.

The presence of larger particles during July 1997 could be linked with the influence of fresh water flow on the mechanisms and behaviour of the turbidity maximum in the Trent. It is well known that high fresh water flow can lead to rapid downstream flushing of fine sediment (Grabeman and Krause, 1989), and previous research in this system has suggested that under these conditions the turbidity maximum region was located downstream of the Burringham observation point (Mitchell et al., 1998). For July 1996, however, antecedent fresh water flow conditions had been much lower prior to the deployment, thus the turbidity maximum region was located further upstream. This points to a system of sediment sorting leading to higher concentrations of coarser particles at the landward end of the turbidity maximum, similar to that suggested by Wolanski and Gibbs (1995) and Li et al. (1999) for the Fly river and the Jiaojiang river estuaries respectively. The larger mobile particles, carried downstream by high fluvial floods, are resuspended during the high velocity flood tide. A similar degree of resuspension of larger particles does not occur during the subsequent ebb tide, leading to their concentration at the landward end of the turbidity maximum under fluvial flow recession conditions. Such a mechanism could help explain these observations in the tidal Trent.

The presence of larger particles and/or flocs may be responsible for the non-linear nature of the relationship obtained in the calibration between SSC and turbidity, measured simultaneously using a Partech IR15C transmissometer. In general an approximately linear calibration was observed, except for around high slack water when for high values of optically measured turbidity, low values of SSC were obtained by gravimetric analysis of pumped samples (see Figure 6). The R^2 value quoted on the figure does not take the two high slack water points into account. This was thought to be due to the settling out of larger particles and flocs below the level of the transmissometer lens, and the concentration of smaller flocs around the level of the lens. This result has important implications for calibrations used in highly turbid systems such as the Trent estuary between SSC and turbidity, particularly for cases where large primary particles with high settling velocity are known to exist. However, for significant portions of the tide, turbidity measurements are potentially adequate for use in long-term monitoring studies. Thus it may be seen that relatively cheap, simple optical devices for measuring turbidity may be used to good effect over long (>1 month) time periods, providing useful data about the turbidity maximum characteristics of an estuary. Particular caution needs to be exercised when using a simple calibration during slack water periods, however.

5. CONCLUSIONS

Preliminary conclusions on the results from primary, deflocculated particle size analysis from two similar deployments during July 1996 and July 1997 in the Trent estuary are outlined below.

1. Analysis of pumped samples taken over a tidal cycle at a fixed point in the water column

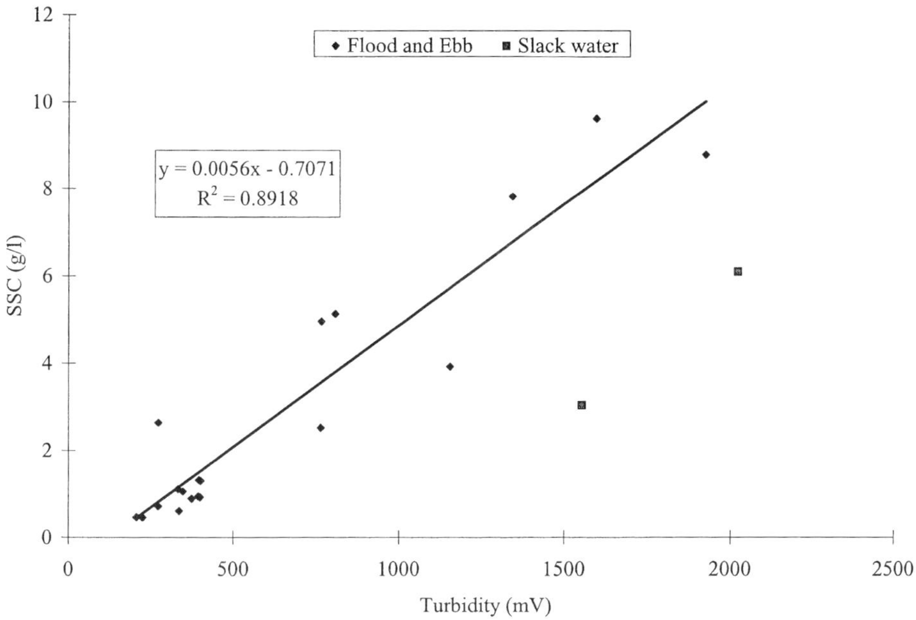

Figure 6 Suspended Solids Concentration (SSC) against Turbidity, Burringham, 21.7.97 (morning tide) with line of best fit. Slack water points not included in best fit calculation

reveals differences between results obtained from the two deployments in flood and ebb tide suspended solids concentrations. It is suggested that this can be explained in terms of the presence of large primary particles (D_{50} = 50 μm) in July 1997 which were not present during the July 1996 deployment. This in turn can be explained in terms of the position of the turbidity maximum region relative to the observation point, which in turn is related to the magnitude of the antecedent fresh water flow entering the estuary. Thus differences in the type of available sediment lead to differences in observed sediment size characteristics and concentration, even though the hydrodynamics were similar on both occasions.

2. Although based on only limited observations, a concentration of larger, heavier primary particles is proposed at the landward tip of the turbidity maximum region for the Trent estuary. This suggests that a sediment sorting regime exists similar to that described by Li et al. (1999) and Wolanski and Gibbs (1995) for the Jiaojiang and Fly estuaries respectively. The high settling velocities, which characterise such particles, lead to a far greater degree of transport on the flood compared with the ebb, thus the passage of these particles downstream is limited, compared with smaller particles.

3. Calibration of a 'transmissometer' type Partech IR15C turbidity sensor against known values of suspended solids concentration obtained by gravimetric analysis of pumped

samples reveals that a generally linear correlation may be obtained for these conditions. However, during high slack water, high concentrations of smaller, less dense particles and flocs lead to anomalously high turbidity readings. This demonstrates the need for continuous reappraisal of techniques for calibration of turbidity sensors, specific to the site and its suspended sediment characteristics at any given time.

4. Further deployments of this kind are needed to investigate this effect more fully, and to investigate similar particle-size sorting effects in other estuaries and for different tidal and fluvial conditions.

ACKNOWLEDGEMENTS

The authors gratefully acknowledge the provision of data by the UK Natural Environment Research Council Land-Ocean Interaction Study. The authors also wish to thank the UK Environment Agency for granting permission to access the culvert platform at Burringham and for providing daily mean flow data. The assistance of personnel at the School of Geography and Environmental Science, University of Birmingham, UK, especially Professor Geoff Petts and Dr John Couperthwaite, is also gratefully acknowledged.

REFERENCES

ten Brinke, W. B. M., 1997. Temporal variation in aggregate size and settling velocity in the Oosterschelde (The Netherlands) *In:* Burt, T. N., Parker, W. R. and Watts, J. (Eds) *Cohesive Sediments.* Wiley, Chichester, U.K., 63-73.

Dyer, K. R., 1986. *Coastal and Estuarine Sediment dynamics.* Wiley, Chichester, U.K.

Dyer, K. R., 1997. *Estuaries: A Physical Introduction.* 2nd Ed., Wiley, Chichester, U.K.

Fennessy, M. J., Dyer, K. R. and Huntley, D. A., 1994. INSSEV: An instrument to measure the size and settling velocity of flocs in-situ. *Marine Geology* (117), 107-117.

Grabeman, I. and Krause, G., 1989. Transport processes of suspended matter derived from time series in a tidal estuary. *Journal of Geophysical Research* (94), 14419-14428.

Guezennec, L., Lafite, R., Dupont, J-P., Meyer, R. and Boust, D., 1999. Hydrodynamics of suspended particulate matter in the tidal freshwater zone of a macrotidal estuary (the Seine, France). *Estuaries* (22), 717-727.

Law, D. J., Bale, A. J. and Jones, S. E., 1997. Adaptation of focused beam reflectance measurement to in-situ particle sizing in estuaries and coastal waters. *Marine Geology* (140), 47-59.

Lawler, D. M., 1995. Turbidimetry and Nephelometry. In: *Encyclopedia of Analytical Science*. Academic Press.

van Leussen, W., 1997. The Kolmogorov microscale as a limiting value for the floc sizes of suspended fine-grained sediments in estuaries. *In:* Burt, T. N., Parker, W. R. and Watts, J. (Eds) *Cohesive Sediments*. Wiley, Chichester, U.K., 45-62.

Li B. G., Eisma D., Xie, Q. Ch., Kalf, J., Li Y. and Xia X., 1999. Concentration, clay mineral composition and coulter counter size distribution of suspended sediment in the turbidity maximum of the Jiaojiang river estuary, Zhejiang, China. *Journal of Sea Research* (42), 105-116.

Mitchell, S. B., 1998. Aspects of sediment transport in the Humber estuary system. Ph. D. Thesis, University of Birmingham.

Mitchell, S. B., West, J. R., Arundale, A. M. W., Guymer, I. and Couperthwaite, J. S., 1998. Dynamics of the turbidity maxima in the upper Humber estuary system, UK. *Marine Pollution Bulletin* (37), 190-205.

Puls, W., Kuehl, H. and Heymann, K., 1988. Settling velocity of mud flocs: results of field measurements in the Elbe and the Weser estuary. In: Dronkers J. and van Leussen W. (Eds.) *Physical Processes in Estuaries*. Springer-Verlag, Berlin, 404-424.

Uncles, R. J., Easton, A. E., Griffiths, M. L., Harris, C., Howland, R. J. M., King, R. S., Morris, A. W. and Plummer, D. H., 1998a. Seasonality of the Turbidity Maximum in the Humber-Ouse Estuary, UK. *Marine Pollution Bulletin* (37), 206-215.

Uncles, R. J., Stephens, J. A. and Harris, C., 1998b. Seasonal variability of subtidal and intertidal sediment distributions in a muddy, macrotidal estuary: the Humber-Ouse, UK. *In:* Black, K. S., Paterson, D. M. and Cramp, A. (Eds) *Sedimentary Processes in the Intertidal Zone*. Geological Society, London, Special Publications, 139, 11-24.

Vanous, R. D., Larson, P. E. and Hach, C. C., 1982. The theory and measurement of turbidity and residue. *In:* R.A. Minear and L.H. Keith (eds) *Water Analysis*, Vol. 1, Academic Press, New York and London, 163-234.

Wolanski E. and Gibbs R. J., 1995. Flocculation of suspended sediment in the Fly River Estuary, Papua New Guinea. *Journal of Coastal Research* (11), 754-762.

On the geometry of cohesive settling flocs

P.D. Scarlatos[a] and H.-S. Kim[b]

[a]Department of Civil Engineering, Florida Atlantic University,
777 Glades Road, Bldg. 36, Boca Raton, Florida 33431, USA

[b]Department of Civil Engineering, Sun Moon University,
100 Kalsan Ri, Tangjeong Myeon, Asan Si, Chungnam, 336-840 Korea

Numerical experiments have been conducted to investigate the aggregating processes of settling flocs within the mobile and stationary suspension layers. The methodology used included the maximum chain collision model for floc aggregation of mobile suspensions and differential settling collision for flocculation of stationary suspensions. The aggregates were analyzed in terms of their fractal dimension and other geometric characteristics using the Feret's diameter and Minkowski's logic. The results indicated a fractal dimension varying between 1.87 and 2.1 for the aggregates and a dimension of 1.65 to 2.45 for the floccules. The results are in agreement with observations from other numerical and experimental studies.

KEY WORDS
aggregation; cohesive sediments; flocculation; fractals; sedimentation

1. INTRODUCTION

Cohesive sediments constitute an important component of many aquatic systems. Due to their small particle size, cohesive sediments, whenever introduced into a liquid medium, demonstrate a complex behavior involving electrochemical reactions, diffusion, sedimentation and non-Newtonian rheology (Russel et al., 1989). As a result, cohesive sediment particles tend to bond together to form aggregates (clusters of single-particles) and floccules (clusters of aggregates). The process of particle bonding depends on the physicochemical and microbiological properties of the water-sediment system as well as the flow regime. The size, shape and strength of the individual aggregates/floccules are strongly affected by the intensity of Brownian motion, differential settling and velocity-shear (Krone, 1984).

Understanding of the geometric characteristics of cohesive particles is very important in quantifying different physical phenomena such as mud re-suspension and settling events in estuarine environments, and dissolved/particulate pollutant exchange between sediments and ambient water (Scarlatos, 1997). In addition, knowledge of particle geometry can lead into design of more effective industrial processes related to colloidal particles (Van de Ven, 1990). However, aggregates and floccules both in dilute and concentrated suspensions present a very complex geometry that can be effectively described only through the usage of fractal geometry (Falconer, 1985; Takayasu, 1990, Vicsek, 1992, Kranenburg, 1994, Winterwerp, 1999).

Depending on the wave and current energy levels available, suspended cohesive sediments in estuaries can be found in a highly dispersed state (mobile suspension), or as a highly flocculated near-bed layer (stationary suspension) (Mehta, 1989). In the mobile suspension particles are either single particles, or small particle number aggregates, that move under the action of Brownian motion, turbulent velocity fluctuations and flow current. Within the stationary suspension, particles are found as large number complex floccules that deposit under their own weight on the consolidating layer.

Based on the above observations, this study was focused on quantification of aggregate formation in the mobile suspension zone and flocculent formation in the stationary suspension zone. Particle aggregation in the mobile suspension zone was simulated using the maximum chain model (Sutherland, 1970). In the stationary suspension zone aggregates were formed after colliding through differential settling (Swamee and Ojha, 1991). The analysis of the resulting particle geometries was accomplished using the Feret's diameter and Minkowski's logic (Mandelbrot, 1983). The fractal geometry was then estimated from the exponential relationship between the perimeter of the particle and the Minkowski's lattice intervals or between the particle size and its radius of gyration (Kim, 1992).

2. MODEL DEVELOPMENT

The model developed for this study was comprised from two independent computational modules: (a) The particle-aggregation module describing particle motion and interaction within the mobile suspension layer and, (b) The aggregate-flocculation module describing particle settling and interaction within the stationary suspension layer. Aggregates formed in the mobile suspension computational module were used as input for the stationary suspension module. The model did not consider the consolidating layer that lies below the stationary suspension zone. Therefore consolidation restructuring and dewatering effects were not included.

2.1. Particle-Aggregation Module

There are various models available for the description of aggregate formation (Sutherland, 1967; Szep et al., 1985; Deutsch, 1985; Okuyama et al., 1981; Wang and Friedlander, 1967; Bunde et al., 1985). All of these models are basically similar to the original Smoluchowski's model, where particles moving at random collide, according to the some collision probability law, to form aggregates (Smoluchowski, 1917). Some special categories of random motion aggregation models include the diffusion-limited aggregation (DLA) model (Witten and Sander, 1981), the diffusion-limited cluster-cluster aggregation (CCA) model (Donnell and Finegold, 1981), the reaction- or chemically-limited aggregation (CLA) model (Jullien and Kolb, 1984), and the ballistic-aggregation (BA) model (Sutherland and Goozard-Nia, 1971).

For this study the maximum chain aggregation model was adapted. This model provides the collision sequence most likely to produce aggregates similar to those observed in nature (Sutherland, 1970). In the maximum chain aggregation model all single particles collide to form doublets, all doublets collide to form quadruplets, all quadruplets to form octuplets, and so on (Sutherland and Goodarz-Nia, 1971). The collision rate c_{ij} for a polydispersed system was defined as

$$c_{ij} = 4\pi D_1 R_1 (\frac{1}{R_i} + \frac{1}{R_j})(R_i + R_j) n_i n_j$$

where c_{ij} is the collision rate, D_1 is the diffusion coefficient of the primary (single) particles, R_1 is the radius of the primary particles, R_i and R_j are the distances from the centroids of aggregates of ith and j^{th} order to their furthest extremity, and n_i and n_j are the concentrations by particle number of the of ith and j^{th} order aggregates. The particles were subjected to a random motion defined by the following transformation matrix (Moran and Kendall, 1963):

$$\begin{bmatrix} \xi & \sqrt{1-\xi^2}\cos\varphi & \sqrt{1-\xi^2}\sin\varphi \\ \sqrt{1-\xi^2}\cos\theta & -\xi\cos\varphi\cos\theta+\sin\varphi\sin\theta & -\xi\sin\varphi\cos\theta-\cos\varphi\sin\theta \\ -\sqrt{1-\xi^2}\sin\theta & \xi\cos\varphi\sin\theta+\sin\varphi\cos\theta & \xi\sin\varphi\sin\theta-\cos\varphi\cos\theta \end{bmatrix}$$

where ξ is a random number ($-1 \geq \xi \geq 1$), and ϕ and θ are random polar and equatorial angles ($-\pi \geq \phi, \theta \geq \pi$) respectively. The three vectors defined by the rows of the above transformation matrix are orthogonal to each other. Particles and clusters rotate until collision occurs. After collision, a new coordinate system is established for each aggregate and the process continues. The simulation did not considered any advective motion of the particles. However, by varying the intensity and other characteristics of the random motion, both Brownian and turbulent fluctuation effects can be simulated effectively.

2.2. Aggregate-Flocculation Module

In the aggregate-flocculation model aggregates simulated in the mobile suspension zone randomly entered the stationary suspension layer after they have reached a certain size. The terminal velocity w, of the arbitrary shaped floccules was defined as (Swamee and Ojha, 1991)

$$w^* = \left[\frac{44.84(v^*)^{0.667}}{(1+4.5\beta^{0.35})^{0.833}} + \frac{0.784}{[\beta^4+20\beta^{20}+(v^*)^{2.4}\exp(18.6\beta^{0.4})]^{0.125}}\right]^{-1}$$

where w^* is the normalized terminal velocity, v^* is the normalized kinematic viscosity and β is the floccule shape factor. More specifically, the normalized terminal velocity is given as

$$w^* = \frac{w}{\sqrt{(s-1)gD_n}}$$

where $s = \frac{\rho_s}{\rho}$ the ratio of sediment over water densities, and D_n is the nominal floccule diameter $D_n = \sqrt[3]{\frac{6V}{\pi}}$, where V is the volume of the floccule. The shape factor is estimated as

$$\beta = \frac{a}{\sqrt{bc}}$$

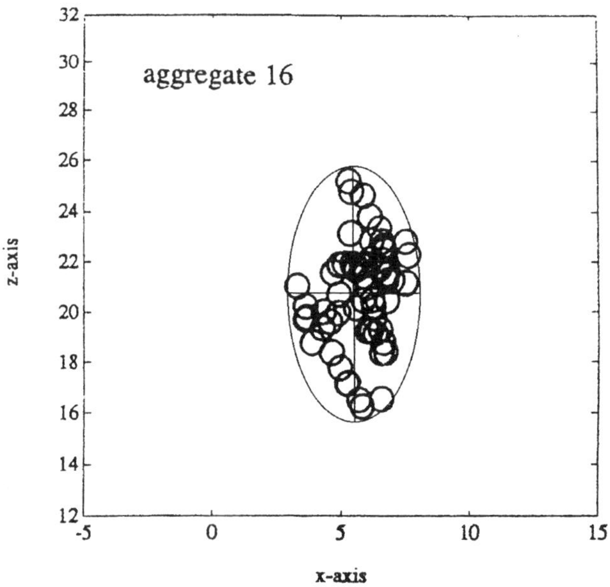

Figure 1. Principal axes of ellipsoid-equivalent shape of a floccule.

where a, b, c are the three principal axes of the floccule ($c \geq b \geq a$). These axes correspond to an ellipsoid that has the same radii of gyration with the floccule area projections along the x-z and y-z planes (z being the gravitational direction) (Figure 1).

The normalized kinematic viscosity is defined as

$$v^* = \frac{1}{R_e}\sqrt{\frac{4}{3C_D}}$$

where $R_e = \sqrt{\frac{wD_n}{v}}$ is the Reynolds number and C_D is the drag coefficient. The equation of the terminal velocity is valid for $v^* \geq 1.8x10^{-4}\sqrt{\beta}$ (Swamee and Ojha, 1991). Once the particles are in contact a new particle is formed and the resulting geometric characteristics and terminal velocity of the new particle are estimated. Again, during this simulation module advective motion was neglected. However, by changing the pattern of aggregate input, the process of either rapid or slow sedimentation can be effectively reproduced. Since no advection or velocity shear was present, differential settling was the main flocculation process (Eisma, 1991; Van Leussen, 1994).

2.3. Analysis of Aggregate/Floccule Fractal Geometry

Once the aggregates and the floccules are formed their geometric characteristics are analyzed. This included estimation of the cluster size (number of primary particles N), projected area A (either in the x-z or y-z planes), anisotropy S, bulkiness B, structure factor F, and radius of gyration about the centroid R_G (Medalia, 1967; 1970; Medalia and Heckman, 1969). The parameters S, B, F and R_G are given as

$$S = \frac{R_a}{R_b}, \qquad B = \frac{4\pi R_a R_b}{A}, \qquad F = SB - 1, \qquad R_G = \sqrt{\frac{I_c}{A}}$$

where R_a, R_b are the radii of gyration about the principal axes a and b, and I_c is the moment of inertia about the centroid of area A.

In addition to the above geometric parameters the fractal dimension δ, of both aggregates and floccules was estimated. For the aggregates the fractal dimension δ was estimated from the general relationship between particle size N, and the radius of gyration (Botet et al., 1984)

$$N \propto R_G^{\delta}, \qquad N \to \infty$$

For the more complex floccule configurations the fractal dimension was calculated according to the relationship

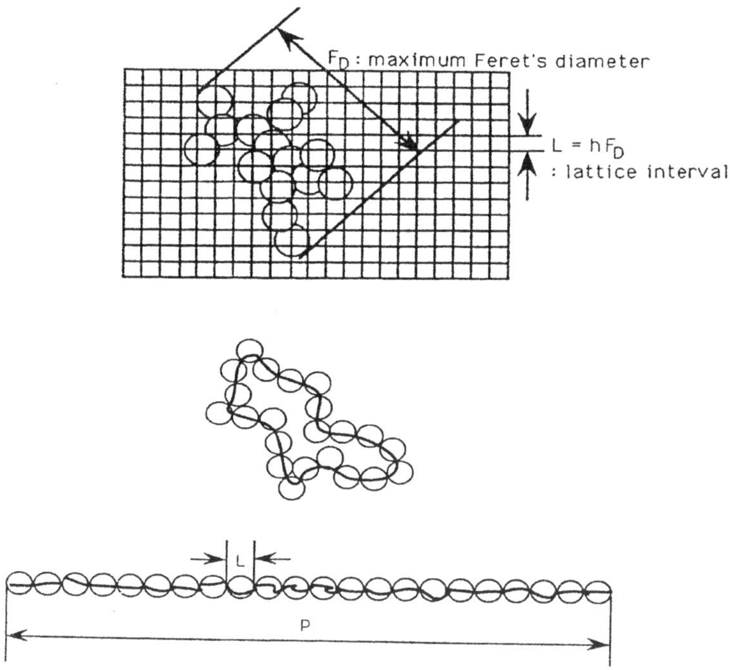

Figure 2. Maximum Feret's diameter, lattice interval and Minkowski's logic.

$$P = P_o L^m, \qquad \delta = 1 + |m|$$

where P is the perimeter of the floccule estimated using the Minkowski's logic, P_o is the exact perimeter of the aggregate, L is the lattice interval, and m is an exponent (Figure 2). Using the maximum Feret's diameter F_D, the perimeter P is estimated as

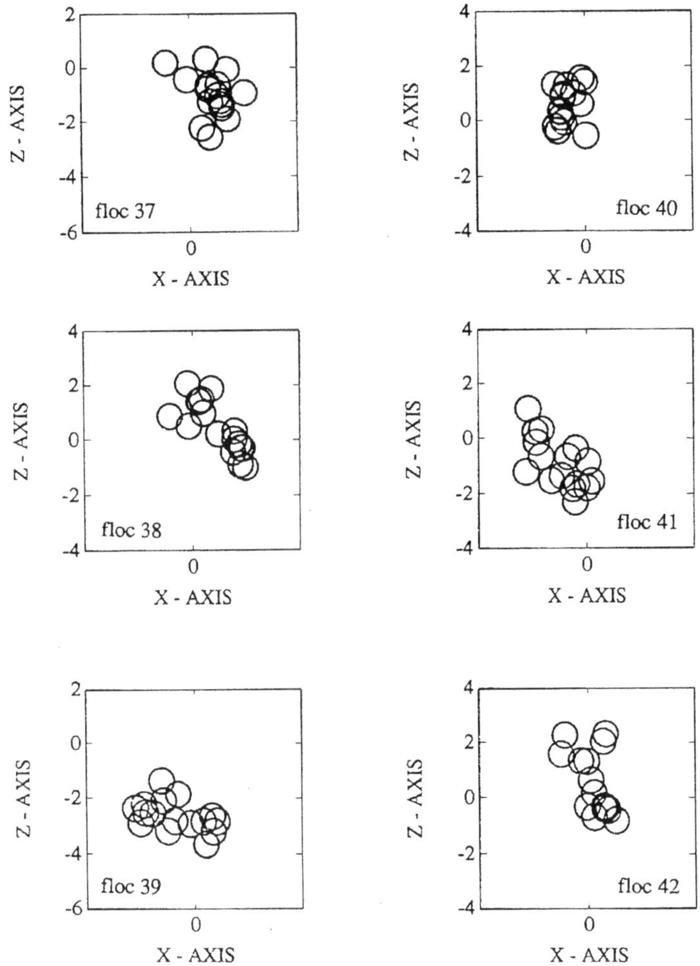

Figure 3. Examples of simulated aggregates in the mobile suspension zone.

$$P = nhF_D = nL$$

where n is the number of lattice intervals and h is a fraction of the Feret's diameter. The maximum Feret's diameter is taken as the maximum length between two extreme points of any projected area of the floccule. As is it expected, by decreasing the size of the lattice interval, a more accurate prediction of the perimeter is accomplished (Kaye, 1989) (Figure 2).

For both computational modules a collision efficiency rate of 10% was assumed. This assumption probably should not have any substantial effect on the procedure since both the random paths of the particles in the mobile suspension zone and their settling motion in the stationary suspension zone were allowed to continue until the aggregates/flocculants have reached a certain particle size. Since, the collision efficiency rate was addressed in a parametric form any other rate than 10% could be easily adjusted in the simulations.

3. RESULTS AND DISCUSSION

In order to investigate the geometry of the aggregates in the mobile suspension zone, fifty particles (each comprised of sixteen primary particles) were analyzed (Figure 3). The average geometric parameters for those and other smaller size aggregates (4- and 8-particle) were compared to the results of another investigation (S-GN) (Sutherland and Goodarz-Nia, 1971). The simulated data of the two studies showed a very good agreement (Table 1).

As it can be seen from Table 1, for increasing aggregate size the aggregate area, bulkiness, structure factor and radius of gyration increase, while anisotropy decreases.

Using the aggregate size N versus the radius of gyration R_G relationship, the fractal dimension of these aggregates was found to be equal to 2.1. This value is close to the value 1.92 estimated for the results of Sutherland and Goodarz-Nia's simulation. However, $\delta = 2.1$, is much higher than the fractal dimension of 1.53 estimated by using a chemically limited cluster-to-cluster aggregation model (Jullien and Kolb, 1984), but close to the value of 1.8 reported from analysis of experimental results (Torres et al., 1991). The fractal dimension of 2.1 is also very close to the dimension of 2.0 reported for cohesive sediments (Winterwerp, 1999).

Table 1
Average geometric characteristics of simulated aggregates

	Simulated results	S-GN (1971)	Simulated results	S-GN (1971)	Simulated results	S-GN (1971)
Aggregate size (N)	4	4	8	8	16	16
Area (A)	2.71	2.63	4.87	5.00	8.89	9.38
Anisotropy (S)	2.27	2.14	2.01	1.96	1.96	1.97
Bulkiness (B)	1.15	1.19	1.34	1.40	1.39	1.55
Structure factor (F)	1.61	1.54	1.69	1.75	1.73	2.09
Radius of gyration (R_G)	1.94	2.09	2.73	2.91	3.64	4.14

Table 2
Example of input data for simulation of the stationary sediment suspension zone

Particle number	x-axis location	y-axis location	a	b	c	β	Injection time	Fall velocity	D_n
1	13.13	6.76	3.12	4.01	5.18	0.68	0.487	1.04	2.18
2	4.38	38.00	3.91	4.77	5.00	0.80	0.269	1.19	1.29
3	35.62	29.25	2.86	4.89	6.35	0.51	0.050	0.90	1.04
4	26.87	20.50	4.25	7.14	7.52	0.58	0.831	0.96	2.29
5	18.12	11.75	3.51	5.33	5.47	0.65	0.612	1.02	2.36
6	9.37	2.99	2.82	4.21	4.83	0.63	0.393	1.00	1.67
7	0.61	34.24	4.67	5.17	6.06	0.83	0.175	1.26	2.32
8	31.86	25.49	3.03	4.04	4.68	0.70	0.956	1.06	2.20
9	23.11	16.74	3.02	3.90	5.17	0.67	0.737	1.04	1.06
10	14.36	7.99	3.69	4.11	5.40	0.78	0.518	1.16	1.91

For the simulation of the stationary suspension zone, the kinematic viscosity of the water-sediment mixture was taken as 1.003×10^{-6} m^2/s, the size of a primary particle as 2 μm and the flocculent and water densities as 1.34 t/m^3 and 0.998 t/m^3 respectively. The "primary" aggregate entering the settling zone was consisting of sixteen primary particles as generated by the previous computational module. This assumption is realistic since the smaller settling floc sizes in an estuarine environment have a diameter of about 30 to 50 μm (Lick and Huang, 1993).

Hindered-settling effects were depicted directly through the simulation since the individual trajectories of the particles were followed. However, the possible effects on floccules settling due to the hydrodynamic wake generated by adjacent falling particles was not included in this study. Possible breakings of very large settling floccules due to hydrodynamic or electrochemical effects were not considered in this study.

As expected the resulting flocculated particles varied both in size and shape but generally resembled other simulated and experimental data (Medalia and Heckman, 1969; Sutherland and Goodarz-Nia, 1971). Also, some of the original aggregates remained unchanged throughout the simulation period. The fractal geometry of the resulting floccules was estimated using different lattice sizes of the Feret's diameter (Figure 4).

The analyses indicated that the fractal dimension of the floccules ranged from 1.65 to 2.45. The small values (1.65 to 2.10) mainly corresponded to the 16-particle aggregates that went through the simulation of the settling process unchanged. This aggregate fractal dimension is slightly smaller than the one estimated by using the particle size versus the radius of gyration relationship.

Primarily the small fractal dimensions correspond to fragile aggregates while the higher dimensions are indicative of stronger estuarine floccules.

Some of the larger particles demonstrated a dual fractal dimension (e.g. 2.15 and 2.35). The larger value is indicative of the structural while the other of the textural fractal dimension of the floccules.

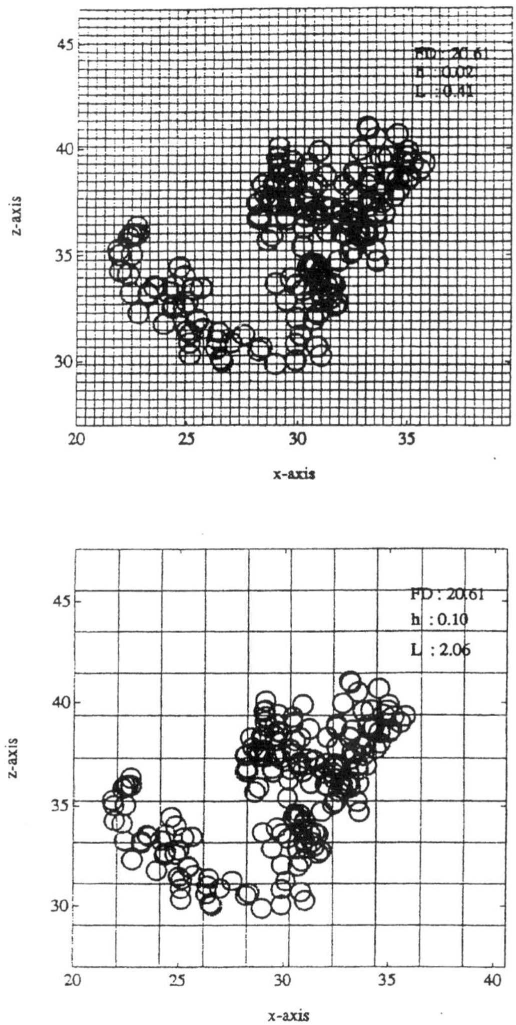

Figure 4. Lattice intervals for fractal analysis of a 256-particle size floccule.

4. CONCLUSIONS

The conclusions of this study are the following:

- The processes of aggregation within the mobile suspension zone can be effectively simulated, under slack tide condition (no advection), by using the maximum chain aggregation model.

- The simulated aggregates in the mobile suspension zone have an average fractal dimension δ, of either 1.87 or 2.1. The lower fractal dimension, corresponds to the one estimated by the aggregate size versus the radius of gyration relationship, while the higher value to the lattice interval and Minkowski's logic analysis. These results are very close to fractal dimension values reported in the literature (e.g., $\delta = 1.53$, 1.80, 1.92, 2.0).
- For increasing aggregate size N, anisotropy S decreases while aggregate area A, bulkiness B, structure factor F, and radius of gyration R_G, increase.
- The processes of flocculation within the stationary suspension zone area can be simulated by randomly introducing aggregates from the mobile suspension and following their collisions through differential settling processes.
- The settling velocity of the arbitrarily shaped floccules is defined as a function of the particle geometry, and flow and fluid/sediment characteristics. The settling velocity along with the collision rates simulates the effect of hindered settling.
- The fractal dimension of the floccules in the settling zone varies from 1.65 to 2.45 with the smaller values being representative of smaller size particles. Also it is believed that the lower fractal dimensions correspond to weak aggregates while the higher fractal dimensions are indicative of stronger estuarine floccules.
- Some complex shaped floccules demonstrated two fractal dimensions: one for structural and another for textural geometry, (e.g., $\delta = 2.15$ and $\delta = 2.35$).

5. RECOMMENDATIONS FOR FURTHER STUDY

The results of this study have been compared primarily with data from aggregates/floccules found in industrial and chemical processes. Thus, a more close comparison with estuarine cohesive sediments is needed.

The simulation excluded advective motion and velocity shear. Thus, inclusion of these two components can provide a more realistic aggregate/floccule formation during the entire tidal cycle.

Since pollutant exchange between ambient water and sediment particles depends on the specific surface of the particles, there may exist relationships between the fractal dimension and other geometric features of the aggregates/floccules versus their adsorption/desorption capacity.

REFERENCES

Botet, R., Jullien, R. & Kolb, M., 1984, Hierarchical model for irreversible kinetic cluster formation, *J. of Physics A: Mathematical & General*, (17), L75-L79.

Bunde, A., Herrmann, H.J. & Stanley, H.E., 1985, The shell model: A growth model with a tunable fraction of forgotten growth sites, *J. of Physics A: Mathematical & General*, (18), L523-L529.

Deutsch, J.M., 1985, Aggregation-disorder transition induced by fluctuating random forces, *J. of Physics A: Mathematical & General*, (18), 1449-1456.

Eisma, D., 1991, Particle size of suspended matter in estuaries, *Geo-Marine Letters*, (11), 147-153.

Eisma, D., 1991, Particle size of suspended matter in estuaries, *Geo-Marine Letters*, (11), 147-153.

Falconer, K.J., 1985, *The geometry of fractal sets*, Cambridge University Press, Cambridge, U.K.

Jullien, R. & Kolb, M., 1984, Hierarchical model for chemically limited cluster-cluster aggregation, *J. of Physics A: Mathematical & General*, (17), L639-L643.

Kanenburg, C., 1994, On the fractal structure of cohesive sediment aggregates, *Estuarine, Coastal and Self Science*, (39), 451-460.

Kaye, B.H., 1989, *A random walk through fractal dimensions*, VCH, Berlin, Germany.

Kim, H.S., 1992, Settling of fine particles, *MS Thesis*, Department of Ocean Engineering, Florida Atlantic University, Boca Raton, Florida, U.S.A.

Krone, R.B., 1984, The significance of aggregate properties to transport processes, in: *Estuarine Cohesive Sediment Dynamics*; A.J. Mehta (Ed.), Springer-Verlag, Berlin, Germany, 66-84.

Leussen, W. Van, 1994, Estuarine macroflocs and their role in fine-grained sediment transport, *PhD Thesis*, Utrecht University, Utrecht, The Netherlands.

Lick, W. & Huang, H., 1993, Flocculation and the physical properties of flocs, in: Nearshore and Estuarine Cohesive Sediment Transport, A.J. Mehta (ed.), *Coastal and Estuarine Studies*, (42), 21-39.

Mandelbrot, B.B., 1983, *The fractal geometry of nature*, W.H. Freeman and Company, San Francisco, U.S.A.

Medalia A.I., 1967, Morphology of aggregates - I. Calculation of shape and bulkiness factors; Application to computer-simulated random flocs, *J. of Colloid and Interface Science*, (24), 393-404.

Medalia, A.I., 1970, Morphology of aggregates - IV. Effective volume of aggregates of carbon black from electron microscopy; Application to vehicle absorption and to die swell of filled rubber, *J. of Colloid and Interface Science*, (32), 115-131.

Medalia, A.I. & Heckman, F.A., 1969, Morphology of aggregates – II. Size and shape factors of carbon black aggregates from electron microscopy, *Carbon*, (7), 567-582.

Mehta, A.J., 1989, Fine sediment stratification in coastal waters, *Proc. 3rd Nat'l Conf. On Dock and Harbour Engineering*, Karnataka Regional Engineering College, Surathkal, India, 487-492.

Moran, P.A.P. & Kendall, M.G., 1963, *Geometrical probability*, Hafner Publishing Company, New York, New York, U.S.A.

Okuyama, K., Kousaka, Y. & Payatakes, A.C., 1981, Evaluation of the effect of nonsphericity of fine aggregate particles in Brownian coagulation, *J. of Colloid and Interface Science*, (81), 21-31.

Russel, W.B., Saville, D.A. & Schowalter, W.R., 1989, *Colloidal dispersions*, Cambridge Monographs on Mathematics and Applied Mathematics, Cambridge University Press, Cambridge, U.K.

Scarlatos, P.D., 1997, Experiments on water-sediment nutrient partitioning under turbulent, shear and diffusive conditions, in: *The Interactions Between Sediments and Water*; R.D. Evans, J. Wisniewski & J.R. Wisniewski (Eds.), Kluwer Academic Publishers, Dordrecht, The Netherlands, 411-425.

Smoluchowski, M. Von, 1917, Versuch einer mathematischen theorie der koagulationkinetik kollider losungen, *Z. Phys. Chem.*, (92), 129-168.

Sutherland, D.N., 1967, A theoretical model of floc structure, *J. of Colloid and Interface Science*, (25), 373-380.

Sutherland, D.N., 1970, Chain formation of fine particle aggregates, *Nature*, (226), 1241-1242.

Sutherland, D.N. & Goodarz-Nia, I., 1971, Floc simulation: The effect of collision sequence, *Chemical Engineering Science*, (26), 2071-2085.

Swamee, P.K. & Ojha, C.P.S., 1991, Drag coefficient and fall velocity of nonspherical particles, *J. Hydraulic Engineering*, (117), 660-667.

Szep, J., Cserti, J. & Kertesz, J., 1985, Monte Carlo approach to dendritic growth, *J. of Physics A: Mathematical & General*, (18), L413-L418.

Takayasu, H., 1990, *Fractals in the physical sciences*, Manchester University Press, Manchester, U.K.

Torres, F.E., Russel, W.B. & Schowalter, W.R., 1991, Floc structure and growth kinetics for rapid shear coagulation of polystyrene colloids, *J. of Colloid and Interface Science*, (142), 554-574.

Van de Ven, T.G.M., 1990, Microrheological aspects of particle interaction in flowing suspensions, in: *Hydrodynamics of Dispersed Media*; J.P. Hulin, A.M. Cazabat, E. Guyon & F. Carmona (Eds.), Elsevier Science Publishers, Amsterdam, The Netherlands, 103-117.

Vicsek, T. 1992, *Fractal growth phenomena*, World Scientific, Singapore.

Wang, C.S. & Friedlander, S.K., 1967, The self-preserving particle size distribution for coagulation by Brownian motion, *J. of Colloid and Interface Science*, (24), 170-179.

Winterwerp, H., 1999, On the dynamics of high-concentrated mud suspensions, *PhD Thesis*, Delft University of Technology, Delft, The Netherlands.

Comparison of flocculation models for applied sediment transport modelling

J R Spearman* and W Roberts*

*H R Wallingford, Wallingford, Oxon, UK.

As part of the MAST3-COSINUS European Project, various flocculation models were tested against settling velocity measurements also undertaken as part of the same project and obtained during a comprehensive field data exercise that place in Calstock in the Tamar Estuary, UK. Based on the ability of each of the flocculation models to reproduce the observed data set conclusions are drawn regarding their applicability in applied modelling. Further investigation is made into the possibility of using suspended sediment profiling and the Rouse profile theory for estimating settling velocity.

Cohesive sediment, flocculation, settling velocity

1. INTRODUCTION

This paper describes work undertaken by HR Wallingford under Task E of the MAST3-COSINUS European Project regarding the parameterisation of flocculation processes for use in the applied modelling of cohesive sediment transport. The aim of the parameterisation task within COSINUS is to simplify the main results of the detailed studies of sediment properties to develop formulations for use in applied engineering models, which can be run with large numbers of grid cells for long prototype time periods. The algorithms developed must therefore be reasonably quick, sparing in the use of storage and robust, whilst as far as possible representing the most important observed features of the physical processes.

The objectives of the study were as follows:

- To examine various available flocculation models and to test how they behave in different types of situation.
- To examine the ability of the flocculation models to reproduce a specific data set, namely the measurements taken at Calstock under Task B of the MAST3-COSINUS Project.
- To draw conclusions regarding the applicability and accuracy of each of the tested models for applied modelling purposes.

2. DESCRIPTION OF MODELS

The following settling/flocculation models were incorporated into a 1DV sediment-turbulence interaction model. The 1DV model allowed the specification of a time varying or constant pressure gradient or depth-averaged velocity and the specification of a constant depth-averaged concentration. The model then calculated the steady-state or time varying sediment concentration profile depending on the input conditions and the choice of flocculation/settling model. The effect of damping on turbulence was represented by the Munk-Anderson damping functions.

Model F1. Constant settling velocity

$$w_s = W_0 \tag{1}$$

Setting the settling velocity to a constant value forms the most basic approach to the assignment of values for settling velocity. Although this approach cannot describe any of the variation in settling velocity that can occur as a result of increased turbulence or concentration, etc, it has the merit of only needing one parameter to calibrate. This approach is most commonly used for applied modelling situations, which often involve little or no data regarding settling velocity in the model area.

Model F2. Power law

$$w_s = kC^m \tag{2}$$

where k and m are empirical coefficients to be determined.

Next to model F1 this model is perhaps the most widely used of the various models consisted here, as it allows some flexibility in the value of settling velocity, based on a readily calculable parameter, but retains some simplicity in that calibration is restricted to two parameters. However, these two parameters can vary considerably from estuary to estuary and from measurement device to measurement device (Whitehouse et al, 2000).

Model F3. Power law with dissipation parameter function

$$w_s = kC^m \frac{1+aG}{1+bG^2} \tag{3}$$

where G is the dissipation parameter given by,

$$G = \sqrt{\frac{\varepsilon}{\upsilon}} \tag{4}$$

ε is the turbulent dissipation rate per unit mass,
v is the kinematic viscosity,
and k, m, a and b are empirical coefficients to be determined.

The dissipation parameter function was first suggested by van Leussen (1994). This model has the benefit of including the effect of increasing turbulence on floc breakup, represented by the dissipation parameter function, as well as the effect of more frequent collision on flocculation, represented by concentration. It has been used successfully to model settling in the Weser Estuary (Malchareck, 1995, summarised in Peltier et al, 1996).

Model F4. Equilibrium floc size model (Winterwerp, 1999)

As part of his doctoral thesis (Winterwerp, 1999), Winterwerp hypothesised a framework for the growth and recession of flocs under the influence of changing conditions within the water column. He developed the following equation governing floc evolution,

$$\frac{\partial D}{\partial t} = K_A C G D^2 - K_B G^{3/2} D^2 (D - D_p) \tag{5}$$

where D_p and D are is the size of an individual particle and of the floc size respectively,
G is the dissipation parameter,
K_A and K_B are empirical constants to be determined which govern the relative importance of the mechanisms of floc break-up and floc aggregation.

This equation for floc size has a stable equilibrium, D_e, given by

$$D_e = D_p + \frac{K_A C}{K_B \sqrt{G}} \tag{6}$$

Winterwerp went on to define a relationship relating settling velocity to floc size,

$$w_s = \frac{\alpha}{18\beta} \frac{(\rho_s - \rho_w)g}{\mu} D_p^{3-n_f} \frac{D^{n_f - 1}}{1 + 0.15 \mathrm{Re}_p^{0.687}} \tag{7}$$

where α and β are coefficients depending on the shape (sphericity) of the particles.
μ is the dynamic viscosity,
n_f is the fractal dimension of the floc,
Re is the Reynolds number.

Winterwerp suggested that giving a sufficiently long length of time, under steady state conditions, the settling velocity of a floc could be related to its size by Equations 6 and 7. He went on to use this assumption in a time dependent model with unsteady conditions. The purpose of testing this model is to examine how useful an assumption of the equilibrium state would be for describing the behaviour of settling flocs in applied modelling.

Model F5. Dynamic floc size model

$$\frac{\partial (CD)}{\partial t} + \frac{\partial}{\partial z}(w_s CD) = \frac{\partial}{\partial z}\left(K_t \frac{\partial (CD)}{\partial z} \right) + C \frac{\partial D}{\partial t} \tag{8}$$

where K_t is the turbulent diffusion coefficient,

$\frac{\partial D}{\partial t}$ is given by Equation 5,

w_s is as specified in Equation 7, but with D a time varying parameter.

Equation 8 is a one-dimensional equation describing the growth and settling of the average floc size (weighted by mass) in the water column. The model represents a simplification of the full 1DV equations developed by Winterwerp (1999). The model describes the tendency of a floc to move towards its equilibrium settling velocity, a goal which it never attains, except under steady state conditions.

By using Equation 6, and dividing throughout by D_e, Equation 5 can be simplified to

$$\frac{\partial x}{\partial t} = \frac{1}{T'} x^2 (1 - x) \qquad \text{where } x = \frac{D}{D_e} \text{ and } T' = \left(K_b G^{3/2} D_e^2 \right)^{-1} \tag{9}$$

which can be solved to give the time scale of evolution, T_R, from initial to equilibrium floc size,

$$T_R = T' \left[\ln\left(\frac{\{D_e - D_0\} D}{\{D_e - D\} D_0} \right) + \frac{D_e}{D_0} - \frac{D_e}{D} \right] \tag{10}$$

where D_0 is the initial particle size and D is the new particle size after time T_R. For the scenario where $D_e >> D > D_0$, T_R can be characterised as,

$$T_R = T_1 \approx T' \left(\frac{D_e}{D_0} - \frac{D_e}{D} \right) \tag{11}$$

For the scenario where $D_e << D < D_0$, T_R can be characterised as,

$$T_R = T_2 = T' \left(\ln\left\{ 1 - \frac{D_e}{D_0} \right\} - \ln\left\{ 1 - \frac{D_e}{D} \right\} + \frac{D_e}{D_0} - \frac{D_e}{D} \right) \approx \frac{T'}{2} \left(\frac{D_e^2}{D^2} - \frac{D_e^2}{D_0^2} \right) + O\left(\frac{D_e}{D} \right)^3 \tag{12}$$

Equations 9,10, 11, and 12 suggest that while over small time periods the evolution of flocs for the case where $D_e << D_0$ occurs rapidly, but as flocs reduce in size the time scale becomes much larger. This contrasts with the evolution of flocs for the case $D_e >> D_0$ where the evolution becomes more rapid with time until $D \approx {}^2/_3 D_E$ whereupon the rate of evolution becomes increasingly slower. In both cases, however, as D becomes very close to D_e the time scale of evolution tends to infinity. The complex behaviour described by Equations 8 and 9 was not found to be well reproduced by approximating $\frac{\partial D}{\partial t}$ as $\frac{(D_e - D)}{T_R}$. Trials with this simplifying approach were found to be prone to instability whereas use of Equation 5 was found to produce much better results. This was further improved by modifying $\frac{\partial D}{\partial t}$ using a relaxation parameter.

The dynamic model is considerably more complex than the other models. In practice with this model the onset of turbulent collapse is very sensitive, and in particular, is sensitive to the resolution of the water column near the bed.

3. COMPARISON OF MODELS WITH MEASUREMENTS OF IN SITU SETTLING VELOCITY

3.1 Field measurements

Calstock is situated at the turbidity maximum of the Tamar Estuary (Plymouth) in the UK. In September 1998 a comprehensive field exercise was undertaken at this location, including measurements of tidal currents, suspended sediment concentrations, and settling velocity at two locations roughly 1km apart. Settling velocity was measured in-situ at the downstream location by Plymouth University at a height of 0.5m above the bed and at the upstream location by HR Wallingford at mid-depth. The depth of water at these locations varies over the range 1-5m. These measurements are discussed in more detail in (Dyer et al, 2000a and Dyer et al, 2000b) but a brief description is included here.

The Plymouth measurements were undertaken using the INSSEV system (Fennessy et al, 1994). This is an in situ device which traps suspended sediment in a horizontal tube, and then allows a portion of the trapped sediment to settle in a quiescent settling column where the settling can be filmed by video camera. The HR Wallingford measurements were undertaken by trapping sediment in a Valeport horizontal tube, manually retrieving the tube, turning the tube vertical (to form a settling column) and filming a small area of the tube with a magnifying video camera. In both cases the resulting videos were further processed using video imaging software to give the settling velocities (by comparing individual video frames a short time apart) and sizes of a large number of individual flocs. Of the two devices Plymouth's INSSEV could be said to be less vulnerable to errors in results caused by the initial delay and disturbance of initially hauling the sediment sample through the water column before filming as in the HR Wallingford device. The number of flocs processed using image-processing software was also much larger for the Plymouth results than the HR Wallingford results.

The flocculation/settling models discussed previously were used to predict the variation in settling velocity seen at Calstock on the 16th (neap tide) and 22nd (spring tide) September. These two data sets represent the tidal cycles with the most measurements taken.

The following models were tested using the measured field data:

- Fixed settling (F1)
- Power law (F2)
- Power law with the dissipation parameter function (F3) $f(G)=\dfrac{1+aG}{1+bG^2}$

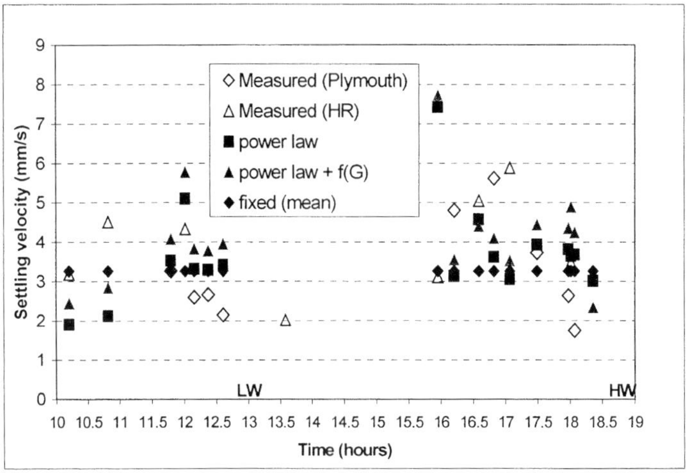

Figure 1. Predicted and measured settling velocity, spring tide of 22 September.

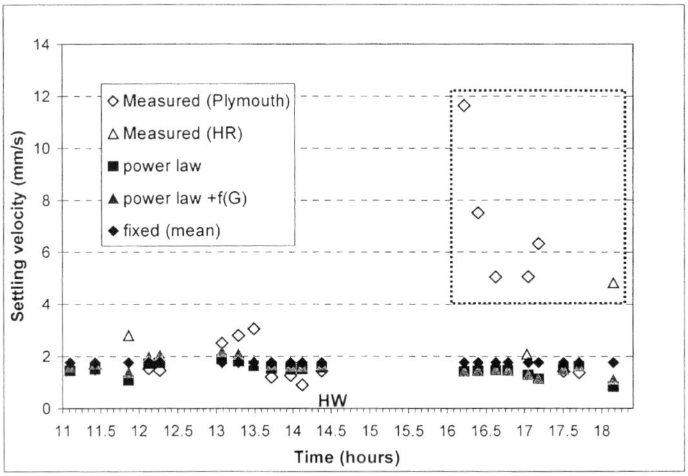

Figure 2. Predicted and measured settling velocity, neap tide of 16 September.

The following models were tested by running the 1DV model under similar conditions:

- Equilibrium model (F4)
- Non-equilibrium model (F5)

3.2 Results of models tested using the measured field data

The comparisons between the settling models and measured field data are shown in Figures 1 and 2. The measurements shown are mean values weighted according to mass, from many different flocs (hundreds, in the case of the Plymouth data) captured at any one time. The highest settling velocities measured during the whole exercise were in excess of10mm/s. The smallest measured were of the order of 0.05mm/s. The Plymouth data set considered only macro-flocs (those with size greater than 160 microns) it being considered that flocs below this size were microflocs or primary particles and therefore would be relatively unaffected by variations in conditions (Dyer et al, 2000). In practice, the exclusion of these much smaller flocs makes only a small difference to the settling velocity when weighting by mass.

The first striking aspect of the measured settling velocities is that they vary over such a large range. W_s is seen to vary between 0.5mm/s and 9mm/s, with the highest settling velocities occurring at peak flood on the 22nd and peak ebb on the 16th. The HR data is seen to agree well with the Plymouth data for the flood tide of the 22nd but the two sets of data have significant differences for the preceding ebb tide. HR only processed its videos of floc settling for the 16th for three points in the tide, and so comparison between data sets for this day is inconclusive. The differences between data sets may be a function of the different location and height of measurement but differences caused by measurement devices cannot be ruled out. The second striking aspect of the data is that the highest settling velocities occur at the start of the neap ebb tide (those lying within the dotted rectangles in Figure 3) when the concentrations within the water column are relatively low (~50mg/l). This phenomenon is observed on other tides from the data set and occurs at times of relatively low concentrations.

Plotting all of the measured settling velocities (using the Plymouth data set which is the largest, but the results are qualitatively similar for the HR data set) gives a pattern of scatter which at first sight appears to negate the traditional idea of dependence of settling velocity on concentration (Figure 3). However, there is evidence to suggest that these fast settling flocs are in fact tourmaline crystals that have been sheared out of flocs in the upper part of the water column (from fresh water flowing over the salt wedge during the more stratified neap tide), and which should be removed from the data set (Dyer et al 2000b). Taking these early ebb measurements out of the data set results in a much stronger relationship with concentration.

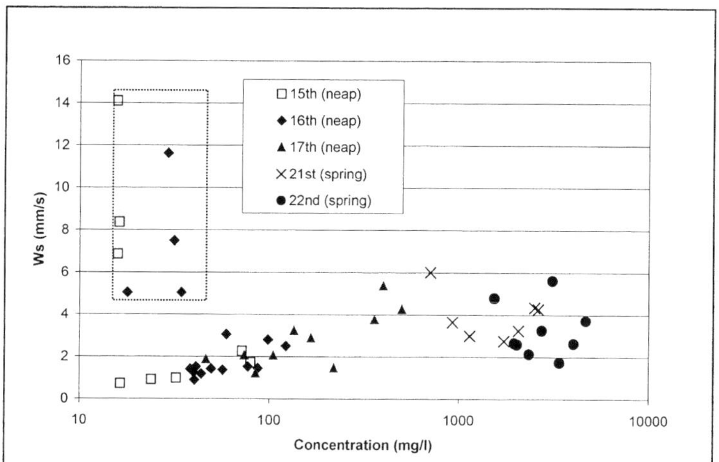

Figure 3. A comparison of the variation in settling velocity with suspended sediment concentration for the Plymouth data set.

From now on these points will be ignored in the discussion that follows, although the presence of these high settling velocities inevitably creates some uncertainty in the results.

The fixed settling velocities shown are the mean settling velocities from each of the tide data sets considered. The fixed settling velocity model used different mean values for spring and neap tides as this is both a sensible applied approach (one would generally expect larger values of spring tides and smaller values on neap tides) and this approach ensured that the number of calibrated parameters (degrees of freedom) for the fixed mean and power law (F2) approaches was the same. The power law (F2) and the Power law with dissipation parameter function (F3) were calibrated using the spring and neap tide data sets combined – these models were not calibrated separately for each tidal data set. The Power law with dissipation parameter function (F3) was used the same calibration for k and m as the Power Law model and the coefficients a and b in the dissipation function were adjusted by trial and error to give the lowest RMS error.

From Figure 3 the best fit power law $w_s \approx 0.715.C^{0.202}$ was derived. Note that although the picture at first glance may appear to suggest a relationship for w_s which at first increases with concentration and then decreases with concentration at higher values, those equations fitting best statistically to the data increased monotonically – ie there is no hard evidence for hindered settling in these results, even though they might suggest so to the naked eye.

The values a and b in the dissipation parameter function, f(G), were calibrated to be taken to be 0.12 and 0.02, respectively. However, except at times of low values of G, the effect of a and b is relatively minor and the dissipation function behaves as 1/G.

The power law model reproduces some of the pattern of spring tide variation of the measurements (Figure 1) but is unable to reproduce the relatively rapid increase and decrease in w_s on the spring flood tide. The prediction of settling velocity does not increase as sharply owing to the relatively low exponent of the power law. The fit to the spring tide data base could be improved by increasing the magnitude of the exponent but thus would be at the cost (see below) of the fit with the neap tide observations. The inclusion of the dissipation parameter function improves the fit to the measurements for this peak – largely due to the low values of G caused by the turbulent damping effect of high concentrations – but results in a poorer fit on the ebb tide. Note that for both the power law models the peak in the values of w_s lags behind the observed peak in w_s by around half an hour. The models predict the fastest settling speed to occur in the decelerating phase of the tide whereas the observations appear to predict the fastest settling at the time of peak currents. On the neap tide, both power law models appear to fit the data well but this is mainly because there is little variation of w_s in the neap tide values. The variation that does exist in the observations is reproduced to a small extent by the models.

Even though the pattern of spring tide variation is broadly reproduced by the models, the absolute values of the predicted settling velocity in some cases vary considerably from the observed values. In order to test this objectively the square root of the mean square of the error for each of the power law models was compared with those produced by the mean neap and the mean spring values of w_s. The data set used for this exercise was the Plymouth

INSSEV data set as it was the larger of the two and judged to be the most reliable. The results are shown in Table 1 below.

The results of the analysis showed that there was very little to choose between taking the mean settling velocity and the power law models in terms of accuracy. Taking a mean value was in fact a better prediction than both the "power law" (F2) and the "power law with dissipation parameter" (F3) models for the spring tide results, although overall the "power law" (F2) model performed slightly more accurately. It is possible that more sophisticated calibration might improve the fit of the "power law with dissipation parameter function" model but the point is made that increased complexity, and even a better physical description of the processes does not necessarily bring better accuracy of results. Note that the errors in the model predictions are less than or of the same order as the differences between the Plymouth and HR measurements. In this context there is even less to chose between models F1, F2 and F3.

Dyer et al (2000) found that the Plymouth University measurements at Calstock were well described by the following formula which used a slightly different formulation of the dissipation parameter, $G' = u^*(u/gH)^{0.5}$. The formula derived took the form:

$$w_s = 0.000567.C + 0.981.G' - 0.0934.G'^2 - 0.243 \qquad (13)$$

This is a site specific rather than generic formula with a form that is not applicable to numerical models in general. However, it is useful to note that the quadratic form of the dependence on G mirrors the initial growth followed by decay of f(G), which essentially assumes that turbulence at first aids aggregation by creating more floc collisions but at some point increased turbulence causes floc disruption. However, Equation 13 implies that at Calstock the turning point from floc growth to floc disruption corresponds to significantly higher values of G than those normally associated with van Leussen's dissipation parameter function.

One possible reason for the increase in settling velocity occurring at spring ebb tide is that the high current speeds erode material which is of a more aggregated state than the rest of the flocs in suspension. This continuing aggregation as a first stage of consolidation has been observed in tests carried out by Oxford University (Lintern, pers.comm.). In this case some of the variation in w_s with G (or G') might in part be coincidental. This point is also made by Winterwerp (1999).

Table 1
RMS mean errors for the different flocculation models (mm/s)

	Fixed mean (F1)	Power law (F2)	Power law with dissipation parameter (F3)	Equilibrium Model (F4)	Dynamic model (F5)
Spring tide 22nd	1.40	1.41	1.46	1.85	1.77
Neap tide 16th	0.66	0.57	0.57	-	-
All measurements	1.26	1.22	1.29	-	-

As found by Malcharek (1995) the dissipation parameter can be a useful tool in describing the variation in settling velocity. However, although formulas based on this parameter may enhance the ability of a model to describe different physical processes, such formulas may not necessarily increase the accuracy of prediction when included in a flocculation model.

3.3 Results of models tested using the 1DV model

The flocculation models were incorporated into the 1DV model and were run with water depth, tidal currents and concentrations similar to those of the Calstock site. Early generic tests with the equilibrium flocculation model (F4) showed that unreasonably high settling velocities could be generated near the bed under some circumstances which were mitigated by the dynamic nature of the F(5) model. However, the same tests showed that for the dynamic model, at low concentrations and low turbulence, the time scale of evolution towards the equilibrium floc size (and settling velocity) could be in excess of a tide which can also lead to problems in some circumstances. The dynamic model is also sensitive to the discretization of the water column near the bed. This is particularly true of the onset of turbulent collapse. Experiments by Winterwerp (1999) using his fully dynamic model showed that the vertical resolution of the water column had to be less than 0.04% of the water depth if results were to become independent of the discretization used although this problem can be overcome to some extent by a varying grid-size near the bed. Here a more applied modelling approach was considered and the dynamic model (F5) used here was run with a time step of 0.02 seconds and a near bed layer thickness of 0.01-0.05cm or approximately 1% of the depth. This layer thickness is suggested by Winterwerp (1999) as being an upper limit for acceptability as beyond this value numerical effects may cause significant errors in the results. The time step used was considerably smaller than the stability requirement, $\Delta t < \Delta z / w_s$. Various different values for the empirical coefficients K_A and K_B were tried to explore their effect on model prediction but the best results were obtained using values very similar to those derived from measurements undertaken by Van Leussen and reported by Winterwerp (1999), namely K_A=14.6 m^2/kg and K_B= 14,000 $s^{1/2}/m^2$. The values used in this study were K_A=16.0 m^2/kg with K_B set to the measured value.

The comparisons between the settling models F4 and F5 and the measured field data are shown in Figure 4. It can be seen that both models reproduce the main features of the measurements reasonably, but quantifying the accuracy of the predictions numerically shows that these results are worse than those of the power law model results (see Table 1). These features are in part a result of the specified depth-averaged concentrations which inform the models how much sediment is being advected past the point of consideration and is really a feature of the uncertainty that advective effects cause in 1DV models, rather than an inherent problem in the 1DV model itself.

The results of the models are similar to those discussed for the power law models above in that the peak settling speed lags behind that of the measurements in the decelerating part of the tide and while it is possible to predict the right order of magnitude for the peak in settling speed it is difficult to find a good fit to the results for both the flood tide and the ebb tide. The inability of the models to reproduce the measurements, in addition to the comments above, can be partly put down to advection and time lag effects occurring upstream or downstream of the measurements, and also to the problems posed by the treatment of eroded sediment.

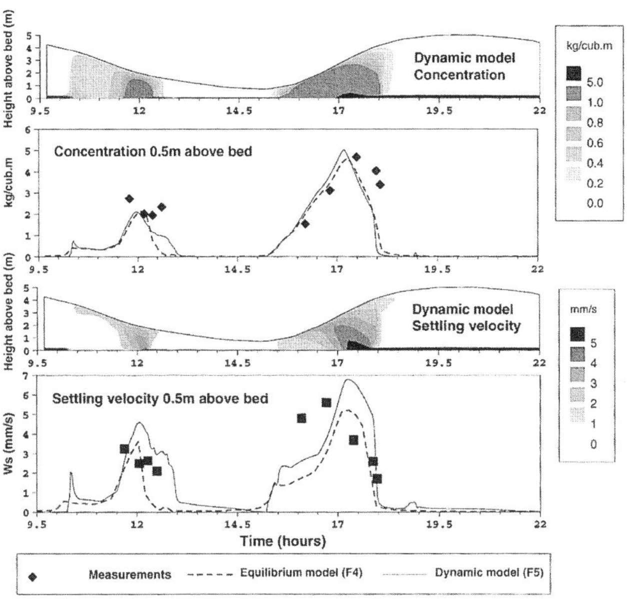

Figure 4. Predicted and measured settling velocity, spring tide of 22 September.

4. DERIVATION OF SETTLING VELOCITY USING SEDIMENT CONCENTRATION PROFILES

4.1 The Rouse profile

A further method for estimating the settling velocity is based on the assumption that the distribution of sediment throughout the water column can be described by the Rouse profile. This approach was used to provide an independent check on the measured in situ settling velocity data but it may also serve as a useful method of describing concentration profiles in 2D applied models.

If it is assumed that the settling flux of suspended sediment is balanced against the upward flux of sediment resulting from turbulent diffusion and a further assumption is made that the eddy diffusivity is of the form,

$$K_s = \kappa u_* z\left(1 - z/H\right) \tag{14}$$

where K_s is the eddy diffusivity,

H is the water depth,

z is the height above the bed,

κ is the von Karman constant,

and u_* is the friction velocity,

then it can be shown that the suspended sediment concentration profile produced is of the form:

$$C(z) = C_a\left(\frac{z\{h - z_a\}}{z_a(h - z)}\right)^{-b} \tag{15}$$

where C_a is a reference concentration at a level z_a above the bed,

b is given by $b = \frac{w_s \sigma_s}{\kappa u_*}$ and σ_s is the turbulent Schmidt number (16)

Toorman (1999) has shown that the effect of stratification is equivalent to a reduction in the value of the von Karman constant by the ratio of the momentum and diffusivity damping functions. In his recent re-examination of the published data for the damping functions (Toorman, 2000) he suggests that this ratio is best described by the relationship, $(1+ \xi \, Ri)^{-n}$ where Ri is the gradient Richardson Number, and ξ and n are empirical coefficients having values 21 and 0.8, respectively.

The Rouse profile exponent under stratified conditions therefore corresponds to a value given by,

$$b = \frac{w_s}{0.41u_*}(1 + \xi Ri)^{0.8} \tag{17}$$

4.2 Measured suspended sediment profiles

The 1998 field exercise at Calstock included a series of suspended sediment concentration measurements made throughout the water column throughout a number of tides by Bangor University. Measurements were made for the 15th, 16th , 17th, 21st and 22nd September simultaneously with the other measurements discussed above. Details of these measurements can be found on the COSINUS web site. The measured suspended sediment concentrations for the 16th and 22nd of September were plotted against the function zH/(H-z) so that the exponent of the Rouse profile b could be determined from a best fit power law. The resulting Rouse profiles are exhibited in Figures 5 and 6.

It can be seen that in general the sediment concentration profiles fit a Rouse profile well. The fit is less good during the formation of the lutocline at peak flood tide where the measured data exhibits a negative density gradient, and less good for the top 25% or so of the water column ($^{zh}/_{h-z} > 10$) where the Rouse profile is known to underpredict sediment concentration in coastal waters (Soulsby, 1997).

It is suggested that the Rouse profile could be used effectively in a 2D mud transport model to characterise the suspended sediment profile with the inclusion of a fluid mud layer near the bed which would exchange sediment with the upper part of the water column.

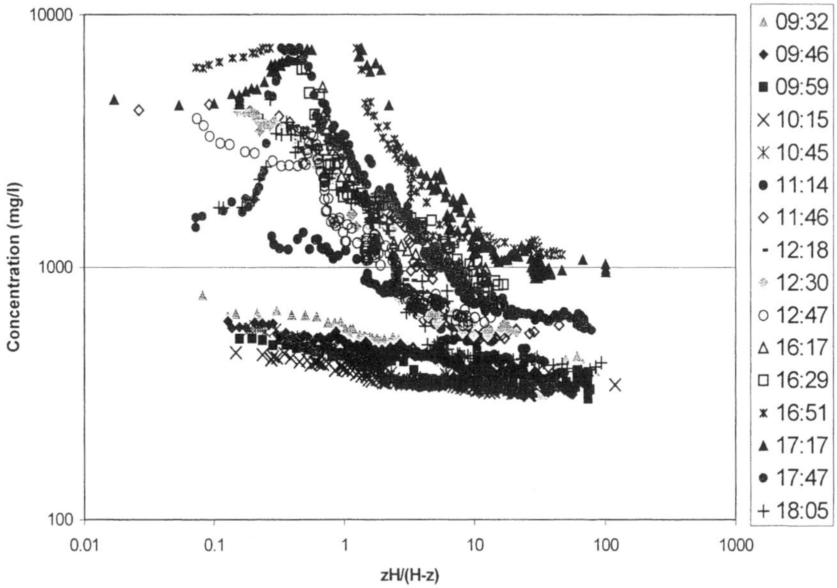

Figure 5. Sediment concentration profiles, spring tide of 22 September.

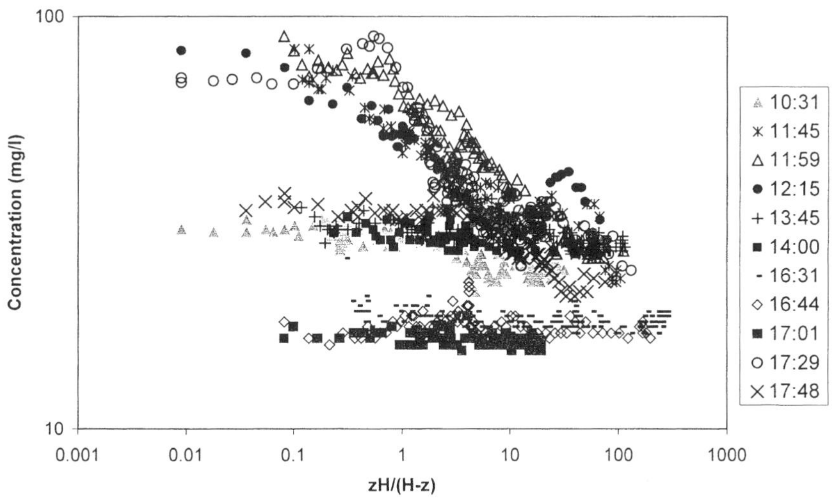

Figure 6. Sediment concentration profiles, neap tide of 16 September.

4.3 Derivation of average settling velocity using the Rouse profile

Using Equation 17 estimates of the average settling velocity throughout the water column were derived. The values of u_* were calculated from the measurements of current speed made simultaneously by Bangor using the equation,

$$U(z) = \frac{u_*}{\kappa} \ln\left(\frac{z}{z_0}\right) \qquad \text{where } z_0 \text{ is the physical roughness length.} \qquad (18)$$

The estimates of settling velocity are compared to the in situ measurements of settling velocity undertaken by Plymouth and HR Wallingford for the spring tide and neap tide data sets in Figures 7 and 8.

For the results of the spring tide on the 22nd September the estimates of settling velocity generated through the assumption of a Rouse profile on the whole reproduce the variation in settling velocity measured by Plymouth well, although there appears to be some discrepancy at LW. However, on the neap tide of the 16th the approach does not reproduce the observations as well. Part of the reason for this is that the smaller settling velocities of the neap tide are more prone to error arising from trying to fit a straight line through the scatter of points. In particular, near the bed there is often a reduced or even negative gradient at peak ebb or flood compared to the upper part of the water column. This accounts for the underestimation of settling velocity between 11:45 and 12:15 and at 17:29. However, the gradients of the sediment concentration profiles are clearly close to zero for all other times, suggesting that the measured values are at times significantly over-predicting the settling velocity or that the Rouse model is deficient under these circumstances. This evidence can be added to the current discussion regarding the use of in situ video techniques since, though only one study of one location in an estuary, it implies that the faster settling observed from such approaches is more valid than the gravitational analysis of settling tubes, which produces smaller settling velocities.

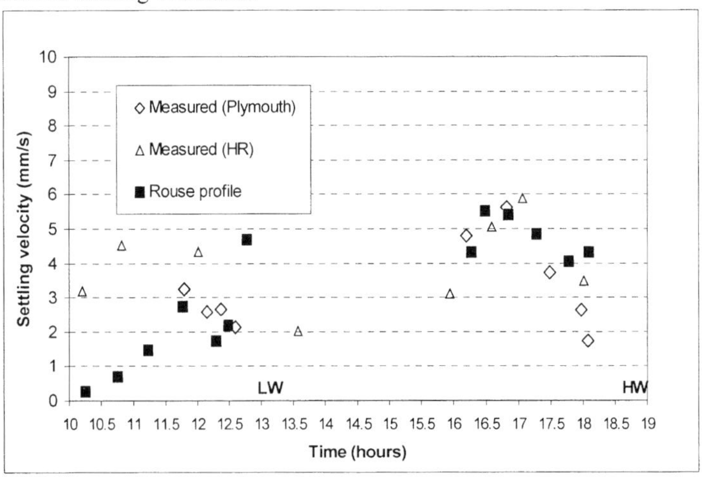

Figure 7. Comparison of measured settling velocity with estimates from Rouse profile approach, spring tide of 22nd September.

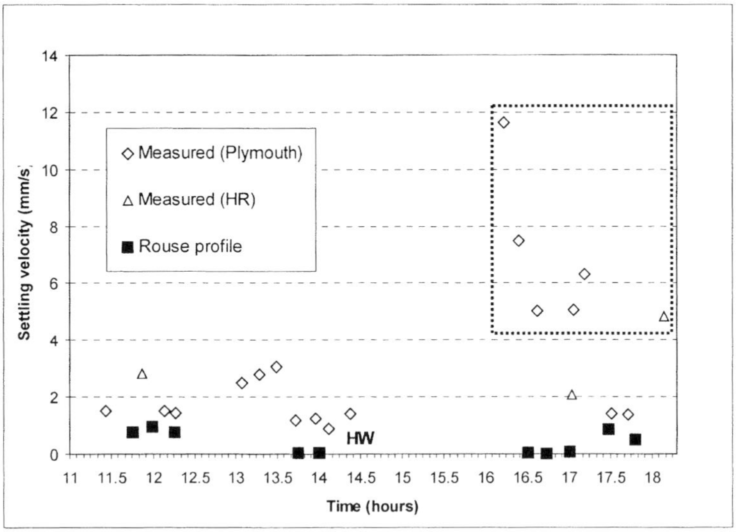

Figure 8. Comparison of measured settling velocity with estimates from Rouse profile approach, neap tide of 16th September.

The predictions using the Rouse profile model show a better accuracy of fit to the Plymouth data than the models tested earlier, even accounting for the poorer fit to the neap tide data. These results would appear to corroborate the general magnitudes of the measured in situ settling velocities (although there is some implied uncertainty regarding some of the neap tide field measurements).

5. CONCLUSIONS

The applied modeller seeks to represent the important mechanisms of the studied system adequately but has no need for added complexity, which leads to greater costs in time and money, if such complexity does not improve the ability of a model to reproduce the observed trends of the real system. This principle is of significance when considering the choice of flocculation model for use in applied 2D or 3D sediment transport modelling of cohesive sediments.

It is apparent that while some flocculation models incorporate more of the physical processes than others, they are not necessarily more accurate for any given situation. The tests undertaken in this study, which were based on a good data set but from a certain height above the bed in a small section of one estuary, have shown that of the models tried, the most accurate, when compared to the in situ measurements, was to use a simple power law model and under some conditions to use a fixed (mean) settling velocity. This result is an obvious product of the type and calibration of models chosen, together with the features of the estuary location and the intricacies of the method of measurement. Other comparisons with other data sets and measurement devices are likely to provide different answers. The development of the power law model, for instance, came from a large number of studies using the Owen

Tube/gravitational analysis method. Samples were trapped in the Owen tube that was then inverted to form a settling column, with samples being drained off from the column over a long period of time and measured for total solids content. Such studies indicated a clear and strong relationship between concentration and settling velocity. More recently video techniques have questioned the accuracy of these measurements owing to circulations within the settling column produced by the sampling process (Dearnaley, 1996). Furthermore, the increasing use of in situ video technology has consistently resulted in estimates of settling velocity many times higher than the gravitational method. The evidence would suggest inherent flaws in the gravitational method although in situ methods are by no means proven in their own right. If this is true, however, then all of the settling velocity measurements made on the basis of gravitational circulation are severely compromised. For those methods (such as the HR method described here) where a sample is manually hauled from the water column before measurements are actually undertaken, further scrutiny should also be made of the effect on settling velocity measurements of the initial few minutes of delay and disturbance to the sample. Even more sophisticated and fully in situ methods (such INSSEV) may have shortcomings which further scientific research may establish.

The important point to be made here is that there is still considerable uncertainty regarding measurement of settling velocities in the field and questions regarding the ability of all models to reproduce such measurements. It is therefore yet to be shown that any particular flocculation model is better than another. In a particular system, for a particular measurement device, a particular model may be shown to give good results. However, to be truly confident about such results the measurements ought to be validated by another measuring device. As for this study it may be that the variation between measurements is of the same order as the variation between measurement devices in which case use of a sophisticated flocculation model is unwarranted. Until the quality and quantity of settling velocity data allows an informed choice to be made, it is suggested that the applied modeller use the simplest of flocculation models and only increase the complexity of the model if there is good evidence for doing so.

The use of Rouse profiles to estimate the settling velocity from field measurements has, in this study, worked well enough to warrant further examination on other detailed data sets. It constitutes another approach to the estimation of settling velocities which can usefully validate other approaches. The method, under circumstances of limited budget, could provide a next best alternative to the in situ measurement approach. Furthermore the use of the Rouse profile can provide a mechanism for parameterising the concentration profile in 2D applied modelling.

The measured in situ data provides a difficult test for the flocculation models considered, in particular the large range of concentrations and flow conditions. It is suggested that similar comparisons be carried out with other detailed data sets in different conditions to gain a better overall understanding of the relative usefulness of the different flocculation models.

6. ACKNOWLEDGEMENTS

This work is co-financed by the European Commission, Directorate XII for Science, Research and Development, through the COSINUS project within the framework of the MAST-3 programme, contract MASC3-CT97-0082.

REFERENCES

Dearnaley, M.P., 1996, Direct measurements of settling velocities in the Owen Tube: A comparison with gravimetric analysis, *Journal of Sea Research*, volume 36, 41-47.

Dyer, K.R., Christie, M.C., Lintern, D.G., Manning, A.J., Roberts, W., and Winterwerp, J.C., 2000a, Measurements and modelling of flocculation and settling, *Proceedings INTERCOH-2000, Elsevier, Coastal and Estuarine Fine Sediment Processes*, ed. J.C. Winterwerp and C. Kraner, this volume.

Dyer, K.R., Bale, A.J., Christie, M.C., Feates, N., Jones, S. and Manning, A.J., 2000b, The dynamics of suspended sediment in an estuarine turbidity maximum, *Proceedings INTERCOH-2000, Elsevier, Coastal and Estuarine Fine Sediment Processes*, ed. J.C. Winterwerp and C. Kraner, this volume.

Leussen, W. van, 1994, Estuarine macroflocs and their role in fine-grained sediment transport, PhD Thesis, University of Utrecht.

Malchareck, A. ,1995, Mathmatiche Modellierung von Strömungen und Stofftransportprozessen in Ästuaren, Dissertation, ,Institut für Stromungsmechanik und Elektronisch Rechnen im Bauwesen der Universität Hannover, Bericht Nr. 44/1995 (in German).

Nezu,I. and Nakagawa, H., 1993, Turbulence in open channel flows, *International Association for Hydraulic Research,* Monograph Series, Balkema, Rotterdam.

Peltier, E., Le Normant, C., Teisson, C., Malcherek, A., Markofsky, M., Zielke, W., Cornelisse, J., Molinaro, P., Corti, S. and Greco, G., 1996, Three dimensional numerical modelling of cohesive sediment transport processes in estuarine environments, Final report to the EC Contract MAS2-CT92-0013, Departement Laboratoire National d'Hydraulique, EDF-DER, Paris.

Soulsby, R.L., 1997, *Dynamics of Marine Sands, A manual for practical applications*, Thomas Telford Publications.

Toorman, E., 1999, Numerical simulation of turbulence damping in sediment-laden flow, Part 1. The Siltman test case and the concept of saturation, Report HYD/ET99.2, Hydraulics Laboratory, Katholieke Universiteit Leuven.

Toorman, E., 2000, Stratification in fine-grained sediment-laden turbulent flows, *Proceedings of the 5th International Symposium on Stratified Flows*, Vancouver, Canada, 10-13 July 2000, IAHR.

Whitehouse, R.J.S., Soulsby, R.L., Roberts, W. and Mitchener, H. J., 2000, *Dynamics of Estuarine Muds, A manual for practical applications*, Thomas Telford Publications.

Winterwerp, J.C., 1999, On the dynamics of high-concentrated mud suspensions, Doctoral Thesis for the Technical University of Delft.

In situ measurements of settling velocity and particle size distribution with the LISST-ST

M. van Wijngaarden[a], J. R. Roberti[b]

[a]Ministry of Transport, Public Works and Water Management
Institute for Inland Water Management and Waste Water Treatment (RIZA),
PO Box 52, 3300 AB Dordrecht, The Netherlands

[b]Ministry of Transport, Public Works and Water Management
Institute for Marine and Coastal Management (RIKZ)

In situ techniques are essential for a proper characterisation of suspended sediment. The Laser In Situ Scattering and Transmissiometry -Settling Tube (LISST-ST) uses laser diffraction to measure both the in-situ size and settling velocity distribution of suspended sediment. The instrument is equipped with a settling tube in which a water sample is captured. By pre-programmed scanning of the sample the LISST-ST keeps up a settling record of the water sample for either 12 or 24 hour. The instruments' facility to measure stand-alone offers a relatively easy way to collect a substantial data set. Size distributions and volume concentrations are calculated from the laser diffraction patterns using an algorithm based on Mie scattering theory. From the time series of the volume concentrations, settling velocities are estimated using a model based on Stokes' law. The instrument was deployed in the Hollandsch Diep and Haringvliet, two fresh water basins in The Netherlands.

Measured settling velocities are in the order of 0.005 mm/s for the finest fractions (<7 μm) and 0.3 mm/s for the larger particles (>96 μm and < 186 μm); corresponding effective densities are in the order of 600 to 30 kg/m^3. Settling velocities in the upstream Hollandsch Diep are higher than those in the downstream Haringvliet; the data indicate that due to flocculation within the system large flocs of a relatively low density are formed in the downstream area. In general small particles have a relatively small volume concentration, but because of their high density, they dominate the mass concentration. The mass flux is dominated by the larger particles, because despite their low density these particles have a relatively high settling velocity.

Keywords
Mud flocs, laser diffraction, settling velocity, particle size

1. INTRODUCTION

The size and settling velocity distribution of suspended sediment are essential parameters in studies on sediment transport. Due to particle collisions mud flocs of variable size and strength are formed. It appears that these rather fragile aggregates, which take up a substantial part of the suspended sediment in natural waters, are easily disrupted by sampling (Gibbs, 1981; Gibbs1982a,b; Gibbs and Konwar 1982,1983). Hence, laboratory analysis of water samples will never lead to reliable estimates for the in-situ transport characteristics of the sediment, such as particle size and density. Therefore in situ techniques are indispensable for a proper characterisation of the suspended sediment.

Several in situ techniques have already been developed, amongst which the settling tube based systems. These systems, such as the well known 'Owen tube', are relatively easy to use in the field, but yield only information regarding bulk settling velocities (Cornelisse, 1996; Pejrup and Edelvang, 1996; Puls and Kühl, 1996). This in contrast with in-situ camera techniques, which are more accurate because they provide both size and settling velocity information. However this type of instruments often require highly skilled, time consuming processing of the data (Van Leussen and Cornelisse, 1993; Fenessy *et al.*, 1994a; Hill *et al.*, 1998).

The instruments of the Laser In Situ Scattering and Transmissiometry (LISST)-series, manufactured by Sequoia Scientific (Seattle, USA), measure floc characteristics through laser diffraction. The LISST-100 measures the in-situ particle size distribution and volume concentration, whereas a version with settling tube, the LISST-ST, determines the settling velocity distribution associated with the in-situ particle size distribution. In November 1999, April 2000 and August 2000, both instruments were deployed in the Hollandsch Diep and Haringvliet, two large fresh water basins in the downstream reaches of the Rhine and Meuse.

This paper presents a characterisation of the suspended sediment in the area measured by the LISST-ST, with special attention paid to the spatial and temporal variability in the settling velocity and density of the particles. Because of the novelty of the instrument the study also aims to describe the methodology followed to obtain the data. A model specifically developed to derive settling velocities and effective densities from the LISST data is presented. The suspended sediment transport in the area will only be globally analysed and a first comparison with existing data and information about suspended sediment transport in the study area will be presented in order to evaluate the order of magnitude and/or trends in the LISST data.

2. IN SITU LASER DIFFRACTION IN THE HOLLANDSCH DIEP AND HARINGVIET

2.1 Field deployment

The rivers Rhine and Meuse form a combined estuary in the southwestern part of The Netherlands (Figure 1). Annual average discharges for the rivers Rhine and Meuse are 2300 and 230 m^3/s respectively. The northern outlet of the estuary to the North Sea is via Rotterdam Waterway. The southern part of the estuary consists of two fresh water bodies: the Haringvliet and Hollandsch Diep. After the completion of the Haringvliet Dam in 1970, this southern branch of the estuary has changed from a dynamic brackish tidal inlet into a semi-stagnant freshwater area. Sluices in the dam are operated using a manipulation programme which guarantees a minimum flow (1500 m^3/s) through the northern outlet of the estuary, the

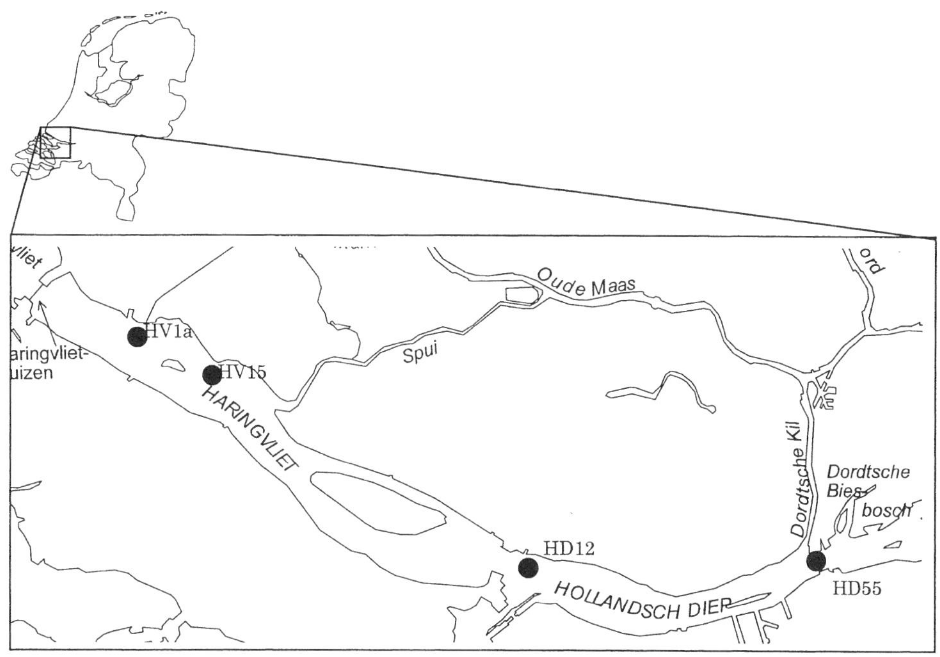

Figure 1. Measurement locations in Hollandsch Diep and Haringvliet, located in the SW of The Netherlands.

Rotterdam Waterway. When the Rhine discharge exceeds 1700 m^3/s (approximately 60 % of the year); the remaining flow is discharged through the Haringvliet during low tide. Due to the construction of the Haringvliet Dam sedimentation was initiated in the eastern part of the Hollandsch Diep basin. Some 75% of the suspended sediments by the rivers Rhine (1.9 Mton/y) and Meuse (0.3 Mton/y) is deposited in the Hollandsch Diep basin (Van Dreumel, 1995). For proper modelling of the cohesive sediment transport in the area, information regarding the settling velocity of the sediment has proven to be a critical parameter (Van Wijngaarden, 1999).

At the locations selected for this study (HD55, HD12, HV15 and HV1a in Figure 1) flow velocities are generally in the order of decimetres per second, the tidal range is limited to several decimetres and salt intrusion is completely absent. During the field deployment in November discharges of Rhine and Meuse were 1700 and 150 m^3/s respectively, in April discharges of both rivers were slightly enhanced to 2500 m^3/s and 325 m^3/s and in August discharges were again lower: 2200 and 115 m^3/s respectively. The first survey started at location HD55 on November the 8th 1999 and finished at location HV1a on November the 22nd 1999. The second survey ran from the 3rd of April 2000 until April the 17th (2000). The third survey ran from the 7th of August until the 20th of August 2000. During the November and August deployments each LISST-ST water sample was left to settle for 12 hours, while during the April deployment a settling time of 24 hours was chosen. The frame was left at each location for 3 to 4 days and water samples were taken at three depths at the start of each deployment and before pick-up of the frame. In the laboratory these samples were analysed on dry weight.

2.2 In-situ particle size distribution measured by laser diffraction

In laser diffraction a laser beam illuminates a water sample over a limited distance. Suspended particles in the sample will scatter part of the incident light over various angles. A fourier lens is used to project the forward scattered light on a 32 logarithmic spaced ringshaped detector, which is placed in the focal plane. The undiffracted light of the laser beam is focused by the fourier lens through a pin hole in the focal plane and measured with a separate detector. Because large particles scatter light over a narrower angular band than small particles the angular light distribution can be used to calculate a particle size distribution. For spherical particles of arbitrary size and refractive index the diffraction of light is described by Mie's theory (1908). Agrawal and Pottsmith (2000) give an extensive description of the LISST optical design and the data analysis used.

Under field conditions the LISST will encounter a mixture of massive primary particles and fragile flocs, while Mie's theory (1908) to describe the diffraction of spherical particles, is applied to process the data. The question arises what influence such a mixture of particles will have on the calculated particle size distribution. Heffels (1995) demonstrates that using ring-shaped detectors the particle shape information will be lost, which will also occur when large particle numbers are involved. Another point often raised is the fact that flocs may not be of solid constitution. If flocs have small pores, light will be diffracted through these holes which will contribute to the diffraction pattern in a similar way as smaller particles would. Still, the mayor diffraction contribution will come of the outer edge of the floc and will determine its final size in the particle size distribution. Finally, it should be discussed that differences in refractive index will have little influence on the position of the peaks in the particle size distribution (Agrawal and Pottsmith, 2000). Considering these points, it can be expected that the LISST particle size distribution will be representative for the actual volume occupied by suspended particles.

2.3 The LISST instruments

The LISST100 is designed to supply continuous information on particle size distribution and volume concentration, while the LISST-ST supplies information on both size and settling velocity distributions. The LISST-ST is an extension of the LISST-100, with a 30 cm height settling tube mounted vertically over the optics' end. Both devices offer pre-programmed and stand-alone data acquisition, with fully computerised processing of the data. Figure 2 displays the LISST-100 (at the back, right hand corner) and LISST-ST (in front) mounted on a frame, which is positioned on the sediment bed during deployment. Although some interference of the frame and the flow cannot be ruled out, the instruments' optics are positioned in an optimal position. The 80 cm long tube houses a datalogger, electronics, power supply and laser source. While water flows through the optics block of the LISST100, the laser beam scans the suspended sediment in the centre of the tube continuously. Although turbulence created at the inflow may affect the measurement, statistical analysis of the collected data has not yet suggested this to be a problem. Nevertheless future research will have to further address this question. In this paper only LISST-ST data will be presented, because its settling data are an innovative source of information. In a later stage the LISST-100 data will be integrated with the LISST-ST data.

The LISST-ST settling tube is equipped with doors positioned at the upper and lower end, which can be programmed to open at a given time interval. Using a motorised propeller a water sample is drawn into the tube through 8 openings of 20 mm diameter. This procedure may disturb the most fragile flocs, simultaneously collected field data from likewise instruments, for example the LISST100, are needed to test the magnitude of this effect.

The LISST-ST samples approximately 20 cm above the frame at the upper end of the settling tube. After sampling, a few seconds are allowed for turbulence to break down before the doors are closed and the water sample is allowed to settle for several hours. It is assumed that after the doors are closed the water flows stops instantly and settling becomes the governing process for suspended particle movement. According to their size and density all particles will start to settle with a specific velocity. During settlement of 12 and 24 hour runs, respectively 72 and 83 laser scans are made in logarithmically scheduled time intervals.

Figure 2. The LISST-ST is in front of the frame with the optics end and settling tube at the right hand side.

2.4 Settling data from the LISST-ST

In the theoretical ideal situation a water sample contains primary particles of single size and density, which have by definition the same settling velocity. Assuming that the particles are initially homogeneously distributed throughout the settling tube, the particles at the top of the settling tube will be the last to pass through by the laser beam at the bottom of the tube. The behaviour of such particles in time will be a step-function, which initially has the value of the total volume concentration of the particles in the sample. This value will instantaneously fall down to zero at the moment in time that the last particles pass by the laser beam, this behaviour is illustrated in figure 3a. From the elapsed fall-time T_s [s], given by the moment the doors close until the downgoing step in the signal, and length L [m] from the top to the laserbeam the settling velocity w_s [m/s] is derived:

$$w_s = \frac{L}{T_s} \tag{1}$$

The LISST-ST also provides the size *d* [μm] of these particles, with which the effective density ρ_{eff} can be calculated using Stokes Law, assuming that particles can be considered spheres, as:

$$w_s = \frac{d^2(\rho_s - \rho_w)g}{18\eta} = \frac{d^2 \rho_{eff} g}{18\eta} \tag{2}$$

With ρ_s the density of the particles [kg/m^3], ρ_w the density of water [kg/m^3], g the acceleration due to gravity [m/s^2] and η the viscosity of water [Pa.s].

The LISST-ST response on single sized particle mixture with two different densities will show a superposition of two different stepfunctions each belonging to a specific effective density. Therefore for single sized particles with a density distribution a gradual decrease from the initial volume concentration to zero is expected, which is illustrated in figure 3b.

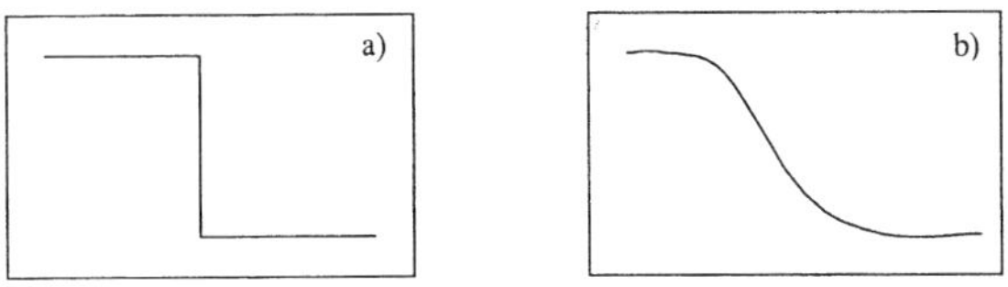

Figure 3. Theoretical settling time histories of suspension with a) single sized, uniform density particles, b) mixture of particles with different densities.

Under natural conditions, suspended particles will be graded both in size and density. Especially for the large size classes detected by the LISST-ST a mixture of single-sized and aggregated particles having very different effective densities is to be expected. The behaviour of such a particle mixture is illustrated in figure 4, presenting the concentration time series derived from a LISST-ST during 12 hours; the log-mean diameter of each size class is presented on top of each subplot. Each subplot shows the figure 3b type of behaviour: a slowly decreasing concentration in time. In figure 4 the particle size distribution is subdivided into 8 size classes instead of 32 size classes to improve the statistical accuracy for the larger size classes, because for an equal volume concentration the particle numbers in a size class will reduce with increasing size with the diameter to the power three. The effect of the reduction in particle numbers is visible in figure 4: the scatter around the initial volume concentration of the larger sizes (49 to 359 μm) is higher than for the smaller sizes (3 to 25 μm).

2.5 Data analysis

For the analyses of the LISST-ST data, as presented in Figure 4a, model based on Stokes law was developed in Matlab to fit the data non-linearly. Within each size class two effective densities are assumed to be present, both having a variable but finite width, with a cosinusoidal distribution on a logarithmic scale. The volume concentrations of each of the two density particle fractions are left variable. A least squares fit of the model to the data was carried out using the Matlab function 'lsqnonlin'. The results are presented by the continuous line in each size class in figure 4; the plusses at the bottom of each of the eight subplots present the absolute mismatch of the fit to the data.

For all data sets of the LISST-ST the fit results seemed reliable at first visual inspection; the use of a 24 hour sampling period did not seem to improve the accuracy of the results of the curve fitting procedure significantly. However further inspection indicated that especially in the size classes with the smallest particles only the lowest calculated density corresponded to a value which could be expected in nature. Towards the larger sizes these two densities converged to the same value. Most likely the in-situ suspension is in a dynamic equilibrium with constant exchanges between the size classes due to flocculation and break-up. Once captured in the settling tube, the equilibrium is destabilised due to differences in settling velocity between the size classes. The fastest disappearing, mostly the larger, size classes break the chain of (de-) flocculation between the size classes. Because the mass from the slow settling classes is moved into these disappearing size classes, the volume concentration in the slower settling classes will start to decrease prematurely. Hence, the density estimated for this stage will be an overestimation. However, this effect will decrease gradually while simultaneously sedimentation of the slower settling particles is initiated. Once the settling of these slower settling particles dominates the decrease in concentration, the estimated density will be correct. This (lowest) estimated effective density is used in the further data analysis.

Mass concentrations c_{eff} were calculated by multiplying the volume concentration $v_{f,i}$ [μl/l] of a size class i with the measured corresponding effective density $\rho_{eff,i}$ [kg/m^3]. The resulting 'effective' mass concentrations $c_{eff,i}$ of all measured size classes are summed to a total effective mass concentration c_{eff}. This c_{eff} has to be calibrated to 'real' dry-mass concentrations c_m [mg/l], because the total volume v_f of an aggregated particle or floc is filled with primary particles *and* water. The mass of this floc, being the product of density and volume, is the summed mass of the water content and the sediment content inside the floc. The mass concentration and volume concentration in a size class can be expressed as:

$$\begin{cases} v_s\rho_s + v_w\rho_w = v_f\rho_f \\ v_s + v_w = v_f \end{cases} \quad (3) \quad \text{combining yields:} \quad v_s = v_f\left(\frac{\rho_f - \rho_w}{\rho_s - \rho_w}\right) = \frac{v_f\rho_{eff}}{\rho_s - \rho_w} \quad (4)$$

with v_s the volume concentration of the primary particles in unaggregated form of the floc, ρ_s the density of these primary particles, v_w the volume concentration of the water content inside that floc, ρ_w the density of water, v_f the volume concentration of the floc, ρ_f the density of the floc and ρ_{eff} the effective density of the floc. It is assumed that the density of the primary particles is constant over all size classes.

The dry mass concentrations are related to the effective mass concentrations measured by the LISST-ST as:

$$c_m = \rho_s v_s = \frac{\rho_s}{\rho_s - \rho_w}\sum_{i=1}^{8} v_{f,i}\rho_{eff,i} = \frac{\rho_s}{\rho_s - \rho_w}c_{eff} = K.\sum_{i=1}^{8} v'_{f,i}\,\rho_{eff,i} \quad (5)$$

with :

$$K = \frac{\rho_s}{\rho_s - \rho_w}k \quad (6)$$

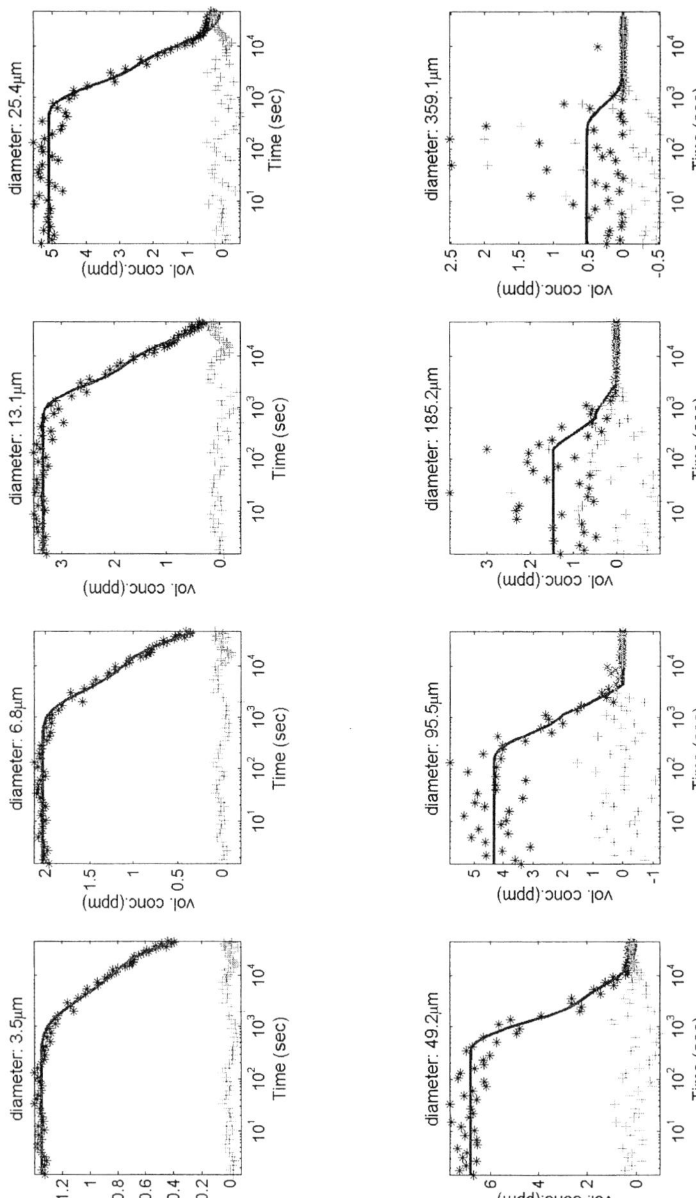

Figure 4 Volume concentration histories of location HD12 (7th run) in August 2000. The stars represent the data points, the line the model fit and the plusses the absolute mismatch of this fit.

In equation 6, k is the volume calibration constant used for all size classes in the calculation of the volume concentration from the light scattering data; consequently $v'_{f,i}$ is the uncalibrated volume concentration.

The equations show that the volume calibration constant k can be set to an arbitrary value, because the calibration constant K derived from a calibration curve incorporates the value of k. This calibration procedure does not require the density of the primary particles and is independent of particle size. From figure 5 the value for K was derived to be 5.6 with a correlation coefficient of 0.66.

Weighted mean values for effective density ($\rho_{eff,w}$), settling velocity ($w_{S,w}$) and diameter (d_w) were calculated for each location. The average values of these variables for all runs were used, with the relative mass concentration of each size class as a weight factor:

$$w_{S,w} = \sum_{i=1}^{8} w_{S,i} \cdot \frac{c_{m,i}}{c_m} \quad \text{and} \quad \rho_{eff,w} = \sum_{i=1}^{8} \rho_{eff,i} \cdot \frac{c_{m,i}}{c_m} \quad \text{and} \quad d_w = \sum_{i=1}^{8} d_i \cdot \frac{c_{m,i}}{c_m} \qquad (7)$$

With d_i the log-mean size of each of the size classes.

Although some smoothing of the data cannot be ruled out, the dynamics in the suspended sediment were expected low enough to justify such an approach. It should be stressed that the procedure followed will provide merely a global insight in sediment transport. Therefore the data will only be used to detect spatial trends and the order of magnitude in the investigated parameters.

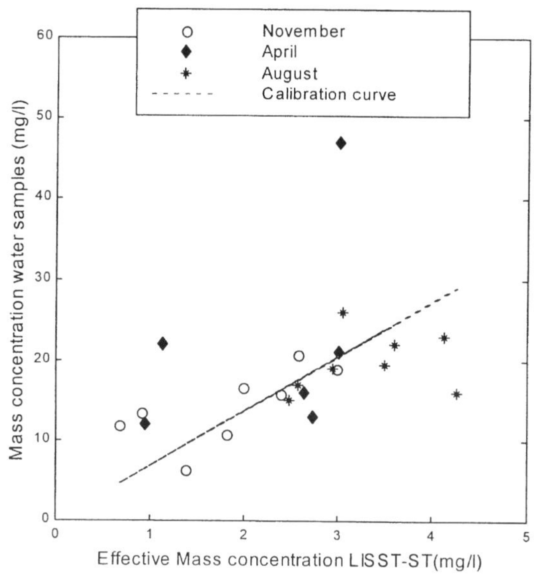

Figure 5. Calibration curve dry mass.

3. MUD FLOCS IN THE HOLLANDSCH DIEP AND HARINGVLIET

3.1 Suspended sediment concentrations from water samples

In Figure 6 the suspended sediment concentrations from water samples taken during the three surveys are presented. Each location is presented twice: before and after the consecutive days of measurement. For each location the results are presented for each depth level. The concentrations are rather low, in the order of 10-20 mg/l. In general suspended sediment concentrations were higher in April than in November and August. At HD55 and HD12 an increase in concentration with depth is observed, most clearly in April and August. In November and April a gradual decrease in concentration from HD55 to HV1a is observed, which points at the deposition of sediment towards the Haringvliet, a familiar feature of the suspended sediment transport in the area (Van Dreumel.1995, Van Eck et al., 1997, Van Wijngaarden, 1999).

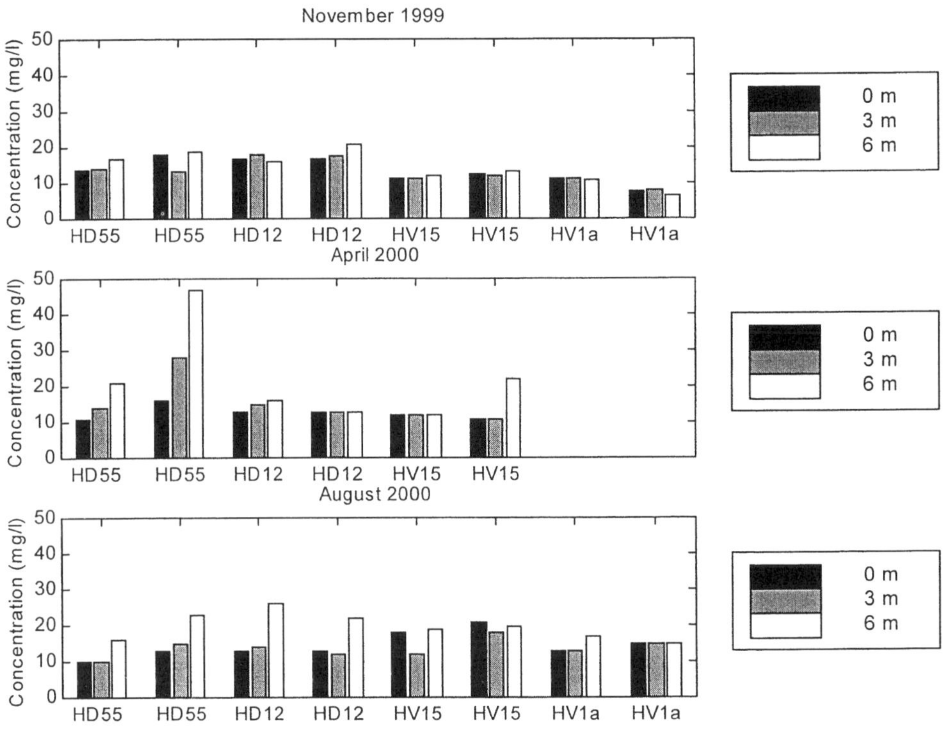

Figure 6. Measured suspended sediment concentrations in water samples (vertical averages) at LISST locations in November 1999 (top figure), April 2000 (middle figure) and August 2000 (lower figure).

3.2 LISST-ST data

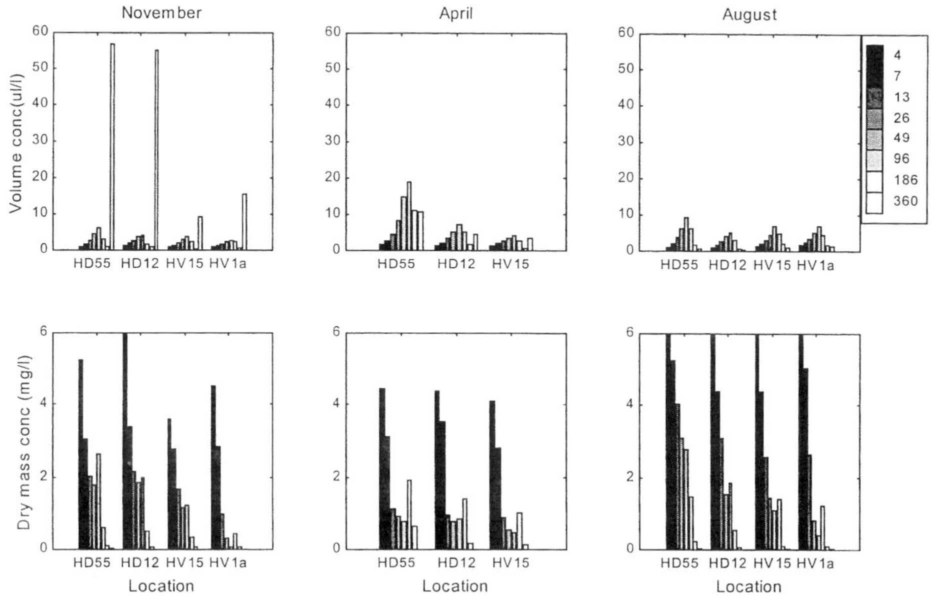

Figure 7. Top half : initial volume concentrations in November (left) and April (middle) and August (right). Lower half : dry mass concentrations November (left) and April (middle) and August (right). The legend applies to all figures, presenting the log-mean diameter (μm) of each size class.

In Figure 7 the volume concentration and the calibrated dry mass concentration distributions are presented. The volume concentrations (Figure 7, top figures) are the initial concentrations at t=0 when the sample has just been drawn in. The November volume concentrations show at all locations a sizeable amount of the largest -360 μm- particles. In both April and August a much smaller fraction of the largest- 360 μm- particles is present in the initial volume concentration. In April middle sized particles (26 to 186 μm) are most abundantly present. Since the density of the larger particles is low when compared to that of the finer particles, the contribution of these larger particles to the dry mass concentration is negligible. In general, the contribution of the finer size classes exceeds that of the larger size classes in the dry mass concentration distributions at each location, most pronounced for the August data (see Figure 7, lower half).

The weighted mean diameters are presented in Figure 8. The weighed mean diameters in November are lower than in April, but comparable to those in August. The figure shows a decrease in mean diameter in the downstream direction for each of the three surveys, illustrating how the suspension is being sorted.

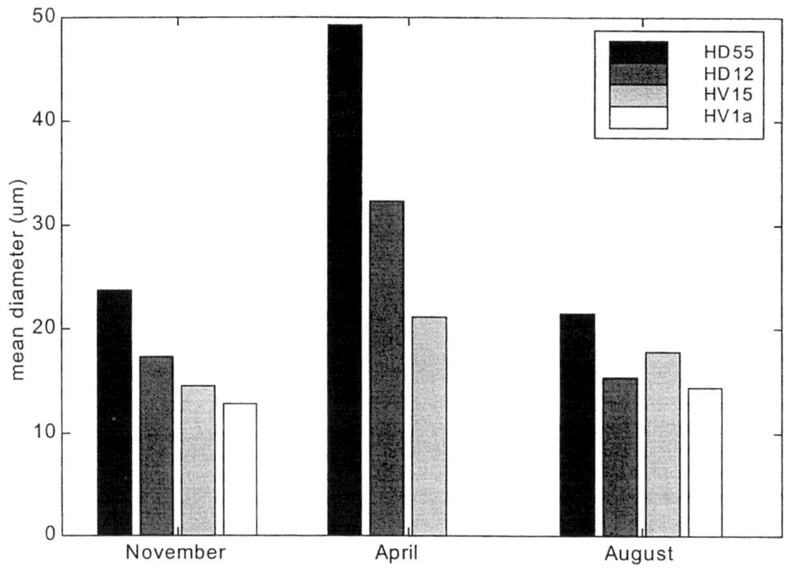

Figure 8. Weighted mean diameter concentration LISST-ST.

In November 1990 and March 1995 floc size and - settlement measurements have been carried out with an in situ underwater video camera, VIS, near location HD55 and HD12 (Verbeek, 1991; Van Wijngaarden, 1999). These data will only be used to compare the order of magnitude of the LISST-data. A direct in-situ comparison was not possible during this study, because the VIS system was not available due to technical problems. From VIS measurements both in 1990 and 1995 a mean diameter of 70 to 80 μm was found near HD55; near HD12 80 to 200 μm flocs were detected. It can be concluded that the LISST-ST diameters are within the same size range, although they are somewhat smaller. An explanation for this can be found in the instrument specifications : the VIS has a size range of 10 to 1000 μm, which will result in relatively larger mean diameter when compared to the LISST-ST data with a size range of 2.5 to 500 μm.

The settling velocity distribution and the corresponding density distribution for each location is presented in Figure 9. In general settling velocities and densities in April are similar to those in November with the exception of the 96 and 186 μm particles, which show higher values in April. Settling velocities in August are throughout the whole range somewhat enhanced. The shape of the settling velocity curves is more or less comparable for the Hollandsch Diep locations (HD55 and HD12) in both November and April with highest values for the 96 and 186 μm size classes. The August data show also a relatively high velocity and density for the largest -360 μm- fraction. During all three surveys in the Haringvliet lower settling velocities are detected for the 26 and 49 μm fractions. These low velocities are a direct result of a lower density of those specific fractions in the Haringvliet when compared to the Hollandsch Diep (see lower half Figure 9). This could be an effect of flocculation between the Hollandsch Diep and Haringvliet, although it is remarkable that this has not affected the three largest size classes.

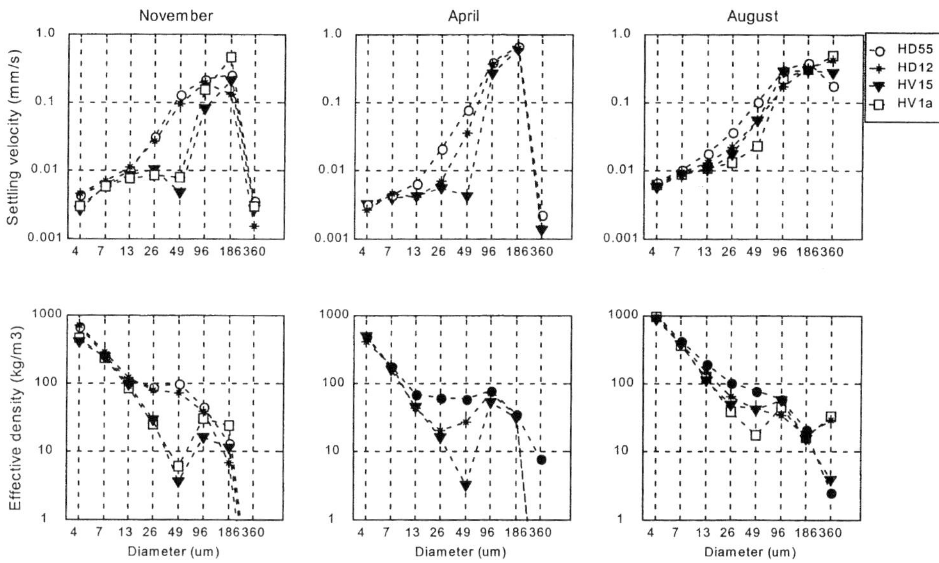

Figure 9. Settling velocity distributions for 8 size class fractions for November (top left hand figure), April (top middle figure) and August (top right figure) and density distributions for 8 size class fractions for November (bottom left figure), April (bottom middle figure) and August (bottom right figure).

The weighted mean settling velocity is presented in Figure 10. The weighted mean settling velocities in November are lower than those in April, but, similar to those in August. This is a result of [a] the finer suspension in November and August and [b] the increased settling velocities for the fractions of 96 and 186 μm in April. A decrease in the downstream direction for all three surveys is observed, which is a result of [a] the settlement of larger particles in the upstream Hollandsch Diep and [b] the fact that settling velocities in the Haringvliet are lower for the middle sized particles.

The settling velocities can also be, although tentatively, compared to those from VIS measurements. In 1990 a mean settling velocity of 0.2 mm/s was measured near HD55 and 0.25 mm/s near HD12. In 1995 a much higher mean settling velocity of 2 mm/s was measured in the Hollandsch Diep. Only settling velocities in the order of 0.2 mm/s fit into the LISST derived settling velocities measured for the larger fractions (see Figure 9).

Mass fluxes, the product of the dry mass of each fraction and the corresponding settling velocity, is presented in Figure 11. Mass fluxes in April are higher than those in November, which is a result of an increase in both concentration and settling velocity of the larger fractions. In November and August the fractions 49 and 96 μm are responsible for maximum flux and in April these are the fractions 96 and 186 μm. The variability in the shape of both the particle size, settling velocity and mass flux distributions, supports the concept to use a settling velocity distribution and not a mean settling velocity to estimate the mass flux

(Burban et. al. (1990)). This is mainly because the larger flocs, in this specific situation not the largest, are responsible for the bulk in mass transport (Eisma et. al.,1990).

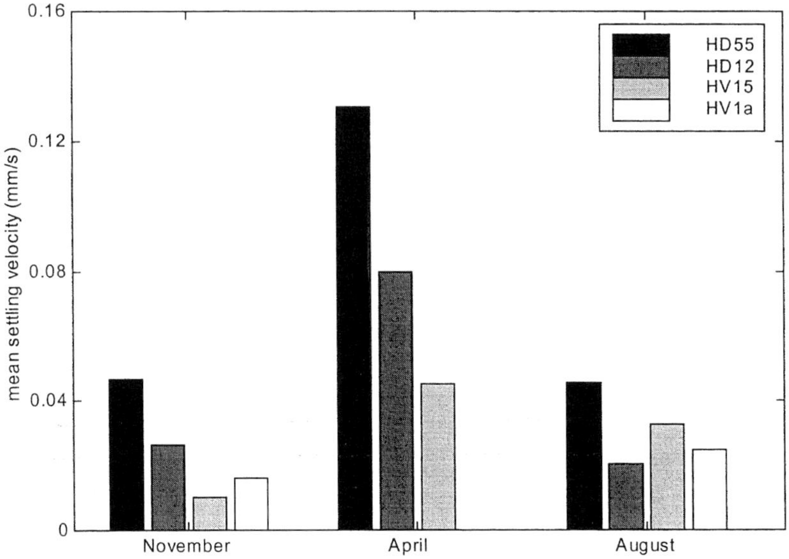

Figure 10. Weighted mean settling velocity.

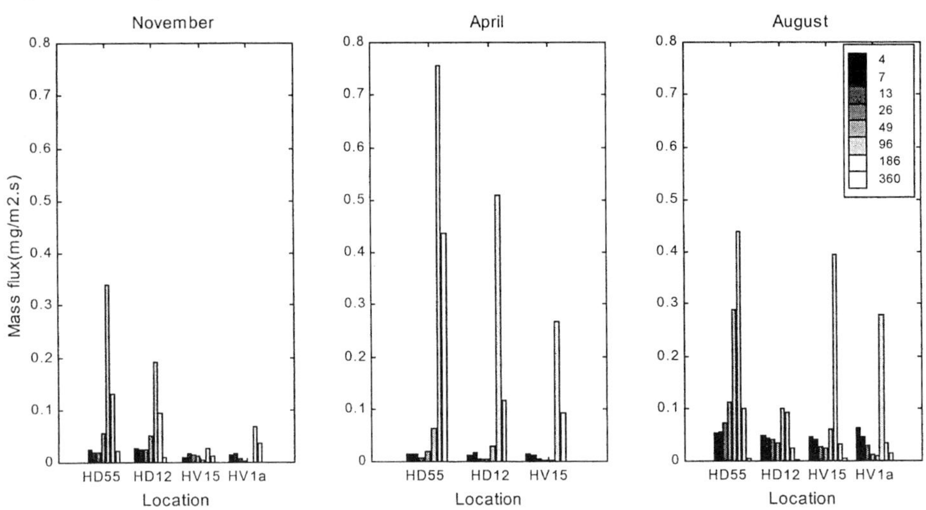

Figure 11. Calculated mass fluxes based on LISST-ST concentrations and settling velocity distributions.

A correlation often investigated is the one between mean diameter and settling velocity as presented in Figure 12. This figure shows a positive correlation between the two parameters, which illustrates that larger particle size fractions will have higher settling velocities than the smaller particle size fractions. Such a trend corresponds to the findings of for instance Fenessy *et al.* (1994a and b);Van der Lee (2000) Burban et al (1990). Figure 12 shows that a similar trend is present in all three seasons, with the April values the highest. The relationship, as presented in figure 12 on a log-log scale, can be described by a straight line with a slope of 1:2. This implies that $w_{s,w} \sim d^2$ and that density is constant, whereas on the other hand the density has been proven to vary over the particle size range (see figure 9). It is most likely that density only varies for individual particle size classes, while such density variations average out when the weighted values (calculated conform eq. 7) are used. When the uncertainty in the relationship from Figure 12 would be taken into account, the density variation would be reflected by the uncertainty in the intercept and the y-axis of the regression line.

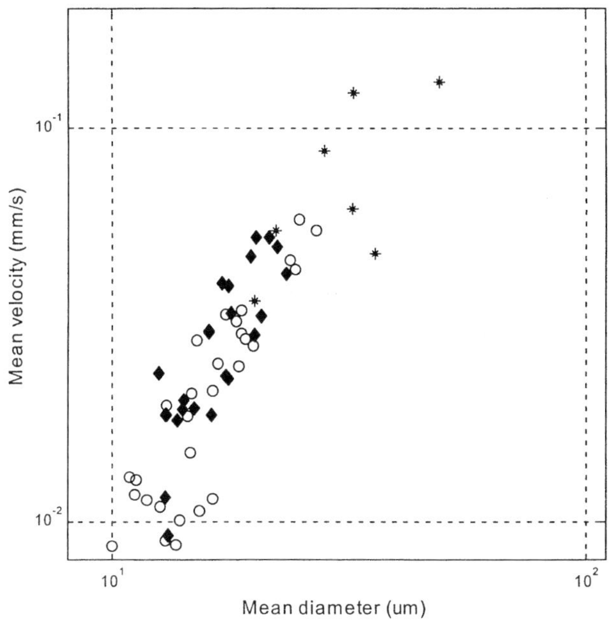

Figure 12. Weighted mean diameter versus weighted mean settling velocity. Circles represent the November data, stars the April data and the diamonds the August data.

4. DISCUSSION AND CONCLUSIONS

The LISST-ST enables the characterisation of suspended sediment in terms of size and settling velocity distribution. Both the spatial and temporal variability in these parameters can be determined. A major advantage was found to be the instruments' facility to measure stand alone, which results in extensive data sets to be collected in a relatively short period of time. A comparison of the LISST-ST data with in-situ video recordings obtained in previous studies points towards similar order of magnitude of the floc sizes an settling velocities for both instruments.

A model based on Stokes' law, specifically programmed to calculate of the settling velocity and density of the particles from the LISST-ST data, produces fair results. The dynamics of the initial suspension complicate this analysis, as destabilisation of the dynamic equilibrium between the particle size classes in the initial in-situ suspension occurs once captured in the settling tube. This effect is accounted for in a model, which was used to estimate a density and settling velocity for each particle size fraction.

When dry mass concentrations are calculated from the volume concentration and density for each size fraction, a final calibration is required. Based on water samples taken during the surveys, a calibration constant, independent of particle size, was derived valid for all locations in the area.

The results of three surveys in the Hollandsch Diep and Haringvliet show that sedimentation and selective transport is the governing process in suspended sediment transport. As in general the smaller flocs have lower settling velocities than the larger flocs, the direct result is that the mean diameter and settling velocity in the downstream area is lower than in the upstream area. Additionally, the results indicate that due to flocculation of suspended sediment between the Hollandsch Diep and Haringvliet middle sized flocs of lower density are formed. Differences between the three surveys were observed, both in particle size and settling velocity distributions. The data illustrate for instance that samples with a comparable size distribution do not necessarily have a similar settling velocity distribution. Seasonal factors, such as biologic activity may play a role in floc formation or strength, but differences in discharge level and/or in the initial sediment composition may also be important. From the data obtained no final conclusions can be drawn; combining these LISST-data with a flocculation model may reveal more about the processes responsible for floc formation and break-up in the area.

Finally, it has been demonstrated that for the calculation of mass transport, the combination of both a settling velocity distribution and a mass concentration distribution is essential. Although the flocs in the upper sizes classes do take up only a small fraction of the mass concentration, their high settling velocity accounts for the fact that they dominate the total mass flux. This is however not always the case for flocs in the largest size class: these aggregates can be of low density and hardly settle out. The use of mean values for concentration and settling velocity would average out such effects.

REFERENCES

Agrawal, Y.C. and Pottsmith, H.C., 2000, Instruments for particle size and settling velocity observations in sediment transport. Marine Geology, **168** pp 89-114.

Burban, P.Y., Xu, Y.J., McNeil, J and Lick, W., 1990, Settling speed of flocs in fresh water and sea Water. Journal of Geophysical Res. **95** (C10) pp. 18,213-18,220.

Cornelisse, J.M., 1996, The field pipette withdrawal tube (FIPITIWU). Journal of Sea Research **36** (1/2) pp 37-39.

Eisma, D. Schuhmacher ,T., Boekel, H., Van Heerwaarden, J., Franken, H., Lann, M., Vaars, A., Eijgenraam, F and Kalf, J., 1990, A camera and image analysis system for in situ observation of flocs in natural waters. Netherlands Journal of Sea Res. **27** pp 43-56.

Fenessy, M.J. , Dyer, K.R. and Huntley, D.A., 1994a, INSSEV: An instrument to measure the size and settling velocity of flocs in situ. Marine Geology **117** pp 107-117.

Fenessy, M.J., Dyer K.R. and Huntley, D.A., 1994b, Size and settling velocity distributions of flocs in the Tamar Estuary during a tidal cycle. Netherlands Journal of Aquatic Ecology. **28** (3-4) pp 275-282.

Gibbs, R.J., 1981, Floc breakage by pumps. Journal of Sedimentary Petrology **51** pp 30-31.

Gibbs, R.J., 1982a, Floc stability during coulter counter size analysis. Journal of Sedimentary Petrology **52(2)** pp. 657-660.

Gibbs, R.J., 1982b, Floc breakage during HIAC light blocking analysis. Environmental Science and Technology **17** pp. 347-375.

Gibbs, R.J. and L. Konwar, 1982, Effects of pipetting on mineral flocs. Environmental Science and Technology **16** pp. 119-121.

Gibbs, R.J. and L. Konwar, 1983, Sampling of mineral flocs using Niskin bottles. Environmental Science and Technology **17** pp. 374-375.

Hill, P.S. , Syvitski, J.P. , Cowan, E.A. and R.D. Powell, 1998, In situ observations of floc settling velocities in Glacier Bay, Alaska. Marine Geology **145** pp 85-94.

Mie, G., 1908, Contributions to the optics of suspended media, specifically colloidal metal suspensions. Ann. der Physik **25** pp 377-455.

Pejrup, M. and K. Edelvang, 1996, Measurements of *In Situ* settling velocities in the Elbe Estuary. Journal of Sea Research **36** (1/2) pp 109-113.

Puls, W. and H. Kühl, 1996, The field pipette withdrawal tube (FIPITIWU). Journal of Sea Research **36** (1/2) pp 37-39.

Verbeek, H., 1991, Measurements with the VIS in the Hollandsch Diep and Lake Volkerak (In Dutch). Department of Inland water Management, Public works, internal document.

Van der Lee, W.T.B, 2000, The settling of mud flocs in the Dollard Estuary. Phd thesis University Utrecht.

Van Dreumel, P.F., 1995, Sand and mud transport in the Rhine-Meuse Estuary, Ministry of Transport, Public Works and Water Management, Directorate Southern Holland, Rotterdam (in Dutch).

Van Eck, G.T.M., Zwolsman, J.J.G., and Saeijs, H.L.F., 1997, Influence of compartimentization on the water quality of reservoirs: lessons learned from the enclosure of the Haringvliet estuary, The Netherlands, Proceedings of the Dix-neuvieme Congres des Grand Barrages, Florence, 647-669.

Van Wijngaarden, M., 1999, A two dimensional model for suspended sediment transport in the Rhine-Meuse estuary (The Netherlands), Earth Surface Processes and Landforms **24**, pp. 1173-1188.

Van Leussen, W. and J.M. Cornelisse, 1993, The determination of the sizes and settling velocities of estuarine flocs by an underwater video system. Netherlands Journal of Sea Research **31** (93) pp 231-241.

Chapter 4:

Processes in and on the bed: consolidation and erosion

Fine Sediment Dynamics in the Marine Environment
J.C. Winterwerp and C. Kranenburg (Editors)

On the erodibility of fine-grained sediments in an infilling freshwater system

T.J. Andersen[a], E. J. Houwing[b] & M. Pejrup[a]

[a]Institute of Geography, University of Copenhagen, Øster Voldgade 10, DK-1350 København K, Denmark.

[b]Institute for Inland Water Management and Waste Water Treatment (RIZA), Rijkswaterstaat, van Leeuwenhoekweg 20, 3316 AV Dordrecht, The Netherlands.

The erodibility of fine-grained sediments from the Hollandsch Diep freshwater system has been measured using a portable EROMES system. The erosion experiments were carried out on natural sediment surfaces sampled with a box-corer. The measured erosion thresholds varied between 0.16 and 0.70 N m^{-2}. The erosion thresholds were positively correlated to the dry density of the bed material ($r^2 = 0.33$) whereas no significant dependence was found on either grain size, organic content or chlorophyll *a* content. The measured erosion rates varied between 0.04 and 1.68 g m^{-2} s^{-1} and a strong positive correlation with bed shear stress was observed. The equivalent settling diameters of the eroded material were generally 5 to 10 times larger than that of the primary particles which shows that the material was eroded as aggregates. However, a strong positive correlation between the two settling diameters ($r^2 = 0.74$) also indicates that the coarser modes of the primary grain size distributions affect size or density of the faster settling aggregates. The lack of correlation between the erodibility and organic content and chlorophyll *a* content indicates that microphytobenthos is not significantly modifying the erodibility of the sediments at the study sites. The macrofaunal community is also poorly developed with few species and small numbers of individuals and it is concluded that the erodibility is mainly controlled by accumulation rate, dry density of the bed material and perhaps maximum current velocity at the sites. This is further supported by the similarity with earlier measurements carried out on settled beds in laboratories.

Keywords: Erosion threshold, erosion rate, sediment settling, Hollandsch Diep, The Netherlands.

1. INTRODUCTION

The marine influence in the Hollandsch Diep freshwater system has been almost completely cut of since the closure of the Haringvliet Dam in 1970 and due to an additional marked decrease in current velocities in the channels, rapid deposition of fine-grained sediments has taken place. These deposits are polluted and are therefore a potential ecological risk if resuspended. The possibility of partial re-opening of the connection to the North Sea has led to increased interest in the erosion potential of the deposited material. Prior to this study, measurements of the erosional characteristics of the deposited sediments from the study area have been undertaken by use of flume-experiments using settled beds (Kuijper et al., 1990, 1991; Winterwerp et al., 1992).

Until now, no measurements have been done on natural, undisturbed sediment beds from the area. However, measurements on undisturbed sediments with natural flora and fauna have generally turned out to be inevitable if more reliable estimates of the erodibility of fine-grained sediments are requested (e.g., Paterson, 1989; Paterson & Black, 1999; Austen et al., 1999, Houwing, 1999; Tolhurst et al., 2000; Andersen, 2001). By use of a portable version of the EROMES erosion device, such measurements were carried out for 7 sites within the Hollandsch Diep - Biesbosch water system during this study. The aim of the present paper is to calculate erosion thresholds, erosion rates and equivalent aggregate size for undisturbed sediments from a freshwater system with high sediment accumulation rates and to compare the results with main physical and biological parameters characterizing the bed sediment.

2. STUDY SITES

The study sites are situated in the Hollandsch Diep freshwater system which is part of the former estuary of the rivers Rhine and Meuse. After the almost complete closure of the Haringvliet Dam in 1970, the estuary was turned into a freshwater system with very low current velocities. The tidal range diminished from about 2 m to 0.2 - 0.35m (van Eck et al., 1997) and the mean current velocities decreased from about 0.80 m s^{-1} to 0.20 m s^{-1}. As a result, large amounts of fine-grained riverine material started to deposit in the area. The ecological changes after the closure are described by Ferguson & Wolff (1984) who found that most marine and brackish macrofaunal

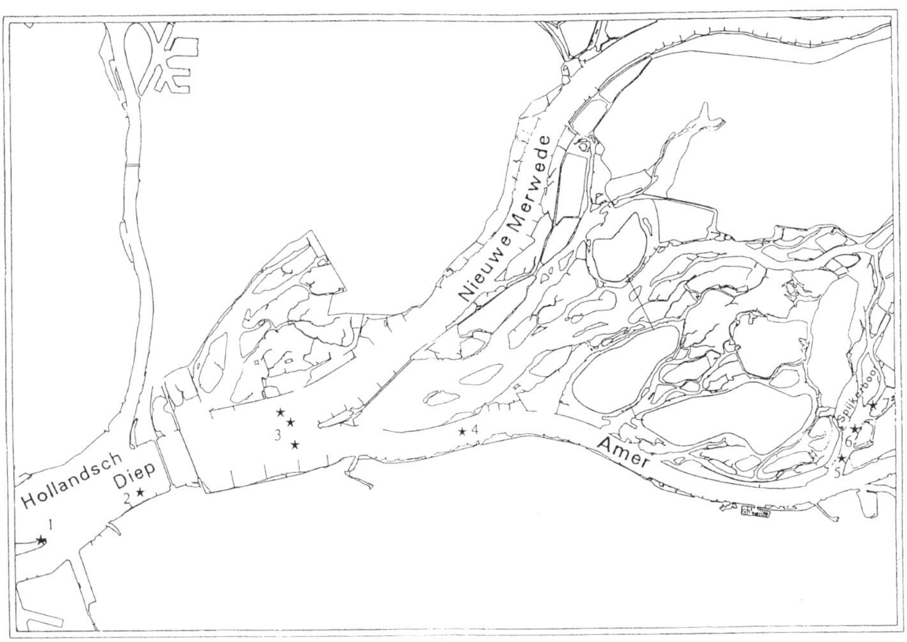

Figure 1. Map showing the 7 study sites within the Hollandsch Diep/Biesbosch, the Netherlands.

species disappeared and these are only slowly being replaced by freshwater species. Erodibility measurements were carried out at 7 positions within the area (fig 1). Station 1 and 2 (water depth 4 and 5.5m respectively) are situated in Hollandsch Diep, station 3 at the junction of Nieuwe Meerwede and the Amer (water depth 3 - 5.5m), station 4 in the Amer (water depth 2.3 m) and station 5, 6 and 7 in the Spijkerboor (water depths 2.5, 1 and 1.6 m respectively).

3 METHODS

3.1. Bed sediment sampling

A large, conventional box corer was used for the collection of undisturbed bed samples. The box measured 25cm * 25cm and the box corer was gently placed on the channel bottom. After retrieval, the eroding unit is placed in the box, avoiding any artificial structures made during the sampling. Erosion experiments were carried out on samples from 7 different positions within the freshwater system with two to four replicas at each site, giving a total of 21 erosion experiments.

3.2 Erosion experiments

The erosion experiments were carried out using a portable EROMES erosion equipment (fig 2). The instrument is similar to the one described by Schünemann & Kühl (1991) and Cornelisse et al. (1997) with the modification that only the erosion chamber itself is used and an OBS-sensor situated in the erosion chamber was used instead of the loop with a transmissometer. Except for the different ways of measuring suspended sediment concentration, the instrument has a configuration similar to the instruments used by Riethmüller et al. (1998) and Austen et al. (1999) but the instrument used in this study is portable and can be used in the field facilitating measurements shortly after sampling. The performance of the EROMES was examined in an

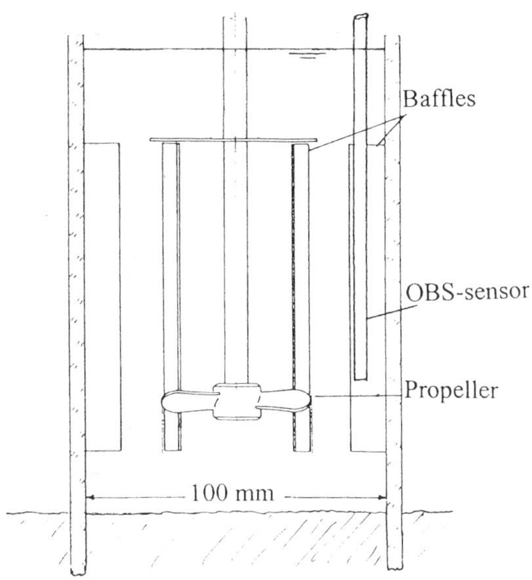

Figure 2. Sketch showing the erosion chamber of the portable EROMES equipment

experimental study of different erosion devices by Gust & Müller (1997). From their results it is seen that the EROMES is able to apply a known bed shear stress although with larger turbulent fluctuations than most of the other tested devices. The higher turbulence levels simulate the bed shear stress under a combination of current and waves (Riethmüller et al., 1998) which is often observed in shallow waters.

The instrument consists of a 100 mm diameter perspex tube which is pushed into the undisturbed bed sediment. The tube is gently filled with local water and the eroding unit is placed on top of the tube. This eroding unit consists of a propeller which generates bed shear stresses and an OBS-sensor which monitors the changing suspended sediment concentration (SSC).

The propeller revolutions are transferred to bed shear stress by use of a calibration based on the onset of erosion of quartz sands with known critical erosion shear stress (Schünemann & Kühl, 1991; Andersen, 2001). Additionally, the bed shear stress has been measured directly by use of a hot-film probe at different radii within the instrument.

The samples for the calibration of the OBS-sensor were withdrawn from the instrument during the experiments with a pipette (2 * 50 ml) and subsequently filtered through pre-weighed Millipore 0.45 μm CEM filters or 0.7 μm Whatmann GF/F filters. The sediment on some of the Millipore filters was later analysed for grain size by use of a laser-sizer model Malvern Mastersizer/E after careful dispersion in 0.01M $Na_2P_4O_7$ and ultrasonic treatment for two min. prior to analysis. The Whatmann filters were later combusted at 550 ^{0}C for two hours in order to determine organic content by loss on ignition (LOI). The calibration of the OBS-sensor was performed individually for each site by finding either the best linear or polynomial fit. The correlation between the OBS-signal and SSC is high and the variance ranges from 0.89 to 0.98.

During each erosion experiment the bed shear stress is increased in steps of 0.1 N m^{-2} every 2 minutes from 0.1 N m^{-2} to 1 N m^{-2}. Additionally, for 8 of the erosion experiments the bed shear stress was maintained at 0.5 N m^{-2} for 10 min. and for 5 experiments at 1.0 N m^{-2} for 10 min. The experiments with longer time-steps showed that the erosion had virtually ceased after 10min at 0.5 N m^{-2} whereas a decreasing erosion rate was found at 1.0 N m^{-2}. The choice of only 2 min. steps for the applied bed shear stress was made in order to facilitate direct comparison with field-data obtained with the instrument at intertidal mudflats (reported in Andersen, 2001). Two minute steps were chosen for the mudflat sites in order to increase the number of erosion experiments spatially and temporally. Additionally, it was found that the erosion rate generally approached zero after 2 to 5 min for the mudflat sites. The erosion thresholds which are found are not very sensitive to the choice of the length of the time-steps when using the procedure applied in this investigation, however, the erosion rates which are found are influenced more by the choice of time-step. When comparing the erosion rates calculated for the first two minutes with the rates found for the full 10 min, a decrease was found due to the decreasing erosion rate with time. For 0.5 N m^{-2} the decrease was 42% on average and for 1.0 N m^{-2} 47%. Consequently, the rates that are reported here are probably slightly higher than the rates found if longer time-steps were applied.

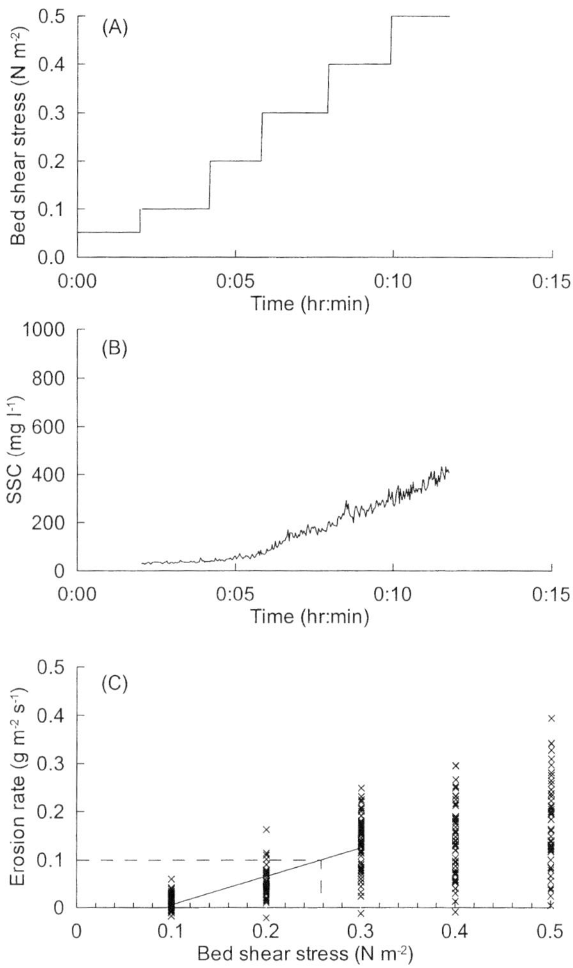

Figure 3a, b & c: An example of the output of an erosion experiment using the EROMES. a = the stepwise increase in applied bed shear stress. b = the variation in SSC during the same erosion experiment. c = a plot of erosion rate versus applied bed shear stress for the same experiment. The full line is the linear regression for the range 0.1 - 0.3 N m^{-2}. The dotted line indicates the chosen level for critical erosion rate. For this experiment an erosion threshold of 0.26 N/m^2 is found.

The excess bed shear stresses which are reported in this paper are calculated as the difference between the applied bed shear stress and the critical bed shear stress for erosion of the surface material (τ_i - $\tau_{cr,\ surface}$). This ignores the fact that the erosion threshold is likely to increase with depth in the sediments but a fairly good correlation was observed between this simple parameter and the erosion rate.

The propeller revolutions and OBS-signal are logged at 0.5 Hz. In order to determine the erosion threshold for each experiment the erosion rate is plotted against the applied bed shear stress. By making a linear fit it is possible to determine the threshold as the bed shear stress at the intercept of this line with a critical erosion rate. The critical erosion rate should not be chosen to be zero as erosion of organic rich fluffy material on top of the bed will give a positive erosion rate before the erosion of the actual bed starts. The choice of the proper critical erosion rate value is performed by examination of the plots of SSC and erosion rate against time. It was found that a critical value of 0.1 g m^{-2} s^{-1} was suitable to discriminate between the erosion of the fluffy top-layer and the erosion of the bed itself. This value is similar to the one which is used for fine-grained, intertidal sediments in the Danish Wadden Sea area (Andersen, 2001). The value 0.1 g m^{-2} s^{-1} corresponds to 360 g m^{-2} h^{-1} or 0.5 to 1 mm h^{-1} for the dry bulk densities found for the bed sediments at the sites. An example of the output on an EROMES erosion test is shown in fig 3. Obviously, the choice of a proper critical erosion rate is not simple. However, for this example a significant increase is observed after increasing the bed shear stress to 0.3 N m^{-2} and using a critical erosion rate of 0.1 g m^{-2} s^{-1} an erosion threshold of 0.26 N m^{-2} is found.

After each erosion experiment the propeller is turned off and the suspended material is allowed to settle. By the continued monitoring of the change in SSC at a known level below the water surface it is possible to calculate the distribution of equivalent settling diameters (ESD) for the eroded material.

3.3 Bed samples

Samples of the surface material have been obtained for all erosion experiments. Each sample consists of four syringe samples of the top 5 mm of the bed, in total 7.8 ml. The samples were weighted before and after drying at 105 ^{0}C to obtain dry bulk density and water content. A subsample was withdrawn from each sample for determination of chlorophyll *a* content by spectophotometry. The chlorophyll *a* was extracted in 10 ml 90 % acetone for 24 hours and the content determinated after Parsons et al. (1984) by measurements of the absorption at 665 nm and 750 nm before and after addition of 200 μl 1M HCl .

Grain size analyses were performed on one sample for each site. Both the equivalent settling diameter of the dispersed material and the volumetric size distribution were measured. The settling diameters were obtained by use of a Sedigraph 5100 and the volume distribution by use of a laser-sizer model Malvern Mastersizer/E. The samples were dispersed in 0.01M $Na_2P_4O_7$ and given an ultrasonic treatment for 2 min. prior to analysis. The organic content of each sample was obtained by loss on ignition at 550 ^{0}C for 2 hours.

4. RESULTS

The results of the erosion experiments on 21 surface samples from 7 positions are listed in table 1 together with data on grain size, dry density, organic content and chlorophyll a content. The erosion thresholds range from 0.20 N m^{-2} to 0.70 N m^{-2} with site averages from 0.24 N m^{-2} to 0.64 N m^{-2}. The erosion rate has been calculated for the 0.5 and 1.0 N m^{-2} increments (time-averaged over 2 minutes) and values of 0.03 - 0.46 g m^{-2} s^{-1} and 0.28 - 1.68 g m^{-2} s^{-1} respectively have been found. For position 2, 4 and 6 the bed shear stress was kept at 0.5 and 1 N m^{-2} for 10 min in order

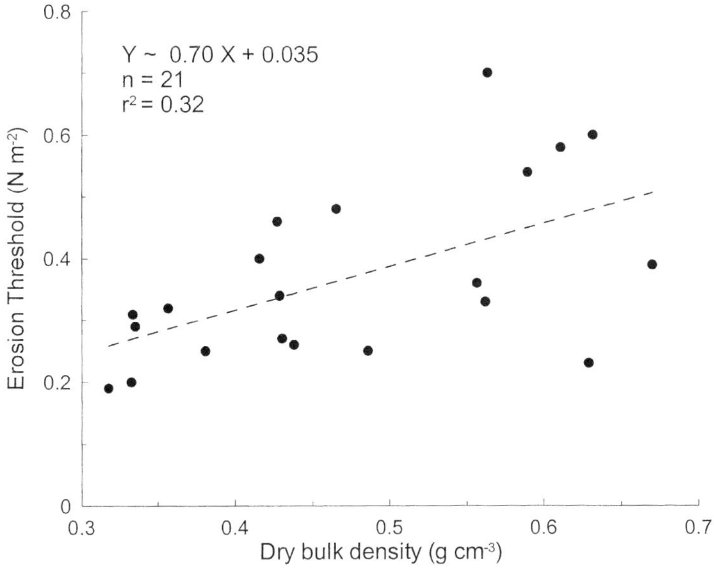

Figure 4. Scatterplot of erosion threshold vs. dry bulk density of the bed material.

to determine the type of erosion. The erosion is of type I (decreasing erosion rate towards a constant SSC) for an applied bed shear stress of 0.5 N m^{-2} and is of type II (constant erosion rate with time) or type I/II (transitional form, Amos et al., 1997) for a bed shear stress of 1.0 N m^{-2}. This is the result of erosion of loosely bound aggregates at lower bed shear stresses and erosion of larger aggregates due to erosion in surface irregularities at higher bed shear stresses.

The texture of the bed material is fine-grained with median equivalent settling diameters of 3 - 20 μm and the sand content is below 12% except for one sample position where 29% sand was found. The equivalent settling diameters of the eroded material were calculated on the basis of the settling of the suspended material after the erosion experiments. The median equivalent settling diameters of the eroded material ($d_{50,\,aggr.}$) are fairly uniform with values from 24 μm to 68 μm with the highest values found on the most sandy position. The dry bulk densities of the bed material vary between 0.32 and 0.63 g cm^{-3}. The organic content is between 5 and 11 % and the absence of biofilms of benthic diatoms was observed in the field and is confirmed by the low chlorophyll *a* contents of the sediments (4 to 21 μg g^{-1} dry wt.). Similarly, only a few macrofaunal species and individuals were observed.

Linear correlation analysis was carried out between all measured variables and a correlation matrix of all the correlation coefficients is given in table 2. A positive correlation was found between the dry bulk density of the bed material and the erosion threshold (fig 4) and the erosion threshold was not significantly correlated to any of the other measured bed sediment properties (primary grain size, organic content, chl *a* content).

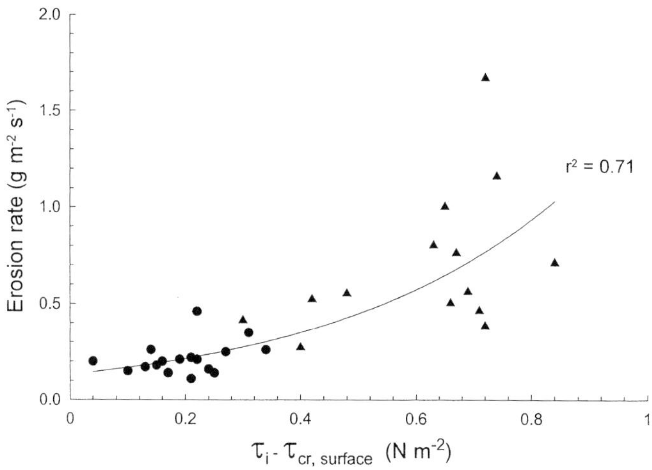

Figure 5. The erosion rate as a function of (τ_i - $\tau_{cr, surface}$). The line shows the best fit to a function of the form $\varepsilon_i = \beta \exp(\alpha(\tau_i - \tau_c))$, $\alpha = 2.4$, $\beta = 0.14$. ●: $\tau = 0.5$ N m^{-2}, ▲: $\tau = 1.0$ N m^{-2}.

The erosion rate at 0.5 and 1.0 N m^{-2} s^{-1} is plotted against the parameter τ_i - $\tau_{cr, surface}$ in fig 5. The best fit to the full dataset is found using an algorithm suggested by Christensen & Das (1973), cited by Mehta et al. (1982):

$$\varepsilon_i = \beta \exp(\alpha(\tau_i - \tau_{cr})) \tag{1}$$

with the constants $\beta = 2.5$ and $\alpha = 0.13$ and ε_i as the erosion rate (g m^{-2} s^{-1}) and τ_i and τ_{cr} as the

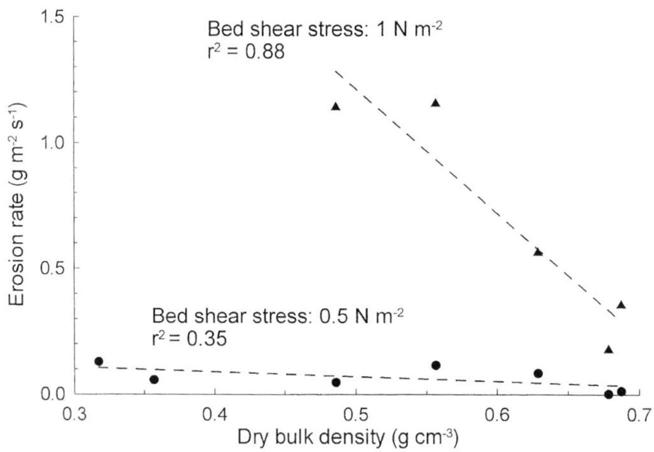

Figure 6. The erosion rate at 0.5 (●) and 1.0 N m^{-2} (▲) as a function of the dry density of the bed material. Lines of the best linear fit are shown.

Table 2.
Correlation coefficients between the erodibility and a range of sediment properties. Numbers in bold indicate highly significant correlations ($P < 0.001$)

	Chl a	Organic content	Dry density	D_{50}, primary grains	D_{50}, aggregates	Erosion rate 1.0 N m^{-2}	Erosion rate 0.5 N m^{-2}
Erosion threshold	-0.28	-0.47	**0.56**	-	0.55	-0.47	**-0.72**
Erosion rate 0.5 N m^{-2}	0.42	0.30	-0.44	-0.51	**-0.65**	**0.69**	
Erosion rate 0.1 N m^{-2}	0.22	0.10	-0.14	-0.32	0.00		
D_{50}, aggregates	-0.35	-0.40	**0.61**	**0.86**			
D_{50}, primary grains	-0.45	**-0.72**	**0.77**				
Dry density	**-0.83**	**-0.84**					
Organic content	**0.74**						

applied and critical bed shear stress respectively (N m^{-2}). There is some scatter in the data but the correlation coefficient is high (0.84) and the correlation is significant.

The erosion rate at 0.5 N m^{-2} is negatively correlated to the median settling diameters of the eroded particles but no correlation is found at 1.0 N m^{-2}. The erosion rate is also correlated to dry density of the bed material but only for higher bed shear stresses (fig 6). Excluding a sample with very high SSC during the erosion experiment, it is found that there is no correlation between $D_{50,\ aggr.}$ and the other measured parameters except primary grain-size and dry bulk density. A positive correlation was found between both the size of the eroded aggregates and primary grain size (fig 7) and dry density of the bed material.

5. DISCUSSION

5.1 Erosion threshold

The fairly low threshold values are typical for soft, fine-grained deposits where surface stabilizing by benthic diatoms or macro-fauna is absent and the thresholds are of the same order as the thresholds found for Danish and German intertidal sites without biological stabilization of the sediments (Riethmüller et al., 1998; Austen et al., 1999; Andersen, 2001).

The erosion thresholds were only found to be highly correlated the dry density of the bed material. Dry densities range from 0.32 to 0.67 g cm^{-3}. These low densities reflect the very fine-grained texture of the material and the high accumulation rate at the sites. A plot of erosion threshold against dry bulk density is shown in fig. 4. The correlation coefficient of 0.57 shows that the correlation is significant ($P < 0.005$) although a lot of scatter obviously is present. Similar correlations for natural sediments have been found by, e.g. Riethmüller et al. (1998) who found that the erosion threshold could be correlated to dry bulk density but only for samples with low contents of chlorophyll *a*. Also Amos et al. (1996) and Amos et al. (1997) reported an increase in erosion threshold with increasing bulk density. In fact, Amos et al. (1996) inferred from their

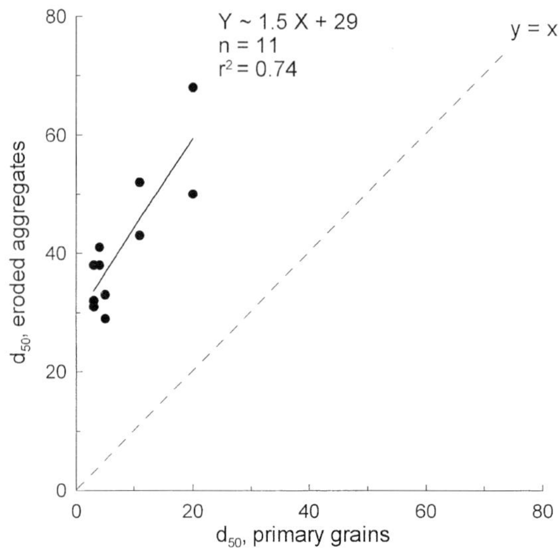

Figure 7. A plot of $d_{50,\ aggregates}$ against $d_{50,\ primary\ particles}$. The solid line shows the best linear fit and the dotted one is the line for which the two are equal.

data set that the deposition rate had a stronger influence on erosion threshold than the sediment texture. However, higher erosion thresholds were found in their study, probably due to lower accumulation rates and the resulting high bulk densities. Data on the deposition of sediment in the present study area is available, but the vertical resolution of the depth soundings is 0.1m which is too low for the detection of small-scale changes over periods of less than a year. Such information would have to be available if predictions of the erodibility of the surface material of cohesive sediments should be based on sedimentation rates alone.

A positive relationship between dry density of the bed material and the erosion threshold is not always found in nature. In fact, this relationship is generally weak or absent due to strong biological mediation of the erodibility (Paterson & Black, 1999; Austen et al., 1999; Andersen, 2001). This is especially true for intertidal sites where microphytobenthos and macrofauna may be very important. When a positive relationship has been found it has mostly been on subtidal sites (e.g. Amos et al., 1996; this investigation) and there seems to be growing evidence that a stronger biological mediation of erodibility is generally found on intertidal sites than on subtidal sites. A possible cause is perhaps the tendency for higher biomass of microphytobenthos in the intertidal zone due to higher solar radiation at these sediment surfaces.

The lowest erosion thresholds were measured at position 1 and 2 situated in the part of the study area with the lowest current velocities. Detailed information on the current velocities at the study sites is not available but a small decrease in current velocity is found when moving westward towards the Haringvliet. Higher current velocities in Amer and part of Spijkerboor is also indicated by the higher sand content of the bed material at these sites. Higher erosion thresholds are observed here and it seems reasonable to conclude that both dry bulk density and current velocity affects the erodibility with higher thresholds at higher dry densities and current velocities. The effect of the current velocity is probably due to winnowing of loose aggregates at sites with higher velocities and perhaps a differences in packing. However, it is not possible to determine this on the basis of the present investigation.

The erosion thresholds found by Winterwerp et al. (1992) for the laboratory experiments on settled beds are of the same order as found in this investigation, although direct comparison of the results is not possible due to differences in the analysis procedure.

Erosion threshold and grain size

The erosion thresholds are not significantly correlated to any of the grain size parameters (median grain size, sand content, clay content, coarse + medium silt) but the number of observations is low. Median grain sizes are 9 to 26 μm (laser analysis) or 3 to 33 μm (Sedigraph analysis), sand content 7 to 27 % (laser sizing) or 2 to 30 % (Sedigraph analysis) and clay content 7 to 13 % (laser sizing) or 18 to 49 % (Sedigraph analysis). The general lack of correlation is probably due to the fairly uniform grain size distributions of the samples.

5.2 Erosion rate

Erosion rate and bed shear stress

The study shows that the erosion rate of the sediments is highly influenced by the applied bed shear stress. The dependence on bed shear stress was also found by Winterwerp et al. (1992) who analysed the results of studies of the erosional behaviour of fine-grained sediments including sediments from Hollandsch Diep and Biesbosch. The experiments were undertaken in laboratory flumes on settled sediments from the sites and the measured erosion rates were fairly similar to the rates found in this study.

It is interesting to see that erosion experiments carried out in the laboratory on settled beds give results which are of the same order of magnitude as experiments carried out on natural samples. The consolidation periods for the laboratory experiments were 1 and 7 days and the erosion rates found for both types of experiments is within the range found on natural samples in this study. This indicates that the erodibility of the sediments at the study site is not significantly affected by benthic macrofauna or microphytobenthos as the effect of this is more or less absent in laboratory experiments.

The dependence of the erosion rate on the parameter $\tau_i - \tau_{cr,\ surface}$ is similar to what is found for Danish intertidal sites with low bioturbation activities. On intertidal sites with high bioturbation activities and strong pelletization of the bed material a much stronger increase in erosion rate with increasing bed shear stress is observed (Andersen, 2001). Consequently, the comparison of the results with results *in situ* from intertidal mudflats also highlights the comparatively low bioturbation activity at the sites.

Erosion rates and dry bulk densities

The erosion rate of the sediments is independent of the erosion threshold but some dependence on dry bulk density is observed for the higher bed shear stresses (fig 6). At 0.5 N m^{-2} no correlation between dry bulk density and erosion rate is observed but at 1.0 N m^{-2} a significant negative correlation is observed ($P < 0.05$). The reason for the difference is that only erosion of the topmost loose aggregates is taking place at 0.5 N m^{-2} (type I erosion) and this erosion is largely independent of the average dry bulk density of the top 5 mm of the bed. On the other hand, at 1.0 N m^{-2}, erosion is proceeding a few mm down into the bed (type II or transitional erosion) where the compaction (and hereby dry bulk density) will be of importance for the erodibility.

Only a limited number of studies of erosion rates on natural sediments have been reported in the literature but the calculated erosion rates are of the same order as the erosion rates reported by Amos et al. (1996), Amos et al. (1997), Widdows et al. (1998) and Houwing (1999). Amos et al. (1997) found that the erosion rate was independent of bed shear stress and largely constant throughout each erosion experiment. This is in contrast to the present investigation where erosion rates at 1 N m^{-2} typically were an order of magnitude higher than the erosion rate at 0.5 N m^{-2}. The reason for this difference is unknown but is presumably related to differences in compaction of the materials as the sediments investigated by Amos et al. (1997) had much lower water contents (and also lower organic contents). For the measured dry bulk densities of approximately 0.5 g cm^{-3} an erosion rate of 1 g m^{-2} s^{-1} corresponds to erosion of 3.6 kg m^{-2} hr^{-1} or 7 mm hr^{-1} but this erosion rate is only observed at high bed shear stresses and the erosion rate will presumably only apply for the top few mm of the bed.

5.3 Settling velocity

The median equivalent settling diameter of the primary grains of the bed material ranged from 2 to 20 μm which strongly suggest that the material was eroded as aggregates, as $d_{50,\ aggr}$ was larger (24 - 68 μm). The tendency for correlation between $d_{50,\ aggr}$ and the primary grain-size of the material indicates that the coarser modes in the primary grain-size distributions could be influencing the settling velocities of the eroded material. This result is in contrast to what is found on mudflats in the Danish Wadden Sea area where $d_{50,\ aggr}$ is found to be independent of primary grain-size and instead mainly controlled by the aggregation induced by benthic macro-fauna (Andersen & Pejrup, in press). Consequently, also the results of the measurements of settling velocity of the eroded material indicate that macro-fauna is not affecting the aggregate structure of the material in the study area as intensely as on intertidal sites. It is possible that this is a general tendency when comparing lacustrine and intertidal sediments. However, this still remains to be shown.

The average settling velocity is 1.1 mm s^{-2} corresponding to a settling diameter of 39 μm. Amos et al. (1996) found mean equivalent settling diameters of 15 μm for the eroded material from sublittoral sediments in a subarctic estuary. The equivalent settling diameters found in this investigation are larger despite the fact that the bed material is more fine-grained. The reason for this apparent paradox is unknown but may be attributed to a more cohesive nature of the clay-rich sediment in the present study area compared to the silty and sandy sediments studied by Amos et al. (1996).

6. CONCLUSIONS

The erodibility of fine-grained sediments from a freshwater system with high accumulation rates and poorly developed macro-faunal community has been investigated. The erosion thresholds ranged from 0.16 to 0.70 N m^{-2} and were correlated to dry bulk density of the bed material. A tendency for higher thresholds at sites with higher current velocities was also observed and it is concluded that the erosion threshold of the bed material is mainly governed by physical properties of the sediment which are a product of sedimentological and hydraulic conditions. The erodibility is apparently not significantly mediated by macrofauna and this is attributed to the generally poorly developed macrofaunal community at the sites.

The erosion rate was correlated to the parameter $\tau_i - \tau_{cr,\ surface}$ in an exponential form: $\varepsilon i = \beta \exp(\alpha(\tau i - \tau cr, surface))$. The rates are similar to the rates found with the same measuring equipment on intertidal sites with low macrofaunal community. When comparing the erosion rates found in this study with erosion rates found on the basis of flume studies with settled beds no significant difference was found. This was slightly surprising as *in situ* measurements generally have shown marked mediation of erodibility by microphytobenthos and macrofauna. Again, the lack of difference between measurements on natural beds and measurements on settled beds points to the conclusion that the erodibility is not significantly mediated by biology at the study sites.

The equivalent settling diameters of the eroded aggregates are at least twice as large as the settling diameters of the primary grains which shows that the material is eroded as aggregates, probably mainly formed due to consolidation of the bed prior to resuspension. However, a significant positive correlation was found between the two settling diameters. This shows that the coarser material of the bed is influencing the aggregate size or density of the aggregates.

REFERENCES

Amos, C. L., Sutherland, T. F. and Zevenhuizen, J. (1996) The stability of sublittoral, fine-grained sediments in a subarctic estuary. Sedimentology 43, 1-19.

Amos, C. L., Feeney, T., Sutherland, T. F. and Luternauer, J. L. (1997) The stability of fine-grained sediments from the Fraser River Delta. Estuarine, Coastal and Shelf Science 45, 507-524.

Andersen, T. J. (2001) Seasonal variation of erodibility at two temperate, microtidal mudflats. Estuarine, Coastal and Shelf Science.

Andersen, T.J. and Pejrup, M. (in press) Biological mediation of the settling velocity of bed material eroded from an intertidal mudflat, the Danish Wadden Sea. Estuarine, Coastal and Shelf Science.

Austen, I., Andersen, T. J. and Edelvang, K. (1999) The influence of benthic diatoms and invertebrates on the erodibility of an intertidal mudflat, the Danish Wadden Sea. Estuarine, Coastal and Shelf Science 49, 99-111.

Christensen, R.W. and Das, B.M. (1973) Hydraulic erosion of remolded cohesive soils. In: National Research Council (ed.) Soil erosion: causes and mechanisms, prevention and control. Highway Research Board Special Report 135, Washington, DC, 8-19.

Cornelisse, J.M., Mulder, H.P.J., Williamson, Witte, G. and Houwing, E.J. (1994) On the development of instruments for in situ erosion measurements. In: Burt, N., Parker, R. and Watts, J. (eds) Cohesive Sediments. John Wiley & Sons Ltd. London, 175-186.

Ferguson, H.A. and Wolff, W.J. (1984) The Haringvliet-project: the development of the Rhine-Meuse estuary from tidal inlet to stagnant freshwater lake. Water, Science & Technology 16, 11-26.

Gust, G, and Müller, V. (1997) Interfacial hydrodynamics and entrainment functions of currently used erosion devices. In: Burt, N., Parker, R. & Watts, J. (eds) Cohesive Sediments. John Wiley & Sons Ltd. London, 149-174.

Houwing, E. -J. (1999) Determination of the critical erosion threshold of cohesive sediments on intertidal mudflats along the Dutch Wadden Sea coasts. Estuarine, Coastal and Shelf Science 49, 545-555.

Kuijper, C., Cornelisse, J. M. and Winterwerp, J. C. (1990) Erosion and deposition characteristics of natural muds. Sediments from the Hollandsch Diep (north of Sassenplaat). Report 27, Rijkswaterstaat, Delft Hydraulics.

Kuijper, C., Cornelisse, J. M. and Winterwerp, J. C. (1991) Erosion and deposition characteristics of natural muds. Sediments from the Biesbosch (Spijkerboor). Report 36, Rijkswaterstaat, Delft Hydraulics.

Mehta, A.J., Parchure, T.M., Dixit, J.G. and Ariathurai, R. (1982) Resuspension potential of deposited cohesive sediment beds. In. Kennedy, V.S. (Ed.) Estuarine Comparisons. Academic Press, New York, 591-609.

Parsons, T.R., Maita, Y. and Lalli, C.M. (1984) A manual of chemical and biological methods for seawater analysis. Pergamon. 173 pp.

Paterson, D.M. (1989) Short-term changes in the erodibility of intertidal cohesive sediments related to the migratory behaviour of epipelic diatoms. Limnology and Oceanography 34 (1), 223-234.

Paterson, D. M. and Black, K. S. (1999) Water flow, sediment dynamics and benthic biology. Advances in Ecological Research 29, 155-193.

Riethmüller, R., Hakvoort, J. H. M., Heineke, M., Heymann, Kühl, H. and Witte, G. (1998) Relating erosion shear stress to tidal flat surface colour. In: Black, K. S., Paterson, D. M. and Cramp, A. (eds.) Sedimentary processes in the intertidal zone. Geological Society, London, Special Publications 139, 283-293.

Schünemann, M. and Kühl, H. (1991) A device for erosion-measurements on naturally formed, muddy sediments: the EROMES-System. Report of GKSS Research centre GKSS 91/E/18, 28 pp.

Tolhurst, T. J., Riethmüller, R. and Paterson, D. M. (2000) In situ versus laboratory analysis of sediment stability from intertidal mudflats, Continental Shelf Research 20 (10-11), 1317-1334.

Van Eck, G. T. M., Zwolsman, J. J. G., and Saeijs, H. L. F. (1997) Influence of compartmentalization on water quality of reservoirs: Lessons learned from the enclosure of the Haringvliet estuary, The Netherlands. In: Dix-neuvieme Congres des Grands Barrages, volume III. Commission Internationale des Grands Barrages, Florence, France, 547-669.

Winterwerp, J. C., Cornelisse, J. M. and Kuijper, C. (1992) Erosion of natural sediments from the Netherlands. Analysis of laboratory experiments. Report 38, Rijkswaterstaat, Delft Hydraulics.

Widdows, J., Brinsley, M. D., Bowley, N. and Barrett, C. (1998) A benthic annular flume for in situ measurements of suspension feeding/biodeposition rates and erosion potential of intertidal cohesive sediments. Estuarine, Coastal and Shelf Science 46, 27-38.

Fine Sediment Dynamics in the Marine Environment
J.C. Winterwerp and C. Kranenburg (Editors)

Gas bubble nucleation and growth in cohesive sediments

Walther van Kesteren[a] and Thijs van Kessel[b]

[a]Marine and Coastal Management Section, WL | Delft Hydraulics, P.O. Box 177, 2600 MH Delft, The Netherlands

[b]same address

Sediment often contains a significant amount of organic material, which can be decomposed by bacterial activity. During this process and under anaerobic conditions that prevail in sediments, mainly methane and carbon dioxide are formed. These compounds will dissolve in the pore water, until the level of saturation is attained.

Experiments show that gas bubble nucleation occurs already at a small oversaturation of methane in pore water. During nucleation, the large solid-liquid interfacial area acts as a catalyst. Bubbles will start to grow, as gas that does not flow out by convective or diffusive transport accumulates in bubbles. As a result of bubble initiation the distance across which transport occurs (to the closest bubble) strongly decreases; the rate of transport therefore increases and finally equals the rate of gas production. The average distance between bubbles is therefore determined by the rate of gas production and generally is in the order of millimetres to centimetres.

From a stability analysis it appears that bubbles in sludge depots may rise once the diameter is tens of centimetres. However, the average bubble diameter remains limited to a few centimetres owing to the large number of bubbles per m^3 and the total volume of gas that can be produced. During bubble growth the amount of deformation energy stored in the grain matrix increases. Crack formation will occur if this amount exceeds the fracture energy and if the stress conditions are suitable. Cracks initially have a plain, closed structure, but may eventually open depending on the depth (i.e. ambient pressure), so that discharge of gas and water can occur.

KEY WORDS
gas production, mud, sludge depots, gas accumulation

1. INTRODUCTION

The decomposition by bacterial activity of organic material present in natural mud results in the formation of carbon dioxide and, under anaerobic conditions, methane. These compounds remain dissolved in the pore water if their concentrations remain below the saturation level. Depending on the balance between the rate of decomposition and the rate of transport of the dissolved reaction products, saturation may or may not be reached. Natural muds generally have a low permeability, strongly limiting advective transport. As diffusive transport of compound dissolved in pore water proceeds very slowly, methane and carbon dioxide production results in the formation of a third phase, a gas phase in addition to the solid phase and liquid phase. This significantly affects the sediment properties and causes the sediment volume to increase. The objective of this study is to investigate, on the basis of laboratory experiments and theoretical analysis, the mechanisms for nucleation and growth of gas bubbles in cohesive sediments, providing means to assess the influence of their presence on sediment properties.

The following topics will be discussed. First bubble nucleation is addressed. Subsequently, the way bubbles grow is elaborated upon. Finally, the occurrence of crack initiation and growth, bubble coalescence and bubble escape is discussed.

2. BUBBLE NUCLEATION

2.1. Solubility

The production of compounds such as CO_2 and CH_4 only results in expansion if molecules that are dissolved in the water phase transit into the gas phase. This will only occur if the water phase becomes oversaturated with CO_2 and CH_4. The solubility of gases in water is given by Henry's law:

$$p_i = k_i\, x_i \tag{1}$$

where p_i is the partial vapour pressure of component i, k_i is Henry's constant for this component and x_i is the mole fraction of i in water.

Equation (1) shows that the vapour pressure is proportional to the concentration of dissolved gas. As soon as the vapour pressure exceeds the liquid pressure, gas bubbles will be formed. Henry's constant is related to the solubility of gas: the larger k_i, the smaller the solubility. An important implication of (1) is that the solubility of gas is proportional to the liquid pressure. For example, in an artificial depot the solubility is already doubled at a depth of about 10 m. The dependency on the temperature is much less pronounced and can easily be quantified.

The influence of a fine pore system on the solubility of gases is not taken into account in the analysis above. However, phase equilibria may definitely change. In pores smaller than 50 nm 'pore condensation' may occur, during which gas molecules adsorb to the surface of the pores. The reduction of vapour pressure is expressed by Kelvin's law:

$$p_i = p_{sat} \exp\left[-\frac{4\sigma V_l}{RTd}\right] \tag{2}$$

where p_{sat} is the saturation pressure, σ the surface tension, V_l the molar volume of the adsorbate, d the pore diameter, R the gas constant and T the temperature (K). In addition, owing to the effect of surface tension, the pressure inside gas bubbles is higher than the ambient pressure[1]:

$$\Delta p = \frac{2\sigma}{r} \tag{3}$$

where r is the bubble radius. Because of these two effects, significantly more gas may be stored in sediment pore water than to be expected on the basis of its saturation pressure (Broekhof, 1969). Figure 1 shows the saturation fraction of CH_4 as a function of the pore diameter, expressed as (Brennen, 1995):

$$x_{CH_4} = \frac{\frac{2\sigma}{r} + p_0}{k_{CH_4} \exp\left[-\frac{4\sigma V_l}{RTd}\right]} \tag{4}$$

Eq. (4) is a combination of (1)–(3) assuming $p_{sat} = p_0 + \Delta p$, with p_0 the ambient pressure. Pore condensation has not been observed during saturation experiments (Van Kessel, 1998a). For both artificial clays (kaolinite, laponite and Westerwalden clay) and a natural mud from Ketelmeer, the solubility of CH_4 did not exceed the value calculated from the water content and Henry's law (1) (for CH_4 at $T = 293.15$ K and $p = 101.325$ kPa, $k_i = 2.806 \times 10^{-5}$).

The solubility of CH_4 was measured as follows. A pressure cell was partly filled with a mud suspension, rinsed with N_2 and depressurised. As a second step, its known headspace volume was filled with pure methane at a pressure of 600 kPa. From the subsequent pressure decrease the amount of CH_4 dissolved in the mud suspension could be calculated.

The absence of any CH_4 storage in addition to the amount according to Henry's law can be explained on the one hand by a limited volume of the meso-pores which are essential for pore condensation, on the other hand by a limited accessibility of these pores. As an example, the pore size distribution Ketelmeer sediment, a natural, normally consolidated mud originating from the Netherlands, is displayed in Figure 2. Its Attenberg limits are: plastic limit PL = 41.7%, liquid limit LL = 83.3% and plasticity index PI = 41.6%. The pore size distribution is obtained using mercury porosimetry, which performs well down to a pore size of a few nm. In addition, the N_2 gas adsorption method was used in the range of 0.35–300 nm. Figure 2 shows that the volume of mesopores (< 50 nm) remains limited. It can therefore be assumed that in most cohesive sediments, the solubility of gas is not influenced by the presence of pores.

---

[1] a further stability analysis for a dynamic system results in $\Delta p = 4\sigma/3r$ (Brennen, 1995)

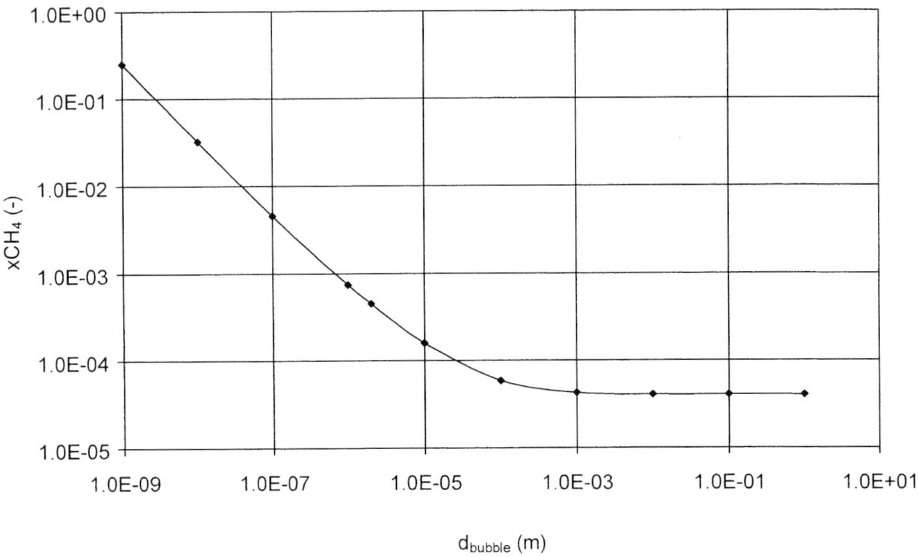

Figure 1. Saturation fraction for CH_4 as a function of bubble diameter.

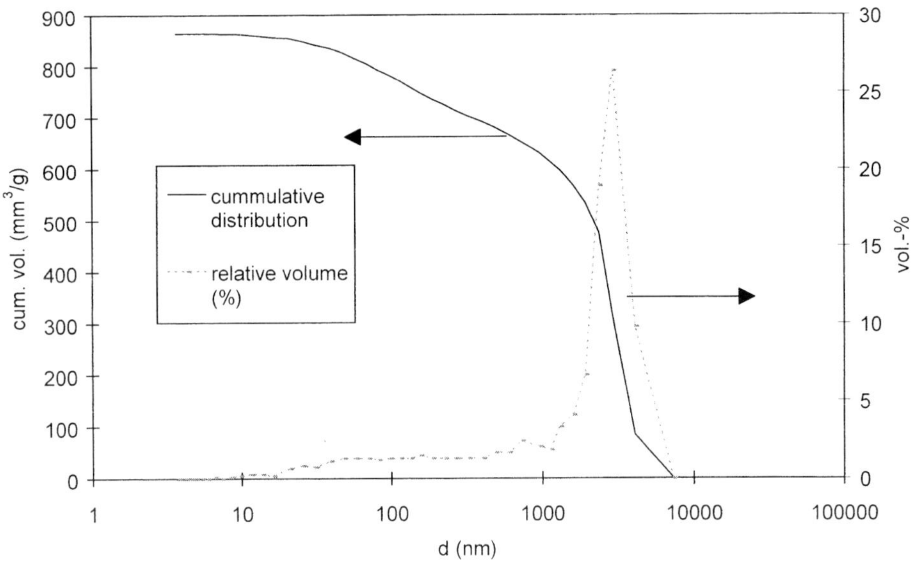

Figure 2. Pore size distribution of Ketelmeer mud, determinded with mercury porosimetry.

2.2. Nucleation

When pore water is oversaturated with gas, bubble nucleation may occur. Nucleation can take place both homogeneously and heterogeneously. For homogeneous nucleation to occur a high oversaturation is needed, as a significant energy barrier (activation energy) has to be overcome to generate micro bubbles (Brennen, 1995). This is caused by the surface tension of water that limits the nucleation process. The excess pressure Δp needed to overcome surface tension is given by Eq. (3). The surface tension of water σ is 0.074 N m^{-1}. From (3) it is clear that the excess pressure (and the oversaturation) increases with decreasing bubble radius.

As a results of this energy barrier, nucleation generally takes place heterogeneously. Small dislocations, on which cavities are stable, are always present on solid surfaces. These are ideal locations for bubble nucleation, as initial cavities in the form of dislocations result in a strong decrease of the energy barrier. Heterogeneous nucleation can therefore occur at a small oversaturation.

As diffusion only proceeds fast over a short distance, a large number of bubbles per litre of sludge is formed. As a result, the degree of oversaturation in the pore water remains limited (Van Kessel, 1998a). The exact number of bubbles n_b and the average bubble distance l_b ($\sim\sqrt[3]{1/n}$) depends on the rate of gas production r_A.

In laboratory experiments, during which pore water was saturated with CH_4 at a high pressure and subsequently the pressure was decreased to create oversaturation, the number of bubbles observed was about 10^4 per litre (Van Kessel, 1998a). At a lower production rate the number is less; assuming $r_A = 0.1$ μmol m^{-3} s^{-1} the calculated average bubble distance is about 0.1 m. Only a few bubbles per litre are then present. The diffusivity of CH_4 in sludge is assumed to be equal to that in water. A lower value leads to a smaller bubble distance.

In the experiments executed, oversaturation was achieved by saturating the sediments tested at a high pressure (0.5 MPa) and subsequent pressure decrease. Both natural mud and artificial clays were tested, one of them being laponite, a transparent clay enabling the observation of bubbles throughout the sample. Pressure, temperature, sediment level, rate of gas outflow and the acoustic impedance of the sample were measured continuously during the experiments to monitor the bubble size and number. Bubble nucleation and growth in laponite was recorded on video tape. Some examples hereof are shown in Figure 3.

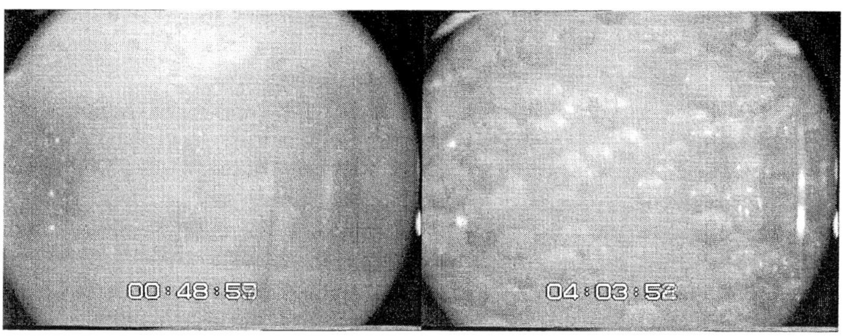

Figure 3. Video images of bubble growth in laponite at two times (displayed in hours and minutes), scale: 1 cm = ⟷

3. BUBBLE GROWTH

3.1. Initial growth

After bubble nucleation the bubbles start to grow as a result of diffusion of dissolved gas towards the bubbles. The rate of bubble growth can be expressed as (Brennen, 1995):

$$r\frac{dr}{dt} = \frac{D[x_\infty - x_i(1 + 2\sigma / rp_\infty)]}{1 + 4\sigma / 3rp_\infty}\left[1 + \frac{r}{\sqrt{\pi D t}}\right] \tag{5}$$

where r is the bubble radius, t is time, D the diffusion coefficient for methane in water, σ the surface tension of water, x_i the CH_4 fraction at the bubble interface, x_∞ the CH_4 fraction far away from the bubble and p_∞ the pore water pressure. At $t = 0$ the concentration is assumed to be x_∞ everywhere. The fraction x_i is calculated according to:

$$x_i = \frac{p_\infty - p_V + 4\sigma / d}{k_{CH_4}} \tag{6}$$

where p_V is the vapour pressure of water, generally negligible. For large t (5) can be written as:

$$d^2 - d_0^2 \approx 8D(x_\infty - x_i)t \tag{7}$$

where d_0 is the bubble diameter at $t = 0$.

From (7) it follows that the rate of bubble growth is proportional to the square root of time. This can be explained by the fact that for an increase dr of the bubble radius, the amount of gas that has to diffuse towards the bubble becomes larger as the bubble radius increases. For example, it only takes one day to form a bubble with a diameter of 10^{-3} m, 100 days for a bubble of 0.01 m and 30 years for a bubble of 0.1 m. However, the maximal bubble size is limited by the total amount of gas produced. Bubbles larger than 0.1 m do not develop, or, on a very long time-scale, will develop at the expense of other bubbles. This is illustrated in Figure 4.

Initially bubble growth proceeds faster, as the effect of surface tension is important for small radii. Because of this, the pressure inside the bubble decreases with increasing radius and the bubble will expand faster. This effect is important for bubbles smaller than 10^{-5} m (Fig. 1). Given the pore size distribution in Fig. 2, it can be concluded that bubbles will therefore attain a diameter equal to the pore diameter soon after their generation.

Figure 5 shows the increase in gas content and average bubble diameter for Ketelmeer mud, a natural mud from the Netherlands. Results are in good agreement with a model for bubble growth, based on Eq. (7). For sufficiently high numbers, the model is insensitive to the number of nucleation sites assumed, as during the growth process large bubbles grow at the cost of small bubbles. The experiments show that the number of nucleation sites is not limiting for bubble growth (Van Kessel, 1998a).

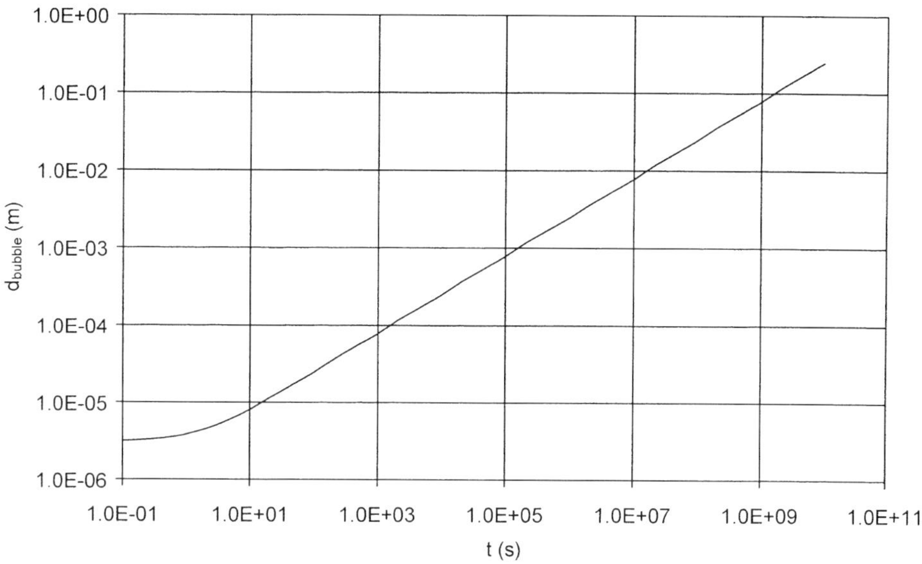

Figure 4. Time scale for bubble growth or shrinkage as a function of bubble diameter according to Eq. 5); $D = 1.5\ 10^{-9}\ m^2/s$; $\sigma/p_\infty = 7.4\ 10^{-7}$ m.

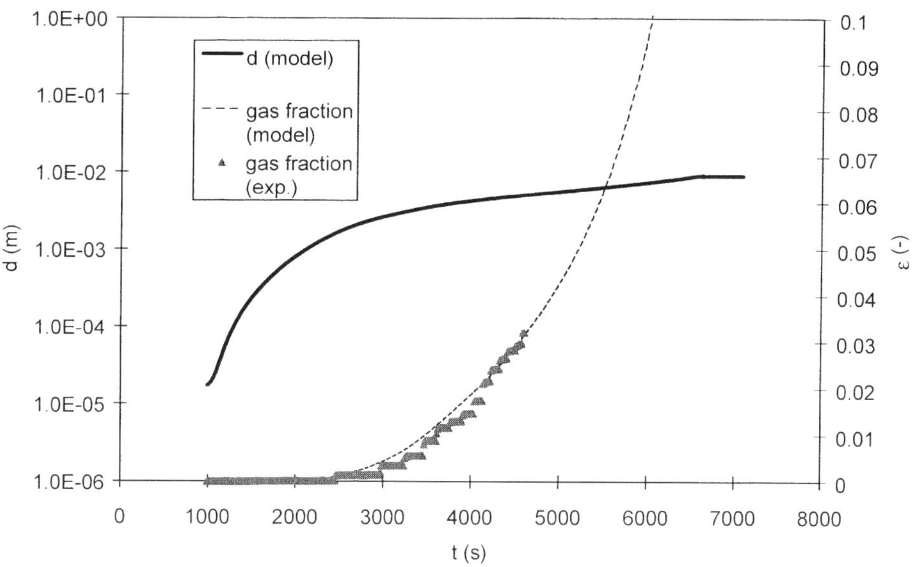

Figure 5. Gas fraction ε and bubble diameter d during pressure decrease; Ketelmeer mud.

3.2. Continued growth

When the bubble becomes larger than the pore diameter, the question is whether gas will push aside water from the pore system or bubbles will deform the grain matrix. This depends on the shear strength of the grain matrix and the diameter of the pores.

The pressure inside a bubble just after nucleation is completely determined by surface tension according to Eq. (3). When the bubble starts to deform the grain matrix, the unambiguous relation between bubble pressure and bubble radius disappears. The following expression now applies (Wheeler *et al.*, 1990a, 1990b):

$$4\sigma/d_{\text{bubble}} \leq \Delta p \leq 4\sigma/d_{\text{crit}} \tag{8}$$

where d_{bubble} is the bubble diameter and d_{crit} is the diameter of the meniscus just bridging the largest distance between two grains at the bubble interface. Note that d_{crit} is independent of d_{bubble} and is in the order of the grain diameter.

However, another independent upper limit exists for the pressure difference Δp between gas bubble and sediment matrix. It follows from plastic cavity expansion theory (Bagualin *et al.*, 1972):

$$\Delta p_{\max} = (4/3)\, c_u\, \{1 + \ln [G/c_u] + \ln [1-(r/r_0)^{-3}]\} \tag{9}$$

in which r is the bubble radius and r_0 the initial bubble radius, σ the surface tension of water, c_u the undrained shear strength and G the shear modulus. The effect of Darcy flow has been neglected, which is a valid assumption for cohesive sediment.

Replacement of pore water only occurs if the pressure limit $\Delta p = 4\sigma/d_{\text{crit}}$ in (8) is smaller than the pressure difference needed to push aside the grain matrix (Eq. 9, limit $\Delta p = (4/3)\, c_u\, [1 + \ln (G/c_u)]$). Replacement of pore water turns out to be possible only in pores larger than about 10^{-4} m for sediment with $c_u = 1$ kPa and $(G/c_u) = 10$, as is shown in Figure 6. For smaller diameters this is unfavourable from a thermodynamic point of view, as the limiting effect of surface tension plays an important part. As the overwhelming majority of pores in sediment like Ketelmeer mud (Fig. 2) is smaller than 10^{-5} m, bubble growth only occurs by deforming the grain matrix.

In view of the undrained shear strength c_u, which ranges between a few and a few tens of kPa in artificial depots, the pressure difference between gas bubble and sediment matrix never exceeds 10^4 to 10^5 Pa (Van Kessel, 1998a).

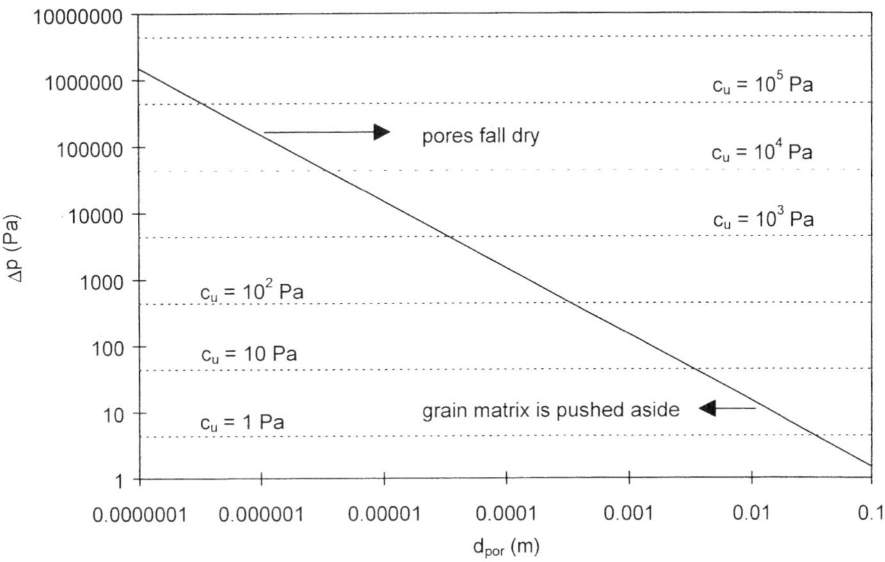

Figure 6. Criterion for the type of bubble expansion; $G/c_u = 10$.

4. GAS ESCAPE

4.1. Bubbles rise

Owing to the large density difference between a gas bubble ($\rho \approx$ 1 kg m^{-3}, pressure dependent) and its surrounding medium, *i.e.* sediment ($\rho \approx$ 1400 kg m^{-3}), a bubble tends to rise. However, the grain matrix can withstand the stresses resulting from the density difference as long as a critical stress, the yield stress, is not exceeded. A criterion for the critical radius r_{crit} at which a bubble just starts to rise is (Nguyen and Boger, 1992):

$$r_{\text{crit}} = fc_u/\rho g \qquad (10)$$

where f is a dimensionless form factor analytically defined to be 7.5. However, from a numerical large strain analysis it follows that $f = 11.6$ (Schotmeyer, 1998).

Based on (10), substituting c_u = 1 kPa and ρ = 1400 kg m^{-3} it follows that r_{crit} = 0.84 m. In the previous section it was demonstrated that such a bubble size will not be found in artificial depots. Bubbles with a diameter of 0.1 m will rise in a sediment matrix having a shear strength of 60 Pa or less. In reality the shear strength is more than an order of magnitude larger, maybe except in the upper few centimetres. The rise and escape of gas bubbles through the grain matrix therefore is an unlikely mechanism for gas transport.

Rising of gas bubbles has also been modelled using the so-called 'Cam Clay'-model in a numerical analysis with DIANA (Cazemier and Visschedijk, 1997). These calculations also do not show a continuous rising of bubbles. Owing to 'hardening' of the material above a bubble, rising becomes more and more difficult. These calculations also showed potential areas for crack initiation in two different areas. The lower one is a so-called Herzian type of crack initiation, which will remain stable. The upper area, however, is a possible location for unstable crack initiation and growth.

4.2. Crack formation

During growth, bubbles may deviate from a spherical shape and grow in the direction of the smallest principle stress. Cracks will form on the boundary between the elastic and plastic deformation zone around bubbles. When the bubbles have reached a critical radius given the fracture toughness of the mud, cracks may create pathways for gas transport after coalescence of adjacent bubbles. However, from fracture mechanics theory it can be derived that cracks can only be formed under the following conditions (Broek, 1983):

1. the ratio between isotropic effective stress and drained cohesion must be less than unity;
2. the length scale of system (*i.e.* bubble size) must be sufficient to allow for stress concentration;
3. the plastic deformation zones around the bubbles must not overlap, as cracks will only form on the boundary between the elastic and plastic zone

In view of the first point, cracks may easily be formed in overconsolidated materials, having a large drained cohesion. In underconsolidated materials however, no cracks will be formed. The second point implies that no cracks are formed around small bubbles. The stress concentration along the tip of shear planes is expressed as (Murdoch, 1993):

$$K_{\mathrm{I}} = 2\sqrt{\frac{r_p}{\pi}}\left[\left(p_{\mathrm{bubble}} - p_{\infty}\right)\frac{2}{\pi}\arcsin\frac{r_{\mathrm{bubble}}}{r_p} + \left(p_{\mathrm{p}} - p_{\infty}\right)\left(1 - \frac{2}{\pi}\arcsin\frac{r_{\mathrm{bubble}}}{r_p}\right)\right] \tag{11}$$

where K_{I} is the stress intensity factor (in N m$^{-1.5}$), r_p the radius of the plastic zone, r_{bubble} the bubble radius, p_{bubble} the pressure in the bubble, p_p the total tangential pressure averaged over the intersection of the plastic zone and p_{∞} the pressure at a large distance from the bubble. Eq. (11) is valid in the case that the grain matrix is deformed (no hydraulic fracture).

When a bubble grows, the stress intensity factor increases and may reach a critical value at which crack initiation and growth will commence. This critical value is called the fracture toughness K_{Ic}. The fracture toughness can be determined experimentally or be derived from the empirical correlation $K_{Ic} = ac_u$, with constant a ranging between 0.13 and 0.21 (Van Kesteren, 1985). However, it must be remarked that fracture toughness is not a material constant, but depends on the type of test used for its determination. A more reliable parameter is the fracture energy required for crack propagation.

Extension tests show a fracture energy for Ketelmeer sediment of about 0.2 N m^{-1}. From these tests a fracture toughness of $K_{Ic} \approx 0.12\ c_u$ is back-calculated, which is close to the empirical correlation. Based on these experiments and on the failure mechanics theory, the critical diameter for crack formation is estimated at 0.25 to 1 cm for Ketelmeer sediment.

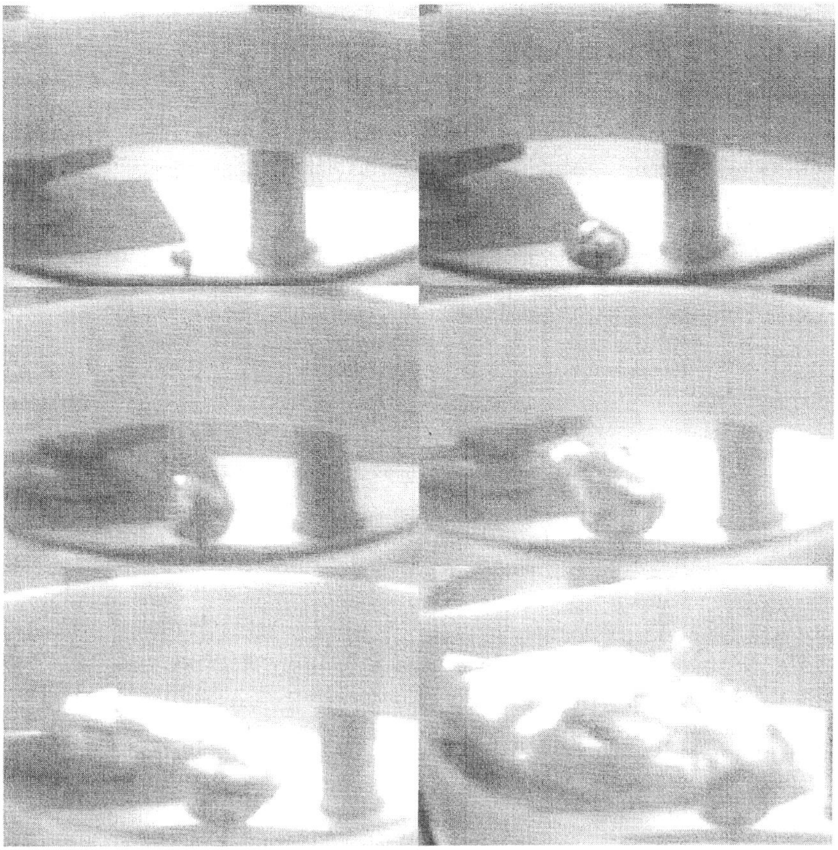

Figure 7. Bubble growth in laponite; times from upper left to lower right 1, 23, 1658, 3846, 4616 and 5662 s after bubble nucleation, respectively; scale: 2 cm = ⟵⟶

Crack propagation is determined by the speed of flow of pore water from the adjacent grain matrix to the zone of fracture. After a fracture plane is formed it is saturated with water. Neighbouring bubbles can communicate by coalescence of fractures. However, exchange of gas does not yet occur. Owing to the flat structure of the fracture, the transport capacity of gas dissolved in pore water and present in bubbles is comparable with that of the largest pores of the surrounding grain matrix. Only if the surrounding grain matrix is pushed aside, a channel can be created along which gas bubbles can be transported. As a result, exchange of gas with surrounding bubbles is possible and an area is created with cracks and cavities filled with both gas and water (Van Kesteren *et al.*, 1998).

The behaviour of growing bubbles has been subject of an experimental study (Van Kessel, 1998b). Under controlled conditions the pressure path inside and around an expanding bubble was measured. Bubble growth was controlled on the one hand by the gas pressure in a hollow needle at the tip of which nucleation occurred, on the other hand by the volume of the triaxial cell, which can be varied with a piston. Pore pressure transducers were mounted at several positions in the cell. The measurements show that the pressure difference calculated with (9) is nearly a factor 2 lower than the measured difference. For example, in Ketelmeer sediment with c_u = 450 Pa a pressure difference is measured of about 4 kPa between an expanding bubble and the surrounding sediment matrix. Based on this experiment, the fracture or deformation energy is estimated at about 0.6 N m^{-1}. From these tests a fracture toughness of $K_{Ic} \approx 0.21\ c_u$ is back-calculated, which is within the range of the empirical correlation proposed by Van Kesteren (1985).

Experiments on a transparent and highly plastic artificial clay, laponite, confirm the expectation that bubbles will deviate from their initial spherical shape during growth (Van Kessel, 1998b). The direction of preference of bubble growth turns out to be the direction of the lowest stress invariant. This direction lies in the horizontal plane (Fig. 7).

5. CONCLUSIONS

Bubble nucleation and growth is not limited by the number of nucleation sites, but controlled by the rate of gas production in relation to the diffusivity. Bubbles larger than a few centimetres in diameter need a very large time to be created and will therefore be scarce. When in equilibrium with the concentration of dissolved gas in the surrounding pore water, bubbles will remain stable unless:

1. the lift force exceeds the matrix strength, or
2. cracks evolve towards the surface, creating pathways for gas transport.

The former process requires a low strength, the latter a high strength. Generally the gas content of natural sediments remains limited to a few percent, but at intermediate strengths (10^2 Pa $< \tau_y < 10^4$ Pa) and at a sufficient rate of gas production (by the decomposition of organic matter) the gas content may exceed 10 or even 20 volume-% *in situ*. The presence of free gas has important consequences for the consolidation and erosion behaviour of sediments. A compressible sediment bed also influences the flow induced (turbulent) bed shear stress.

REFERENCES

Bagualin, F., Jézéquel, J.F., Le Mee, E., Le Méhauté, A. (1972). Expansion of cylindrical probes in cohesive soils. *J. Soil Mech. Found. Div., Proc. ASCE*, Vol. 98, nr. SM 11, pp. 1129–1142

Brennen, C.E. (1995) Cavitation and bubble dynamics. Oxford University Press, Oxford.

Broek, D. (1983). Elementary engineering fracture mechanics. 3rd ed. Martinus Nijhoff Publ. The Hague, Netherlands.

Broekhoff, J.C.P. (1969). Adsorption and Capillarity. Ph.D. thesis Delft University of Technology.

Cazemier, W. and M.A.T. Visschedijk. (1997) Gas bubbles in sediments. TNO report no. 97-NM-R1066 dm11

Kessel, T. van. (1998a) Bubble initiation and bubble growth in sediment layers. WL report no. Z2314 dm18

Kessel, T. van. (1998b) Crack formation in sediments: experimental research. WL report no. Z2312.40 dm21

Kesteren, W.G.M. van, Cornelisse, J.M. and Kessel, T. van (1998) Crack formation and gas formation in sediments. WL report no. Z2312. dm15

Kesteren, W. van (1985). Summary and evaluation plastic clay cutting. Report no. R1233-3 (in Dutch).

Murdoch, L.C. (1993). Hydralic fracturing of soil during laboratory experiments. Part 1 Methods and observations, Part 2 Propagation, Part 3 Theoretical analysis. *Géotechnique* v 43, no. 2 pp. 255–287.

Nguyen, Q. and Boger, D. Measuring the flow properties of yield stress fluids. *Ann. Rev. Fluid Mech.* 24 (1992), pp. 47–88.

Schotmeyer, G.J. (1998). Stability of growing gas bubbles in sediments. GeoDelft report no. 373350/18.

Wheeler, S.J. Sham, W.K. Thomas, S.D. (1990a). Gas pressure in unsaturated offshore soils. *Canadian Geotechnical Journal* v 27 n 1 Feb 1990, pp. 79–89.

Wheeler, S.J. (1990b). Movement of large gas bubbles in unsaturated fine-grained sediments. *Marine Geotechnology* 9(2), pp. 113–129.

Fine Sediment Dynamics in the Marine Environment
J.C. Winterwerp and C. Kranenburg (Editors)

Erosion properties of mud beds deposited in laboratory settling columns

D.G. Lintern[a], G.C. Sills[a], N. Feates[b] and W. Roberts[b]

[a]Department of Engineering Science, Oxford University. Parks Road, Oxford, OX1 3PJ, UK

[b]HR Wallingford Ltd, Howbery Park, Wallingford, OX10 8BA, UK

Work has been conducted at Oxford University using H.R. Wallingford's Instrument for measuring Shear stress In Situ (ISIS). The first several experiments show that the equipment is working well in a column setup. Each bed tested is initially eroded by a distinct critical shear stress, or erosion threshold, which may depend on the properties of the mud and the length of consolidation time. The beds seem to erode in a laminar fashion, until at a very high shear rate a large (2 to 3 cm) liquified section of the surface gives way. Contrary to what might be expected, the beds become less resistant to erosion as they mature. It is proposed that the development of a biochemical surface layer, often containing gas, and the feeding and soil reworking by marine worms contribute to this weakening over time.

KEY WORDS:
ISIS; COSINUS; erosion; consolidation; shear stress; Tamar; mud

1. INTRODUCTION

Many factors affect the erosional properties of cohesive sediment, including its grainsize, maturation time, aerial exposure, biostabilisation, changes in sedimentation rate and bioturbation rate. The results presented here focus on the effects of consolidation time and biological activity, together termed maturation, in sedimentation experiments. Bed properties such as bulk density and pore pressures are measured accurately and the bed is carefully observed during erosion.

A longer consolidation time may be expected to lead to increased resistance to erosion, due to structural collapse of the electrically charged clay aggregates and due to bed cementation, or gelling. Krone (1999), for instance, reviews papers by Roberts *et al.* (1998) and Zreik *et al.* (1998). Both of these papers show that resistance to erosion is directly related to bed density, where the changes in density occur through time with structural alteration of aggregates.

Biological activity is also generally thought to lead to increased resistance of the beds to erosion. Stabilisation by biofilm (mainly diatom) exudates has been suggested for many years, but was first experimentally examined by Holland *et al.* (1974). Burrows of benthic organisms were traditionally thought to increase bed strength (eg. Luckenbach, 1986; Meadows and Tait, 1989); however this is still a topic of debate, since evidence to the contrary has been found. An excellent review of biological effects on sediment stability is given by Paterson and Black (1999), in which the authors propose evolutionary theories to explain why these organisms would require that they decrease the stability of the beds. These will be discussed in a later section.

The Instrument for measuring Shear stress In Situ (ISIS) is used to erode the beds. ISIS was originally designed by H.R. Wallingford to make erosion measurements in situ - its strength is in its ease of transport, since the whole device may be carried and placed on an exposed sediment surface by two people. The creators have developed ISIS further into their present SedErode device. The design and construction of the ISIS is described in Williamson and Ockenden (1996), and shown in Figure 1. Essentially it consists of a circulating mass of water which is pumped up through the bottom of the bell head, through a flow meter and nephelometer (turbidity probe) and then back out between the sides of the ISIS shaft and the column walls. The bell head is mathematically designed to create an even shear force across the entire bed surface. This shear force may be increased either by moving the head closer to the bed, or by increasing the pump speed.

Laboratory settling column tests are well established as a technique in the examination of soil bed development (eg. Sills, 1998). While it is recognised that direct comparison to field situations is difficult, the laboratory technique allows single parameters to be altered, thereby giving a better understanding of the processes. The facility to measure density accurately and non-destructively (see, for example, Sills, 1998), makes it possible to obtain a density profile of the bed prior to erosion. The authors are unaware of any previous publications which describe erosion experiments in settling columns, in which so much information is known about the beds.

Much of this work has been carried out as part of the EU funded MAST III COSINUS project, a main goal of which is to improve estuarine sedimentation models by incorporating the influences of concentrated benthic suspensions. An important aspect of the project is the collection of data to calibrate the models. In particular mass fluxes of sediment from the bed to the water column and from the water column to the bed are important. In the context of the project, the ISIS is used to measure the erodibility properties of sediment beds due to different flocculation, sedimentation, consolidation and biological conditions. The purpose of this paper is to describe the results obtained using novel techniques, such as the ISIS used in a laboratory column, x-ray density profiles and pore water pressure measurements.

2. MATERIALS AND METHODS

Mud beds have been formed by placing a well-mixed slurry in Perspex columns (10 cm ID) with a well mixed initial slurry. This method has been used in studies of the behaviour of very soft soils. The stress history of the soil is similar to that of dredged slurry prior to disposal, where the subsequent consolidation and erosion conditions are often of major interest. These beds have been left to consolidate for a number of weeks, and in some cases primary consolidation was nearly complete by the time the erosion measurements were made. Density measurements were made throughout the consolidation stage and after erosion by an x-ray method detailed in Sills (1998).

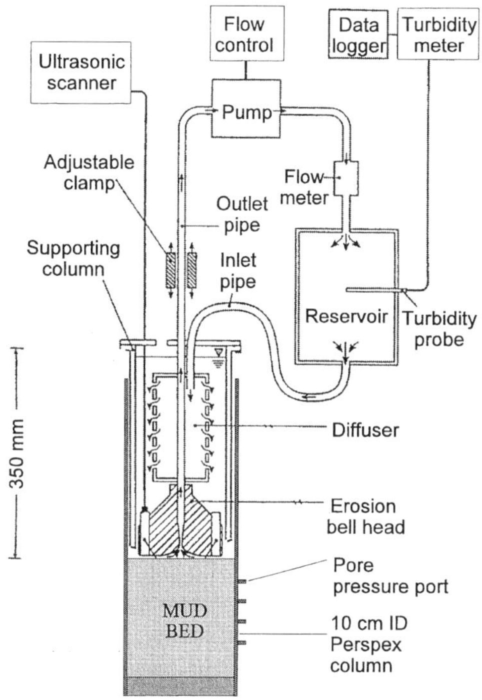

Figure 1 Schematic diagram of ISIS showing the major components of the system. Adapted from Williamson and Ockenden, 1996.

The ISIS has been altered slightly from its field-ready state so that it fits tightly into the columns. Erosion tests are conducted at the end of settling and consolidation experiments within these columns. The ISIS pump is initially started at a very low flow rate, much lower than would be expected to erode any cohesive bed. As the water circulates throughout the system, trapped air bubbles are released in bubble traps and are replaced with water until the entire system is filled with water. After this the ISIS head is lowered to within 6 mm of the bed. According to hot film anemometry calibration tests carried out at the University of Plymouth, this range gives a predictable and even shear stress across the bottom surface of the head (Williamson and Ockenden, 1996). The pump speed, and thus bed shear stress, is increased at two-minute time intervals, until the end of the experiment.

Data from the flow meter and nephelometer (turbidity probe) are collected at 1 second intervals. Flow meter results are converted to flow rates, based on calibrations made during the development of ISIS and are then converted to bed shear stresses based on the hot film anemometry calibration tests. Results from the nephelometer are converted to concentration based on previous calibration tests using Tamar mud (Feates *et al.*, 1999). The results are described in terms of bed shear stress (τ), critical bed shear stress of erosion or erosion

Table 1.
Experimental conditions and results.

Name	Consolidation interval (days)	Initial density (Mg/m^3)	τ_c (N/m^2)	τ_f (N/m^2)	M $\tau_{0.1}$ (mg/m^2)	M $\tau_{0.2}$ (mg/m^2)	M $\tau_{0.5}$ (mg/m^2)
ISIS1	7	1.137	0.177	na	0.045	0.201	0.615
ISIS2	13	1.164	1.3	3.6	0.067	0.071	0.178
ISIS3	19	1.164	0.01	3.0	0.107	0.178	0.517
ISIS4	32	1.164	na	na	na	na	na
ISIS5	5	1.080	0.1	1.225	0.062	0.071	0.196
ISIS6	58	1.162	0.073	na	0.196	0.272	complete
ISIS7	30	1.110	0.075	0.425	0.076	0.187	complete
TAM3	15	1.131	0.65	2.05	0.036	0.058	0.134
TAM5	43	1.113	0.065	0.50	0.107	0.357	complete

threshold, τ_c, and mass of sediment eroded per unit bed area. The erosion threshold is the bed shear stress at which the first signs of erosion occur. This is a partially subjective measurement, based on a rapid increase in mass of eroded sediment, followed by a change in the slope of increasing eroded mass over increasing shear stress.

3. RESULTS AND DISCUSSION

The series of tests reported here investigates the effects of time and the effects of biological activity on erosion properties. The test names and conditions are given in Table 1. In each test, the initial slurry concentration was always above the structural density of the mud. That is, the mud particles were so densely packed that they formed a structural framework and effective stresses existed from the outset (Sills, 1998). Therefore the beds were undergoing consolidation from the start of the experiment until erosion took place.

High resolution video observation gives an indication of how the erosion takes place. In the initial stages of the experiment aggregates are rather gently lifted off the bed, but often these are too heavy to be circulated up the ISIS. Continually increasing the flow rate causes larger aggregates (several mm) to be eroded, until a shear rate is reached where thin layers of the bed are eroded. Increasing the flow rate even further causes the upper bed to liquify. Layers continue to erode from the bed until, at a very high bed shear stress, a massive failure of the top several centimetres occurs. This has been called the failure threshold, τ_f. Figure 2 shows the bed surface immediately before, during and after a massive failure of the bed surface. The white flakes are a combination of biochemically altered mud and algal material consisting mainly of diatoms,

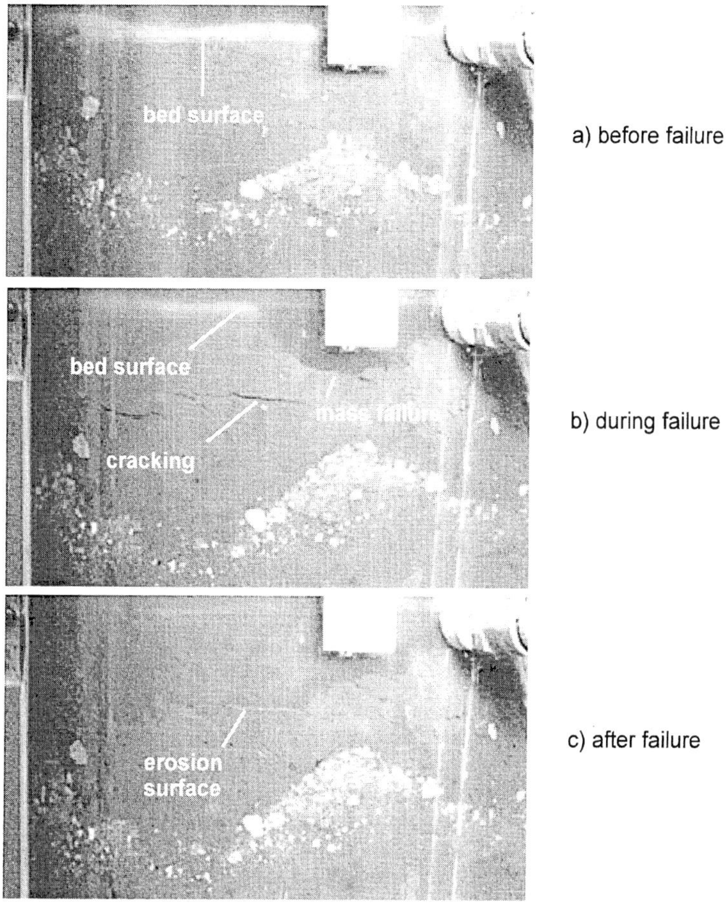

Figure 2 Photographs of bed before during and after massive failure

which are commonly found on top of the beds in column tests. During liquification these migrated down the column wall.

Figure 3 shows data from the first erosion experiment, ISIS1, which was allowed to consolidate for one week before the erosion test was made. The figure shows the experiment duration time on the horizontal axis. Ever-increasing the pump speed at set time intervals is represented in the stepped nature of the shear stress profile.

There is a distinct shear stress at which the mass of eroded sediment circulating throughout the ISIS is increased, representing the first significant piece of bed to be eroded. This can be seen to occur on Figure 3a at a time of approximately 19 minutes. At this time the corresponding shear stress is 0.177 N/m^2, as shown by following the dotted line to the the right hand axis of the

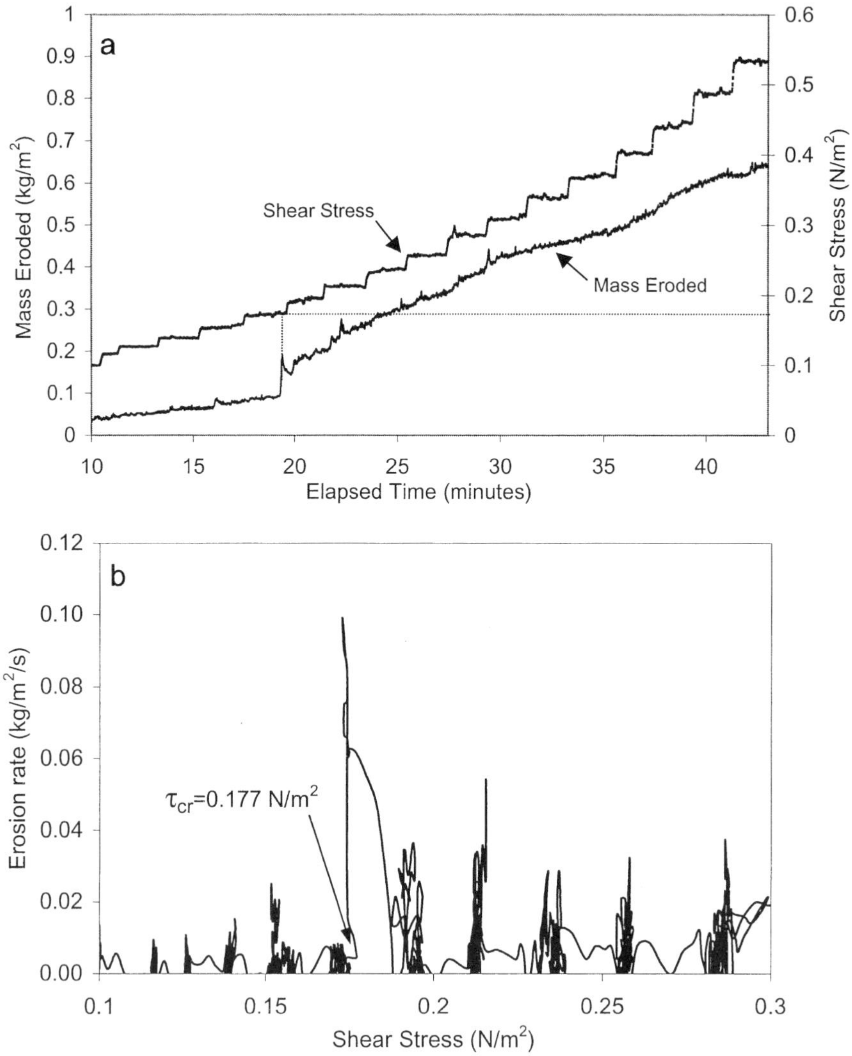

Figure 3 Various methods to represent the critical erosion threshold, τ_c, for experiment ISIS1.

Figure. After this, mass is eroded at a greater rate than before, given increasing shear stress. The critical shear stress, τ_c, is defined as the shear stress at which this occurs.

The critical shear stress may also be represented by other methods of analysis. Figure 3b shows the erosion rate, or mass of sediment eroded per unit bed area per unit time, and Figure 4a shows the accumulated mass of sediment eroded per unit bed area. Both are plotted against the applied shear stress. Figure 4a clearly indicates a significant increase in eroded mass at a shear stress of

0.177 N/m^2, followed by an increased slope in the graph. In Figure 3b all values above zero indicate that sediment mass is actively being eroded. The data plots rather chaotically since both erosion rate and shear stress may increase or decrease during the sampling interval of one second. However it is again clear on Figure 3b that the first significant and sustained erosion for ISIS1 occurs at a shear stress of 0.177 N/m^2.

Although a calculation of erosion rate is possible, the testing procedure used in these experiments does not allow an analysis to be made in terms of erosion rate. A rather short interval of two minutes was used for increasing the shear stress, which differs from more commonly reported laboratory techniques of fewer shear stress levels over much longer time periods. The latter (common) technique allows the bed state to reach a steady state before the shear stress is increased. There are several reasons why the former technique was chosen in this case. First, there may be time dependant changes taking place on the scale of hours which may alter the properties of the bed during the experiment. Most important of these, which are widely known, are daily temperature fluctuations, the further consolidation and cementation in the young beds, and the production and release of gas in the older beds- a single pock mark generated from escaping gas could influence the results drastically. Shorter time intervals were also suggested by the developers of ISIS as it is the interval used for both the ISIS and the SedErode in their field sampling projects. To allow appropriate statements to be made about the 'erosion rate', the bed must have sufficient time to respond fully at each stress interval. Rather the increasing erosion in the present experiments is driven by the history of the stress application in the experiments. This stress history is standardised as far as possible for the tests, with an interval of two minutes, allowing the data from the various experiments to be compared to one another. There are a few exceptions to the two minute rule where x-ray measurements or pore pressure measurements were made. These measurements were infrequent, nevertheless caution is necessary when comparing results which make use of time, or rates. The large majority of mass is eroded at the instant the shear stresses are increased. That is the erosion is driven by the increase in shear stress, and the data may be appropriately presented in terms of shear stress and mass eroded. This method is consistent with the methods used commonly in the field for the SedErode (Feates *et al.*, 1999), the ISIS (Williamson and Ockenden, 1996), and is also similar to methods used with other in situ devices such as the University of St. Andrew's Microcosm (up to ten minute intervals) and their cohesive strength meter (5 second intervals). It may be argued that beds in natural tidal environments rarely actually reach a state of equilibrium, so increasing the shear stress before the bed has time to come to such an equilibrium is acceptable, so long as the measuring technique is consistent throughout compared experiments (Trevor Black, Gatty Marine Laboratory, University of St. Andrews, personal communication).

The tests following ISIS1 went further to investigate the effect of consolidation time. Five columns were prepared using Tamar mud. Three of these, ISIS2, ISIS3, and ISIS4, were sedimented from an initial slurry density of 1.164 Mg/m^3 and height 190 mm, and allowed to mature for 13, 19, and 32 days. The remaining experiments ISIS5, ISIS7 and TAM5 all had slightly lower initial slurry densities and consolidated for 5, 15, and 43 days. Density profiles showed that all columns settled at similar rates with consistent density changes throughout the beds during consolidation.

Plotting mass eroded against applied shear stress gives an indication of the erosion reactions

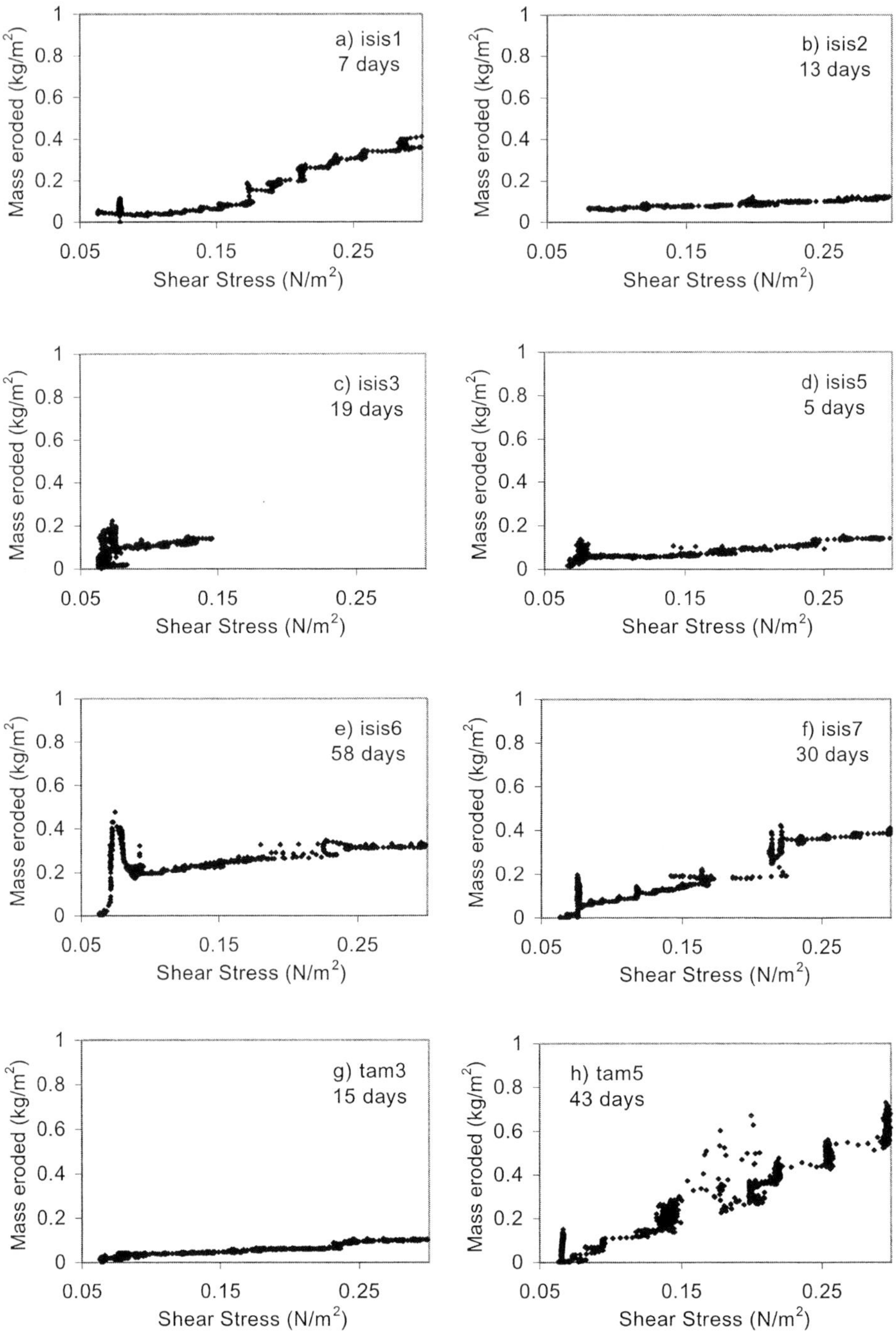

Figure 4(a-h) Cumulative mass eroded given increasing shear stress for each of the beds

of each of the beds given increasing shear forces (Figure 4). It must be emphasized that this is not an attempt to calculate erosion rate. It does, however, allow a comparison of the different behaviours of the beds of different initial conditions, which are subjected to very similar stress/time histories during the erosion tests. It can be seen in Figure 4 that these reactions are drastically different for the different beds. In addition to having different slopes the curves have different shapes, and in most experiments the slopes at lower shear stresses are different from those at higher shear stresses.

In Figures 3b and 4a-h the horizontal axes are adjusted to show the shear stresses up to 0.30 N/m^2 only, since at shear rates higher than this the beds in a few experiments start to massively erode. In some experiments there is a peak at very low shear stresses, showing that these beds are very fragile at the outset of the experiment. This is particularly noticeable in ISIS6, the most mature of the beds. In fact, the more mature experiments all have larger initial peaks in their mass erosion than the younger experiments.

Figure 4 shows that the mass eroded at any given shear stress varies between experiments. One way of analysing these data further is to compare the mass eroded at a set of arbitrarily chosen shear stresses: 0.1, 0.2 and 0.5 N/m^2. The mass of the sediment circulating in the ISIS at these specified shear stresses are given in Table 1 ($M\tau_{0.1}$, $M\tau_{0.2}$, $M\tau_{0.5}$). These have been plotted in Figure 5, against the length of consolidation time for each experiment. At the lower shear stress levels of 0.10 and 0.20 N/m^2 the mass eroded increases with consolidation time. That is, the older beds are less resistant to erosion than the young beds. This relationship continues into the higher shear stresses. The more mature beds undergo massive failure much sooner than the immature beds, and by a shear stress of 0.50 N/m^2 the experiments of the more mature beds are complete (thus the missing data values in table 1). These results are, at first sight, surprising since the normal expectation is that increased consolidation time leads to stronger beds due to bed restructuring and cementation (Krone, 1999). This point is discussed later.

X-ray measurements have been made before, during and after most of the erosion tests. A typical set of density profiles (TAM3) shows how the initial slurry settles and becomes more dense (Figure 6). Figure 7a shows the density before and after the erosion of ISIS2. The erosion has led to lower densities throughout the upper bed. This may be due, in some small part, to the decreased bed height (apparent in the figure) which would decrease the total stress loading the upper bed. However, this unloading would cause only a small swelling, given the irreversible behaviour of soil consolidation. It is therefore most likely that this density reduction is a consequence of liquifaction or dilation, which has affected all of the bed except the bottom 4 cm. This reduced density due to entrainment and maintenance of water during and after erosion is reported in the literature, and has important consequences in erosional processes. For example Mehta (1991) refers to the type of erosion where the bed is first fluidised, and then destabilises causing the fluid mud to be entrained and mixed.

Results to determine how pore water pressures respond to the erosion are also available for one experiment. When the water level in the column is kept constant the pore pressures rise overall during the testing. This might indicate that the force of the water pressure coming down the column walls is having an effect deeper in the bed, and is the likely cause of the liquefaction of the upper bed. The use of x-ray and pore pressure measurements deserves further attention in

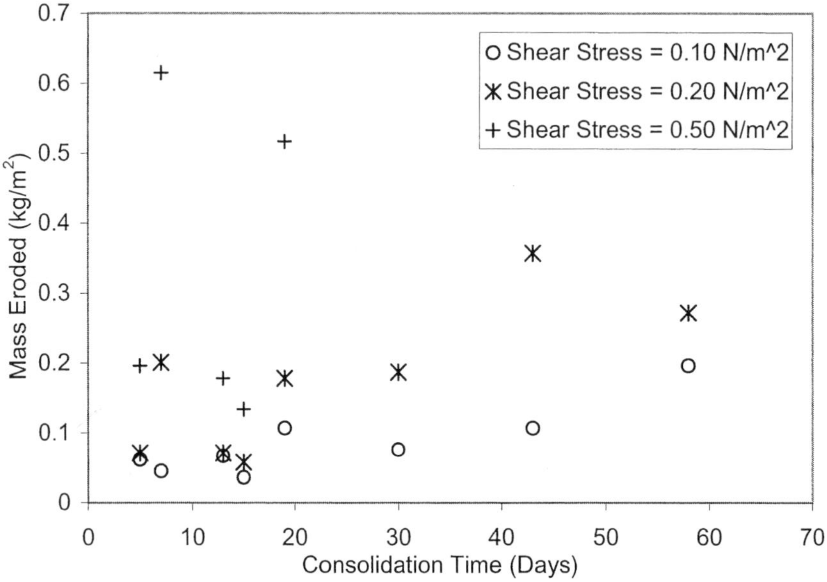

Figure 5 The cumulative mass eroded from the beds at specified shear levels during the experiment show an overall increase with increasing consolidation time for the lower shear stresses

the examination of soil erosion, since together they may be used to provide very accurate information about density, effective stresses and void volume in the beds.

An overall increase in eroded mass, given increasing bed shear stress, has been shown to be distinct for each of the experiments (Figure 4). The density profiles for the experiments may help reveal some aspects of the bed surface which may account for these differences. During the consolidation process, it has frequently been observed that there is not a sharp density change between water and bed. Rather the top surface on the density profiles becomes more well-rounded as a biological layer forms. Figure 6 illustrates this; the density in the earlier profiles changes abruptly between the bed surface and immediately below the bed surface, whereas at a time of 170.5 hours this change is much less abrupt. This has been observed repeatedly throughout the years in column experiments at Oxford. Microscopic analysis of this layer has revealed (as expected) an abundance of dormant diatoms and other organisms (John Hamer, University of Wales, Bangor, personal communication.). The weakening of the mature beds may be attributed to the lower resistance to erosion of this biochemical ‘fluff’ layer. Similar features have been described in nature. For example, Amos *et al.* (1997) describe an easily eroded fluff layer on beds of the Fraser River Delta in Southwestern British Columbia. Erosion of this fluff layer is also the likely cause of the large peak in mass eroded at the outset of the more mature experiments (Figure 4). The small peak seen in the younger beds at the onset of erosion may be due to a surface layer of finer particles that settle out of the overlying water after the main body of the soil.

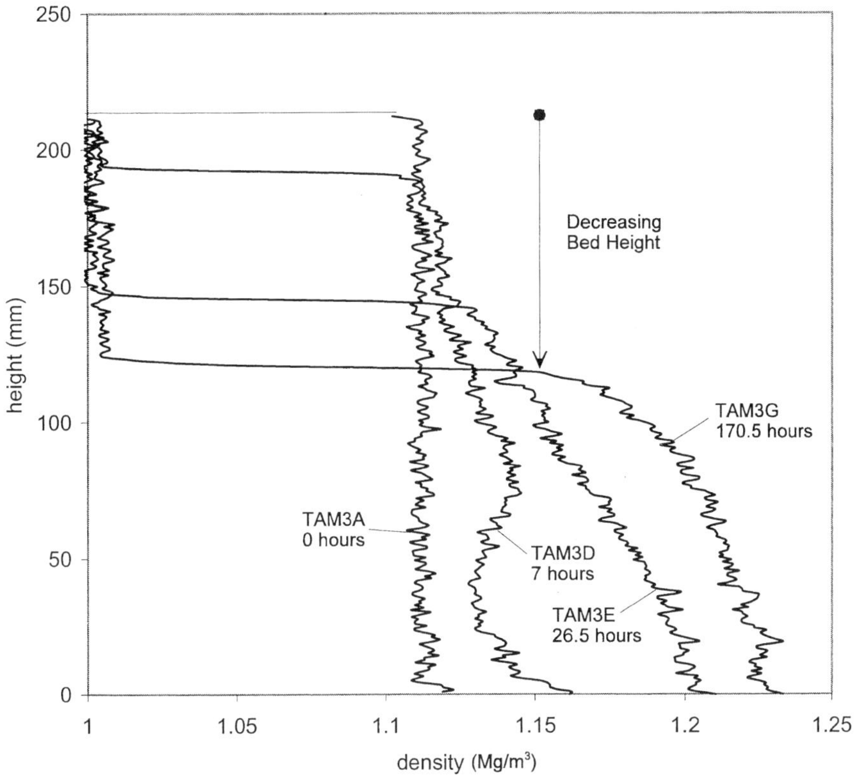

Figure 6 Density profiles for experiment tam3, showing the bed height decreasing and the density of the bed increasing during consolidation.

X-ray profiles have also revealed another interesting feature about the bed surfaces. The profiles shown in Figure 7b are taken from experiments of different length using the same mud, at a similar initial slurry, though with slightly different initial heights. The two more mature profiles, ISIS3D and ISIS4E, show a density minimum at around 7 mm below the surface, whereas the third profile for TAM1A is more typical and does not show this density minimum. This minimum has been seen on occasions in various experiments at Oxford, and for consolidation or soil mechanics research it is rather insignificant. However it becomes considerably more important when the research is focusing on the properties of the bed surface. Close visual observation of the surface reveals that there is a gaseous sublayer corresponding to the low density layer in the more mature experiments. Figure 8 shows a photograph of this gaseous layer forming just below the surface. It should be emphasized that this photograph is of a bed in which there was visibly more gas, and a greater reduction of density than occurred in the erosion experiments. It therefore represents and extreme example, but nevertheless a relevant one. It is likely that this gaseous sub-layer, which shows up so distinctly on the density

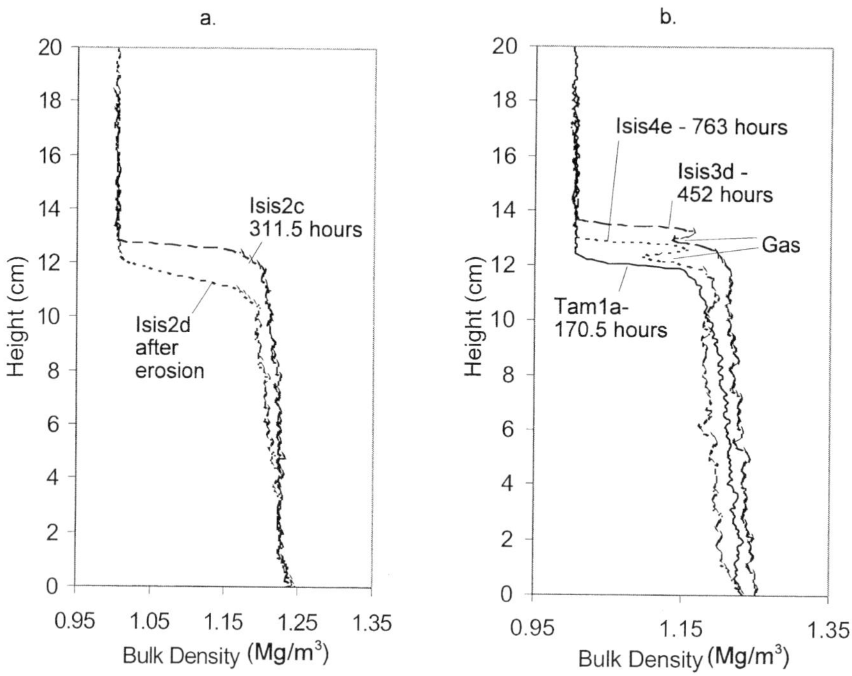

Figure 7 Density profiles showing a) reduced bed density in upper bed due to erosion, and b) a gas layer of reduced density below the surface in more mature beds.

profiles of the more mature beds, is also weakening the bed at this level, and contributes to the larger (bulk) erosion processes.

Figure 4 showed the progression of erosion to complete failure. In Figure 2b there is a rather large piece of bed missing, marked 'mass failure'. In the initial experiments it was believed that these pockets of higher erosion were caused by turbulent vortices at higher flow rates. However on much closer observation using a video technique it has since been found that these pockets occur where a feeding benthic worm is present. The worms' feeding behaviour may weaken the surrounding bed. This worm feeds at the surface, with its head anchored in the sediment, and its body an oscillating cantilever in the water column. As it feeds it ejects pellets which settle to the bed surface. Thus after several weeks the entire bed surface surrounding the worm is composed of these pellets. The constant oscillation, combined with the sediment surrounding a worm being deficient in living organic matter and being composed of faecal pellets rather than fine-grained cohesive material, may hinder the processes of consolidation and cementation. There is supporting evidence in the literature for this conclusion. Paterson and Black (1999) report on a study by Widdows *et al.* (1998) in which benthic invertebrates destabilise the sediment leading to increased erosion. They postulate that this erosion leads to the periodic renewal of food surrounding the organisms.

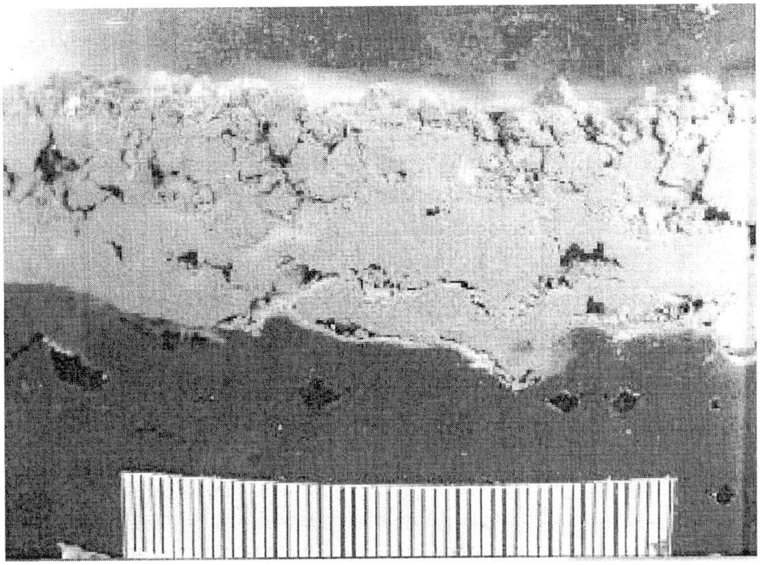

Figure 8 Low density layer at top of bed is characterised by a different colouration, and often gas bubbles. The scale is in millimetres. Photo courtesy of Ramon Gonzalez, Department of Engineering Science, Oxford.

4. CONCLUSIONS

The tests illustrate that overall an increased maturation time reduced the resistance of the beds to erosion. This would not be expected in a pure geotechnical sense, where increased consolidation and cementation would expect to increase the strength of the bed. However, it is consistent with results reported elsewhere and may be attributed to the biological activity of the beds. The deposition of an algal 'fluff' layer may reduce the strength, as might the replacement of fine-grained cohesive sediment with faecal pellets by a benthic worm. The latter was more noticeable at higher shear stresses, where areas around worm burrows were more deeply scoured. The production of gas slightly below the surface may also have an effect on the erosion properties. Each of these will be studied in more detail in future experiments.

As pointed out many times in the literature, the study of the interaction between benthos and sediment stability is in its infancy, and requires closer investigation using improved techniques. The equipment at Oxford University for measuring bed density and interstitial pore water pressures has been applied for the first time in the examination of the beds before, during and after erosion. These techniques have revealed some interesting features of the beds, and warrant further development, either with the ISIS or with a more established flume type arrangement.

ACKNOWLEDGMENTS

Funding for this research has been provided by The European MAST III COSINUS project and by the Engineering and Physical Sciences Research Council, UK. The collaboration between Oxford University and HR Wallingford was established and maintained by Dr. Bill Roberts and Mr. Nigel Feates of HR Wallingford. Dr. Carl Amos, University of Southampton and Dr. David Paterson and Trevor Tolhurst, St. Andrews University, provided helpful comments.

REFERENCES

Amos, C. L., Feeney, T., Sutherland, T.F. and Luternauer, J.L., 1997. The stability of fine-grained sediments from the Fraser River Delta. Estuarine, Coastal and Shelf Science 45: 507-524.

Been, K. and Sills, G.C., 1981. Self-weight consolidation of soft soils: an experimental and theoretical study. Geotechnique 31: 519-535.

Feates, N.G., Hall, J.R., Mitchener, H.J., and Roberts, R., 1999. Cosinus field experiment, Tamar Estuary. HR Wallingford Report TR82, March 1999.

Holland, A.f., Zingmark, R.G. and Dean, J.M, 1974. Quantitative evidence concerning the stabilisation of sediments by marine benthic diatoms. Marine Biology 27: 191-196.

Krone, R.B., 1999. Effects of bed structure on erosion of cohesive sediments. Journal of Hydraulic Engineering, ASCE, 125(12): 1297-1301.

Luckenbach, M.W., 1986. Sediment stability around animal tubes. The roles of hydrodynamic processes and biotic activity. Limnol. Oceanogr. 31: 779-787.

Meadows, P and Tait, J., 1989. Modification of sediment permeability and shear strength by two burrowing invertebrates. Marine Biology 101: 75-82.

Mehta, A.J., 1991. Review notes on cohesive sediment erosion. Coastal Sediments 91 proceedings. Published by American Society of Civil Engineers. Seattle, Washington, June 25-27: 40-53.

Paterson, D. M. and Black, K. S. (1999). Water flow, sediment dynamics and benthic biology. Advances in Ecological Research. D. Raffaelii and D. Nedwell. Oxford, Oxford University Press. 29: 155-193.

Roberts, J., Jepson, J., Gottard, D. and Lick, W., 1998. Efects of particle size and bulk density on erosion of quartz particles. Journal of Hydraulic Engineering, ASCE, 124(12): 1261-1267.

Sills, G.C., 1997. Consolidation of cohesive sediments in settling columns. Cohesive Sediments. Fourth Nearshore and Estuarine Cohesive Sediment Transport Conference INTERCOH 94. Burth, R. Parker and J. Watts (eds.). Wallingford, UK.

Sills, G. C. (1998). Development of structure in sedimenting soils. Phil. Trans. R. Soc. Lond. A 356: 2515-2534.

Widdows, J., Brinsley, M., and Elliot, M., 1998. Use of an in situ flume to quantify particle flux (biodeposition rates and sediment erosion) for an intertidal mudflat in relation to current velocity and benthic macrofauna. In Sedimentary Processes in the Intertidal Zone. Black, K.S., Paterson, D.M. and Cramp, A. (Eds). Geological Society, London. Special publication no. 139: 85-97.

Williamson, H.J. and Ockenden, M.C., 1996. ISIS: An instrument for measuring erosion shear stress in situ. Estuarine, Coastal and Shelf Science 42: 1-18.

Zreik, D.A., Krishnappen, B.G., Germaine, J.T., Madsen, O.S. and Ladd, C.C., 1998. Erosional and mechanical strengths of deposited cohesive sediments. Journal of Hydraulic Engineering, ASCE, 124(11): 1076-1085.

Fine Sediment Dynamics in the Marine Environment
J.C. Winterwerp and C. Kranenburg (Editors)

Strength modelling of consolidating mud beds

L.M. Merckelbach[a*], C. Kranenburg[a], J.C. Winterwerp[ba]

[a]Delft University of Technology, Faculty of Civil Engineering and Geosciences, P.O. Box 5048, 2600 GA Delft, The Netherlands

[b]WL|Delft Hydraulics,
P.O. Box 177, 2600MH, Delft, The Netherlands

Key words: consolidation, mud, shear vane, shear strength, fractal, scale invariance

The research presented focusses on the self-weight consolidation and strength evolution of soft underwater mud beds. A series of experiments was carried out in 1.5 m tall consolidation columns for a maximum duration of 95 days. Segmented settling columns were designed and built in order to provide well-defined samples of the bed, the strength of which was measured using a high-precision rheometer equipped with a miniature vane.

The process of consolidation was modelled as a one-dimensional process using the Gibson equation [9] written in a Eulerian reference frame and with the particle volume fraction as the dependent variable. The integration of the consolidation equation requires boundary conditions, initial conditions and constitutive equations for effective stress and permeability. The strength evolution is taken into account by a failure criterion which relates strength to effective stress.

Constitutive equations for effective stress and permeability and a failure criterion were derived based on the concept of a scale invariant bed structure. The bed is assumed to be formed by aggregates. The properties of the aggregates are assumed to be determined by the clay fraction only. The averaged size of the aggregates determines the effective stress, permeability and shear strength. During consolidation the averaged size of the aggregates reduces, so that effective stress and shear strength increase and permeability decreases. The results of numerical modelling of the experiments agree well with measurements.

1. INTRODUCTION

Soft mud beds are a common appearance in many coastal and estuarine areas. Such mud beds are usually formed from deposition and initially have a weak, fluid like structure. Freshly deposited layers are subjected to self-weight consolidation, during which a significant strength may develop. The modelling of the behaviour of soft mud layers is required in hydraulic engineering applications, such as the estimation of erosion and sedimentation in (estuarine) navigation channels and harbour basins, the definition of navigable depth, gravity currents of mud on a sloping bed or mud flows forced by pressure gradients.

*Present address: WL|Delft Hydraulics, p.o. box 177, 2600MH, Delft, The Netherlands

The process of consolidation of soft mud is usually modelled by means of the large strain Gibson equation [9]. The properties of the soil are accounted for by constitutive equations for effective stress, defined as the total stress minus the pore water pressure, and permeability. Den Haan proposes a generalised power law for soil compression, including a number of formulations that are frequently used in practice, [7]. The disadvantage of this kind of formulations is that the empirical parameters do not reflect the underlying physics, which makes predicting the values of empirical parameters difficult. Formulations that are similar in a mathematical sense are also used for the permeability constitutive equation [6,21], suffering from the same shortcomings. An overview of more sophisticated models of Darcy flow is given in [10]. In some of these models viscosity and aspects as shape and geometry are accounted for.

Although the shear strength of soils is extensively studied in soil mechanics, relatively few studies report on the strength modelling in very soft soils, i.e. effective stress levels below 5 kPa or so. One of the reasons is that soft muds show too less consistence to be tested on (shear) strength in a classical triaxial test. Instead, the shear vane test may be used as an alternative. Shear vane tests on soft soils in conjunction with effective stress measurements have been reported in [8,2].

The aim of present paper is to develop a model for the shear strength of soft soils. The model is based on a failure criterion that relates shear strength to parameters related to the consolidation process, such as effective stress. The present paper presents a constitutive model based on the concept of a scale invariant structure of the bed. The constitutive model encompasses constitutive equations for effective stress and permeability to be used in a consolidation model and a failure criterion to be used in a strength model. The constitutive model is applied on numerical consolidation and strength models and the results are compared with laboratory experiments.

2. CONSTITUTIVE MODEL

2.1. Bed structure model

Generally, mud is a mixture of particles of various sizes, ranging from clay particles as small as 2μm or even smaller, to sand particles as large as 63 μm or larger. Due to their cohesive character, clay particles tend to form aggregates. Krone noted that clay particles (primary particles) form flocs, which can join to form floc aggregates, [12]. The floc aggregates can join to form larger aggregates, and so on. Experimental work done by Krone suggests a floc structure that is more or less independent of the scale considered. Also work by other researchers [5,13,11,22,3,20] indicates that the concept of scale invariant structure is a useful approximation to describe the aggregates' structure.

Assuming that the network structure that forms the bed is built from large, scale invariant aggregates, the concept of fractal geometry can be applied to describe the structures of these aggregates. This means that the aggregates are regarded as self-similar or fractal structures and have geometrical properties that are truly scale invariant [14]. Then, the number N of clay particles in a network structure or aggregate is proportional to the spherical diameter R_a of the (self-similar) clay structure

$$N \propto \left(\frac{R_\mathrm{a}}{R_\mathrm{p}}\right)^D, \tag{1}$$

where R_p the spherical diameter of the clay particles and D the fractal dimension. The fractal dimension, the value of which is in the range of [1,3], is a measure of how efficient the space in the structure is filled with (clay) particles. The higher the fractal dimension, the more efficient the filling is.

The volume fraction of the clay particles ϕ_{clay}, defined as the ratio of the volume of clay particles to the total volume, is given by

$$\phi_{\mathrm{clay}} \propto \frac{N R_\mathrm{p}^3}{R_\mathrm{a}^3}. \tag{2}$$

Combining (1) and (2), and assuming that R_p is constant, it follows that

$$\phi_{\mathrm{clay}} \propto R_\mathrm{a}^{D-3}. \tag{3}$$

2.2. Constitutive equations for effective stress and permeability

2.2.1. Effective stress

Simply put, self-weight consolidation can be seen as a process during which excess pore water pressure is transferred to effective stress. The maximum stress that can occur, is assumed to be determined by the number of inter-particle bonds per unit area times the bond strength. If the effective stress exceeds the maximum stress that can be realised, the network will restructure so as to increase the number of inter-particle bonds per unit area. This is achieved by reducing the size of the scale invariant bed structure, i.e. the bed structure continually breaks up in smaller, more compact structures with more inter-particle bonds per unit area. The break-up process is accompanied by the release of pore water, which is subsequently expelled. As a result of the released pore water being expelled, the bed compacts. Consequently, the effective stress in a consolidating soft mud bed is equal to the maximum compressive stress that can be realised by the network in the current state.

Mitchell shows experimental evidence indicating that the number of bonds per unit area is proportional to the effective stress, [18]. Thus, the effective stress σ' may be written as

$$\sigma' = aS, \tag{4}$$

where S is the number of bonds per unit area and a the strength of a single bond (in N). Strictly, (4) is not valid if creep plays a significant role, since creep may be considered as an increase in the number of bonds at a constant level of effective stress. A simple model to include the effect of creep (or overconsolidation) is given in [16]. The effect of creep will not be considered herein.

According to the definition, the concept of scale invariance of the clay structure implies that the geometrical properties of the clay structure are equal at all scales[2]. Therefore, if the strength of a structure is determined by the number of critical bonds, and the number of critical bonds depends only on the geometrical structure, then the number of critical bonds is independent of the length scale of the structure. The number of bonds per unit area per structure then reads

$$S = \frac{s}{R_\mathrm{a}^2}, \tag{5}$$

[2]In real cases this is only true for scales between a few times the size of primary particles (R_p and the size of the structure (R_a).

where s is the scale invariant number of critical bonds per structure.

Combining (4) and (5) gives the relation between effective stress and the length scale of the structure

$$\sigma' = a\left(\frac{s}{R_{\mathrm{a}}^2}\right). \tag{6}$$

Using (3), equation (6) can be written as

$$\sigma' = K_\sigma \phi_{\mathrm{clay}}^{\frac{2}{3-D}}, \tag{7}$$

where K_σ is an empirical proportionality coefficient. This result relates the effective stress to the volume fraction of clay particles according to a power law. It is remarked that in a mathematical sense (7) is equivalent to Butterfield's natural compression law [4].

2.2.2. Permeability

Inertia effects can commonly be neglected during self-weight consolidation processes. Therefore, the fluid flow–particle interaction can be modelled reasonably well by the generalised Darcy law

$$(1-\phi)(v_{\mathrm{f}} - v_{\mathrm{s}}) = -k\frac{1}{\rho_{\mathrm{w}} g}\frac{\partial p_{\mathrm{e}}}{\partial z}, \tag{8}$$

where v_{f} is the velocity of the fluid, v_{s} the velocity of the solids, k is the permeability, ρ_{w} the density of the fluid, z the vertical coordinate and p_{e} is the excess pore water pressure defined as the difference of the pore water pressure and the hydrostatic pore water pressure.

For reasons of continuity, (8) can be written as

$$v_{\mathrm{s}} = k\frac{1}{\rho_{\mathrm{w}} g}\frac{\partial p_{\mathrm{e}}}{\partial z}, \tag{9}$$

in which it is assumed that the consolidating layer is on top of an impervious base.

For pore water to be expelled from the bed during consolidation, it is required that at least part of the pores are inter-connected. The system of connected pores is modelled by a network of tiny tubes, in which the flow is assumed laminar (Poiseuille flow) in accordance with Darcy's law. The classical formulation of Poiseuille flow [19, p.157] relates the fluid velocity to the pressure gradient as

$$\frac{\partial p}{\partial s} = -\frac{\eta}{l^2}v, \tag{10}$$

where p is the pressure, η the dynamic viscosity, v the flow velocity, l an equivalent diameter of the tube and s the coordinate along the centre-line of the tube. To apply Poiseuille flow to pore water flow, (10) needs to be modified. Similar to the generalisation of the Darcy law, the water flow velocity relative to the particle velocities needs to be taken into account, yielding

$$v_{\mathrm{s}} \propto \frac{\partial p_{\mathrm{e}}}{\partial z} l^2. \tag{11}$$

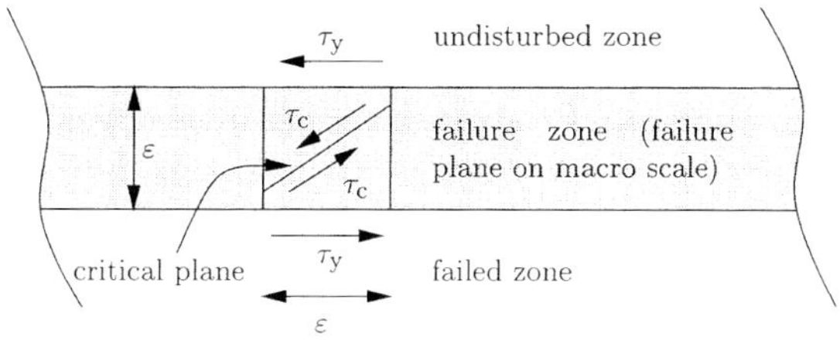

Figure 1. Definition sketch of τ_y and τ_c.

Combining (9) and (11) yields

$$k \propto l^2. \tag{12}$$

The assumption of scale invariance implies that the size of the largest connecting pores scales as R_a [3,11]. Letting l be the equivalent diameter of these pores, l is proportional to R_a, so that with (3), equation (12) becomes

$$k = K_k \phi_{clay}^{-\frac{2}{3-D}}, \tag{13}$$

where K_k is a empirical coefficient which takes account of effects due to viscosity, pore water density and shape.

It is noted that it is sufficient to consider the flow through the largest pores only, since the flow rate Q of Poiseuille flow scales as $Q \sim v_s l^2 \sim l^4$, see (11).

2.3. Failure criterion

The undrained strength of a material may be regarded as the resistance against failure. If a mud layer is failing locally, for example during a vane test or plug flow on a sloping bed (both failure mechanisms are undrained), then two regions can be distinguished: a region where the material is still undisturbed and a region where the material failed. The transition is marked by the plane of failure. However, the orientation of the critical plane, which is the plane in which the maximum shear stress occurs, is not necessarily parallel to the plane of failure, but depends on the normal stresses (effective stresses). A conceptual sketch is shown in Figure 1, in which the critical plane (on micro scale) is assumed to be within the failure zone with an infinitesimal thickness, so that on macro scale, the failure zone is considered a failure plane. The transition from no-flow to flow can be characterised by a yield stress τ_y, acting in the plane of failure. However, the shear stress in the critical plane τ_c represents the strength of the bed.

The shear rate of the material in the undisturbed zone is assumed to be equal to zero. The material in the failed zone, on the other hand, may flow, as is the case in for example the shear vane test and mud flow on a sloping bed. The shear rate at the transition, i.e. in the failure plane, is also assumed equal to zero. As a result, both τ_y and τ_c are shear rate independent.

In case of shear deformation, the aggregates move relative to adjacent aggregates, expressed by a shear strain between aggregates. At the same time, the aggregates themselves are deformed as well, expressed by a shear strain within the aggregates. The motion between aggregates results in stresses generated by bonds between particles of adjacent aggregates. The deformation of aggregates results in stresses generated by bonds between particles within a single aggregate. It seems fair to assume that the shear strain due to motion between aggregates is equal to the shear strain due to the deformation of the aggregates. Then, it can be shown that the (total) critical shear stress is a linear summation of the critical shear stresses as a result of inter- and intra-aggregate bonds, respectively

$$\tau_c = \theta \tau_{aa} + (1 - \theta) \tau_{pp}, \tag{14}$$

with $0 < \theta < 1$.

2.3.1. Shear stress resulting from bonds between particles of adjacent aggregates

The shear stress generated by inter-aggregate bonds, τ_{aa}, is assumed proportional to the number of inter-aggregate bonds per unit area. The number of inter-aggregate bonds per aggregate is, in accordance with the concept of scale invariance, proportional to R_a^{D-1} so that

$$\tau_{aa} \propto R_a^{D-3}. \tag{15}$$

From (3) it follows that

$$\tau_{aa} \propto \phi_{clay}. \tag{16}$$

2.3.2. Shear stress resulting from inter-particle bonds within aggregates

The process of shearing of the individual aggregates is similar to the compression of the aggregates: an aggregate will break up under shear if the critical shear stress is reached. The shear stress generated by inter-particle bonds within the aggregate, τ_{pp}, is assumed to be proportional to the number of inter-particle bonds per unit area. According to the concept of a scale invariant structure, the number of inter-particle bonds within an aggregate is constant. Then, it follows that τ_{pp} varies linearly with the effective stress:

$$\tau_{pp} \propto \sigma'. \tag{17}$$

2.3.3. Shear stress at failure

According to (14) the critical stresses generated by inter-aggregate bonds (τ_{aa}) and critical stresses generated by inter-particle bonds within the aggregates (τ_{pp}) give the failure criterion

$$\tau_c = c_1 \phi_{clay} + c_2 \sigma', \tag{18}$$

where c_1 and c_2 are empirical coefficients in which the coefficients θ and $(1-\theta)$ are already included.

The traditional failure criterion for soils is the Mohr-Coulomb failure criterion, given by

$$\tau_c = c' + \sigma' \tan \phi', \tag{19}$$

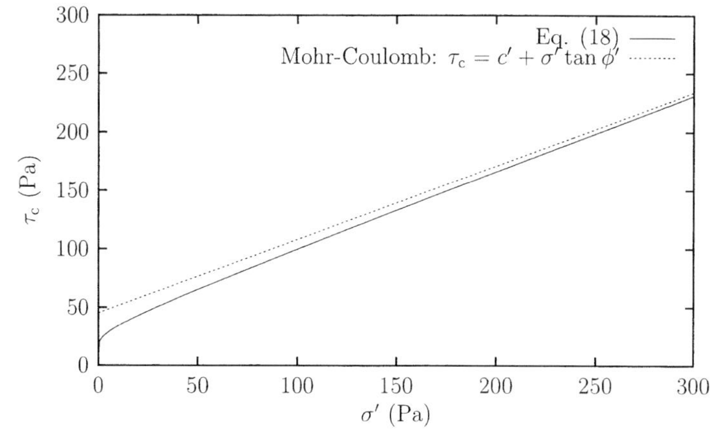

Figure 2. Schematic plot of the failure criterion. The values of the parameters used in (18) are $c_1 = 300$ Pa and $c_2 = \tan \phi' = 0.63$, and those used in (7) are $K_\sigma = 2 \times 10^9$ Pa and $D = 2.75$. The value of the Mohr-Coulomb cohesion is $c' = 45$ Pa.

where c' represents the shear resistance at zero effective stress (true cohesion) and ϕ' is the friction angle. Equating (18) and (19) gives

$$c' = c_1 \phi_{\text{clay}} \tag{20}$$

and

$$\tan \phi' = c_2. \tag{21}$$

Figure 2 shows the relationship between σ' and τ_c. The behaviour of this relationship can be explained as follows. From (7) it follows that the zero effective stress is found for a clay volume fraction equal to zero. Consequently, it follows from (20) that, if the level of effective stress approaches zero, then the true cohesion approaches zero as well. If, on the other hand, effective stress has developed a substantial level, then changes in effective stress are accompanied by relatively small changes in the clay volume fraction, which follows from (7). Consequently, for substantial levels of effective stress, the failure criterion given by (18) resembles that of the traditional Mohr-Coulomb failure criterion. However, for relatively low levels of effective stress; the failure criterion is non-linear, yielding zero strength at zero effective stress.

3. VERIFICATION WITH LABORATORY EXPERIMENTS

3.1. Experimental set-up

The constitutive equations and the failure criterion were verified with measurements obtained from laboratory experiments. The measurements are labelled CTxx, where xx refers to the day on which the measurements were taken (xx=0 refers to the start of the experiment). The experimental techniques used are described in detail in Merckelbach [16]. Only a brief description is given herein.

A natural mud, obtained from the Caland-Beer Channel, Port of Rotterdam, The Netherlands, with a particle size distribution of 40/45/15 % (clay/silt/sand) was used. The mud was diluted until a bulk density of 1080 kg/m^3 and allowed to consolidate from a suspension in settling columns for 95 days. The initial height of the suspension was 1.54 m.

To obtain accurate strength measurements, the settling columns were designed, such that a lower section consisted of removable segments. After a preselected time of consolidation, a settling column was dismantled by first removing the upper part of the column and then segment by segment, starting from the top, see Figure 3. Subsequently, shear strength measurements were carried out on the well-defined samples thus obtained using a high-precision rheometer (Physica, UDS200) equipped with a miniature vane (height : diameter=2.0 cm : 1.0 cm). Since this procedure is destructive for the bed, four segmented columns were identically set up to monitor the strength development with time (after 9, 24, 58 and 95 days).

Density profiles were measured with a conductivity probe after the shear vane tests. Furthermore, a conventional column was set up for a non-destructive γ-ray densimeter so that density profiles could be measured during the course of the experiment. Pore water pressures were also measured in all columns during the course of the experiments: pressure ports along the column (provided with a filter only letting through water) were connected via tubes to a device that houses a single pressure transducer. This device was designed to connect the transducer to one pressure port at a time.

3.2. Determination of parameters

3.2.1. Effective stress

The effective stresses were calculated from the measured density and pore water pressure profiles. Total stresses were calculated by integrating the density profile. Subtracting the pore water pressures from the total stresses gives the effective stresses.

A selection of available effective stress data is presented in Figure 4. For stress levels above 20 Pa, an approximate linear relationship is observed on log-log scales, which implies a power law relationship. The data are fitted by a power law (solid line). Error bands are also shown (dotted lines). The two error bands are given by the translations of the power law fit over $(+\tilde{\sigma}', -\tilde{\phi})$ and $(-\tilde{\sigma}', +\tilde{\phi})$, respectively, in which $\tilde{\sigma}' = 30$ Pa and $\tilde{\phi} = 2.0 \times 10^{-3}$. Almost all data fit within the error bands.

The empirical parameters in (7) are estimated from the linear approximation of the data in Figure 4. The fractal dimension D, obtained from the slope of this approximation, is determined at 2.75; the proportionality coefficient K_σ is determined at 2.1×10^{12} Pa.

3.2.2. Permeability

The local permeability can be estimated by assuming the validity of Darcy's law. To that end, measured pore water pressures are required and an estimate for the (local) settling rate. The latter can be obtained from two consecutive density profiles. This procedure is described in [1]. The accuracy of this method depends largely on the accuracy of the density measurements. Unfortunately, the accuracy of the density measurements

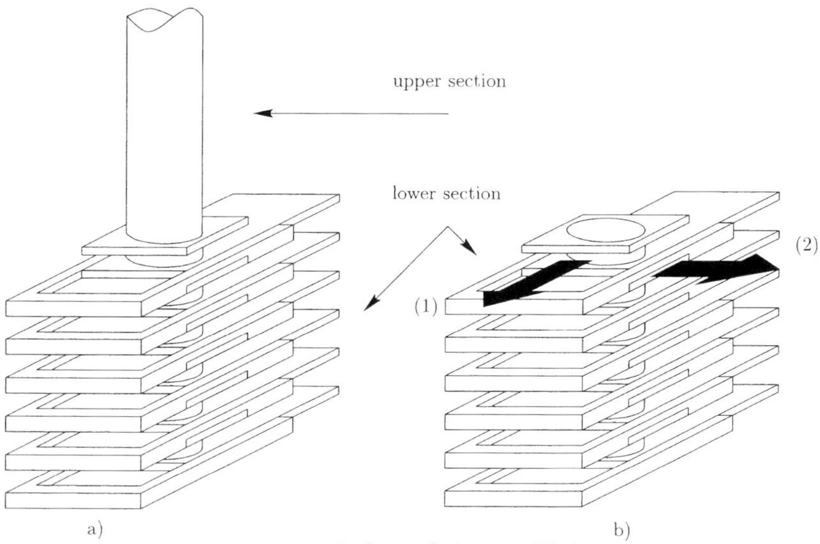

1. Set-up during consolidation.

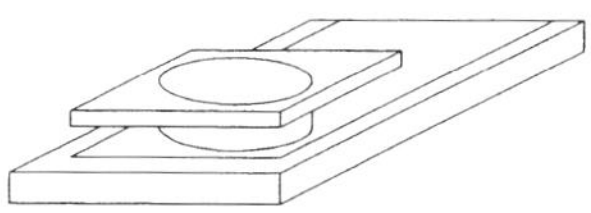

2. Lower section after removal of upper section. The column section of the uppermost segment is pushed on to the base plate (1), then, the segment is slid aside (2).

3. The removed segment with the sample, ready for the shear vane test.

Figure 3. A schematic picture of the dismantling of a segmented column.

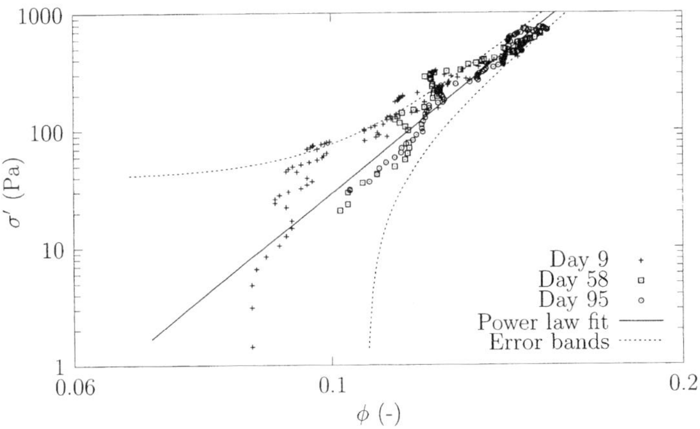

Figure 4. Effective stress against particle volume fraction.

(with the γ-ray densimeter) did not meet the requirements.

Alternatively, the permeability may be estimated by an analytical expression for the initial stage of the settlement of the interface. This method is based on the assumptions that i) Gibson's consolidation equation is valid, ii) the permeability is a power law function of the clay volume fraction, and iii) the soil can be considered homogeneous. A detailed derivation and verification of this expression is given in [16,17]. Here only the resulting expression for the initial settling is given:

$$h(t) = \left(\zeta \frac{2-n}{1-n}\right)^{\frac{1-n}{2-n}} \left(K_k \frac{\rho_s - \rho_w}{\rho_w}(n-2)\right)^{\frac{1}{2-n}} t^{\frac{1}{2-n}}, \tag{22}$$

where h is the settlement of the interface, ζ is the Gibson height (the product of the initial height and the initial volume fraction of solids), ρ_w the density of the pore water, ρ_s the density of the solids, $n = \frac{2}{3-D}$ and t time.

The interface settlement of measurement CT95 is shown in Figure 5. Using the analytical expression (22) for $t < 2 \times 10^6$ s, the unknown parameters in (13) are estimated at $K_k = 4.4 \times 10^{-18}$ m/s and $D = 2.75$.

3.2.3. Undrained shear strength

Undrained shear strength profiles were measured after various times of consolidation. The strength is represented by the peak shear strength measured at a constant rotation rate of 1 rpm (undrained conditions). Figure 6 shows the measured peak shear strength against effective stress. The results show that for a given time, the peak shear strength increases approximately linearly with effective stress, and that the peak shear stress increases with time for a given level of effective stress. The gain in peak shear strength with

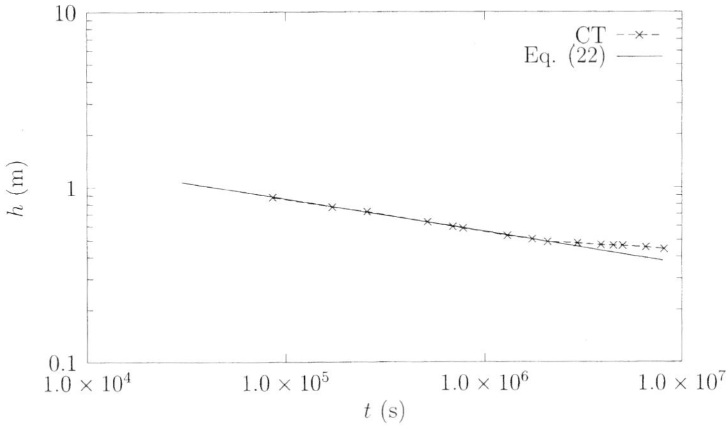

Figure 5. Settlement of the interface versus time for measurement CT95.

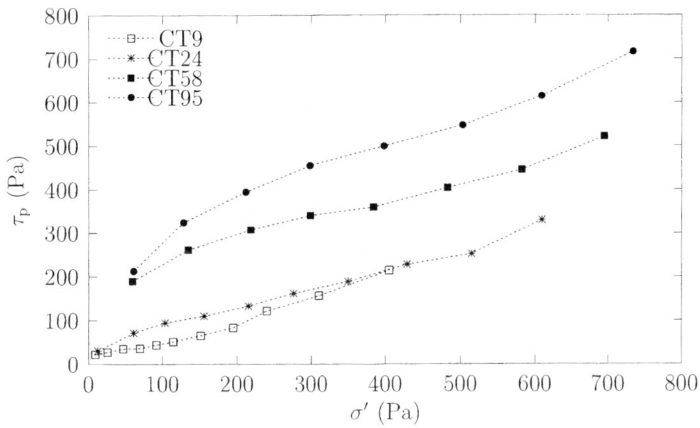

Figure 6. Undrained peak shear stress against effective stress.

time seems to be independent of the level of effective stress, since the curves of the data points are more or less parallel. Furthermore, for low levels of effective stress ($\sigma' < 100$ Pa) the data curves tend to bend down towards the origin.

Interpreting these observations in relation to the failure criterion (18), it follows that i) c_2 is approximately constant, irrespective of the level of effective stress and time, ii) c_1 is time dependent, but independent of the level of effective stress.

Since it is well known that the magnitude of the measured peak shear stress depends on the testing conditions such as the vane's geometry, the rotation rate, the parameters c_1 and c_2 implicitly take into account rate effects and effects due to thixotropy. The value of c_2 is obtained from the averaged slope of the curves, yielding $c_2 = 0.63$. The value of c_1 is estimated as follows. For relatively high levels of effective stress, the volume fraction of solids is about 0.1 (Figure 4), so that the clay volume fraction at these stresses, $\overline{\phi}_{\text{clay}}$, is about 0.05 for the given particle size distribution. From the results shown in Figure 6 it can be shown that the offset of the linear approximation of the data points (equal to $c_1\overline{\phi}_{\text{clay}}$) increases by about 3.3 Pa/day. Consequently, it follows that $\dot{c}_1 = 66$ Pa/day.

3.3. Comparison of the model results with measurements

The process of consolidation was modelled by the Gibson equation [9], but written in an Eulerian coordinate frame rather than the usual material coordinate reference frame, and with the volume fraction of solids as dependent variable rather than the void ratio. The consolidation equation reads [15]

$$\frac{\partial \phi}{\partial t} - \frac{\rho_{\text{s}} - \rho_{\text{w}}}{\rho_{\text{w}}} \frac{\partial}{\partial z}\left(k\phi^2\right) - \frac{1}{\rho_{\text{w}} g} \frac{\partial}{\partial z}\left(k\phi \frac{\partial \sigma'}{\partial z}\right) = 0. \tag{23}$$

Equation (23) was numerically integrated together with the initial conditions (Section 3.1), the constitutive equations (7) and (13) and the empirical parameters K_σ, K_k and D as found in Section 3.2. Computed density profiles after 8, 24, 58 and 95 days of consolidation are compared with measured density profiles in Figure 7. It is seen that the computations are in a fair agreement with the measurements. The observed differences are partly due to relatively inaccurate density measurements. It is noted that the parameters used in the numerical modelling are determined from the same experimental data set. A true validation requires the modelling of an experiment with independently determined parameters. In [16] more experimental data is shown that indicates that for the same material the fractal dimension is independent of the initial conditions, but this is not necessarily true for the coefficients K_σ and K_k.

Strength profiles were calculated from the failure criterion (18) with the effective stresses computed by means of the consolidation model and the $\dot{c}_1$ and c_2, as determined in the previous section. The computed peak shear stresses are presented together with measured peak shear stresses in Figure 8. A fairly good agreement between the modelled and measured strength profiles can be observed. The absolute deviation between measured and modelled peak shear stresses is up to 100 Pa for the lower part of the bed on day 9 and day 24, but with respect to the absolute values of the peak shear stresses, this is an error of maximally 33%, which is considered fair.

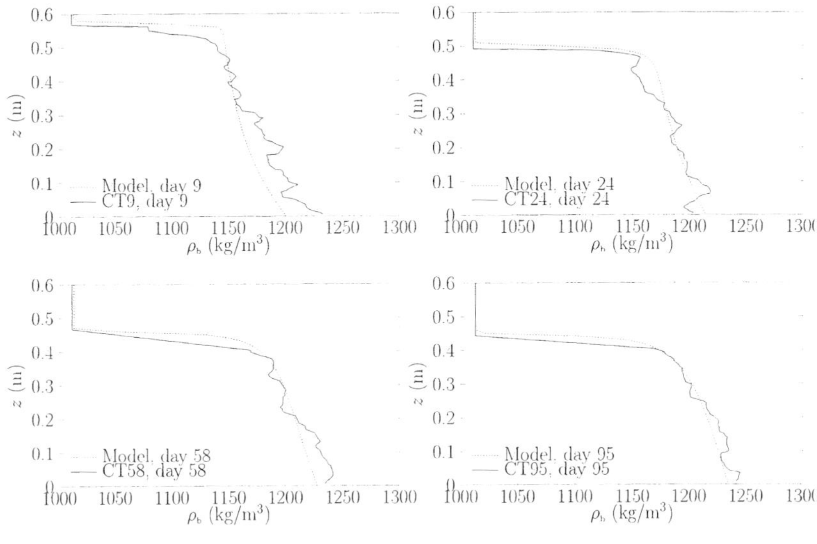

Figure 7. Measured and computed density profiles.

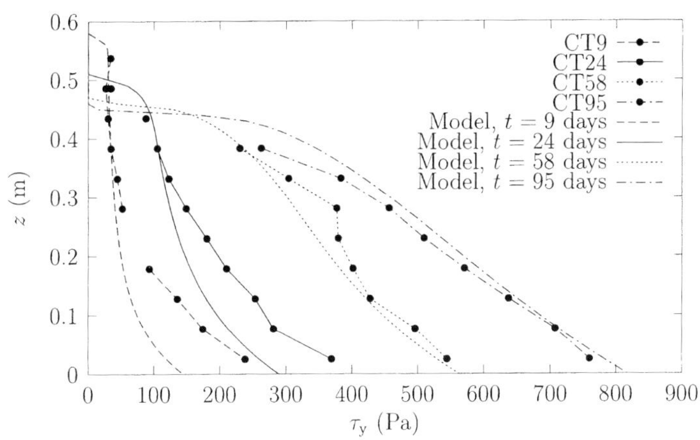

Figure 8. Measured and modelled evolution of the strength of the bed.

4. CONCLUSIONS

Constitutive equations for effective stress and permeability were derived assuming a scale invariant bed structure. The resulting expressions relate the effective stress and permeability to the clay volume fraction according to power laws. The exponents are determined by the fractal dimension only. The fractal dimension was determined for the present experiments at 2.75 for both the effective stress relationship and the permeability relationship.

A failure criterion was derived also assuming a scale invariant bed structure. The shear stress at failure is linearly related to the effective stress and also linearly, but time dependent to the clay volume fraction. For relatively high levels of effective stress ($\sigma' >$ 100 Pa for the current experiments) the failure criterion resembles the traditional Mohr-Coulomb failure criterion with true cohesion, but for low levels of effective stress the failure criterion in ($\sigma' - \tau$) space is non-linear, yielding zero strength at zero effective stress.

A series of consolidation experiments is described. The empirical parameters for the effective stress relationship were determined from measured effective stress data. The empirical parameters for the permeability relationship were determined from the initial settling of the interface, using an analytical expression.

The new constitutive equations are applied to the Gibson consolidation equation. Computed density profiles compared well to measured density profiles. Computed strength profiles using computed effective stress profiles also compared well to measured peak shear stress measurements.

In conclusion, the concept of a fractal bed structure allows the derivation of useful constitutive equation. In future research the theory will be extended to identify the influence of physical parameters, such as the size of clay particles and viscosity, and to include the effect of a varying particle size distribution.

ACKNOWLEDGEMENTS

This research was jointly financed by The Netherlands Foundation of Technology (STW) and the Commission of the European Communities, DG XII, MAS3-CT97-0082, MAST-COSINUS.

REFERENCES

1. K. Been. *Stress-strain behaviour of a cohesive soil deposited under water.* PhD thesis, Oxford University, 1980.
2. R.K. Bowden. *Compression behaviour and shear strength characteristics of a natural silty clay sedimented in the laboratory.* PhD thesis, Oxford University, 1988.
3. L.G.B. Bremer, V. Van Vliet, and P. Walstra. Theoretical and experimental study of the fractal nature of the structure of casein gels. *Journal of the Chemical Society, Faraday Transactions I*, 85:3359–3372, 1989.
4. R. Butterfield. A natural compression law for soils (an advance on $e - \log(p')$. *Géotechnique*, 29(4):469–480, 1979.

5. S. Chen and D. Eisma. Fractal geometry of in situ flocs in the estuarine and coastal environments. *Netherlands Jounal of Sea Research*, 32(2):173–182, 1995.
6. J.M. Cornelisse, C. Kuijper, and J.C. Winterwerp. Analyse konsolidatieproeven en toepassing konsolidatiemodel met ketelmeerslib. Technical report, Rapport van Rijkswaterstaat en WL|Delft Hydraulics, 1993.
7. E.J. Den Haan. The formulation of virgin compression of soils. *Géotechnique*, 42(3):465–483, 1992.
8. D. McG. Elder. *Stress–strain and strength behaviour of very soft soil sediments.* PhD thesis, Oxford University, 1985.
9. R.E. Gibson, G.L. England, and M.J.L. Hussey. The theory of one-dimensional consolidation of saturated clays. *Géotechnique*, 17:261–273, 1967.
10. A.A Hannoura and F.B.J. Barends. Non-darcy flow; a state of the art. In A. Verruijt and F.B.J Barends, editors, *Flow and transport in porous media*, volume 143, pages 37–51, 1981.
11. C. Kranenburg. The fractal structure of cohesive sediment aggregates. *Estuarine, Coastal and Shelf Science*, 39:451–460, 1994.
12. R.B. Krone. A study of rheological properties of estuarial sediments. Technical Report 63-8, Hydraulic Engineering Laboratory and Sanitary Engineering Laboratory, University of California, Berkeley, 1963.
13. B.E. Logan and J.R Kilps. Fractal dimensions of aggregates formed in different fluid mechanical environments. *Water Resources*, 29(2):443–453, 1995.
14. B.B. Mandelbrot. *The fractal geometry of nature.* Freeman, New York, 1982.
15. L.M. Merckelbach. Consolidation theory and rheology of mud; a literature survey. Technical report, no. 9–96. Hydromechanics Section, Delft University of Technology, The Netherlands, 1996.
16. L.M. Merckelbach. *Consolidation and strength evolution of soft mud layers.* PhD thesis, Delft University of Technology, The Netherlands, 2000.
17. L.M. Merckelbach and C. Kranenburg. A method to determine empirical coefficients for effective stress and permeability from simple laboratory experiments. *In preparation*, 2001.
18. J.K. Mitchell. *Fundamentals of Soil Behavior.* John Wiley & Sons, 1976.
19. H. Rouse. *Elementary mechanics of fluids.* John Wiley and Sons, New York, 1946.
20. D.L. Turcotte. *Fractals and chaos in geology and geophysics.* Cambridge University Press, Cambridge, 1992.
21. B.G.H.M Wichman. *Consolidation behaviour of gassy mud: theory and experimental validation.* PhD thesis, Delft University of Technology, 1999.
22. J.C. Winterwerp. *On the dynamics of high-concentrated mud suspensions.* PhD thesis, Delft University of Technology, 1999.

Fine Sediment Dynamics in the Marine Environment
J.C. Winterwerp and C. Kranenburg (Editors)

Description of Vertical exchange processes in numerical mud transport modelling

Ole Petersen[a] and H. Jacob Vested[a]

[a]DHI Water & Environment, Agern Alle 11, DK-2970 Hørsholm, Denmark

The physical background for entrainment and erosion of cohesive sediment is briefly discussed followed by suggestions for parameterisations in both two-dimensional horizontal (2DH) and three-dimensional (3D) sediment models. The 2DH parameterisation uses a 2D-layer approach. The 3D parameterisation is based on the application of a coupled turbulence-sediment model. In order to investigate the feasibility of the proposed parameterisations with respect to functionality and the ability to predict reasonable values these are tested on data collected during the Tamar Field experiment. The results of the parameterisations are discussed in relation to the Ariathurai-Partheniades equation for erosion. A direct comparison of results from the turbulence-sediment model and the entrainment expression of Kranenburg and Winterwerp (1997) shows a good agreement. For a more comprehensive comparison of field data with model results, reference is made to Petersen *et al.* (2000).

KEY WORDS
models, cohesive sediments, entrainment, concentrated benthic suspension layers

1. INTRODUCTION

The interaction between a suspension of mud and the deposits on the bed in terms of resuspension and deposition is a key process in the quantification of mud transport and has been the subject of many experimental and theoretical studies. For cohesive sediments, a common understanding is that a gradual transition may exist, when characterised e.g. by the density of the layer between the material suspended in the water column and the immobile bed. Several definitions of distinct suspension layers exist, e.g. Winterwerp (1999) distinguishes between low concentrated mud layers (0.01-0.1 g/l), high concentrated mud suspensions (0.1-1 g/l) and fluid mud (10-100 g/l). In each layer different physical processes are prominent, and the mechanisms responsible for resuspension are therefore also different. Concentrated near-bed suspensions in the order of 5-10 g/l are sometimes referred to as Concentrated Benthic Suspension (CBS) layers, see also Berlamont and Toorman (2000). This terminology is used for the regime in which there is a significant interaction with the flow, but the sediment/water mixture still behaves as a Newtonian fluid.

For settled beds and denser suspensions, the classical theory depicts a gradual wearing of the sediment interface. The rate of resuspension depends on the excess bed shear stress above a critical value, characteristic for the bed surface, as:

$$E = E_o \left(\frac{\tau_b}{\tau_c} - 1 \right), \tau_b > \tau_c \tag{1}$$

where E is the rate of resuspension, E_o is an empirical constant, τ_b is the shear stress acting on the bed and τ_c is the critical shear stress characterising the strength of the mud layer. This equation is referred to as the Ariathurai-Partheniades equation, Ariathurai (1974). Experimental verifications of this relation have usually considered consolidation times larger than 12 h (Mehta, 1989; Zriek *et al.* (1998)).

More recently Mehta (1989) and Winterwerp and Kranenburg (1997) suggested to apply an analogy with density stratified flows to high concentrated mud layers. The suspension is assumed to behave like a denser fluid, and the resuspension is seen as an entrainment process, mainly controlled by the flow and the vertical density gradient induced by the mud suspension. Interactions between the turbulence and the buoyancy jump across the lutocline may start an entrainment process where the lutocline is raised due to entrained water from above and the suspension is diluted. Kranenburg and Winterwerp (1997) have studied this situation theoretically and Bruens *et al.* (1999) experimentally. The situation is in analogy to the ocean mixed layer deepening that arises due to wind induced turbulence or in inhomogeneous topographies where a turbulent tidal current enters above a pool of quiescent fluid mud. Based on experiments in a rotating flume, Kranenburg and Winterwerp (1997) demonstrated that the entrainment rate under certain conditions could be approximated using.

$$w_e / u_* = \left(\frac{con_1}{con_2 + R_{io}} \right)^{0.5} \tag{2}$$

where w_e is the ascent of the lutocline, u_* is the shear velocity and R_{io} a Richardson number,

$$R_{io} = \frac{g \Delta\rho H}{\rho u_*^2},$$

where H is the depth, $\Delta\rho = \rho_b - \rho_u$ is the density difference between upper and lower layer, g is the gravitational acceleration and ρ_b and ρ_u are the bulk density of lower and upper layer, respectively. con_1 and con_2 are constants. They also derived more elaborate expressions taking mud properties and viscous effects into account.

It is seen from Eq. (1) that the erosion rate depends on τ_b or (u_*) and ρ_b, if the latter parameter can be related to τ_c. The entrainment rate according to Eq. (2) depends on ρ_b, u_* and H, thus the structure of the relation (1) and (2) is similar. It does however appear more appropriate to describe the resuspension of dense suspensions as entrainment rather than erosion, and furthermore, the entrainment relation avoids the necessary τ_c-ρ_b relation. This may therefore represent a simplification of (1) under certain conditions.

The objective is to investigate to which extend the vertical exchange process can be described as an entrainment process in numerical mud transport models. In applied modelling of mud transport, evidently in 2DH models, but also in 3D models where the resolution to resolve thin suspension layers may not be available, mathematical descriptions that parameterise resuspension processes are essential. In order to investigate the feasibility of the parameterisations for vertical exchange processes with respect to functionality and the ability to predict reasonable values, the data set gathered in the Tamar Estuary as part of the COSINUS project has been applied.

2. THE TAMAR ESTUARY DATA SET

Inspection of the Tamar measurements of the vertical profiles of concentrations of Suspended Particulate Matter (SPM) indicates that it is possible to identify a lower CBS layer. The concentration in this layer is in the order of 2-5 g/l. In Figure 1, selected examples of the measured vertical profiles during neap and spring tide are shown. Notice that the concentration scale is logarithmic. Following the observations taken under spring conditions and going from falling to rising tide, the development of the profile can shortly be described as follows: While the flow increases towards low tide, the lower layer is entrained and the sediment is kept in suspension. When the flow ceases at low tide, the sediment settles out from the upper layer and the lower layer is rising in thickness. This lower layer is entrained into the upper layer as the flow increases. Towards high tide, as the flow ceases, the sediment settles and the lower layer is formed again and a lutocline can be identified about one metre above the bed. It is further observed that the thickness of the lower layer is greater at high tide than at low tide. The concentration of the upper layer is in the range from 0.5-1.0 g/l. During neap, the development of a lutocline can hardly be identified and the vertical profile is smoother. The concentrations during neap are in the order of 0.01 to 0.1 g/l. For a thorough discussion of sediment dynamics and the Tamar data, see Dyer *et al.* (2000).

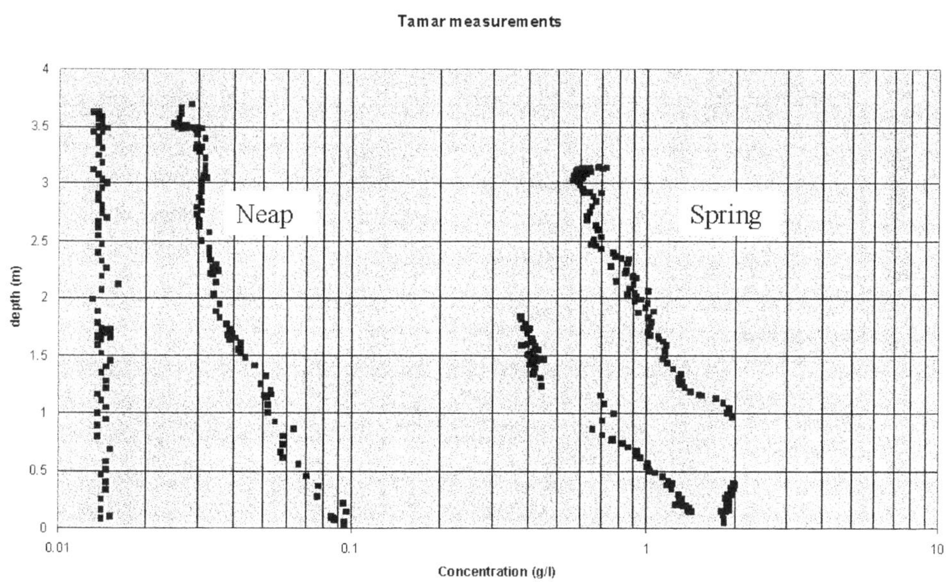

Figure 1. Examples of the vertical variation of the measured concentration profiles during spring and neap conditions in the Tamar Estuary.

3. PARAMETERISATIONS OF ENTRAINMENT IN 2D HORIZONTAL MODELS

A common approach (Mehta *et al.*, 1989) for definition of the vertical sediment and bed processes is shown in Figure 2. These processes can be parameterised (referred to as *old*) for use in 2DH models, e.g. Brøker *et al.* (1994). The horizontal transport of the depth averaged sediment concentration is computed from the water fluxes and the horizontal dispersion. It is thus necessary to establish relations between the depth averaged sediment concentrations and the local vertical variations of SPM profiles.

The erosion processes follow Ariathurai-Partheniades (Ariathurai, 1974) for dense consolidated beds and deposition follows Krone (1962). To compute the deposition flux at the bed, an expression is needed that relates the average concentration to the concentration at the bed. The vertical SPM or concentration profile can for example be assumed to follow the parameterisation of Teeter (1986). This gives an analytical expression for the ratio of near-bed concentration to the depth averaged, assuming a parabolic distribution of vertical eddy viscosity with variable bed flux conditions, i.e. erosion and deposition and a constant settling velocity.

Limitations of the above parameterisation are: The CBS layer should be part of the water column rather than the bed. The exchange of sediment between the suspension layer and the CBS layer is better described as an entrainment approach than as a surface erosion process. The variation of the settling velocity in the vertical is not accounted for, here in particular, its decrease in the CBS layer due to hindered settling. Movements of the CBS or fluid mud layer due to external pressure gradients are not included. Alternatively, one could define an upper and lower sediment layer as being within the water column and apply an entrainment description. This is conceptually shown in Figure 2 (referred to as *new*).

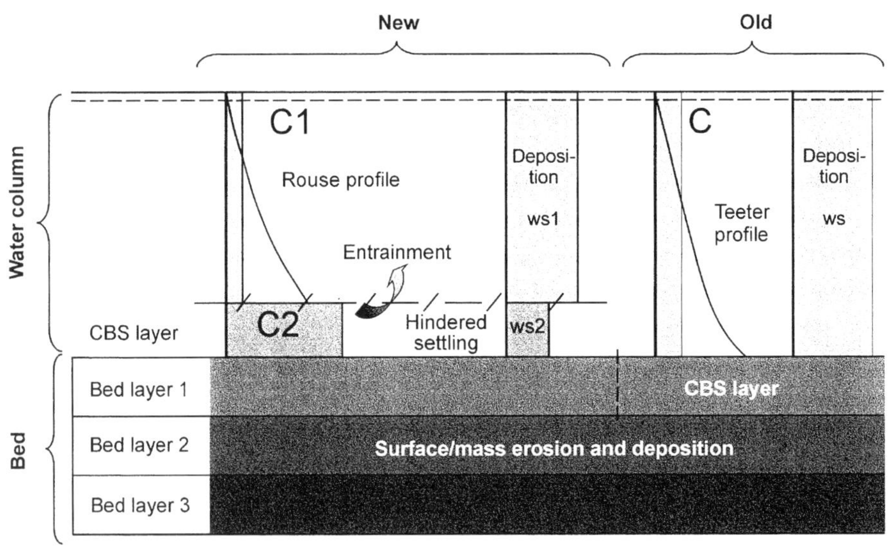

Figure 2. New and old parameterisation of cohesive sediment processes in 2D vertical depth averaged models.

The entrainment model is based on a two layer concept, where the turbulence mixing capacity in the upper layer is significantly higher than in the lower, thus sediments mixed across the interface will quickly be fully mixed into the turbulent layer. In this way, the interface can stay relatively sharp for a long period.

Applying the sediment mass balance gives for the upper suspended sediment layer:

$$\frac{Dh_1c_1}{Dt} = E - D \qquad (3)$$

where h_1 and c_1 are the thickness and concentration of the upper layer. D/Dt is the rate of change along a streamline. E is the entrainment flux and D the deposition flux.

For the lower CBS layer the sediment mass balance reads:

$$\frac{Dh_2c_2}{Dt} = D - E \qquad (4)$$

where h_2 and c_2 are the thickness and concentration of the CBS layer. In the above equations interactions with the solid bed below the CBS layer have been neglected. The deposition flux D can be given by (Mehta, (1989)):

$$D = p_D w_s c_1 \qquad (5)$$

and the entrainment flux E by:

$$E = c_2 w_e \qquad (6)$$

where p_D is the probability of deposition defined as $(1-\tau_b/\tau_d)$, τ_d is the critical shear stress for deposition, w_s is the fall velocity, c_2 is the CBS layer concentration and w_e is the entrainment velocity, which as shown by Kranenburg and Winterwerp (1997) can be related to a bulk Richardson number through Eq. (2). Hereby it is assumed that the concentration of the CBS layer c_2 is constant. The two-layer system is thus solved for the thickness of the CBS layer and the upper layer concentration. The applicability of Eq. (2) for tidal flow, as is the case here, is discussed in Chapter 4.2.

The CBS layer is subject to movements due to gravitational forces caused by density differences or steep bottom gradients (sliding). In addition the movement due to external pressure, gradients can be very important. This is described using an advection scheme for the CBS layer. The advection speed of the CBS layer is determined as a certain fraction of the mean current speed:

$$U_{CBS} = facU_{mean} \qquad (7)$$

The factor depends on the vertical profile of the horizontal velocity and the vertical sediment profile. It will also depend on the density stratification and layer thickness. Thus, in reality it will be time and space varying. For the present purpose it is assumed constant.

It should be mentioned that Odd and Cooper (1989) have applied a two-layer approach incorporated into a depth integrated two-dimensional suspended mud transport model. This approach goes back to 1972. The present proposed approach is fundamentally similar to this work, but simpler as it does not involve the solution of the 2D equations for the conservation of mass and momentum of fluid mud. In addition, the present paper concerns the CBS layer and not the fluid mud layer, which Odd and Cooper assumed to have non-Newtonian rheological properties (Bingham fluid).

3.1. 2DH simulation of the Tamar Estuary by application of Ariathurai-Partheniades erosion equation

A 2DH Finite Element Model (FEM) has been applied to simulate the entire Tamar Estuary. Figure 3 shows the layout of the grid and the location of the measurement stations. The model boundaries are tidal elevations and freshwater discharge. The model has been calibrated to reproduce the measured water level variations and velocities at Calstock.

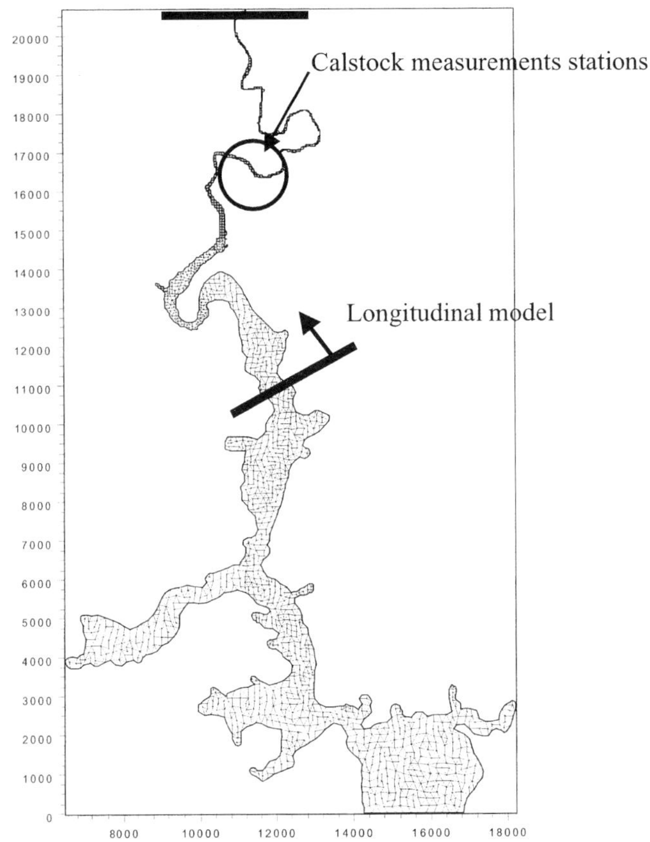

Figure 3. The Tamar Estuary Finite Element Model grid. The location of the two measurement stations (within the circle) and the extension of the longitudinal two-layer model are also indicated.

Table 1.
Model parameters applied in the Ariathurai-Partheniades erosion and Krone deposition equations

W_s Settling velocity	0.003 m/s
τ_d critical deposition	0.1 N/m^2
τ_c critical erosion	0.1 N/m^2
E_o Erosion constant	0.0002 kg/m2/s

In a discussion about operational modelling of cohesive sediment transport it is relevant to investigate to which extend the widely used Ariathurai-Partheniades equation can reproduce the observed sediment behaviour. This parameterisation is referred to as *old* in Figure 2. The 2D FEM model was therefore run for the neap-spring period of the Tamar experiments. The initial conditions for the modelling were established by assuming a distribution of sediment over the bed of a length of 10 km located at about the expected position of the turbidity maximum (around Calstock). This stock of sediment was not limited in depth. The initial sediment concentrations in the water column were put equal to zero. The model parameters are shown in Table 1.

In Figure 4, the simulated time series of the suspended sediment concentrations are shown. The levels of concentration are in accordance with the observations, especially one notices the important increase in sediment concentration from neap to spring caused by the increasing tide. Also seen is the strong difference in concentration from rising to falling tide, caused by the strong tidal asymmetry. In Figure 5 is shown the horizontal distribution of the sediment at rising tide illustrating the horizontal distribution of the turbidity maximum. Two maxima are seen due to the change in currents along the estuary.

The measured averaged SPM concentrations are in the order of 0.05 to 0.5 g/l for neap and spring, respectively. The results of the simulation are in the same order of magnitude. This indicates that the Ariathurai-Partheniades equation does work very well on the Tamar Estuary when properly calibrated. This is in accordance with the experience from many estuarine studies undertaken in practice, and also the reason for the success for this formulation of the erosion process. The Ariathurai-Partheniades equation can, as it scales with the square of the friction velocity, describe both the erosion of a hard bed as well as the resuspension from a soft bed, and in practice it is very difficult to distinguish these phenomena.

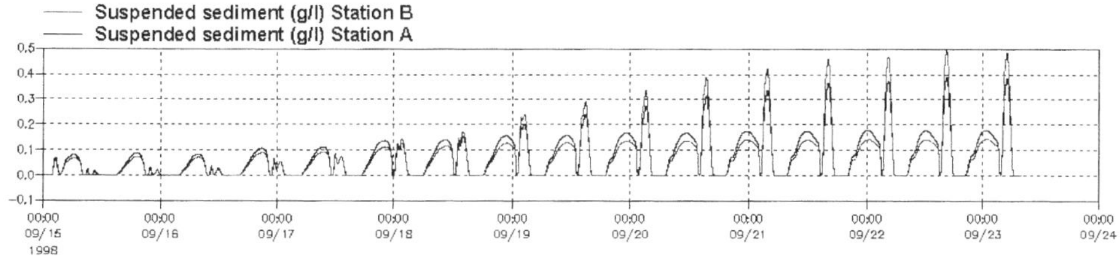

Figure 4. Time series of simulated suspended sediment concentration from neap to spring at Calstock. The location of Stations A and B is shown in Figure 5.

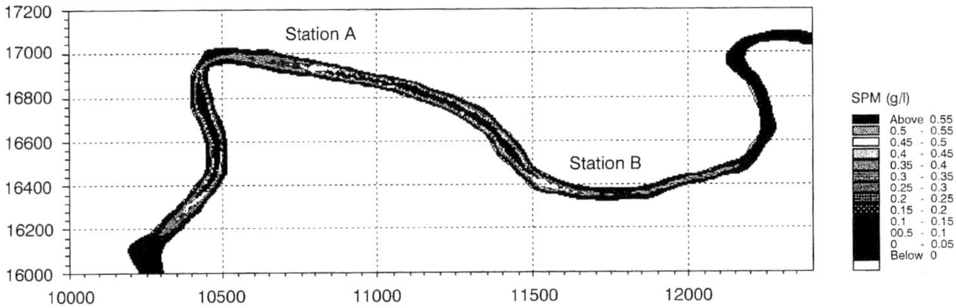

Figure 5. Horizontal distribution of sediment concentration during spring at rising tide showing the turbidity maximum.

The calibration of the model was facilitated by advice on the threshold values of erosion and deposition on the basis of the measured values of the bottom shear stress, Christie (2000). The value of the settling velocity was based on the estimate of Spearman and Roberts (2000). The only remaining parameter to adjust was the rate of erosion E_o. Furthermore, it is emphasised, although trivial, that an accurate determination of the hydrodynamic conditions is crucial for reproducing the sediment phenomena.

3.2. Test of the two-layer approach with inclusion of advective effects

From the observations in the estuary it is known that a CBS layer exists during spring. This makes it relevant to test the two-layer approach under realistic conditions. In order to do this a 1D longitudinal model has been set up for a stretch of the Tamar Estuary, c.f. Figure 3. The imposed hydrodynamic boundary conditions are taken from the FEM model of the entire estuary. The bathymetry and extension of the 1D longitudinal model is similar to the 2D vertical longitudinal model applied in Petersen et. al. (2000). The two-layer model described above (referred to as *new* in Figure 2) has been implemented and the behaviour of the upper layer concentrations and the CBS layer is calculated for a spring period consisting of six uniform tidal cycles (3 days).

The additional input data specified for the two-layer description is listed in Table 2.

Table 2.
Model parameters applied in the two-layer description

C_2, CBS layer concentration	5 g/l
Initial condition, thickness of CBS layer along the entire length (16 km)	0.1 m
W_s, Settling velocity	0.003 m/s
τ_d, Critical shear stress for deposition	0.5 N/m^2
Fac, Advection factor for CBS layer	0.4

The simulation presented in Figures 6 and 7 shows that during rising tide there is strong increase in upper layer concentration due to entrainment of sediment from the CBS layer. The upper layer sediment is subject to advection and is moved upstream in the estuary. Sediment settles preceding slack water, i.e. as soon as the velocity starts to decrease. Thereupon, at falling tide, the upper layer concentration is lower. The CBS layer flows with the tide downstream, and the cycle repeats itself. This behaviour is in agreement with the results of the detailed study presented in Petersen *et al.* (2000). Though the simulation was not an attempt to reproduce the Tamar observations, it is noted that the concentrations in the upper layer are within the range of the measured values (0.5-1g/l). The applied settling velocity and critical shear stress is in agreement with the observed values (Spearman and Roberts (2000) and Christie (2000)). After two tidal cycles, the initial uniform distribution of the CBS layer forms a CBS layer of a length of about 4 km, which is a realistic figure for spring tide. During calibration it was found that the advection of the CBS layer with the mean tidal current was decisive to reproduce the expected behaviour. The simulation is only describing the interaction between the CBS layer and the upper layer. In order to describe neap-spring variations, erosion of the solid bed below the CBS layer should be taken into account as well as the consolidation of the CBS layer.

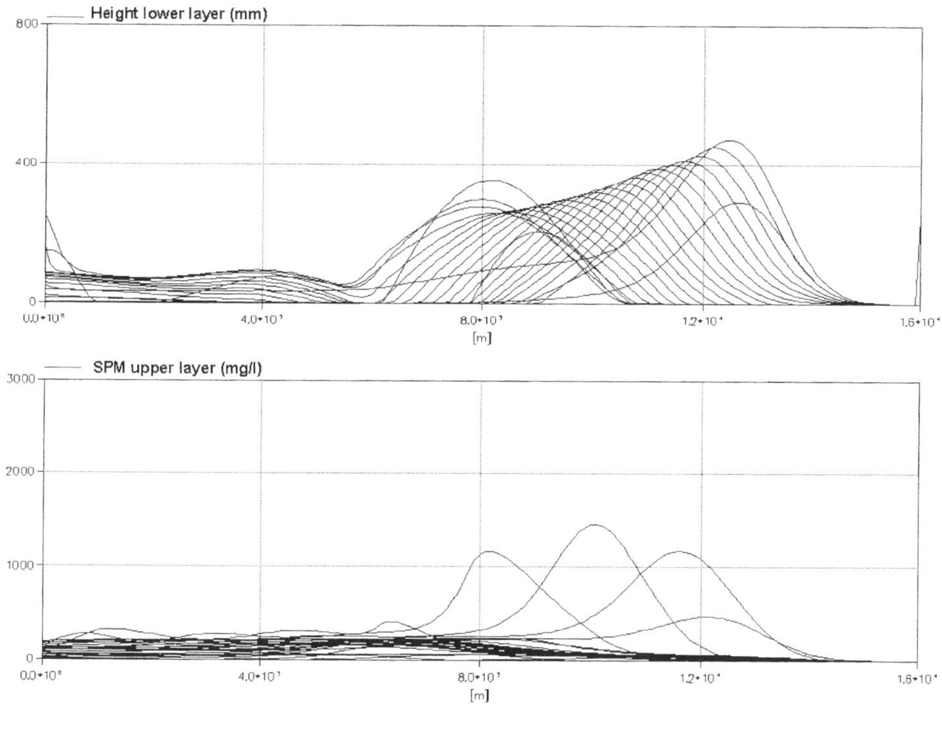

Figure 6. Longitudinal variation of lower layer thickness and upper layer concentration during the sixth tidal cycle.

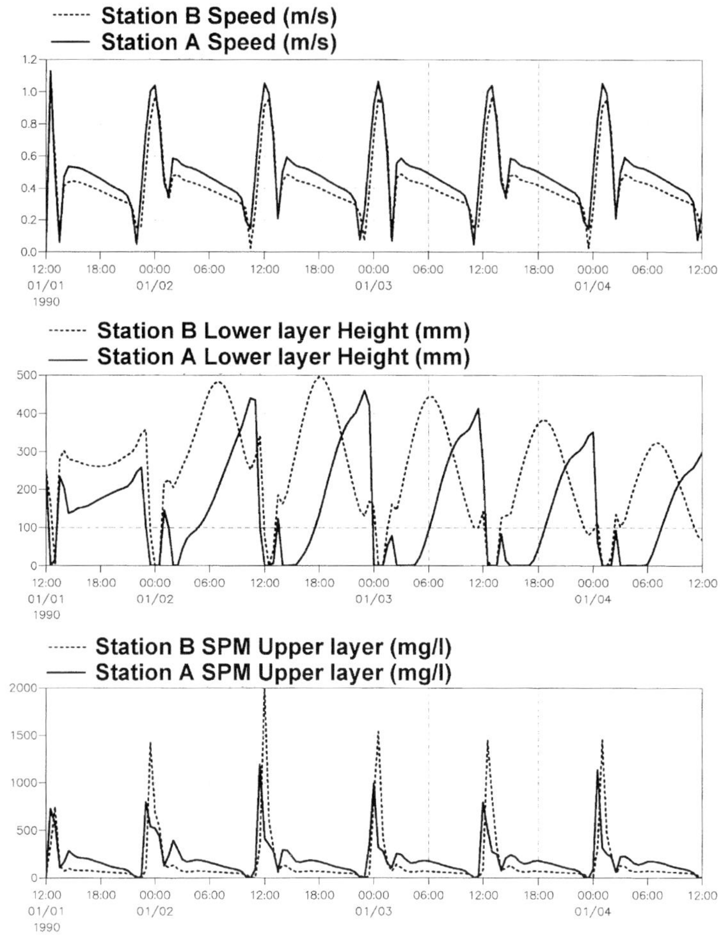

Figure 7. Time series showing the variation of the velocity, the lower layer thickness and upper layer concentration, respectively, at two positions (Stations A and B) in the middle of the estuary.

4. PARAMETERISATIONS OF ENTRAINMENT IN 3D MODELS

In 3D models for estuarine flows, the horizontal grid size is usually an order of magnitude higher than the vertical grid size. This implies that it may be reasonable to separate the horizontal and vertical eddy viscosity and diffusivity parameterisations. The parameterisations of the vertical eddy viscosity and diffusivity is normally based on a mixing length approach or a k-ε type model. The latter model has shown to be successful in describing the mixing and entrainment across density stratifications due to salinity or temperature differences, see e.g. Burchard *et al.* (1998b).

Considering a 1D vertical turbulence model that solves the equation for one velocity component, suspended sediment and a k-ε model, assuming horizontal homogeneity of all terms, see e.g. Burchard *et al.* (1998a), the equations for momentum, suspended sediment and turbulence are:

$$\partial_t u = \partial_z(\nu_T \partial_z u) + \partial_x P \tag{8}$$

$$\partial_t c = \partial_z(\nu_T / \sigma_T \partial_z c) + \partial_z(w_s c) \tag{9}$$

$$\partial_t k = \partial_z(\nu_T / \sigma_T \partial_z k) + \nu_T(S^2 - N^2/\sigma_T) - \varepsilon \tag{10}$$

$$\partial_t \varepsilon = \partial_z(\nu_T / \sigma_T \partial_z \varepsilon) + \frac{\varepsilon}{k}(\nu_T(c_1 S^2 - c_3 N^2/\sigma_T) - c_2\varepsilon) \tag{11}$$

where u is the velocity, c the suspended sediment concentration, k the turbulent kinetic energy and ε the dissipation respectively. t is time and z is the vertical co-ordinate. P is the hydrostatic pressure, ν_T is the eddy viscosity given by $\nu_T = C_u k^2/\varepsilon$, w_s the mean fall velocity of the suspended sediment, S is the r.m.s. velocity gradient, N the buoyancy frequency $N^2 = g/\rho \partial_z \rho$, σ_T the sediment Schmidt number. c_1 and c_2 and c_3 are empirical constants. The density is linked to the sediment concentration by an equation of state as:

$$\rho = \rho_w + \beta\, c \text{, where the constant } \beta = 1 - \rho_w/\rho_s\,. \tag{12}$$

where ρ_s is the density of the sediment grains.

At the fixed bed standard rough bed conditions are used for the u, k and ε equations with a bed roughness of k_s, while a no flux boundary is used for the sediment.

The sediment erosion of the fixed bed is assumed to follow Eq. (1). In order to describe the formation of a lutocline, it is necessary to take into account the variation of the settling velocity in the water column. This is done by applying a simplified hindered settling description where it is assumed that:

$$w_s = \begin{cases} w_0 & for\ c < c_h \\ \alpha w_0 & for\ c > c_h \end{cases} \tag{13}$$

where α is a constant (<1). The concentration c_h can be interpreted as the concentration at which the suspension changes from increasing flocculation to the hindered settling regime.

4.1. Application of the 1D vertical k-ε model to the Tamar

In order to validate the application of a 1D vertical k-ε model for the purpose of simulating the vertical distribution of CBS layers and dense sediment suspensions, the k-ε model described above has been applied to simulate the sediment dynamics in the Tamar Estuary. The model parameters are summarised in Table 3.

Table 3.
Model parameters applied in the 1D vertical turbulence sediment model

	Neap	Spring
H	3.8 m	3.8 m
w_s (above lutocline)	0.0045 m/s	0.0045 m/s
E_o	1.3 10^{-3} kg/m2/s	1.3 10^{-3} kg/m2/s
τ_c	0.1 N/m2	0.4 N/m2
τ_d	1.0 N/m2	0.6 N/m2
I (Surface gradient)	1.3 10^{-4}	4.0 10^{-4}

In addition to the parameters above, it has been assumed that hindered settling occurs at the level of 2 g/l. This is a necessary condition to form a CBS layer with a level of concentration as observed. The coefficient α in Eq. 13 is equal to 0.05 providing a discontinuity in w_s at c=2 g/l. The applied critical shear stress for deposition allows for simultaneous erosion-deposition. This is done as it simplifies the calibration in the present situation.

The water surface level is assumed constant. The numerical solution is obtained on an equidistant grid using 2nd order explicit time marching numerical scheme. The vertical grid size is 0.1 m and the time step is 2 s. The model is calibrated to simulate approximately the observed current variations while it describes the distribution of sediment within the column during the tidal cycle.

It should be noted that the effect of gravitational circulation due to salinity differences is not taken into account in the present calculations. The occurrence of the salinity intrusion is not coinciding with the formation of the CBS layer and the formation of the CBS layer cannot as such be explained as an advective process, only.

A cyclic behaviour is observed. The results are presented as vertical profiles during the tidal cycle in Figure 8 for neap and spring as compared to examples of measurements. The comparison is very good showing that the model does indeed capture the range of the measured data. During neap the SPM profile is very smooth, whereas during spring, at slack water, a lutocline is clearly formed. One of the findings of the simulations was the importance of the hindered settling.

In the simulation, the deposition is adjusted via the settling velocity defined as αw_0 just above the bed, such that the suspended mass concentration in the profile corresponds to the measured data. As the measured data do not include the bed, this adjustment is empirical. Varying the coefficient α makes this adjustment to the settling velocity, such that the settling velocity is in the hindered settling regime. A sensitivity analysis of the effects of different α has been prepared. Table 4 gives the relation between different settling velocities and the concentration 1m above the seabed for the sensitivity analysis.

From Table 4 it is seen that the bottom concentration measured 1 m above the bed is not very sensitive to the value of α. This is very encouraging. The description of the settling velocity through inclusion of hindered settling is found to be decisive in reproducing the dynamics of the observations. It is of course important to have a good estimate of the concentration c_h at which the hindered settling starts. This emphasises the importance of measuring close to the bed.

Table 4.
Sensitivity of concentration 1 m above bed to the reduction of settling velocity

	αw_0			
	α=0.0	α=0.01	α=0.05	α=0.1
Concentration 1 m. above bed	1.46 g/l	1.39 g/l	1.36 g/l	1.33 g/l

The 1DV parameterisation of the resuspension process has been embedded into a full 3D model and has been applied to a stretch of the Tamar, see Petersen *et al.* (2000).

4.2. Entrainment in an idealised tidal flow

In order to verify that the entrainment concept is applicable for non-steady tidal flow also, where the turbulence is generated near the bed, a 1DV model comparison between the two-layer entrainment approach and the k-ε model is made. The model set-up is made with reference to the SILTMAN case used for model comparison in the COSINUS project, see also Violeau *et al.* (2000). The simulated rates of entrainment are compared with values measured in steady flow situations by Kranenburg and Winterwerp (1997). The model describes a 16 m deep water column where the flow is forced by a periodically varying horizontal pressure gradient, which can be given as $d_xP = 1.5\ 10^{-4}\ (0.2 + sin\ (2\ \pi\ t/12.5\text{h})$, corresponding to a M_2 tidal component, superposed on a net flow. A range of settling velocities and concentration levels were simulated. Other relevant parameters are the sediment Schmidt number $\sigma_T = 0.7$, the bed roughness $k_s = 0.05$ m and a kinematic viscosity $\nu = 1.0\ 10^{-6}\ \text{m}^2/\text{s}$. The numerical solution is obtained on an equidistant grid with 0.5 m resolution using 2nd order explicit time marching.

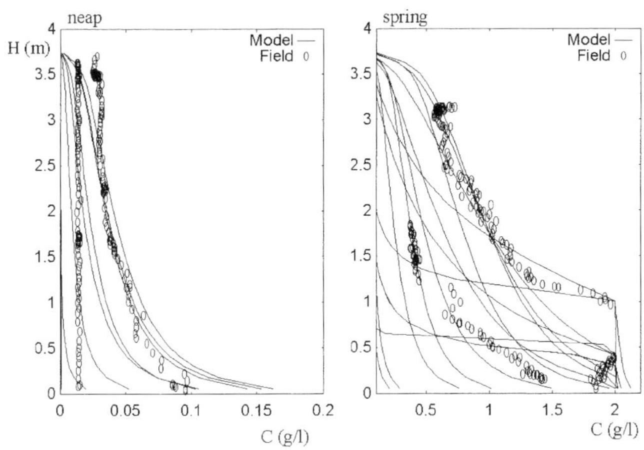

Figure 8 Vertical profiles, comparison between k-ε model and Tamar data for neap (left) and spring (right), respectively.

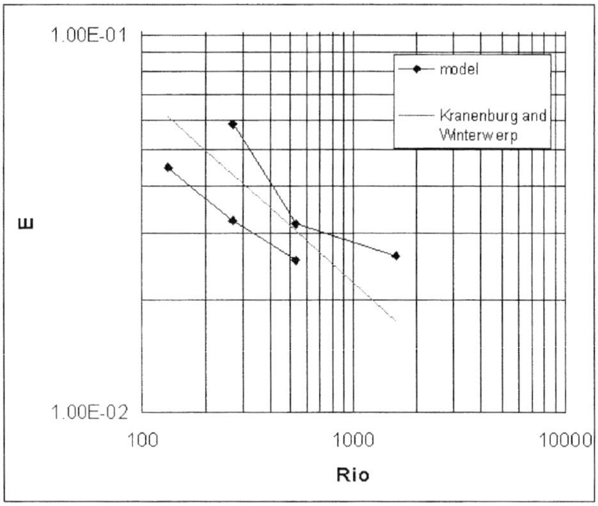

Figure 9. Comparison between model estimated and entrainment formula of Kranenburg and Winterwerp (1997).

The entrainment rates, estimated from the simulations, have been plotted as a function of the Richardson number in Figure 9. The relation suggested by Kranenburg and Winterwerp (1997), Eq. (2), is also plotted. This shows that over a range of R_{io} of 100-1000, the simulation appears to follow the experimentally derived relationship reasonably, say, within a factor of 2. As suggested by Kranenburg and Winterwerp (1997), con_1 is 0.5 and con_2 is 5.6 in Eq. (2). The simulations shown have been made with a grid with a relatively coarse (Δz = 0.5 m) vertical resolution. Trial runs with Δz down to 0.12 m show that the concentration gradients do become steeper across the lutocline. However, neither do the concentration profiles depend significantly on the resolution, nor does the entrainment rates change with the resolution. From an applied point of view this is assuring, as apparently it is possible to resolve the CBS layer entrainment with a resolution that can be used also in operational models. This supports the application of 3D models to describe the CBS dynamics without further parameterisation, provided a turbulence model that includes the buoyancy effects is used. The results shown here are in general agreement with other studies, e.g. numerical models by Winterwerp (1999) on CBS layers and Burchard and Petersen (1998a) on salinity stratified layers or laboratory experiments using salinity stratification by Kranenburg (1984) or Turner (1973).

5. DISCUSSION AND CONCLUSIONS

Parameterisations of entrainment of cohesive sediments and vertical exchange processes have been presented for 2DH and 3D numerical mud transport models. For 2DH models, a comparison have been made with the often-used Aritahurai-Partheniades equation. In order to investigate to which extent the vertical exchange process can be described as an entrainment process, the Tamar Estuary data set has been applied. From observations it is known that a CBS layer exists during spring tides.

The Ariathurai-Partheniades equation was able to qualitatively reproduce the behaviour and observed range in sediment concentration from neap to spring. In the present set-up this description does not distinguish between erosion from a soft or a solid bed. The calibration was facilitated by independent estimates of the threshold values of erosion and deposition and measurements of the settling velocity. The calibration was thus an adjustment of the rate of erosion.

For 2DH models, the simplest parameterisation of a CBS layer is to apply a two-layer description distinguishing between an upper suspension layer and a lower CBS layer. This requires an expression for the rate of entrainment between the two layers. The entrainment formula of Kranenburg and Winterwerp (1997) was applied. The two-layer entrainment description was embedded in the 2DH model and then applied for spring tides where a CBS layer exists. The two-layer model was able to reproduce the observed dynamical behaviour of the CBS layer qualitatively. The associated levels of concentrations in the upper layer and the thickness of the lower layer were within the range of the observed.

The Ariathurai-Partheniades equation can, as it scales with the square of the friction velocity, describe both the erosion of a hard bed as well as the resuspension from a soft bed. In practice it is very difficult to distinguish between these phenomena. For example, in the transition from neap to spring, erosion of a more dense bed takes place. The advantage of the two-layer description lies in the definition of the water column and bed. Movements of the CBS layer due to density and external pressure gradients causing sliding and advective transport, respectively, can be included.

In conclusion, the Ariathurai-Partheniades equation is very attractive for its applicability. However, this is on the expense of simplifications of the physical processes. The calibration of the equation is also difficult. The threshold values for erosion and the associated rate of erosion in Ariathurai-Partheniades equation can only be measured indirectly in nature. The two-layer entrainment description demands a good estimate of the amount of sediment that can be exchanged between the upper water column and the CBS layer. This requires knowledge of the thickness and level of concentration of the CBS layer. These are measurable quantities in an estuarine environment.

For parameterisation of entrainment in 3D models it was found that a combined turbulence sediment description based on a k-ε model could be applied. The vertical variation of the settling velocity is decisive. In the upper part of the water column, the settling velocity is determined by floc aggregation. As the concentration increases towards the bed, hindered settling becomes important. By introduction of a level of concentration c_h , at which hindered settling starts, it was possibly to reproduce the formation of a lutocline forming at slack water. It has been shown in a 1DV description that the parameterisations can reproduce the observed behaviour of the cohesive sediment in the Tamar Estuary. A complete quantitative comparison between model results and observations has been made in Petersen *et al.* 2000.

The entrainment formula according to Kranenburg and Winterwerp (1997) applied in the two-layer description is applicable for flow conditions that may results from wind action or turbulence produced at other locations than the bed. By a comparison between 1DV simulations and derived results for entrainment rates it is argued that the formula also can be used for tidal flow conditions and the bed generated turbulence.

The parameterisation of entrainment in the turbulence-sediment model requires knowledge of the near bed concentration defined as the value at which hindered settling becomes important. This knowledge can be obtained from in situ measurements. In practice, this is difficult due to the problems connected with making field measurements close to the bottom. However, further emphasis on the measurements of sediment profiles will be a major step forward. This should be complemented with longer time series in order to address the spring neap behaviour.

The entrainment description is relevant to use when a pronounced CBS layer exists and significant horizontal movements of sediment are related to this phenomenon. This may be the case if the near bottom concentration values are in the order of 1-10 g/l Mehta (1989) or Winterwerp (1999). However, the concentration changes with the hydrodynamic conditions and the supply of sediment. Thus for practical applications, the entrainment description must be embedded into a modelling concept that can describe the variation from neap to spring. This also requires the inclusion of consolidation processes and the erosion of a denser layer. The obvious next step will therefore be to combine the entrainment description with the erosion model for solid beds in order to be able to simulate the complete neap-spring cycle.

Theoretically, a method to determine the near bed sediment properties is crucial. However, this is essentially the known problem of determining the growth of strength of the newly deposited bed, a problem not yet fully resolved (Winterwerp (1999)).

ACKNOWLEDGEMENTS

The paper is part of Task E, Applied Modelling of MAST III - COSINUS Project. The authors would like to express acknowledgement to colleagues in the COSINUS project group for help with data and many fruitful and inspiring discussions. The comments and suggestions of the two reviewers are also highly appreciated. This work was co-financed by the European Commission, DG XII, MAST III - COSINUS Project, Contract No. MAS3-CT97-0082.

REFERENCES

Arithurai, C.R., (1974), A finite element model for sediment transport in estuaries, *Ph.D Thesis, University of California, Davis.*

Berlamont, J.E. & E.A. Toorman (editors) (2000), COSINUS Final Scientific Report, Hydraulics Laboratory, K.U. Leuven.

Bruens, A., Kranenburg, C. and Winterwerp, J. (1999), Experiments on CBS dynamics, *Book of abstracts, MAST III - Cosinus, 2nd annual general meeting.*

Brøker I., Johnsen J., Lintrup M. and Møller J.S. (1994), The spreading and dredging of soils during construction of the Denmark Sweden link, Proceedings *Coastal Engineering, ASCE, Kobe Japan.*

Burchard, H. and Petersen, O. (1998a), Model of turbulence in the marine environment, *Journal of Marine Systems,* (21), 29-53.

Burchard, H., Petersen, O. and Rippeth, T. P. (1998b), Comparing the performance of of the Mellor-Yamada and the k-ε two-equation turbulence models, *Journal of Geophysical Research*, (103) C5, 10543 - 10554.

Christie M. (2000), Tamar data for COSINUS model validation. Personal communication.

Dyer, K. R., Bale A.J., Christie M.C., Feates N., Jones S. and Manning A.J. (2000), The Dynamics of Suspended Sediment in an Estuarine Turbidity Maximum, Proceedings *INTERCOOH-2002, Elsevier, Coastal and Estuarine Fine Sediment Processes*, ed. J. C. Winterwerp and C. Kranenburg, this volume.

Kranenburg, C. (1984), Wind induced entrainment in a stably stratified fluid, *Journal of Fluid Mechanics,* (145), 253-273.

Kranenburg, C. and Winterwerp, J. C. (1997), Entrainment of fluid mud layers, I: Entrainment model, *J. Hydraulic. Engineering,* ASCE, (123) 6, 504-511.

Krone R.B. (1962), Flume studies of the transport of sediment in estuarial processes, Final report Hydraul. Eng. Lab. Sanit. Eng. Res. Lab., University of California, Berkeley.

Mehta, A. J. (1989), On estuarine cohesive sediment suspension behaviour, *Journal of Geophysical Research*, (94) C10, 14303-14314.

Mehta, A. J., Hayter E.J., Parker W.R. Krone R.B. and Teeter A.M. (1989), Cohesive sediment transport I: Process description, *Journal of Hydraulic Engineering,* (115) 8, 1076-1093.

Odd, N.V.M., and Cooper A.J. (1989), A two-dimensional model of the movement of fluid mud in a high energy turbid estuary, *Journal of Coastal Research*, 5.

Petersen, O., Vested H.J., Manning, A., Christie, M.C. and Dyer, K.R. (2000), Numerical modelling of mud transport in the Tamar Estuary, Proceedings *INTERCOOH-2002, Elsevier, Coastal and Estuarine Fine Sediment Processes*, ed. J. C. Winterwerp and C. Kranenburg, this volume.

Spearman J. and Roberts W. (2000), Parameterization of flocculation models for applied sediment transport modelling, Proceedings *INTERCOOH-2002, Elsevier, Coastal and Estuarine Fine Sediment Processes*, ed. J. C. Winterwerp and C. Kranenburg, this volume.

Teisson, C. (1991), Cohesive suspended sediment transport: feasibility and limitations of numerical modelling, *Journal of Hydraulic Research*, (29) 6.

Teeter, A.M. (1986), Vertical transport in fine grained suspension and newly deposited sediment. Estuarine Cohesive Sediment Dynamics, A. J. Mehta, ed., *Lecture notes on coastal and estuarine studies, Springer Verlag*, (14), 170-191.

Turner, J. S. (1973), *Bouyancy effects in fluids,* Cambridge University Press.

Violeau, D., Bourban S., Cheviet C., Markofsky M., Petersen O., Roberts W., Spearman J., Toorman E., Vested H.J. and Weilber H. (2000), Numerical simulation of cohesive sediment transport, Intercomparison of several numerical models, Proceedings *INTERCOOH-2002, Elsevier, Coastal and Estuarine Fine Sediment Processes*, ed. J. C. Winterwerp and C. Kranenburg, this volume.

Winterwerp, J.C. and Kranenburg, C. (1997), Entrainment of fluid mud layers. II: Experiments and model validation. *J. of Hydraulic Engineering,* ASCE, **123**(6), 512-519.

Winterwerp, J.C. (1999), *On the dynamics of high-concentrated mud suspensions.* Brinkman & Ammerlaan, Delft. PhD thesis Delft University of Technology.

Zriek, D.A., Krishnappan, B.G., Germaine, J.T., Madsen, O.S. and Ladd, C.C. (1998), Erosional and mechanical strengths of deposited cohesive sediments. *Journal of Hydraulic Engineering,* ASCE, **124**(11), 1076-1085.

Fine Sediment Dynamics in the Marine Environment
J.C. Winterwerp and C. Kranenburg (Editors)

Simulation of biogenic sediment stabilisation by heterotrophic bacteria in an annular flume

Joern Prochnow [a], Christoph Schweim [a], Juergen Koengeter [a]

[a] Institute of Hydraulic Engineering and Water Resources Management, Aachen University of Technology, Kreuzherrenstrasse, 52065 Aachen,

The stabilising effect of assemblages of blue-green algae and diatoms on cohesive sediments in shallow estuarine waters has been investigated in detail for a number of years now. These biofilms can decrease the roughness and hence the erodibility of fine materials. In the case of sediments in reservoirs the occurence of phototrophic biogenic stabilisation is unlikely because the amount of sunlight penetrating the upper layers of the water body is too small to enable phototrophic growth. On the other hand information on heterotrophic sediment stabilisation is still lacking. This paper describes an experimental setup suitable for the simulation of heterotrophic stabilisation of cohesive sediments in the lab. It features annular flume tests and the use of surrogate materials (china clay, activated sludge, silica flour). The use of the FISH-technique (in situ hybridisation with fluorescent-labelled genetic probes) for the identification of more or less active bacteria in sediments is discussed

KEYWORDS: Biogenic Stabilisation, EPS, Annular Flume, Activated Sludge, FISH

1. INTRODUCTION

The biogenic influence on the erodibility of cohesive deposits in lowland rivers and reservoirs can not be neglected. Both, negative and positive effects of biota on transport processes of fine-grained sediments have been investigated and described (Baumgartner, 1992; Eckmann & Nowell, 1984; Grant & Gust, 1987; Humann, 1996; Heinzelmann & Wallisch, 1991; Montague, 1986; Spork et al., 1997). In most of these publications the authors focussed on stabilisation effects induced by assemblages of photoautotrophic microorganisms like blue-green algae and diatoms. Furthermore, most attention has been paid to microbial communities in shallow estuarine waters. The work presented in this paper is conversely engaged in sediment stabilization caused by heterotrophic microorganisms in rivers and artificial lakes.

1.1. Basics on biogenic stabilisation

The most important feature of micro-organisms, as far as stabilisation of sediments is concerned, is the exudation of extracellular polymeric substances (EPS). These high molecular-weight mucous secretions range from tight capsules to amorphous EPS-species that form the matrix associated with sediment particles, detritus and other surfaces (Decho,

1990). Most EPS are composed of polymer chains with high molecular weight that have been characterised as being predominantly polysaccharide with lesser amounts of amino acids, uronic acids and glycoproteins (Decho, 1993; Pavoni et al., 1972). In the following the key functions of EPS to the microbial cell are summarised. The exopolymers create a microenvironment in which nutrients and exoenzymes are concentrated. They serve as protection against grazing and chemical hazards. In particular, EPS are important in the attachment of microbial cells to surfaces. The physical properties of the EPS matrix change with its water content. By absorption of water the less hydrated gel-like state can be transformed to a loose slime and finally to a colloidal solution (Decho, 1994). EPS play a major role in biogenic stabilisation of cohesive sediments, because the organic coating improves the cohesion of sediment particles and aggregates. This cohesive effect has been described by Vos et al. (1988). The microbial exopolymers can form a dense biofilm on the sediment surface and moreover fill up inter-particle voids. These processes result in the formation of a compact, elastic sediment showing a significant reduction of bottom roughness and an increase in the threshold of sediment motion (Fuehrboeter, 1983). The ultrastructure of these microbial assemblages and the binding mechanisms have been described elsewhere (Grant et al., 1986; Hoagland et al., 1993).

Filamentous organisms like cyanobacteria, fungi and bacteria can build inter-particle networks that extend to deeper sediment layers. This "network effect", as described by Vos et al. (1988) is often caused by the mucilage threads left behind by motile organisms migrating through the sediment and not by the cells themself.

Biogenic stabilisation is a dynamic process because the growth rate and EPS-production of a given microbial community depend on seasonal, climatic and biochemical parameters. Underwood & Paterson (1993) have reported seasonal changes in the sediment stability of intertidal sediments. In case of nutrient depletion or a high carbon to nitrogen ratio the amount of EPS secreted can increase significantly and the composition of the secreted substances changes (Decho, 1993). Furthermore high loads of hazardous substances can encourage EPS-secretion. The EPS-matrix is exposed to microbial breakdown, affecting the composition and molecular weight of the polymeric chains. On the other hand high-MW (MW = molecular weight) and low-MW compounds of the DOM (Dissolved organic matter) readily bind to exopolymers, especially in case of microbial communities in nutrient-poor environments (Lock et al., 1984).

1.2. Autotrophic and heterotrophic organisms: Differences in stabilisation effects

When comparing biogenic stabilisation effects caused by autophototrophic organisms and heterotrophic bacteria it has to be considered that both phenomena can take place in the same environment, as far as the trophogenic layer, in which a sufficient amount of sunlight penetrates the water body, is concerned. In these shallow waters the EPS-matrix secreted by algae is inhabited by dense assemblages of bacteria breaking down the algal exopolymers (Prochnow, 1998). In the tropholytic layer of a water body only hetrotrophic growth takes place. This means that the structure and composition of the EPS secreted differs from the polymers exudated in the trophogenic layer. On the other hand it can be neglected that the basic principles of EPS exudation and particle binding are very similar because the cellular mechanisms of EPS-Production and the composition of the material secreted are analogous.

Only limited information on biogenic effects on sediment transport caused by heterotrophic organisms is available (Dade et al. 1990; Skafel & Krishnappan, 1998). This may be due to the fact that consistent investigations in laboratories require an environment free of bacteria and germs or at least accurate information on the background level of microbial growth. Especially in larger scale experiments this demand is difficult to meet. But one has to realise that cohesive sediments in streams and reservoirs are frequently found in the tropholytic zone of the water bodies. That is why focussing on hetrotrophic biogenic stabilisation phenomena is considered to be rather promising.

Most hetrotrophic microorganisms associated with cohesive mud deposits generally develop into small chains and clumps. First the bond formed holding the organisms together is weak and the EPS hull enclosing the cells is almost not present. Furthermore most of the bacteria are motile after mitosis has taken place. In the later stages of development the secretion of exopolymers increases and the composition of the exudates changes. In contrast to algal exopolymers the concentration of nutrients from the relatively nutrient-poor water is an important attribute of bacterial EPS (Lock et al., 1984) In particular detrital DOM and POM (Particulate organic matter) is absorbed and utilised by the bacteria. While small-MW DOM compounds are directly taken up by the cell to a large extend, high-MW compounds absorb to the matrix substances and are hydrolysed by exoenzymes (Pearl, 1975). Especially nitrogenous compounds such as proteins and amino acids are concentrated within the EPS-hull (Joyce & Dugan, 1970).

While photoautotrophic growth depends mainly on the amount of light energy caught by the organisms, hetrotrophic bacteria need to utilise carbon and nitrogen sources. Furthermore, the biochemical pathways used for the breakdown reactions depend on the availability of dissolved oxygen. This leads to a stratification of the sediment (Doelman, 1995). In the upper layer aerobic processes like respiration and nitrification take place, but anaerobic degradation processes like denitrification, sulphate reduction and methanogenesis may occur within a distance of a few millimeters.

A survey into stabilisation effects caused by heterotrophic organisms consequently has to encompass monitoring of dissolved oxygen levels and the availability of carbon and nitrogen sources. As POM and DOM are consumed by the organisms both parameters can not be used to correlate the biological status of the bed to hydraulical parameters like the critical shear stress of erosion. On the contrary one has to obtain information on the taxonomic structure of the microbial community, its activity and growth rates.

2. MATERIAL & METHODS

2.1. Simulation of biogenic influences in the lab

We decided to use a surrogate material rather than collecting samples of natural cohesive sediments from the field. Investigations into soft sediment erodibility have revealed the problems involved in sampling procedures, subsequent storage of the samples and changes in chemical parameters, for instance the dissolved oxygen concentration (Skafel & Krishnappan, 1998). In laboratory studies physical and chemical parameters like water quality, temperature and light penetration can be held constant. In the complex field of biogenic influences on sediment transport the application of a simplified experimental set-up seems to be reasonable. It has to be accepted that not all aspects of biogenic stabilisation can be simulated in the lab. Therefore, the results obtained from these laboratory surveys have to be evaluated cautiously.

2.2. Natural sediments and substitutes

Soft sediments in natural rivers and lakes are composed of cohesive materials like clay and silt, the non-cohesive sand fraction and organic compounds like organisms, EPS and detritus. An artificial sediment must contain materials representing the most important properties of fine-grained deposits (Spork, 1997).

A replacement for the silt and clay fraction that has successfully been used in various investigations into soft sediment transport (Parchure, 1984; Mehta & Partheniades, 1973; Prochnow et al., 1999a) is Kaolinite. We used a china clay (Chinafill K75, Amberger Kaolinwerke, Germany) with low water content (1%), moderate loss on ignition (11.4%) and a uniform grain size distribution (99,9% < 15 microns, d_{50} = 3.6 microns). Small amounts (up to 25% on the basis of the dry weight of the fractions) of silica flour (Quarzwerke Frechen, Germany) were added to simulate the non-cohesive sand fraction.

Activated sludge was chosen as a replacement for the organic fraction of natural cohesive sediments. A local wastewater treatment plant (Klaerwerk Aachen-Soers, Germany) provided thickened sludge from the centrifuges. It was stored in a tank with a capacity of 1 m^3. The dry weight of the material amounted to 6.5 to 8 gl^{-1}. To ensure that the properties of the activated sludge remained constant, the content of the tank was aerated and stirred at a high velocity continuously. It is noteworthy that as a result of the permanent stirring limitations in floc size occurred. Organic matter (Acetol 30 = industrial product, high ethanol content, Rethmann Lippewerk Recycling GmbH, Germany) was added at intervals of three days to provide nutrients for the microbial community.

Because of the high dissolved oxygen level in the aeration basins aerobic and facultative aerobic bacteria are dominating in the sludge community. On the other hand organisms that are usually associated with soft deposits in natural waters like *Lamprocystis rosea-pericina*, *Spirillum undulans*, *Zoogloea ramigera* are frequently found in activated sludge samples (Lorenz & Lorenz, 1995). Furthermore, the EPS production in the aeration basins is supposed to be small because of high nutrient levels and accelerated growth. As soon as the sludge enters the secondary clarifiers the oxygen concentration drops and nutrient supply is interrupted. That is why EPS-secretion increases and anaerobic metabolism is favoured. During the storage of the sludge sample the volume and dry weight remained nearly constant. Monitoring of the sludge by light microscopy revealed a dense floc-structure and a prevalence of filamentous bacteria and fungi.

The resulting mixture used in the flume tests showed a grain size distribution and levels of DOC and POC very similar to natural deposits available from reservoirs (Spork, 1997) proving the suitability of the substitute material. The composition of the artificial sediment was diversified in order to determine the effect of higher and lower biological and organic loads.

2.3. Flume tests

The annular flume at the Institute of Hydraulic Engineering and Water Resources Management of Aachen University of Technology used in this investigation (Fig. 1) has been described in detail elsewhere (Spork et al., 1994; 1996; Spork, 1997; Prochnow, 1998; Schweim et al. 2000).

It comprises of a ring-shaped channel with a width of 0.25 m, a mean diameter of 3.25 m and a lid of the same size touching the water surface.

Fig. 1: The annular flume at Aachen University of Technology, top lid view

Flume and top lid move in opposite directions generating an endless flow in tangential direction. This type of flow is most suitable for investigations into transport processes of cohesive sediments. The speed ratio of both parts can be altered in order to ensure uniform bottom shear stresses in case of erosion tests or minimal secondary currents in case of deposition experiments. The flume resides in an air-conditioned room, because differences in temperature are supposed to modify flocculation characteristics of the suspension (Parchure, 1984; Burt, 1986).

A mixture of the substitute materials was suspended using de-ionised water. The concentration of solids amounted to 10g/l . Small amounts of Sodium Nitrate were added to suit the salinity of the suspension to the concentration level of natural waters again. This procedure had to been chosen to overcome the problem of fluctuant metal loads in the tap water (Prochnow, 1998). The suspension was mixed thoroughly at high speeds for at least 30 min to ensure a uniform floc size and to allow formation of organic-mineral aggregates to take place. Then the flume was stopped and most of the solids deposited over a 2-hour period. Subsequent consolidation processes continued for up to 12 days at a constant temperature of 15° C in the dark.

In order to determine the threshold value for erosion to commence, the bottom shear stress was increased stepwise. A sample of the suspension was withdrawn at the end of each shear stress step, dried at 105° C and weighted to calibrate the turbidity-meter recording continuous readings of the suspension load. In addition these samples were analysed for DOC and POC-content.

2.4. Biological and hydraulic parameters

The transport characteristics of cohesive deposits can be described by the critical bottom shear stress, the characteristic bottom shear stress and averaged erosion rates (Partheniades, 1984; Parchure, 1984). While the critical bottom shear can be derived easily from concentration-time plots, the characteristic shear stress is calculated by plotting the final concentration of solids in suspension in each shear stress step against the bottom shear stress as proposed by Spork (1997). The average erosion rate was calculated for each shear stress step. The values below and above the characteristic bottom shear stress are averaged

Various biological parameters have been utilised to correlate the biological status of the cohesive bed to its hydrodynamic properties. In numerous publications on biogenic stabilisation the photopigment-content of autotrophic organisms has been employed (Grant & Gust, 1987; Madsen et al., 1993; Rhese, 1992). This parameter can not be used in case of heterotrophic organisms. Other established biological parameters are the DOC and POC-contents (Humann, 1996; Prochnow, 1999). Particulate species of organic carbon and nitrogen and the C to N ratio were determined by means of an elemental analyser (Carlo Erba Strumentatione, Model No. 110) as described by Kirsten (1983). DOC-content was analysed following the experimental procedure proposed by Dubois et al. (1956) and altered by Liu et al. (1973). Counts of bacteria and other inhabitants of the sediment have been applied by Skafel & Krishnappan (1998) and others. But one has to realise that the staining and counting techniques available (Trolldenier, 1993; Schuller, 1991) do not permit to quantify microbial activity. Even a differentiation of dead and living cells is not possible. All procedures involving growing of the organisms are not suitable because 90% of the species in a natural cohesive sediment can not be grown in the lab (Castenholz, 1994). That is why the application of recent methods, utilising the progresses in the field of molecular genetics, i. e. improved in-situ-hybridisation techniques (Amann et al., 1995; Christensen & Poulsen, 1994; Hahn et al. 1992), is considered and will be elucidated later on.

3 RESULTS

3.1 Flume tests

Results of the preliminary flume tests are given in Fig.2-Fig.6. Fig. 2 shows the concentration-time plots of three selected experiments. Sediments containing activated sludge are compared to abiotic sediments. The stabilising effect generated by the sludge microbes is easy to recognise. The differences in the initial concentration show the changes in the turbidity of the sediment-free water body caused by free bacteria that became more and more abundant while consolidation took place. While the abiotic sediment shows a significant gap after reaching the critical bottom shear stress (erosion commences) and before reaching the characteristic bottom shear stress (substantial erosion), the biogenic stabilised sediments show a rapid increase in the suspended sediment concentration as soon as the threshold value is exceeded. Once the surface layer, containing most of the microbial biomass, is removed the unprotected remainders can be easily eroded. This finding again reflects the stratification of the populated sediment. The final concentrations for all three tests are very similar. Therefore it can be concluded that significant stabilisation processes in deeper layers of the sediment did not take place.

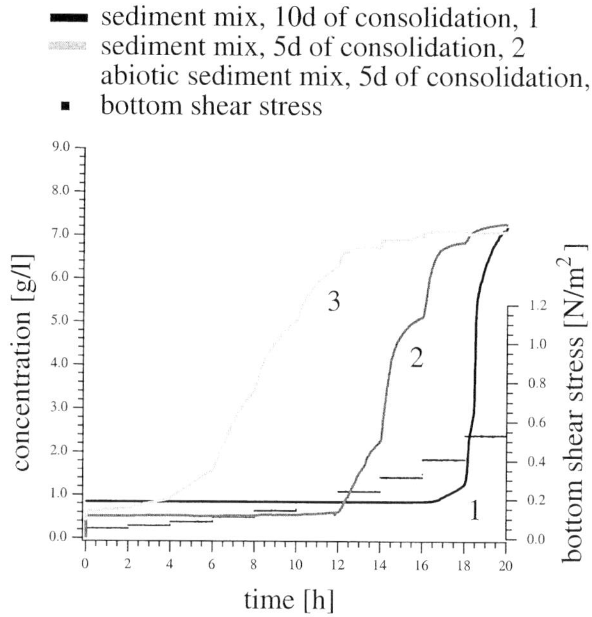

Fig. 2: concentration of suspended solids in annular flume experiments: sediment mix: mixture of activated sludge (25%), china clay (50%) and silica flour (25%); abiotic sediment mix: mixture of china clay (75%) and silica flour (25%)

Figure 3 displays the dependency of the critical bottom shear stress of erosion and the consolidation-time. The scattering of the values indicates that fluctuations in the properties of the activated sludge in terms of number and activity of bacteria and composition of the microbial community occurred. However, a trend towards increasing critical bottom shear stresses can be observed when averaging all flume tests. These findings correspond to the values for the critical bottom shear stress that were obtained in tests with blue-green algae and diatoms using the same experimental set up (Prochnow, 1999; Spork, 1997). The missing discrepancy in the two series with different sediment mixes indicates that a higher portion of cohesive materials in the mixture does not result in an increase in the threshold of sediment motion. Similar findings were obtained in case of the characteristic bottom shear stress (Fig. 4). Especially the results for the sediment containing no non-cohesive material show a decrease of this parameter with increasing consolidation time. It can be concluded that

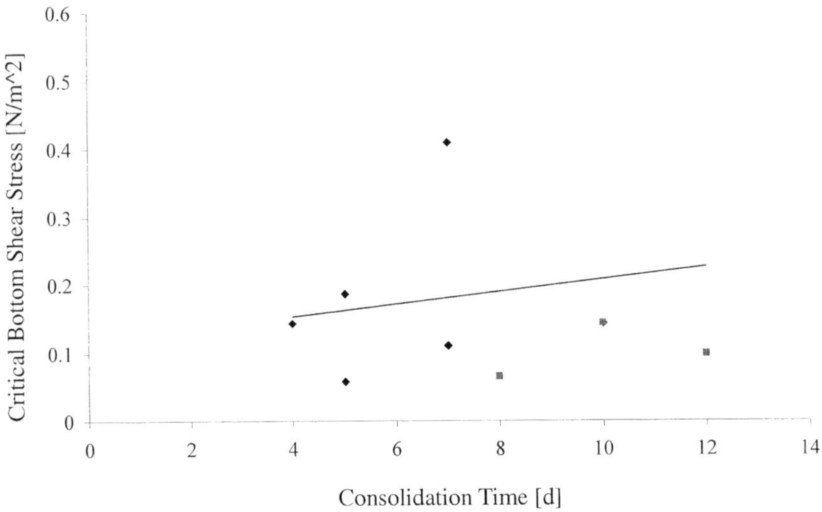

Fig 3: Changes in the critical shear stress of erosion with consolidation time; diamonds: sediment mix: mixture of activated sludge (25%),china clay (50%) and silica flour (25%); squares:sediment mix: mixture of activated sludge (25%), china clay (75%)

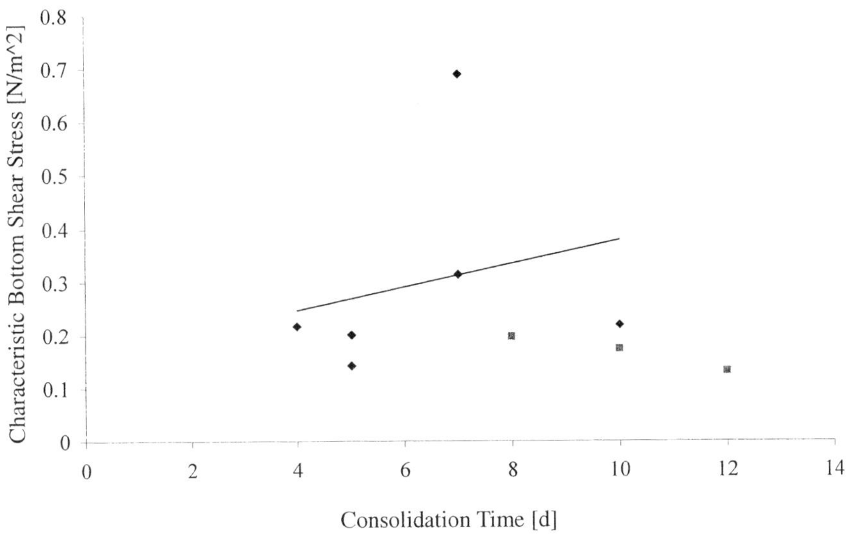

Fig 4: Changes in the characteristic shear stress of erosion with consolidation time; diamonds: sediment mix: mixture of activated sludge (25%), china clay (50%) and silica flour (25%); squares: abiotic sediment mix

microbial metabolism induces changes in the cohesiveness of particles. In particular the formation of gas bubbles leads to a looser sediment structure.

The dependency of the averaged erosion rates (below and above τ_{ch}) and the consolidation-time is shown in figures 5 and 6. The lower values obtained in case of longer consolidation periods, particularly in the series of tests without non-cohesive materials, gives a clue that further experiments should encompass a extension of consolidation and growth periods.

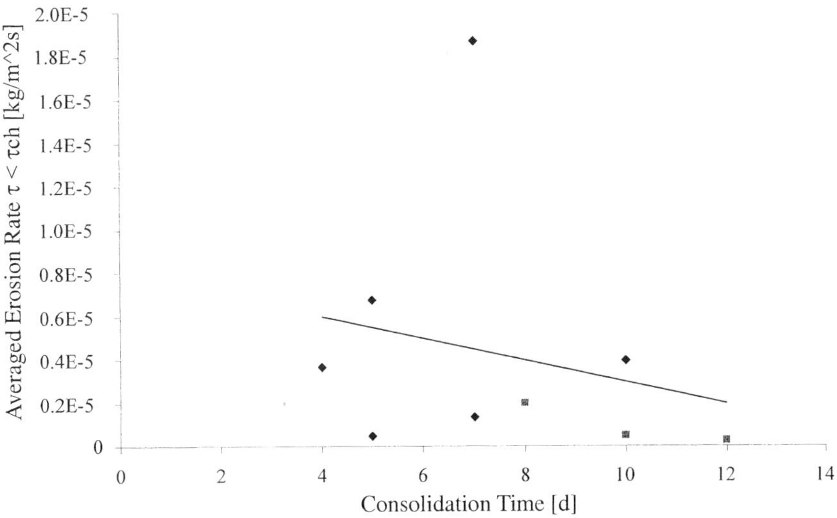

Fig 5: Changes in the averaged erosion rate (below τ_{ch}) with consolidation time; diamonds: sediment mix: mixture of activated sludge (25%), china clay (50%) and silica flour (25%); squares: sediment mix: mixture of activated sludge (25%), china clay (75%)

Compared to erosion rates and the threshold of erosion found in case of natural cohesive sediments from artificial lakes (Spork, 1997) there is no significant difference to the mixture of surrogate materials we used. On the other hand the fluctuations in the activated sludge composition requires further investigations.

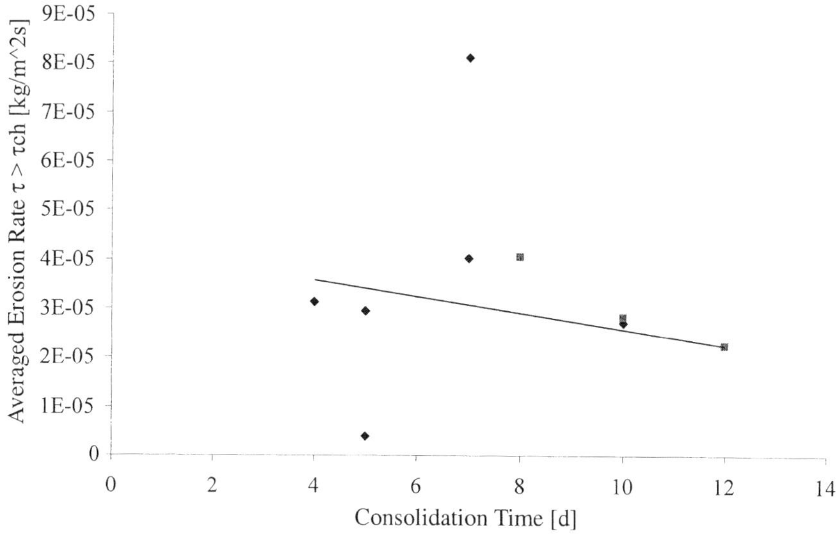

Fig 6: Changes in the averaged erosion rate (above τ_{ch}) with consolidation time; diamonds: sediment mix: mixture of activated sludge (25%), china clay (50%) and silica flour (25%); squares: sediment mix: mixture of activated sludge (25%), china clay (75%)

3.2 Biological parameters

A proper evaluation of the findings for the hydraulic parameters requires biological parameters to be correlated to them. The suitability of biological parameters has been discussed in the opening chapter.

Analysis of the samples of the suspended solids for POC and DOC-content (data not shown) revealed no dependency of hydraulic and biological parameters. One has to take the microbial uptake of dissolved and particulate species of organic carbon into account. Due to the rapid decrease in the dissolved oxygen level, numerous aerobic bacteria will show reduced activity and finally die off or turn to dormancy, all the more the longer the consolidation period lasts. This means that the consumption of carbon and nitrogen sources is not constant, it decreases in the beginning and rises as the microbial community adapts to the new conditions. The duration of both phases of growth has not been determined yet.

Further investigations will focus on the taxonomic structure and the activity of the sediment inhabitants. A reliable method to gain accurate information on the composition of a microbial community is FISH (in-situ hybridisation using genetic probes labelled with a fluorescent marker). Whole cells are hybridised, which means that the expensive DNA-isolation and purification techniques are not needed. Furthermore the methodology is especially well suited in case of nutrient rich environments like activated sludge (Wagner et al., 1994; Weiss et al., 1996) and allows a differentiation of organisms showing a more or less active metabolism (Christensen et al., 1999).

As the uronic acids in the EPS matrix are supposed to be the compound that is responsible for the coherence of mineral and organic particles (Leppard, 1997) this fraction of the exopolymers will be determined in future investigations. Furthermore the dissolved oxygen levels will be monitored to gain additional information on the growth characteristics of the activated sludge microbes.

4. CONCLUSIONS

From the present stage of the presented work it can be concluded that the experimental set up is suitable for investigations into biogenic stabilisation effects caused by heterotrophs. The limited data available indicate that simulation of these phenomena in the laboratory requires a set of suitable biological parameters for accurate evaluation.

On the other hand the scattering in the hydraulic parameters suggests that the properties of the activated sludge component in the sediment mixture have to be controlled more thoroughly. It seems that the length of storage of the material influenced the experimental results. Consequently the procedure of storage has to be improved in terms of constant aeration rates and constant nutrient levels. The amount of activated sludge added to the sediment mixture has to be varied in order to meet the natural levels of particulate and dissolved organic carbon.

However, the results presented are in concordance with data available from other publications on biogenic sediment stabilisation (Prochnow et al., 1999; Spork, 1997; Humann, 1996).

The use of POC and DOC as biological parameters requires further investigation. It has to be taken into account that in natural cohesive sediments and in the surrogate mixture we used only a small portion of the organic carbon species will contribute to sediment stabilisation. Furthermore detailed information on the composition of natural POC and DOC-pools has to be obtained.

The utilisation of artificial sediments and an annular flume facilitates consistent investigations into biogenic stabilisation phenomena in the laboratory. The problems associated with sampling procedures and storage of natural cohesive deposits can be overcome. Additional effort has to be spent on holding up constant activated sludge properties.

ACKNOWLEDGEMENT

The authors would like to thank the Deutsche Forschungsgemeinschaft (German Research Foundation, DFG) for funding the research project over the last four years.

REFERENCES

Amann, R.I.; Ludwig, W.; Schleifer, K.-H. (1995): Phylogenetic Identification And In-Situ Detection Of Individual Microbial Cells Without Cultivation. Microbiol. Rev. 59, Pp. 143-169

Baumgartner, B.(1992): Oberflächenrauheit Und Erosionsstabilität Biogen Verfestigter Sandsohlen In Flußstauhaltungen, Teil 2: Vorkommen Und Aufwuchsdynamik Benthischer Diatomeen Einer Main-Staustufe In Abhängigkeit Von Strömungsgeschwindigkeit, Wassertiefe Und Nährstoffparametern, DFG-Abschlußbericht, Technische Hochschule Darmstadt

Burt, T. N. (1986): Field Setteling Velocities Of Estuarine Muds; In: Mehta, A.J. (Ed) Estuarine Cohesive Sediment Dynamics; Lecture Notes On Coastal And Estuarine Studies, Vol. 14; Springer, Berlin Heidelberg New York, Pp. 126-150

Castenholz, R. W. (1994): Microbial Mat Research: The Recent Past And New Perspectives, In: Stal, L.J.; Caumette, P. (Eds.) Microbial Mats, Proceedings Of The NATO Advanced Research Workshop On Structure, Development And Environmental Significance Of Microbial Mats, Arachon, France; NATO ASI Series, Vol. G35 Springer, Berlin Heidelberg New York, Pp.3-18

Christensen, H.; Poulsen, L.K. (1994): Detection Of Pseudomonas In Soil By RNA Targeted In-Situ Hybridization Technique. Soil Biol. Biochem. 26, Pp. 1093-1996

Christensen, H.; Hansen, M.; Sorensen, J. (1999): Counting And Size Classification Of Active Soil Bacteriae By Fluorescence In Situ Hybridization With An Rrna Oligonucleotide Probe. Appl. Environ. Microbiol. 65 No. 4, 1753-1761

Dade, W. B.; Davis, J. D.; Nichols, P. D.; Nowell, A. R. M.; Thisle, D.; Trexler, M. B.; White, D. C. (1990): Effects Of Bacterial Exopolymer Adhesion On The Entrainment Of Sand; Geomicrobiology Journal, Volume 8, Pp. 1-16

Decho, A. W. (1990): Microbial Exopolymer Secretions In Ocean Environments: Their Role(S) In Food Webs And Marine Processes; Oceanogr. Mar. Biol. Annu. Rev. 28, Pp. 73-153

Decho, A. W. (1993): Exopolymers In Microbial Mats: Assesing Their Adaptive Roles; In: Stal, L.J.; Caumette, P. (Eds.) Microbial Mats, Proceedings Of The NATO Advanced Research Workshop On Structure, Development And Environmental Significance Of Microbial Mats, Arachon, France; NATO ASI Series, Vol. G35 Springer, Berlin Heidelberg New York, Pp.214-219

Decho, A. W. (1994): Molecular-Scale Events Influencing The Macroscale Cohesiveness Of Exopolymers; In: Krumbein, W.E.; Paterson, D.M.; Stal, L.J. (Eds.) Biostabilization Of Sediments; Bibliotheks- Und Informationssystem Der Universität Oldenburg, 1994, Pp.135-148

Doelman, P. (1995): Microbiology Of Soils And Sediments, In: Biogeodynamics In Soils And Sediments, Springer, 1995, Pp. 31-52

Dubois, M.; Gilles, K. A.; Hamilton, K. J.; Rebers, P. A.; Smith, F.(1956): Colorimetric Method For Determination Of Sugars And Related Substances; Analytical Chemistry, 28, Pp. 350 - 356

Eckmann, J. E.; Nowell, A. R. M. (1984): Boundary Skin Friction And Sediment Transport About An Animal Tube Mimic; Sedimentology 31, Pp. 851-862

Fuehrboeter, A. (1983): Über Mikrobiologische Einflüsse Auf Den Erosionsbeginn Bei Sandwatten; Wasser Und Boden, 1983, Heft 3, Pp.106-116

Grant, J.; Bathman, U. V.; Mills, E. L. (1986): The Interaction Between Benthic Diatom Films And Sediment Transport; Estuarine, Coastal And Shelf Sience 23, Pp. 225-238

Grant, J.; Gust G. (1987): Prediction Of Coastal Sediment Stability From Photopigment Content Of Mats Of Purpur Sulfur Bacteria; Nature, 330, Pp. 244-246

Hahn, D; Amann, R.I.; Ludwig, W.; Akkermans A.D.L.; Schleifer, K.-H. (1992): Detection Of Micro-Organisms In Soil After In Situ Hybridization With Rrna Targeted Fluorescently Labeled Oligonucleotides. J. Gen. Microbiol. 138, 876-887

Heinzelmann, C. H.; Wallisch, S. (1991): Benthic Settlement And Bed Erosion, A Review; Journal Of Hydraulic Research Vol. 29, No.3

Hoagland, K. D.; Rosowski, J. R.; Gretz, M. R.; Roemer, S. C. (1993): Diatom Extracellular Substances: Function, Fine Structure, Chemistry, And Physiology; J. Phycol. 29, Pp. 537-566

Humann, K. (1996): Der Einfluß Des Mikrophytobenthos Auf Die Sedimentstabilität Und Die Schwebstoffbildung Aus Sedimenten Im Elbe-Ästuar, Dissertation, Universität Hamburg

Joyce, G. H.; Dugan, P.R. (1970): The Role Of Floc-Forming In BOD Removal From Waste Water, Dev. Ind. Microbiol., 11, Pp. 377-386

Kirsten, W.J. (1983): Organic Elemental Analysis; Academic Press, London

Leppard, G. G. (1997): Colloids And Surfaces A: Physicochemical And Engineering Aspects, 120, 1.

Liu, D; Wong, P. T. S.; Dutka, B. J. (1973): Determination Of Carbohydrat In Lake Sediment By A Modified Phenol-Sulfuric Acid Method; Wat Res 7, Pp. 741-746

Lock, M. A.; Wallace, R. R.; Costerton, J. W.; Ventullo, R. M.; Charlton, S. E. (1984): River Epilithon: Towards A Structural-Functional Model, Oikos, 42; Pp. 10-22

Lorenz, P.; Lorenz, P. (1995): Einführung In Die Biologisch-Mikroskopische Belebtschlammanalyse, Quelle & Meyer, Wiesbaden

Madsen, K. N.; Nilson, P.; Sundbeck, K. (1993): The Influence Of Benthic Microalgae On The Stability Of A Subtidal Sediment; Journal Of Experimental Biology 170, Pp. 779-787

Mehta, a. J.; Partheniades, E. (1973): Depositional Behaviour Of Cohesive Sediments, Technical Report No. 16, Coastal And Oceoanographic Engineering Laboratory , University Of Florida, Gainesville (FL), March, 1973

Montague, C. L. (1986): Influence Of Biota On Erodibility Of Sediments, In: Mehta A.J. (Ed) Estuarine Cohesive Sediment Dynamics; Lecture Notes On Coastal And Estuarine Studies, Vol. 14 Sringer, Berlin Heidelberg New York, Pp.251-269

Parchure, T. M. (1984): Erosional Behaviour Of Deposited Cohesive Sediments; Dissertation, University Of Forida, Gainesville

Partheniades E. (1986): A Fundamental Framework For Cohesive Sediment Dynamics. In: Mehta A.J. (Ed) Estuarine Cohesive Sediment Dynamics; Lecture Notes On Coastal And Estuarine Studies, Vol. 14 Springer, Berlin Heidelberg New York, Pp. 219-250

Pavoni, J. L; Tenney, M. W.; Echelberger, W. F. (1972): Bacterial Exocellular Polymers And Biological Flocculation, Journal WPCF, Vol. 44, No.3, Pp. 414-431

Prochnow, J.V. (1998): Laboruntersuchungen Zur Beeinflussung Der Erosionsstabilität Und Flockung Kohäsiver Sedimente Durch Biofilmbildende Diatomeen Und Cyanobakterien; Diplomarbeit, Biologie I, Abteilung Ökologie Des Bodens Sowie Institut Für Wasserbau Und Wasserwirtschaft, RWTH Aachen

Prochnow, J.V.; Spork; V. Jahnke, J.; Schweim, C.(1999): Using Dissolved And Particulate Carbon For The Prediction Of Soft Sediment Erodibility, Proceedings Of The XXIV General Assembly Of The European Geophysical Society, April 19th-23rd, The Hague, The Netherlands

Prochnow, J.V.; Spork, V.; Jahnke, J.; Schweim, C. (1999a): Stabilizing Effect Of Biota On Cohesive Sediments, Proc. IAHR Symposium On River, Coastal And Estuarine Morphodynamics, Genova, Italy, University Of Genova, Department Of Environmental Engineering, 725 - 734

Rhese, C. (1992): Oberflächenrauheit Und Erosionsstabilität Biogen Verfestigter Sandsohlen In Flußstauhaltungen, Teil 3: Aufwuchsdynamik Von Navicula Seminulum Und Kultivierung Repräsentataiver Diatomeenfilme Auf Sandsohlen, DFG-Abschlußbericht, Technische Hochschule Darmstadt

Schuller, E. (1991): Schwermetalle, Mikrobielle Biomassen Und Enzymaktivitäten In Oberböden Von Altlasten, Wasser Und Abfall, 47

Scchweim, C.; J. Zhou; V. Spork; J.V. Prochnow; J. Koengeter (2000): Large Eddy Simulation Of A Lid-Driven Rotating Annular Flume Flow; Fourth International Conference On Hydrodynamics, Iowa, 23-27 July 2000, Accepted For Publication

Skafel, M. G.; Krishnappan, B. G. (1998): A Laboratory Investigation Of Depositional Characteristics Of Mud From An Inland Harbour Using A Rotating Circular Flume, Water, Air And Soil Pollution, 112: Pp. 1-19

Spork, V.; Ruland, P.; Schneider, B.; Rouvé, G. (1994): A New Rotating Annular Flume For Investigations On Sediment Transport; International Journal Of Sediment Research, 9(3), 141-147

Spork V.; Cueppers, A.; Koengeter, J. (1996): The Turbulent 3D Flow Field In A Rotating Annular Flume; Proc. 8Th International Symposiom On Applications Of Laser Techniques To Fluid Mechanics, Lisboa, Portugal

Spork, V.(1997): Erosionsverhalten Feiner Sedimente Und Ihre Biogene Stabilisierung; Dissertation; Mitteilungen Des Instituts Für Wasserbau Und Wasserwirtschaft, RWTH Aachen; 114

Spork V.; Jahnke, J.; Prochnow, J.; Koengeter, J. (1997): Stabilizing Effect Of Benthic Algae On Cohesive Sediments; International Journal Of Sediment Research, 12, No.3, Pp. 399-406

Trolldenier, G. (1993): Fluoreszensmikroskopische Zählung Von Bodenbakterien, In: Bodenbiologische Arbeitsmethoden; Schinner, F.; Öhlinger, R.; Kandeler, E.; Margesin, R. (Eds.), 2. Aufl., Springer, Berlin Heidelberg New York, Pp. 16-21

Underwood, G. J. C.; Paterson, D. M. (1993): Seasonal Changes In Diatom Biomass, Sediment Stabilityand Biogenic Stabilization In The Severn Estuary; Journal Of The Marine Biological Association Of The U.K., 73, Pp. 871-887

Wagner, M.; Erhard, R.; Manz, W.; Amann, R.I.; Lemmer, H.; Wedi, D.; Schleifer, K.-H. (1994): Development Of A Rrna-Targeted Oligonucleotide Probe Specific For The Genus Acinetobacter And Its Application For In Situ Monitoring In Activated Sludge. Appl. Environ. Microbiol. 60,792-800

Weiss, P.; Schweizer, B.; Amann, R.I.; Simon, M. (1996): Identification In Situ And Dynamics Of Bacteria On Limnetic Organic Aggregates (Lake Snow). Appl. Environ. Microbiol. 62, 1998-2005

Vos, P. C.; De Boer, P. L.; Misdorp, R. (1988): Sediment Stabilization By Benthic Diatoms In Intertidal Sandy Shoals; Qualitative And Quantutative Observations; In: De Boer, P.L.; Van Gelder, A.; Nio, S.D.(Eds.), Tide-Influenced Sedimentary Environments And Facies, D. Reidel Publishing Co., Pp. 511-526

Fine Sediment Dynamics in the Marine Environment
J.C. Winterwerp and C. Kranenburg (Editors)

The influence of an extracellular polymeric substance (EPS) on cohesive sediment stability

T. J. Tolhurst[a], G. Gust[b] and D. M. Paterson[a].

[a] Gatty Marine Laboratory, The University of St. Andrews, St Andrews, Fife, UK KY16 8LB
[b] Technical University of Hamburg-Harburg, Ocean Engineering 1, Lämmersieth 72DE-22305 Hamburg

Extracellular polymeric substances (EPS) are secreted by a variety of benthic organisms including diatoms and bacteria. These substances are often considered the primary mechanism by which intertidal benthic organisms increase sediment stability. The stabilisation is caused by physico-chemical interactions between clay minerals and EPS and physical strengthening and gluing by EPS strands. The effects of an EPS (xanthan gum) on the erosion characteristics (threshold and rate) of cleaned cohesive sediments were investigated in the laboratory for the first time using two erosion devices, and compared to *in situ* measurements on natural sediments. The addition of EPS was found to increase the erosion threshold and to decrease the erosion rate of the cohesive sediment, even at the lowest EPS content. Low-temperature scanning electron microscopy (LTSEM) images of the sediment revealed differences in the microstructure and EPS distribution between the five laboratory treatments and natural sediments. This study supports the contention that EPS secretion is partly responsible for the increases in sediment stability caused by many intertidal benthic organisms.

KEYWORDS
EPS, xanthan gum, cohesive sediment, erosion threshold, stabilisation, LTSEM.

1. INTRODUCTION

The contribution of biota to intertidal mudflat sediment dynamics is well-documented (Paterson 1994a and 1997). Many intertidal organisms have been shown to increase sediment stability (defined here as both increased erosion threshold and reduced entrainment rate), including unicellular microalgae such as diatoms (Paterson 1989, Tolhurst *et al.* 1999), bacteria (Grant and Gust 1987) and macrofauna (Meadows *et al.* 1990). The mechanisms of these effects remain at present poorly understood. Most intertidal mudflat organisms secrete extracellular polymeric substances (EPS); these consist of various long chain molecules, the majority of which are polysaccharides (Decho 1990). EPS is often considered the principle way in which benthic organisms such as diatoms stabilise intertidal mudflats (Paterson 1994); although this would appear to be a secondary effect of these secretions, the primary functions include facilitating movement and desiccation resistance (Hoagland et al 1993).

The work of Holland et al. (1974) on the relationship between diatom EPS production and stability indicated the possible importance of EPS in sediment dynamics. Surrogate measures that reflect microbial biomass (e.g. pigment content) have been shown to correlate with sediment stability (Grant and Gust 1987, Paterson *et al.* in press). Correlative information also exists showing that as EPS content increases so does sediment stability (Paterson 1988, Dade et al. 1990, Humann *et al.*, in press). Visualisation of the sediment fabric by low-temperature electron microscopy revealed how extracellular polymeric substances form strands between sediment grains (Chenu and Jaunet 1992, Chenu 1993, Paterson 1995, Defarge, 1997). Moreover, it has been shown that EPS fundamentally affects the physico-chemical properties of cohesive sediments by increasing the molecular attractive forces between clay particles. For example, EPS acts as a catalyst in the flocculation process promoting flocculation, leading to larger flocs which settle more quickly (Eisma *et al.* 1982, Decho 1990). The presence of polysaccharides (EPS) has been shown to increase the tensile strength of clay minerals, by the formation of polymer bridges between the clay particles that increase the strength of inter-particle bonds (Chenu and Guérif 1991). Thus, EPS can affect sediment stability in two ways, by physically binding both cohesive and non-cohesive sediment particles together and by molecular electrochemical interaction with cohesive clay minerals. At present it is not possible to determine the relative importance of these two mechanisms.

Underwood and Smith (1998) have proposed a relationship between the biomass of microphytobenthos and the colloidal EPS content of the sediment, which can explain a mechanistic relationship between measures of chlorophyll *a* and EPS mediated sediment stability. However, the influence of EPS *per se* has not been established for natural cohesive intertidal sediments as it has for sandy sediments (Dade et al. 1990). This constitutes a significant gap in knowledge as most natural cohesive sediments contain microbiota that may have a measurable effect upon the sediment dynamics through EPS production. Indeed there is evidence that diatom blooms can alter estuary wide sediment dynamics (Kornman and De Deckere 1998).

Whilst some studies have shown a positive relationship between EPS content and sediment stability (Amos *et al.* 1998, Tolhurst 1999, Paterson *et al.* in press), there are no abiotic intertidal sediments which could serve as control sediments to which these data can be compared. The influence of EPS is therefore difficult to assess at present as many other factors, such as bulk density, bioturbation, water contents and exposure influence the sediment stability (Humann *et al.*, in press). The goal of this study was to establish such a link by using a controlled sediment-EPS system to assess the influence of EPS without confounding factors, thus identifying the relative contribution of physical and microbiologically mediated stability parameters of natural sediments.

2. MATERIALS AND METHODS

2.1. Erosion devices

Two erosion devices were used in this study, the Microcosm system (Gust 1991, Gust and Muller 1997) and the CSM device (Paterson 1989, Tolhurst *et al.* 1999). The Microcosm (30 cm diameter erosion chamber) generates a spatially homogeneous bottom stress field with a fine structure close to that of natural flows (by a combined suction/stirring action). Both the erosion threshold and the erosion rate can be calculated. The CSM is a vertical jet device that measures the erosion threshold and the Suspension index (S_i) a relative semi-quantitative erosion rate (derived from the slope of the erosion data). Its rapid deployment time facilitates replication.

It is important to note that the results from these two devices are not directly comparable. There are fundamental hydrodynamic differences in the erosion generation mechanism, the Microcosm erodes by the tangential component of the bottom wall shearing stress, whilst the CSM utilises a vertically impinging jet and thus generates a normal and tangential stress component at the interface. The selected threshold criteria were slightly different as well, being 0.05 kg of sediment eroded per m^2 for the microcosm and 0.02 kg m^{-2} for the CSM. It was not the goal of this study to establish a quantitative link between the results of these two devices. Instead, they were used to establish the trend in sediment stability under EPS addition. Two devices were used because the microcosm does not provide the same dynamic range in bottom stresses as the CSM, whilst the CSM does not provide the fine shear stress resolution of the Microcosm. Thus, the microcosm was used to validate the data collected with the CSM.

2.2. Microcosm: Time averaged erosion rate (ε)

As erosion progresses, the change in suspended particulate matter (SPM) in the enclosed water body is used to calculate the erosion rate. The SPM was measured in the microcosm directly by water samples collected from the chamber every ten minutes and indirectly every minute by a nephelometer (turbidity sensor) calibrated to these SPM samples. The erosion rate was calculated for each nephelometer reading giving a time averaged erosion rate over a minute.

$$\varepsilon = (1 / A) \times [(SPM_2 - SPM_1) / (t_2 - t_1)] \quad (1)$$

ε = time averaged erosion rate (kg m^{-2} s^{-1})
A = area (m^2)
SPM = suspended particulate matter (kg/microcosm volume)
t = time (seconds)

2.3. CSM test settings

For the CSM measurements, six replicate measurements were made on each treatment. It is possible to vary the jet pulse duration and the pressure step increments of the CSM, three different tests were used in this study. Test S3: jet fired for 0.3s starting at 0 kPa, incrementing by 3.45 kPa per step, up to 34.5 kPa, then incrementing by 6.9 kPa per step up to 207 kPa. Test S7 jet fired for 1s pressure starting at 0 kPa, incrementing by 2 kPa per step up to 82 kPa. Test S9 jet fired for 1s pressure starting at 0 kPa incrementing by 3.45 kPa per step up to 34.5 kPa, then incrementing by 6.9 kPa per step up to 207 kPa.

2.4. Sediment samples

Cohesive sediment was collected from the Eden estuary Scotland (02°51'00", 56°22'00"), and brought back to the laboratory, sieved and cleaned by adding a solution of 30% v/v hydrogen peroxide. The sediment was then dried in an oven and powdered. Varying amounts of a commercially extracted EPS (Xanthan gum) were added to 350 ml of distilled water and mixed thoroughly with a magnetic stirrer. NaCl was added to give a salinity of 32. When it was fully dissolved the EPS solution was added to 650 g of the dry, cleaned sediment. This was mixed manually and then with an electric mixer, to distribute the EPS throughout the sediment. The result was a porewater-sediment system with a final water content of 35% m/m.

The sediment was left to stand for 30 minutes after mixing, and re-mixed before use. The sediment was poured directly into the Microcosm chamber and levelled with a specially built straight edge, resulting in a flat surface (roughness height < 0.1 mm). For the CSM runs, the sediment was poured to a depth of 1 cm into plastic petri dishes (5 cm diameter) and levelled by tapping gently. A smooth level surface was essential; irregularities acted as micro-sites for erosion by increasing bed roughness and generating isolated wakes which increased the stress on the bed.

The laboratory experiment was designed to mimic natural contents of EPS. Field data revealed an upper range of the natural contents of colloidal carbohydrates of 10 μg of carbohydrate per mg of dry sediment (Tolhurst 1999). Consequently, zero (as a control), 1.25, 2, 5, and 10 μg of EPS per mg of dry sediment were used to mimic the range of naturally occurring carbohydrate contents. Frozen sediment samples were taken from both devices using the Cryolander (Wiltshire *et al.* 1997) to identify the sediment structure using LTSEM (Paterson *et al.* 1986).

To determine the effects of naturally occurring organic matter (including EPS) on sediment stability and structure, field measurements of sediment stability were taken with the CSM in the Eden estuary during February 2000. Cryolander sediment samples were collected from the measurement site, for analysis of carbohydrate, water content and LTSEM. LTSEM images from sediments with a thick diatom biofilm from the Eden (September 1999) and Ems Dollard (July 1996) provide a comparison for periods of significant biotic activity. A sample of the surface 1 cm of the sediment was brought back to the laboratory, this sample was homogenised and split into two treatments to remove the organic matter. The first was cleaned in hydrogen peroxide as outlined above, the second was placed in a furnace at 550°C. The two dry sediment samples were then reconstituted to the original water content so the stability and structure could be investigated.

3. RESULTS

3.1. The Microcosm.

Three replicate measurements of sediment stability were made for each content of EPS using the Microcosm. The operational settings generated maximum shear velocities (U_*) of 1.73 cm s^{-1} and the erosion threshold of the sediment was set to an SPM of 0.05 kg m^{-2}. As EPS content increased, so did the erosion threshold (Table 1). At an EPS content of 1.25 μg mg^{-1}, there was erosion in all three runs, at 2.0 μg mg^{-1} erosion occurred in only one run and at 5.0 μg mg^{-1} there was no detectable erosion for the friction velocity range 0-1.73 cm s^{-1} (Figure 1). The stability of the sediment with EPS was always higher than that without. The erosion rates decreased for each of the four treatments with increasing EPS content (Table 2).

Table 1
Microcosm erosion thresholds, a, b and c are replicate measurements.

	Threshold U* (cm s^{-1})			
EPS (μg mg^{-1})	a	b	c	average ± se
0	1.46	1.51	1.51	1.5 (±0.017)
1.25	1.73	1.73	1.64	1.7 (±0.03)
2	no erosion	1.73	no erosion	-
5	no erosion	no erosion	no erosion	-

Table 2
Microcosm peak erosion rates, a, b and c are replicate measurements.

	Peak erosion rate kg m^{-2} s^{-1}		
EPS (μg mg^{-1})	a	b	c
0	0.0012	0.0015	0.0015
1.25	8.8×10^{-4}	3.81×10^{-5}	7.6×10^{-4}
2	no erosion	4.0×10^{-4}	no erosion
5	no erosion	no erosion	no erosion

3.2. The CSM

Six replicate measurements of sediment stability were made for each EPS content using the CSM. The erosion threshold generally increased with increasing EPS content (Table 3). The relationship between eroding pressure and EPS content was approximately linear; however, an exponential growth function gave a better fit to the data (Figure 2).

The most significant effect of increasing EPS content was not on the thresholds, but on the erosion profiles (Figure 3). The S_i values (relative measure of the average erosion rate after Tolhurst *et al.* 1999) decreased with increasing EPS content (Figure 4) showing that EPS content affects the erosion rate (Note that the S3/9 S_i values are not directly comparable to the S7 S_i values due to the different pressure step increments).

Table 3
The erosion thresholds and S_i values measured with the CSM. A low S_i value indicates a low erosion rate.

	Average Threshold (Nm^{-2})			S_i		
eps (μg mg^{-1})	s3	s9	s7	s3	s9	s7
0	0.64	0.43	0.58	11.96	12.85	23.85
1.25	0.64	0.64	0.73	7.86	8.32	8.98
2	0.85	0.64	0.70	4.84	5.39	6.77
5	1.25	1.00	1.18	2.22	2.72	3.24
10	2.15	1.87	2.03	1.21	0.75	1.39

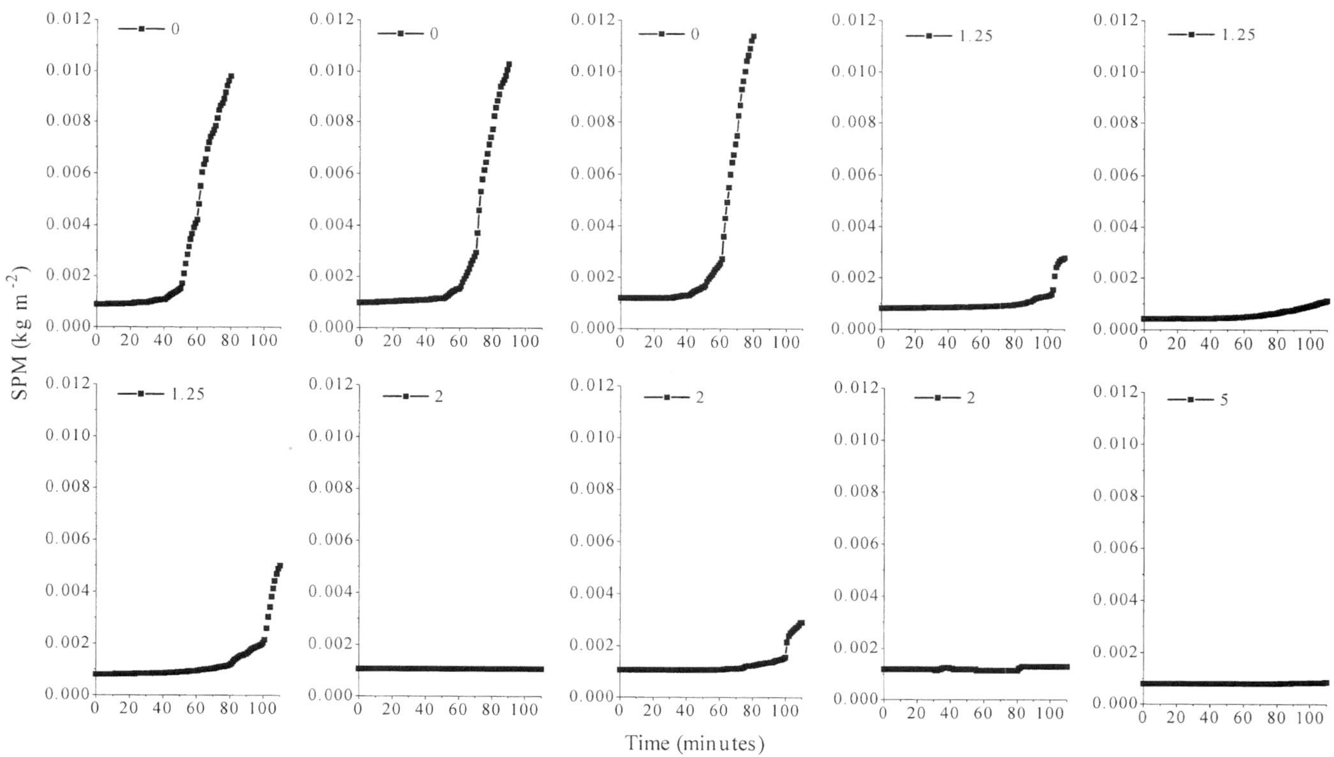

Figure 1 Replicate SPM concentration curves from cleaned cohesive sediment with different EPS contents, measured with the Microcosm. As EPS content increases relative SPM is reduced.

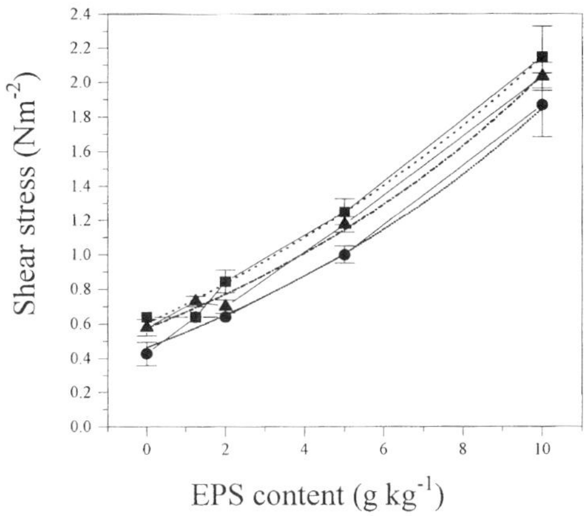

Figure 2. The erosion thresholds of cleaned cohesive sediment with different EPS contents, measured with the CSM using three different test settings (Sand 3 = filled square, Sand 7 filled circle and Sand 9 filled triangle). Best fit curves (dotted lines) were fitted using the exponential growth model y0+Ae^(x/t). As EPS content increases the erosion threshold increases. Error bars are standard error, n = 5.

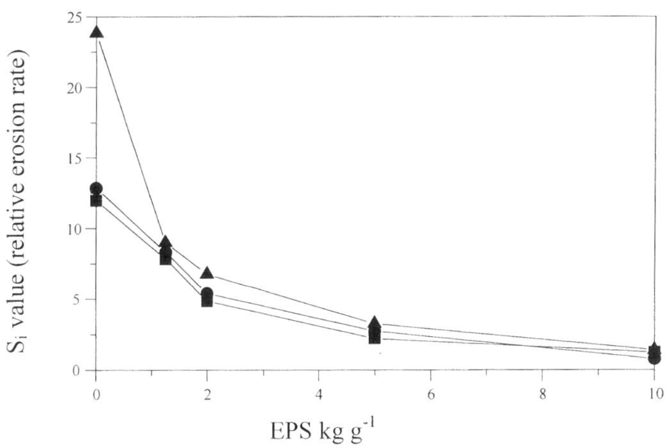

Figure 4. The decrease in S_i value (erosion rate) with increasing EPS content. Note the values from the three different tests are not directly comparable, but nevertheless show the same trends. Sand 3 = squares, Sand 9 = circles and Sand 7 = triangles.

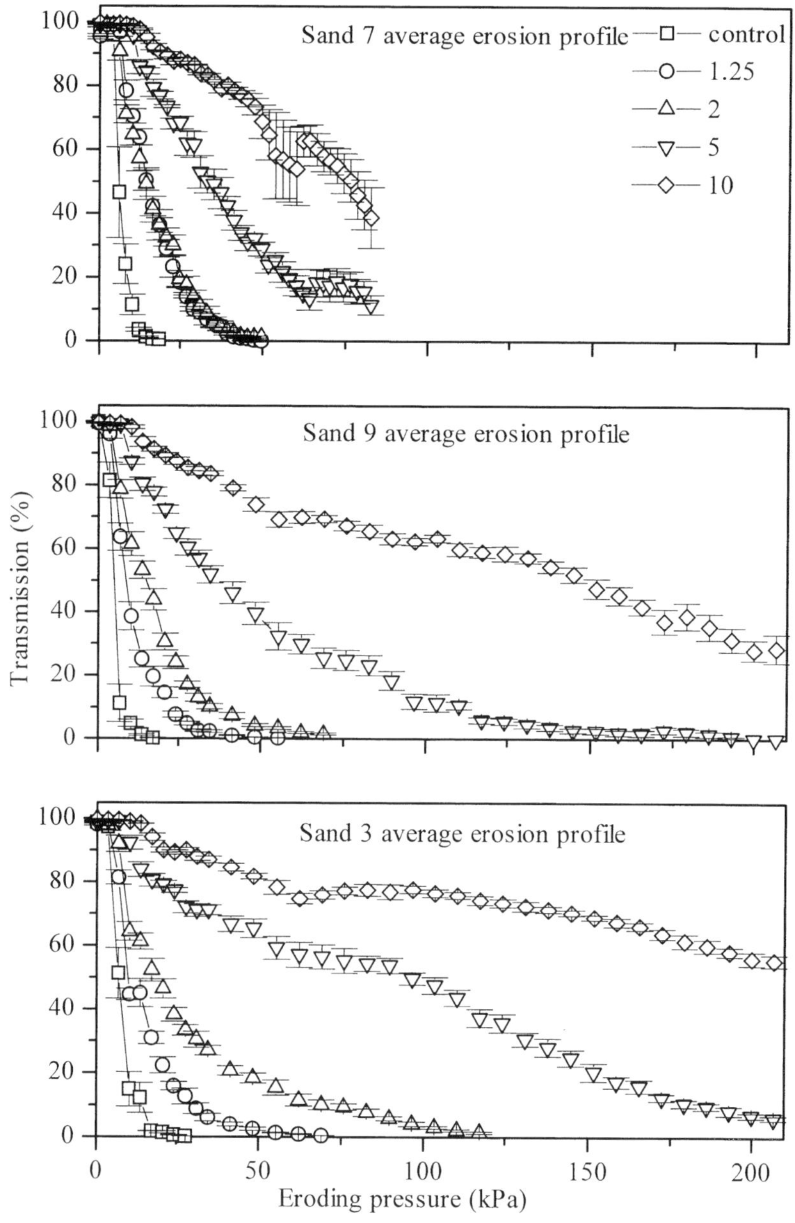

Figure 3. CSM erosion profiles for sediment of different EPS contents (tests Sand 3, 7 and 9). As EPS content increases, so does the erosion threshold. More significantly, the slope of the erosion profile changes, such that as EPS content increases, the erosion rate decreases.

3.3. Field measurements

The field measurements of sediment stability were made during the winter (February 2000) in the Eden estuary (Figure 5), a time when biotic activity is supposed to be minimal. Five replicate measurements were taken with the CSM, the average threshold was 2.42 Nm^{-2} (standard error 0.11). The average colloidal carbohydrate content was 5.7 μg glucose equivalents per mg of sediment (standard error 1.6), the average total carbohydrate was 6.42 μg glucose equivalents per mg of sediment (standard error 0.98), the average water content was 57% (standard error 1.7). When reconstituted to the same water content, the cleaned sediment samples were both fluid, so no measurements of stability were possible.

3.4. The LTSEM images

Low-temperature scanning electron microscopy was used to visualise the sediment microstructure. Images were obtained from each of the laboratory treatments and are compared to images from natural sediments *in situ*.
The LTSEM images, taken from Cryolander samples reveal the microstructure of natural sediments with and without diatom biofilms (Figure 6). The sediment has a typical open card-house structure, and where thick biofilms occur strands of EPS are found (Figure 6 c and d). Light microscopy reveals that diatom EPS (stained with Alcian blue/yellow) naturally forms strands (Figure 7a).

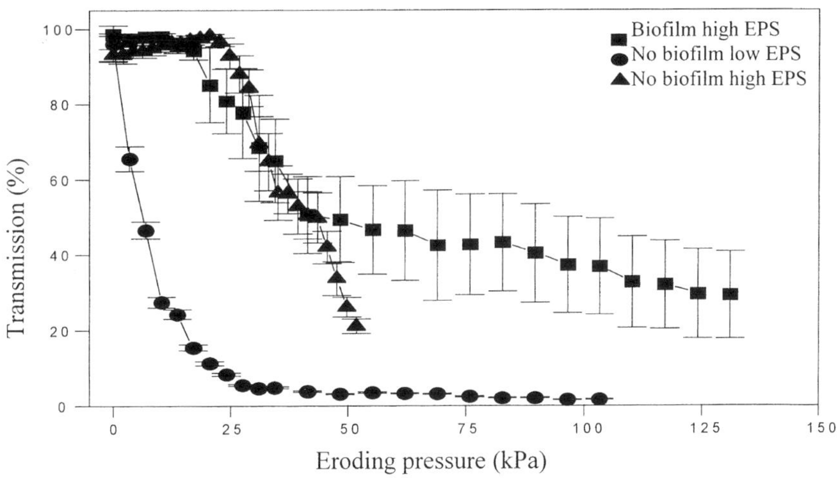

Figure 5. (Including data redrawn from Tolhurst et al. 1999). Field data shows that the erosion profiles from natural sediments with a diatom biofilm (which have high contents of EPS) (squares), and the field data from this study (triangles) where there was no diatom biofilm, but the colloidal carbohydrate contents were relatively high are similar to the cleaned sediment with high contents of EPS (compare to figure 4). Areas without a diatom biofilm and low contents of EPS (circles) are similar to the profiles from sediment with a little or no added EPS (compare to figure 4). Note the x axis scales of the two figures are not the same.

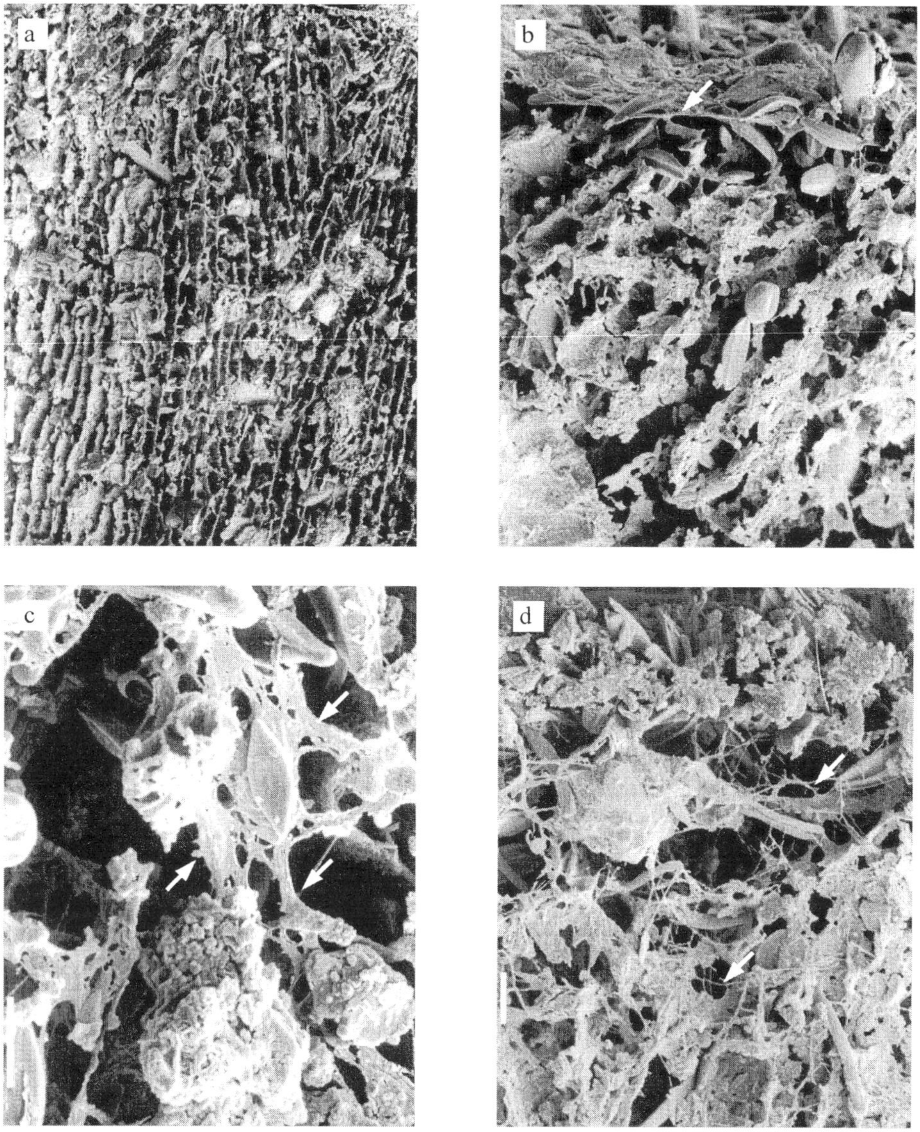

Figure 6. LTSEM images of sediment fracture faces (vertical section through the sediment), surface at top of images. a: Natural sediment from the Eden estuary during February 2000 when no biofilms were present has a typical open card structure, scale = 100 μm. b: during September 1999 thick biofilms occur in the Eden estuary at the sediment surface (marked by white arrow), scale = 10 μm. c: thick strands of EPS are found within this biofilm (white arrows), scale = 10 μm. d: EPS strands have also been found in the Ems Dollard July 1996, scale = 10 μm.

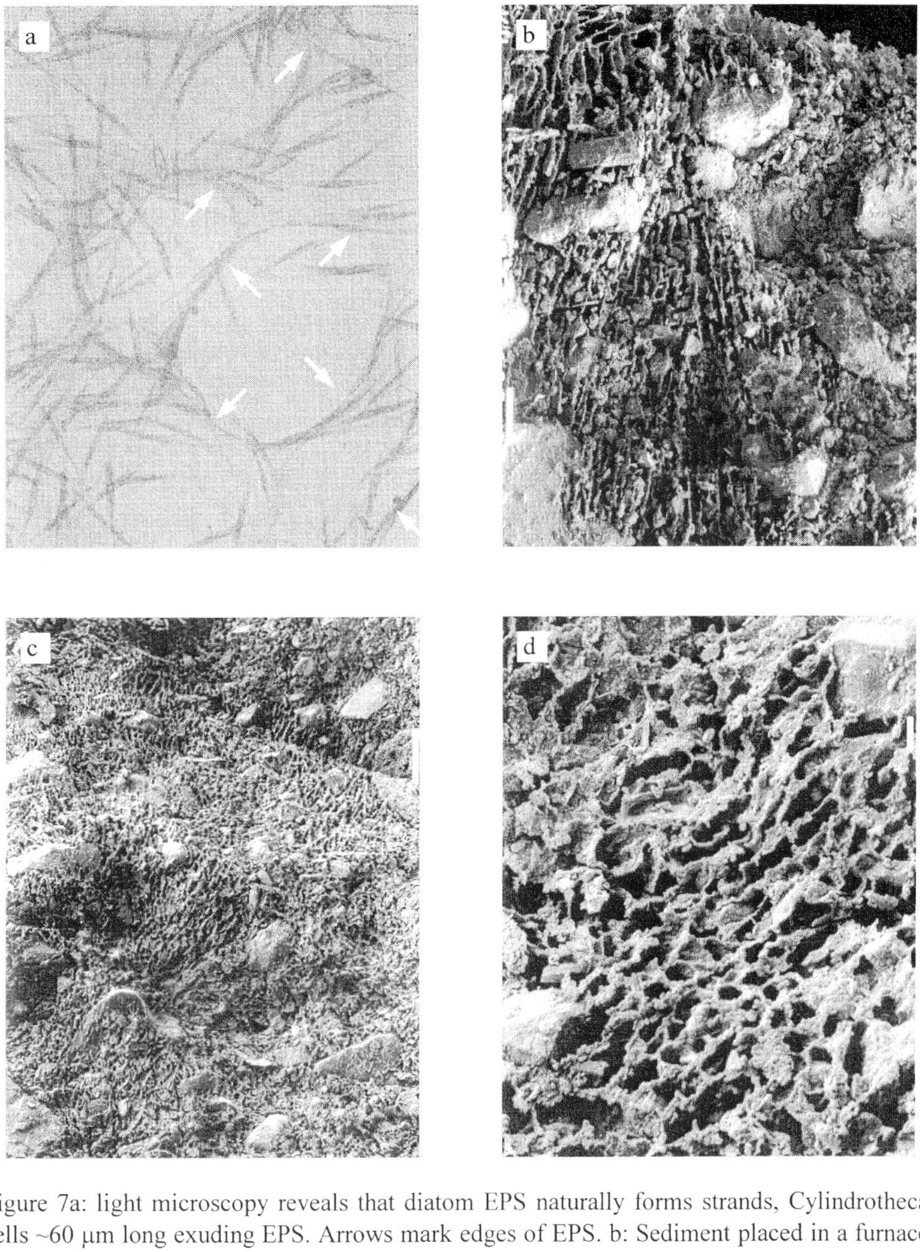

Figure 7a: light microscopy reveals that diatom EPS naturally forms strands, Cylindrotheca cells ~60 μm long exuding EPS. Arrows mark edges of EPS. b: Sediment placed in a furnace at 500°C to burn off organic matter and subsequently reconstituted shows a similar open card house structure to the field samples, scale = 100 μm. c and d: The treated cleaned sediment control with no added EPS, showing the typical open card house structure, scale = 100 μm and 10 μm respectively.

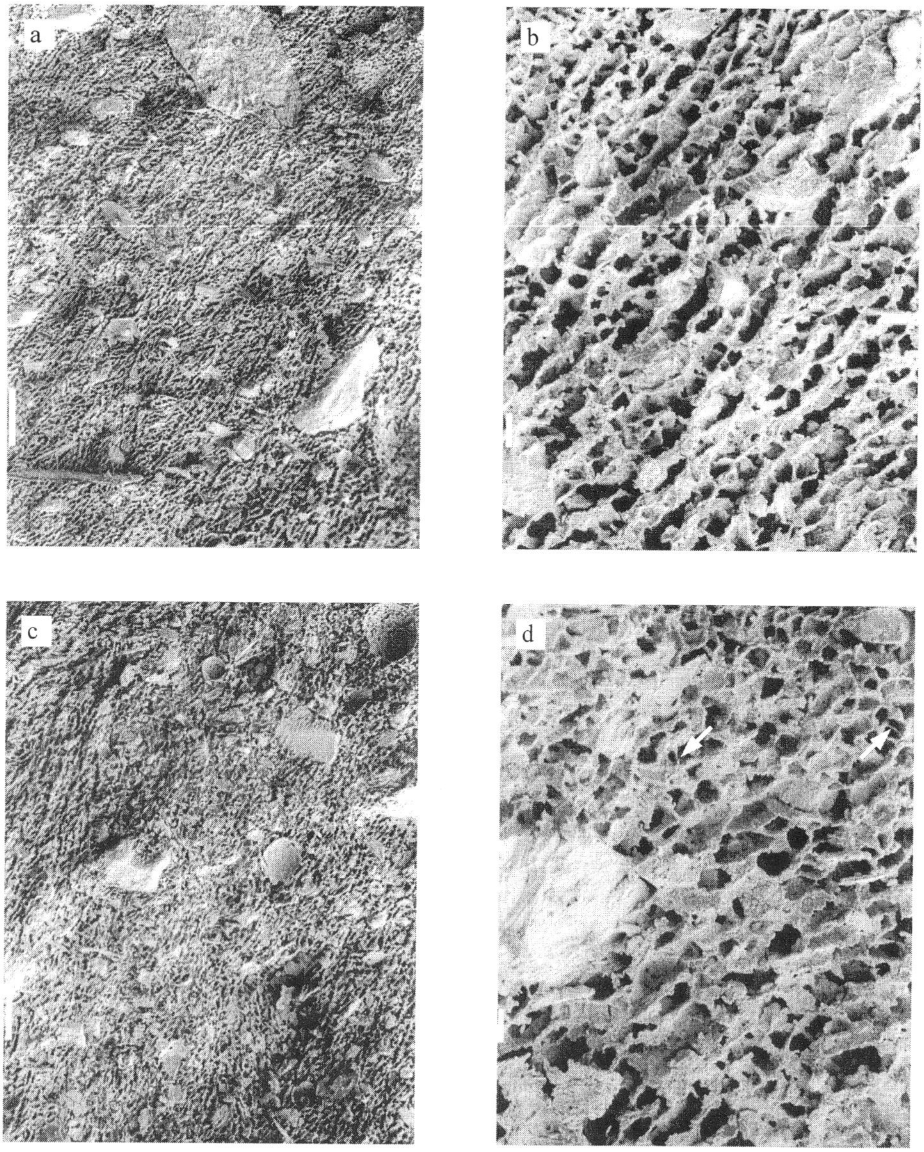

Figure 8. a and b: The treated sediment with 1.25 μg mg^{-1} of added EPS, the structure is a little less open than the control, scale = 100 μm and 10 μm respectively. c and d: The treated sediment with 2.5 μg mg^{-1} of added EPS, the structure is similar to that of the 1.25 μg mg^{-1} treatment, a few small strands of EPS are found between the sediment particles (white arrows), scale = 100 μm and 10 μm respectively.

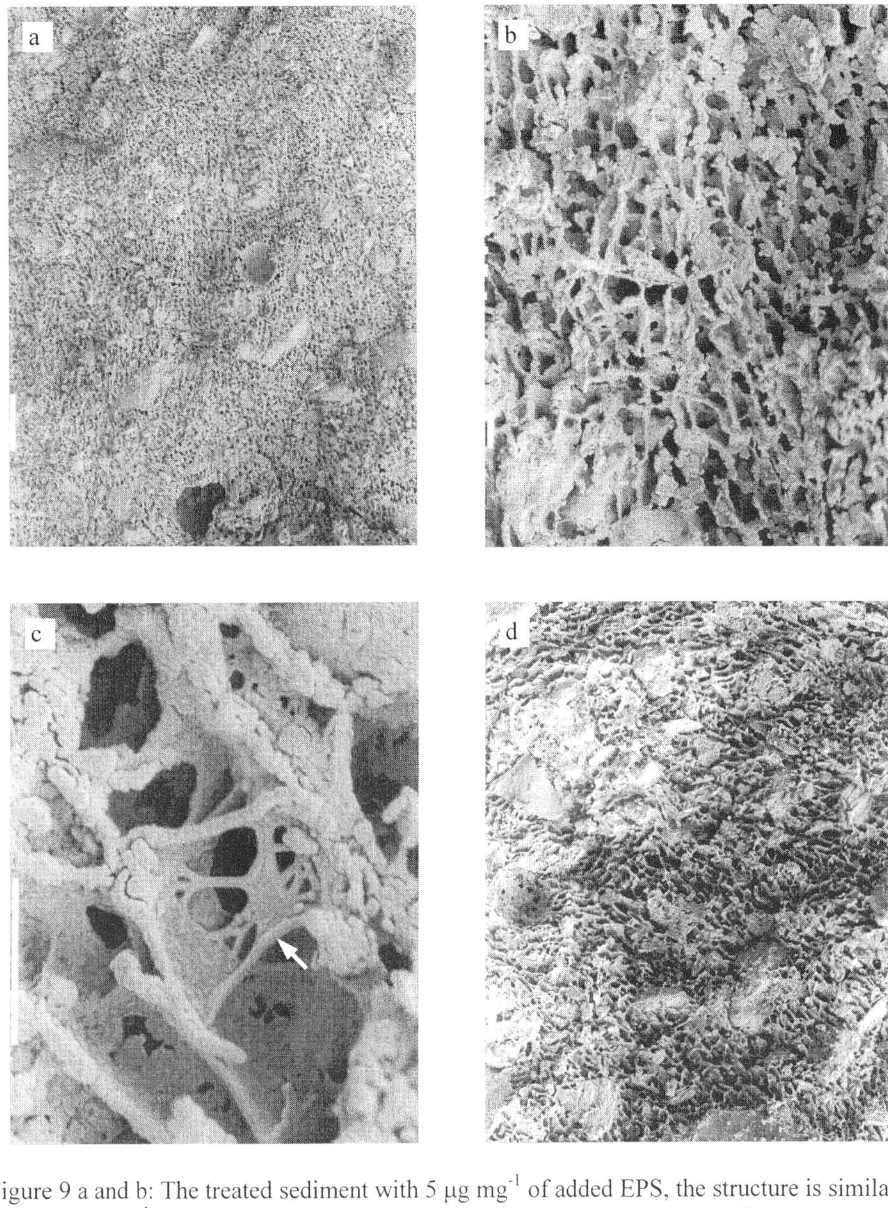

Figure 9 a and b: The treated sediment with 5 μg mg^{-1} of added EPS, the structure is similar to the 2.5 μg mg^{-1} treatment, scale = 100 μm and 10 μm respectively. c: Close examination reveals strands of EPS connecting the sediment particles, scale = 10 μm. d: The 10 μg mg^{-1} treatment has many strands of EPS connecting the sediment particles, resulting in a quite different structure to the other treatments, scale = 100 μm.

Sediment placed in a furnace at 500°C to burn off organic matter and subsequently reconstituted shows a similar structure to the field samples (Figure 7 b). The addition of EPS at high contents to the cleaned sediment alters the microstructure (Figures 8-10). Close examination of the sediment microstructure reveals that at high EPS contents (5 and 10 μg mg^{-1}) strands of EPS form between the sediment particles similar to those found in natural sediments with a biofilm (Figures 9 and 10).

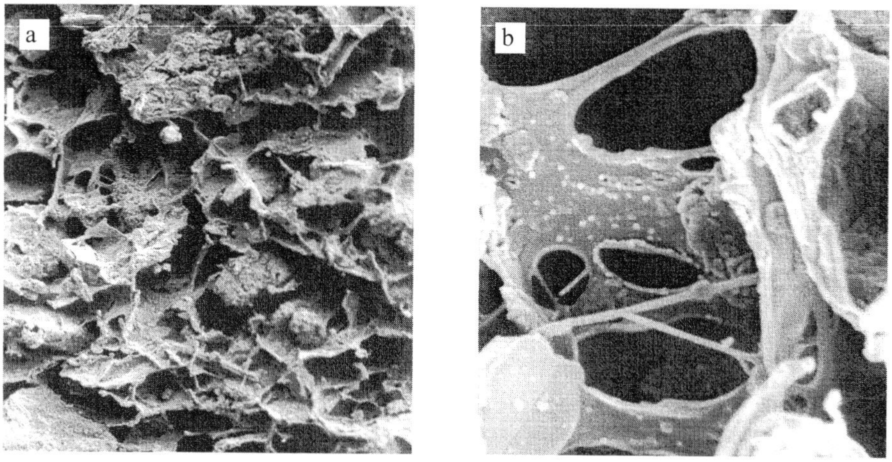

Figure 10. a and b: Close up view of the 10 μg mg^{-1} treatment shows the quite different structure to the other treatments, scale10 μm. f: Close examination reveals that the strands coalesce in places to form sheets, similar to those found associated with diatoms under the light microscope (compare to Figure 7a), scale = 1 μm.

4. DISCUSSION

The Microcosm data show addition of EPS to cleaned sediment significantly increased the critical erosion threshold and reduces the erosion rate (the sediment is more stable), even at low EPS content (Tables 1 and 2, Figure 1). The same trend was found with the CSM (Table 3). The increase in critical erosion threshold at low EPS contents is not accurately resolvable with the CSM (Figure 3), due to the operational procedure which generates a relatively large increase in eroding stress with each pressure step. At higher EPS contents, the Microcosm system used could not generate sufficient bottom stress to initiate erosion (Table 1), but the CSM was capable of exceeding the threshold stress of the sediment.

The increase in threshold friction velocity with EPS measured by the CSM was approximately linear, but an exponential growth model fitted the data better for the contents tested (Figure 2). The increase in erosion threshold with increasing EPS content was relatively high; the average threshold was 0.55 Nm^2 for the controls and 2.02 Nm^2 for 10 μg mg^{-1} of EPS. The erosion profiles from the control and low contents of EPS (1.25 μg mg^{-1}) were similar to those measured on natural sediments with few diatoms and low carbohydrate contents. Field measurements from areas with a diatom biofilm (which have a high EPS content) and field measurements from the Eden during periods when there was no biofilm, but colloidal carbohydrate contents were still high, are similar to the laboratory profiles with a high EPS content (5-10 μg mg^{-1}) (compare Figure 4 to Figure 5).

The high EPS content treatments erode less rapidly than the natural biofilm after the threshold is exceeded (compare Figure 4 to Figure 5). It has been demonstrated that EPS content in the field is highest at the sediment surface (top 250 μm) and decreases rapidly with depth (Taylor and Paterson 1998). This results in a decrease in sediment stability with depth (on the μm scale). In this study, the contents of EPS in the laboratory treatments were homogenous throughout the sediment, as the authors could not devise a practical way of mimicking the change in EPS content with depth on the μm scale. This difference appears to have had an effect on the sediment entrainment results, in particular the erosion rate. It seems likely that in this study, the erosion rate has been reduced more than under natural conditions due to the presence of high EPS contents with depth. However, the fundamental role of EPS in mediating cohesive sediment erosion was confirmed.

The LTSEM images show that the sediment treatments are similar in structure, although there is some variation in the interstitial pore spaces (Figures 7 and 8). At contents of less than 5 μg mg^{-1} little EPS is visible indicating it has adsorbed onto the sediment in a physico-chemical fashion. This type of EPS incorporation alters the surface properties of the sediment particles and increases stability by strengthening physico-chemical inter-particle bonding forces. This seems especially likely as the sediment was cleaned with hydrogen peroxide, which removes cations from the clay minerals. At higher contents strands begin to form physical connections between the grains, more strands were visible at 5 than at 10 μg mg^{-1} (Figures 9 and 10). Images from the Eden and Ems Dollard show that EPS from diatoms forms strands, as the diatoms move through the sediment. Under bloom conditions, EPS production is high and these strands are very thick (Figure 6 c and d). They ramify through the sediment gluing particles together, helping to dissipate stress and significantly increasing the elasticity of the sediment. Sediment with a thick biofilm deforms elastically, whereas sediment without a biofilm deforms plastically (Paterson 1989).

Cleaned sediment remains liquid when reconstituted to the same water content as the natural sediment, this may be due to the segregation of the water molecules among the phases within the sediment matrix. The EPS found naturally in sediments not only increases physical and chemical stability, but also binds water molecules reducing the free water available in the sediment. In this study, field sediments (water content of 57%) were approximately 4.5 times as stable as cleaned sediment (water content of 35%). Thus, electrochemical binding of clay minerals would appear to be less important in natural sediments when compared to the stabilising effects of EPS and other, at present unclear factors (possibly including tube structures and other organic substances).

The cleaned sediment with added EPS is approximately half as stable as *in situ* sediment with a similar amount of EPS, despite having a considerably lower water content. There are a number of possible reasons for this. As hydrogen peroxide is an acid, it is possible that the chemical composition of the clay minerals is being altered by removal of cations resulting in weaker interparticle bonding. However, the loss of stability was also found in samples heated to 500°C, to remove organic material. A possible explanation is that the natural sediment contained a suite of various organic molecules (including proteins and lipids) that together were more effective stabilisers than the Xanthan gum used in the laboratory manipulation. This possibility requires further investigation. Alternatively, it may be that the way in which EPS is naturally secreted by organisms (in strands and as burrow coatings) that significantly increases its stabilising properties and that mechanical mixing does not have the same effect (Dade 1990).

Biostabilised sediments often have high stability yet also high water contents and low bulk density. Physically, high water contents and low bulk densities should result in low stabilities, not high ones. The unusual properties of biostabilised sediments have been partly attributed to a high EPS content, which can increase both the stability and the water content, whilst reducing the bulk density; in effect overriding the abiotic physical factors (Chenu 1993, Tolhurst 1999). Our sediment EPS treatment is not an ideal mimic of natural sediments, however it is clear that organic material in natural sediments may be more important in determining the erosion dynamics of cohesive intertidal sediments than was previously thought.

5. CONCLUSION

The addition of EPS to cleaned sediment increases the erosion threshold and decreases the erosion rate. The increase in stability occurs by two mechanisms. EPS binds to the surface of sediment particles increasing the strength of the physico-chemical interparticle attractive forces. EPS also occurs as strands that connect sediment particles, increasing stability by physical binding and increased elasticity. Sediments cleaned of organic matter lose all of their structural integrity becoming liquid when reconstituted to the same water content as before treatment. This provides evidence that cohesive intertidal mudflats exist in their present form largely due to the integrated action of biota, whether by removal of free water or by direct binding action. The contention that EPS is partly responsible for the increase in sediment stability caused by certain biota is supported by this data.

ACKNOWLEDGEMENTS

The authors would like to thank all members of the Sediment Ecology Research Group and in particular to Irvine Davidson for help with LTSEM, and two anonymous reviewers for their comments on the manuscript.
This work was funded by a St. Andrews university Ph.D. studentship to T.J. Tolhurst and the European Union project CLIMEROD (EU MAS3-CT98-0166).

REFERENCES

Chenu, C. and Guérif, J., 1991, Mechanical strength of clay minerals as influenced by an adsorbed polysaccharid,. *Soil Sci. Soc. Am. J.* (55), 1076-1080.

Chenu, 1993, Clay- or sand-polysaccharide associations as models for the interface between micro-organisms and soil: water related properties and microstructure, *Geoderma*, (56), 143-156.

Dade, B.W., Davis, J.D., Nichols, P.D., Nowell, A.R.M., Thistle, D. Trexler, M.B. and White, D.C., 1990, Effects of bacterial exopolymer adhesion on the entrainment of sand. Geomicrobiology Journal, (8), 1-16.

Decho, A.W., 1990, Microbial exopolymer secretions in ocean environments: Their role(s) in food webs and marine processes, *Ocean. and Marine Biology Ann. Review,* (28), 73-153.

Défarge C., 1997, Cryoscanning electron microscopy and high resolution scanning electron microscopy of organic matter and organomineral associations in modern microbial sediments, *Geomat. Pet. Sed.*, (324), 2a, 553-561

Eisma, D., Cadee, G.C., Laane, R. and Kalf, J., 1982, Preliminary results of Aurelia and Navicula cruises in the Rhine and Eems Estuaries, January-Febuary, 1982. *Mitteilungen – Geologisch Palaontolgis Chen Institut der Univ. Hamburg.*

Grant, J. and Gust, G., 1987, Prediction of coastal sediment stability from photopigment content of mats of purplesulphur bacteria, *Nature,* (330), 6145, 244-46.

Gust, G., 1991, Fluid velocity measuring instrument, U.S. Patent No. 4,986. 122.

Gust, G. and Muller, V., 1997, Interfacial dynamics and entrainment functions of currently used erosion devices, In Burt et al., Eds. Proc. 4th Nearshore & Estuarine Cohesive Sediment Transport Conference INTERCOH '94, July 1994, UK. CRC. 149-174.

Hoagland, K.D., Rosowski, J.R., Gretz, M.R. and Roemer, S.C., 1993, Diatom extracellular polymeric substances: function, fine structure, chemistry and physicology, *J. Phycol.,* (29), 537-66.

Holland, A.F., Zingmark, R.G. and Dean, J.M., 1974, Quantitive evidence concerning the stabilization of sediments by marine benthic diatoms, *Marine Biology,* (27), 191-196.

Humann, K., Kies, L. and Gust, G., The effects of benthic algae on the stability of cohesive sediments in an intertidal region. In press.

Paterson, D.M., Crawford, R.M. and Little, C., 1986, The structure of benthic diatom assemblages: A preliminary account of the use and evaluation of low-temperature scanning electron microscopy, *Journal of Experimental Marine Biology and Ecology,* (96), 279-289.

Paterson, D.M., 1989, Short term changes in the erodibility of intertidal cohesive sediments related to the migratory behaviour of epipelic diatoms, *Limnol. Ocean.* (34), 1, 223-234.

Paterson, D.M., 1994, Microbial mediation of sediment structure and behaviour, In: NATO ASI series, Vol. 635 *Microbial Mats.* Stal, L. J. and Gaumethe, P. Eds., Springer-Verlag.

Paterson, D.M., 1995, The biogenic structure of early sediment fabric visualised by low-temperature scanning electron microscopy, *J. Geol. Soc.* (152), 131-140.

Paterson, D.M., 1997, Biological mediation of sediment erodibility: ecology and physical dynamics, In Burt, N., Parker, R. and Watts, J. (Eds.) "Cohesive Sediments." John Wiley & Sons Ltd.

Taylor, I. and Paterson, D.M., 1998, Microspatial variation in carbohydrate concentrations with depth in the upper millimetres of intertidal cohesive sediments, *Est. Coastal Shelf Sci.,* (46), 359-370.

Tolhurst T.J., 1999, Microbial mediation of intertidal sediment stability, Unpublished Ph.D thesis, St.Andrews University.

Tolhurst T. J., Black K. S., Shayler S. A., Mather S., Black I., Baker K. and Paterson D.M., 1999, Measuring the *in situ* erosion shear stress of intertidal sediments with the Cohesive Strength Meter (CSM), *Estuarine Coastal and Shelf Science* (49), 281-294.

Underwood G.J. and Smith D.J., 1998, Predicting epipelic diatom exopolymer concentrations in intertidal sediments from sediment Chlorophyll a, *Microb. Ecol.*(35), 116-125.

Wiltshire, K.H., Blackburn, J. and Paterson, D.M., 1997, The Cryolander: A new method for in situ sampling of unconsolidated sediments minimising the distortion of sediment fabric, *Journal of Sedimentary Research,* (67), 5.

Chapter 5:

Field observations on mud dynamics and insturmentation

Fine Sediment Dynamics in the Marine Environment
J.C. Winterwerp and C. Kranenburg (Editors)

The seasonal dynamics of benthic (micro) organisms and extracellular carbohydrates in an intertidal mudflat and their effect on the concentration of suspended sediment

E.M.G.T. de Deckere[a,*], B.A. Kornman[b,**], N. Staats[c], G.R. Termaat[b,***], B. de Winder[c,***], L.J. Stal[a] & C.H.R. Heip[a]

[a]Netherlands Institute of Ecology, Centre for Estuarine and Coastal Ecology, P.O.box 140, 4400 AC Yerseke, The Netherlands

[b]Institute for Marine and Atmospheric Research Utrecht, Utrecht University, Department of Physical Geography, P.O. Box 80115, 3508 TC Utrecht, The Netherlands

[c]Laboratory for Microbiology, ARISE, University of Amsterdam, Nieuwe Achtergracht 127, 1018 WS Amsterdam, The Netherlands

The Ems-Dollard estuary, situated in the north of the Netherlands near to the German border, is covered for a large part by intertidal flats. Currents and wind- induced waves exert a shear stress on these flats, resulting in resuspension of the sediment. Fluctuations of the suspended sediment concentration in the Dollard are strongly affected by erosion due to wind-induced waves and by settling during calm weather conditions. However it is believed that benthic processes influence the amount of sediment that can be resuspended from the intertidal flats. Therefore suspended sediment concentration above the Heringsplaat, an intertidal flat in the Ems-Dollard estuary, was measured during two seasons and related to the dominant benthic biological features. The benthos and the abiotic sediment characteristics were monitored from 1995 till 1997, at two stations. At both stations a peak of chlorophyll-*a* concentration was found in spring, caused by a diatom bloom. Meiobenthos was dominated by nematodes, macrobenthos by the polychaete *Marenzelleria viridis* and the amphipod *Corophium volutator*.

The seasonal pattern was studied most intensively during 1996. Monthly sampling was done on the Heringsplaat. The flat was covered with ice during March. This period was followed by a clear water phase during spring and an increase of the silt content at the edge of the Heringsplaat. During this period an increase was found of chlorophyll-*a* concentration and of the carbohydrate concentration. This carbohydrate secretion stabilized the sediment. This finally resulted in a stable diatom mat at the end of the spring and a two-centimeter thick layer of silt. During June a high increase of nematodes and *Corophium volutator* was found. Nematode density was about 5000 ind per 10 cm^2 during June and July and *Corophium* density reached almost 100.000 per m^2. The diatom population decreased rapidly during June. It is assumed that this is caused by high grazing pressure of the nematodes and of *Corophium*. The reduced diatom concentration and the increased

[*] Present address: University of Antwerp, Department of Biology, Universiteitsplein 1, 2610 Wilrijk, Belgium.
[**] Present address: National Institute for Coastal and Marine Management/RIKZ, P.O.Box 9039, 4330 EA Middelburg, The Netherlands.
[***] Present address: Ministry of Public works, Directie Zeeland, P.O.Box 5014, 4330 KA Middelburg, The Netherlands

bioturbation activity of *Corophium* resulted in reduced sediment stability. The silt content of the sediment surface at the edge of the Heringsplaat reduced and at the same time the turbidity in the water column increased.

KEY WORDS
sediment stability, suspended sediment, macrobenthos, *Corophium*, Dollard estuary

1. INTRODUCTION

Estuaries are highly dynamic systems that are continuously subject to morphological changes. Erosion, transport and deposition of sediment trigger these changes. Several studies have been carried out in order to understand the physical processes that cause sediment transport and deposition in estuaries (Dyer, 1988; Postma, 1967). Erosion, on the other hand, depends as well on physical processes, induced by waves and currents (De Jonge & van Beusekom, 1995), as on biological processes (Kornman & de Deckere, 1998; Nowell et al., 1981). Organisms living in or on the sediment will alter the sediment properties, which will result in either stabilization or destabilization. Stabilization of the sediment is mainly a result of the excretion of extracellular polysaccharides by diatoms and/or bacteria (Dade et al., 1990; Paterson, 1988). Bioturbation by meio- and macrobenthos results in a destabilization of the sediment by an increase of the water content and/or microtopography (Davis, 1993; de Deckere et al., 2001; Nowell et al., 1981) and indirectly by grazing on stabilizing organisms (Gerdol & Hughes, 1994a).

The Dollard, a part of the Ems-Dollard estuary, consists of intertidal flats for 80 %. These flats are regularly covered with diatom mats, which increase the erosion resistance of the sediment (Kornman & de Deckere, 1998). Meiobenthos consist for almost ninety percent of nematodes, most species found are diatom feeders (Bouwman, 1983). A lot of nematodes are known to build tubes in the surface layers of the sediment (Jensen, 1996), thereby increasing the water content. Protruding tubes will enhance micro turbulence in the boundary layer of the sediment (Eckman et al., 1981). This will be more pronounced by tubes of the amphipod *Corophium volutator*, which is one of the dominant species in the Dollard (Essink et al., 1998). *Corophium* destabilises the sediment directly by changing the microtopography (unpublished results) and indirectly by grazing on diatoms (Gerdol & Hughes, 1994a). *Corophium* can affect the sediment dynamics also by active resuspension of fine sediment particles (de Deckere et al., 2000).

The suspended sediment concentration in the Ems-Dollard estuary is strongly affected by erosion due to wind-induced waves and by settlement during calm weather conditions (De Jonge & van Beusekom, 1995). It is hypothesised that benthic destabilisation and stabilisation processes affect the amount of sediment that can be suspended. Therefore suspended sediment concentration above the Heringsplaat, an intertidal flat in the Ems-Dollard estuary, was measured during two seasons and related to the dominant benthic biological features.

2. MATERIAL AND METHODS

2.1. Study area

The Ems-Dollard estuary is situated in the northeast part of the Dutch Wadden Sea on the border between the Netherlands and Germany (Fig. 1). The Dollard is the upper reach of the estuary, approximately 100 km^2, and consists for ± 85 % of intertidal flats. The tidal prism of this area is $115 * 10^6$ m^3 and the tidal range is 3 m (De Jonge, 1992). Measurements were performed on the Heringsplaat, an ebb dominated mesotidal flat located in the central part of the Dollard (Fig. 1). The sediment at this flat varies between sandy mud and muddy sand. Maximum current velocities during ebb are approximately 0.3 $m.s^{-1}$.The maximum significant wave height above the flat is about 0.5 m, due to a combination of the fetch of a few kilometers and the relative shallow water depth (max 1.5 m).

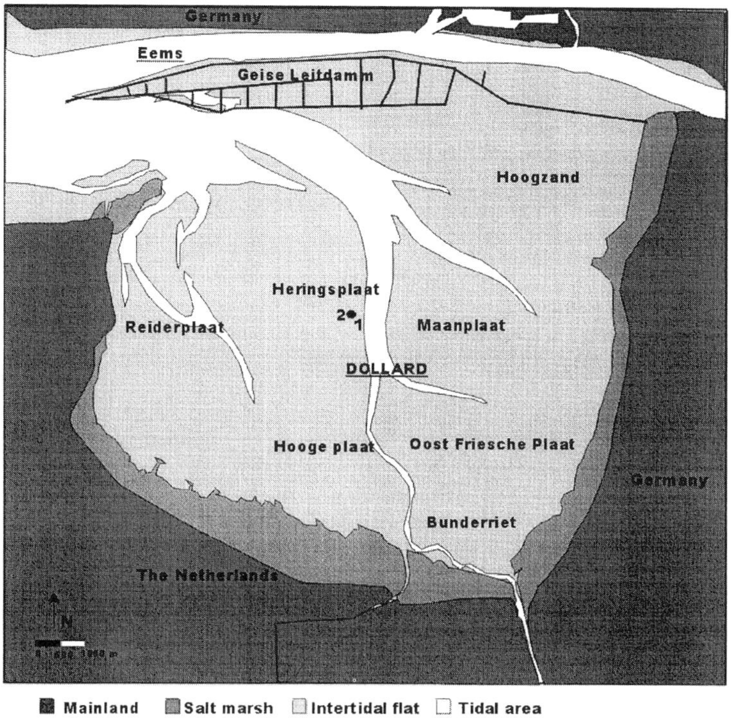

Figure 1. Map of the Dollard estuary with the measuring-bridge situated in the middle on the intertidal flat Heringsplaat (•) and the two sampling stations 1 and 2.

2.2. Sampling

Two plots of 100 m^2 were monitored from 1995 till 1997. Station 1 was located 100 m from the channel and 100 m south of a huge measuring-bridge (Figure 2)(Kornman & de Deckere, 1998). Station 2 was situated at the higher part of the flat, 250 m from the channel. Sampling was done seasonally during 1995 and with monthly intervals from spring till autumn in 1996 and 1997. Undisturbed cores were taken for chlorophyll-*a* and carbohydrates (0-5 mm, ø = 1.8 cm, n = 5 (n=number of cores)), grain size (0-5 mm, ø = 3.6 cm, n = 3) and *Corophium* and nematodes (0-30 cm, ø = 12 cm, n = 3). The samples for chlorophyll-*a*, carbohydrates and grain size were immediately frozen in the field, samples for benthos were fixed with borax buffered formaldehyde (4%). The suspended sediment concentration in the water column was measured continuously with 3 MEX turbidity sensors during 1996 and 1997 at the bridge (type BTG RD-20/10). The data of the three MEX sensors were averaged and finally an average value was calculated per day. Wind velocity and wind direction were measured at the top of this bridge. The wind velocity data were also averaged per day. The suspended sediment concentrations measured above the Heringsplaat were compared to data measured at the entrance of the Dollard in the main channel "Groote Gat" (Figure 1).

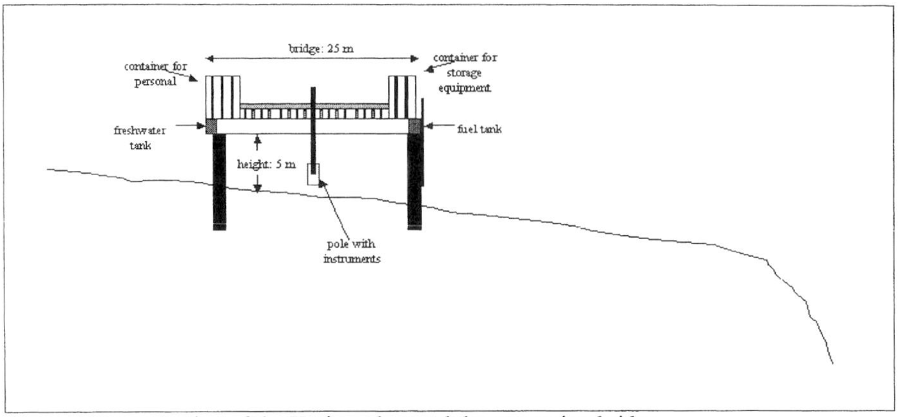

Figure 2. Cross-section of the Heringsplaat and the measuring-bridge

2.3. Analysis

Chlorophyll-*a* was extracted from freeze-dried sediment (100 mg) by N,N-dimethylformamide. Samples were incubated for 1hour and centrifuged for 5 minutes at 20,000 g. The absorption was measured spectrophotometrically at 665 nm (E_n). Subsequently 15 µl HCl (5 M) was added to the extract and absorption was measured again (E_a). The chlorophyll-*a* concentration can then be calculated according to the following equation: chlorophyll-a ($g\ l^{-1}$) = 2.3 * (E_n - E_a) / 72.114 (de Winder et al., 1999).

Carbohydrates were determined using the phenol-sulfuric acid assay (Dubois et al., 1956). Carbohydrates were extracted from the sediment in two steps. About 200 mg of freeze-dried sediment was extracted with 1 ml of distilled water for 1 hour at 30°C. The sample was centrifuged for 5 min. at 20,000 g. The supernatant contained the colloidal carbohydrates. Subsequently the pellet was extracted for 4 h. by 1.5 ml of a 0.1 M Na_2-EDTA-solution at 20°C. The supernatant contained the EDTA-extractable fraction.

Grain size was analysed with a Malvern Particle Sizer 3600 EC. *Corophium* and nematodes were counted under a binocular after sieving the sediment respectively over a 500-µm and a 38-µm sieve.

3. RESULTS

Both stations showed an increase of chlorophyll *a* in spring during 1995 and 1996 (Figure 3a), but contents were higher at station 1, which was situated at the edge of the flat. A pronounced peak of chlorophyll *a* was observed at this station in the summer of 1996. This coincided with an extensive diatom mat on the surface of the sediment. This mat was not found at station 2 neither during 1995 and 1997 at station 1. Chlorophyll *a* did not show any seasonal dynamics in the surface layer of the sediment during 1997.

The carbohydrate fractions, the colloidal as well as the EDTA extractable, had higher contents at station 1 than at station 2 till August during 1995 and 1996. The seasonal dynamics of the colloidal and EDTA extractable carbohydrates showed a similar trend as the chlorophyll *a* dynamics (Figure 3b and 3c). The EDTA extractable fraction showed some differences. An extreme peak was found for this fraction at the beginning of May 1995 and there was not a peak at the end of June 1996. The EDTA extractable carbohydrates remained also higher throughout the winter. A significant relation however was found between chlorophyll *a* and respectively the colloidal and EDTA extractable

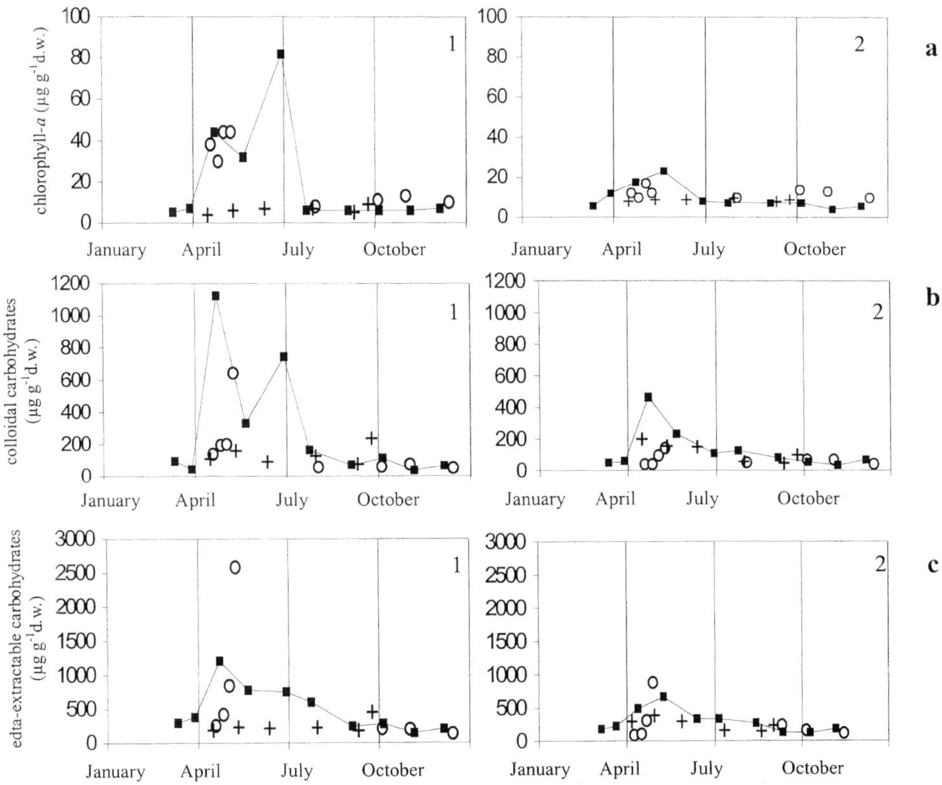

Figure 3. Time series of (**a**) chlorophyll-*a* content, (**b**) colloidal carbohydrate content and (**c**) EDTA-extractable carbohydrate content, in the upper layer of the sediment (0-0.5cm) at plot 1 (left column) and plot 2 (right column) during 1995(o), 1996(■) and 1997 (+).

fraction of carbohydrates (ANOVA: $F_{18,28} = 4.62$; $p < 0.001$ and $F_{18,26} = 3.29$; $p < 0.01$), indicating that the EPS production is strongly affected by the algal community.

The macrofaunal community at the two stations consisted out of 7 up to 12 species (Table 1), no significant difference was found between the plots. In spring the benthos was not dominated by one of the species. *Corophium volutator*, *Hydrobia ulvae*, *Macoma balthica*, *Marenzelleria viridis*, *Nereis diversicolor* and oligochaetes were the most common species, but they were far outnumbered by huge densities of *Corophium volutator* in the summer. The enormous increase of *Corophium* during the summer was found every year (Figure 4a). Highest density occurred in the beginning of July 1996, namely 82000 ind.m^{-2}. Density decreased rapidly at station 2, but remained high at station 1. This station was situated in the zone, where a second diatom bloom was observed. Nematodes showed also a peak at the end of June (Figure 4b). Nematode density varied between 400 up to 2300 ind.10 cm^{-2}, but increased up to 5300 ind.10 cm^{-2}.

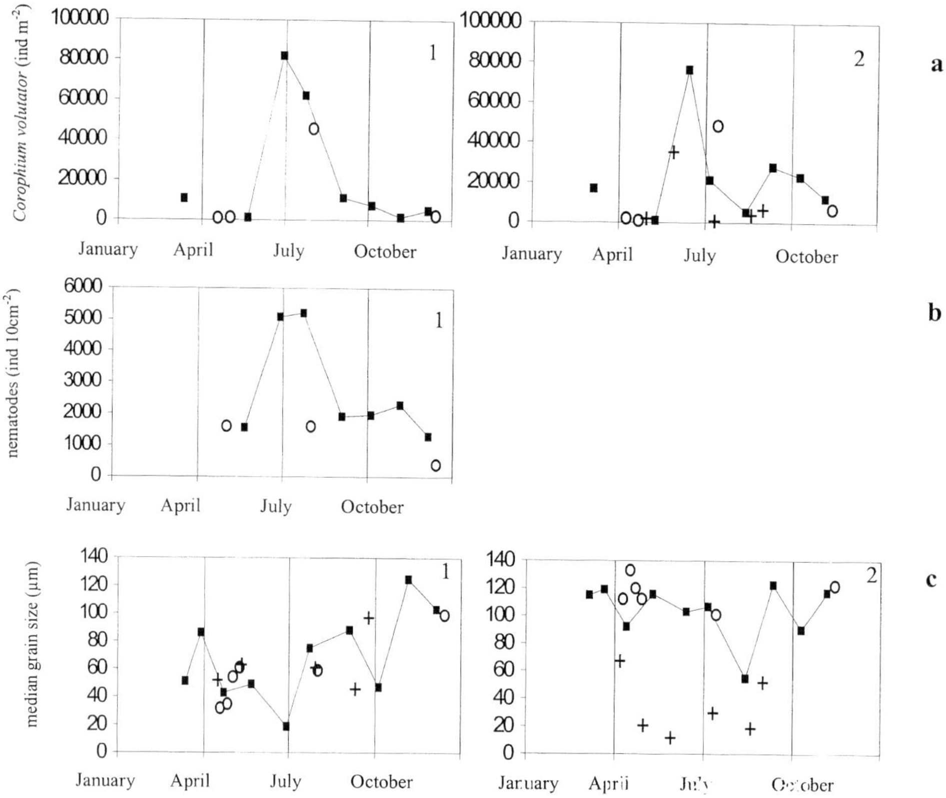

Figure 4. Time series of (**a**) *Corophium volutator* density, (**b**) nematode density and (**c**) median grain size of the sediment in the upper layer of the sediment (0-0.5cm), at plot 1 (left column) and plot 2 (right column) during 1995(o), 1996(■) and 1997 (+). Nematodes were not sampled in 1997, neither in plot 2.

The sediment at station 1 was sandy silt. The median grain size varied between 20 and 80 µm in spring (Figure 4c). A decrease was observed from April 1996 to July 1996. A layer of at least two centimeter of fine sediment settled at this site (Figure 5), but this layer eroded during the rest of the summer. Median grain size in the top half centimeter of the sediment increased up to between 50 and 130 µm. Station 2 was sandier during 1995 and 1996. The median grain size varied around 110 µm, with a small decrease in the autumn of 1996. Fine sediment settled down at station 2 during the spring of 1997. The median grain size decreased to ± 20 µm, and increased slowly during summer.

The suspended sediment concentration in the Dollard showed a similar seasonal trend above the intertidal flat "Heringsplaat" and in the channel "Het Groote Gat", situated at the entrance of the Dollard (Figure 6). The concentration above the intertidal flat was significantly higher. Sediment is eroded here directly by wind waves. The eroded sediment is diluted in the water column, which results in lower concentrations in the channels. Regression analysis showed a low correlation between the daily average wind speed and the daily average suspended sediment concentration during 1996 ($r^2 = 0.24$). No relation was found at all for 1997. This suggests that besides wind other factors, such as the sediment bed characteristics, are important for the suspended sediment concentration. Analysis of covariance showed that the suspended sediment concentration during 1996 depended both on wind as well as on the dominant biological feature in the sediment ($F_{2, 159} = 0.071$, $p<0.001$). Therefore the data of 1996 were divided into three periods, during two of these periods one group of organisms was clearly more abundant and affecting the sediment surface than the other organisms. The first period was characterised by a diatom mat on the surface of the sediment (chlorophyll *a* > 20 μg.g^{-1} d.w. sed.), from 2 April till 19 June. The mat was still present at the end of June, but the abundance of *Corophium volutator* increased enormous. *Corophium volutator* was abundant in high densities (> 25000 ind.m^{-2}) at both stations during the second period, which was from 19 June till 9 August. The rest of the year no dominant features were present in the sediment.

Table 1
Macrobenthos density (N.m^{-2}) at station 1 and 2 at the Heringsplaat in 1995

Species	Station 1				Station 2			
	19/4	3/5	2/8	14/12	19/4	3/5	2/8	14/12
Arenicola marina							35	
Corophium volutator	580	1275	45647	2293	1971	928	48583	6806
Crangon crangon							35	
Eteona spec.			70				70	50
Heteromastus filiformis	464	696	772	1243	348	116	210	149
Hydrobia ulvae	1044	2087	2210	2238	1044	2667	2771	1492
Hydrobia ventrosa	464	696	491		696	1160	491	
Macoma balthica		1507	1157	497	464	580	456	448
Marenzelleria viridis	2667	2783	1754	1592	2667	3479	3964	2636
Mya arenaria		116	35		116			
Nereis diversicolor	348	812	491	497	1391	464	175	149
Oligochaeta spec.	1623	1507	982	895	580	696	631	647
Pygospio elegans		928	1403		812		1929	

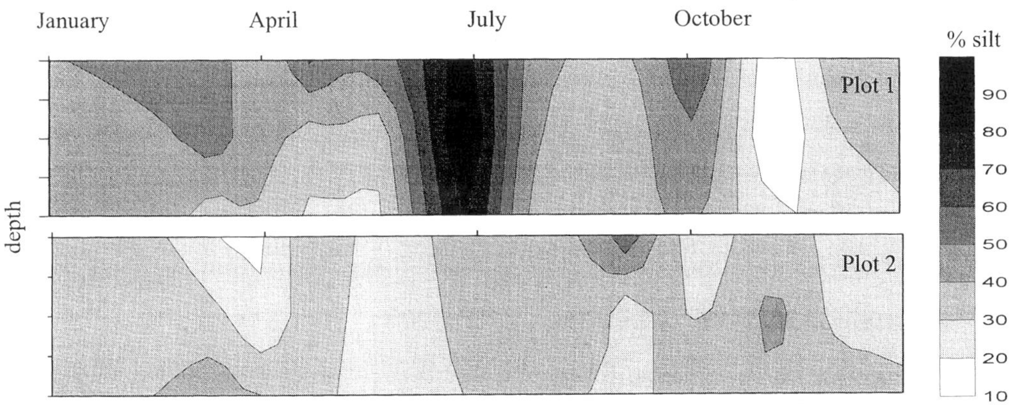

Figure 5. Percentage of silt in the upper two centimeter of the sediment in plot 1 and plot 2 during 1996.

This results in a good relation between the suspended sediment concentration and the wind speed when *Corophium* density was low and a diatom mat was absent (Figure 7, $r^2 = 0.62$). A good relation was also found for the second period, when *Corophium* was present at high densities ($r^2 = 0.67$). Suspended sediment concentrations were significantly higher during this period than the other periods. The steepness of the relation, however, was similar to the steepness of the relation for the period when *Corophium* density was low and a diatom mat was absent. This indicates similar erosion rates in these periods, but an extra input of sediment into the water column when high *Corophium* densities are present, which cannot be related to the wind speed variability. No significant relation was found between the suspended sediment concentration and the wind speed in the first period, when a diatom mat was present ($r^2 = 0.02$). However the suspended sediment concentrations remained low during this period, clearly indicating a reduced erosion rate.

The results indicate that benthic organisms do have an effect on the amount of sediment that will resuspend when the sediment bed is exposed to wind- and wave stress. The suspended sediment concentration is lower when the tidal flat is covered with a diatom mat binding sediment particles together with mucus, while more sediment will be resuspended when high densities of *Corophium volutator* are abundant.

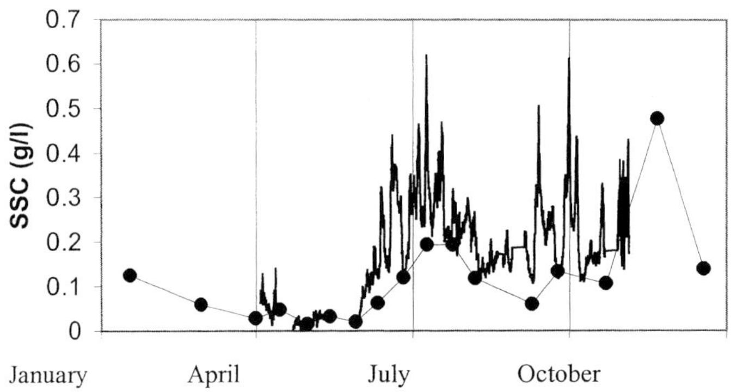

Figure 6. The seasonal variation of the suspended sediment above the Heringsplaat (solid line) and in the channel "Het Groote Gat" (line with dots) during 1996.

4. DISCUSSION AND CONCLUSIONS

The suspended sediment concentration in the Dollard estuary will depend on the wave energy caused by wind as is shown in Figure 7, but benthic processes in the intertidal flats affect the relation. This was most pronounced during 1996, when a diatom mat was formed on the edges of the Heringsplaat. The suspended sediment concentration remained low as long as the mat was intact, but increased as soon as the abundance of the amphipod *Corophium volutator* started to increase. This coincided with the disappearance of the diatom mat.

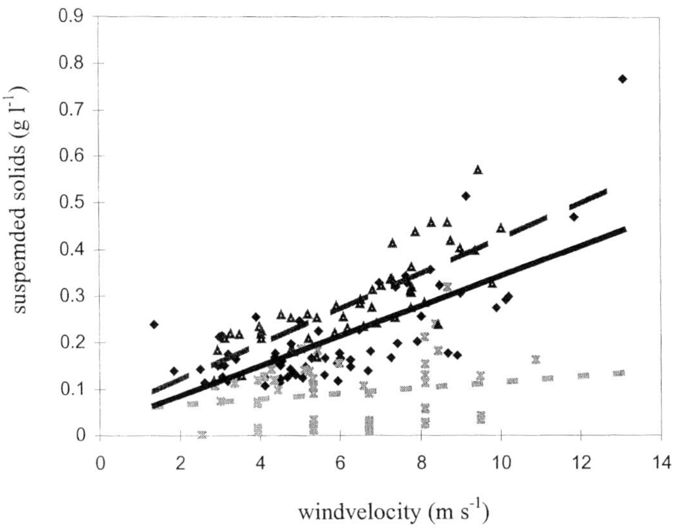

Figure 7. Suspended sediment concentration above the Heringsplaat during 1996 versus the wind velocity. The data were divided over three periods. One period that is dominated by *Corophium* (Δ), one by diatoms (X) and during the rest of the season (♦) no dominating feature was found.

It was observed during the field study that 1996 differed from the other two years. The water column was extremely clear after a severe winter and calm weather (Staats et al., 2001). This resulted in the development of an algal bloom both in the water column and in the sediment. Finally a diatom mat was formed at the end of the spring on the edge of the Heringsplaat (Staats et al., 2001). The dominant diatom species in this mat was largely made up of the genus *Nitzschia* (Wiltshire et al., 1998). A diatom mat was observed also on the muddier flats at the upper reaches of the Dollard estuary in this period (personal observations). High chlorophyll *a* contents were found in the spring of 1997 at the upper reaches, but no algal mat was observed (unpublished results). The occurrence of a diatom mat seems to depend on the severity of the winter (Gillbricht, 1964; Staats et al., 2001). The sediment on the intertidal flats seemed to be protected from erosion by the algal mat. The suspended sediment concentration remained low. This was due to an increased shear stress for erosion (Kornman & de Deckere, 1998), which most likely is a result of increased carbohydrate contents of the sediment. These carbohydrates are produced by diatoms and bind sediment particles together. The binding capacity can be ascribed mainly to the EDTA-extractable fraction. The colloidal carbohydrates on the other hand will dissolve every time when water covers the flat and will have less effect on the sediment stability. The algal bloom disappeared during spring, but was followed by a second bloom at the edges of the Heringsplaat. This time a diatom mat was formed. Fine sediment particles were trapped to this mat, thereby increasing the silt content of the sediment. However the increase of silt can also be due to the increase of small *Corophium*, who probably collect fine particles out of the sediment column to build their tubes (Jensen, 1996).

The disappearance of the diatom mat at the end of June is most likely a result of the increased grazing pressure by the increasing number of benthos. A typical estuarine community, such as *Corophium volutator*, *Hydrobia ulvae*, *Macoma balthica*, *Nereis diversicolor* and oligochaetes, dominates the benthos at the Heringsplaat. Since the early nineties there is also an increase observed of the spionid *Marenzelleria viridis*. This typical community is to be expected to have a destabilising effect on muddy sediments, thus enhancing erosion of fine sediments at a lower shear stress (de Deckere et al., 2001). Destabilisation occurs directly by an increase of the microtopography, thereby enhancing microturbulence resulting in an increased shear stress (Eckman & Nowell, 1984). Indirect destabilisation is due to grazing and reduction of microphytobenthos. Microphytobenthos is known to stabilise the sediment by secretion of carbohydrates. On the other hand the benthos can enhance the amount of suspended material by ejecting sediment particles into the water column (de Deckere et al., 2000). The secretion of faecel pellets, which will erode more easily, was also observed in the field for *Marenzelleria*. The erosion of faecal pellets can also strongly effect the erosion rate. This was shown for the faecel pellets of *Hydrobia ulvae* at two microtidal mudflats in the Danish Wadden Sea (Andersen, 2001).

Benthic diatoms form a significant part of the diet of *Corophium volutator* (Creach et al., 1997; Gerdol & Hughes, 1994b). They can selectively pick out diatoms, but they can also feed on bacteria or organic films, like colloidal or EDTA-extractable carbohydrates. An ingestion rate of 1.5 ng chl ind^{-1} h^{-1} was found in laboratory experiments (Gerdol & Hughes, 1994b). This was equivalent to approximately 4000 small diatoms. Considering the density of *Corophium* of ± 80000 ind m^{-2} at the end of June, a grazing pressure of 120 μg chl m^{-2} h^{-1} could be expected. However this grazing rate will be an overestimate, because *Corophium* were starved before the experiment started. At the same time an increase was reported for nematodes. Nematodes were the most abundant meiobenthic species in the Heringsplaat. Up to 90% of the species found in the Dollard estuary are categorised as diatom feeding species (Bouwman, 1983; Riemann & Schrage, 1978). Feeding rates found for nematodes vary between 40 diatoms per day up to 7 diatoms per hour (Admiraal et al., 1983). This means a daily consumption of approximately 10 ng C ind^{-1} d^{-1}. Blanchard (1991) found similar rates, but, contrary to the previous author, he concluded that nematodes could become food limited because of high grazing pressure. Despite the inaccuracy of the reported grazing rates, it seems likely that the diatom mat collapsed during July due to the high grazing pressure by the benthos.

The resuspension of both sandy as well as muddy sediments in estuaries can be strongly related to the wind-induced waves (De Jonge & van Beusekom, 1995; Freire & Andrade, 1999). Our results show that benthic processes affect this relation. The clear water phase in spring was most likely not a result of this, but consequently a diatom bloom at the intertidal areas restricted the resuspension of the sediment. This confirms the hypothesis that diatoms stabilise the sediment by mucus secretion (Paterson, 1989). The increased suspended sediment concentration in the summer confirmed both the direct as well as the indirect effect of the benthic population. The indirect effect by grazing on the diatoms showed a decrease of the sediment stability, but the direct effect seemed much more related to a direct input of suspended sediment into the water column than reduced sediment stability.

The results of this study demonstrate clearly the impact of benthic organisms, both of microphytobenthos and of macrobenthos, on the suspended solids concentration. The relation is not straightforward and will also depend on climatic and hydraulic conditions, but the effect can clearly be distinguished in the field. Therefore it is recommended to include information about benthic organisms when studying the behaviour of sediments in tidal areas, especially for the prediction of resuspension.

ACKNOWLEDGEMENTS

The authors wish to thank the crew of the R.V. "NAVICULA" and Willem van der Lee, skipper of the R.V. "GEOS". 'Meetdienst Noord' of the Ministry of Transport, Public Works and Water Management provided the data of the "Groote Gat". This study was financially supported by the Dutch organisation NWO-GOA as a part of the BOA research theme on tidal areas and by the European Community grants MAS3-CT95-0022 INTRMUD. This is publication 2812 of the Netherlands Institute of Ecology, Centre of Estuarine and Coastal Ecology, Yerseke.

REFERENCES

Admiraal, W., L.A. Bouwman, L. Hoekstra, & K. Romeyn, 1983, Qualitative and quantitative interactions between microphytobenthos and herbivorous meiofauna on a brackish intertidal mudflat, *Internationale Revue der gesamten Hydrobiologie.*, (68), 175-191.

Andersen, T.J., 2001, Seasonal variation in erodibility of two temperate, microtidal mudflats, *Estuarine, Coastal and Shelf Science*, (53), 1-12.

Blanchard, G.F., 1991, Measurement of meiofauna grazing rates on microphytobenthos: is primary production a limiting factor?, *Journal of experimental Marine Biology and Ecology*, (147), 37-46.

Bouwman, L.A., 1983, Systematics, ecology and feeding biology of estuarine nematodes, BOEDE Publications and reports 3, Ph.D.-thesis, Agricultural University, Wageningen, 173 p.

Creach, V., M.T. Schricke, G. Bertru, & A. Mariotti, 1997, Stable isotopes and gut analyses to determine feeding relationships in saltmarsh macroconsumers, *Estuarine, Coastal and Shelf Science*, (44), 599-611.

Dade, W.B., J.D. Davis, P.D. Nichols, A.R.M. Nowell, D. Thistle, M.B. Trexler, & D. C. White, 1990, Effects of bacterial exopolymer adhesion on the entrainment of sand, *Geomicrobiological Journal*, (8), 1-16.

Davis, W.R., 1993, The role of bioturbation in sediment resuspension and its interaction with physical shearing, *Journal of experimental Marine Biology and Ecology*, (171), 187-200.

de Deckere, E.M.G.T., T.J. Tolhurst & J.F.C. de Brouwer, 2001, Destabilisation of muddy intertidal sediments by benthos, *Estuarine, Coastal and Shelf Science*, (53), 665-669.

de Deckere, E.M.G.T., J. van de Koppel, & C.H.R. Heip, 2000, The influence of *Corophium volutator* abundance on resuspension, *Hydrobiologia*, (426), 37-42.

De Jonge, V.N., 1992, Physical processes and dynamics of microphytobenthos in the Ems estuary (The Netherlands), Ph.D.-thesis, University of Groningen, 176 p.

De Jonge, V.N. & J.E.E. van Beusekom, 1995, Wind- and tide-induced resuspension of sediment and microphytobenthos from tidal flats in the Ems estuary, *Limnology &Oceanography*, (40), 766-778.

de Winder, B., N. Staats, L.J. Stal, & D.M. Paterson, 1999, Carbohydrate secretion by phototrophic communities in tidal sediments, *Journal of Sea Research*, (42), 131-146.

Dubois, M., K.A. Gilles, J.K. Hamilton, P.A. Rebers, & F. Smith, 1956, Colorimetric method for determination of sugars and related substances, *Analytical Chemistry*, (28), 350-356.

Dyer, K.R., 1988, Fine sediment particle transport in estuaries. In Dronkers, J. (ed.), *Physical processes in estuaries*, Springer-Verlag, Berlin : 295-310.

Eckman, J.E. & A.R.M. Nowell, 1984, Boundary skin friction and sediment transport about an animal-tube mimic, *Sedimentology*, (31), 851-862.

Eckman, J.E., A.R.M. Nowell, & P.A. Jumars, 1981, Sediment destabilization by animal tubes, Journal of Marine Research, (39), 361-374.

Essink, K., J. Eppinga, & R. Dekker, 1998, Long-term changes (1977-1994) in intertidal macrozoobenthos of the Dollard (Ems Estuary) and effects of introduction of the North American spionid polychaete *Marenzelleria* cf. *wireni*, *Senckenbergiana Maritima*, (28), 211-225.

Freire, P. & C. Andrade, 1999, Wind-induced sand transport in Tagus estuarine beaches, *Aquatic Ecology*, (33), 225-233.

Gerdol, V. & R.G. Hughes, 1994a, Effect of Corophium volutator on the abundance of benthic diatoms, bacteria and sediment stability in two estuaries in southeastern England, *Marine Ecology Progress Series*, (114), 109-115.

Gerdol, V. & R.G. Hughes, 1994b, Feeding behaviour and diet of Corophium volutator in an estuary in southeastern England, *Marine Ecology Progress Series*, (114), 103-108.

Gillbricht, M., 1964, Einwirkungen des kalten Winters 1962/63 auf die Phytoplanktonwicklung bei Helgoland, *Helgoländer Meeresuntersuchungen*, (10), 263-275.

Jensen, P., 1996, Burrows of marine nematodes as centres for microbial growth, *Nematologica*, (42), 320-329.

Kornman, B.A. & E.M.G.T. de Deckere, 1998, Temporal variation in sediment erodibility and suspended sediment dynamics in the Dollard estuary. In Black, K., D.M. Paterson, & A. Cramp (eds), *Sedimentary Processes in the Intertidal Zone*, Geological Society, London, Special Publications, 139 : 231-241.

Nowell, A.R.M., P.A. Jumars, & J.E. Eckman, 1981, Effects of biological activity on the entrainment of marine sediments, *Marine Geology*, (42), 133-153.

Paterson, D.M., 1988, The influence of epipelic diatoms on the erodibility of an artificial sediment, Proceedings of the 10th Diatom Symposium, 345-355.

Paterson, D.M., 1989, Short-term changes in the erodibility of intertidal cohesive sediments related to the migratory behavior of epipelic diatoms, *Limnology &.Oceanography*, (34), 223-234.

Postma, H., 1967, Sediment transport and sedimentation in the estuarine environment. In Lauff, G. H. (ed.), *Estuaries*, American Association of Advanced Scientific Publications, 158-184.

Riemann, F. & M. Schrage, 1978, The mucus-trap hypothesis on feeding of aquatic nematodes and implications for biodegradation and sediment texture, *Oecologia*, (34), 75-88.

Staats, N., E.M.G.T. de Deckere, B. de Winder & L.J. Stal, 2001, Spatial patterns of benthic diatoms, carbohydrates and mud on a tidal flat in the Ems-Dollard estuary, *Hydrobiologia*, (448), 107-115.

Staats, N., E.M.G.T. de Deckere, B.A. Kornman, W. van der Lee, R. Termaat, J. Terwindt & B. de Winder, 2001, Observations on suspended particulate matter (SPM) and microalgae in the Dollard estuary, The Netherlands: importance of late winter ice cover of the intertidal mudflats, *Estuarine, Coastal and Shelf Science*, (53), 297-306.

Wiltshire, K.H., T.J. Tolhurst, D.M. Paterson, I. Davidson, & G. Gust, 1998, Pigment fingerprints as markers of erosion and changes in cohesive sediment surface properties in simulated and natural erosion events. In Black, K.S., D.M. Paterson, & A. Cramp (eds), *Sedimentary processes in the intertidal zone*, Geological Society, London, 99-114.

Fine Sediment Dynamics in the Marine Environment
J.C. Winterwerp and C. Kranenburg (Editors)

Interaction of Submerged Vegetation, Hydrodynamics and Turbidity; Analysis of Field and Laboratory Studies

E.J. Houwing[a], I.C. Tánczos[b], A. Kroon[c] and M.B. de Vries[b]

[a]Institute for Inland Water Management and Waste Water Treatment (RIZA), Rijkswaterstaat, P.O Box 52, 3300 AK Dordrecht, The Netherlands

[b]Delft Hydraulics, P.O. Box 177 Delft, The Netherlands

[c]Institute for marine and Atmospheric research (IMAU), Utrecht University, P.O. Box 80.115, 3508 TC Utrecht, The Netherlands

Both field studies and laboratory experiments were carried out in order to identify relevant processes that cause the phenomenon of a clear water phase above submerged vegetation fields, as commonly observed in lakes in The Netherlands.

Results from the field study revealed that an increase in the turbidity level of lake waters is due to local wind induced wave activity. Advective transport of suspended sediment is shown not to contribute to changes in the turbidity level. Resuspension of bed material by waves is likely confined to a so-called 'fluffy layer'.

Results from the laboratory study showed that submerged vegetation decreased the eddy diffusivity by affecting both the turbulent kinetic energy and the sizes of the turbulent structures. However, this did not result in an increase in sedimentation within the vegetation field. Waves were effectively damped by the vegetation. This effect is a function of plant morphology (stiffness and plant length). Results from the laboratory experiments therefore corroborate the findings from the field study: the phenomenon of a clear water phase above submerged vegetation canopy is most likely due to the dampening effect of waves by the vegetation, which inhibits local resuspension of the sediment bed.

KEY WORDS
submerged vegetation, plant-flow interaction, cohesive sediment, dissipation, turbulence

1. INTRODUCTION

Since the 1950's, human impact has caused a dramatic decline in the number of clear water lakes in The Netherlands. This was mainly caused by a removal of submerged vegetation and an increase in nutrient loading of the surface water by direct discharge of wastewater. Since the 1980's, numerous restoration projects have been carried out to induce the switch back from turbid to clear water (see for overview Hosper, 1997; Meijer, 2000). In these restoration programmes, the return of macrophytes seems crucial for a stable clear water state (e.g. Perrow et al., 1997) as it tends to enhance water transparency (Van den Berg, 1998; Scheffers et al., 1994). However, an abundance of vegetation, especially in the upper water layer, is a problem for recreational activity in the lakes. Better understanding of the dominant processes responsible for the occurrence of clear water is needed to optimise maintenance strategies for these lakes.

The interaction between submerged vegetation and hydrodynamics have been extensively investigated (see also Petryk and Bosmajian, 1975; Dawson and Charlton, 1988; Gurnell and Midgley, 1994; Nepf, 1999). The effects can be attributed to:

- a reduction in the current velocity (e.g. Gambi et al., 1990; Petticrew and Kalff, 1992; Kutija and Hong, 1996),
- a redirection or even a blocking effect of the current (Gambi et al., 1990; James and Barko, 1990) and
- dissipation of wave energy (for instance Dubi and Torum, 1997; Verduin and Backhaus, 2000; Mendez et al., 1999).

As a result of these processes, vegetation canopies enhance sedimentation in and above themselves (e.g. Fonesca and Fisher, 1986; James and Barko, 1990). This is thought, amongst others, to be of major importance for the occurrence of clear water patches in turbid lakes in The Netherlands (Van den Berg, 1998).

However, both field and laboratory research reveals the existence of a two-layer velocity profile. Although in the vegetation canopy the current velocity decreases rapidly as function of depth, in the free water zone above the vegetation a logarithmic velocity profile still exists (Pethick et al., 1992; Shi et al., 1996). Gambi et al. (1990) reported an increase of the current velocity by 10 to 20% in the free water layer just above the canopy, depending on the initial current velocity. Shi et al. (1996) found that this strong increase in current velocity hampered sediment particles to enter the canopy. In this respect, it may be doubtful that the clear water above submerged vegetation is a result of enhanced sedimentation. The dissipation of wave energy by vegetation could be more important (Mendez et al., 1999). The decrease in wave induced shear stress may inhibit resuspension of bottom material reducing local turbidity.

A study is carried out in the field and in the laboratory with the objectives to quantify the effects of vegetation canopy on current velocities, wave propagation and sediment transport. In this paper, the interaction between vegetation and processes of resuspension and sedimentation is considered separately from the results of field and laboratory experiments and their contribution to a clear water phase is discussed.

2. EXPERIMENTS

2.1. Field study

The field measurements were carried out in the Gouwzee, The Netherlands. The Gouwzee is a shallow, relatively turbid lake located along the western shore of the Markermeer (Fig. 1). The area is about 2100 ha and mean water depth ranges from 1.8 metres from the southern part of the lake to 2.2 metres in the northern part. The bed consists mainly of clay and fine silt. As a result the Gouwzee is highly susceptible to wind and wave disturbance, resulting in relatively easy resuspension of sediments and high turbidity levels. Suspended sediment concentrations can increase from 20 mg l^{-1} in fair weather to values over 300 mg l^{-1} during storm conditions. Monitoring carried out in years prior to the study had indicated submerged vegetation to be concentrated at the southern end of the lake.

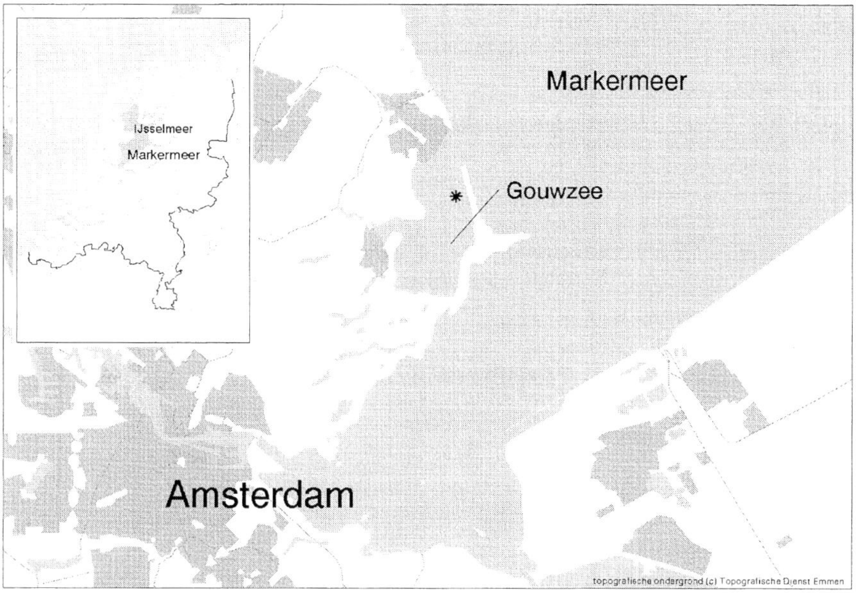

Figure 1. The Gouwzee and the measurement location (*).

A tripod system was positioned in the Gouwzee in order to measure current velocities, wave heights and the resulting turbidity in the water as function of the wind climate (Fig.1). The tripod system contained an electromagnetic flow meter (EMF, Delft Hydraulics), a pressure transducer (Keller, No. 56) and turbidity sensors (BTG MEX-3; BTG Bonnier Technology Benelux BV). The EMF measured the water flow along two perpendicular horizontal axes. The accuracy of the flow meter was 0.01 m s^{-1}. The pressure transducer measured fluctuations in water pressure due to wave propagation. The accuracy of the transducer was 50 Pa. Significant wave heights and corresponding wave periods were calculated from the pressure data. The turbidity sensor measured an increase in the suspended sediment concentration by a decrease in the light extinction. The output from the turbidity sensors was calibrated to concentrations (in mg l^{-1}) with in-situ water samples.

The tripod system was stationed for apr. 3 weeks. The local water depth was 1.8 m. Current velocities and turbidity were measured at a height of 0.10 m above the bed. Burst mean data values were obtained from 10 minute time series, at one hour intervals. The data were recorded at a frequency of 4 Hz and stored in a pc-unit.

Continuous wind measurements were carried out at a station on the dike and obtained from the Royal Dutch Meteorological Institute.

The time averaged value of the wave induced bed shear stress ($\tau_{b,w}$) is related to the instantaneous fluid velocity just outside the boundary layer (U_δ) and reads as:

$$\tau_{b,w} = \frac{1}{2}\rho f_w \left(\hat{U}_\delta\right)^2 \tag{1}$$

where

$$f_w = \textit{wave friction coefficient} = 0.09\left(\frac{\hat{U}_\delta \hat{A}_\delta}{\upsilon}\right)^{-0.2} \tag{2}$$

For a current the overall time-averaged bed shear stress ($\tau_{b,c}$) can be defined as:

$$\overline{\tau}_{b,c} = \frac{1}{8}\rho f_c \overline{U}_{10}^2 \tag{3}$$

with U_{10} = current velocity measured at 0.10 m above the bed. The friction factor follows from the Chézy-coefficient (C) and the acceleration of gravity (g):

$$f_c = \frac{8g}{C^2} \tag{4}$$

yielding for hydraulic rough flow (Van Rijn, 1993):

$$f_c = \textit{current friction coefficient} = 0.24\log\left(\frac{12h}{k_s}\right)^{-2} \tag{5}$$

where k_s = Nikuradse roughness parameter and h = water depth.

2.2. Laboratory study

Sediment is kept in suspension by turbulence generated by waves or currents. Therefore not only the intensity but also the size of the turbulent eddies is important. Their product determines the mixing capacity and is expressed in the so called eddy diffusivity. Wave induced shear stress is able to resuspend bed material but is not capable of keeping sediment in suspension.

One goal of the laboratory experiments was to study the effect of submerged vegetation on the turbulent intensity and eddy diffusivity generated by currents. A further goal was to study the energy dissipation of waves by vegetation.

The experiments were conducted in the Tidal Flume at Delft Hydraulics. Experiments were carried out separately with waves or a current. Plants were attached to the bottom of the 130 m long flume over the full width (= 1 m) of a 3 to 6 metre section.

Water velocity and wave heights were measured at several locations relative to the vegetation. Velocity was measured with a Electro Magnetic Flow meter (EMF) and an immersible Laser Doppler Velocity Meter (LDA) with measuring accuracies of 0.01 m s^{-1} and 1 10^{-4} m s^{-1}, respectively. Wave heights were recorded with a conducting wave height meter (GHM) with an accuracy of approximately 1 10^{-4} m.

For the experiment with a current only, fine sediment (china clay) with a fall velocity of approximately 0.1 mm s^{-1} was added. Suspended sediment concentration profiles were measured with an Optical Silt Concentration Meter (OSLIM).

Two types of natural vegetation were used: *Calitriche hamulata* and *Ceratophillum demersum* (Fig. 2). Experiments were also carried out with artificial vegetation of the type *Egeria densa*. It is available in 10 cm long strips, which can be attached to each other to form a complete plant of the desired length.

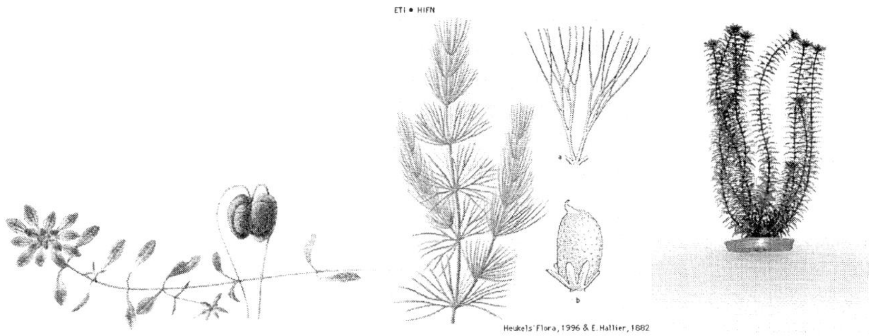

Figure 2. a) *Calitriche hamulata*, b) *Ceratophillum demersum* and c) *Egeria densa*.

The experiments were carried out with values for hydraulic parameters typically found in Dutch shallow lakes. Current velocities ranged from 0.05 to 0.20 m s^{-1} at a water depth of 0.8 m. Regular waves were generated with a height of 0.05 m and a period of 1 s in water depth of 0.4 m. Plant length ranged between 0.10 to 0.50 m.

The total turbulent kinetic energy per mass unit k [m^2 s^{-2}] is calculated according to:

$$k = 0.5 \times \left(u'^2 + v'^2 + w'^2\right) \tag{6}$$

with u',v' and w' the turbulent intensity in the x, y and z-direction, respectively (Tennekes & Lumley ,1972).

Fourier analysis was used to calculate the turbulent energy spectrum. The eddy diffusivity Γ can be obtained by (Bendat & Piersol (1971) and Uittenbogaard (1995)):

$$\Gamma(L) \approx L\left[\int_0^{1/L} E(1/L)d(1/L)\right]^{1/2} \tag{7}$$

Where
E= turbulent energy spectrum (m^2 s^{-1})
L= length (mm)

3. RESULTS

3.1. Field measurements

Field data from the period 29 May to 9 June were analysed (Fig. 3). The measurement period started with fair weather (winds up to 2 m s^{-1}). Both the waves and current velocities were hardly detectable (<0.05 m and 0.02 m s^{-1} resp.). Suspended sediment concentrations were ~50 mg l^{-1}. When wind velocities increase wave heights increase as well. Current velocities on the other hand remain low (<0.05 m s^{-1}), while suspended sediment concentrations increase. For example on The 3rd of June, wave heights up to 0.2 m and concentrations of 300 mg l^{-1} were recorded. It appears that the increase in concentration lags the increase in wave height, which suggests that a certain threshold has to be overcome before resuspension occurs. The wave induced shear stress was calculated according to Eq. 1. Erosion of the bed occurred at a critical shear stress of 0.01 N m^{-2}. These values are very low compared to erosion thresholds found for other fresh water systems in The Netherlands (varying between 0.16 and 0.7 N m^{-2}, reported in Anderson and Pejrup, 1999) and correspond to thresholds found for "fluffy" layers, 0.02-0.08 N m^{-2} (Gust & Morris, 1989; Maa et al., 1998).

There appears to be a decrease in measured sediment concentrations in time for similar values of excess bottom shear stress (Fig. 3). It is not clear if this is due to the presence of the tripod or instead maybe the composition of the bottom material changes locally as result of repeated resuspension.

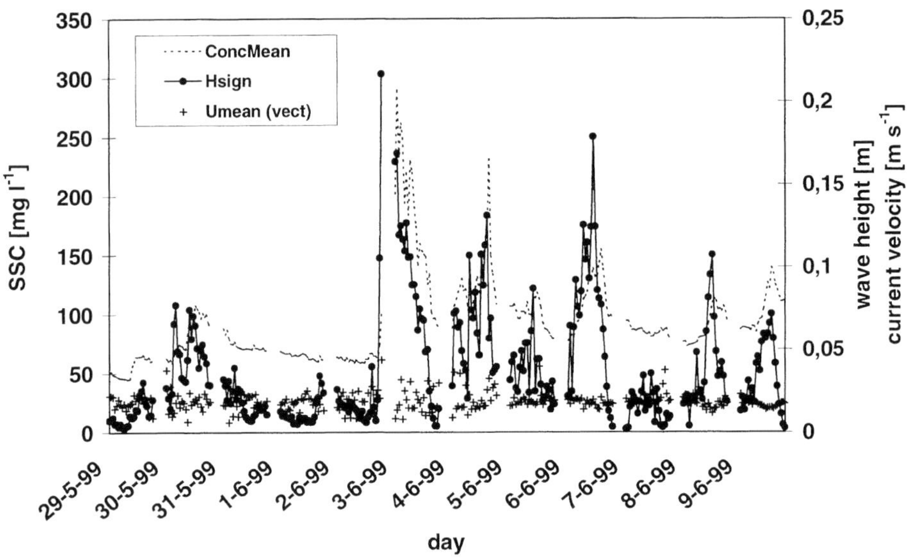

Figure 3. Significant wave height, mean suspended sediment concentration and mean current velocity measured in the Gouwzee.

In Figure 4 the average of the instantaneous suspended sediment concentration is plotted as a function of the mean value of local significant wave height. The correlation is clearly visible although some scatter is present. No correlation between the mean current velocity and the suspended sediment concentration could be found.

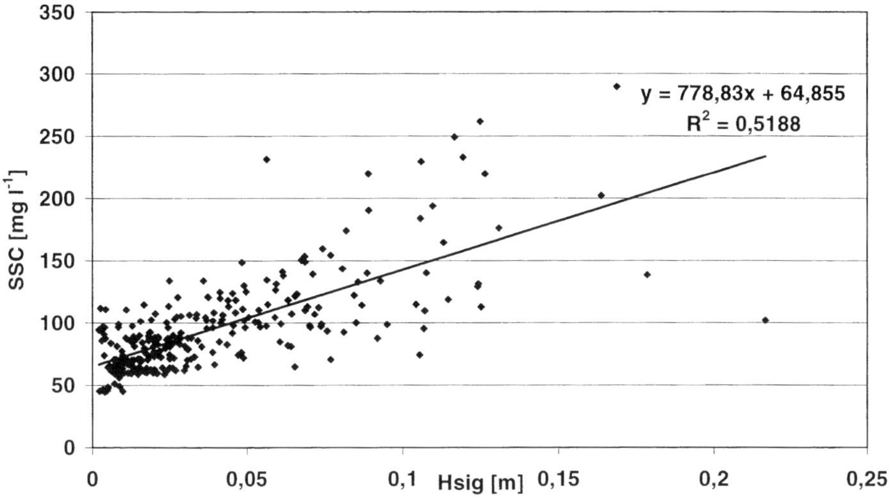

Figure 4. The mean suspended sediment concentration as function of significant wave height.

3.2. Laboratory measurements

Current only

When not otherwise indicated, velocities are in the x-direction over the length of the flume. All distances are given in metres with respect to the beginning of the vegetation (x= 0). Results of experiments with artificial vegetation 0.50 m long, a current velocity of 0.05 m s^{-1} and a water depth of 0.80 m are indicated in Figure 5. Average velocities are given at different positions in the flume. Before the flow reaches the vegetation canopy (x=-2) the velocity structure is relatively undisturbed and the profile approximates the logarithmic shape.

At the margins of the field (x=0) the effect of the vegetation can be detected in the lower part of the profile. Within the vegetation canopy it decreases rapidly for positions of 1/3 (x=0.7), halfway (x=1.5) and to approximately zero at the end of the vegetation canopy (x=2.1). Above the plants the velocity increases to almost twice the depth average value.

At a height of 0.50 m (the top of the plants) maximum values of the turbulent kinetic energy occurred (see Eq. 1). Inside the canopy energy drops by approximately 30-50%.

Lengthscales of turbulent structures decreases from approximately 0.20 m outside to values of 0.02 m within the canopy at 0.12 m from the bottom (WL | Delft Hydraulics, 1998). As a result eddy diffusivity will decrease by a factor of 10 - 20 (Eq. 7).

A constant sediment concentration of approximately 200 mg l^{-1} over the depth was generated and maintained in the flume. Figure 6 shows concentration profiles at several positions within and outside the vegetation canopy as recorded with the OSLIM. The influence of the vegetation on turbidity appears to be minimal. Only in the lower 0.30 m of the water column can a significant decrease in concentration be observed. However, the drop in concentrations is limited to a value of 10 mg l^{-1} only and appears not to be enough to improve the transparency of the water.

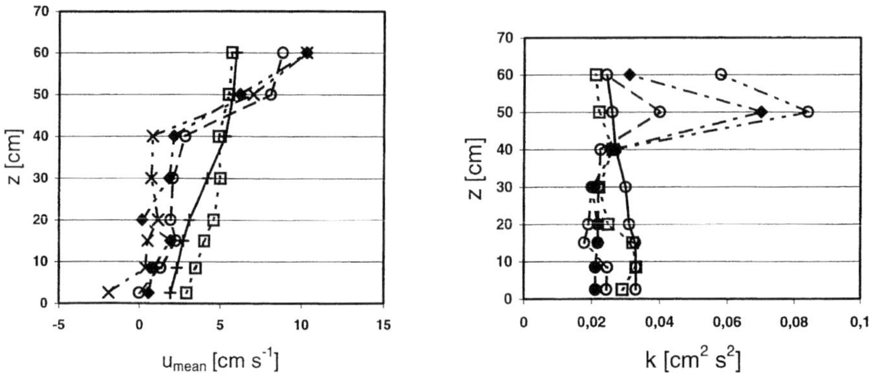

Figure 5. Experimental results for artificial vegetation and current of 5 cm s^{-1}. a) average velocity and b) kinetic energy. Measurement locations are at xp=-2 m (•), x=0 (+), x=0.7 (o), x=1.5 (♦) and x=2 m (x), see text.

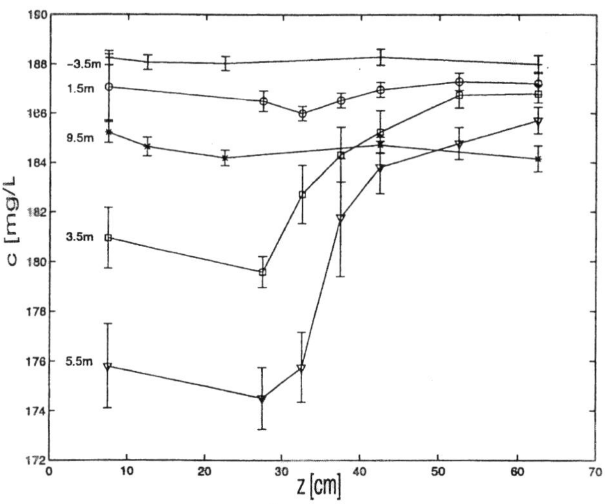

Figure 6. Sediment concentrations profiles at several locations in the flumes at a free current velocity of 5 cm s^{-1}.

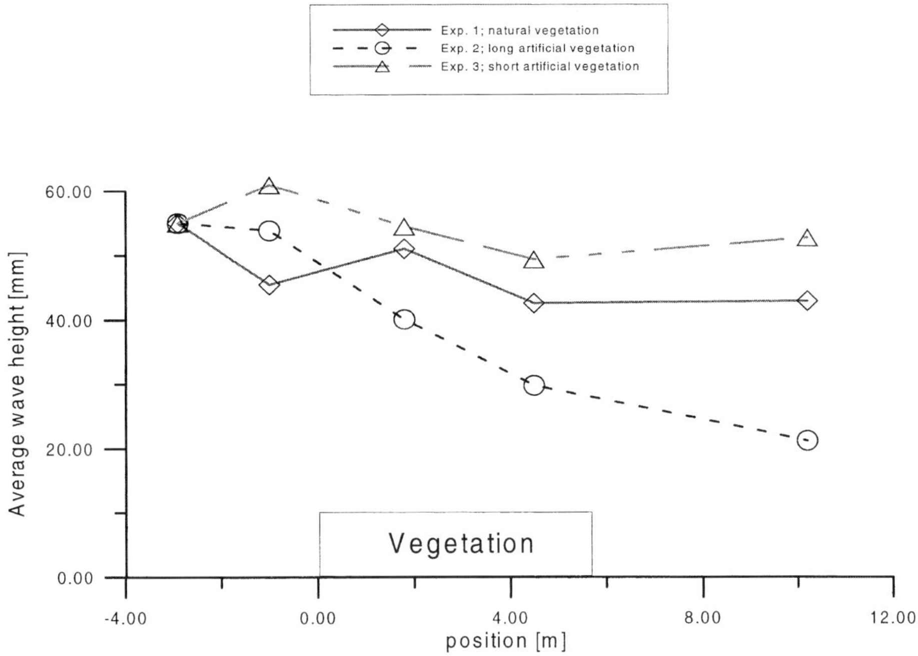

Figure 7. Change in wave height when passing the vegetation canopy.

Waves only

The measured change in average wave height over three types of vegetation is shown in Figure 7 for:

1. Short artificial vegetation 0.1 m high,
2. Natural vegetation 0.5 m high, and
3. Long artificial vegetation 0.5 m high .

All values of wave height were normalised relative to the height measured 3 m before the vegetation canopy, which was set to be 0.055 m. The canopy was almost 6 m long and the water depth was fixed at 0.4 m.

For the long natural vegetation the drop in measured wave height was approximately 20%. For the less flexible long artificial vegetation this was as high as 50% while for the short plants only a 5 % reduction was measured.

When entering the canopy of short artificial vegetation, waves experienced a drop in water depth together with an increase in bottom roughness. This resulted in an initial increase in height followed by a loss of energy due to the higher bottom friction. The net result was a damping that lies close to zero. However, over the plants themselves a decrease of approximately 5% was measured.

A return current was observed in the flume immediately ahead of the vegetation field. This is probably a result of wave set-up due to the presence of the plants. It is likely that this is an artifact of the experimental set-up as the plants were able to block the complete water column over the full width. Therefore no strong conclusions should be drawn from the increase in wave height observed immediately in front of the short vegetation field (see Fig. 7).

4. DISCUSSION

The field measurements show a correlation between local instantaneous wave height and suspended sediment concentrations. No correlation between current velocity and sediment concentration was recorded. This indicates that local turbidity levels are caused by local wave activity.

Increase in suspended sediment concentration occurred at very low bottom shear stresses (<0.01 N m^{-2}) . These values correspond to values reported for fluffy layers, see for instance Maa et al. (1998).

The laboratory experiments show that submerged vegetation causes a decrease in flow velocity and turbulent kinetic energy in the lower part of the water column. Moreover the vegetation restricts the size of turbulent eddies. Consequently the eddy diffusivity or turbulent mixing capacity of the flow decreases by a factor of 10-20 within the vegetation canopy. Due to the experimental set-up the water was pushed through and over the canopy as the vegetation covered the full width of the flume. Therefore, it can be expected that in the field situation this decrease in diffusivity might be even higher, because current velocities will be lower.

However, no significant effect on turbidity was measured. Suspended sediment concentrations did not decrease strongly enough to improve the clarity of the water (Fig. 6). Model calculations with a 1-DV advection diffusion model, adapted to incorporate the presence of vegetation (Winterwerp & Uittenbogaard, 1997; WL I Delft Hydraulics, 1998) indicate that even 50 m of vegetation will not result in a significant decrease in the turbidity level (Fig. 8).

Results of the wave experiments make clear that the degree of wave damping depends on plant characteristics such as length and flexibility. A decrease in height of only 5 % was measured for short artificial vegetation, to values as high as 50 % for the long artificial plants. For long natural (flexible) vegetation a reduction in wave height of 20% was found. These results correspond to values reported by Mendez et al. (1999) from experiments of Asano et al. in 1988. According to this study a canopy of 8 m long consisting of artificial seaweed of 25 cm in a water depth of 45-52 cm caused a decrease in wave height of 20-30 % with an initial height of 6-9 cm.

For a shallow wind swept lake like the Gouwzee, wave height can be considered as the main parameter that determines the degree in turbidity of the water system. The field measurements showed that erosion of the sediment bed by waves is most likely restricted to a fluffy layer characterised by a relatively low critical shear strength (<0.01 N m^{-2}). Figure 3 shows that wave heights of 0.1 - 0.15 m are sufficient for a significant increase in turbidity in the Gouwzee. We also see that the turbidity substantially decreases for wave heights below 0.05 m. Based on Eq. 1 and assuming that linear wave theory can be applied in this situation, a critical wave height for resuspension of bed material can be calculated to be 0.07 m, which seems to correspond with the field observations. A reduction in wave height of about 50 -70% is thus needed to obtain values for the wave induced shear tress which will lie below the critical value for erosion.

Based on the laboratory results and assuming a constant decrease over length a vegetation canopy of less than 100 metres would be sufficient to reduce turbidity. This is a rather cautious estimate based on a wave damping coefficient of 5-10% over 5 metres of vegetation cover. Increasing this coefficient to 20%, the transition of turbid to clear water would be achieved within a field width of 30 m. This latter is confirmed by field observations made by Van den Berg et al. (1998).

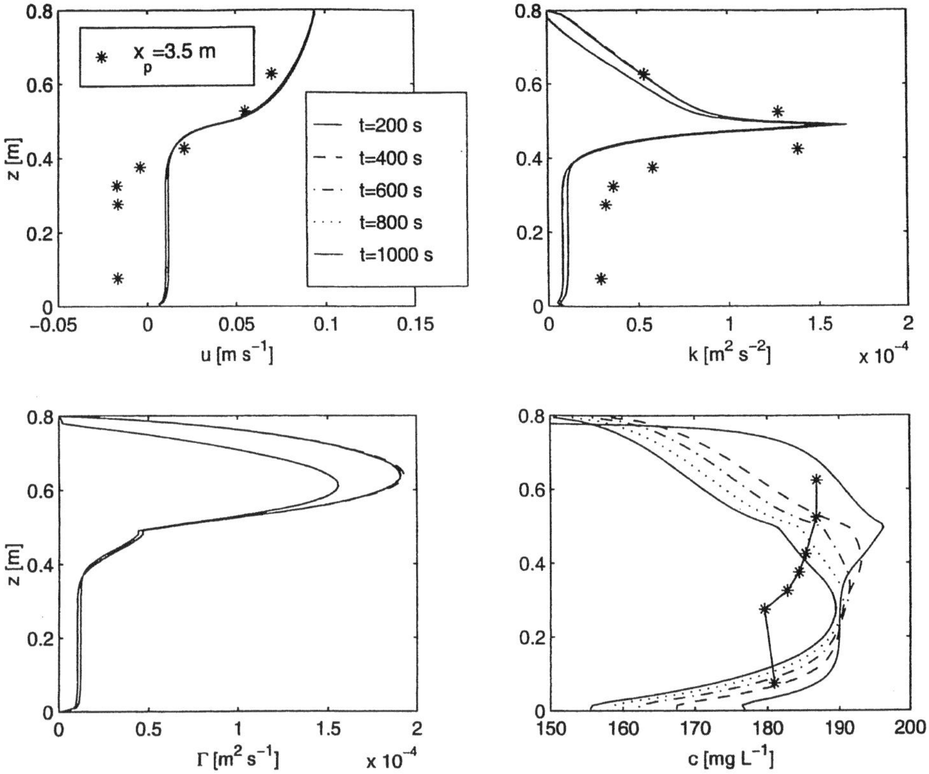

Figure 8. Calculated (lines) and measured (*) results with vegetation for a velocity of 4 cm s^{-1}. a) velocity (m s^{-1}), b) kinetic energy (m^2s^{-2}), c) eddy diffusivity (m^2s^{-1}) and d) sediment concentrations (mg l^{-1}).

Results from the field study revealed that, in relatively shallow wind swept lakes, turbidity is mainly caused by resuspension from a so called 'fluffy layer' induced by wind waves. The laboratory experiments showed that within a submerged vegetation canopy a reduction in turbidity can be associated with wave damping and a restriction in the local resuspension of bed material. Brouwer (1998) discussed the fact that clarity of the water was more strongly dependent on biomass than on percent coverage in the Gouwzee. At increasing coverage of vegetation, visibility in the water column gradually improved, but not until biomass reached its peak in August did visibility increase such that the bottom could be seen. This fact might be explained by the results from the laboratory experiments, where a positive correlation between the damping effect on waves and the vegetation length was found.

Although advective transport seemed not to be a dominant process, the blocking effect on the current by the vegetation could not be investigated during this study due to the settings of the laboratory experiments. It can be argued that this blocking effect will also increase with increasing biomass. Little is known about two-dimensional effects of an underwater vegetation canopy on water movement and associated turbidity levels. In future, this aspect will be investigated more thoroughly by means of a 2-Dh model-study.

The vegetation is capable of improving its own light climate, but its ability to influence water clarity outside the vegetation is limited to only a short distance (see also Van den Berg et al., 1998; Brouwer, 1998). Implementation of submerged vegetation in management to improve lake water clarity will need to take into consideration lake morphology, hydrology and the life-cycle of vegetation species. The model-study under construction will be a useful tool to identify the relevant processes as a function of extension and growth of the submerged vegetation and will be used as a decision supporting system for implementation of new management strategies.

References

Anderson, T.J. and Pejrup, M., 1999, Determination of erosion thresholds and erosion rates on fine-graines freshwater sediments in The Netherlands using a portable EROMES-system, *Internal report*, Institute of Geography University of Copenhagen, Denmark, 16 p.

Bendat, J.S. and Piersol, A.G., 1971, Random data: analysis and measurements, Wiley & sons, 1971.

Brouwer, L., 1998, Charophytes and turbidity: the impact of *Nitellopsis obtusa* vegetation on water clarity in the Gouwzee, *RIZA work document 98.151x*, Lelystad, The Netherlands.

Dawson, F.H. and Charlton, F.G., 1988, Bibliography on the hydraulic resistance or roughness of vegetated watercourses, *Freshwater Biological Association*, Occasional Publications No. 25. River Laboratory, Warehem, Dorset, UK.

Dubi, A. and Torum, A., 1997, Wave energy dissipation in kelp vegetation, *Proceedings of the Twenty-Fifth Coastal Engineering Conference*, ed. B.L. Edge, 142-156.

Fonesca, M.S. and Fisher, J.S., 1986, A comparison of canopy friction and sediment movement between four species of seagrass with reference to their ecological restoration, *Marine Ecology-Progress Series* (29), 15-22.

Gambi, M.C., Nowell, A.R.M. and Jumars, P.A., 1990, Flume observations on flow dynamics in *Zostera marina* (eelgrass) beds, *Marine Ecology-Progress Series,* (61), 159-169.

Gurnell, A.M. and Midgley, P., 1994, Aquatic weed growth and flow resistance: influence on the relationship between discharge and stage over a 25 year river gauge station record, *Hydrological progress,* (8), 63-73.

Gust, G. and Morris, M.J., 1989, Erosion thresholds and entrainment rates of undisturbed in situ sediments, *Journal Coastal Research,* (5), 87-99.

Hosper, S.H., 1997, Clearing lakes: an ecosystem approach to the restoration and management of shallow lakes in the Netherlands, *Ph.D. thesis*, landbouwuniversiteit Wageningen, 168 p.

James, W.F. and Barko, J.W., 1990, Macrophyte influence on the zonation of sediment accretion and composition in a north-temperate reservoir, *Archive fur Hydrobiologie,* (120) 2, 129-142.

Kutija, V. and Hong, H.T.M., 1996, A numerical model for assessing the additional resistance to flow introduced by flexible vegetation, *Journal of Hydraulic Research*, (34) 1, 99-114.

Maa, J.P.-Y., Sanford, L. and Halka, J.P., 1998, Sediment resuspension characteristics in Baltimore Harbour, Maryland, *Marine Geology*, (146), 137-145.

Meijer, M.-L., 2000, Biomanipulation in the Netherlands; 15 years of experience, *Ph.D. thesis*, Landbouwuniversiteit Wageningen, 208 p.

Méndez, F.J., Losada, I. and Méndez, M.A., 1999, Hydrodynamics induced by wind waves in a vegetation field, *Journal of Geophysical Research*, (104) C8, 18,383-18,396.

Nepf, H.M., 1999, Drag, turbulence, and diffusion in flow through emergent vegetation, *Water Resources Research*, (35) 2, 479-489.

Pethick, J., Leggett, D. and Husain, L., 1992, Boundary layers under salt marsh vegetation developed in tidal currents and waves, *Vegetation, erosion processes and environments*, ed. J.B. Thornes, John Wiley & Sons Ltd, London, 113-124.

Petryk, P. and Bosmaijan, G., 1975, Analysis of flow through vegetation, *Journal of the Hydraulics Division, Proceedings of the American Society of Civil Engineers (ASCE)*, (101) HY7, 871-884.

Petticrew, E.L. and Kalff,J., 1992, Water flow and clay retention in submerged macrophytes beds, *Canadian Journal of Fisheries and Aquatic Sciences*, (49), 2483-2489.

Perrow, M.R., Meijer, M.-L., Dawidowicz, P. and Coops, H., 1997, Biomanipulation in shallow lakes: state of the art, *Hydrobiologia*, (342/343), 355-365.

Scheffers, M., Van den Berg, M.S., Breukelaar, A., Coops, H., Doef, R.W. and Meijer, M.-L., 1994, Vegetated areas with clear water in turbid shallow lakes, *Aquatic Botany*, (49), 193-196.

Shi, Z. Pethick, J.S., Burd, F. and Murphy, B., 1996, Velocity profiles in a salt marsh canopy, *Geo-Marine letters*, (16), 319-323.

Táncos, I.C., De Vries, M. and Cornelisse, J.M., 1998, Interaction of submerged vegetation and physical processes; 1998 (in Dutch), *WL|DELFT HYDRAULICS, report Z22473/Z2633*, Delft, The Netherlands.

Tennekes, H. and Lumley, J.L., 1972, A first course in turbulence. *The MIT Press*, Cambridge.

Uittenbogaard, R.E., 1995, The importance of internal waves for mixing in a stratified estuarine tidal flow, *Ph.D. Thesis*, Delft University.

Van den Berg, M.S., 1998, Charophyte colonization in shallow lakes; processes, ecological effects and implications for lake management, *Ph.D. thesis*, Vrije Universiteit Amsterdam, 137 p.

Van den Berg, M.S., Coops, H. Meijer, M.-L., Scheffer, M. and Simons, J., 1998, Clear water associated with a dense Chara vegetation in the shallow and turbid Lake Veluwemeer, The Netherlands, *The structuring role of submerged macrophytes in lakes*, ed. E. Jeppese, M. Sondergaard and K. Christoffersen. Springer-Verlag, New York, 339-352.

Van Rijn, L.C., 1993, Principles of sediment transport in rivers, estuaries and coastal seas, Aqua publications, Amsterdam, 658 p.

Verduin, J.J. and Backhaus, J.O., 2000, Dynamics of plant-flow interactions for the seagrass *Amphibolis antartica*: Field observations and model simulations, *Estuarine, Coastal and Shelf Science*, (50), 185-204.

Winterwerp, J.C. and Uittenbogaard, R.E., 1997, Sediment transport and fluid mud flow, *WL|DELFT HYDRAULICS, report Z2005*, Delft, The Netherlands.

Fine Sediment Dynamics in the Marine Environment
J.C. Winterwerp and C. Kranenburg (Editors)

Sedimentation in a coastal mangrove system, Red River Delta, Vietnam

B.M. Janssen-Stelder[a], P.G.E.F. Augustinus[a] and W.A.C. van Santen[a]

[a] Institute for Marine and Atmospheric Research Utrecht (IMAU)
Dept. of Physical Geography, Utrecht University
P.O. Box 80.115, 3508 TC Utrecht, The Netherlands

The mangroves in the Red River Delta in Vietnam are exposed to impacts of climatic change and anthropogenic forcing. Therefore, their ecological function as well as their function as coastal and riverbank defence are threatened. An enhancement of the process knowledge can help to predict further development of the coastal mangroves in the Red River Delta, which can be applied in management issues. In this paper field measurements are used to identify the dominant hydrodynamic processes on the banks of an estuary where bare mudflats grade into mangrove areas. Furthermore the resulting sedimentation pattern is described. The measuring results show that there is no set-up of the water level caused by the mangroves. The mangroves do reduce the current velocity significantly. This leads to a fining of the sediment from the riverbank into the mangrove forest. Sedimentation rates are high in the frontage of the pioneer zone and low in the back of the mangrove forest. The mangroves appear to colonise the mud banks when these reach a critical height.

KEYWORDS
Mangroves, sedimentation, hydrodynamics, Red River Delta, Vietnam

1. INTRODUCTION

1.1 General introduction

Mangroves are trees, which form tidal forest ecosystems in sheltered saline to brackish environments. They thrive in the tropics, but can extend into the temperate zone in areas where frost is rare and of short duration (Augustinus, 1995). Mangrove forests have many functions. They serve as a reservoir and refuge for many animals and plants and in particular have a nursery and shelter function for several kinds of fishes and crustaceans (e.g. shrimps). They support commercial and recreational fisheries and deliver several direct and indirect services to the local population, especially in the field of potentially renewable products, like firewood (Hamilton and Snedaker, 1984). Furthermore, they have an important function in coastal and riverbank defence.

Impacts of climatic changes (e.g. sea level rise) and anthropogenic forcing (e.g. overexploitation, reduced sediment supply due to dam building) threaten the sustainability of the mangrove system of the Red River Delta in Vietnam. A research programme was formulated to study the manner in which mangroves in the Red River Delta environment facilitate the deposition of muddy sediments. An enhancement of the process knowledge can help to predict further development of the mangrove covered estuary banks in this area, which can be applied in management issues. In this paper field measurements are used to describe the sedimentation processes. The aim of the study is to identify the dominant hydrodynamic processes on the banks of an estuary where bare mudflats grade into mangrove areas and discover to what sedimentation pattern this leads.

2.1 Characteristics of mangroves

Mangroves are adapted to grow under environmental conditions such as high salinities and waterlogged, often muddy soils. Their most important adapted feature is the root system (Figure 1). In addition to wind stress, mangroves must cope with strong (tidal) currents and high waves. Normal taproots cannot develop, due to the anaerobic conditions of loose and waterlogged soils. Mangrove trees therefore need special roots for fixation and for aeration. There are basically two root systems in which the requirements for stability and aeration are combined. In the root system of the *Avicennia* type stability is obtained by a star shaped network of cable roots, which radiate out from the trunk at a depth of 20 to 50 cm (Figure 1). From these cable roots, anchor roots shoot downward, while vertical aeration roots, pneumatophores, are pushed up. The height of the pneumatophores is in the order of 25 cm, depending on the species. The rooting of the *Rhizophora* type consists of a system of prop roots, arching out from the trunk and anchoring at a depth of about 30 cm (Figure 1). The prop roots contain lenticles for aeration. Most mangrove species have nutritive roots allowing normal respiration in aerobic conditions. In the Red River Delta the mangrove vegetation is dominated by *Aegiceras corniculatum* and *Kandelia candel*, which appears in a dwarf form. In this case the shrub itself has the same function as the dense network of prop roots or pneumatophores and an aboveground root system is absent.

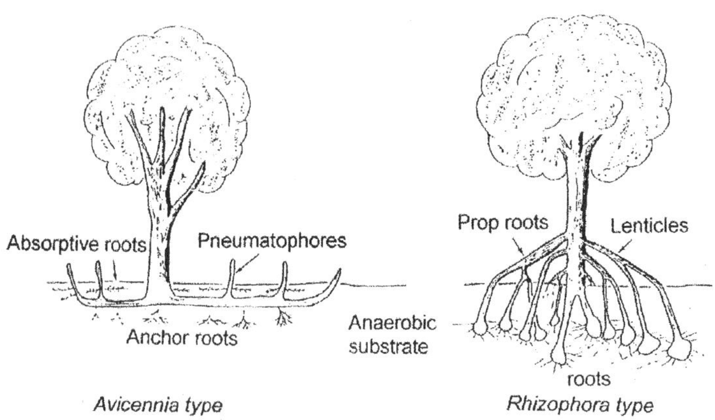

Figure 1: Mangrove root systems.

Due to the aboveground root system, mangroves are able to influence hydrodynamics. Mangrove forests along estuaries and coasts are periodically inundated during high tide. The dense and extensive network of trunks and aboveground roots induces an increased bed roughness and currents and waves are altered (Bunt and Wolanski, 1980; Wolanski et al., 1980). The reduction of currents and the attenuation of waves result in sedimentation as the dominant process in mangrove forests (e.g. Augustinus, 1978; Walsby and Torckler, 1992).

1.3 Study area

The Red River Delta is located about 110 km southeast of Hanoi, in the north of Vietnam (Figure 2). The estuary contains the last significant remnant of the mangrove/mudflat ecosystem along the coast of Vietnam. Large parts of the mudflats and mangroves have recently been impounded for shrimp ponds. Part of the mangroves along the estuary of the Red River mouth is now part of the Xuan Thuy Reserve and has RAMSAR status. Mangroves are protected in this area. A larger variety of species are beginning to return to the area. A natural community of *Soneratia caeseolaris*, *Bruguira gymnorhiza, Aegiceras corniculatum* and *Kandelia candel* occupies the area.

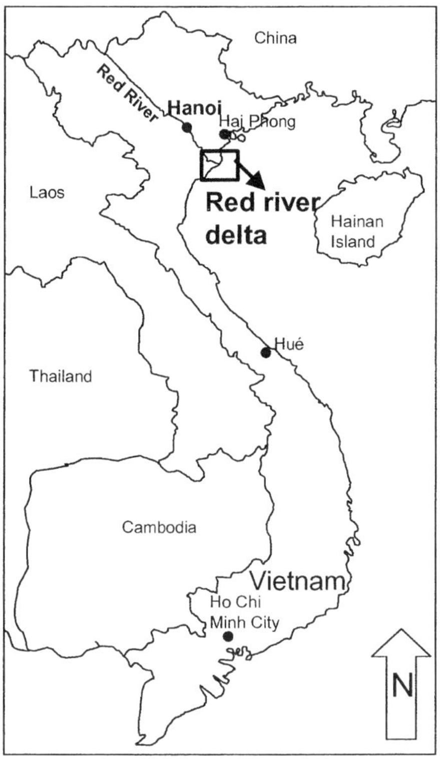

Figure 2: Location of the Red River Delta.

The Red River delta has a diurnal tide with a tidal cycle of about 25 hours. The spring-neap tidal cycle has a duration of two weeks (Figure 3). During neap tide the tidal curve has an irregular, non-sinusoidal shape. The mean tidal amplitude varies between 1.50 m and 1.80 m, with a maximum of 3.30 m and a minimum of 0.25 m. The area has a tropical monsoon climate with a pronounced maritime influence. Prevailing winds are north to east in the dry season (November-March) and east to southeast in the wet season (April-October). Typhoons and tropical cyclones frequently occur between July and October, causing severe erosion along the coast.

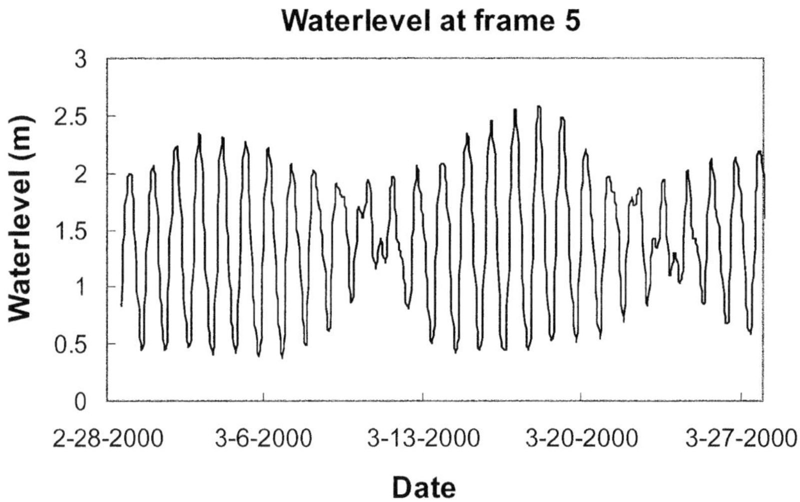

Figure 3: Tidal curve during the measuring period in the mouth of the Red River.

Both accretion and erosion take place in the estuary of the Red River delta, but the accretion rates are twice as high as the erosion rates. Accretion in the delta has resulted in a horizontal extension of the mudflats of 500-600 m per year and a vertical sedimentation of 10 cm/y (Kinh, 1992).

The study area used for the sedimentation rate measurements was located in the RAMSAR mangrove area, in the mouth of the Red River (Figure 4). It is a mudflat with a small gradient (less than 2%) bordered by the Red River, two small creeks and a small dike. The bed sediment of the mudflat consists mainly of silt smaller than 20 μm. Under the mangrove vegetation the silty bed layer varies in thickness between 60 and 75 cm and overlies sandy deposits. On the mudflat, mangrove vegetation is developing from the pioneer zone close to the river to a dense vegetation cover, mostly of *Aegiceras corniculatum* and *Kandelia candel* of up to 2 m in height close to the small dike (Figure 5).

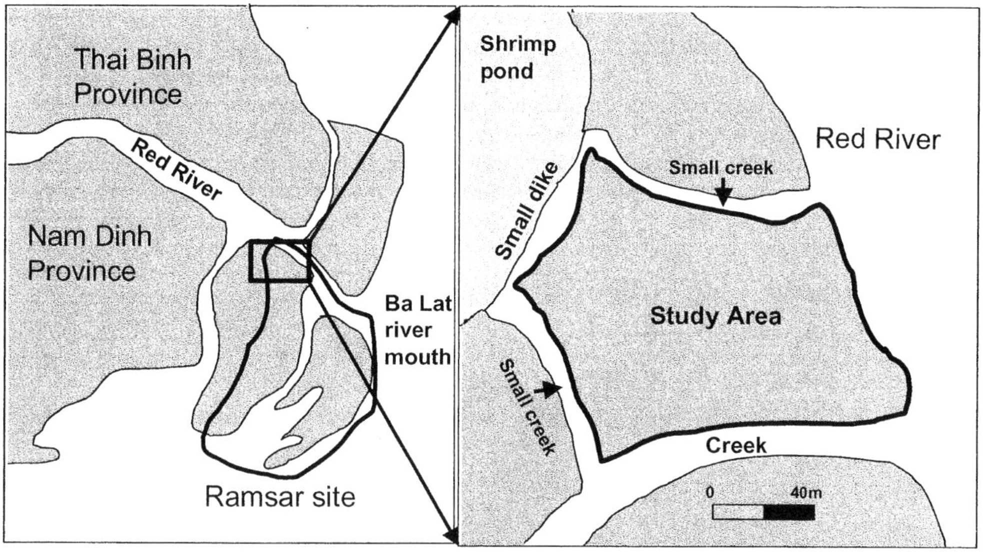

Figure 4: Location of the study area.

Figure 5: Study area.

2. METHODS

In spring 2000 a field study was carried out. The measurements from this period are assumed to be representative for the dry season. A station for meteorological measurements was built at the coast about 30 km from the river mouth. Wind speed and direction gathered by a cup-anemometer at this station were used in the analysis. At three locations within the study area (Figure 6) frames were built to measure the hydrodynamic conditions. A pressure sensor was used to measure water depth. Current velocity and current direction were measured by an electromagnetic flow meter. At the frames an optical backscatter meter measured the suspended sediment concentration. The measurements were continuous in bursts of 18 minutes per hour at a measuring frequency of 2 Hz. Three measuring frames were placed on a transect from the river to the small dike. The sites were located in the river, in the pioneer zone on the mudflat and in the densely vegetated zone (Figure 6). In order to describe the study area accurately, height data were gathered by levelling. These data were used to make a digital elevation model.

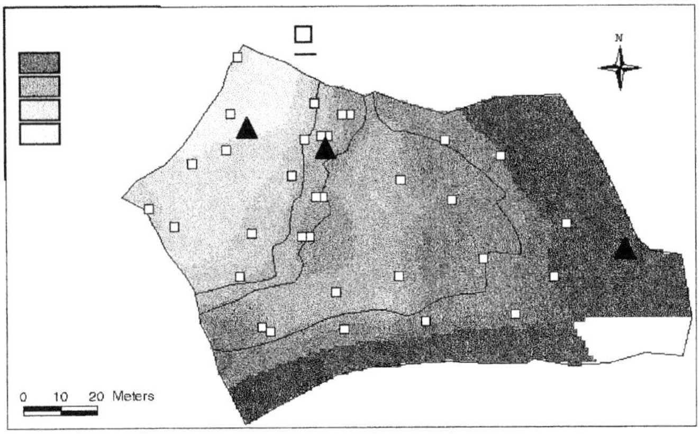

Figure 6: Elevation differences, location of the measuring frames and location of the sediment traps within the study area.

A special type of sediment trap was used to measure sedimentation rates in the mangrove system. Data on accretion rates in mangrove forests are scarce. Measurements are usually done with rods or stakes but the results show a great deal of variation, which is partly caused by the method. Rods only measure at a one point and scour effects make it difficult to indicate the measuring surface. To increase the measuring accuracy, experimental sediment traps with a larger surface were used in this study, made of carpet measuring an area of 40x40 cm. The traps were placed on the surface of the substrate during low water at 33 locations (Figure 6). During high tide, when the area is inundated, sediment is deposited on the carpets. After the measuring period of 21 days, the traps were removed and the amount and grain size distribution of the deposited sediment was analysed. In this way the net sedimentation rates were calculated for the different locations. The hourly sedimentation was calculated by dividing the amount of sediment by the time of submergence of the area. During low tide the area is completely emerged. The data from the pressure sensors on the measuring frames were used to calculate the submergence period for different zones in the area.

3. RESULTS

3.1 Wind, hydrodynamics and suspended sediment concentrations

Easterly winds dominated in the measuring period of the month of March (Figure 7). In that period the average wind speed was low at values of 2.7 ± 1.4 m/s. The maximum wind speed was 7.7 m/s.

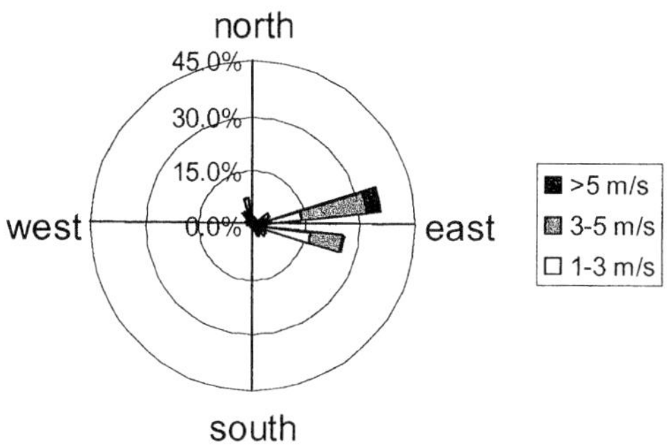

Figure 7: Wind distribution at Quat Lam, Red River Delta, Vietnam for the month March.

The water levels in the study area were tide dominated during the measuring period, wind set-up hardly occurred. The water level above local bed level at measuring Frames 5, 6 and 7 are shown in Figure 8. Due to the gradient on the mudflat, Frame 5 shows the highest water levels, followed by Frame 6 and 7. The bed elevation of Frame 6 and 7 is respectively 1.45 m and 1.70 m higher than the bed level at Frame 5, which was located in the river. The differences in the water level are caused by the differences in bed elevation. There is no phase lag in the water level at the three measuring locations or set-up of the water level caused by the mangrove vegetation.

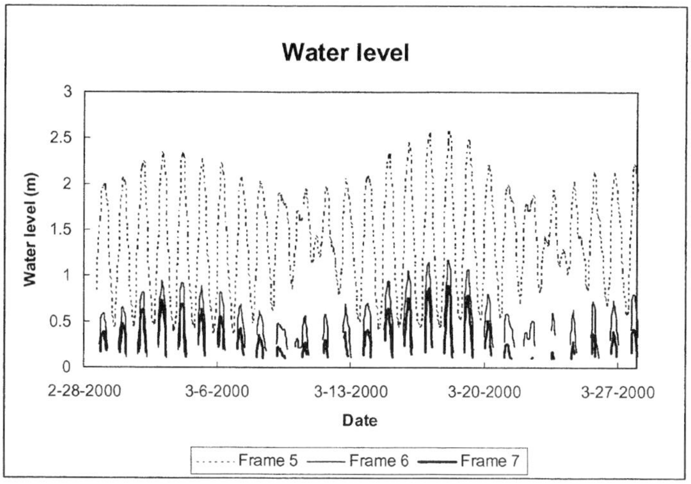

Figure 8: Water level at the different measuring locations during the measuring period.

As an example, the current velocities at the measuring frames during four tidal cycles are shown in Figure 9 and 10. The current velocity in the river (measured by Frame 5) ranged from 0.02 m/s to 0.50 m/s (Figure 9). Minimum velocities occurred at slack water, timed at one hour after high water and 2 to 3 hours after low water. The current velocity during ebbing tide always exceeded the velocities during flooding tide. Similar patterns occur on the mudflat at the locations of Frames 6 and 7. However, the velocities are much smaller and range from 0.01 m/s to 0.04 m/s (Figure 10). These values come close to the measuring accuracy of 0.01 m/s. However, a consistent pattern appeared in the current velocities so the values can be used for interpretation.

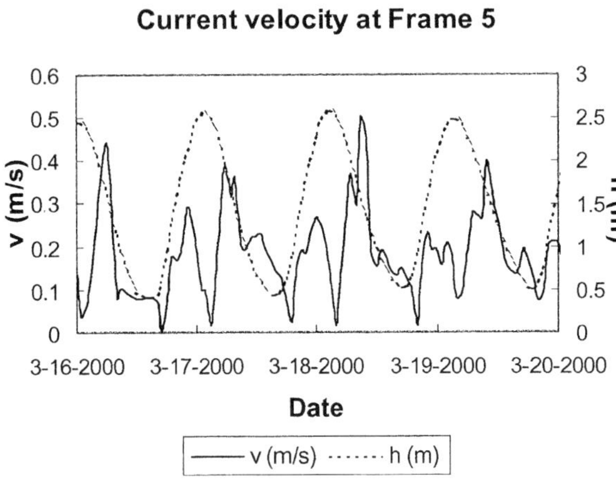

Figure 9: Current velocity and water level at frame 5 during four tidal cycles.

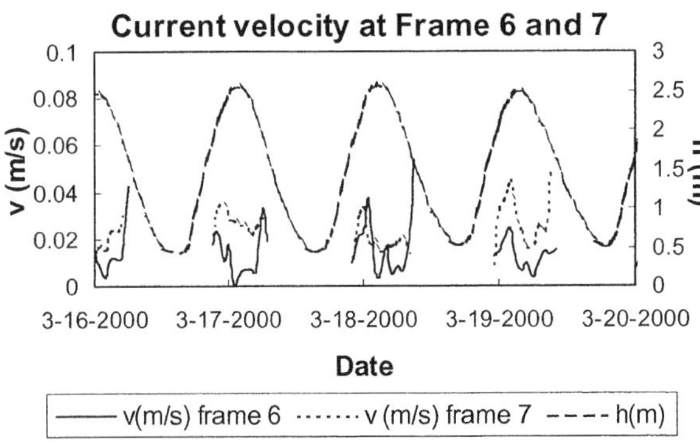

Figure 10: Current velocity at Frame 6 and 7 and water level at Frame 5 during four tidal cycles.

The average suspended sediment concentration (SSC) measured in the river at Frame 5 was 53.3 ± 78 mg/l. Figure 11 shows the large variation in SSC. Therefore the spread around the average SSC is very high. Peaks of 800 mg/l were measured (Figure 11, Table 1). The peaks in the SSC occurred predominantly at low current velocities (< 0.10 m/s) and at relatively shallow water depths of about 1 m. Apparently, during these conditions there is an optimal bed shear stress for suspending sediment. The average SSC seems to increase from Frame 5 towards Frame 6 and 7, with highest concentrations at Frame 7 (Figure 11, Table 1). The reason for this will be treated in the discussion. At both Frames 6 and 7 the peaks in SSC occurred during periods when the water was shallow (about 0.5 m) combined with low current velocities (<0.05 m/s) directly after a period of high current velocities (0.05-0.10 m/s). At Frame 7 peaks in SSC also occurred during peaks in current velocity (>0.10 m/s). The combination of shallow water and high current velocities seems to lead to high amounts of suspended sediment. During these conditions most of the sediment can be entrained. The lag between the high current velocities and the increase in SSC indicates advective transport of sediment.

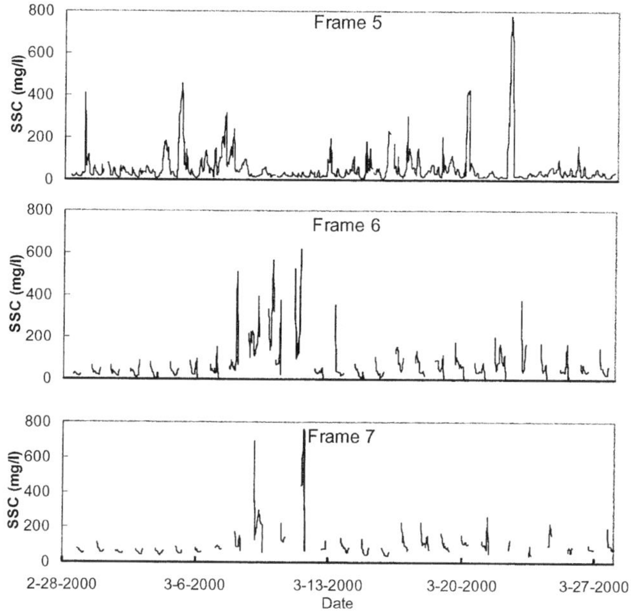

Figure 11: Suspended sediment concentrations in mg/l measured by the OBS-sensors at the three

Table 1: Values of the SSC at the three measuring frames

Frame #	*Average SSC (mg/l)*	*Standard deviation SSC (mg/l)*	*Peak value SSC (mg/l)*
5	53.3	78	800
6	73.7	89	600
7	106.0	99	750

3.2 Sedimentation

The dry weight of the sediment collected on the sediment traps in the study area ranged from 34.3 g to 681 g. On one of the traps in the northeastern corner of the study area a significantly larger amount of 937 g was collected. However, this trap was situated in a depression close to a small creek that bordered the area. Therefore, the trap was submerged for a significantly larger amount of time compared to the other traps. The water stagnated on the trap and much sediment could settle. This trap is not representative for the sedimentary conditions in the study area and therefore the trap was excluded from further analysis.

The average amount of sediment collected in the study area was 317 ± 207 g, which equals an average sedimentation of 0.8 ± 0.5 mm in the study area over a period of 21 days. In the pioneer zone and at locations in the open mangrove front, sedimentation rates are high (1.5-3.4 g/hr) (Figure 12). The higher elevation of the mudflat compared to the river, in combination with the presence of the pioneer vegetation, leads to wave and current reduction and the settling of sediment. Low sedimentation rates of 0-0.5 g/hr occurred in the densely vegetated part of the study area. Most sediment has settled in the area in front of the denser mangroves.

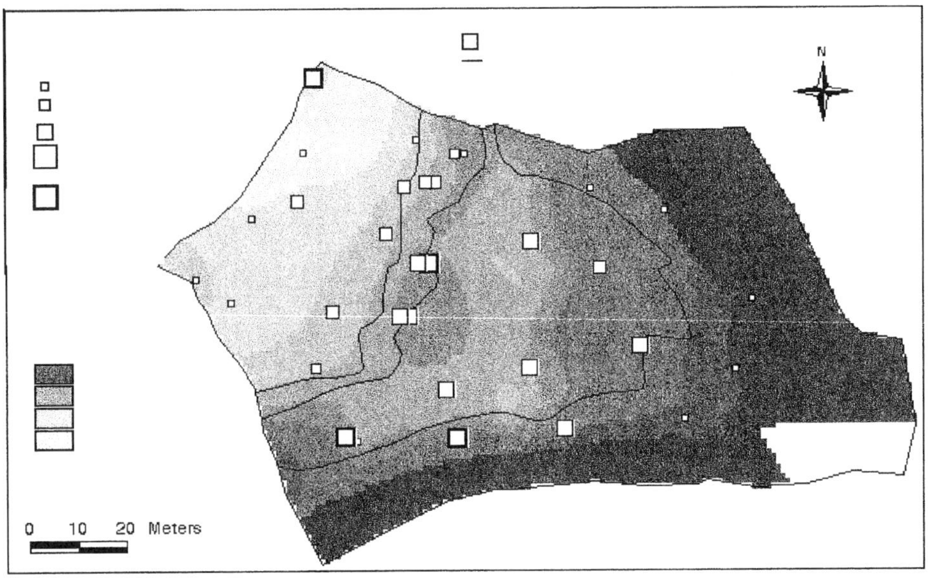

Figure 12: Sedimentation rates and elevation differences in the study area.

The area very close to the river in the eastern part of the study area has eroded. At the end of the measuring period the traps were located at a higher elevation than its surroundings. The current velocities in the river were too high to cause sediment to settle directly along the river.

The sediment in the study area consists mainly of silt (2-50 μm). The average proportion of silt found on the sediment traps is 71.5 %. The majority of the silt (66 %) is finer than 20 μm. The proportion of silt is constant throughout the study area. The average proportion of clay (< 2 μm) and sand (> 50 μm) collected on the sediment traps is both 14 %. The maximum proportions of clay were found in the north of the study area, close to the dike, in the densely vegetated mangroves. The finest fraction only settles in areas where water stagnates, such as in the north of the study area. The maximum proportions of sand were found at the riverside and decreased in the vegetated area. The coarsest fraction settles at the start of the flooding tide when the water submerges the mudflat and current velocities decrease.

4. DISCUSSION

The physical processes of the interaction between currents and waves in combination with the availability of sediment influence the amount of sedimentation in mangrove forests. The mangrove trees obstruct the flow and therefore they stimulate sedimentation. In the study area in the Red River delta, sedimentation occurred particularly in the frontage of the mangroves during the measuring period of one month in the spring. Assuming that this month is representative for the dry season, it appears that in this period the mangroves colonise the intertidal mud banks when these reach a critical height. This is opposite the conclusion of Furukawa and Wolanski (1996) who argue that mangroves create their own environment. However, measurements carried out in the wet season are needed to verify whether the mangroves in this area colonise the intertidal mud banks throughout the year.

During the study period wind speeds were low and set-up of the water level hardly occurred. Currents, more than waves, were responsible for the sediment transport from the river towards the mangrove area. Hydrodynamic conditions were calm and sufficient sediment was available for deposition. Therefore sedimentation dominated in the area. During the wet season when rougher hydrodynamic conditions occur, waves are expected to play a significant role. Then the frontage of the mangroves may have an important role in wave breaking. This could result in sedimentation further landward in the mangrove forest.

During the measuring period the largest concentrations of suspended sediment were found in the river (Frame 5) when water depths were around 1 m. At these shallow water depths sediment is resuspended. In the mangroves at Frame 6 and 7, peaks in SSC occurred after the peaks in SSC in the river. The river is the only sediment source for the mangrove forest, so SSC peaks in the mangrove forest occur when river water containing a high sediment concentration floods the area. Peaks in the SSC occurred during ebbing tide, directly after the maximum current velocity.

The average SSC increased from Frame 5 in the river towards Frame 6 and 7 in the mangroves (Table 1). This was unexpected since the river is the only sediment source for the mangrove area. There is no sediment input from the landward side because a dike borders the area. There is also no sediment input from the coast, because even during high tide the measured current direction in the river is directed downstream. This could imply that part of the sediment has been entrained locally from the mudflat leading to high SSC in the mangroves. However, the hydrodynamic energy in the mangroves is very low so resuspension is unlikely. Possibly a timelag effect can explain the apparent increase in concentrations. This is confirmed when examining the peak values of the SSC, which are quite similar for the entire study area. Furthermore, the highest peak values occurred in the river. The presence of a timelag effect makes it very difficult to draw conclusions concerning sediment transport in the study area.

The experimental sediment traps were found to work well in this sedimentary environment. After only one flooding period the traps were completely covered with a thin layer of sediment. In the remaining time of the measuring period the bed at the location of the traps looked exactly like the surroundings. This made them representative for the surrounding area. The only disadvantage of the sediment traps is that they cannot be used in an eroding area like the river levees.

5. CONCLUSIONS

One aim of the study was to identify the dominant hydrodynamic processes on the banks of an estuary where bare mudflats grade into mangrove areas. Current reduction was the main process identified during the measuring period in the spring of 2000 (dry season). The results showed that the presence of the mangroves did not influence the height of the water level. Due to very low wind velocities during the measuring period there was no significant influence of waves. The measurements did show that the mangroves caused a large reduction in current velocity. The mean current velocities in the river of (0.20 m/s) were up to ten times larger than mean current velocities on the mudflat and in the mangrove forest (0.02 m/s). This resulted in a fining of the bed sediment from the riverbank into the mangrove forest.

Another aim of the study was to describe the sedimentation pattern in the study area. Sedimentation rates were high in the pioneer zone and at locations where the vegetation became somewhat denser. In this frontage of the mangroves most hydrodynamic energy was reduced and sedimentation could take place. The mangroves seem to colonise the intertidal mud banks when these reach a critical height. Low sedimentation rates occurred in the back of the study area where sediment supply is lacking. Erosion occurred on the mudflat edge where current velocities are highest.

REFERENCES

Augustinus, P.G.E.F., 1978, The changing shoreline of Surinam. *Ph.D. thesis*, Utrecht University.

Augustinus, P.G.E.F., 1995, Geomorphology and sedimentology of mangroves. In: *Geomorphology and sedimentation in estuaries*, ed. G.M.E. Perillo. Amsterdam, Elsevier Science Publishers, 333-357

Bunt, J.S. and E. Wolanski, 1980, Hydraulics and sediment transport in a creek mangrove swamp system. *7th Australasian Hydraulics and Fluid Mechanics Conference*, Brisbane, 492-495.

Hamilton, L.C. and S.C. Snedaker, eds., 1984, Handbook for mangrove area management. *United Nations Environment Programme and East-West Center, Environment and Policy Institute.*

Furukawa, K. and E. Wolanski, 1996, Sedimentation in a mangrove forest. *Mangroves and Salt Marshes* (1), 3-10.

Kinh, N.K., 1992, Management and monitoring of the Red River estuary. *Asian Wetland Symposium*, Otsu, Japan, 1-11.

Walsby, J. and D. Torckler, 1992, Forests in the sea, N.Z. *Geography* (15), 40-65

Wolanski, E., Y. Mazda and P. Ridd, 1980, Mangrove hydrodynamics. In: *Coastal and Estuarine Studies (41), Tropical Mangrove Ecosystems*, eds. A.I. Robertson, D.M. Alongi, American Geophysical Union, Washington DC, 43-62.

Fine Sediment Dynamics in the Marine Environment
J.C. Winterwerp and C. Kranenburg (Editors)

A Preliminary Study on Using Acoustic Waves to Measure High Resolution Marine Sediment Bed Structure

Jerome P.-Y. Maa[a] and D.-Y. Lee[b]

[a]Assoc. Prof., School of Marine Science, Virginia Institute of Marine Science, College of William and Mary, Gloucester Point, VA, 23062, U.S.A.

[b]Principal Research Scientist, Korea Ocean Research and Development Institute, P.O. Box 29, Ansan 425-600, Korea

Using the chirp technique with high frequency (210 to 760 khz) supersonic waves, we have explored the possibility of measuring high resolution bulk density profiles of marine sediments. Results from laboratory experiments on four different sediments clearly show the gradient of possible bulk density profiles. Coarse granular sediment beds consolidate fast and have a relatively uniform density within the bed. A sharp gradient of bulk density near the sediment surface can be observed in the very early stages of consolidation. Clayey sediment beds show very different rates of consolidation among each other. The density gradient is usually rather uniform near the surface, however, it increases significantly near the bottom.

KEY WORDS
acoustic waves, chirp signal, sediment density structure, measurement device.

1. INTRODUCTION

Recent *in-situ* studies on the erosion behavior of marine sediments have revealed an interesting bed structure (Amos and Droppo, 1996; Maa *et al.*, 1998; Maa and Kim, in press). In general, at the water-sediment interface, there is always a fluff layer with a thickness of the order of millimeters. The sediment in this layer is soft and may not be consolidated at all. When applying a bed shear stress, the sediment in this layer can be easily dispersed into the water column. Further down into the bed, however, the bed may have been under various stages of consolidation. Inasmuch as the alternation of different erosional and depositional environments, marine sediment beds usually have a layered structure. For this reason, during an erosion experiment, one can see different bed responses even when the applied erosion bed shear stress is a constant (Maa and Kim, in press). While this deductive conclusion can be obtained from the measured changes of suspended sediment concentration during erosion experiments and from X-ray photographs of a undisturbed sediment core (Dellapenna, 1999), a direct *in-situ* measurement of sediment structure at fields on an erosion resistant profile has never been attempted.

Since the bulk density of a sediment bed reflects the degree of consolidation, bulk density usually has a good correlation with the erosion resistance. This leads to an attempt to obtain the bed bulk density profile first, and then working on the erosion resistant profile.

Nuclear probes were used frequently in estuaries and near coasts during the late 70's and 80's for measuring the bulk density of sediment beds (Hirst *et al.*, 1975; Kirby, 1988). However, there is always a chance that the instrument could be lost during field operations. Under the new U.S. regulation, users have to recover the probe at their expense in order to maintain a safe environment. For this reason, the use of this technique can be expensive, and thus, it is urgent to find an alternative method.

Experimental data (Hamilton, 1969) have revealed that sediment mechanical properties (*i.e.*, grain size, porosity, and bulk density) can be identified using acoustic waves. For example, sound velocity correlates reasonably well with grain size and porosity for marine sediments. Recent studies have also provided a reasonable analytical model for predicting sound speed and attenuation in marine sediments (Buckingham, 1997).

To explore the idea of using supersonic waves to identify sediment properties with high resolution, two laboratory experiments using an intrusive type of measurement (*i.e.* requiring the insertion of sensors into the sediment) have been attempted (Maa *et al.*, 1997). We found a clearly noticeable difference (with a 2 mm resolution) in supersonic wave attenuation caused by consolidation of cohesive, fine-grained sediments.

Based on the strength of reflected acoustic waves, caused by the difference in acoustic impedance (the product of bulk density and sound speed), the chirp technique is a non-intrusive approach to identify sediment properties (LeBlanc *et al.*, 1992). The chirp technique was first developed for RADAR systems to increase the signal/noise ratio. This technique has been implemented successfully on low frequency (10 - 20 Khz) acoustic sub-bottom profiler for identification of large area sediment properties (LeBlanc *et al.*, 1992). Because of the low frequency operation, however, the resolution is limited to approximately 10 to 20 cm. In this study, we adopted the chirp technique but changed the operation frequency to around 500 khz in order to obtain a high resolution (on the order of millimeters) for revealing the bulk density information near the water-sediment interface. We will explain the chirp technique, test the selected instrument configurations on several well defined interfaces, and present the experimental results to identify the density profile for four selected sediments.

2. CHIRP TECHNIQUE

The chirp technique uses a frequency and amplitude modulated signal as the source wave form (Fig. 1a) to excite an acoustic transducer. After the wave form, the signal remains at zero for a preselected duration. The total duration can be easily changed by adjusting the length of the zero signal period. The wave form is stored in a micro-processor that controls the chirp device. When taking auto-correlation of this wave form, a pulse type wave form (Fig. 1b) can be produced which is just like that used in the traditional acoustic industry. After receiving the echo waves

(Fig. 2a) for a preselected total duration, a cross-correlation of the source and the echo time series produces a pulse-like echo time series (Fig. 2b).

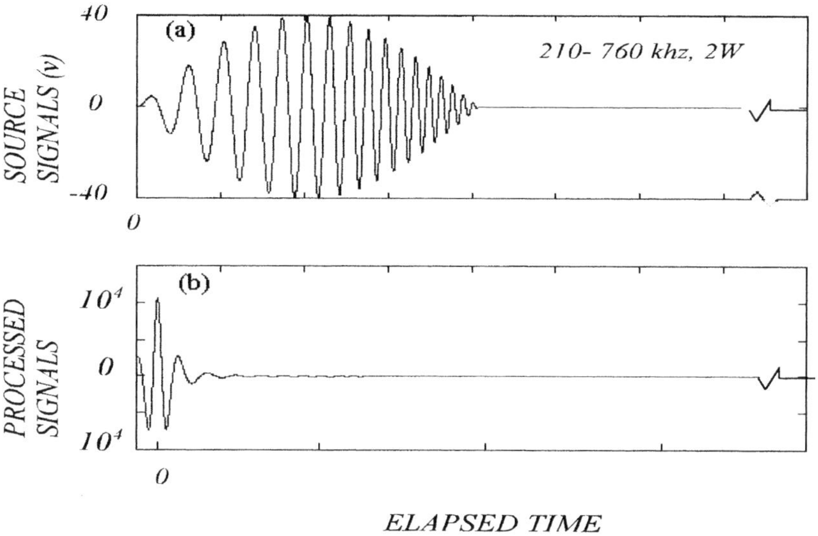

Fig. 1. Chirp Wave Train to Provide a Constant Energy Pulse. (a) Original Wave Form; (b) After Auto-correlation.

There are three reasons to use the chirp technique instead of the traditional pulse type wave form: (1) to provide a constant source of acoustic energy; (2) to increase the efficiency of converting electric energy to acoustic energy; and (3) to increase the signal/noise ratio for a longer distance operation or to reduce power requirement. Although acoustic pulses (Fig. 1b) are the standard wave form used in the acoustic industry, the pulse energy is hard to maintain as a constant. The pulse type wave form is selected as the standard for the acoustic industry because it provides clear points for identifying the duration of wave propagation. The echo strength, however, is not used because it is difficult to reproduce pulses with identical peak energy. When converting the electric energy to acoustic energy, a pulse type signal is not efficient because it is far away from operating at an acoustic transducer's resonant mode. Each transducer has its own design frequency band and the conversion of electric energy to acoustic energy is efficient only when the input signal frequency is the same as the transducer's design frequency. For example, the highest efficiency is achieved using a tone-burst (a given number of sine waves, Maa *et al.*, 1997) that oscillates at the transducer's design frequency. A chirp wave train can be designed to operate around a transducer's design frequency, and thus, has a much higher efficiency.

When the received echo signals are amplified, the noise is amplified as well. After cross-correlation, however, the noise is depressed because there is no correlation with the source wave form. An example of an echo chirp wave train with a low signal/noise ratio is given in Fig. 2c, and the processed echo waves (Fig. 2d) show a significant improvement on the signal/noise ratio.

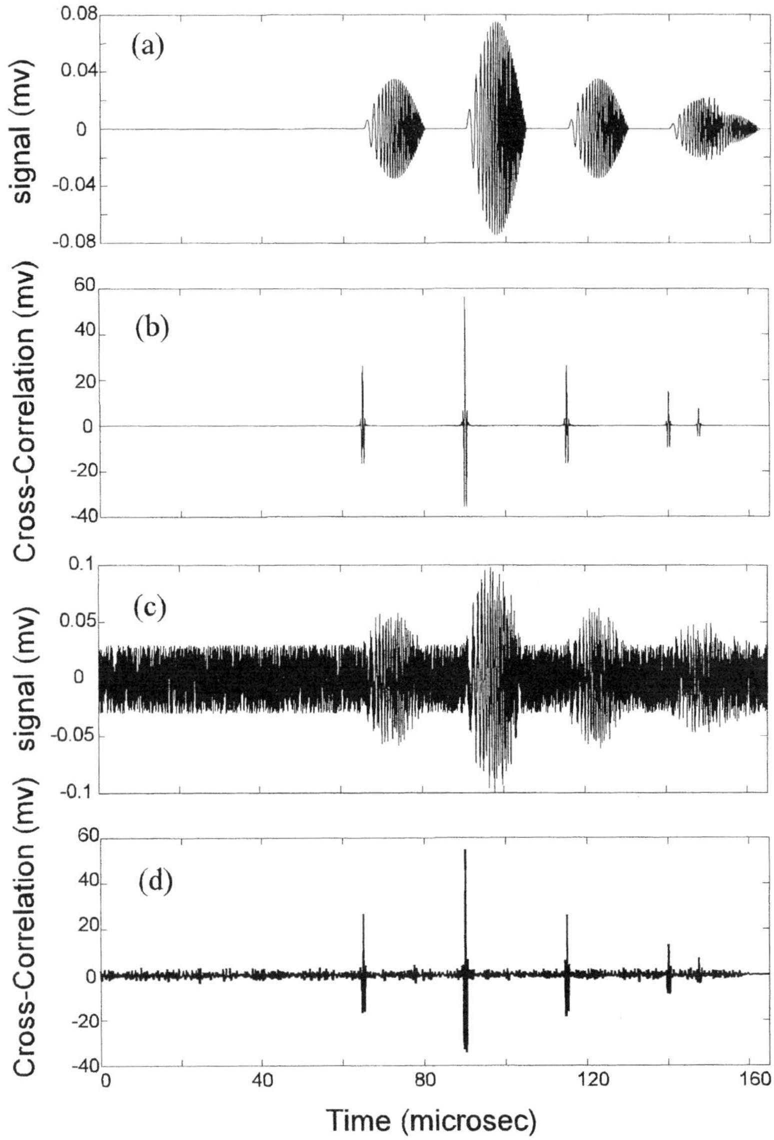

Fig. 2. Example of Echo Wave Trains. (a) A Clean Echo Represents Five Interfaces. The last two echoes are superimposed; (b) The Processed Echo Wave Train Showing the Five Interfaces; (c) An Echo Wave Train with Low Signal/noise Ratio; (d) The Processed Echo Wave Train Showing the Depression of Noises.

In this study, the chirp wave form was generated according to the following equation

$$y(i) = \sin\left(\frac{i\pi}{n}\right)\sin\left(\frac{2(i-1)\pi}{T}\right) \tag{1}$$

where T is the wave period varied as T = 220 - 0.07(i-1), i = 1 to n, and n is the total number of signal data points. In this study, we selected n = 1700. Notice that the wave form generated does not exactly follow the chirp specification (LeBlanc *et al.*, 1992) which use a shape of Gaussian function to modulate the wave amplitude. We used a low frequency sine function, the first sine function on the right hand side of Eq. 1, to modulate the wave amplitudes. As shown in Figs. 1 and 2, the pulses generated by auto- and cross-correlation are as good as those suggested by using the Gaussian function.

3. INSTRUMENTS

The numerical time series, y(i), generated from Eq. 1 was sent to a Stanford Research Systems' arbitrary function generator (model DS340), and checked on a digital oscilloscope (LeCroy, Model 9310). The electronic signal was then fed into a 3 watt power amplifier (EIN, model 403LA) and the power signal was sent to the transmitter, a Panametrics transducer (model V389, 0.5 MHz, sensor diameter = 3.81 cm, with a spheric focus distance of 9.65 cm) that was placed 10 cm above the water-sediment interface using a Nsite miniature echo sounding device (Model NDSW-500).

Another Panametrics transducer (model V301, 0.5 MHz, sensor diameter = 2.54 cm) was used as the receiver. Although the transmitter can also be used as a receiver, we used a separate transducer in order to simplify the system design.

The echo signal was fed directly into the digital oscilloscope and recorded using the built-in math function to average 10 echoes for data smoothing. Because the electronic signal was repeating itself at a rate of 100 Hz, the math average only takes about 0.1 second to accomplish. For high resolution on the echo time series, the echo waves were recorded using a screen window of 50 μs. At least five consecutive windows were recorded to cover a total duration of 250 μs, which roughly translates to a one-way wave travel distance of 20 cm.

Four sediment samples were prepared for the test: (1)sandy sediment with a grain size between 1.5 to 2 ϕ; (2) sandy sediment with a grain size between 3 to 3.75 ϕ; (3) kaolinite; and (4) redart. The sediment samples were first mixed with tap water in 2 liter beakers and then allowed to consolidate for selected times.

4. CHECKING EXPERIMENTS

Before applying the acoustic signal to sediment beds, we tested it on selected clear interfaces and examined the performance of the selected wave form. In the first test, the wave train was sent down to impinge on two plexiglass plates (with a thickness of 25 mm) that were staked together. The first interface was a water-plexiglass interface, and the second was a plexiglass-plexiglass interface. Below the bottom plexiglass, there was a plastic container. It is hard to see any clear interfaces from the raw echo wave train (Fig. 3a). The processed wave train (*i.e.*, after cross correlation of the chirp source wave form given by Eq. 1 and the measured echo wave train), however, clearly showed the three interfaces marked in Fig. 3b. The processed echo pulse at elapsed time = 170 μs was caused by the plastic container. At the water-plexiglass interface, there should have been a clear jump in acoustic impedance, from 1.48 x 10^6 Pa.s/m in water to 3.26 x 10^6 Pa.s/m in plexiglass (speed = 2750 m/s, density = 1190 kg/m^3). The processed echo pulses were not very sharp and some degree of spreading was observed. It is not clear yet what causes this spreading.

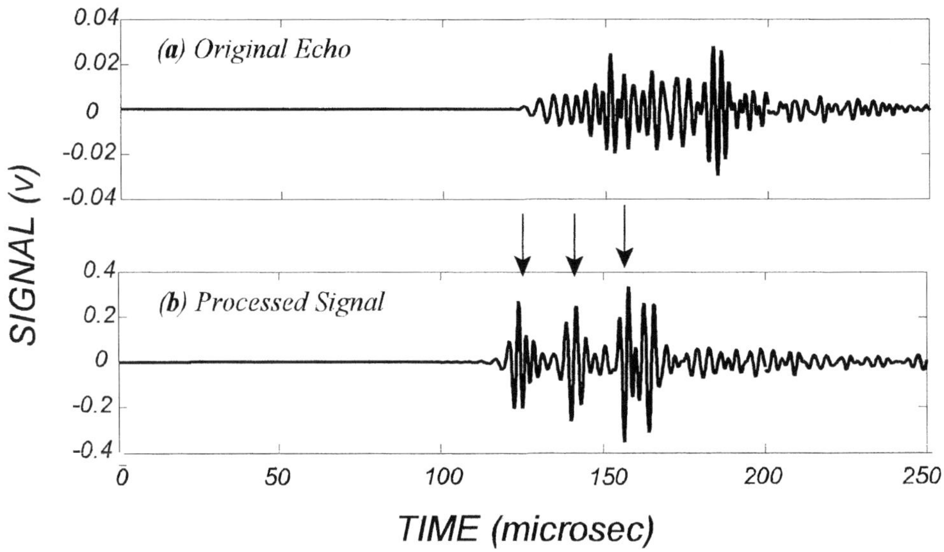

Fig. 3. Performance Test of the Selected Wave Train on Two 2.5 cm Thick Plexiglass Plates Stacked together.

Another test replaced the two plexiglass plates with one 2.4 cm thick PVC plate. The results, however, show a better pulse shape with little spreading (Fig. 4a) which may indicate that the signals used were reasonably good. The acoustic impedance also jumped from 1.48 x 10^6 Pa.s/m in water to 3.28 x 10^6 Pa.s/m (speed = 2380 m/s, density = 1380 kg/m^3) in the PVC plate.

The last test replaced the 2.4 cm thick PVC plate with a 6.35 mm thick PVC plate in order to check the resolution. This plate was hung in the water. The processed signal (Fig. 4b) is capable of showing the PVC plate, but it may also demonstrate the possible limit of resolution using the selected wave form. For a better resolution, higher frequency are needed.

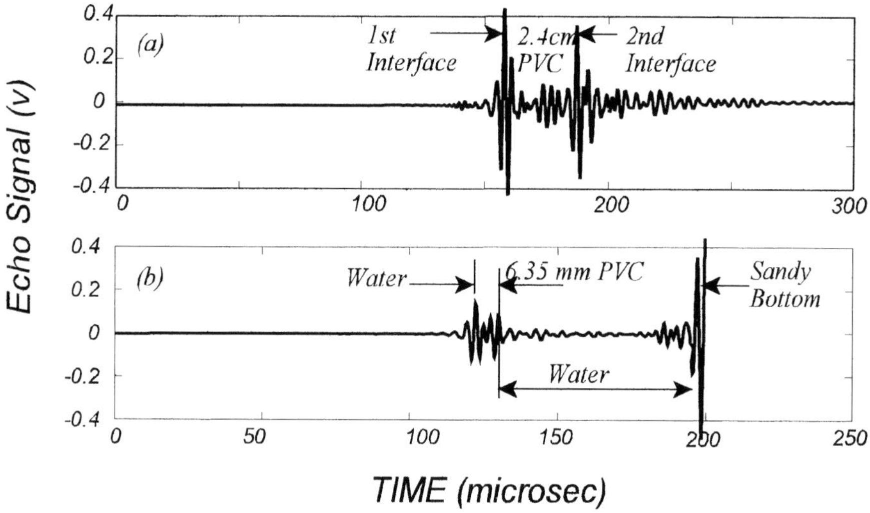

Fig. 4. Processed Wave Echo from PVC Plates. (a) Thickness = 2.5 cm; (b) Thickness = 6.35 mm hanged in the water.

5. RESULTS

Since the echo wave strength will be used to identify the sediment density, the hardware and software design must be able to catch the maximum echo wave amplitude. Implementing this requirement is not a simple issue, especially for field operations using a towed fish. To avoid this requirement during the early stages of instrument design, the distance from the transducer to the sediment bed was set at 10 cm. Because of this reason, the results presented are not absolute values, and only the relative echo strengths are given.

Sediment beds with different grain sizes have different attenuation rates on acoustic waves (Maa *et al.*, 1997). The attenuation is small for clays and high for granular material.

At this stage, we did not attempt to correct the signal amplitude based on the distance from the transducer. This correlation, however, is needed for future applications to identify the sediment density gradient below the water-sediment interface.

5.1. Sand with size from 1.5 - 2 ϕ

For this medium sand bed (grain size from 0.25 to 0.355 mm), we expected consolidation to be completed quickly. Thus, we took only 2 measurements at 10 minutes and one day after mixing. In our first measurement (Fig. 5a), the processed echo indicated that below the water-sediment interface (at 127 μs), the acoustic impedance increased significantly and reached a plateau at 148 μs. Considering the change of acoustic wave speed was minimal (about 1780 m/s), we may conclude that within the top 2 mm (*i.e.*, 2 mm $\approx$ 0.5*(148-127)μs * 1780 m/s), the bulk density increased sharply. The density remained about the same for the rest of the sand column until it was close to the bottom at 200 μs, which was about 1 mm above the glass beaker (at 212 μs). The sediment bed thickness was about 7.56 cm (*i.e.*, ≈ 0.5* (212-127)μs * 1780 m/s).

After one day, the top 2 mm layer disappeared (Fig. 5b), and left a relatively strong echo at 124 μs which reflected the fact that the sediment was well consolidated. The total thickness of the bed was reduced to 7.38 cm, a change of 1.8 mm.

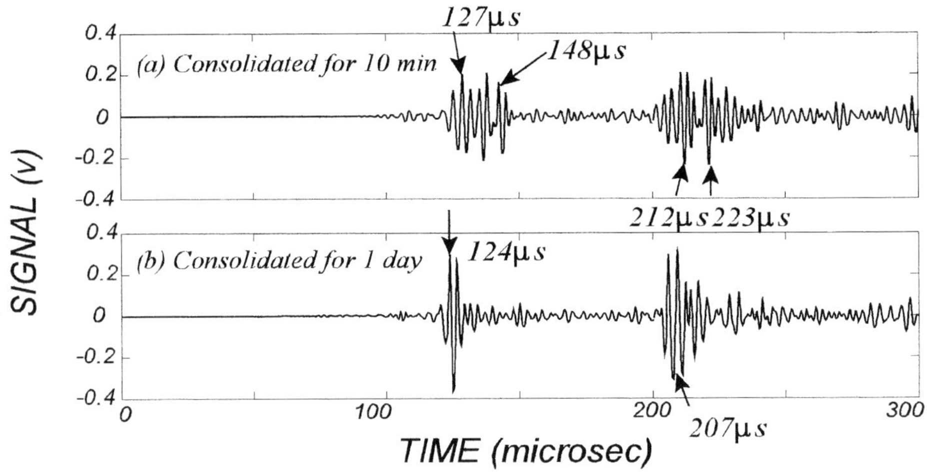

Fig. 5. Processed Echo Signals from Medium Sand with Size Between 1.5 to 2 ϕ. (a) Ten minutes after mixing; (b) One day after mixing.

5.2. Sand with size from 3 - 3.75 ϕ

For this very fine sand (grain size from 0.075 to 0.125 mm), two measurements were taken two days (Fig. 6a) and nine days (Fig. 6b) after mixing. The total thickness of this sediment bed (4.8 cm) did not change in seven days, and the echo wave amplitude increased only slightly at the water-sediment interface (elapsed time = 126 μs). The echo amplitude at the sediment-glass interface (elapsed time = 182 μs for two days consolidation), however, increased significantly after seven days (elapsed time = 179 μs for nine days consolidations). This may reflect better development of bed rigidity, which causes less dissipation of acoustic energy. Because most of the acoustic energy was reflected at the sediment-glass interface, there was not much energy for reflection at the second glass-air interface. The small pulses in the processed echo wave train between the first two interfaces indicate that there must be a mild and continuous increase of bulk density in this region.

5.3. Kaolinite

Newly deposited clay slurry is usually hard to detect using acoustic waves because of the small difference in acoustic impedance (from 0.02 to 1 x 10^6 Pa.s/m). Acoustic waves travel in clay with a speed (~ 1450 m/s) slightly less than that in water (~ 1480 m/s) regardless of the duration of

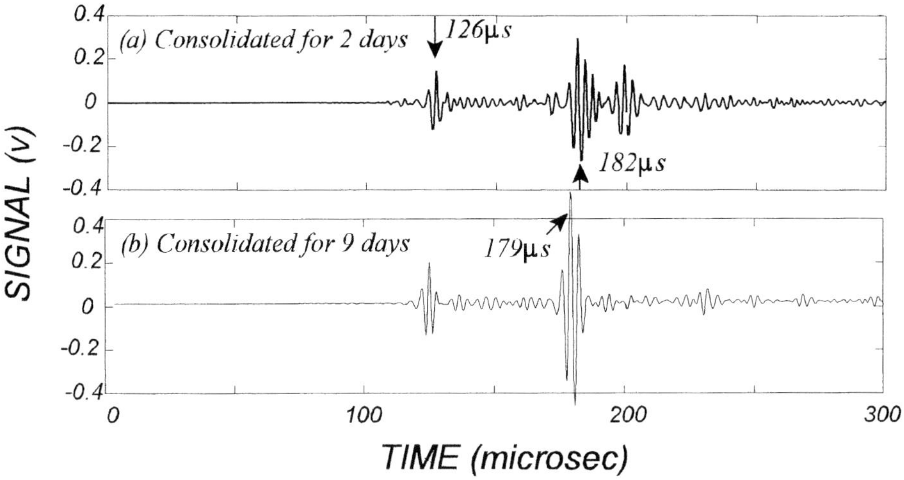

Fig. 6. Processed Echo Signals from Very Fine Sand with Size between 3 to 3.75ϕ. (a) Two days after mixing; (b) Nine days after mixing.

consolidation (Maa *et al.*, 1997). The acoustic wave attenuation, however, increases with the duration of consolidation because of higher viscosity in a more consolidated bed. We did not test for fresh and new clay slurry because we need more power to drive the transducer and combine a good quality pre-amplifier to boost the weak signal. For this reason, only two measurements were taken for the kaolinite bed with three and 10 days consolidation.

At the elapsed time = 125 μs, the water-kaolinite interface was detected for the three days kaolinite bed (Fig. 7a). Within the next 27 μs (*i.e.*, 0.5 * 27 μs * 1450 m/s = 19.5 mm), there was little change in the kaolinite bulk density. A continuous increase of bulk density (because of a near constant wave speed) was observed for the next 30 μs (about 21 mm). The glass beaker was located at elapsed time = 200 μs (*i.e.*, a total bed thickness of about 54 mm).

After another seven days, a similar sediment bulk density profile was observed (Fig. 7b). The top 55 μs (*i.e.*, ~ 40 mm) kaolinite bed had no density change. The rest of the kaolinite bed (*i.e.*, from 176 μs to 190 μs, about 10 mm), however, had a rather large gradient of bulk density. The total bed thickness reduced slightly to 50 mm.

5.4. Redart

This particular kind of clay has very low echo waves and high attenuation. After consolidation for three days, the water-redart interface was barely identifiable (at 131 μs, Fig. 8a). For the next 11 μs (*i.e.*, ~ 8 mm), there was no change in density. Below this level, the bulk density of the redart bed increased continuously to the bottom (elapsed time = 201 μs). The total bed thickness was 50 mm.

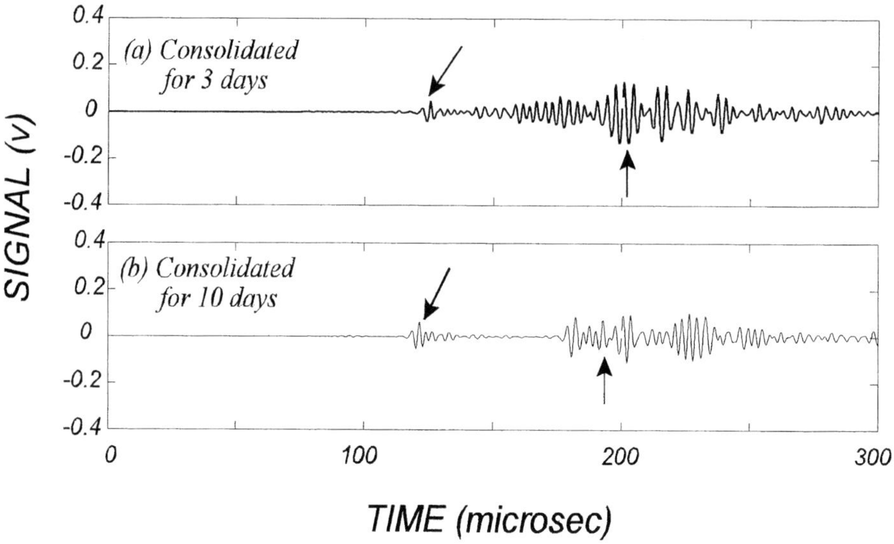

Fig. 7. Processed Echo Signals from Kaolinite Bed. (a) Three days after mixing; (b) Ten days after mixing.

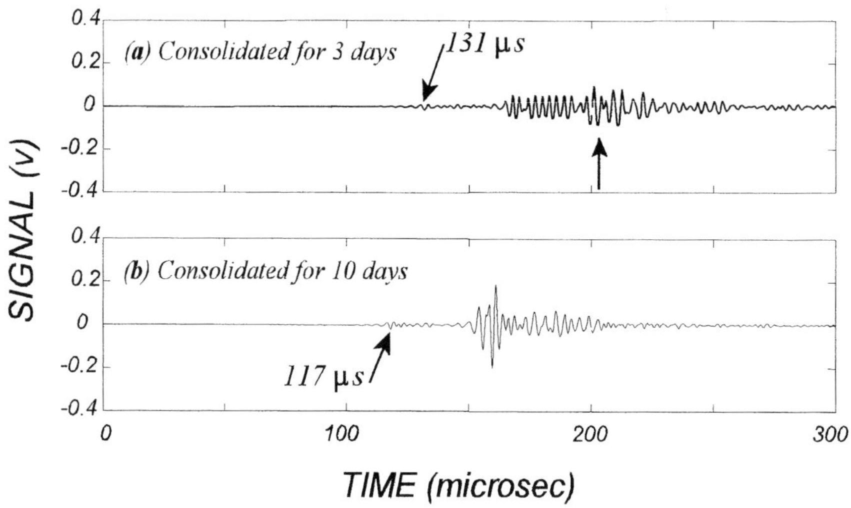

Fig. 8. Processed Echo Signals from Redart Bed. (a) Three days after mixing; (b) Ten days after mixing.

After another seven days, the water-redart interface was still barely identifiable at elapsed time = 117 μs (Fig. 8b). The flat response for the next 11 μs indicated that the bulk density did not change in this region. Only at the last 4 μs (3 mm) did the bulk density increase significantly. The total bed thickness decreased to about 31 mm.

6. DISCUSSION AND CONCLUSIONS

In this study, the chirp technique was elaborated and implemented in the laboratory with a high operation frequency, around 500 khz. Although the selected chirp wave form and the acoustic transducers are not perfect, the results indicate that it can be used to identify the change of acoustic impedance with depth.

To translate the acoustic impedance gradient information to bulk density profile requires knowing the local acoustic wave speed and local attenuation rate. Since there are three variables (bulk density, wave speed, and attenuation coefficient) involved in one measurement, theoretically, it is impossible to find the exact solution. An approximate solution, however, can be obtained if a data base that provides wave speed and attenuation rate based on local bulk density information is available. Even if the data base is not accurate, an estimated bulk density profile with reasonable accuracy can be obtained because the range of these two variables is small (*e.g.*, acoustic wave speed is about 1450 m/s for clay and 1780 m/s for sand).

Although the actual procedure has not been established yet, we envision that the acoustic impedance for the top sediment layer can be calculated according to the echo amplitude at the water-sediment interface. Using an iteration approach, the possible best combination of bulk density and wave speed for the top sediment layer can be estimated first and followed by the wave attenuation rate. The echo wave amplitude can then be adjusted for the next layer and the process repeated for the next layer of sediment. In such cases, the accuracy will be degraded with the number of layers because of the accumulation of errors. Nevertheless, an approximate bulk density profile can be established.

Acquisition of the maximum echo signals for a moving device is difficult. If the instrument is fixed near the bottom, *e.g.*, mounted on a tripod, then it will not be too difficult to accomplish this objective. Details on the design of hardware and software, however, are waiting further effort.

The measured possible density gradient for medium sand at the first 10 minutes after mixing is interesting. This gradient occurred right below the water-sediment interface. Cohesive sediments (kaolinite and redart), however, showed the gradient starts at the bottom of the bed.

Because of the instrument setup, absolute amplitude of the echo pulses at the water-sediment interface cannot be obtained yet. The relative amplitudes of the four sediment samples (Fig. 9) show a clear difference among these beds. The relationship shown in Fig. 9 is widely known; this time, we have a number to demonstrate the relative difference.

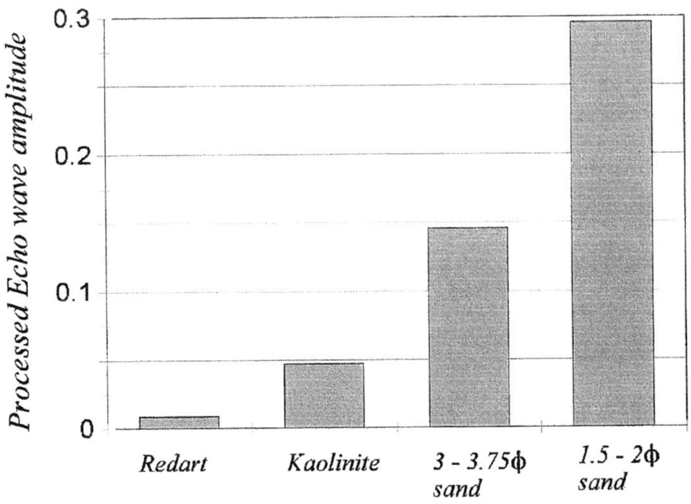

Fig. 9. A Comparison of Echo Wave Amplitudes at the Water-sediment Interface.

Although we are not able to provide numbers on the bulk density profile at this time, the suggested approach using the chirp technique to obtain high resolution information on sediment mechanic properties in a timely manner is promising. Once the data base that correlates the bulk density, wave speed, and attenuation coefficient has been established, an estimated bulk density profile will be available.

7. ACKNOWLEDGMENTS

Sincere appreciation goes to the Seed-grant Program of the Virginia Water Resource Research Center for partial support for this study. This is VIMS contribution No. 2394.

REFERENCES

C.L. Amos and I.G. Droppo, The Stability of Re-mediated Lakebed Sediment, Hamilton Harbour, Lake Ontario, Canada, Geological Survey of Canada, Open File Report #2276, (1996).

M.J. Buckingham, "Theory of Acoustic Attenuation, Dispersion, and Pulse Propagation in Unconsolidated Granular Materials Including Marine Sediments," J. Acoustical Society of America, 102(5), 2579-2596 (1997).

T. M. Dellapenna, Fine-scale Strata Formation in Biologically and Physically Dominated Estuarine System within the Lower Chesapeake and York River Subestuary, Ph.D. Dissertation, School of Marine Science, Virginia Institute of Marine Science, College of William and Mary, pp273 (1999).

E.L. Hamilton, Sound Velocity, Elasticity, and Related Properties of Marine Sediments, North Pacific, TP 144, Naval Undersea Research & Development Center, San Diego, CA (1969).

T. J. Hirst, M. Perlow, Jr., and A. F. Richards, "Improved In Site Gamma-ray Transmission Densitometer for Marine Sediments," Ocean Engineering, 3(1), Pergamon Press, 17-27 (1975).

R. Kirby, "High Concentration Suspension (Fluid Mud) Layers in Estuaries," in Physical Processes in Estuaries, Eds. J. Dronkers and W. van Leussen, Springer-Verlag, 463-487 (1988).

L.R. LeBlanc, L. Mayer, M. Rufino, S.G. Schock, and J. King, "Marine Sediment Classification Using the Chirp Sonar," J. Acoustical Society of American, 91(1), 107-115 (1992).

J.P.-Y. Maa and S.-C. Kim, "A constant erosion rate model for fine sediment in the York River, Virginia," submitted to Environmental Fluid Mechanics, (in print).

J.P.-Y. Maa, L. Sanford, and J.P. Halka, "Sediment Resuspension Characteristics in the Baltimore Harbor," Marine Geology, 146, 137-145 (1998).

J.P.-Y. Maa, K.-J. Sun, and Q. He, "Ultrasonic Characterization of Marine Sediments," Marine Geology, 141, 183-192 (1997).

Fine Sediment Dynamics in the Marine Environment
J.C. Winterwerp and C. Kranenburg (Editors)

AN UNUSUAL TURBIDITY MAXIMUM

Bruce W. Nelson

Department of Environmental Sciences, University of Virginia
36 University Circle, Charlottesville, VA 22903

Turbidity maxima develop upstream from an arrested salt-wedge and near the limit of salinity intrusion under partially-mixied conditions in Sungai Selangor, a tropical, mesotidal estuary in Malaysia. The TM has peak surface suspended sediment concentrations of 300 mg./l. on neap tides and >2,000 mg./l. on spring tides. The TM occurs in a "muddy reach" where discontinuous "fluid mud" patches form on the bottom during neap tides. Tidal range and current speed determine the amount of sediment entrained, but the vertical density gradient, which varies greatly with changes in freshwater discharge and tidal range, determines the amount of sediment that reaches the surface layer. A "lag" in sediment transport associated with the vertical density gradient needs to be considered in modelling sediment transport in estuaries.

Key Words: estuaries, sediment, turbidity maximum, Malaysia

1. INTRODUCTION

The peak in suspended sediment concentration that occurs near the limits of salinity intrusion in estuaries was observed first in Europe and in the Chesapeake Bay region in 1950-60 (Wellershaus, 1981). The "turbidity maximum", or TM, migrates up and downstream with the ebb and flood tides, with neap and spring tides, and with variations in river discharge. Much observational and theoretical development was summarised by Dyer (1997, p159): "The residual vertical gravitational circulation produces the broad background of turbidity in partially mixed estuaries." The net non-tidal estuarine circulation accumulates, or "traps", sediment in the bottom layer near the null point. Mathematical modelling, summarized by Jay & Musiak (1994), confirms the importance of this mechanism. Postma (1967) and Allen et al. (1975, 1980) suggested that the net landward flux of sediment by "tidal pumping" due to the dominance of flood currents over ebb currents in an asymmetric tidal cycle may enhance sediment trapping. Dyer (1988, p306) concluded that tidal pumping was a major factor in generating and supporting the TM. In addition, Geyer (1993) showed that turbulence suppressed by stratification probably contributes to sediment trapping in salt-wedge systems. Geyer's model calculations indicate that suppression of turbulence by stratification is much more effective in trapping sediment than in prior models in which a uniform diffusivity was assumed. All of the trapping occurs near the bottom, so the TM becomes much more localized on realistically short time scales. Strong stratification during ebb tide favors this trapping mechanism. The parameters used in most models generate a TM that is averaged over depth and over a tidal cycle or much longer periods, but they usually do not include erosion and deposition of sediment from the bed (Jay and Musiak, 1994).

However, Wellershaus (1981) early observed that the TM was associated closely with local accumulations of muddy sediment on the bottom in the Weser estuary. The TM observed is formed by tidal resuspension of sediment from these bottom muds. Thus, he distinguished between the longitudinal convergence of sediment trapped in a zone along the channel and the vertical distribution of sediment produced by deposition and resuspension. Riethmuller, et al (1988) analyzed the tidal dynamics of the turbidity zone in the Weser over flood and ebb cycles during a neap period (A= 2.3-2.5 m.). Their field observations and modelling show that the TM originates partly from the null zone, as model calculations suggest. But the null zone is close to a reach where extensive mud deposits lie on the bed, where high turbidity exists in the water column throughout the tide, and where sediment erosion and resuspension are dominate. The observed turbidity is mainly due to resuspension and the rapid rate of particle settling observed.

Kirby (1988) reviewed the history of observations on highly concentrated fine-grained sediment suspensions in estuaries that began in the 1950's. These studies revealed the discontinuous vertical structure of sediment profiles formed by the settling of highly concentrated, floccular suspensions. These observations led to the concepts of "mobile suspensions", "stationary suspensions", the "lutocline", and "settled mud". Much of the observational data comes from high tidal estuaries, such as the Severn, in which the tides efficiently mix salinity and temperature vertically and determine the sediment dynamics. Laboratory studies by Mehta (1989) and co-workers linked the consolidation history of muddy bed sediment to its deposition and erosion, i.e., the dynamics of "cohesive sediment" transport. Stratified sediment concentration profiles arise from floccular settling behavior of suspensions more concentrated than about 300 mg./l. So far, the analytical expressions for the behavior of these suspension are depth averaged, and Mehta (1988) remarks: "with less than adequate emphasis placed on the evolution of the vertical structure of the suspension as observed in nature". Odd (1988, p509-10) discussed how vertical stratification, due both to salt and to sediment, may affect the vertical turbulent exchange. But generally the dynamics of mud erosion and deposition have been analyzed without reference to vertical stratification of the water column due to salinity.

So neither field observations nor modelling studies very clearly portray how vertical stratification affects the behavior of cohesive sediments and formation of the turbidity maximum that is so characteristic of many estuaries. The purpose of this paper is to supply some qualitative field observations that show how important a factor it is. The data come from a tropical setting where sediment production is high and where a large range of tidal amplitudes coupled with major seasonal variations in freshwater discharge give rise to great variety in sediment behavior in stratified systems.

2. GEOGRAPHICAL SETTING

Sungai Selangor flows down the western slope of the Main Range in Peninsular Malaysia near 3° N. latitude. It drains uplands that rise to 1500-1900 m. These are underlain by deeply weathered Mesozoic granitic rocks and clayey, tropical soils. Tropical

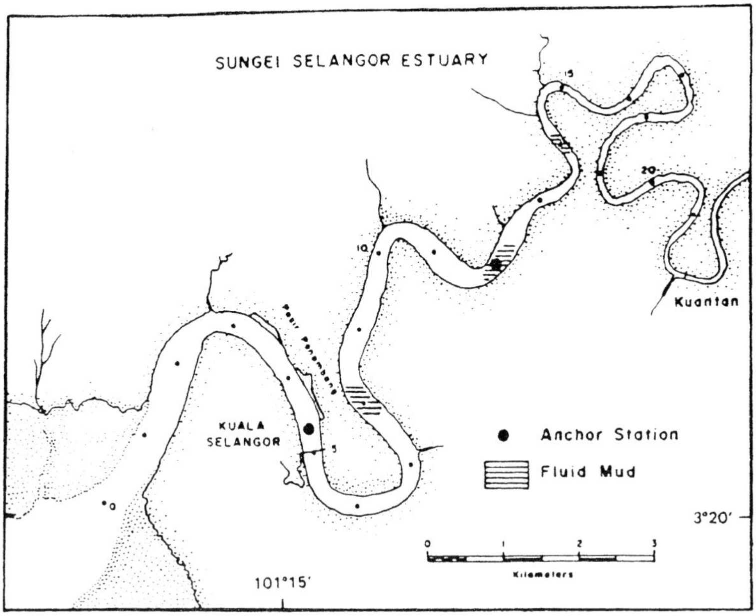

Fig. 1 Sg. Selangor Estuary, Malaysia

rain forest (hill dipterocarp) still occupies much of the upper catchment. Its middle reaches pass through hilly country with elevations to 250 m., which are underlain by tropical soils developed on metamorphosed Paleozoic rocks (schist and phyllite with included masses of limestone). Some tributaries contain remnants of tin mining activity, and palm oil and rubber plantations occupy the interfluvial areas. A few areas are subject to early stages of urbanization and industrial development. The river meanders across a coastal plain of Quaternary sediments that is 30-35 km. in width before flowing into the Melaka Straits some 70-80 kilometers from the Main Range. The catchment area above the coastal plain is about 1450 $km.^2$ and the average discharge is 53.1 $m.^3/s$. Seasonal variations in rainfall cause the flow to exceed 122 $m.^3/s$. or to fall below 23 $m.^3/s$. about 10 percent of the time. The monsoon periods produce 27 percent of the annual discharge in November-December and 21 percent in April-May; only 8.1 percent occurs in the July-August dry season. Where the catchment is under forest, rubber, or oil palm, suspended sediment concentrations in streams are <200 mg./l. (often <100 mg./l.). Stream loads increase wherever tin mining or urban development occur. Suspended sediment concentrations at the head of the estuary are 200-500 mg./l.

To a considerable degree still, mangrove forest occupies the shoreline of the tidal portion of the coastal zone. The estuarine portion of Sg. Selangor extends about 20 km. upstream from the mouth (Fig. 1). It has a meandering, funnel-shaped form whose average channel depth determined by echo sounding surveys is 5-6 m. Depths in the meanders increase to as much as 9 m., and an entrance shoal of <5 m. separates the estuary from Klang Straits. The channel cross section decreases up to a highway bridge located 5 km.

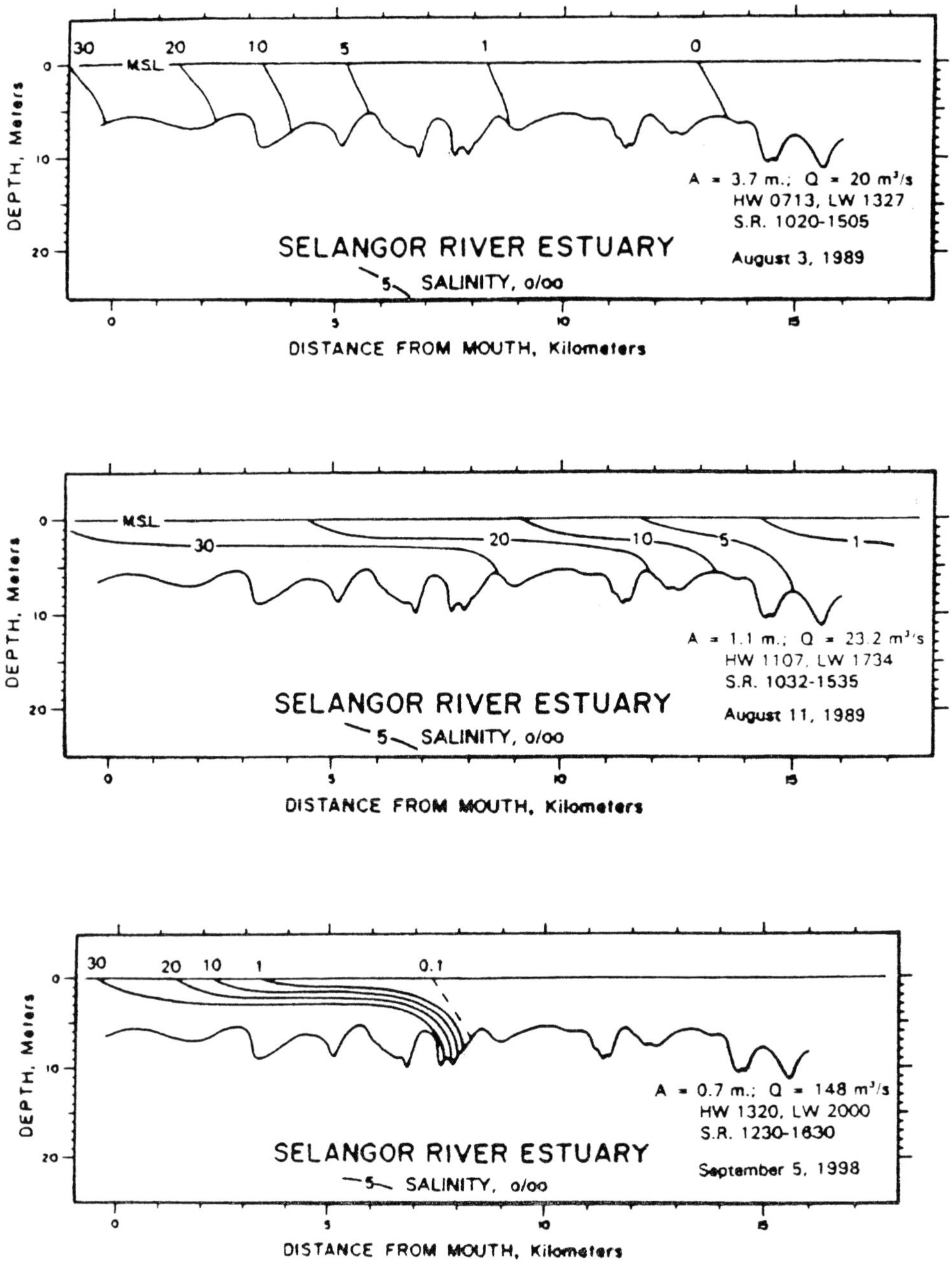

Fig. 2 Longitudinal Salinity Distributions in Sg. Selangor

above the mouth. The bottom consists of consolidated muds along most of the length, but from 8 to 12 km. above the mouth patches of "fluid mud" rest on the bottom during neap tides (Fig. 1). The "fluid mud" has much lower salinity and dissolved O_2 than the estuarine water above it (Nelson, et al. 1994). The most abundant clay mineral in these silty clay sediments is kaolinite, but gibbsite is usually present as well.

3. OBSERVED TIDAL CONDITIONS & SALINITY DISTRIBUTIONS

The reference tidal station is at Port Klang, 40 km. to the south, where the average tidal range is 2.7 m., mean neaps are 1.3 m., and mean springs are 4.1 m. Temperature, salinity, suspended sediment and other variables were measured from a small boat using a salinometer and discrete samples between 1987 and 1990 over tidal ranges of 0.7 to 5.2 m. and freshwater inputs of 20 to 148 $m.^3/s$. The water column is usually isothermal at 28-30° C. During spring tides and low flow the estuary is well mixed (Fig. 2A). The 1 o/oo isohaline is located 8 km. from the mouth at low water and more than 15 km. at high water, a tidal excursion of 10 km.; the 30 o/oo isohaline penetrated 9 km. upstream. As freshwater discharge increases, the intrusion length decreases: to 6 km. at 41 $m.^3/s$ and 2 km. at 86 $m.^3/s$. During neap tides and low flow the estuary is well stratified (Fig. 2B) during ebb tide; the salinity difference between surface and bottom near mid-estuary is 15 o/oo. During neap tides and very high inflows (Fig. 2C), the Selangor becomes a salt-wedge estuary. The salt wedge is arrested 8 km. from the mouth and doesn't break up until near the end of ebb tide. Under these conditions the salinity difference between surface and bottom in the lower estuary is about 30 o/oo. Little upstream motion occurs in the saline layer, while the surface flow with an average suspended sediment concentration of 109 mg./l. exceeds 1 m./s.

4. OBSERVED TURBIDITY DISTRIBUTIONS

A turbidity maximum is nearly always observed in the Selangor estuary, except when a salt-wedge forms. Under the low flow and neap tidal conditions of Fig. 2B, the peak concentrations observed at the surface during a mid-ebb slack run were comparatively low, i.e., 300 mg./l. (Fig. 3). The TM occurs near the limit of saline intrusion where the mean salinity is 1.5 o/oo (0.5 o/oo at the surface; 3.5 o/oo at 5 m. depth). This is quite analogous to turbidity maxima that have been observed in microtidal estuaries, such as Chesapeake Bay and its tributaries (Nelson, 1960; Schubel, 1969). However, under spring tidal conditions at low flow (Fig. 2A) when there is very little vertical stratification, the surface suspended sediment distribution along the estuary is quite different (Fig. 4). Mean sediment concentrations in the TM near the limits of salt intrusion (S = 0.4 o/oo) are 1,500 mg./l. (2,300 mg./l. near the bottom). A 3.4-fold change in tidal range has increased sediment concentration 5-fold. The TM is located over the patch of "fluid mud" 12 km. above the mouth (Fig. 1). The peak surface concentrations that occur in the lower and middle estuary are of two kinds. First, those between the mouth and location S-6 are related to changes in channel configuration. They are downstream from channel meanders and probably are generated by helical flow around the bends. Peak concentrations increase upstream from about 1,488 mg./l. at S-3.5 [8,200 mg./l. on the bottom] to about 3,550 mg./l. at S-6.5 [11,640 mg./l. on the bottom!] Second, the strongest peak of 5,780 mg./l. at S-8 was collected nearly at low water. It appears to be a mobile suspension sampled above the "fluid mud" patch at that location. In any case, these data show that the tidal dynamics and the processes of erosion or resuspension determine the details of the turbidity distribution in a muddy estuary.

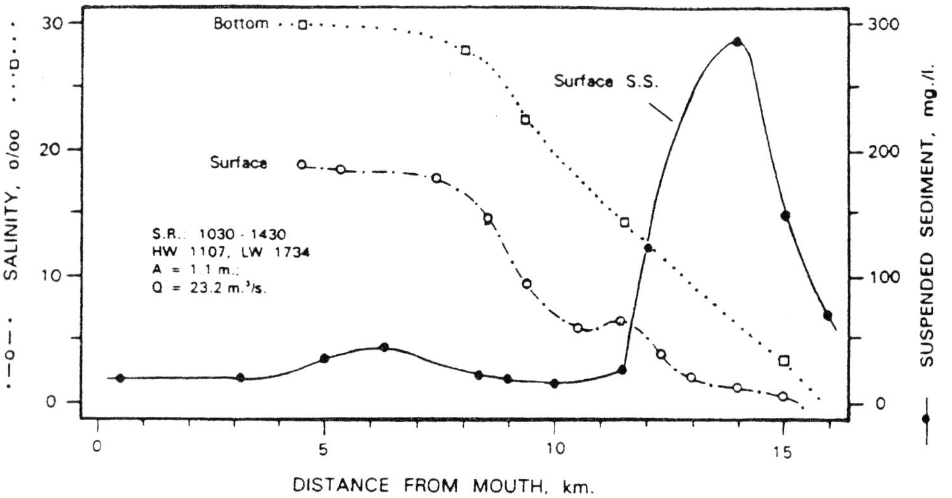

Fig. 3 Longitudinal Salinity & Turbidity at Neap Tide; 8/11/89

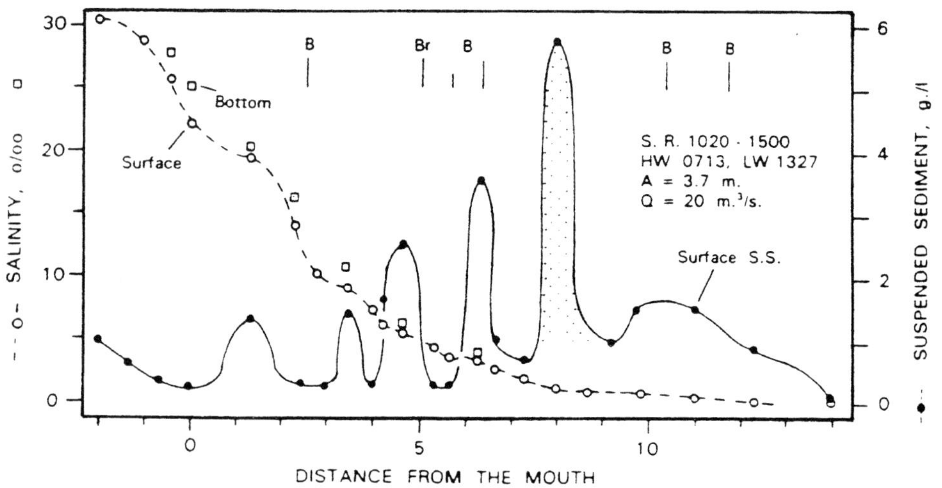

Fig. 4 Longitudinal Salinity & Turbidity at Spring Tide; 8/3/89

In order to study the process of resuspension in more detail, a series of half-tidal cycle observations were made from high water to low water at several stations. Fig. 5 shows how sediment was resuspended from the bottom and then appeared in the surface layer during a large amplitude tide of 5.2 m. and a higher than average freshwater discharge of 75.5 m.^{3}/s. The station lies above the fluid mud patch 12 kilometers above the mouth (Fig. 1). At HWS the surface to bottom salinity difference was 14.4 o/oo and the depth 6.5 m. The surface turbidity was low (105 mg./l.) and consisted of reddish brown silty clay. The bottom 1.5 m. consisted of a 6,210 mg./l., olive gray suspension. By 1000 hours the surface sediment was olive-gray in color and the concentration exceeded 500 mg./l. The peak turbidity shown at 1015 hours was 2,500 mg.l., and the surface to bottom salinity difference was 6 o/oo. By 1030 it was 2 o/oo, and by 1045 the differential was 1.3 o/oo. This very sharp peak in turbidity was probably caused by a mobile suspension, which was brought to the surface during the peak of ebb tide when the current broke up the vertical salinity gradient. The broader, more modest peak of 1,794 mg./l. at 1145 hours was the normal turbidity maximum. This appeared when the mean salinity was <0.5 o/oo, and there was no vertical salinity gradient. The sediment concentration fell gradually to a value of 610 mg./l. at 1545 hours. The salinity was 0.6 o/oo and the sediment was reddish-yellow brown again, i.e., river borne. The depth at LWS was 2.0 m. This yellow-brown sediment was present in the water column at HWS and afterwards until the stratification broke down and olive-gray clay from the bottom layer was incorporated into the surface flow.

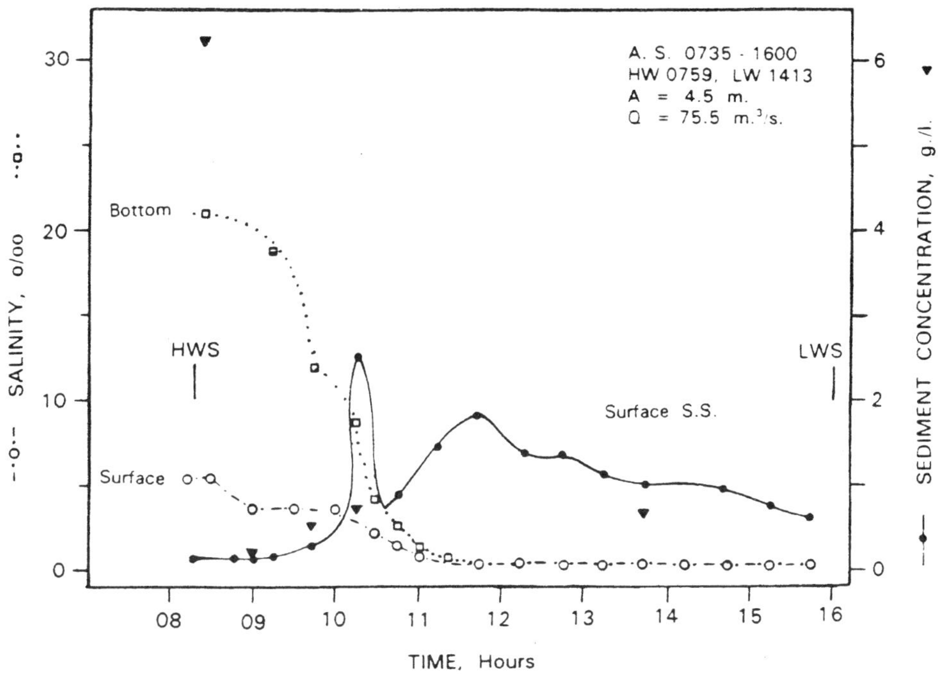

Fig. 5 Ebb Tide Salinity & Turbidity at S-12; 8/30/88

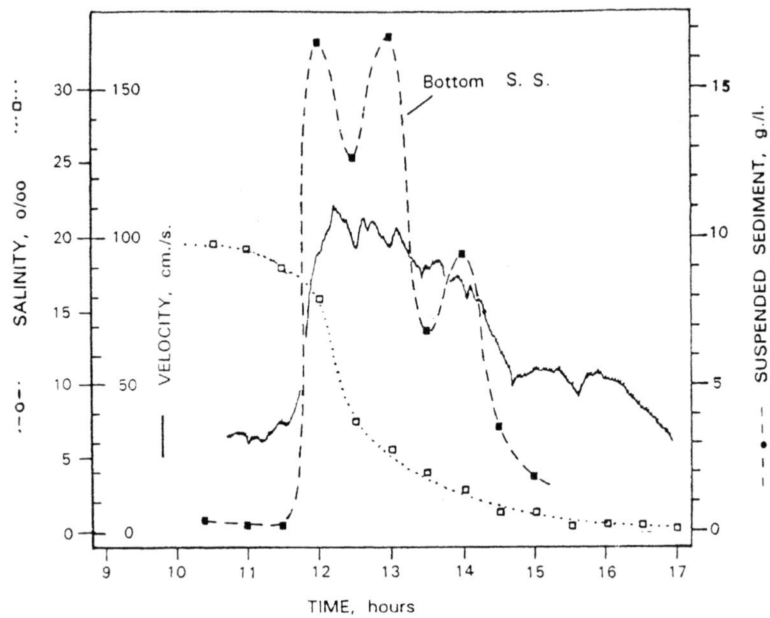

Fig. 6A Ebb Tide Salinity & Bottom Turbidity at S-12; 7/24/89

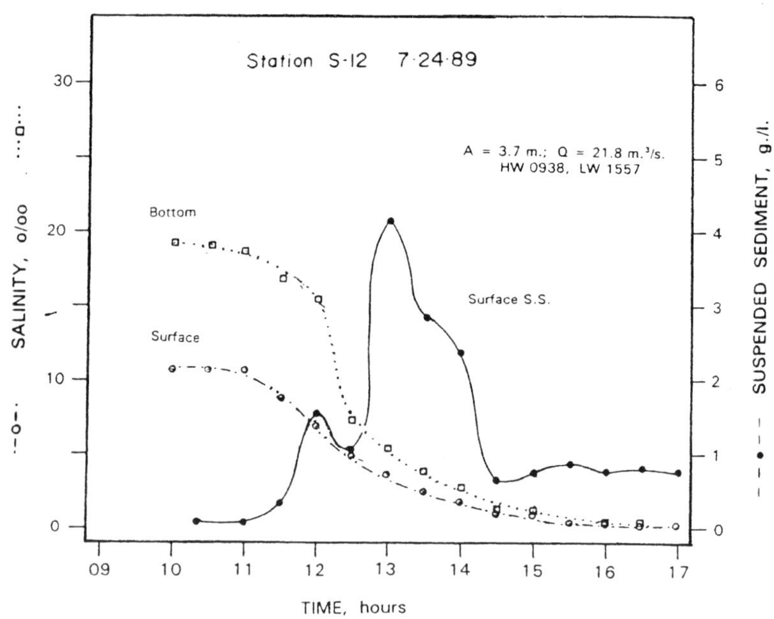

Fig. 6B Ebb Tide Salinity & Surface Turbidity at S-12; 7/24/89

Fig. 6A shows the impact of less intense tidal flow at the same location, i.e. over the uppermost patch of "fluid mud" at S-12 when the tidal amplitude was 3.7 m. The initial depth at HWS was 5.7 m. and Q = 22 $m.^3/s$. Velocity, measured 2.5 m. above the bottom, rose quickly to a peak of 100 cm./s. about 1.5 hours after HWS and then fell gradually throughout the ebb. Sediment was entrained from the bottom abruptly when the current reached 0.5 m./s. at 1200 hours. The concentration in the bottom water peaked at 16,500 mg./l., 2-3 hours after HWS and then fell abruptly during the last half of the ebb tide. Meanwhile, the bottom salinity, initially 20 o/oo dropped gradually to 15 o/oo at 1200 hours. Then as sediment was resuspended from the bottom, the bottom salinity dropped abruptly to 7.3 o/oo at 1230. Such a rapid drop in salinity was not anticipated; a steady ebb flow should bring water of lower salinity from upstream gradually to this station. As noted above, the "fluid mud" at this location had a pore water salinity of <5 o/oo. The coincident drop in bottom water salinity and resuspension of sediment suggests that the reduction in salinity was due to admixture of low salinity "fluid mud" to the flow. The fluid mud must have been deposited initially at the end of an earlier ebb tide, or perhaps during some earlier freshet.

Fig. 6B shows synchronous developments in the surface layer at 1 m. depth. Here the surface salinity is 10.5 o/oo and it falls gradually during the ebb. Starting at HWS, the low concentration sediment samples were yellow-red in color, which is indicative of river-borne soil material. As the turbidity increases, the color changes to olive-gray, an indication that material from the bottom is reaching the surface layer. This occurs abruptly in two pulses, first to 1,574 mg./l. at 1200 hrs and then to 4,116 mg/l. at 1300 hrs. Then it falls to <1,000 mg./l. as the current slackens. These peaks correspond to those of 16,500 mg./l. in the bottom water described above. The first upward flux of sediment at 1200 hours seems to be inhibited by the stratification, i.e. the surface to bottom salinity difference of 8.4 o/oo. The main peak in surface turbidity occured at 1300 hours when the differential was 1.7 o/oo. It lags the initial resuspension from the bed by one hour. In any case, maximum turbidity occurs in the surface only after the vertical salinity gradient has been reduced to <2 o/oo. There also was a depression of the dissolved oxygen in the bottom water, which probably was related to the resuspension of the anoxic "fluid mud". The oxygen dynamics in this system were described by Nelson, et al (1994).

Fig. 7 is an example of how stratification influences the turbidity above a consolidated mud bottom. The observations at this station began one hour after HWS when the depth was 5.0 m. The tidal range was 4.2 m. and the discharge 86.5 m./s. At HWS the top to bottom salinity difference was 19 o/oo. The suspended sediment concentration varied from <100 mg./l. at the surface, to 6,342 mg./l. at 4 m. overlying a mobile suspension of 19,786 mg./l. at 5 m. This mobile suspension began to disperse in the bottom layer when the current exceeded 1 m./s. at 0930. The near bottom turbidity varied from 2,200-6,400 mg./l., peaked at 1045 hours, and gradually decreased to <1,000 mg./l. at 1230 hours. The surface turbidity rose as the salinity differential fell to <10 o/oo and reached 1,000 mg.l. at 1015 hours. The maximum of 2,156 mg./l. occurred at 1100, but only **after** the salinity difference between surface and bottom fell to about 0.5 o/oo and 1.5 hours after the peak flow velocity of 1.6 m./s. Thus, a lag of 90 minutes occurred between resuspension of sediment in the bottom layer and the appearance of maximum turbidity at the surface.

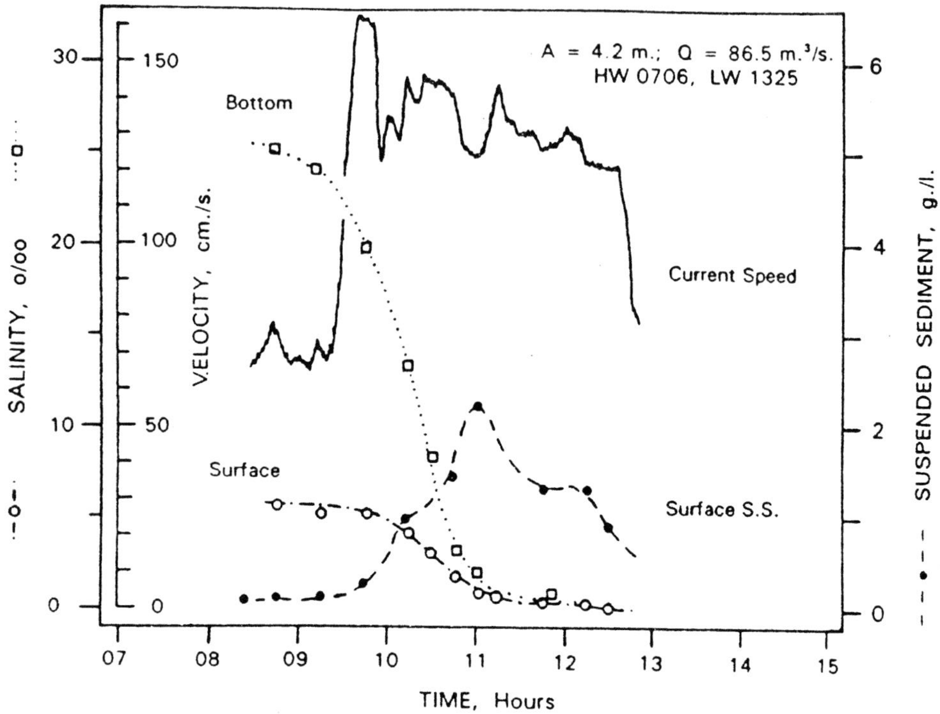

Fig. 7 Ebb Tide Salinity & Turbidity at S-6.3; 9/13/88

In the lower estuary, high concentrations of sediment usually are not found at the surface unless the stratification is broken down by channel geometry or an obstruction. For example, Fig. 8 shows evolution of the turbidity maximum during an ebb tidal cycle at station S-4.7 on July 23, 1989 when A = 4.0 m., Q = 22.5 m.^{3}/s., and initial depth at HWS = 5.2 m. This station is located 0.3 km. downstream from a bridge (see Fig. 1). From 0940 to 1230 hours the boat was located in mid-channel between the bridge piers. The salinity difference between surface and bottom was very small at HWS, and the suspended sediment concentration was <25 mg./l. As the tide ebbed, surface velocities reached 1 m./s. or more, and a moderate stratification developed, as measured by dS = 4-6 o/oo (Fig. 8). This level of stratification is similar to Fig. 2A. The bridge supports upstream did generate turbid swirls that became visible by 1130 hrs., but there seems to have been little effect on the stratification. This is in accord with the results of Miller and Valle-Levenson (1996). Resuspended sediment appeared in the bottom water as the current peaked at 1200 hrs. However, the surface turbidity remained compratively low, 492 mg./l., while at 4 m. depth the concentration was 4,612 mg./l. Even this modestly stratified water column inhibited the upward flux of suspended sediment from the bottom layer.

At 1250 the boat was repositioned directly downstream from one of the bridge piers, where observations were made from 1300 to 1600 hours. Sediment concentrations in the

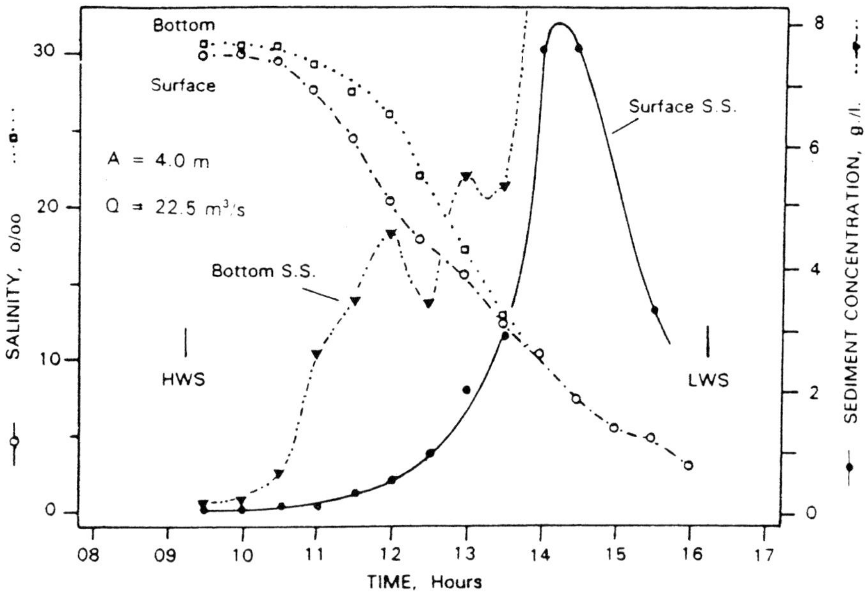

Fig. 8 Ebb Tide Salinity & Turbidity at S-4.7; 7/23/89

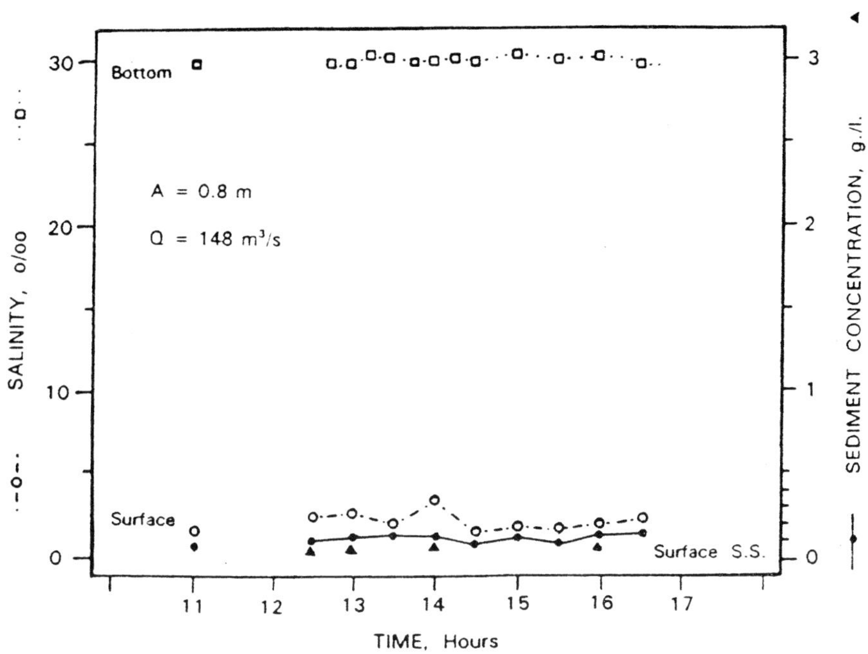

Fig. 9 Ebb Tide Salinity & Turbidity at S-4.7; 9/5/98

surface layer reached a maximum 7,500 mg./l. at 1400-1430. The surface TM peak occured more than four hours after HWS when the salinity was 10 o/oo! By that time the current had begun to slacken. Coincident with the TM at the surface, concentrations in the bottom layer reached 10,700 mg./l. Turbulence from the bridge pier apparently had eliminated any stratification (Fig. 8), because surface-to-bottom salinity differences were <1 o/oo. The extreme contrast between the surface and bottom turbidities from 1000-1230 and those from 1330-1530 show that even modest stratification profoundly affects the vertical exchange of suspended sediment. In the lower estuary where large differentials between surface and bottom salinities usually occur, a greater than average spring tide and artificially generated turbulence were needed to eliminate the stratification and permit sediment to rise upward into the surface layer.

When salt-wedge conditions exist at this station, such as those shown in Fig. 2C, the tidal energy is too weak to break up the stratification and the surface water is isolated from the bottom water by a salinity difference of 28 o/oo throughout the ebb cycle (Fig. 9). There is little resuspension of sediment in the saline bottom water. Small amounts of suspended sediment, inheritied from the river discharge, flow out in the low-salinity surface layer without any augmentation by resuspension from the bottom.

5. DISCUSSION

Wellershaus (1981) showed that high suspended sediment concentrations (500 mg./l.) associated with the salt intrusion result from the resuspension of sediment that was trapped in the convergence zone. Geyer explained this as follows: "The position of this pool of sediment upstream of the region where it is trapped during the ebb may be explained by landward advection during the flood followed by settling." Fig. 2 shows that the limit of salt-wedge intrusion in the Selangor estuary is 8 km. under the high discharge, low tidal amplitude conditions that create a salt-wedge. Under lower flow conditions and higher tidal amplitudes, the salinity intrusion may extend upstream as far as 16 km. The pools of "fluid mud" extend from 8 to 14 km. upstream. Thus, the "fluid mud" patches must be formed, not by the arrested salt-wedge, which is relatively uncommon, but by the partially mixed salinity regime that exists during periods of average and higher tidal energies and average and smaller freshwater inflows.

Riethmuller, et al. (1988) located the TM in the Weser 4 km. downstream from the null point between the 3-5 o/oo isohalines. This was a zone of little stratification as the salinity difference between top and bottom was <1 o/oo. Suspended sediment concentrations ranged from 100 mg./l. at the surface to 800 mg./l at the bottom under these neap tide conditions. However, concentrations > 200 mg./l. were restricted to the lowest 4-5 m. in a total depth of 13-15 m. Thus, even during neap tides this mesotidal system was well mixed vertically where the maximum surface turbidity occurred.

In the macrotidal Gironde estuary Allen, et al. (1980) observed high suspended sediment concentrations just landward of the salt intrusion during spring tides, but fluid mud layers formed there during neap tides. The fluid mud increased in thickness throughout the neap period, but it was re-eroded during the next spring period. The Selangor estuary presents a similar picture of "fluid mud" formation during neaps and resuspension during springs, but the patches of "fluid mud" are scattered along the intrusion length. The observations at S-12, for example, made just following maximum spring tides, show that the TM occurs at a salinity of 3 o/oo when the vertical salinity difference is very small.

Miller and Valle-Levinson (1996) studied the extent of water column destratification from tidal flow around supports of the Chesapeake Bay Bridge where the tidal range is 1 m., the peak tidal currents are 1 m./s., and the water depth is 10 m. The surface to bottom salinity difference was 6-8 o/oo. Stratification varied a little between flood and ebb tide as expected, but it remained nearly constant along the station transects spaced 75, 350, and 1,300 m. from the bridge pilings. The stratification was not significantly modified by the pilings. They found the mixing energy due to the vertical pilings was much less than that due to frictional interaction with the bottom. "The peak effect of bottom stress is several times greater than that due to the pilings and its mean effect is an order of magnitude greater." Since, the mean depth in the Selangor is only 5 m., the data in Fig. 8 need to be interpreted primarily in terms of bottom-generated turbulence. The parallel rise in surface and bottom turbidity after 1300 hours is likely due to the decreasing depth late in the tidal cycle and the progressively greater impact of turbulence generated from the bottom. It is clear that this parallelism is supported by a zero vertical salinity gradient and a related high diffusivity.

The turbidity maxima described above are "unusual" in the sense that they occur in a small estuary with moderate tidal range and current speeds. Macrotidal estuaries, such as the Gironde and the Severn, generate higher turbidities. The relatively high values at mesotidal energies encountered in the Selangor must be related to the abundant supply of fine-grained sediment that flows into this estuary. It also may be that the shallow depths reflect a system that is closer to an "equilibrium estuary" as defined by Dyer (1988, p308) which may be expected to produce a heavy sediment flux to the sea.

But the fundamental feature to emphasize is that most estuarine sediment resuspended from the bottom cannot rise to the surface under even modestly stratified conditions. Geyer (1993) showed that particle settling velocities and density stratification affect particle sizes in the turbidity maximum. Stratification in a salt-wedge estuary suppresses the turbulence within the halocline and decreases the quantity of sediment that can remain suspended in the surface layer. The vertical distribution of sediment depends on the relative magnitude of the particle fall velocity compared to the turbulent motion of the fluid (the Rouse profile). For fine sediment or strong turbulence (small values of R_o), the distribution of sediment is nearly uniform vertically, but for coarse sediment and weak turbulence (large values of R_o) the sediment and sediment fluxes are confined to the bottom layer. This especially affects silt-sized particles which sink below the halocline and are advected towards the toe of the salt wedge. Such reasoning helps explain the vertical distribution of sediment in the Weser and here in the Selangor. But the observations above also show that the vertical flux of sediment must be extremely sensitive to even very limited stratification, such as occurs in partially mixed estuaries. In the Selangor, very little suspended sediment appears in the surface layer (let alone drop out from it), unless the salinity gradient is less than about 1 o/oo in 4-5 m. And this is in a system dominated by turbulence generated along the bottom. It would appear that the vertical salinity gradient needs to be included in future models that hope realistically to portray the cohesive sediment dynamics of estuaries. West and Shiono (1988) analyzed the interaction of bottom generated turbulence on the vertical density gradient in shallow estuaries such as the Selangor, and their work would be a good basis for quantitative treatment of the cohesive sediment dynamics.

The observations above show that a definite time lag occurs between the resuspension of "fluid mud" from the bottom and the appearance of significant quantities of suspended sediment in the surface layer. The maximum turbidity in the Selangor consistently occurs over a short range where the surface-to-bottom salinity difference is less than one or two parts per thousand. The time lags observed are on the order of one to three hours, a significant fraction of the ebb tidal cycle. Such lags must have a large effect on the net transport of sediment, since the surface layer is mainly responsible for the downstream flux in the estuarine circulation. It would appear that this factor should shorten the length over which one observes a turbidity maximum. Maximum turbidities have been observed late in the ebb cycle in other estuaries, even though the ebb current peaks earlier. The effects of this late arrival or "stratification lag" should be evaluated as a factor in the net transport of sediment.

REFERENCES

Allen, G. P., G. Sauzay, P. Castaing, and J. M. Jouanneau, 1977, Transport and deposition of suspended sediment in the Gironde estuary, p63-81 in M. Wiley [Ed.], *Estuarine Processes*, v2, Acadamic Press, New York

Allen, G. P., J. C. Salomon, P. Bassoulet, Y. du Penhoat, & C. de Grandpre, 1980, Effects of tides on mixing and suspended sediment transport in macrotidal estuaries, *Sedimentary Geology*, v26, p69-90

Dronkers, J., 1986, Tide-induced residual transport of fine sediment, p228-244 in J. vanden Kreeke [Ed.], *Physics of shallow estuaries and bays, Lecture Notes on Coastal and Estuarine Studies* #16, Springer-Verlag, Berlin, 280p

Dyer, K. R., 1997, Estuaries, a physical introduction, 2nd Ed., John Wiley & Sons, Chichester, U.K., 195p.

Dyer, K. R., 1988, Fine sediment particle transport in estuaries, p295-310, in J. Dronkers & W. van Leussen [Eds.], *Physical Processes in Estuaries*, Springer-Verlag, Berlin, 560p

Geyer, W. R., 1993, The importance of suppression of turbulence by stratification on the estuarine turbidity maximum, *Estuaries*, v16, n1, p113-125

Jay, D. A. & J. D. Musiak, 1994, Particle trapping in estuarine tidal flows, *J. Geophysical Research*, v99, nC10, p20,445-20,461

Kirby, R., 1988, High concentration suspension (fluid mud) layers in estuaries, p463-487, in J. Dronkers & W. van Leussen [Eds.], *Physical Processes in Estuaries*, Springer-Verlag, Berlin, 560p

Kirby, R. and Parker, W. R., 1983, The distribution and behaviour of fine sediment in the Severn estuary and Inner Bristol Channel, *Can. J. Aquatic Sciences*, v40, Sup. 1, p83-95

Lewis, R., 1997, Dispersion in estuaries and coastal waters, John Wiley & Sons, New York, 312p.

Mehta, A. H. [Ed.]. 1989, On estuarine cohesive sediment suspension behavior, *J. Geophysical Research*, v94, n10, p14303-14314

Mehta, A. J., 1988, Laboratory studies of cohesive sediment deposition and erosion, p427-444, in J. Dronkers and W. van Leussen [Eds.], *Physical Processes in Estuaries*, Springer-Verlag, Berlin, 560p

Miller, J. L. & A. Valle-Levinson, 1996, The effect of bridge piles on stratification in lower Chesapeake Bay, *Estuaries*, v19, n3, p526-539

Nelson, B. W., 1960, Recent sediment studies in 1960, *Mineral Industries Journal*, Virginia Polytechnic Institute, v7, n4, p1-4 [cited in R. H. Meade, 1972, Transport and deposition of sediments in estuaries, *Geological Society of America, Memoir 133*, p100]

Nelson, B. W., A. Sasekumar, & Z. Z. Ibrahim, 1994, Tidal effects on dissolved oxygen in two Malaysian estuaries, *Regional Seminar on Ecology and Conservation of Southeast Asian Marine and Freshwater Environments, Kuala Lumpur, November 4-6, 1991, Hydrobiologia*, v285, p7-17.

Odd, N. V. M., 1988, Mathematical modelling of mud transport in estuaries, p503-531, in J.Dronkers & W. van Leussen [Eds.], *Physical Processes in Estuaries*, Springer-Verlag, Berlin, 560p

Postma, H., 1967, Sediment transport and sedimentation in the estuarine environment, p158-179 in G. Lauff [Ed.], *Amer. Assoc. Adv. Sci.*, Pub. No. 83

Riethmuller, R., *et al*, 1988, Hydrographic measurements in the turbidity zone of the Weser estuary, p 332-344 in J. Dronkers & W. van Leussen [Eds.], *Physical Processes in Estuaries*, Springer-Verlag, Berlin, 560p

Schubel, J. R., 1968, Turbidity maximum of the northern Chesapeake Bay, *Science*, v161, p1013-1015

Wellershaus, S., 1981, Turbidity maximum and mud shoaling in the Weser estuary, *Archiv. Hydrobiology*, v92, n2, p161-198

West, J. R. and K. Shiono, 1988, The structure of turbulence in partially mixed estuaries, p196-210 in J. Dronkers and W. van Leussen [Eds.], *Physical Processes in Estuaries*, 560p

Near bed sediment transport in the Itajaí-açu River estuary, southern Brazil.

C.A.F. Schettini

Centro de Ciências Tecnológicas da Terra e do Mar CTTMar-UNIVALI
C.P. 360, Itajai SC 88.302-202, Brazil.

The objective of this study is to assess the near bed sediment transport in the Itajaí-açu River salt-wedge estuary, especially regarding two fortnight periods: the first being dominated by tidal currents, and the second being dominated by river flood currents. The data acquisition system consisted of an acoustical current meter moored in a tripod, with an optical backscatter turbiditymeter and a pressure sensor. The tripod was deployed at the channel thalweg 10 m deep and at 4 km from the estuarine mouth, throughout a 75-day period. Daily river discharge data were provided by the National Power Agency. During the period of data acquisition two discharge peaks greater than 1,000 $m^3.s^{-1}$ occurred within a seven-day interval, the first being in neap tide and the second in spring tide condition. During the high discharge events, the near bottom current speed peaks reached 1.0 $m.s^{-1}$, while during normal conditions they are usually lower than 0.7 $m.s^{-1}$, even during spring tides. Furthermore, the current during the river flood events was unidirectional towards the sea, lasting 49 and 35 hours, respectively. The tidal signal could still be observed, however, as the oscillation of the seaward current. Strong ebb current asymmetry followed such periods. The yield shear stress exceeded 2 Pa during the current peaks, and the bottom suspended sediment concentration increased from 0.02 to more than 0.25 $kg.m^{-3}$. The mean seaward sediment transport during the events was 303 $kg.m^{-2}$ and 509 $kg.m^{-2}$ per hour, respectively. Comparatively, during the previous fortnight period with low river discharge, the mean sediment transport was 32 $kg.m^{-2}$ per hour landward. These results show that critical sporadic high discharge events play an important role in the sediment dynamics in the Itajaí-açu estuary, furnishing higher critical bottom shear stress than tidal currents as well as unidirectional seaward flow. During the tide-dominated period, erosion appears to take place only during the flood of spring tide, contributing to estuarine basin infilling. On the other hand, during river the flood-dominated period, erosion appears to take place as a function of river discharge only, promoting intense seaward sediment transport.
KEY WORDS: episodic events; near bed sediment transport; Itajaí-açu River estuary

1. INTRODUCTION

Man often tries to fit the environment to his needs in the way of economic growth and social development. Estuaries are environments which are frequently cited as, although having been adequate to meet the demands placed upon them in the past, have become increasingly stressed due to the greater demands placed upon them as a result of area growth and development. In this context, navigational conditions are of prime importance and this lead the searching for better understanding of sediment dynamics in estuaries.

The sediment in many estuaries consists mainly of fine sediment, which are clay and silt size particles with a smaller content of sand and organic matter. Because of the characteristics of clay particles, these sediment present cohesive properties that originate from the surface electro-chemical forces of broken bonds (Raudkivi, 1990; Drever, 1988). These particles are supplied mainly by rivers, but they can also be provided by the inner shelf (Meade, 1969; Schettini & Carvalho, 1998). They occur as aggregates rather than single particles (Krone, 1978; Eisma; 1986), and once in the estuary, they are in a remarkably complicated environment where they undergo a repeated cycle of erosion, transport and deposition (Postma, 1967; Mehta et al., 1982; Nichols, 1984; Dyer, 1986 & 1995). Such processes are still not fully understood, and are generally described by semi-empiric models obtained mainly through laboratory experiments (Partheniades 1984; Mehta, 1988; van Rijn, 1993; Winterwerp et al., 1993). *In situ* observations of the cohesive sediment dynamics are very complicated, since all concerning variables are acting with different weights to produce a given response, under varying conditions. The deployment of an *in situ* flume such as the "sea carousel" produces excellent results (Amos et al., 1992), but its use involves costs and logistics so that its deployment is not possible everywhere.

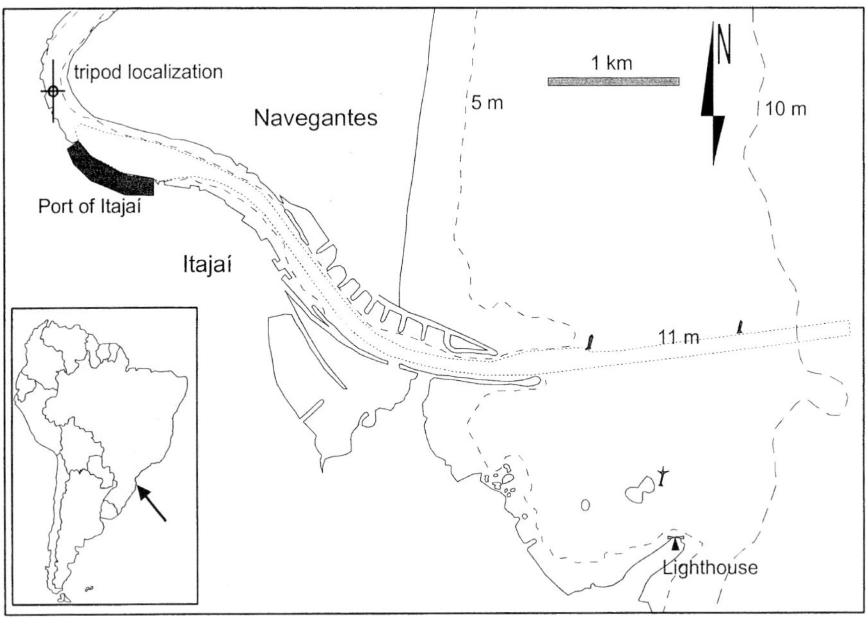

Figure 1: The localization of the estuary of the Itajaí-açu River, with the position where the tripod was deployed.

Beyond the complexity of estuarine hydrodynamics and cohesive sediment behavior, field observations must also cope with episodic events that significantly change the 'standard' patterns. Ingram et al. (1986), Kirby et al. (1993) and Nichols (1993), give some examples. These events, rather than episodic, are cyclic. As one would expect, they will repeat within a given time scale. Such time scale can vary from hours to years depending on which processes are being observed, leading to important implications for deposition and erosion of cohesive sediment in estuaries (Mehta et al., 1989). This study assesses the near bed sediment dynamics of the Itajaí-açu River salt wedge estuary (Figure 1), based on a 75-day time series of near bed currents and turbidity. During the period, two episodic events occurred within a seven-day interval, changing the physical setting of the estuarine dynamics. In an attempt to observe erosional events during the time series, an effort to estimate critical bed shear stress for erosion is presented as well.

2. SETTING

The Itajaí-açu River estuary is a narrow salt-wedge system, where the mean channel was 6-7 m deep in former times. This condition was conducive to the early development of harbor activities at the beginning of the 20th century. However, the harbor activities continue to develop, and at present, the port access channel and turning basin must be maintained at 11 m deep along the first 4 km of the lower estuary (Figure 1). Siltation occurs promptly after dredging, requiring a continuous program to maintain the desirable depth.

The bottom sediment in the lower estuary consists mainly of clay size material, > 70 %, with a lesser content of sand and silt, and 10 ± 3 % of organic matter. Ponçano (1982 & 1987) pointed out that the sedimentary facies distribution changes significantly after a high discharge event. The percentage of well-sorted fine sand increases after prolonged periods of low discharge, suggesting near bottom sediment transport landward. After a high discharge event, the sediment becomes muddy again, and the sand, if present, is riverborne. After a prolonged drought, suspended sediment is exported to the inner shelf through the upper layer, and imported through the lower layer, with net transport landward (Schettini & Carvalho, 1998).

River discharge is the main driving agent for estuarine hydrology and hydrodynamics, with the tides playing a minor role. The mean river discharge measured daily at the Indaial lymnimetric station since 1934 is 230 ± 280 $m^3.s^{-1}$, with a minimum and maximum observed of 16 and 5,390 $m^3.s^{-1}$, respectively. The discharge does not present a regular seasonal pattern; it is usually low, below 200 $m^3.s^{-1}$, with sporadic pulses greater than 800 $m^3.s^{-1}$. The river discharge pulses occur all year long as a response to cold front passage or extra-tropical storms. The highest discharges ever observed were related to El Niño–Southern Oscillation years.

The mean suspended sediment load (SSL), monitored daily at the Indaial station since 1998, is 2,039 tons per day. It, too, is highly variable as it is a function of river discharge, and can be related to it by $SSL\ [ton.day^{-1}] = -5{,}137 + 28.8\ River\ Discharge\ [m^3.s^{-1}]$, with r^2 = 0.77. From November 1998 to November 1999, the SSL presented a minimum and maximum of 29 and 83,709 tons per day, with total sediment delivery of about 760,362 tons. The day with maximum SSL accounted for 11 % of the total SSL for the entire year. During this period, there were only 9 days in which the river discharge exceeded 1,000 $m^3.s^{-1}$, accounting for 41 % of the total SSL.

The suspended sediment concentration (SSC) in the estuary varies as a function of the SSL, ranging from less than 25 $mg.l^{-1}$ to more than 300 $mg.l^{-1}$. The distribution of the SSC follows the distribution of salinity: the upper layer of fresh and/or brackish more turbid waters; and the lower layer of salt and less turbid waters. The maximum turbidity zone is not observed in this estuary, and fluid mud is only observed in the turning basin and access channel, due to the action of the water-injection dredging system used.

Local tide is mixed semi-diurnal with a mean range of 0.8 m, varying between 0.3 to 1.2 m during neap and spring tide cycles, respectively. The tide can be observed up to 70 km upstream, presenting syncronous behavior (e.g., Nichols & Biggs, 1985). The meteorological effects on the sea level can induce surges of about 1 m above the astronomical signal, and account for 30 % of the sea-level variability (Truccolo, 1998).

The estuary presents highly stratified distribution of salinity, and it is of Type 4 – Salt Wedge, according to the circulation-stratification classification scheme of Hansen & Rattray (1966; Schettini et al., 1996). The salt-wedge intrusion in the estuarine basin is non-linear and inversely related to the river discharge. Under near mean river discharge, 300 $m^3.s^{-1}$, the salt-wedge intrusion can enter up to 18 km upstream; when the discharge exceeds 1,000 $m^3.s^{-1}$, the salt wedge is completely flushed out of the estuary (Döbereiner, 1985). The river discharge can explain about 70 % of the salt wedge displacements, and can be empirically related by *Salt Wedge Intrusion [km] = 1.72 + 32.69 exp(-2.17e-3 River Discharge [*$m^3.s^{-1}$*])* (Schettini & Truccolo, 1999).

3. DATA ACQUISITION

The data used in this study were acquired with the deployment of a Famouth™ 3D-ACM acoustic current meter fixed in a tripod in the estuarine thalweg. The tripod was situated about 4 km from the estuarine mouth, just upstream of the turning basin of the Itajaí Port (Figure 1). The instrument recorded North-South and East-West velocity components, hydrostatic pressure, temperature and turbidity at a point 1 m above the bottom. The turbidity was measured by a Seapoint™ optical backscattering sensor, previously calibrated in the laboratory with local sediment. The instrument-operating mode was averaging bursts of 5 minutes at 5.5 Hz., every 20 minutes for 75 days with external power supply, from September 8 to November 21, 1999.

Yield bed shear stress was calculated by the quadratic stress law $\tau = c_D \cdot \rho \cdot U_{100}^2$, using a drag coefficient c_D for mud bed of 0.0022 (Dyer, 1986), and constant density ρ = 1025 $kg.m^{-3}$. The application of the quadratic stress law is not recommended for stratified flows, as is the case in the Itajaí-açu River estuary. However, the region where the instrument was deployed is sea-dominated most of time with a thickness of about 5 to 7 m of homogeneous coastal salt water with more or less constant salinity of 30 ‰. In the present case the obtained values of τ must be considered as an approximation.

Daily Itajaí-açu River discharge information was obtained from the Brazilian Power Agency (ANEEL) for the Indaial lymnimetric station. This station is 90 km from the estuarine mouth, representing 71 % of the total river drainage, and is the closest station of estuarine mouth without tidal influence. The SSC was also monitored at this station for the purpose of SSL calculations.

4. RESULTS

Figure 2 presents the hourly time series of river discharge, water level, longitudinal velocity component (positive values mean landward, negative values mean seaward) and SSC. The bold line represents the low-frequency non-tidal oscillations.

The river discharge during the sampling period well exemplifies the random behavior of low discharge with peaks of higher discharge. The probability of occurrence of a river discharge peak greater than 1,000 $m^3.s^{-1}$ is about 0.025. The occurrence of the two peaks in a matter of a week as occurred on days 25 and 31, presenting 1,156 and 1,167 $m^3.s^{-1}$, respectively, was a rare situation.

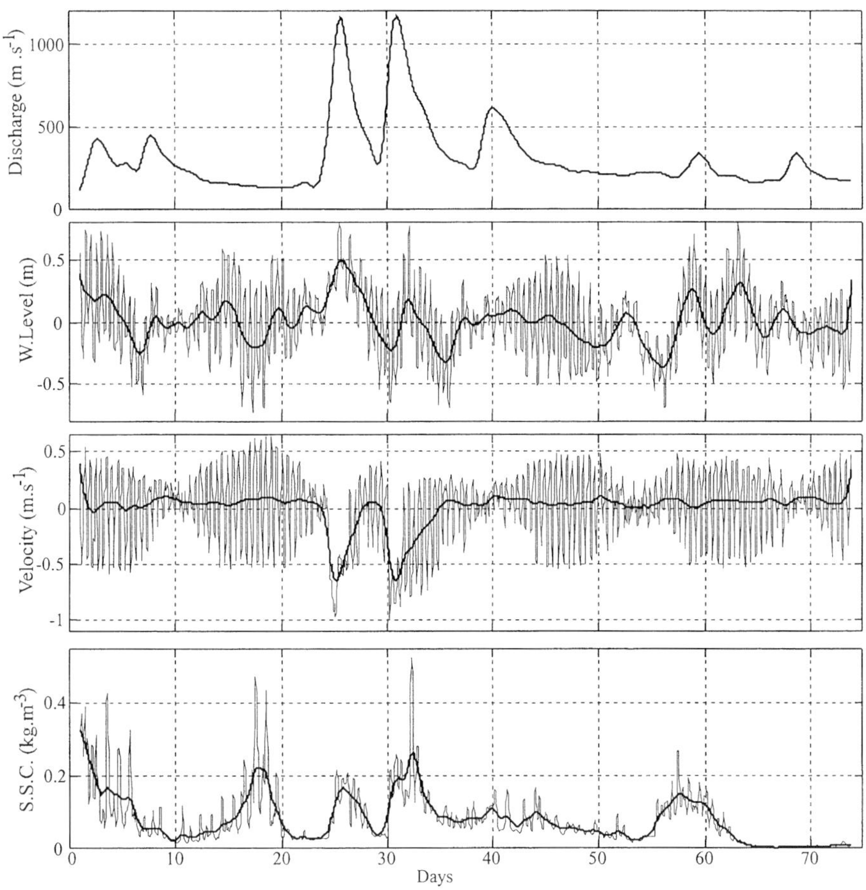

Figure 2: Time series of daily river discharge at the Indaial lymnimetric station, and hourly water level, longitudinal velocity component and suspended sediment concentration. The bold lines in the latter three variables are non-tidal low-frequency signals.

The lower estuary water level response to river discharge pulses has not been thoroughly assessed so far. However, from Figure 2 it is possible to ascertain that only the most expressive peaks present some importance. Even so, despite the same magnitude of the peaks on days 25 and 31, different signals in the low-frequency water level were observed. The remaining non-tidal oscillations in the water level were caused mainly by local effects of the wind blowing on the coastal sea (Truccolo, 1998).

The response of the near bed current to the river discharge peaks was much clearer than its response to the water level. As the two higher flood waves flushed out the salt water from the estuarine basin, the higher seaward velocity peaks of about 1 $m.s^{-1}$ followed. The other smaller river discharge peaks did not generate a noticeable response on the currents. The non-tidal low-frequency sea level oscillations also appeared to have almost no influence. The residual non-tidal currents during low discharge periods were always landward. The salt wedge under these conditions is well developed and can reach up to 30 km upstream from the estuarine mouth. The river advection and tidal mixing induce salt water entrainment to the upper seaward flowing layer, and landward bottom current is expected to respect volume conservation (Officer, 1976; Dyer, 1997).

The SSC varied greatly during the entire monitored period, ranging from 0.015 to 0.550 $kg.m^{-3}$. The higher peaks before day 10, as well as after day 55, were caused by mechanical perturbation due to the dredging activity in the harbor area. Between these days there was no dredging activity. The disappearance of the turbidity signal after day 60 was due to biological activity blocking the sensor. Perturbations of the SSC signal could also have originated from ship maneuvering in the harbor, where resuspended sediment plumes with concentrations higher than 1 $kg.m^{-3}$ occur. Due to the proximity of the harbor turning basin to the monitoring site, such plumes could have reached it during periods of flood currents.

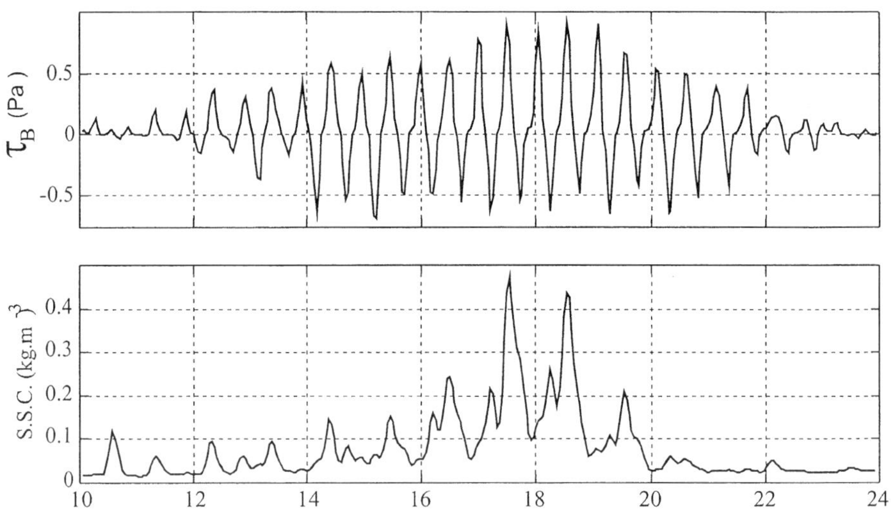

Figure 3: Yield shear stress and suspended sediment concentration during the fortnight period before the high river discharge peaks, which can be considered typical for low river discharge periods (positive values mean landward, and negative, seaward).

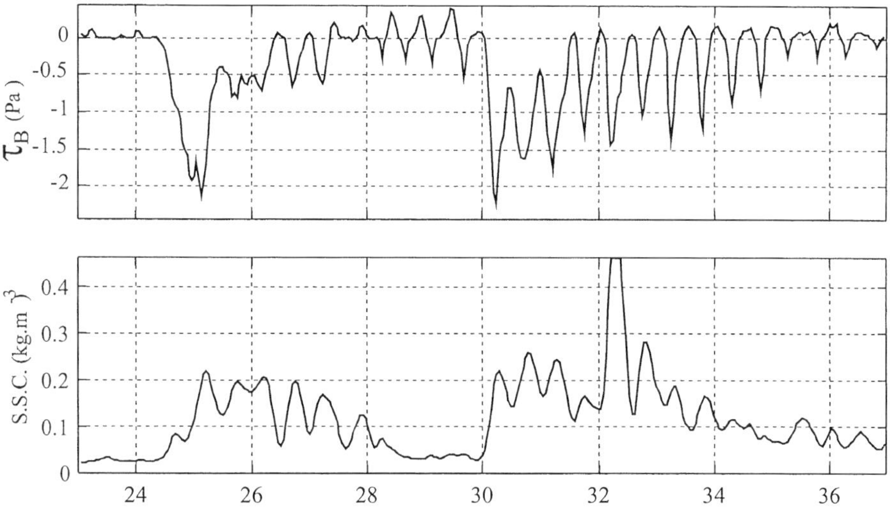

Figure 4: Yield shear stress and suspended sediment concentration during the fortnight period when the observation of two high river discharge peaks greater than 1,000 $m^3.s^{-1}$ occurred, on days 25 and 31 (positive values mean landward, and negative, seaward).

The fortnight period from day 10 to 23 presented a tide-dominated condition of the SSC related to the yield bed shear stress (Figure 3). As the tidal range increases, the same happens with the yield bed shear stress, providing shear excess for erosion (Mehta et al., 1982; Mehta, 1988; van Rijn, 1993). However, a delay in the response of the SSC to the increasing shear occurred, when the maximum SSC is observed at the end of spring tide period. This behavior can be explained by the progressive reduction of the bed shear strength, which reaches its smaller value at the end of spring tide. As the shear excess diminishes after the spring tide, the bottom sediment consolidates recovering its initial shear strength, decreasing the erosion rate. The asymmetry of the SSC peaks is noticeable between day 16 and 20, following the tide inequalities. It can be deduced that during maximum spring tide currents the yield bed shear stress exceeds the critical bed shear stress for erosion of lower sediment layers.

During the following fortnight period, from day 23 to 36, the discharge peaks promptly furnished much more shear excess than had been observed during the previous tide-dominated fortnight period (Figure 4). Furthermore, the periods with shear excess were much longer than a tidal cycle: the first lasted 49 hours and the second lasted 35 hours of continuous seaward flow. The mean yield bed shear stresses were 0.84 and 1.15 Pa, respectively, with a maximum of 2.29 and 2.70 Pa, respectively. The mean shear values were high enough to provide shear excess for erosion of estuarine beds in the consolidation process, and the peak shears were high enough to erode even moderately compacted deposits (van Rijn, 1993). Along with the increase in bed shear stress, the salinity dropped from an average of over 30 ‰ during salt-wedge presence to 0 ‰. Owen (1975), and Winterwerp (1989), reported the reduction of the critical bed shear stress for erosion associated with the decrease in salinity, and this phenomenon probably also contributes to the augmentation of erosion rates during high river discharge events.

5. CURRENT ASYMMETRY AND NEAR BED SEDIMENT TRANSPORT

Figure 5 presents the frequency distribution of yield bed shear stress for the periods of days 10 to 23 and days 23 to 36. Positive shear stress means seaward and negative means landward. The frequency distribution from the former fortnight period resembles a quasi-symmetrical shape, with a slight trend landward, with a mean of about 0,05 Pa. This period can be stated as typical for low river discharge conditions. Considering a critical bed shear stress for surface erosion of the top sediment layer of 0.20 Pa (e.g., van Rijn, 1993), we have that it occurs at 21 % of the time landward, and 14 % seaward. It is clear that such unbalance of 7 % will result in the net transport of sediment towards the tip of the salt wedge. During prolonged low river discharge when the salt wedge can reach up to 30 km upstream, such a mechanism can significantly increase the estuarine trapping efficiency, and can even generate periods of importation of sediment from the inner shelf (Schettini & Carvalho, 1998).

During the next fortnight period, stronger asymmetry occurred in response to the river discharge pulses, with a mean of about -0.32 Pa. Bed shear stress higher than 0.20 Pa occurred less than 3 % of the time, while values smaller than -0.20 Pa accounted for 40 % of the time. The asymmetry can be distinguished well by the non-tidal low-frequency velocity (Figure 2). The asymmetry induced by the first peak was shorter and the tidal signal back to normality in a matter of two days. The second peak was as short as the first, but the discharge did not diminish in the same way, presenting a step at 600 $m^3.s^{-1}$, and taking about 5 days to back down. However, not only the magnitude of the peak is important to the estuarine response, but the behavior of the discharge after that as well. The low-frequency velocity component could be linearly related to the discharge during such events, and for this particular period the relationship was $Velocity\ [m.s^{-1}] = 0.26 + 8\times10^{-4}\ River\ Discharge\ [m^3.s^{-1}]$, with $r^2 = 0.88$. It is noteworthy that this linear relationship does not apply for smaller discharge peaks like that of day 40, or perhaps it applies but only upstream of the salt-wedge tip.

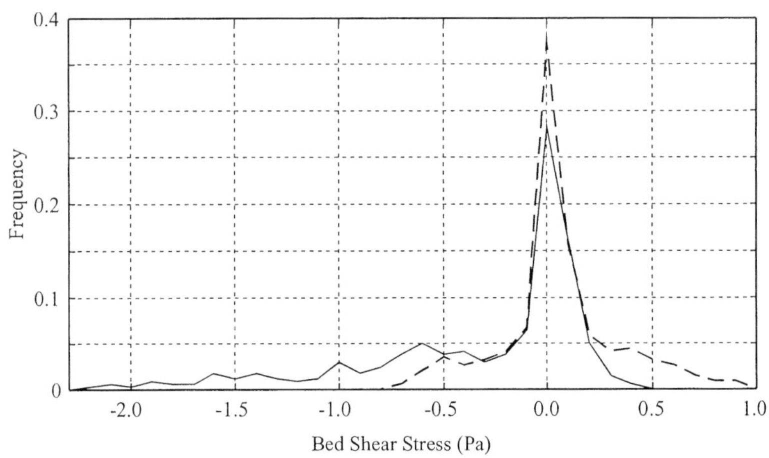

Figure 5: Frequency distribution of the yield shear stress during the fortnight period from day 10 to 23, traced line, and for the period from day 23 to 36, full line.

The implication of current and bed shear stress asymmetry on the estuarine bottom sediment transport is obvious. During the typical fortnight period with low river discharge, the net near bottom sediment transport was $+9\times10^{-3}$ kg.s^{-1}.m^{-2}, while for the high river discharge fortnight it was -44×10^{-3} kg.s^{-1}.m^{-2}. Specifically during the periods of unidirectional seaward flow, the mean transports for the first and second peaks were -80×10^{-3} and -140×10^{-3} kg.s^{-1}.m^{-2}, respectively.

For the fortnight period that succeeded the high river discharge events, the mean transport was again positive, about $+6\times10^{-3}$ kg.s^{-1}.m^{-2}. This value is not much smaller than the value observed during the tide-dominated fortnight period, which preceded the high river discharge events. After the episodic events, the bed surface sediment were eroded leaving a sediment layer with higher bed shear strength exposed. Observing the SSC time series for this period, the relationship between the peaks of SSC with erosion events due to shear excess is unclear. However, most of the sediment that were transported landward during this period were probably not from the bed but from inner shelf suspension. Along the inner shelf there is a near bed high turbidity zone associated with the river plume dispersion and nearshore processes (Schettini et al., 1998), which could be the source for this material.

5.1. Erosion Events

One basic requirement to understand the cohesive sediment dynamics in estuaries is to have a good knowledge of bottom sediment properties, as well as hydrodynamics. The bed shear strength is the most important property, and other properties can be empirically related to it. Notwithstanding, at present it is not possible to measure it directly *in situ* (Parker & Kirby, 1982). Other properties such as bulk density and dry density are easier to measure, from which the bed shear stress for erosion can be estimated (van Rijn, 1993).

A great problem in the assessment of a reasonable value for the critical bed shear for erosion from field observations of SSC is how to identify its source of the latter: how much is from erosion, and how much is from advection. The erosion rate is defined as the gain of mass of suspended sediment by unit of time, considering the absence of sediment advection. It is linearly related to the excess shear stress either over the shear strength (Mehta, 1984) or the critical bed shear stress for erosion (van Rijn, 1993) in homogeneous sediment; or non-linearly related in the case of increasing shear strength with depth in non-homogeneous sediment. In both cases it is necessary to have further knowledge of the shear excess and dimensional coefficients determined in laboratory to adequately determine the erosion rate.

Figure 6 presents the relationship between yield bed shear stress and SSC for the periods of days 10 to 23 and days 23 to 36. The SSC was directly related to the yield bed shear stress, but no useful coefficients could be drawn from such crude information: for a given yield shear stress there was one order of magnitude of variation in the SSC. To estimate a critical bed shear stress for erosion, a correlation analysis between the yield shear stress and erosion rate was performed. The latter was given by the derivative of the SSC per unit of time and unit of volume. The negative values were excluded, since erosion and deposition are mutually exclusive processes. The correlation was performed recursively by selection of the pairs according to the increasing shear.

Figure 7 presents the results of the correlation analysis for both the fortnight periods. It is reasonable to state that in a system where it is not possible to define the origin of the SSC, the increase of yield shear stress augments the probability that the observed SSC was originated by erosion instead of advection. The correlation coefficients showed in Figure 7 are not very high, but they present an interesting behavior. The correlation coefficient for the tide-dominated period does not increase as the yield shear stress increases: it stays under 0.2 and presents a steep increase when the shear stress reaches 0.45 Pa, remaining at this level after that. The same was not observed for the river-dominated period, where a better correlation was observed with all data, indicating the dominance of advection in the control of the SSC.

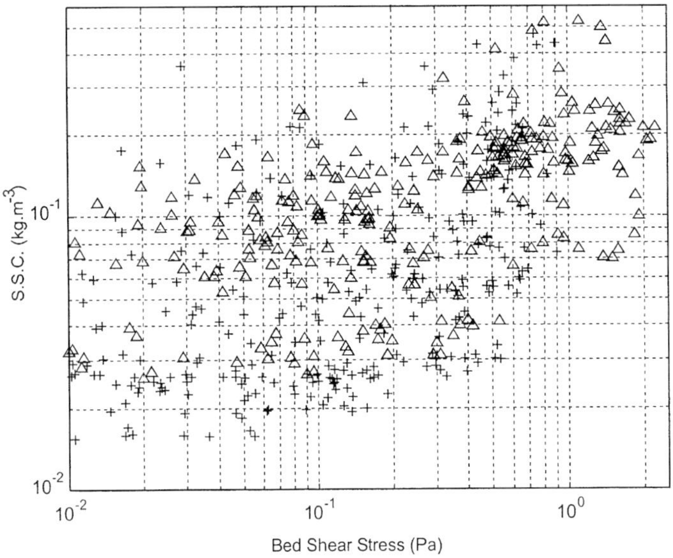

Figure 6: Relationship between yield bed shear stress and suspended sediment concentration: crosses for the 10-23 period; triangles for the 23-36 period.

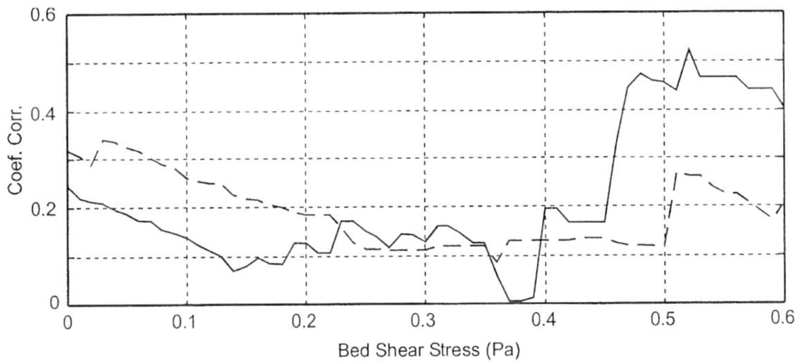

Figure 7: Results of the correlation analysis with progressive selection by shear increase. Full line for 10-23 period; dashed line for 23-36 period.

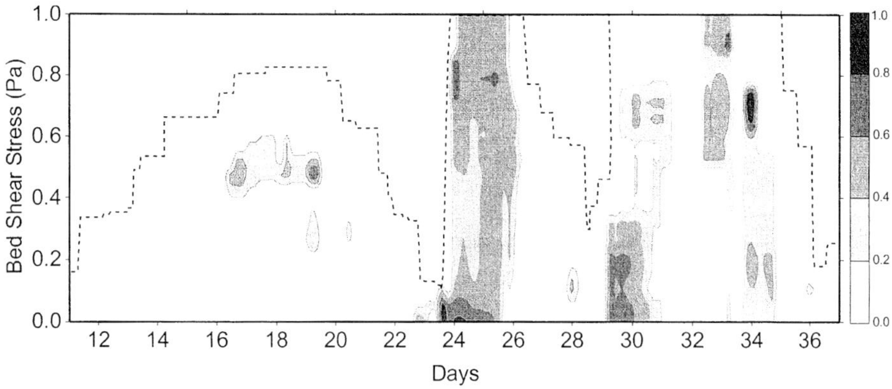

Figure 8: Results of the correlation analysis with progressive selection by shear increase performed with a running window of 48 hours. The dashed line represents the maximum yield shear stress for a given 48-hour period.

The significance of the correlation coefficient decays as the number of pairs diminishes as the occurrence of higher yield shear stress decreases. However, although it could be merely coincidence, it is noteworthy that a steeped behavior was observed around the value of 0.4 – 0.5 Pa, which is the erosion threshold of lower layers for several types of cohesive sediment (e.g., van Rijn, 1993). Therefore, this approach is not able to indicate critical shear stress for erosion of surface layer, probably because of the masking effect of suspended sediment advection. On the other hand, it suggests a reasonable value for the critical shear stress for erosion of lower layers, and erosion events as well.

Figure 8 presents a similar correlation analysis, which was performed using a 48-hour window running from day 10 to 36. During the tide-dominated fortnight period, the better correlation coefficients were obtained during the maximum yield shear stress period, with better values for 0.4 – 0.5 Pa, as previously shown in Figure 7. For the river-dominated fortnight period, during the first high river discharge event, erosion took place at very low yield shear stress. This could be because the bottom during the neap tide periods is formed by loose flocs structure deposited under low shear, so, it could be easily eroded even with a low increment in shear (Partheniades, 1984; Mehta, 1988). At the beginning of the second high river discharge event, erosion seems to have taken place again at low yield shear stress, but not continuously as it was observed in the first event. Apparently, during this period erosion occurred when the shear reached was about 0.6 Pa, as the bed was then formed by exposed sediment with a longer period of consolidation and higher shear strength (Mehta et al., 1982).

6. CONCLUDING REMARKS

Estuaries are remarkably complex environments. Cohesive sediment processes present great complexity. Thus, cohesive sediment dynamics in estuaries means a synergy of complexity. However, through the observation of a long (in the sense of, much greater than a tidal cycle) time series, it is possible to discern the quintessence of some obvious processes.

The near bed cohesive sediment dynamics in the Itajaí-açu River estuary present a phenomenological duality: (1) long-term tide-dominated, and (2) short-term river discharge events.

During the long-term tide-dominated period with low river discharge, the bed processes are dominated by tides, and either semi-diurnal cycles or fortnight lunar cycles play an important role. Semi-diurnal inequalities can produce stronger differences in erosion processes in an hourly time scale, while fortnight lunar cycles will act in a daily to weekly time scale. During such periods, the near bed currents present asymmetry landward, resulting in net sediment transport in the same direction. Apparently, during long-term low river discharge periods, only the surface sediment layer participates in the erosion-deposition tidal cycle, with the lower layers being affected only during the spring tide. The critical bed shear stress for erosion of lower layers was estimated between 0.4 and 0.5 Pa, which can be generated only during spring tide current peaks.

During the short-term river-dominated mode, the bed processes are dominated by river discharge energetic events, no matter what the tidal range is like. During such events, when 1,000 $m^3.s^{-1}$ is a reference value, the near bed currents experience extreme values, much higher than those observed as a tidal response. The bed shear stress can exceed 2 Pa, or even more depending on the discharge, and erosion of old, strongly compacted layers takes place.

ACKNOWLEDGEMENTS

This work was carried out at the request of HAM Dredging for the purpose of observation of near bed conditions in the Itajaí-açu River estuary, which partially sponsored the mooring. The author would like to thank Reinier, Nanne, Dermeval and Ernest, for their help in the instrument deploy and recovery, and special thanks to Ewerton and Paulo for the unpleasant dives in the muddy waters for instrument care.

REFERENCES

Amos, C. L., Grant, J., Daborn, G. R. and Black, K. S. 1992. Sea carousel - a benthic annular flume. *Estuarine, Coastal and Shelf Science* 34: 557-577.

Döbereiner, C.E. 1985. *Comportamento hidráulico e sedimentológico do estuário do Rio Itajaí - SC.* Rio de Janeiro, INPH, Relatório 700/03, 34p.

Drever, J.I. 1988. *The geochemistry of natural waters.* 2nd Ed., New Jersey, Prentice Hall, 275p.

Dyer, K.R. 1986. *Coastal and estuarine sediment dynamics.* New York, John Wiley and Sons, 342p.

Dyer, K.R. 1995. Sediment transport processes in estuaries. In: Perillo, G.M.E. (Ed.) *Geomorphology and sedimentology of estuaries.* New York, Elsevier, p423-449.

Dyer, K.R. 1997. *Estuaries: a physical introduction.* New York, John Wiley and Sons, 195p.

Eisma, D. 1986. Flocculation and de-flocculation of suspended matter in estuaries. *Netherlands Journal of Sea Research*, (1/3) 20:183-199.

Hansen, D.V. & Rattray, M. 1966. New dimensions on estuarine classification. *Lymnology and Oceanography*, 11:319-326.

Ingram, R.G.; D'Angleja, B.F.; Lepage, S. 1986. Changes in current regime and turbidity in response to a freshwater pulse in the Eastmain estuary. *Estuaries*, 9:320-325.
Kirby, R.; Bleakley, R.J.; Weatherup, S.T.C.; Raven, P.J.; Donaldson, N.D. 1993. Effect of episodic events on tidal mud flat stability, Ardmillan Bay, Strangford Lough, Northern Ireland. In: Mehta, A.S. (Ed.) *Nearshore and estuarine cohesive sediment tranport.* Washington, American Geophysical Union, p378-392.
Krone, R. 1978. Aggregation of suspended particles in estuaries. In: Kjerfve, B. (Ed.), *Estuarine transport processes.* Columbia, USC-Press, p177-190.
Meade, R.H. 1969. Landward transport of bottom sediment in estuaries of the Atlantic Coastal Plain. *Journal of Sedimentary Petrology*, 39:222-234.
Mehta, A.J. 1984. Characterization of cohesive sediment properties and transport processes in estuaries. In: Mehta, A.J. (Ed.) *Estuarine cohesive sediment dynamics.* Berlim, Springer-Verlag, p290-325.
Mehta, A.J. 1988. Laboratory studies on cohesive sediment deposition and erosion. In: Dronkers, J. & van Leussen, W. (Eds.) *Physical processes in estuaries.* Berlim, Springer Verlag, p427-445.
Mehta, A.J.; Parchure, T.M.; Dixit, J.G.; Ariathurai, R. 1982. Resuspension potential of deposited cohesive sediment beds. In: Kennedy, V.S. (Ed.) *Estuarine comparisons.* New York, Academic Press, p591-609.
Mehta, A.J.; Hayter, E.J.; Parker, W.R.; Krone, R.B.; Teeter, A.M. 1989. Cohesive sediment transport. 1: processes description. *ASCE Journal of Hydraulics Engineering*, 115:1076-1093.
Nichols, M.M. 1984. Effects of fine sediment resuspension in estuaries. In: Mehta, A.J. (Ed.) *Estuarine cohesive sediment dynamics.* Berlim, Springer-Verlag, p5-42.
Nichols, M.M. 1993. Response of coastal plain estuaries to episodic events in the Chesapeake Bay region. In: Mehta, A.S. (Ed.) *Nearshore and estuarine cohesive sediment tranport.* Washington, American Geophysical Union, p1-20.
Nichols, M.M. & Biggs, R.B. 1985. Estuaries. In: Davis Jr., R.A. (Ed.) *Coastal sedimentary environments.* New York, Springer Verlag, p77-186.
Officer, C.B. 1976. *Physical oceanography of estuaries and associated coastal waters.* New York, Wiley, 465p.
Owen, M.W. 1975. *Erosion of Avonmouth mud.* Hydraulics Research Station, Report INT 150.
Parker, W.R. & Kirby, R. 1982. Time dependent properties of cohesive sediment relevant to sedimentation management – european experience. In: Kennedy, V.S. (Ed.) *Estuarine comparisons.* New York, Academic Press, p573-589.
Partheniades, E. 1984. A fundamental framework for cohesive sediment dynamics. In: Mehta, A.J. (Ed.) *Estuarine cohesive sediment dynamics.* Berlim, Springer-Verlag, p219-250.
Ponçano, W.L. 1982. *Sedimentação atual na área de interesse ao Porto de Itajaí – SC.* São Paulo, IPT, Relatório 17.502, 56p.
Ponçano, W.L. 1987. Reconhecimento sedimentológico do estuário do Itajaí-açu (SC). *Revista Brasileira de Geociências*, (1) 17:34-41.
Postma, H. 1967. Sediment transport and sedimentation in the estuarine environment. In: Lauff, G.H. (Ed.). *Estuaries.* Washington, AAAS Publication No. 83, p158-179.
Raudkivi, A.J. 1990. *Loose boundary hydraulics.* 3rd Ed., New York, Pergamon Press, 600p.
van Rijn, L.C. 1993. *Principles of sediment transport in rivers, estuaries and coastal seas.* Amsterdam, Acqua Publications.

Schettini, C.A.F.; Carvalho, J.L.B.; Jabor, P. 1996. Comparative hydrology and suspended matter distribution of four estuaries in Santa Catarina State, Southern Brazil. Workshop on Comparative Studies of Temperate Coast Estuaries, Bahia Blanca, *Proceedings*...IADO, p29-32.

Schettini, C.A.F. & Carvalho, J.L.B. 1998. Suspended sediment balance in the estuary of Itajaí-açu River during a low discharge period. *Anais da Academia Brasileira de Ciências,* (2) 70:325-334.

Schettini, C.A.F.; Kuroshima, K.N.; Pereira FO., J.; Rörig, L.R.; Resgalla JR. 1998. Oceanographic and ecological aspects of Itajaí-açu river plume during a high discharge period. *Anais da Academia Brasileira de Ciências*, (2) 70:335-351.

Schettini, C.A.F. & Truccolo, E.C. 1999. Dinâmica da intrusão salina no estuário do Rio Itajaí-açu. In: Congresso Latino Americano de Ciências do Mar, 8, Trujillo, *Resumenes ampliados*...Tomo II, UNT/ALICMAR, p639-640.

Truccolo, E.C. 1998. *Maré meteorológica e forçantes atmosféricas locais em São Francisco do Sul, Sc.* Florianópolis, UFRS, MSc Thesis, 100p.

Winterwerp, J.C. 1989. *Flow-induced erosion of cohesive beds.* Delft, Delft Hydraulics, Report 25.

Winterwerp, J.C.; Cornelisse, J.M.; Kuijper, C. 1993. A laboratory study on the behavior of mud from the Western Scheldt under tidal condition. In: Mehta, A.S. (Ed.) *Nearshore and estuarine cohesive sediment transport.* Washington, American Geophysical Union, p295-313.

Fine Sediment Dynamics in the Marine Environment
J.C. Winterwerp and C. Kranenburg (Editors)

Field study and modelling on the characteristics of bed mud formation processes at the Rokkaku River

R.Watanabe[a], T.Kusuda[b], H.Yamanishi[c] & K.Yamasaki[a]

[a]Department of Civil Engineering, Fukuoka University, Nanakuma 8-19-1, Jonan-ku, Fukuoka 814-0180, Japan

[b]Department of Urban and Environmental Engineering, Kyushu University, Hakozaki 6-10-1 (SUIKO), Higashi-ku, Fukuoka 812-8581, Japan

[c]Institute of Lowland Technology, Saga University, Honjyou 1, Saga 840-8502, Japan

Field observations of bed mud formation prosesses were continuously carried out at a site 11km upstream from the river mouth of the Rokkaku River in Japan from November 28, 1994 to February 4, 1995. A large experimental reservoir was constructed along a mudbank in the midstream of the Rokkaku River. From the observations, the maximum sedimentation rate on the flats per one tide cycle reached about 0.01m at spring tide. Based on the observations, modelling the bed mud formation process in the reservoir was performed. The process of the bed mud formation consists of three stages: deposition-consolidation, sweep-out and evaporation processes. The simulation results explain the formation process of the bed mud layers in the reservoir very well, so that, the process in tidal rivers can be explained in terms of this method by composing shear stress in the bed itself and that generated by the flow.

key words
muddy tidal river, estuary, bed mud, sedimentation rate, intertidal mudflats

1. INTRODUCTION

Cohesive sediments on the bottom and the banks of tidal rivers play a major role in narrowing the cross section. From an engineering standpoint, the estimation of the deposition rate is of importance for planning and maintaining river channels, because of high cost of dredging. Since plenty of areas have suffered from the siltation problem, many field studies have been already performed on mudflat formation to solve this problem. In the Amazon River, the deposition rate of fine-grained sediments from fluid mud (10-400 kg•m^{-3}) was about 1 cm•day^{-1} for half a year (Allison et al. 1995). The thickness of the bed mud during this period reached about 1.5 m. In the Rokkaku River in western Japan, whose tidal range is 5 m at the river mouth, the maximum sedimentation rate for one tidal cycle reaches about 10 mm (Futawatari and Kusuda 1993). In the Jobaru River in western Japan, the accumulated bed mud was recorded to be about 2.2 m thick

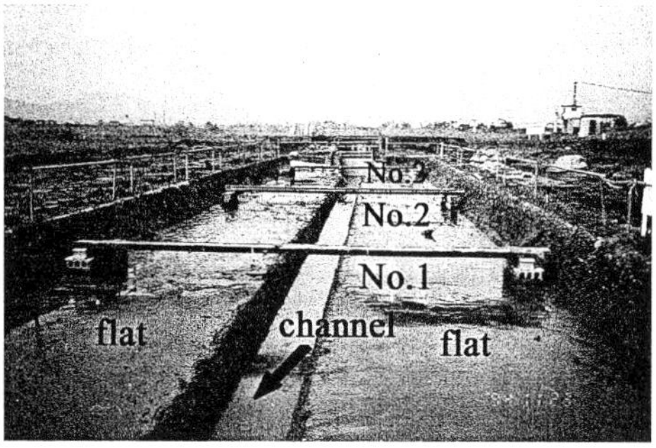

Photo 1. Experimental reservoir.

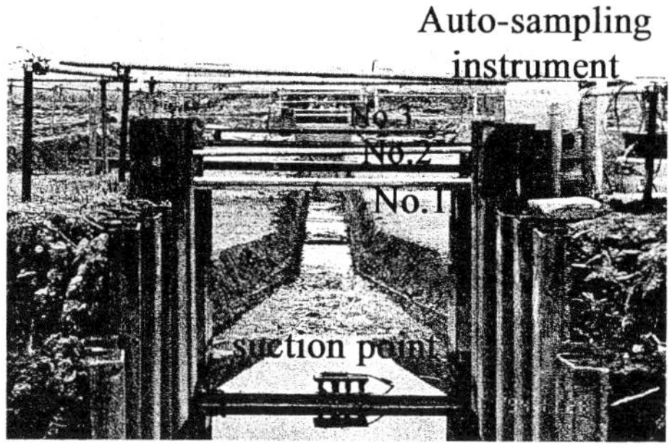

Photo 2. Reservoir inlet.

for 431 days. Roberts and Whitehouse (2001) considered the concept of equilibrium mudflat morphology and developed modelling methods for a long-term simulation of the morphology, however, they have not described yet in detail mechanisms of the mudflat formation. The fundamental elements involved in the process, such as sedimentation and consolidation, have not been fully accounted for yet, so that studies are necessary to explain the mudflat formation process. Toorman and Huysentruyt (1997) proposed a constitutive equation for effective stress in self-weight consolidation, their model clearly indicated the consolidation process in the water column. Teisson et al., (1993) investigated the cohesive sediment transport process and proposed several models for the sedimentation process. Their models, however do not include the formation process of mudflat such as bed mud level variation. This study aims to make the formation process of mudflats

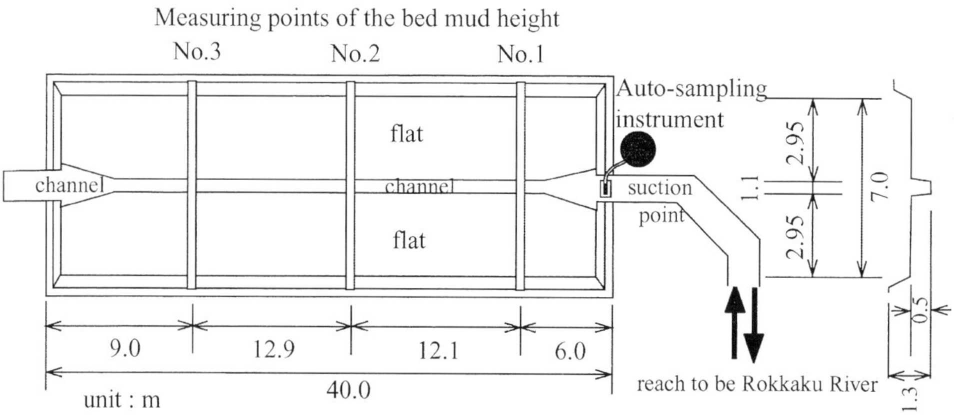

Figure 1. Outline of the reservoir.

experimentally clear by use of a large experimental reservoir under natural tidal action and to simulate the sedimentation process to explain the mechanism of the process.

2. FIELD OBSERVATION

Field observations were performed at a site 11km upstream from the river mouth of the Rokkaku River in Japan. The tide reached a weir at the 29km upstream point. A large experimental reservoir was constructed along a mudbank in the midstream of the Rokkaku (refer to Photos.1 and 2). Figure 1 outlines the reservoir, 40m long, 7m wide, and 1.3m deep, with a channel (bottom) and two flats (they are located 0.5m above the bottom), into which river water comes in and out with tide. Part of suspended solids coming into the reservoir with water settle down in it. Water in the reservoir was sampled every half-hour at the channel entrance, using an automatic sampling instrument whose

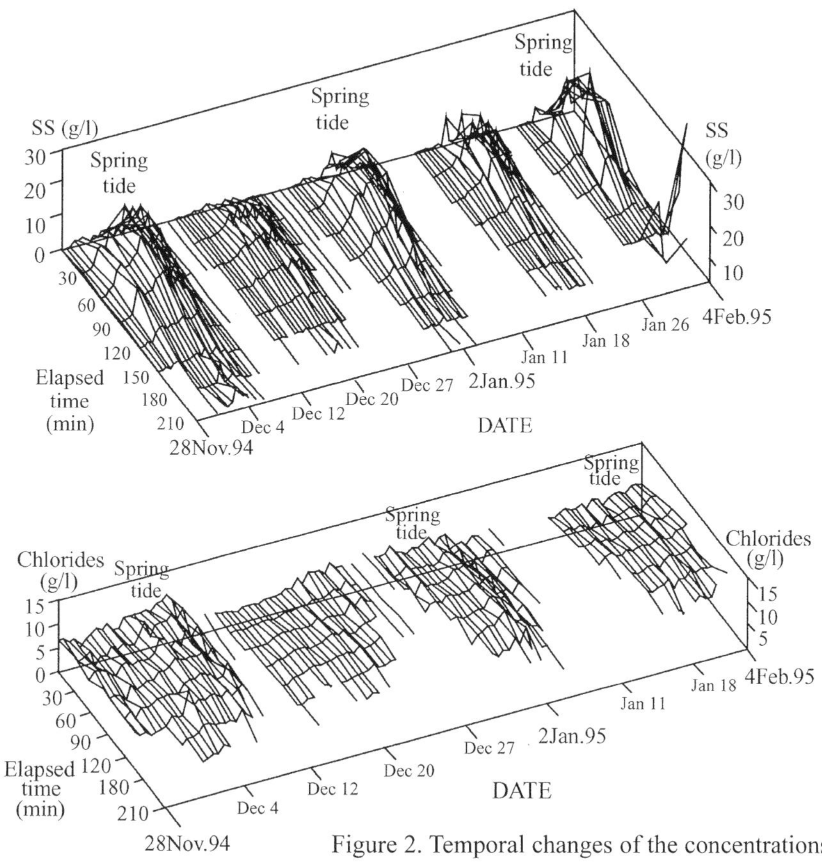

Figure 2. Temporal changes of the concentrations of suspended solids and chlorides at the entrance.

Table 1. Critical angle of the deposited mud on flats.

Point	Left hand side	Right hand side
No.1	44.8	46.8
No.2	42.2	40.7
No.3	42.3	49.2

suction inlet was set 0.2m above the bottom, and the concentrations of suspended solids and chlorides were measured. The heights of the bed mud surface in the channel and flats were measured at three lines perpendicular to the channel. These were set at 6.0m, 18.1m and 31.0m from the entrance. At the neap tides, the surface of the mud bed was exposed to the air, because of the bottom of the flume was higher than the water levels. Measuring the heights of the mud bed and sampling of inflowing water were continuously carried out from November 28, 1994 to February 4, 1995. Using 1000ml syringes, bed materials were sampled at fixed periods. After freezing the samples, the water contents in the mud were calculated.

Figure 2 shows the temporal changes of suspended solids concentration and chlorides of inflow. The concentrations of suspended solids and chlorides showed periodic variations, becoming lower during the neap tides and higher during the spring tides. The minimum and maximum values of the concentrations of suspended solids and chlorides in tidal cycles take place at neap tides and spring tides, respectively. Figure 3 illustrates the observation results of the heights of the bed mud surface in the channel and flats. The bed mud surface in the channel and on the flats rose with time. In the channel, the maximum height of the deposited mud reached about 50cm thick for the period of the observation (refer to No.3 point). On the flats, the maximum height of the deposited mud recorded about 20cm (refer to No.1 point). This result indicates quite a possibility that the sedimentation rate in the channel is faster than that on the flats, due to greater immersion time, greater water depth and hence potentially larger supply of sediment. In this figure, at the front part of the bed mud on the flats, the angle of the mudbank attained the critical angle for deposition. Table 1 indicates the critical angle of the deposited mud on the flats at the end of the observation (on February 4, 95). From the results, the critical angle of the bed mud is about 45 degree. On the other hand, the experimental results in the flume indicate the critical angle at the edge part of the mud bank was 50 degrees (Watanabe 2000).

Figure 4 indicates the temporal changes of the bed mud surface at No.1 and the amount of rainfall in this area. The maximum sedimentation rate on the flats per one tide cycle reached about 0.01m at spring tide. By rainfall, the deposited mud on the surface is washed away by the surface flow. After heavy rain, the level of the bed mud surface declines abruptly. The bed mud surface lowers during neap tides due to evaporation. Figure 5 illustrates the water content profiles in the bed mud on the flats at a 200cm point from the edges. The water content at the surface range from about 600 to 1000 %, and the water content at bottom indicate about 300 %. Figure 6 shows the water content profile at No.3 point in the channel. The water content at the surface and at the bottom indicate almost the same values as on the flats.

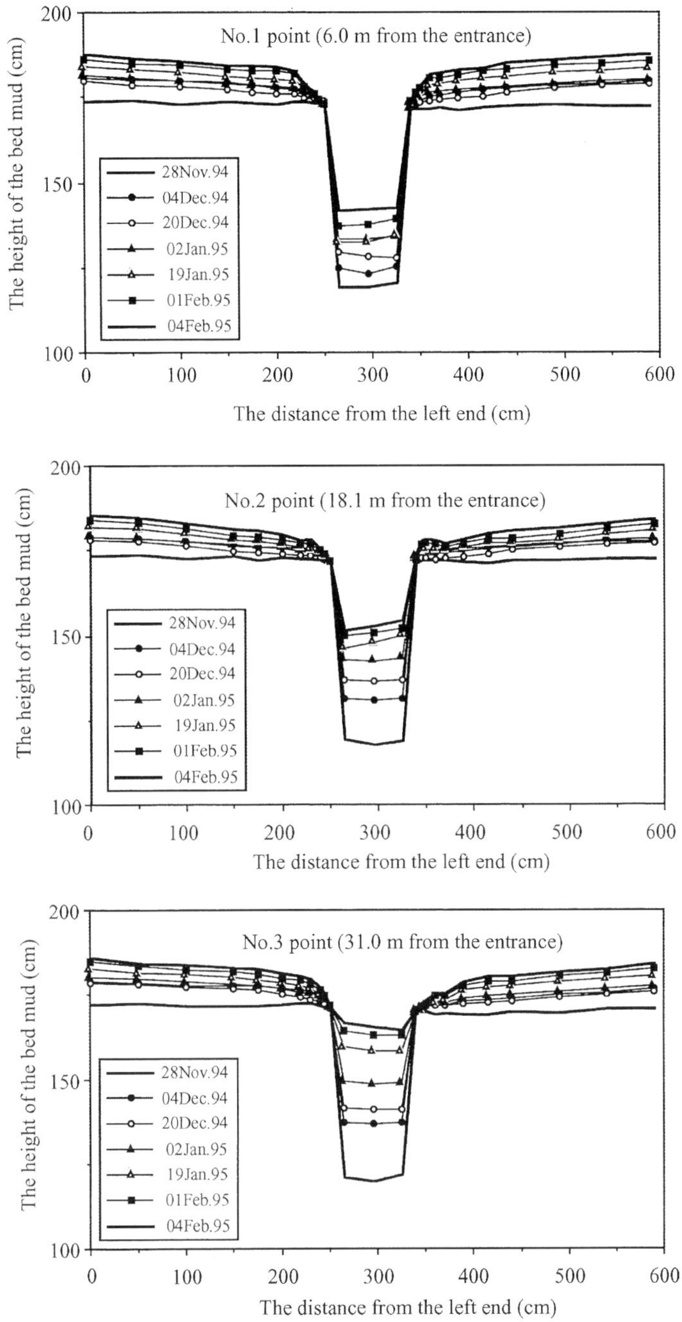

Figure 3. Observation results of the bed mud heights in the reservoir.

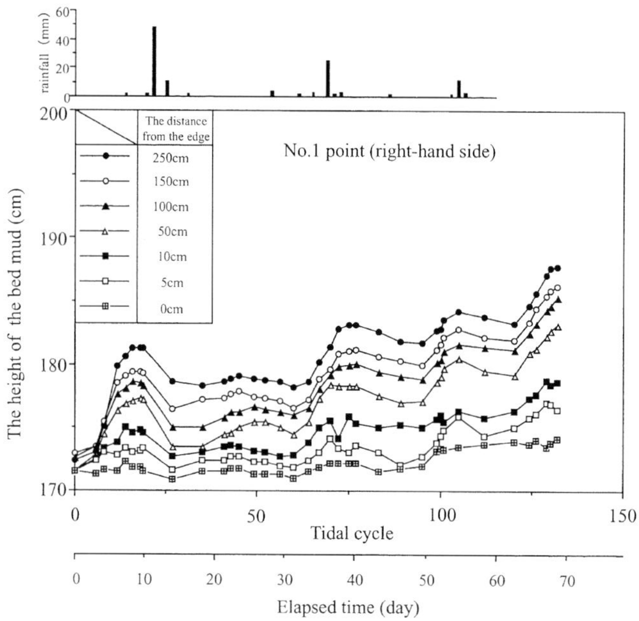

Figure 4. Temporal changes of the bed mud surface.

3. MODELLING OF THE MUDBANK FORMATION PROCESS AND SIMULATION RESULTS

Based on the field observations, a simulation model on the formation process of bed mud in the reservoir was developed. In tidal rivers, water level changes periodically according to the tide so that tide and river flow control the bed mud formation process. The process consists of three stages: deposition-consolidation, sweep-out and evaporation (refer to Figure 7). In the first stage, deposited sediments are consolidated on the flats. In the second stage, the edge part of the bed mud layer collapses due to emerging into the air and the surface of the bed mud flows down as a thin surface layer of fluid mud. In the third stage, the surface of the bed mud is exposed to the air and water in the bed mud layer evaporates. Repeating these processes, a bed mud layer grows on the bed. In the deposition-consolidation stage, deposited sediments form consolidated bed mud. Kinetic and mass conservation equations for the consolidation process are to be established for the solid and liquid phases of bed mud. Assumptions to constitute the governing equations are as follows:

1) Suspension is homogeneous in the overlying water;
2) No compression occurs at the solid phase of bed mud;
3) Pore water is drained vertically;
4) Bed mud is consolidated vertically;
5) The Reynolds numbers of the movements of liquid and solids are smaller than unity; and

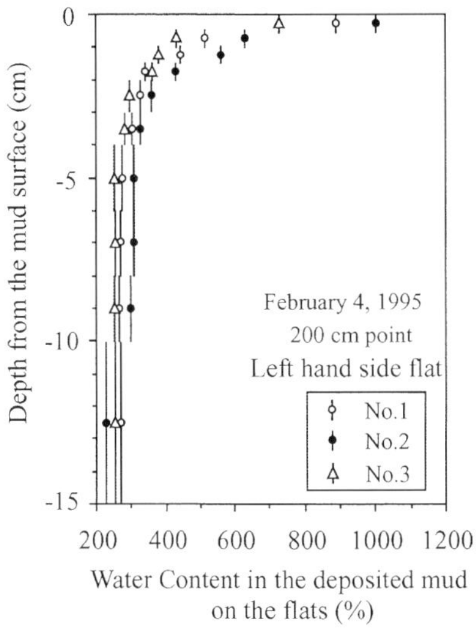

Figure 5. Water content in the mud on the flats.

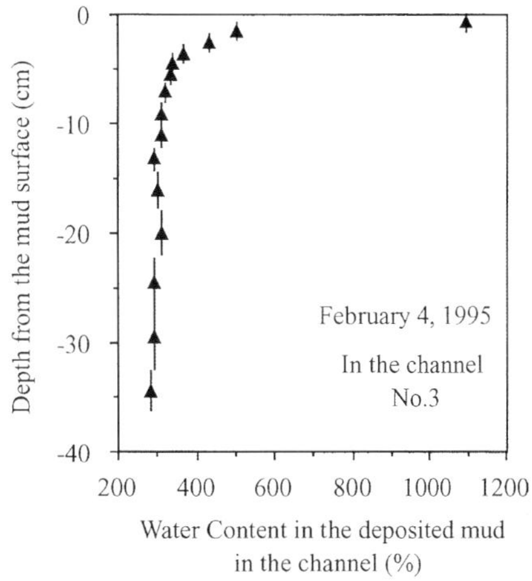

Figure 6. Water content in the mud at the channel (No.3 point).

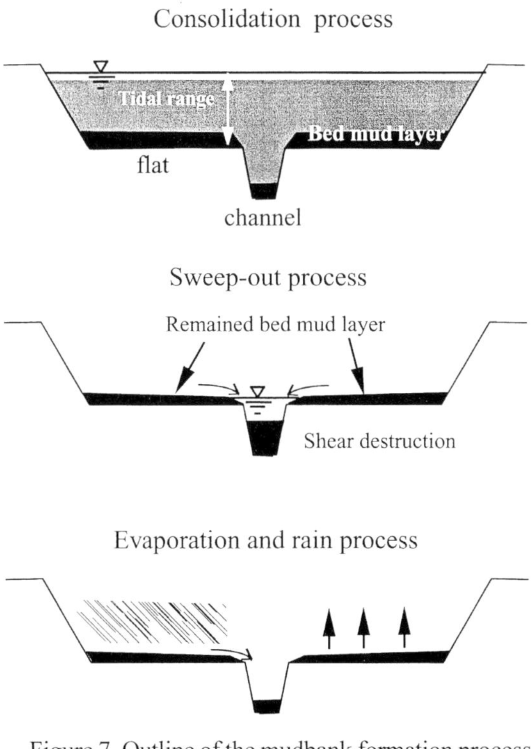

Figure 7. Outline of the mudbank formation processes.

6) The wall effects on the movements are little as to be neglected.
Since the effective stress and the excess pore pressure support the submerged weight of the solid particles, the kinetic equations of the solid and liquid phases are expressed as follows (Gibson et al., 1967):

$$\text{Solid phase} \quad : \frac{\partial P_s}{\partial z} - \frac{\mu}{K}\left[V_l - \frac{\varepsilon}{1-\varepsilon}V_s\right] + (1-\varepsilon)(\rho_s - \rho_l)g = 0 \qquad (1)$$

$$\text{Liquid phase} \quad : \frac{\partial P_l}{\partial z} + \frac{\mu}{K}\left[V_l - \frac{\varepsilon}{1-\varepsilon}V\right] = 0 \qquad (2)$$

The mass conservation equations of the solid and liquid phases are derived as follows (Gibson et al., 1967):

$$\text{Solid phase} \quad : \frac{\partial (1-\varepsilon)}{\partial t} + \frac{\partial V_s}{\partial z} = 0 \qquad (3)$$

$$\text{Liquid phase} \quad : \frac{\partial \varepsilon}{\partial t} + \frac{\partial V_l}{\partial z} = 0 \qquad (4)$$

where μ : the viscosity of liquid, ε : the porosity, P_s : the effective pressure, P_l : the excess pore water pressure, K : the coefficient of permeability, V: the volumetric flux, ρ : the density and g : the acceleration of gravity. Suffixes s and l mean the solid and liquid phases, respectively.
Figure 8 expresses relationships between the coefficient of permeability and $\varepsilon^3/(1-\varepsilon)^2$. Figure 9 indicates the relationships between the effective pressure and $\varepsilon_c - \varepsilon$. The coefficient of permeability and the effective pressure based on experimental results are expressed as follows (Kusuda, et al., 1980):]

$$\rho_l\, g\, K = \alpha\left\{\varepsilon^3 / (1-\varepsilon)^2\right\}^{0.76} \qquad (5)$$

$$\frac{P_s}{\rho_l\, g} = \beta(\varepsilon_c - \varepsilon)^m \qquad (6)$$

in the above equations, the coefficients of α, β, and m are given as functions of time, such as Eqs.7 to 9. The critical solid fraction ε_c above which the effective pressure appears, is set to be equal to 0.012 in this model (Kusuda, et al., 1980). Through theoretical considerations and experimental results about self-weight consolidation, it is well known that the deformation of network structure and the interparticle contact force vary depending on consolidation time. In this model, the above phenomena are approximated as a creep process as follows:

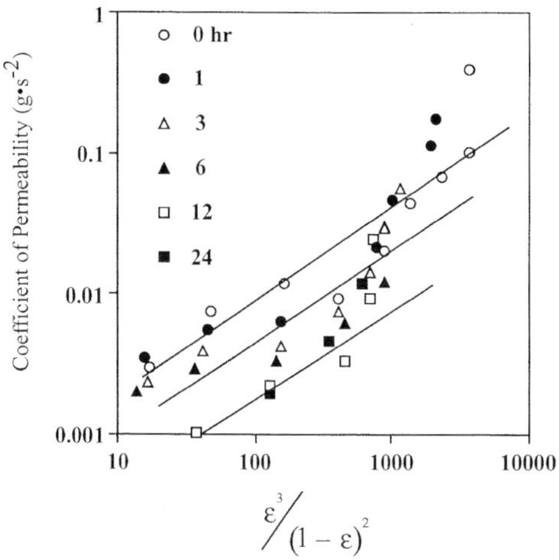

Figure 8. Relationship between $\varepsilon^3 / (1-\varepsilon)^2$ and the coefficient of permeability.

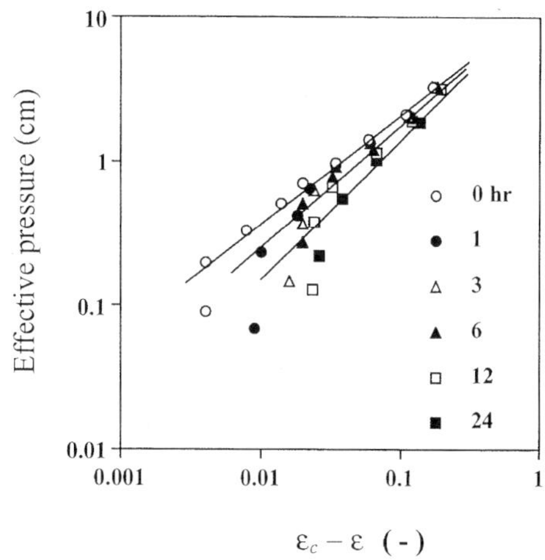

Figure 9. Relationship between $\varepsilon_c - \varepsilon$ and the effective pressure.

$$\alpha = \alpha_0 \, exp\left[-t/\lambda\right] \tag{7}$$

$$\beta = \beta_0 \, exp\left[-t/\lambda\right] \tag{8}$$

$$m = m_0 + \left(m_\infty - m_0\right)\left\{1 - exp\left[-t/\lambda\right]\right\} \tag{9}$$

where, t means the elapsed time, λ, α_0, β_0, and m_0 are equal to 100, 0.00059, 0.081 and 0.76, respectively.

In the simulation of the consolidation process, the coordinate (z, t) is converted to a moving coordinate (w, t) for computation simplicity, in which w is the accumulated dry mass per unit area. The computation is carried out using a finite difference method with Δt = 0.1 seconds and Δw = 0.1 kg•m^{-2}. The initial solid volumetric fraction of the suspension on the flats was obtained from the suspended solids concentration profiles in the reservoir. Figure 10 presents experimental (circles) and simulated (solid lines) results for the suspended solids concentration profiles in the reservoir. This figure indicates that numerical solutions for the suspended solids concentration profiles agree well with the observation results. Based on these results, we can estimate the suspended solids concentrations on the flats and channel.

In the edge collapse and surface sweep-out stages, the edge part of the bed mud layer collapses on emerging to the air, and then the surface of the bed mud flows down as a thin surface water layer (refer to Figure 7). Because of its low water content, fluidity at the surface of the layer is higher than that in the other layers. We assume that destruction in the edge collapse and surface sweep-out stages is caused by shear stress.

Probably the sliding surface formed by shear destruction is quite irregular, depending on the homogeneity of the bed mud. If the bed mud is homogeneous and enough bed mud exists on a bed, the most critical sliding surface will be like an arc. However, the water content profiles in the bed mud were not uniform (refer to Figures 5 and 6). In the model developed here, a plane sliding surface is assumed for the shear destruction process (refer to Figure 7), in which the shear strength at the bottom is assumed to be greater than that at the surface. In other words, the water content at the bottom is smaller than that at the surface (refer to Figures 5 and 6), so that the yield shear stress at the bottom of the bed mud layer is larger than that at the surface.

From the aforementioned point of view, it is reasonable that the plane failure is assumed in the shear destruction process in the bed mud layers.

Figure 11 illustrates the outline of the calculation procedure for the shear destruction process. This figure shows a situation of the drainage and sweep-out processes. In the simulation of the collapse stage, the procedure for calculation is as follows: at first, we set a water level, and then assumed several plane sliding surfaces from a contact point with the water level. Next the mass of a portion of the bed mud is calculated and finally the mass is converted to shear stress acting over a plane sliding surface. If the shear stress is larger than the shear strength, the portion of the bed mud is collapsed. If the shear stress is less than the shear strength, the portion of the bed mud remains as it is.

Figure 12 shows a relationship between solid volumetric fraction and yield shear stress. The yield shear stress is expressed as a function of solid volumetric fraction in the

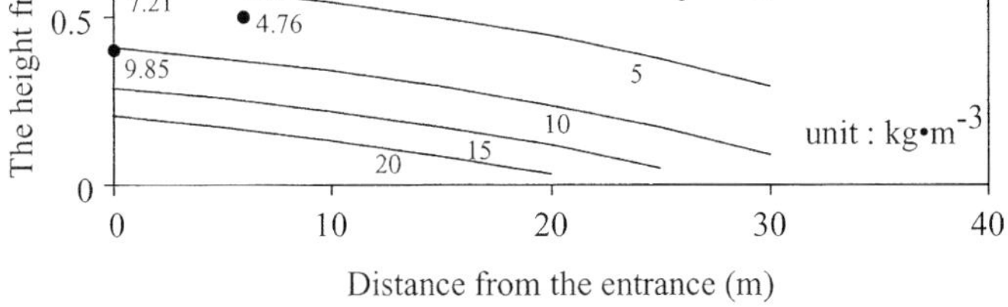

Figure 10. Suspended solids concentration profiles in the reservoir.

cohesive sediments as follows (Umita et al., 1986, Teisson et al., 1993):

$$\tau_y = 230 \times \left(1 - \varepsilon\right)^{1.7} \tag{10}$$

where, τ_y : yield shear stress (Pa), $1-\varepsilon$: solid volumetric fraction in the bed mud. After repeating the above procedure on other water levels, a final profile of the bed mud layer is determined for a tidal cycle. Computation was carried out by the finite difference method with Δx of 0.04 m and Δz of 0.002 m.

Figure 13 gives a comparison between simulated and field observations on the flats. In this figure, the solid and broken lines indicate simulated results, and the symbols indicate experimental results. This figure indicates that the simulated profiles of the bed mud

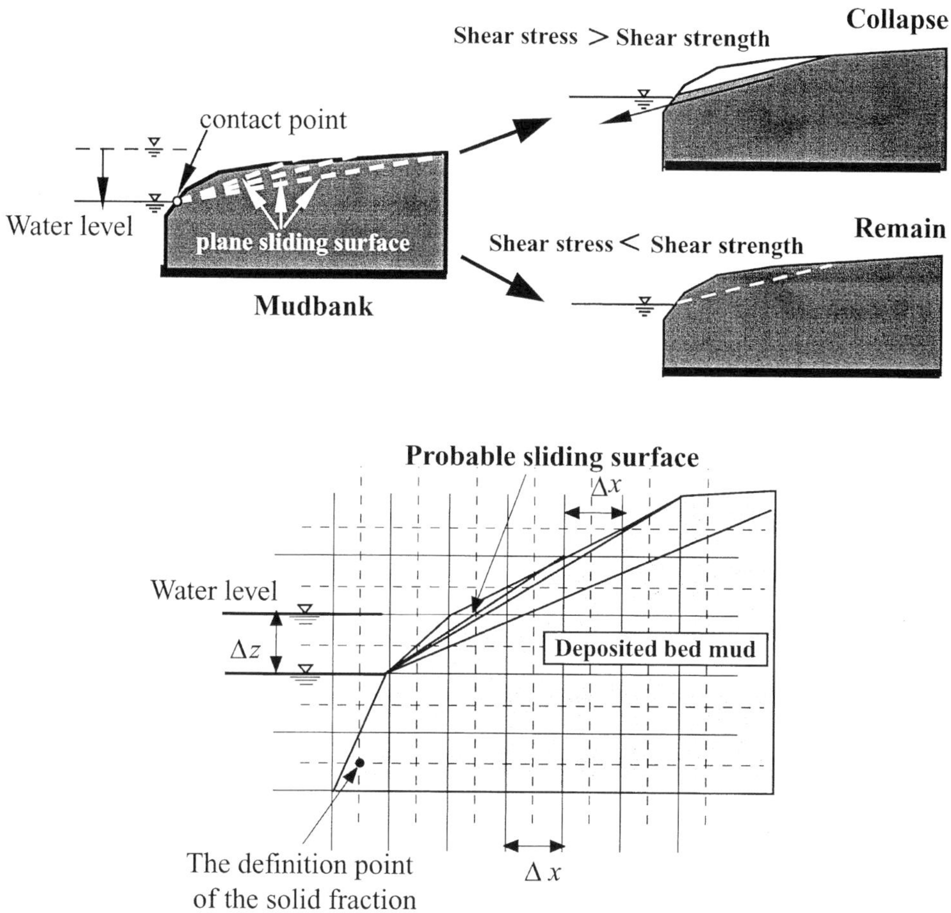

Figure 11. The outline of the calculation procedure for the shear destruction process.

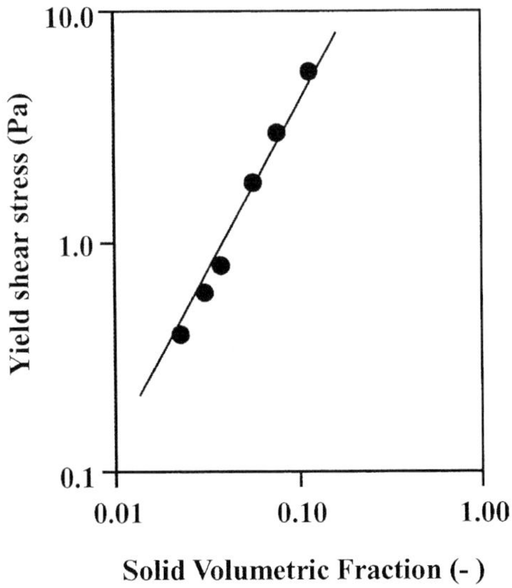

Figure 12. Relationship between solid volumetric fraction and yield shear stress.

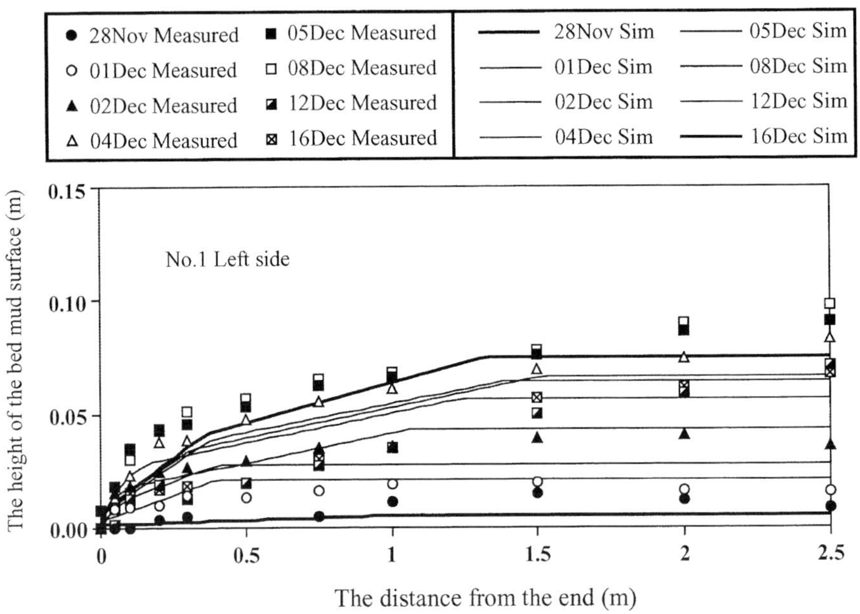

Figure 13. Comparison between simulation and observation results in the reservoir.

layer on the flats agree well with the field observation results. However around the edge region of the bed mud, the simulated results are lower than the experimental ones. This results from that the water contents in the bed mud at the edge region in the observations are higher than those in the simulation. As a result, we underestimated the yield shear stress in the bed mud layers.

The above results explain the formation process of bed mud layers in the reservoir very well, so that, the process in tidal rivers can be explained in terms of this method by composing shear stress in the bed itself and that generated by the flow.

4. CONCLUSION

The formation process of the bed mud layer in a reservoir was observed from November 28, 1994 to February 4, 1995. From the field observations, characteristics of the bed mud deposition became clear. The maximum sedimentation rate per tidal cycle was about 0.01 m at a spring tide. Since the critical angle of the mudbank is about 45 degrees, we can determine its shape by using this value. Therefore, engineers can avoid unnecessary dredging based on the estimated shape of river mudbanks. By using the field observation data, the formation process was analyzed theoretically. The model developed on the consolidation and collapse stages is available for simulating the mudflat formation process in tidal rivers.

REFERENCES

1. G.J. Kynch, A theory of sedimentation, *Transactions of Faraday Society* Vol.48 (1952) 166-172.
2. R.E. Gibson, G.L. England and M.J.L. Hussey, The theory of one-dimensional consolidation of saturated clays 1.Finite Non-Linear Consolidation of Thin Homogeneous Layers, *Geotechnique* 17 (1967) 261-273.
3. C. Teisson, M. Ockenden, P.Le Hir, C. Kranenburg and L. Hamm, Cohesive sediment transport processes, *Coastal Engineering* 21 (1993) 129-162.
4. T. Futawatari, and T. Kusuda, Modeling of suspended sediment transport in a tidal river, In: A.J.Mehta (ed), *Coastal and Estuarine Studies* No.42.AGU, Washington D.C (1993) 504-519.
5. T. Umita, T. Kusuda, T. Futawatari, Y. Awaya and M. Onuma, A model of erosion of soft cohesive sediments, *Third international symposium on river sedimentation.* The University of Mississippi (1986) 1658-1667.
6. T. Kusuda, K. Koga and Y. Awaya, Gravity thickening of sludge, *Proceedings of Japan Society of Civil Engineers* Vol.294 (1980) 59-71.
7. K. Lee, and G.C. Sills, The consolidation of a soil stratum, including self-weight effects and large strains, *International Journal for Numerical Methods in Geomechanics* Vol.5 (1981) 405-428.
8. K. Been and G.C. Sills, Self-weight consolidation of soft soils :an experimental and theoretical study, *Geotechnique* 31 No.4 (1981) 519-535.
9. T. Kusuda, R. Watanabe, T. Futawatari and H. Yamanishi, Fluid-mud movement on an inclined bed. In: A.J.Mehta (ed), *Coastal and Estuarine Studies* No.42.AGU, Washington D.C (1993) 281-294.

10. M.A. Allison, C.A. Nittrouer, G.C. Kineke, Seasonal sediment storage on mudflats adjacent to the Amazon River, *Marine Geology* 125 (1995) 303-328.
11. W. Roberts and R.J.S. Whitehouse, Predicting the profile of intertidal mudflats formed by crossshore tidal currents, In: W.H.McAnally and A.J.Mehta (eds), *Coastal and Estuarine Fine Sediment Processes.* ELSEVIER (2001) 263-285.
12. E.A. Toorman and J.E. Berlamont, Mathematical Modeling of Cohesive Sediment Settling and Consolidation, In: A.J.Mehta (ed), *Coastal and Estuarine Studies* No.42 .AGU, Washington D.C. (1993) 167-184.
13. E.A. Toorman and H. Huysentruyt, Towards a new constitutive equation for effective stress in self-weight consolidation, In: N.Burt, R.Parker and J.Watts (eds), *Cohesive Sediments.* John Wiley & Sons, New York (1997) 121-132.
14. R. Watanabe, Fluid mud transport and bed mud formation processes in muddy tidal rivers, *PhD thesis*, Kyushu University (in Japanese) (2000) 52-55.

Chapter 6:

Numerical modelling

Fine Sediment Dynamics in the Marine Environment
J.C. Winterwerp and C. Kranenburg (Editors)

Numerical simulation of cohesive sediment transport in the Loire estuary with a three-dimensional model including new parameterisations

C. Cheviet, D. Violeau, M. Guesmia

Laboratoire National d'Hydraulique et Environnement (EDF), 6 quai Watier, 78400 Chatou, France

3D simulations of cohesive sediment transport in the Loire estuary (France) have been carried out, using mainly the TELEMAC-3D software. In addition to hydrodynamics, three main features regarding sediment processes have been implemented into the code : turbulence damping, flocculation and entrainment of fluid mud. The results are in good agreement with the observations.

KEY WORDS

Cohesive sediment transport, 3D numerical simulation, fluid mud, turbulence damping, flocculation, Loire estuary.

1. INTRODUCTION

The main goal of the MAST3-COSINUS European Project was to improve the knowledge regarding the behaviour of cohesive sediment transport in estuaries, and to better understand the specific processes which are involved (see Violeau et al. 2000). As a validation of this work, numerical models have been provided with operational techniques to take into account the most important of these phenomena, namely damping of turbulence by suspended matter, entrainment of CBS (Concentrated Benthic Suspension) layer, and flocculation. For this purpose, the numerical models have been tested on a real estuary. The main contribution of the *Laboratoire National d'Hydraulique et Environnement* (LNHE) was to reproduce the behaviour of the turbidity maximum in the Loire estuary (France).

The above-mentioned cohesive processes have been first tested in the case of a schematic estuary, in order to examine their effects. In a second step, these processes have been introduced into the 3D Loire model, designed in order to reproduce hydrodynamical and sediment processes in the Loire estuary, thanks to the software TELEMAC-3D, developed by the LNHE. *In situ* measurements have also been performed. The results of the numerical model have been examined and compared to the data collected, regarding water surface elevation and sediment concentration. Comparison to the *in situ* measurements has shown that the numerical simulation of cohesive sediment transport has improved significantly, thanks to the progress made within COSINUS.

2. PARAMETERISATION OF COHESIVE SEDIMENT PROCESSES

Three main sediment processes have been parameterised for this work, starting from the work done within Tasks A through D of the COSINUS project : damping of turbulence, entrainment, and flocculation.

One of the most important effects of suspended particles is the damping of turbulence due to buoyancy effects. Suspended sediment particles cause damping of turbulent energy in the flow. Traditionally, this effect is parameterised by the use of semi-empirical damping functions, which represent a decrease of both eddy viscosity and eddy diffusivity, depending on the Richardson number *Ri* (Le Normant 1997). In the case of the PML (Prandtl Mixing Length) model, we have

$$\begin{aligned} \nu_t &= \ell^2_{m0} \cdot F_t(Ri)\left|\frac{\partial u}{\partial z}\right| \\ K_t &= \frac{1}{\sigma_0}\ell^2_{m0} \cdot F_s(Ri)\left|\frac{\partial u}{\partial z}\right| \end{aligned} \tag{1}$$

where ℓ_{mo} is the mixing length ; σ_o is the neutral turbulent Schmidt number, with a value of 0.7, based on experimental data (Taylor 1973). A favoured set of functions is provided by the so-called Munk-Anderson damping functions :

$$\begin{aligned} F_t &= (1+10\cdot Ri\ )^{-1/2} \\ F_s &= (1+10\cdot Ri/3)^{-3/2} \end{aligned} \tag{2}$$

However, in the framework of Task A of the COSINUS project, Toorman proposed new damping functions, based on extended experimental data (Toorman 2000) :

$$\begin{aligned} F_t &= (1+100\cdot Ri)^{-1/3} \\ \sigma_s &= \sigma_0(1+21\cdot Ri)^{0.8} \end{aligned} \tag{3}$$

σ_s being the Schmidt number : $\sigma_s = \sigma_0 F_t / F_s$. In addition, it has been proven that the traditional method used to calculate the shear velocity u_* is no longer valid when considering sediment. High density gradients at the bed result in an apparent reduction of the bottom roughness, with consequently higher transport than expected when the model would not account for these buoyancy effects in the bed boundary conditions. A new bottom treatment method has been proposed, which leads to the correct bed shear stress. According to Toorman, the log-law, in the case of sediment-laden stratified flow, should be modified to yield to

$$u = \frac{u_*}{\kappa_0}\ln\frac{z}{\alpha z_0} \tag{4}$$

where α is the apparent roughness correction factor. This parameter is directly related to the mixing length damping function through a differential equation, too complex to be solved analytically. Numerical experiments indicate that α can be parameterised by the following function :

$$\alpha = \exp\left[-\left(1+7.7\frac{w_s}{u_*}\right)\left(1-\exp(-4.25\cdot Ri^{0.6})\right)\right] \tag{5}$$

where w_s is the settling velocity of the sediment. Finally, the calculation of the shear velocity also requires the mixing length damping function. For more details about this model, see the companion paper (Toorman 2000).

Interactions between the suspended matter and the deposition on the bed, in terms of resuspension and deposition, are also a key process in the quantification of sediment transport. The classic theory assumes that the rate of resuspension depends on the bed shear stress τ_b, compared to a critical stress τ_c which is characteristic of the bed surface (Partheniades' model). However, in the framework of Task D of the COSINUS project, Kranenburg & Winterwerp (1997) have developed another model. They suggest an analogy with density stratified flows, considering resuspension as an entrainment process, controlled mainly by the flow and the density gradient caused by the mud suspension. They proved that the entrainment rate, under certain conditions, can be approximated by

$$E = C_b w_e \tag{6}$$

where C_b is the concentration of the bed, and w_e an entrainment velocity, related to u_* by

$$w_e = u_*\left(\frac{0.5}{5.6+Ri_0}\right)^{1/2} \tag{7}$$

In this equation, Ri_0 is an overall Richardson number, depending on the density jump across the lutocline $\Delta\rho$, the bed friction velocity u_* and the water depth H. More details and test cases can be found in Petersen & Vested (2000).

Flocculation is one of the main processes specific to cohesive sediments, which can produce aggregates called « flocs ». The work done in the framework of Task B of the COSINUS project has pointed out that the main influence of flocculation on cohesive sediment transport processes is the change of settling velocity. Various formulations exist, such as constant formulation (w_s = *constant*), constant formulation with hindered settling, power laws ($w_s = kC^m$), power law with dissipation parameter function. In addition, Winterwerp (1999) has developed a framework for the growth and recession of flocs under the influence of changing conditions in the flow, suggesting that the settling velocity could be expressed through the fractal characteristics of the flocs. All these equations involve more or less calibration parameters. Because of the increasing complexity of the methods previously described, Spearman & Roberts (2000) suggested testing the simplest models first, then introducing complexity if the results were not in good agreement with *in situ* measurements. This methodology is justified by the fact that, although some flocculation models incorporate more physical processes than others, they are not necessarily more accurate. As a matter of fact, some tests on real cases have shown that the model with constant settling velocity sometimes gives the best results.

In order to test the parameterisation of turbulence damping and entrainment previously described, these processes have been introduced into a 2DV schematised estuary designed by Le Hir (1997). Le Hir reproduced numerically a well developed turbidity maximum at spring tide and its partial deposition as fluid mud at neap tide, characteristic features of sediment transport in the Loire estuary. The estuary is 106 km long, having a constant width of 600 m. The bathymetry is representative of the Loire navigation channel (see *figure 1*). The hydrodynamic boundary conditions are related to typical hydrodynamic conditions of the Loire estuary : the tidal water level is imposed at the mouth as a sinusoidal function corresponding to a spring tide. At the upstream boundary, the river discharge is prescribed to be 300 m^3/s. The initial mass of sediment is imposed as 8 cm of available bed material between x = 10 km and x = 70 km, and there is no initial suspended matter (for more details about this schematic estuary, see Cheviet et al. 2000a & 2000b). In a first step, turbulence damping has been modelled through the classic Munk-Anderson damping functions (*figure 1a*), and erosion has been taken into account through the classical Partheniades' method. In a second step, the above-mentioned developments have been introduced : first the entrainment model only (*figure 1b*), then Toorman's damping of turbulence only (*figure 1c*).

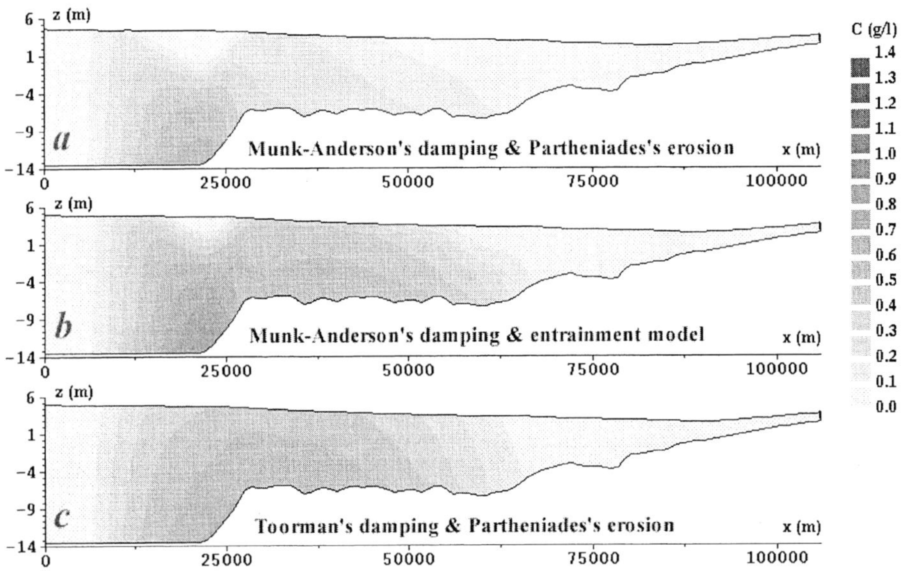

Fig. 1 - Schematic estuary. Sediment concentration 2 hours before high tide for 3 runs.

Figure 1 shows the vertical distribution of sediment two hours before high tide, in the three cases mentionned. The most important effect is caused by the change in the parameterisation of the damping of turbulence, which leads to a higher transport rate near the bed between $x = 5$ and 65 km. This is mainly the effect of the apparent reduction of the bottom roughness (eq. (4)), with consequently higher transport and lower erosion rates than expected. The parameterisation of entrainment processes seems to have lower consequences, particularly when turbulence damping is taken into account. The effect is mainly to increase the sediment rate near the bed, at the instant considered in *figure 1*.

3. DESCRIPTION OF THE LOIRE ESTUARY

The Loire estuary is located in the south of French Brittany, in the area of Nantes and Saint-Nazaire. It can be divided into two parts : the internal estuary (up river from Saint-Nazaire) and the external estuary (down river). The internal estuary can be divided into several parts (see *figure 2*) :

- the « transition zone », up to 17 km up river from Saint-Nazaire, where large sand banks and marshy areas are located ;
- the mid section, between 17 km and 36 km up river from Saint-Nazaire, where are located the large marshy areas, the islands and the branches of the Loire river ;
- the « embanked » section, between Nantes and the point located 36 km up river from Saint-Nazaire ;
- the up river section, up river from Nantes, in which large quantities of sediment (about 70 million m^3) have been dredged between 1910 and 1993.

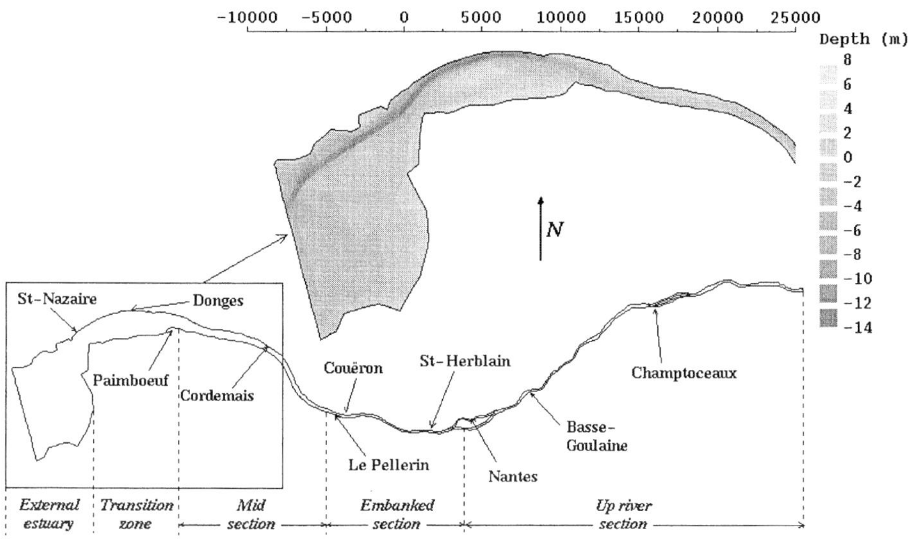

Fig. 2 - General picture of the Loire estuary
and zoom on the down river section.

The bottom level inside the channel is about -14.10 m in the area of Saint-Nazaire (referred to the lowest surface elevation), and -13.25 m between Saint-Nazaire and Donges. From Donges to Nantes, the lowest level is -5.50 m. The tidal signal in the Loire estuary is a semi-diurnal one (the period is 12 hours and 24 minutes), with surface levels varying from +6.25 m (high tide) to +0.20 m (low tide) for a strong spring tide. During the lowest neap tides, the levels change from +3.65 m (high tide) to +1.80 m (low tide). The mean level is about +3.0 m, when the effects of storm surges are negligible.

The river discharge shows large fluctuations : the highest value observed since 1866 is 6000 m^3s^{-1} (December 1910), while the lowest is 48 m^3s^{-1} (August 1949). Available data proves that the mean discharge is about 825 m^3s^{-1}. Sand is predominant up river from Nantes, while cohesive sediments are mainly present in the internal estuary. The sediment carried by the river discharge is due mainly to suspended matter (mud, 90 % of the particles having a diameter smaller than 40 μm). Under the combined effects of tide and river discharge, the suspended matter has an oscillating motion, with an amplitude depending on the tidal characteristics and river flow (for a mean tide and a discharge between 380 and 480 m^3s^{-1}, the amplitude of particle motion is about 17.5 km). Concentrations in the up river part oscillate between 20 and 50 $mg.l^{-1}$. The turbidity maximum (which can grow up to several $g.l^{-1}$) is located between Paimboeuf and Saint-Nazaire when the river discharge is higher than 1000 m^3s^{-1}, but can be confined to the external estuary when strong swellings occur. Its length extends 20 to 45 km, depending on the river flow.

During a neap tide, the suspended matter can fall down onto the bed, making the so-called « fluid mud » layer. Such a fluid mud layer often appears in the area of the turbidity maximum, mainly in the channel. The concentration of this layer is between 50 and 200 $g.l^{-1}$. Its motion is negligible : a few hundreds of meters during one tidal period. The thickness of this mud layer is from 0.5 m to 2.5 m, and its length extends from 1 to 10 km. For more details about the sediment characteristics of the Loire estuary, see Le Normant (1995).

4. IN SITU MEASUREMENTS

Gallenne (1974) was the first to make extensive measurements of turbidity in the Loire estuary ; his PhD thesis is still a reference document regarding the sediment behaviour in this estuary. Extended data has also been collected by Migniot (1993). Even if some of it seem a little old in the context of Loire estuarine evolution, main trends are highlighted and magnitudes for suspended matter are given for different fluvial and tidal conditions.

For this study, field measurements were carried out by the LNHE in the Loire estuary, for different tidal and fluvial conditions. The computed case corresponds to a situation encountered on the 3rd of July 1998, with low river discharge conditions (nearly 300 m^3/s this day and the previous ones) and a mean neap tide. Four water samples were taken 3 km up river Cordemais, at high water level. The choice of this period has been made in order to find a well developed fluid mud, which allows for a reasonable amount of measurements. All the samples have been taken at several depths along the water column (surface, mid-depth, fluid mud and bottom sediments). Aside from LNHE's field measurements, the most recent data has been collected for a study realised by BCEOM for the account of a French Water Agency

(Agence de l'eau Loire-Bretagne 1997). Samples for a low river discharge and a neap tide were taken on the 15th of November 1995, and have been used to calibrate and validate the numerical model, as well as Gallenne and Migniot's data (see *table 1*, section 5).

5. 2D-NUMERICAL MODELS

Because of stratification and vertical exchanges, the numerical modelling of sediment behaviour in estuaries requires a 3D model. However, to obtain more accurate results and to reduce computational time, a 2D model has first been used, in order to provide the 3D model with good initial conditions. Actually, two 2D numerical models have been used: TELEMAC-2D, to compute the tidal currents, and SUBIEF-2D, for sediment transport under currents. Both are developed by the LNHE. TELEMAC-2D computes the non-steady free surface flows in shallow water environments, solving the Saint-Venant equations (Hervouet & Van Haren 1994) through a finite elements method. It gives, at each node of the mesh, the depth-averaged velocity and the water depth at each time step. On the other hand, SUBIEF-2D is used for two-dimensional sediment or water quality applications (Moulin & Gailhard 1995). It includes a prediction model for suspended cohesive sediment, including erosion and deposition processes. SUBIEF-2D solves in finite elements a vertically averaged equation of advection and dispersion, including the exchanges with the bottom. Thus, hydrodynamics and sediment transport are uncoupled, velocity and water depth fields being data resulting from the TELEMAC-2D computation, then used to compute the sediment behaviour using SUBIEF-2D.

The area considered is shown on *figure 2*. The Western boundary extends about 10 km offshore, and the river boundary is located at Montjean, about 120 km up river from Saint-Nazaire. The mesh contains 2309 nodes and 3969 elements. The element size, in the external estuary, is about 500 m, while the smallest element length is about 200 m near Nantes. The bathymetry shows an internal channel and a large area of tidal flats, which corresponds to the so-called « down river section » (see above). The initial conditions for TELEMAC-2D are zero velocity and a free surface that has been deduced by a linear extrapolation from six points of measurements along the estuary; the river discharge is prescribed at the upstream boundary. When a steady state is reached (after 3 tidal cycles), a tidal period is extracted regarding currents and water depths, to be used by SUBIEF-2D. Initial conditions for sediment are imposed through a 20 cm thick layer of mud, extended over 65 km in the middle of the modelled area. For this 2D model, a constant settling velocity has been prescribed, $w_s = 1.5$ mm/s. An input concentration of 0.035 $g.l^{-1}$ is imposed at the mouth during flood, and 0.05 $g.l^{-1}$ is prescribed on the upstream boundary. Simulations with SUBIEF-2D were carried out over 10 tidal cycles, in order to obtain a well established sediment distribution.

A run has been carried out for a river discharge of 1085 m^3s^{-1} and a spring tide. The turbidity maximum oscillates between St-Nazaire and Le Pellerin, and is partially expelled into the external estuary at low water level. The maximum of sediment rates occurs at high water level, reaching 1 $g.l^{-1}$ at Donges. If we consider the turbidity distribution in the channel established with the help of measurements done in 1981 during a spring tide with a low river discharge (Migniot, 1993), the turbidity maximum seems more spread out at high water level than at low water level, with, at high water level, a very slow decrease in the sediment concentrations in the up river part. Consequently, localising the exact maximum upstream extension of the turbidity maximum is difficult. The same remark applies for numerical results. We must also remember that the quality of 2D models results can be questioned because of the assumption of vertical homogeneity. Indeed, for high river flows, salt stratifications appear that are not reproduced by the model. They generate a residual circulation between surface and bottom and hence they contribute to the creation of the turbidity maximum.

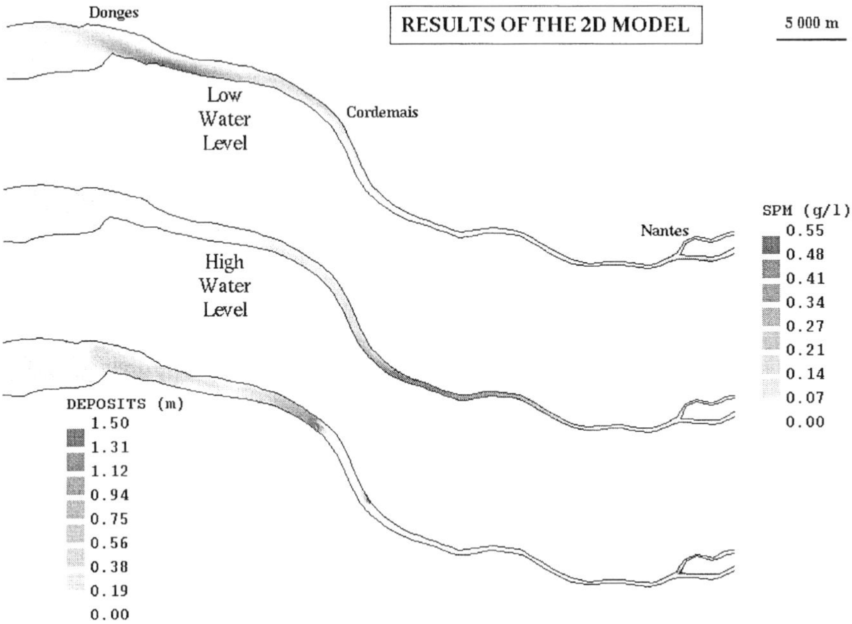

Fig. 3 - 2D model (neap tide). Location of the turbidity maximum at high and low tide, and position of the deposits.

With respect to fluid mud and deposition, the computed values reach a high thickness between two points located respectively 6 km and 15 km up river from St-Nazaire. Gallenne (1974) noted that in this area, characterised by a high water depth, the observed fluid mud is virtually always present before and after swellings. Computed deposit distribution shows that the highest values (reaching locally 1.80 m) don't occur in the navigation channel but on lateral mudbanks. The observed values are overestimated because we haven't introduced any sliding model for the fresh deposition in SUBIEF-2D.

A condition of neap tide with a low river discharge (300 m^3s^{-1}) has been simulated in order to provide the 3D model with initial conditions (see *figure 3*). Sediment modelling has been qualitatively validated through Migniot's and Gallenne's observations.

Table 1 - 2D model. Comparison of the computed sediment characteristics with observations and other computations.

Extension of the fluid mud layer (km up river from St-Nazaire)			Centre of gravity of the fluid mud layer (km up river from St-Nazaire)		
Computed results	DHI 1999 (comput.)	Gallenne 1974 (measur.)	Computed results	DHI 1999 (comput.)	Gallenne 1974 (measur.)
11 to 27 Lavau-Cordemais	15 to 32 Carnet-Martinière	12 to 30 Carnet-Sardine	22	24 Cordemais	26 Cordemais
Turbidity maximum	Computed results	Measurements			
Maximum upstream extension	km 47	km 50 (Migniot)	km 46 (DHI 1999)		
Centre of gravity	km 25	km 20 (Migniot)	km 25 (CSEEL)		
Concentrations	maximum of 0,8 g/l	bottom : 1,5 g/l mid-depth : 0,15 g/l surface : 0,075 g/l (BCEOM 1999)			
Total suspended mass	70 000 t	< 90 000 t (Migniot)	50 000 t (CSEEL)		

Extension, oscillation, ranges for concentrations values, localisation and thickness of the fluid mud, as well as total sediment mass in the estuary, deposited and suspended total masses (see table 1). The results of a study carried out by DHI Water & Environment (Agence de l'Eau Loire-Bretagne 1999) has been also compared with ours.

6. 3D-NUMERICAL MODEL

The 3-dimensional numerical model used for this work is TELEMAC-3D, developed by the LNHE (Janin et al. 1997 and Le Normant 1997). It reproduces three-dimensional flows both in seas and rivers, reproducing the surface motion (tidal signal, river flow). It can take into account a large variety of processes acting upon the flow, such as bottom friction, wind stress, the Coriolis force, turbulent mixing, and variations of water density resulting from salinity and suspended matter (among others). TELEMAC-3D solves the 3D-Navier-Stokes equations through a finite element method. For the present work, it has been used with a hydrostatic pressure assumption ; turbulent mixing processes are reproduced through a mixing length model. The hydrodynamic equations solved are the following :

$$\begin{cases} \dfrac{\partial u}{\partial t}+u\dfrac{\partial u}{\partial x}+v\dfrac{\partial u}{\partial y}+w\dfrac{\partial u}{\partial z}=-\dfrac{1}{\rho_0}\dfrac{\partial p}{\partial x}+\dfrac{\partial}{\partial x}\left(\nu_{tH}\dfrac{\partial u}{\partial x}\right)+\dfrac{\partial}{\partial y}\left(\nu_{tH}\dfrac{\partial u}{\partial y}\right)+\dfrac{\partial}{\partial z}\left(\nu_{tZ}\dfrac{\partial u}{\partial z}\right)-\gamma v \\ \dfrac{\partial v}{\partial t}+u\dfrac{\partial v}{\partial x}+v\dfrac{\partial v}{\partial y}+w\dfrac{\partial v}{\partial z}=-\dfrac{1}{\rho_0}\dfrac{\partial p}{\partial y}+\dfrac{\partial}{\partial x}\left(\nu_{tH}\dfrac{\partial v}{\partial x}\right)+\dfrac{\partial}{\partial y}\left(\nu_{tH}\dfrac{\partial v}{\partial y}\right)+\dfrac{\partial}{\partial z}\left(\nu_{tZ}\dfrac{\partial v}{\partial z}\right)+\gamma u \\ p=\rho_0 g(Z_S-z)+\rho_0 g\displaystyle\int_z^{Z_S}\dfrac{\Delta\rho}{\rho_0}dz \\ \dfrac{\partial u}{\partial x}+\dfrac{\partial v}{\partial y}+\dfrac{\partial w}{\partial z}=0 \end{cases} \qquad (8)$$

where Z_S is the free surface elevation, u, v, w the velocity components, p the pressure, g the acceleration due to gravity, ν_{tH} and ν_{tz} the velocity diffusion coefficients (including molecular and turbulent diffusion), ρ_0 a reference density and $\Delta\rho$ the variation in density. Coriolis effects are taken into account through $\gamma = 2\Omega\sin\lambda$, λ being the latitude (here + 49°) and $\Omega = 7.29\ 10^{-5}$ s^{-1}. The equations for salinity S and sediment concentration C are

$$\frac{\partial S}{\partial t}+u\frac{\partial S}{\partial x}+v\frac{\partial S}{\partial y}+w\frac{\partial S}{\partial z}=\frac{\partial}{\partial x}\left(K_{tH}\frac{\partial S}{\partial x}\right)+\frac{\partial}{\partial y}\left(K_{tH}\frac{\partial S}{\partial y}\right)+\frac{\partial}{\partial z}\left(K_{tz}\frac{\partial S}{\partial z}\right) \tag{9}$$

$$\frac{\partial C}{\partial t}+u\frac{\partial C}{\partial x}+v\frac{\partial C}{\partial y}+w\frac{\partial C}{\partial z}+\frac{\partial (w_s C)}{\partial z}=\frac{\partial}{\partial x}\left(K'_{tH}\frac{\partial C}{\partial x}\right)+\frac{\partial}{\partial y}\left(K'_{tH}\frac{\partial C}{\partial y}\right)+\frac{\partial}{\partial z}\left(K'_{tz}\frac{\partial C}{\partial z}\right) \tag{10}$$

in which w_s is the settling velocity, K and K' the eddy diffusivities for salinity and sediment. In the framework of the mixing length model, ν_{tz} and K_{tz} are expressed by eq. (1). Bottom friction and wind stress are considered through the boundary conditions.

A first model of the Loire estuary was developed a few years ago by Le Normant, with a fine grid, showing the ability of TELEMAC-3D to reproduce correctly both hydrodynamics (surface elevation) and sediment transport (turbidity maximum) in the Loire estuary (Le Normant 1995). The present model, completely new, runs on a coarse grid, which enables us to considerably reduce the computational costs, and uses up to date software versions. In addition, the developments made within the COSINUS project and tested in the case of the schematic estuary (see section 2) have been introduced in the present 3D model. In particular, Toorman's model for damping has been used (equations (3) to (5)), as well as the entrainment model for the interactions with the bed (equations (6) and (7)). The sensitivity to the settling velocity have been tested, the best results being provided by a power law, which is in good agreement with settle measurements in the estuary (Migniot 1993).

Many 3D models have shown problems with the sliding of deposition on the bed. When the deposition thickness is high, the numerical models can underestimate the suspended matter ; when sediment is deposited on tidal flats, it is no longer eroded. Actually, the sediment can slide if the bed slope is strong enough, re-introducing suspended sediment by diffusion into the water column. A model has been developed in TELEMAC-3D, to take into account this phenomenon : when the bed slope is higher than a critical slope, the bed is smoothed, through a finite element process involving a sliding parameter to be calibrated. This method prevents from high deposition on the tidal flats at low tide.

The horizontal mesh used is the same as for the 2D model, while 6 σ-layers are used for the vertical discretization. Initial conditions have been chosen several tidal periods after the beginning of the 2D simulations, with a sediment distribution consistent with observations. After too many tidal periods, the suspended matter would be expelled out of the estuary (for a high river discharge) or deposited (for a neap tide). A run with salinity and no sediment has been performed over several tides in order to stabilise hydrodynamics and salt distribution (this run begins with a linear extrapolation of salinity between points of measurements). Once a steady state is reached, this computation is continued with introduction of sediment.

Suspended matter distribution and deposits are initially prescribed as the results of the 2D model, assuming a vertically homogeneous initial distribution is used for suspended matter. Salinity is imposed as a 35 g/l input at the downstream boundary during flood and freely goes out during ebb. At the upstream boundary, fresh water enters the estuary. Boundary conditions for sediment are the same as for the two dimensional model, input concentrations being imposed as a constant over the depth.

7. RESULTS

Free surface levels show a very good agreement with the data, as it can be seen of *figure 4*, for a mean tide and a river discharge of 1085 m^3s^{-1}. Regarding sediment, one tidal condition has been tried, corresponding to a neap tide, with a low river discharge of 300 m^3s^{-1}.

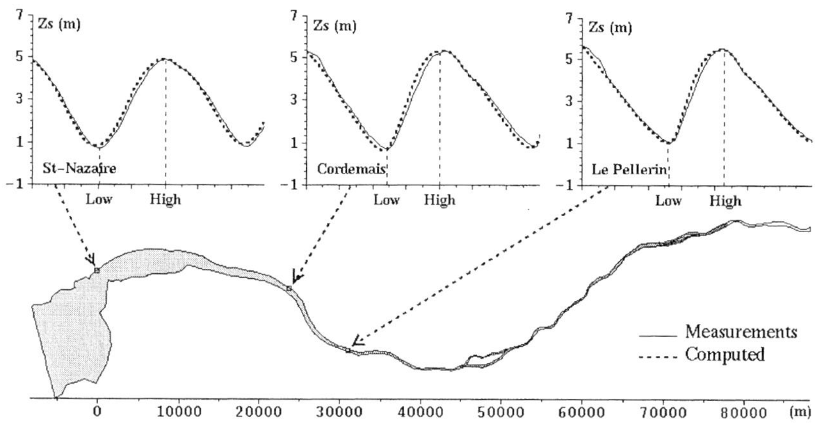

Fig. 4 - 3D model. Surface elevation : comparison between computations and data.

The centre of gravity (tidally averaged position) of the computed turbidity maximum, located at Cordemais, its maximum upstream extension, located 47 km upstream from St-Nazaire, as well as the total suspended mass in the estuary (70 000 tons in average) are in very good agreement with the observations (Migniot 1993 and Gallenne 1974). The longitudinal profile for the turbidity maximum has been compared and validated by comparison to Migniot's plots. Two extreme situations, at high tide and low tide, as well as fluid mud distribution at the end of computation, are plotted on *figure 5*. Our maximum value for fluid mud layer thickness reaches 2.27 m. This value is confirmed by Gallenne's observations which showed that the fluid mud layer could reach a maximum thickness of 3 m after neap tides. The time-and-depth-averaged concentration has a maximum of 0.75 g/l, which is confirmed by Migniot.

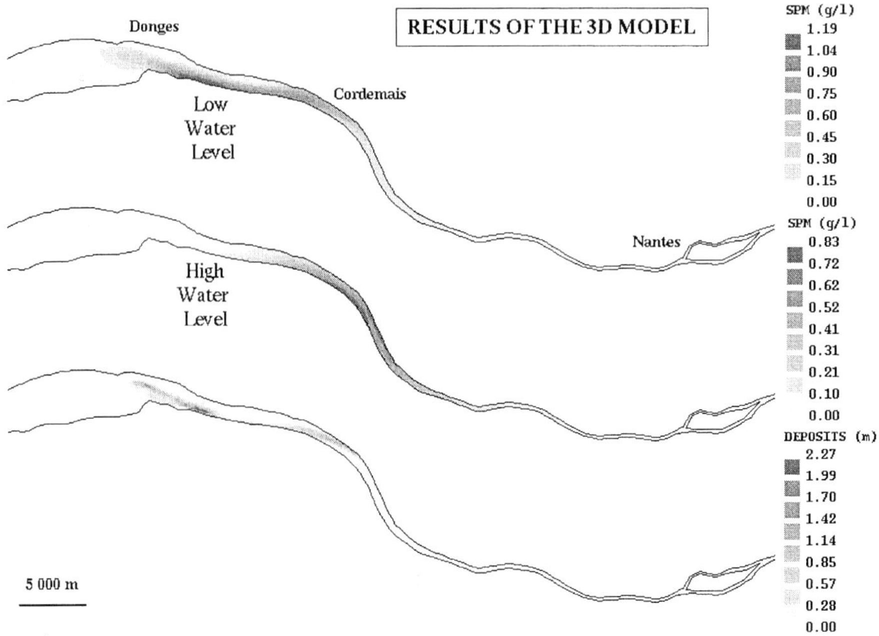

Fig. 5 - 3D model. Location of the turbidity at high and low tide, and position of the deposits.

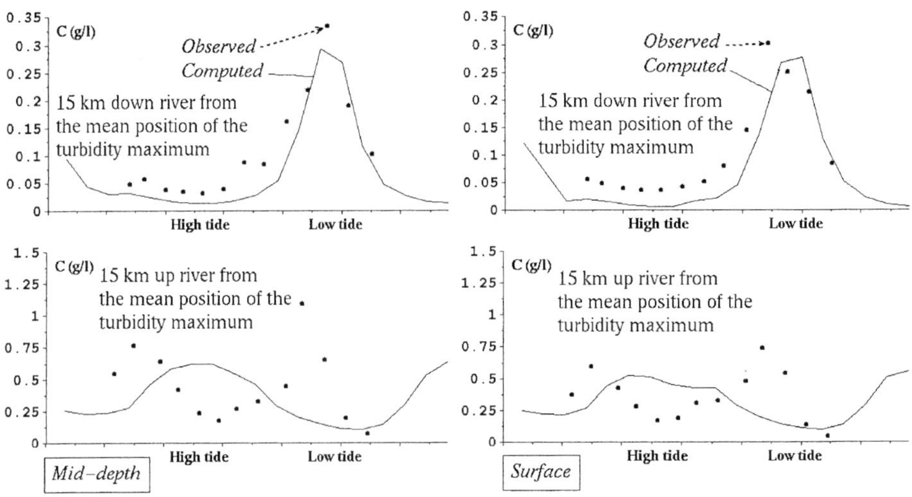

Fig. 6 - 3D model. Time series of concentration at the surface and mid-depth. Comparison between computations and data.

A quantitative validation for the obtained results has been done with the help of recent measurements. Those measurements were undertaken for a river input reaching 285 m^3s^{-1}, while the river input was increasing (reaching values between 250 and 278 m^3s^{-1} during the previous eleven days). Thus, the observed turbidity maximum, whose location is strongly depending on the chronological account of tidal and fluvial conditions, is still localised upstream in comparison to the general observations of Migniot and Gallenne.

A comparison has been done at points situated 15 km down river and 15 km up river from the centre of gravity of the turbidity maximum, at the surface and mid-depth (*figure 6*). Sediment movements are well reproduced 15 km down river from the centre of gravity, with a maximum value of 0.3 g/l. The discrepancies are more visible on the other plot (15 km up river from the centre of gravity), with an apparent dephasing between computed and measured concentrations. Actually, the phase is correct but a strong deposition of suspended matter at high water level slack (which leads to a strong decrease of the observed sediment rates, from 0.7 g/l to 0.2 g/l) doesn't seem to be reproduced by the model, even if computed results show a slight increase of suspended matter rates 3 hours after high tide, which is related to erosion of fresh deposits. Another explanation to the shape of the measurements could be the existence of a temporary second peak of turbidity within the turbidity maximum.

It should be noticed that the location of the turbidity maximum, in the Loire estuary, is very sensitive to the history of tidal conditions before the date when observations are done. The time-dependency of the observed river flow should also play a role, though a constant flow is considered here. However, comparing the general pattern of the computed turbidity maximum with observations, it's obvious that sediment prediction is in progress thanks to the new developments considered here. Further applications should include extensive simulations, covering several days in order to validate the time behaviour of the SPM in the estuary.

8. CONCLUSIONS

The TELEMAC-3D software has proven its ability to correctly reproduce the sediment behaviour in an estuary, thanks to the developments made within the COSINUS European project regarding cohesive sediment processes (damping of turbulence, entrainment of Concentrated Benthic Suspension, flocculation process). After testing the sensitivity of the model to the main parameters involved, it is possible to calibrate it. This calibration was done using the different models provided by the COSINUS project, regarding the above-mentioned processes. In particular, the choice of the settling velocity formulation is crucial. These developments allowed a good prediction of the sediment behaviour. The motion of the turbidity maximum, as well as the values of concentration reproduced by the model, are in good agreement with the measurements. However, a more precise prediction would require a long term simulation, in order to take into account the time dependency of both river flow and tidal conditions.

ACKNOWLEDGEMENTS

This work was co-financed by the European Commission, Directorate XII for Science, Research & Development, through the COSINUS project within the framework of the MAST-3 programme, contract MASC3-CT97-0082.

REFERENCES

Agence de l'Eau Loire-Bretagne, 1997, La Loire estuarienne. Etude de la modélisation prospective. Tranche n°1, rapport d'études n°2 : modélisation mathématique (données, construction et calage).

Agence de l'Eau Loire-Bretagne, 1999, Modélisation prospective de la Loire estuarienne. Tranche n°2, phase 1, rapport : construction des outils Mike 11.

Cheviet, C., Violeau, D., Le Normant, C., 2000, MAST3-COSINUS European Project - Intercomparison of the Results of Several Numerical Models on a Schematic Estuary Case. *Report No* HP-72/2000/026/A, *Electricité de France / LNHE.*

Cheviet, C., Violeau, D., Guesmia, M., 2000, MAST3-COSINUS European Project - 3D-modelling of cohesive sediment transport in the Loire estuary (France). *Report No* HP-72/2000/048/A, *Electricité de France / LNHE.*

Gallenne, B., 1974, Les accumulations turbides de l'estuaire de la Loire. Etude de la crème de vase. *Doctoral Thesis, University of Nantes.*

Hervouet, J.M. and Van Haren, L., 1994, TELEMAC-2D Version 3.0 - Principle note. *Report No* HE-43/94/052/B, *Electricité de France / LNHE.*

Janin, J.M., Marcos, F., Denot, T., 1997, Code TELEMAC-3D Version 2.2 - Note théorique. *Report No* HE-42/97/049/B, *Electricité de France / LNHE.*

Kranenburg, C. and Winterwerp, J.C., 1997, Entrainment of Fluid Mud Layers. I : Entrainment Model. *J. Hydraulic. Engineering, ASCE,* **123**(6), 504-511.

Le Hir, P., 1997, Fluid and sediment « integrated » modelling application to fluid mud flows in estuaries, *Cohesive Sediments, proc.* INTERCOH '94.

Le Normant, C., 1995, Modélisation numérique tridimensionnelle des processus de transport des sédiments cohésifs en environnement estuarien. *Doctoral Thesis for the Institut National Polytechnique de Toulouse, report No* HE-42/95/028/A, *Electricité de France / LNHE.*

Le Normant, C., 1997, Description de la bibliothèque sédimentologique de TELEMAC-3D Version 2.2. *Report No* HE-42/97/070/A, *Electricité de France / LNHE.*

Migniot, C., 1993, Bilan de l'hydrologie et de l'hydrosédimentaire de l'estuaire de la Loire au cours des deux dernières décennies. *Agence pour la Protection de l'Environnement de l'Estuaire de la Loire, Port Autonome de Nantes Saint-Nazaire.*

Moulin, C. and Gailhard, J., 1995, SUBIEF Water Quality Computation Software Version 3.1. User's Manual. *Report No* HE-43/95/074/B, *Electricité de France / LNHE.*

Petersen, O. and Vested, H.J., 2000, An Operational Description of Vertical Exchange Processes in Numerical Mud Transport Modelling, *COSINUS report. DHI 2000, Second Draft version.*

Spearman, J.R. and Roberts, W., 2000, Parameterisation of flocculation models for applied sediment transport modelling. *Proceedings* INTERCOH-2000, *Elsevier, Coastal and Estuarine Fine Sediment Processes*, ed. J.C. Winterwerp and C. Kranenburg, this volume.

Taylor, J.S., 1973, Buoyancy Effects in Fluids *Cambridge University Press.*

Toorman, E. A., 2000, Parameterisation of Turbulence Damping in Sediment-Laden Flows. *Report No* HYD/ET/00.2, *Hydraulics Laboratory, Katholieke Universiteit Leuven.*

Violeau, D., Bourban, S., Cheviet, C., Markofsky, M., Petersen, O., Roberts, W., Spearman, J., Toorman, E.A., Vested, H.J., Weilbeer, H., 2000, Numerical Simulation of cohesive sediment transport : intercomparison of several numerical models, Proceedings INTERCOH-2000, *Elsevier, Coastal and Estuarine Fine Sediment Processes*, ed. J.C. Winterwerp and C. Kranenburg, this volume.

Winterwerp, J.C., 1999, On the Dynamics of High-Concentrated Mud Suspensions, *Doctoral Thesis for the Technical University of Delft.*

Fine Sediment Dynamics in the Marine Environment
J.C. Winterwerp and C. Kranenburg (Editors)

3D application of the *continuous modelling* concept to mud slides in open seas

P. Le Hir and F. Cayocca

Institut Français de Recherche pour l'Exploitation de la Mer (FREMER)
Centre IFREMER de Brest, BP70, 29280 Plouzané, France

Most turbidity current models are vertically integrated, and use a parametric entrainment rate to simulate the growth and dilution of the dense layer. The *continuous modelling* concept integrates all physical processes related to high-concentrated suspensions (stratification-induced turbulence damping, hindered settling and molecular viscosity increase or even visco-plastic behaviour). It is applied in a three-dimensional frame to simulate turbidity currents on a slope. A sensitivity analysis has been carried out in a 2DV configuration. It shows the front celerity is dependent on the initial mass, not so much on the slope, whereas the velocity within the body of the turbidity current is controlled by the slope and vertical mixing processes. The resulting entrainment rate is in agreement with flume experiments. A vertical recirculation is observed at the head of the turbidity current. The contribution of bed erosion is pointed out: such an erosion is generated by the density current and then enhances the density gradient and thus the turbidity current. The application of the 3D model to the 1979 Nice slide and turbidity current is under process. Preliminary results are discussed.

KEYWORDS : turbidity current, sediment transport, mathematical modelling.

1. INTRODUCTION

Depending on the material nature and on the seafloor slope, under water slumps or slides are likely to generate high-density turbidity currents that represent a main type of offshore mass movements. Such events can be illustrated by two well-known cases : the 1929 Grand Banks slumps which traveled more than 700 km from the source area and the 1979 Nice slump generated in shallow waters on a very steep slope; the latter was induced by overloading with discharged materials during the extension of the Nice airport (Mulder and Cochonat, 1996). According to these authors, the high-density turbidity currents contrast with mass flow (like debris flow for instance) by their turbulent regime and their rather low volume concentration (below 0.09, following the Bagnold limit).

Most turbidity current models are "integral", meaning they consider vertically averaged characteristics of the turbibity current such as its mean concentration and velocity, over a variable height (e.g. Parker et al., 1986; Graf and Altinakar, 1995; Bradford et al., 1997; Mulder et al., 1997). Such one-dimensional or two-dimensional models consume less computing time, but cannot describe vertical gradients around the interface between the turbidity current and the overlying water. Furthermore an entrainment law has to be parameterized in order to close the problem (e.g. Parker et al., 1987). A nice exception to these depth-integrated models is provided by Assier-Rzadkiewicz et al.(1997), who develop a free-surface non-hydrostatic model, two-dimensional in the vertical plane (2DV). They simulated submarine landslides and the induced surface waves, but turbulence was not explicitely accounted for, nor its damping at the interface between mud and water. In contrast, the work from Stacey and Bowen (1988) should be mentioned: these authors simulated the vertical profile of a turbidity current in a one-dimensional vertical frame, neglecting longslope gradients.

On the other hand, our knowledge on processes controlling high-concentrated mud suspensions has recently made progress, although applied to fluid mud formation and behaviour in navigation channels or estuaries (e.g. Winterwerp, 1999). Aiming to integrate these processes in a single modelling frame, Le Hir et al. (2001) proposed the concept of *continuous modelling*, which consists in considering the water + sediment complex as a whole. By accounting for stratification-induced turbulence damping, hindered settling in high concentrations and non-newtonian behaviour of the mud, the so-called continuous modelling enables the simulation of realistic density and velocity vertical profiles.

The present paper aims to evaluate the possibility of applying the continuous modelling concept to sediment slumps and turbidity currents over continental slopes. First a 2DV case on a uniform slope is considered, then the 3D simulation of the Nice sliding is attempted.

2. STRUCTURE OF A TURBIDITY CURRENT

It is classical to distinguish the head and the body of a turbidity current, according to a sketch drawn on figure 1. The head is driven by the pressure gradient downslope arising from the density difference between the head of the density current and the ambient water in front of it (Stacey and Bowen, 1988).

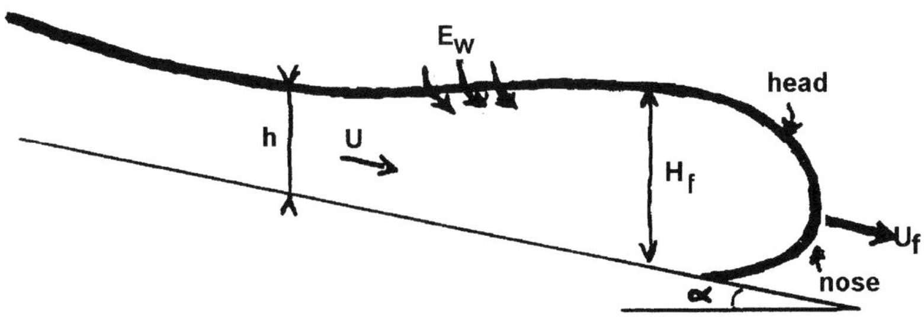

Figure 1 Definition sketch of a turbidity current on a slope.

Following Graf and Altinakar (1995), the velocity of the head ("front velocity", U_f) can be deduced from the Bernouilli equation applied across the head :

$$U_f = (2g'h)^{0.5} \quad (1)$$

where $g' = g(\rho_f-\rho_w)/\rho_W$, the reduced gravity
ρ_f the density of the turbidity current
ρ_w the water density
h the height of the turbidity current far upstream of the head

From experiments, Graf and Altinakar (1995) also expressed U_f as a function of the front height H_f :

$$U_f= 0.75\ (g'\ H_f)^{0.5} \quad (1')$$

The body of the turbidity current is also driven by density gradients, but can be approximated by a steady uniform flow, the density-induced acceleration being compensated by the bottom friction and some entrainment of the overlying water. The turbidity current can be described through integral scales U, C and h, that represent respectively the mean velocity, the mean concentration and a representative thickness of the turbidity current. These integral scales are related to each other by the following relations (e.g. Parker et al., 1987):

$$hC = \int_z cdz \quad (2a), \qquad hC^2 = \int_z c^2 dz \quad (2b), \qquad hU = \int_z udz \quad (2c)$$

Dividing (2b) by (2a), C and then h can be deduced, whereas U can be evaluated from (2c) and h.

The entrainment represents the rate of overlying water progressively included in the turbidity current and can be related to integral scales by considering a mass balance equation (e.g. Parker et al., 1987):

$$\frac{dUh}{dx} = E_w U \quad (3)$$

where E_w is the non-dimensional coefficient of entrainment. Generally E_w is experimentally related to a global Richardson number.

A simple equation is often used to characterize the motion (e.g. Stacey and Bowen, 1988):

$$U = \left[\frac{g' h \sin\alpha}{E_w + C_d}\right]^{0.5} \quad (4)$$

where C_d is a drag coefficient, resulting from a quadratic bottom friction $\tau = \rho\ C_d\ U^2$.

From equations (1) and (4) it can be seen that the front celerity should not depend on the bottom slope, whereas the velocity within the turbidity current depends on both the bed slope α and the bottom friction. According to Graf and Altinakar (1995), for reasonable drag coefficients and non negligible bottom slopes ($\alpha > 0.7\,^{o}$), the front velocity is always lower than the body velocity, the larger the bottom slope, the higher the difference. Accordingly, and for the sake of mass conservation, the head of the turbidity current is higher than the body ($H_f > h$), as represented on figure 1.

3. THE CONCEPT OF CONTINUOUS MODELLING

The concept of continuous modelling (Le Hir et al., 2001) consists in considering the sediment and water compartments as a whole, without empirical exchanges between them. Then erosion or entrainment of sediment by overlying flow results from viscous or turbulent mixing. Assuming the pressure is hydrostatic and density variations only contribute to the pressure term (Boussinesq assumption), governing equations are:

- continuity:

$$\frac{\partial \varsigma}{\partial t} + \frac{\partial \int_{bottom}^{\varsigma} u_i dz}{\partial x_i} = 0 \qquad (5)$$

where ς is the free surface elevation and u_i the horizontal velocity along the x_i direction

- horizontal momentum balance for the mixture:

$$\frac{\partial u_i}{\partial t} + u_j \frac{\partial u_i}{\partial x_j} + w \frac{\partial u_i}{\partial z} \mp f u_j = -\frac{1}{\rho_0} \frac{\partial \int_z^{\varsigma} \rho g dz}{\partial x_i} + \frac{\partial}{\partial z}\left(-\overline{u_i' w'} + \frac{T_{iz}}{\rho_0} \right) \qquad (6)$$

where ρ is the density of the mixture, fu_j the Coriolis acceleration, w the vertical velocity, u_i' and w' the turbulent fluctuations of velocity and T_{iz} the "viscous" shear stress.

- sediment mass conservation:

$$\frac{\partial c}{\partial t} + \frac{\partial c u_i}{\partial x_i} + \frac{\partial}{\partial z}\left[c(w + W_s) \right] = \frac{\partial}{\partial z}\left(K_z \frac{\partial c}{\partial z} \right) \qquad (7)$$

where c is the mass concentration of sediment, W_s the settling velocity and K_z the vertical diffusivity.

The set of equations (5, 6, 7) has to be closed by a classical state equation that relates ρ and the sediment concentration c, and by a turbulence model specifying $-u'_i w'$ and K_z. The key of the continuous modelling is the possibility to simultaneously account for stratification-induced turbulence damping, hindered settling in concentrated environments and non-newtonian behaviour of the sediment mixture. A relatively high vertical resolution is required to describe the vertical structure and behaviour of the mixture.

In order to reduce computing time, the adopted turbulence closure is a mixing length model with empirical damping functions. Details on the formulations of the turbulence model, as well as the settling dependence on sediment concentration are given in Le Hir et al. (2001).

The viscous shear stress is related to the sediment concentration and the shear rate by means of the following expression which empirically accounts for a visco-plastic behaviour, very close to the Bingham type (Le Hir et al., 2001):

$$\frac{T_{iz}}{\rho} = \left\{ \nu_0 + k_1 C^{k_2} \left(1 + \frac{k_3}{k_4 + \sqrt{\sum \left(\frac{\partial u_i}{\partial z} \right)^2}} \right) \right\} \frac{\partial u_i}{\partial z} \tag{8}$$

The continuous model should be suitable for applications to underwater slides and turbidity currents, as it accounts for density gradients forcing. The entrainment of quiescent water by the turbidity current then results from the exchange of momentum by turbulent viscosity within the lutocline and does not need any additional formulation.

In the following, the SAM-3D finite differences model is used to solve equations (5-6-7). This sediment transport model includes a bed module where consolidation can be accounted for (Cugier and Le Hir, 2000). Exchanges between the water column (within which the continuous modelling is applied) and the bed are parameterized by the Krone law for deposition and an empirical erosion law.

4. APPLICATION TO A 2DV CONFIGURATION

The purpose of the 2DV calculations is to describe the ability of the model to simulate a turbidity current and to run a brief sensitivity analysis in a simplified geometry.

4.1. Description of the computed domain

Along the x direction the slope of the bed is uniform and the grid space is 50 m. The boundaries up and down the slope are far from the area of interest (3 km long), in order to avoid any sediment flux across them. The water level is fixed at the boundaries, which will induce unrealistic reflection patterns of the free surface. However these disturbances are unlikely to significantly affect the motion of the slide. In addition the model is hydrostatic, which is probably not valid for the wave induced by the slide. Hence results about the water level will be disregarded. The domain is vertically split into 1 m thick layers. The bottom is thus constituted of steps of varying height depending on the slope. These steps are likely to generate artificial bed roughness : other simulations with smaller grid spacing (not reported here) show that the effect of these steps on the flow and the mean bottom shear stress are similar to the effect of bedforms of the same scale. Last a transverse dimension is introduced in the 3D model, in order to avoid any lateral effect, but results will only be considered in the central plane.

4.2. Initialization

An unstable sediment mass is given *a priori*, at the head of the sloping bed. This sediment is assumed to be constituted of an homogeneous mixing of particles and pore water, with a given visco-plastic behaviour. It is located in the lower layers of the "water column", the rest of which is clear. The whole domain is initially motionless. The horizontal density gradient between this mixture and clear water at the front constitutes a forcing which is likely to generate the slide and subsequent water movements in the surroundings.

For the first run (our "reference"), the bed slope is 3°, the initial mixture is 10 m high, 500 m long and 300 kg.m^{-3} concentrated. The material viscosity is set to be the same as the water viscosity. The computing time step is 1 second.

4.3. Comments on the "reference" run

A first look on the slide is given on figure 2, with no vertical distortion, but results are easier to analyze in a distorted representation (figure 3, and all following pictures).

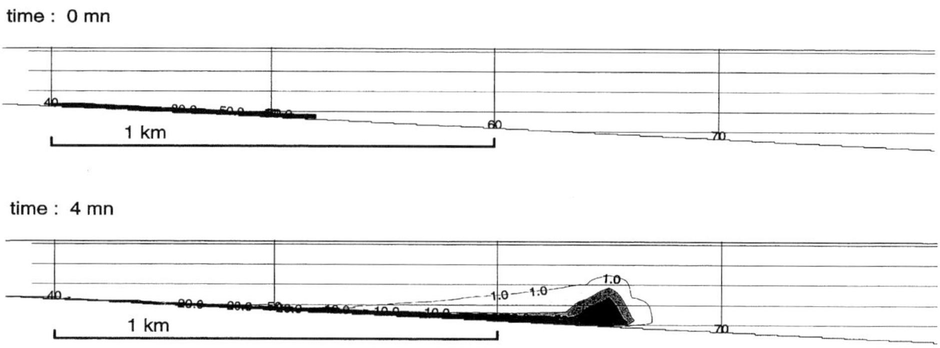

Figure 2 2D simulation of an underwater slide and the induced turbidity current.
Same parameters as on figure 3.

Figure 3 shows that (i) the initial sediment rapidly slides along the slope, and that (ii) induced velocities are large enough for turbulent exchanges to take place at the interface between the sliding sediment and surrounding water. As a consequence, particulate matter partly diffuses in the environment. This dispersion can be seen as a water entrainment or a dilution of the mixture around the interface. The rising of particles at the head is related to vertical advection induced by water continuity. The observed front velocity is about 2.6 m.s^{-1}, but decreases with time, because of a progressive reduction of the density gradients due to mixing. Probably for the same reason, this front velocity is a little smaller than the solution of equation (1') : U_f = 3.2 m.s^{-1} for g'=1.8 m.s^{-2} and H_f = 10 m [application of equation (1) is difficult because of some ambiguity for setting h].

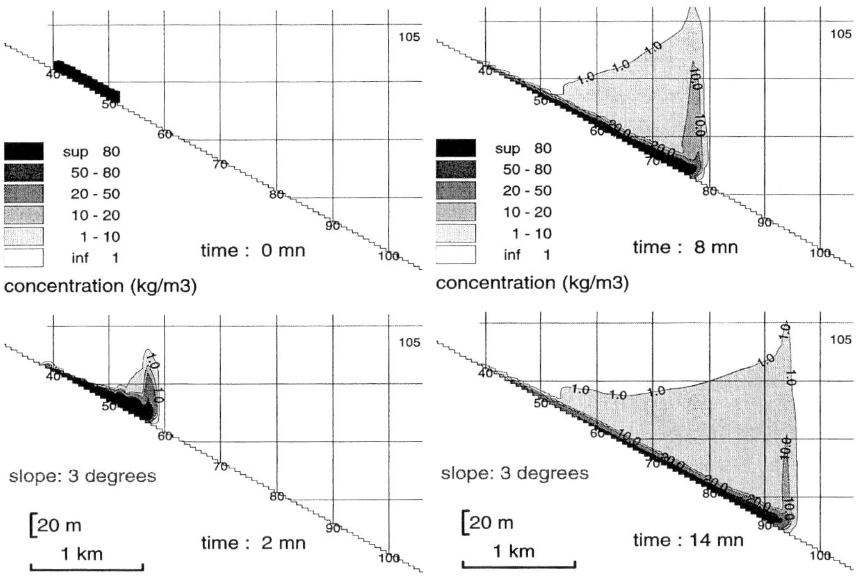

Figure 3 : 2D simulation of a turbidity current: **reference test**.
Bed slope = 3°. Material behaviour: see text.

The vertical structure of the turbidity current (called TC, in the followings) is illustrated in figure 4. At the head of the turbidity current (left hand side of figure 4), the concentration profile presents a maximum 4 m above the bed, in agreement with the nose which is often depicted at the front of a TC. This concentration maximum seems to be induced by advection, the velocity being maximum few metres above the bed. Also, the return flow in the upper layers appear rather strong and guaranties continuity. Behind the head, within the body of the TC (right hand side of figure 4), the concentration increases down to the bed as expected, while the current remains maximum 5 m above the bed, in the middle of the lutocline. The velocity decay below is induced by friction.

The structure of the computed turbidity current is qualitatively very close to observed profiles of density currents in a flume, as shown by figure 5 which illustrates an experiment from Ellison and Turner (1959). In particular the maximum of velocity clearly appears within the pycnocline (the experiment concerned a salt/fresh water interface on an inclined bed).

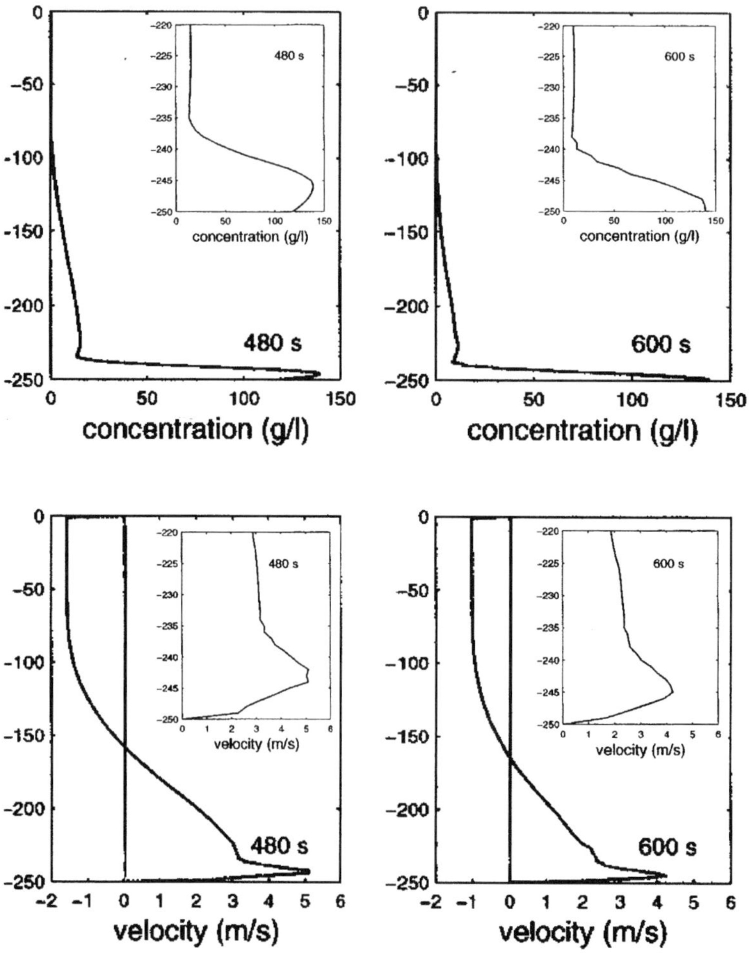

Figure 4 Vertical structure of the simulated turbidity current (TC), reference test.
Left: concentration and velocity profiles through the head of the TC
Right: concentration and velocity profiles within the body of the TC
On top-right corner of each graph is plotted a zoom of the bottom layers (vertical units are metres)
Profiles are drawn at the same location, corresponding to the front location at 480 s, but situated within the TC 120 s later (t = 600 s).

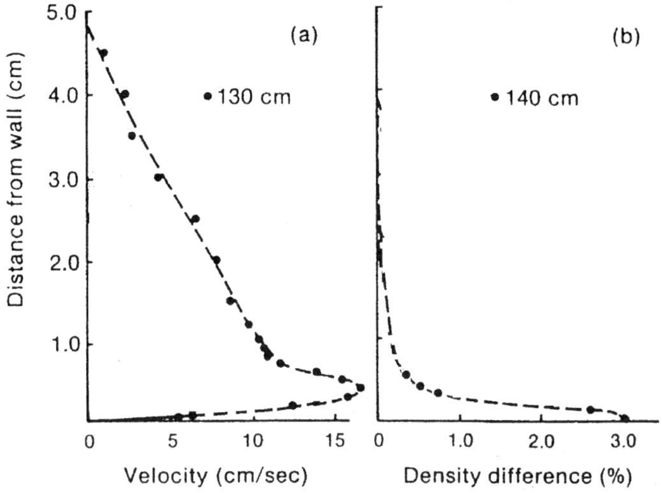

Figure 5 Observed profiles of velocity (left) and concentration (right) for a density current flowing down a slope of 14° (from Ellison and Turner, 1959)

4.4. Effect of turbulence damping

For the reference test, the mixing length was reduced in the stratified zone, according to empirical damping functions detailed in Le Hir et al. (2001). Switching off the turbulence damping process and comparing with the reference test will allow us to assess its effect. With no damping, the mixing (or entrainment) between the turbidity current and the water is increased (figure 6). The induced reduction of concentration at the head then softens the forcing so that the front celerity is slightly reduced (the distance between the respective fronts is about 50 m after 10 minutes).

Differences on velocity and concentration profiles are more sensitive, as shown in figure 7. First the increase of vertical mixing is confirmed when turbulence damping is not accounted for (thin curves on figure 7), and velocity profiles become smoother. Furthermore, depth-averaged concentration is reduced at the front (left side of figure 7), which means that the latter is horizontally spread. One main consequence of the turbulence damping process is the sensitivity of the entrainment rate (see § 4.8). The practical importance of this parameter implies that damping functions of the turbulent closure have to be carefully calibrated, which requires profile measurements at the nature scale for this type of flow. Alternative turbulence closure should also be checked against experimental results at the same scale.

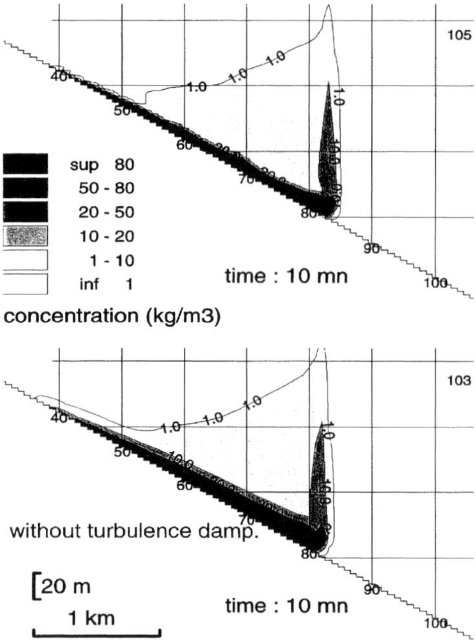

Figure 6 Effect of turbulence damping on the simulated turbidity current.
Top: with turbulence damping (reference test)
Bottom: without turbulence damping.

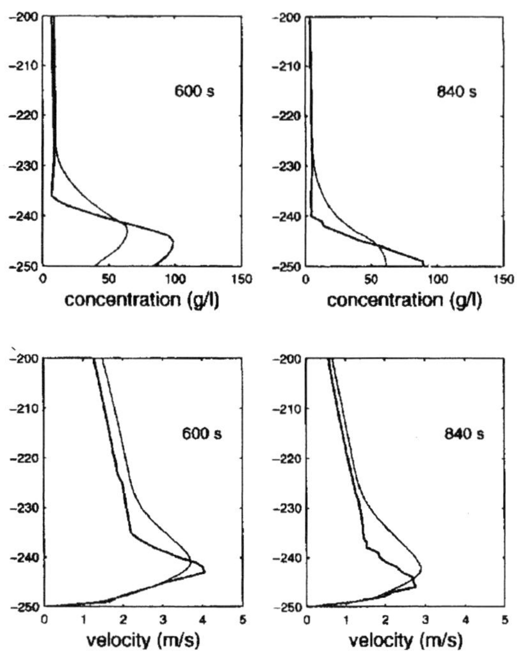

Figure 7 Effect of turbulence damping on the vertical structure of the turbidity current. *Left:* at the head *Right:* within the body. *Thick (respect. thin) curves:* with (respect.without) turbulence damping

4.5. Effect of the bottom slope

The role of the bottom slope is illustrated in figure 8. It can be seen that the front celerity is nearly unchanged for a slope of either 2 or 5 degrees, while vertical mixing increases with the bottom slope, which is due to an increase of velocity within the body of the TC, also perceptible on figure 8 as the tail of the TC is shifted downslope when the bottom slope is larger.

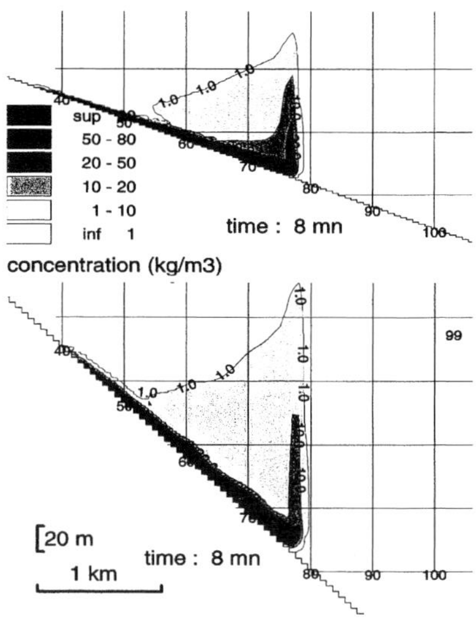

Figure 8 Effect of the bottom slope on the turbidity current
Top: slope $\alpha = 2°$
Bottom: $\alpha = 5°$

4.6. Influence of initial characteristics of the unstable sediment

Three simulations (numbered 2-4) were conducted in order to test the influence of the initial geometry or concentration of the movable sediment. Figure 9 shows the configuration of the TC after 12 mn. The general shape remains unchanged, but the front velocity varies noticeably. The initial mass of simulations 2 to 4 is twice the initial mass of the reference test, and is either twice as thick, twice as long or twice as concentrated. The three simulations exhibit similar patterns of front propagation, about 25% faster than the reference test, even in the case of an initial length of 1000 m if we account for the upstream location of the extension from 500 m to 1000 m. These "experimental" results show that due to a rapid spreading of the sliding mass, at first order, the celerity of the front can be related to the initial movable mass.

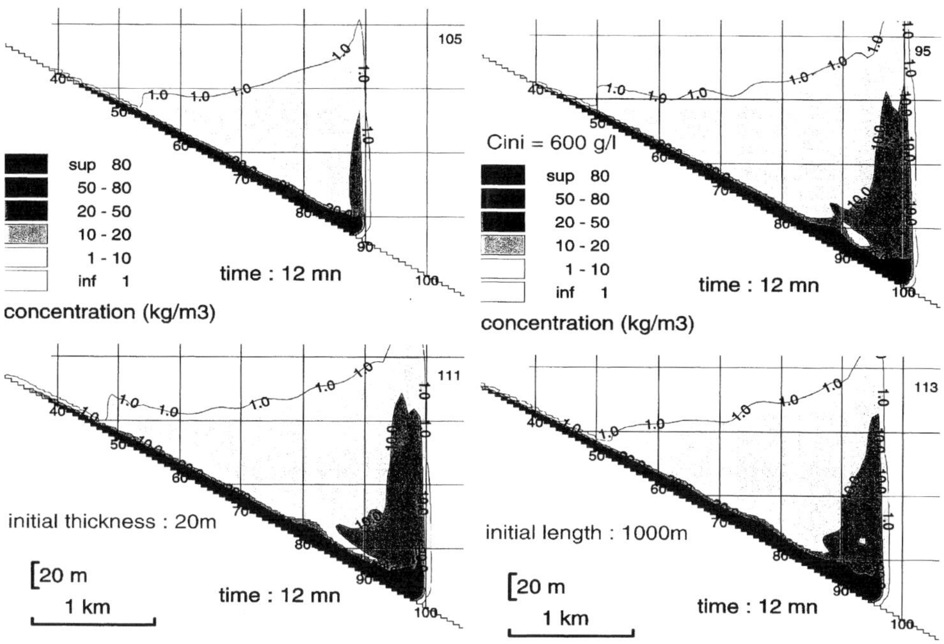

Figure 9 Effect of initial characteristics of the sliding sediment on the turbidity current
Top/left: initial concentration. 300 kg.m^{-3}, length = 500 m, thickness = 10 m (reference test)
Bottom/left: initial concentration 300 kg.m^{-3}, length = 500 m, thickness = 20 m
Top/right: initial concentration 600 kg.m^{-3}, length = 500 m, thickness = 10 m
Bottom/right. initial concentration 300 kg.m^{-3}, length = 1000 m, thickness = 10 m

4.7. Influence of the sediment behaviour

The sedimentological characteristics of the material are also likely to influence the TC, in particular the settling velocity for the suspensions and the rheological behaviour for dense layers.

In the reference case, the settling velocity varies with concentration, in order to account for flocculation processes, but also for hindering effects. Thus the settling velocity increases from 0.5 to 2 $mm.s^{-1}$ when the concentration (dry density) varies from 0 to 20 $kg.m^{-3}$, and then rapidly decreases due to hindered settling, reaching 0.02 $mm.s^{-1}$ for 100 $kg.m^{-3}$ (Le Hir et al., 2001). Under these conditions the settling velocity remains practically below 1 $mm.s^{-1}$ for the reference test. An alternative simulation has been run, with a constant - and unrealistic- settling velocity of 10 $mm.s^{-1}$. The resulting turbidity current (not presented) is thinner, with a reduced vertical mixing, but the front celerity is only 6% slowed down.

One viscous rheological behaviour, has been tested, selecting $k_1 = 10^{-8}$, $k_2 = 3$ and $k_3 = 0$ in equation (8). With such a parameterization, a mixture concentrated 200 $kg.m^{-3}$ would have a molecular viscosity ν of 0.08 $m^2.s^{-1}$. This rheological behaviour has been applied for an initial concentration of 600 $kg.m^{-3}$. Results are presented on figure 10. Due to more friction within the TC, the velocity of the body is reduced, which affects somewhat the front celerity. As a consequence of this velocity reduction, turbulent mixing is reduced at the water interface, as shown on figure 10.

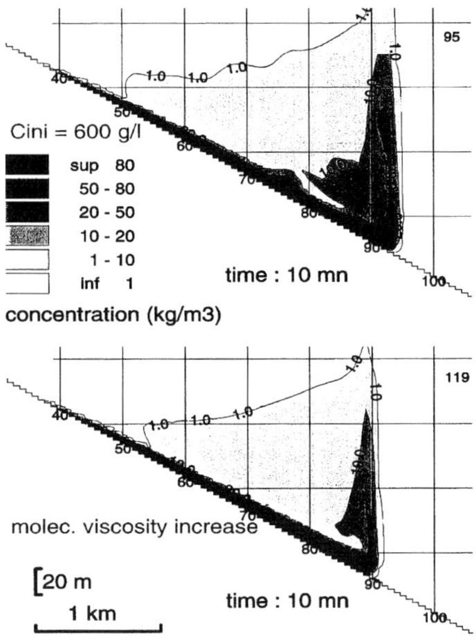

Figure 10 Effect of a molecular viscosity increase in concentrated layers on the turbidity current
Top: init. Conc. = 600 $kg.m^{-3}$, $\nu = 10^{-6}$ $m^2.s^{-1}$
Bottom: init.conc.= 600kg.m^{-3}, ν variable

4.8. Entrainment rate

Although entrainment is simulated as a mixing process in the continuous modelling, the entrainment rate can be computed by using equation (3), after estimating the integral scales C, U and h. A global Richardson number can also be evaluated, following for instance Graf and Altinakar (1995): $R_i = g'h\cos\alpha/U^2$.

Computations of C, U, h and R_i in the body of the TC turned out to be unstable according to time and/or space. The integral scales were also proved to vary depending on the method used to compute them. For instance the mean height h is different when deduced from hU and hU^2 or from hC and hC^2 ; also, the vertical integration in (2) has to be stopped at the level the velocity is zero (below the return flow). Lastly, it should be reminded that our simulations are typically unsteady, whereas the litterature on entrainment processes often considers steady flows. Nevertheless several couples (E_w, R_i) could be deduced from the different computations, and are plotted on figure 11, together with measurements from laboratory.

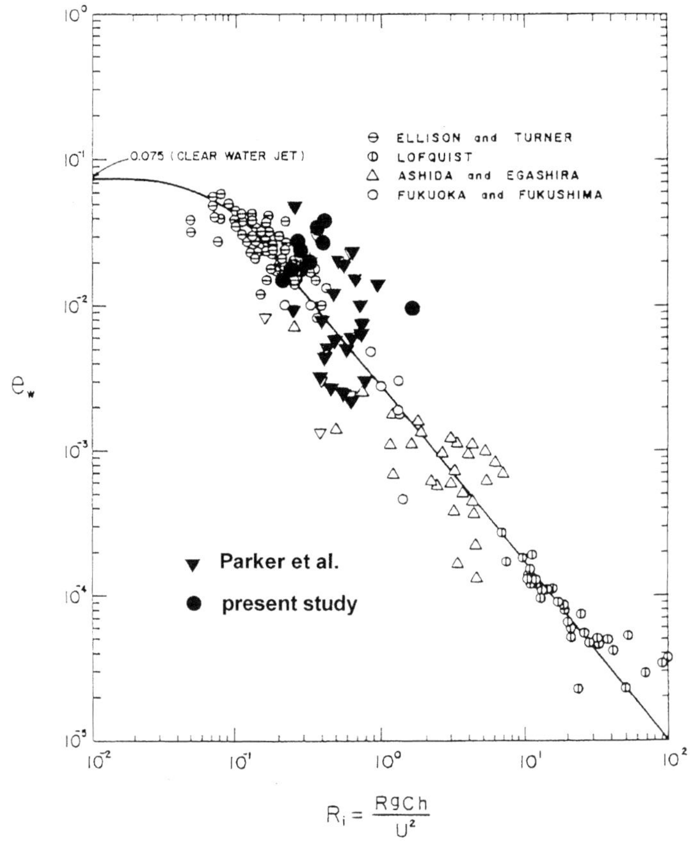

Figure 11 Water entrainment coefficient as a function of Richardson number (data from litterature in Parker et al., 1987). *Open symbols are related to density currents, black ones to turbidity currents (triangles: Parker et al. 1987; circles: present study).*

Simulations fit fairly well these observations, but the range of variations of R_i remains rather narrow. It is noticeable that the points of Parker et al. (which are the only measurements plotted on figure 11 that concern turbidity current, the rest being related to other density currents) exhibit the same features, i.e. a narrow range of global Richardson number and an entrainment rate in the order of 10^{-2}. The range of variation of the latter is rather large, in relation with the aforementioned ambiguities in the determination of the integral quantities.

Nevertheless, the relative agreement of the model results with the classical relationship between the entrainment rate and the Richardson number (represented on figure 11) suggests that vertical eddy diffusion is likely to properly reproduce the entrainment process, provided stratification-induced turbulence damping is accounted for.

4.9. Bed erosion, induced by turbidity currents

Sedimentologists have long known the erosive role of turbidity currents. Typically large velocities of turbidity currents (several tens of meters in the field) exert bottom stresses which are likely to erode the bed and enhance the sliding mass (*e.g.* Parker, 1982, Mulder et al., 1997).

Use of the bed module within the SAM-3D model allowed us to add this erosion process to the "reference" test, as illustrated in figure 12. The erosion law is representative of consolidated beds (Partheniades law):

$$E\ (kg.m^{-2}.s^{-1}) = E_0(\tau/\tau_e - 1),$$
with $E_0 = 0.1$ kg.m^{-2}.s^{-1} and $\tau_e = 1$N.m^{-2}.

The effect of erosion on the TC is noteworthy, as the concentration growth increases the density forcing and then accelerates the front celerity. Such a process is considered as a major one : for instance Mulder et al. (1997) estimate that during the Nice TC (next section), the volume of erosion was larger than the scale of the initial slide volume.

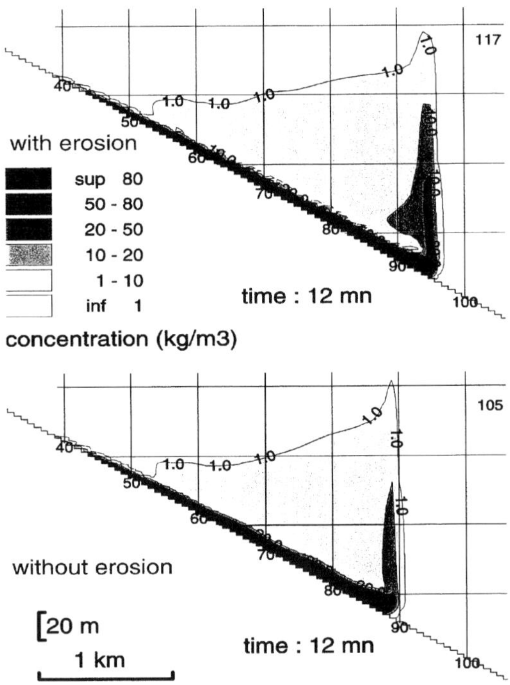

Figure 12 Effect of bed erosion on a turbidity current.
Top : simulation with erosion. *Bottom* : reference simulation (without erosion)

5. SIMULATION OF THE 1979 NICE TURBIDITY CURRENT

On 16 october 1979, a large slide occurred in coastal shallow water (15-50 m) during the landfilling operations for the Nice airport extension (south of France). The initial slide was estimated 8.10^6 m^3, but it caused progressive erosion and quickly transformed into a debris flow and turbidity current over more than 100 km. The estimation of the total turbidite volume is in the order 10^8-10^9 m^3. Several papers document this event, among them Gennesseaux et al. (1980) and Mulder et al. (1997).

As an application of the continuous modelling concept, the SAM-3D model has been used to attempt to simulate the Nice turbidity current. The modelled area extends 6 km x 3 km, from the coast down to 400 m depth. The horizontal grid size is 100 m, the vertical one is 1 m. Preliminary results are illustrated in figure 13, showing the channelling of sliding materials within the canyons. Quantitative results can be discussed from a cross-section along a canyon (figure 14).

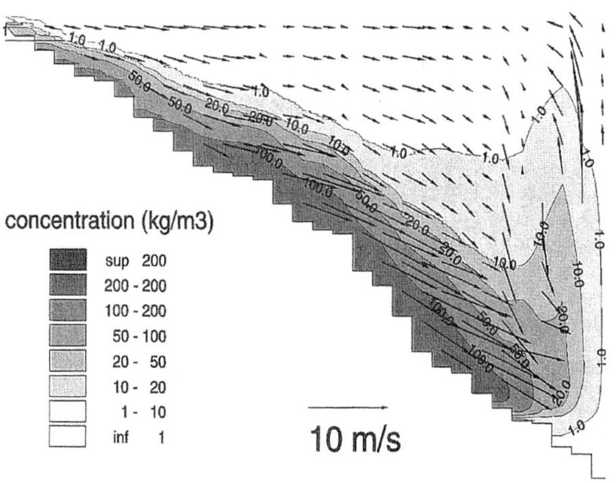

Figure 14 3D simulation of the 1979 Nice turbidity current.
Cross-section along a canyon, 3 min 28 sec after initiation.

The shape of the turbidity current is similar to those relative to the 2D case on a regular slope, but the local thickness of the TC increases when the local slope reduces. The representation of velocity vectors helps to interpret the shape of the head, where vertical recirculation is very strong (attention to the vertical distortion in the representation !). Far behind the head, the absence of return flow is probably due to a 3D circulation. Velocities are quite large - in the order of 10 $m.s^{-1}$ - but remain lower than estimated velocities in the field. This underestimation may be due to a bad initial concentration, to an underestimation of the erosion and/or to an excessive bottom roughness generated by the bottom discretization. Nevertheless, this first computation demonstrates the possibility to simulate large scale turbidity currents in a 3D geometry. A calibration procedure has to be carried out.

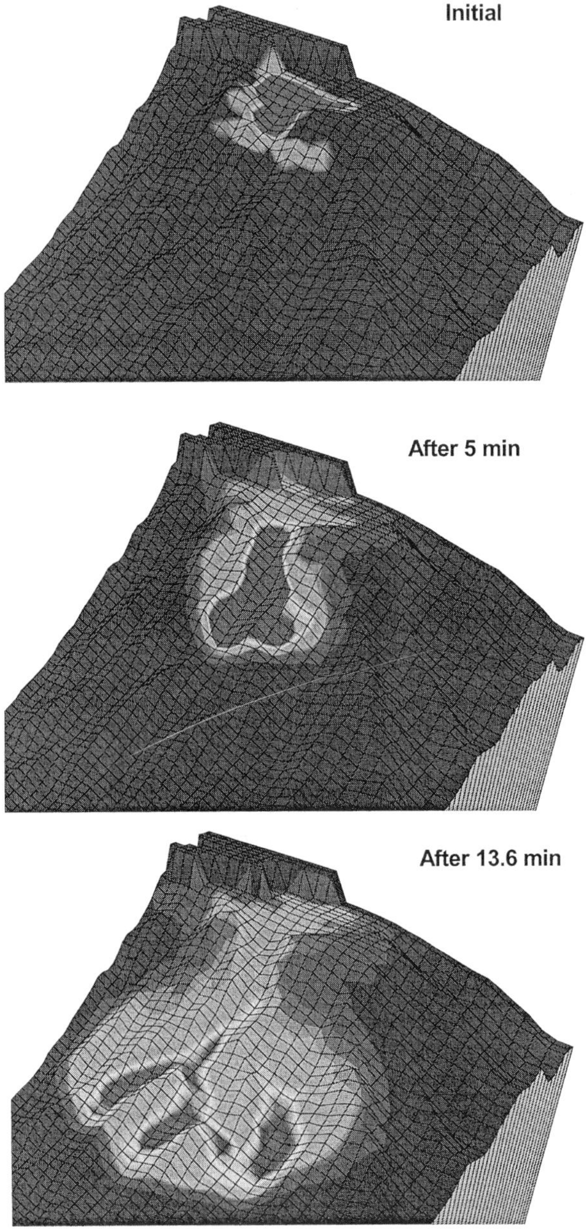

Figure 13 3D representation of the simulated 1979 Nice event.
On pictures, the covered area is approximately 3 km x 3 km.
Grey scale represents the vertically integrated sliding mass

6. CONCLUSION

The concept of continuous modelling has been used within a 3D model in order to simulate a turbidity current on a slope.

Many characteristic features of the turbidity current have been reproduced through computation, including the shape of the head resulting from vertical recirculation.

A sensitivity analysis has shown that the front celerity strongly depends on the initial movable mass, not so much on the bottom slope. On the other hand, within the body of the turbidity current, velocities increase with the bottom slope. Vertical profiles of concentration and velocity depend on the mean flow intensity, in relation with turbulent diffusion. Turbulent mixing can be dampened by stratification, so that the interface between the dense flow and the overlying water can be maintained. For the applied turbulence model and damping functions, turbulence damping appears to have little influence on the front celerity. An entrainment rate is deduced from the increase of the integral height of the turbidity current. When using a mixing length turbulence closure and empirical damping functions, the relationship between the entrainment rate and a global Richardson number fits the classical relationship deduced from flume measurements. However more sophisticated turbulence closures should be used to confirm this relationship.

In addition, the rheological behaviour of the sediment is likely to influence the profile of the turbidity current. Only viscous behaviours have been tested so far, but the model will be used for non-newtonian behaviour in the future. Such studies should help investigating the initiation of the movement.

The importance of the bed erosion induced by the strong velocities of a turbidity current has been pointed out. By increasing the sliding mass and the density gradient, such an erosion enhances the turbidity current and accelerates it (thus increasing the erosion proces).

The fully 3D simulation of the 1979 Nice slide has been attempted. Preliminary results are promising, but the vertical discretization of the model generates an artificial roughness hard to avoid. The difficulty lies in the need for a high vertical resolution in the bottom layers, for several km along a steep slope. Under these conditions, reduced (σ) coordinates do not seem to be more suitable, which may suggest a practical limitation of this method to model such processes.

Lastly, validations of the model are required, but difficult to achieve at the scale of the field. Therefore it will be necessary to run simulations at the scale of flume experiments for which measurements are available.

AKNOWLEDGEMENTS

The authors thank the reviewers for their fruitful comments and suggestions.

REFERENCES

Assier Rzadkiewicz, S., Mariotti, C. and Heinrich, 1997, Numerical simulation of submarine landslides and their hydraulic effects, *Journal of Waterway, Port, Coastal & Ocean Engineering,* Jul/Aug 1997, 149-157.

Bradford, S., Katopodes, N. and Parker, G., 1997, Characteristic analysis of turbid underflows, *Journal of Hydraulic Engineering,* (123) 5, 420-431.

Cugier, Ph. and Le Hir, P., 2000, Three dimensional modelling of suspended matters in the eastern "baie de Seine" (English Channel, France), *Comptes Rendus Académie des Sciences, Paris, Earth and Planetary Sciences,* (331), 287-294.

Ellison, T.H. and Turner, J.S., 1959, Turbulent entrainment in stratified flows, *Journal of Fluid Mechanics,* (6) 423-448.

Gennesseaux, M, Mauffret, A. and Pautot, G., 1980, Les glissements sous-marins de la pente continentale niçoise et la rupture de câbles en mer Ligure (Méditerranée occidentale), *Comptes Rendus Académie des Sciences, Paris,* (290), 959-962.

Graf, W. and Altinakar, M., 1995, Courants de turbidité, *La Houille Blanche,* (7), 28-37.

Le Hir, P., Bassoullet, Ph. and Jestin, H., 2001, Application of the continuous modeling concept to simulate of high-concentration suspended sediment in a macrotidal estuary, In *Coastal and Estuarine Fine Sediment Processes,* W.H. McAnally and A.J. Mehta (Eds), *proceedings in Marine Science (Elsevier),* No 3, 229-248.

Mulder, T. and Cochonat, P., 1996, Classification of offshore mass movements, *Journal of Sedimentary Research, (*66) 1, 43-57.

Mulder,, T., Savoye B. and Syvitski, 1997, Numerical modelling of a mid-sized gravity flow: the 1979 Nice turbidity current (dynamics, processes, sediment budget and seafloor impact), *Sedimentology* (44), 305-326.

Parker, G., 1982, Conditions for the ignition of catastrophically erosive turbidity currents, *Marine Geology,* (46), 307-327

Parker, G., Fukushima, Y. and Pantin, H., 1986, Self-accelerating turbidity currents, *Journal of Fluid Mechanics,* (171), 145-181.

Parker, G., Garcia, M., Fukushima, Y. and Yu, W., 1987, Experiments on turbidity currents over an erodible bed, *Journal of Hydraulic Research,* (25) 1, 123-147.

Stacey, M. and Bowen, A., 1988, The vertical structure of density and turbidity currents: theory and observations, *Journal of Geophysical Research,* (93) C4, 3528-3542.

Winterwerp, H., 1999, On the dynamics of high-concentrated mud suspensions, *PhD Thesis, Report 99-3, Delft Univ. Tech.,* 172 p.

Fine Sediment Dynamics in the Marine Environment
J.C. Winterwerp and C. Kranenburg (Editors)

The influence of fresh water distribution on SPM transport in the Dutch coastal zone

J. M. de Kok

National Institue for Coastal and Marine Management/RIKZ
P.O. Box 20907, NL-2500 EX The Hague

The rivers Rhine and Meuse - interconnected in the Dutch delta - flow into the North Sea through two outlets: the Rotterdam Water Way and the Haringvliet. The distribution of fresh water between the Rotterdam Water Way and the Haringvliet can be controlled by the management of sluices in the Haringvliet.

A change in the management of the Haringvliet sluices can cause a different distribution of river run off over the two outlets and consequently a different salinity pattern around the mouths of Haringvliet and Rotterdam Water Way. This results also in a different transport pattern of cohesive sediment in the area, as it is strongly influenced by the salinity distribution.

A case study with a high resolution 3D baroclinic tidal model combined with a cohesive sediment transport model was conducted to assess the intrusion of cohesive sediment in the Rotterdam Water Way. The model results indicate a reduction of 30 - 60 % of the deposition of cohesive sediment in harbours and shipping channels as a result of a different distribution of fresh water over the two outlets.

KEYWORDS : cohesive sediment, SPM transport, ROFI, numerical modelling

1. INTRODUCTION

The rivers Rhine and Meuse, with a total year average run off of about 2200 m^3/s, meet in the Dutch delta. Their waters flow into the North Sea through two outlets: the Rotterdam Water Way and the Haringvliet (Figure 1).

The Haringvliet is a former estuary, but nowadays it is divided into two parts, separated by sluices. At present these sluices are opened only to discharge fresh water into the western part, but no sea water can enter the eastern part. The western part is referred to as the "mouth of the Haringvliet". It is connected to the North Sea by a number of tidal gullies and has an estuarine character. The eastern part is referred to as "the Haringvliet ". It contains only fresh river water and tidal motion is virtually absent.

The most seaward part of the Rotterdam Water Way, together with the adjacent fairways is called the Maasmond (see Figure 2). The Maasmond is the entrance to a vast harbour area with intensive shipping traffic, including the largest bulk carriers. The Maasmond is a completely artificial water way with a width of 1000 m and a maintenance depth of 25 m below mean sea level. Deposition rates of sands and cohesive sediments, both coming from

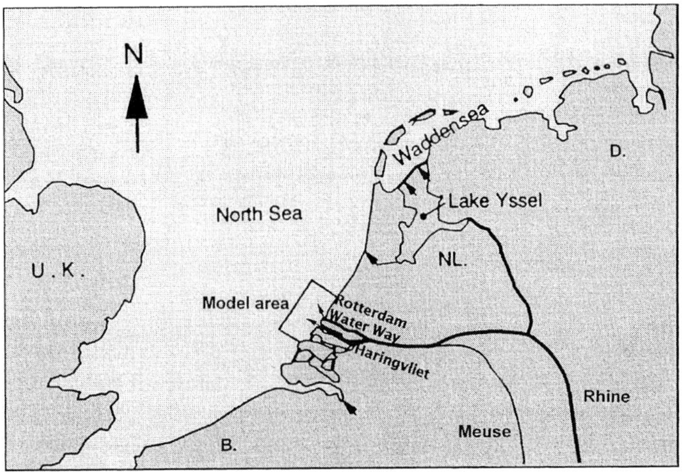

Figure 1 Schematic map of the Dutch Delta.

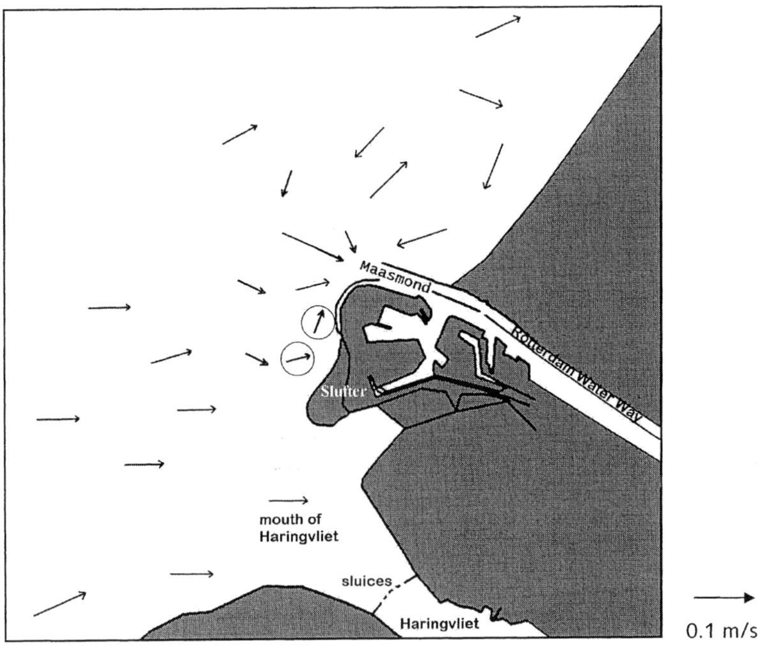

Figure 2 Schematic view of computed residual near bed currents near the Maasmond and the Mouth of the Haringvliet. North-west of the Slufter the circulation is directed towards the Maasmond. T0 scenario, year averaged river outflow and wind and tidal conditions.

the sea, are very high. Yearly an amount of more than 6 million tonnes dry weight of sediment has to be dredged away from fairways and harbour basins. Deposition of sand takes place mainly in the most western part of the Maasmond. The dredged amounts of sand and cohesive sediment are roughly equal to each other.

Cohesive sediments are intruding farther into the Maasmond than sands (Spanhoff and de Kok, 1991). Marine cohesive sediments can still be found in the nose of the salt wedge in the Rotterdam Water Way, some 25 km from the sea. It is estimated that near the turbidity maximum half of the deposited sediment is of marine origin, the other half being supplied by the river.

During the winter and spring seasons the bottom of the Maasmond is covered with a fluid mud layer with a thickness between 2 and 4 m. Its density varies from 1030 kg/m^3 at the top until 1300 kg/m^3 near the sand bed. The upper part of the layer is very mobile with flow velocities above 0.2 m/s.

The Maasmond is sheltered from incoming waves and alongshore tidal currents by breakwaters with a length of more than 4 km. Incoming tidal currents can reach velocities of more than 0.7 m/s near the lutocline, where internal friction is minimal as a result of the damping of turbulence (Winterwerp et al., 1998).

Outside the breakwaters tidal current velocities near the bottom are much higher. Together with wave activity they prevent the formation of a mud layer. However, around the turn of the tide benthic suspensions with concentrations of up to 10 g/l are often observed. Wave activity and hindered settling effects keep the sediment flocs in suspension. At mid depth suspended particulate matter (SPM) concentrations vary between 10 and 500 mg/l, depending on wave activity and current velocity.

However, simultaneously observed SPM concentrations showed large differences over distances of less than 2 km. Observed diameters of aggregates near the bed are between 100 and 200 10^{-6}m, depending on tidal phase. Observed primary grain diameters are between 6 and 10 10^{-6}m.

2. ESTUARINE CIRCULATION AND SPM FLUXES

The general residual circulation along the Dutch coast is driven by wind, tides and the effects of river outflow. The main flow is directed towards the North-East (de Kok, 1996, 1997). This current supplies most of the cohesive sediment in the Dutch coastal area.

The net alongshore SPM flux is estimated at at least 10^7 tonnes a year (van Alphen, 1990), but more recent estimates are several times higher (McManus and Prandle, 1997).

A substantial part of it is entering the Maasmond and the mouth of the Haringvliet as a result of the presence of horizontal and vertical salinity gradients and the associated density gradients in the coastal water (de Kok et al., 1999) . These salinity gradients are caused by the outflow of fresh water of the rivers Seine, Scheldt, Meuse and Rhine into the sea.

Along the entire continental coast of the North Sea cross shore density gradients exist, which are maximal in the outflow area of Rhine and Meuse (Figure 3). The density gradients cause baroclinic pressure gradients, which are increasing in downward direction, leading to an estuarine cross shore circulation with a shoreward near bed residual current (De Ruijter et al.,

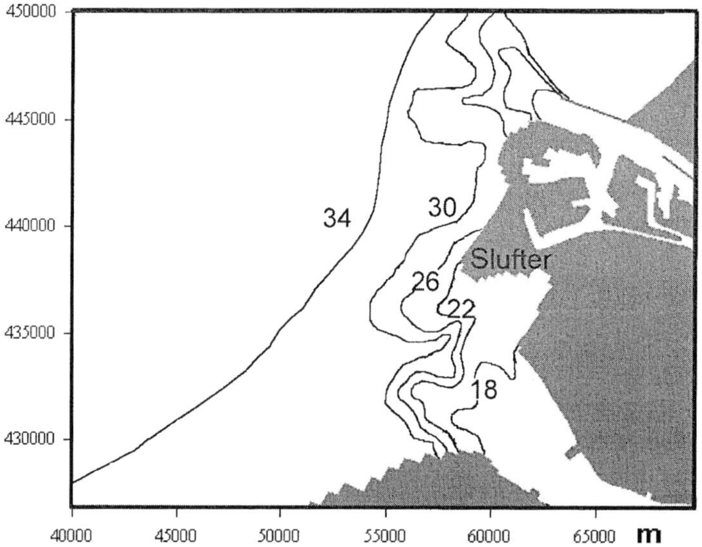

Figure 3 Computed salinity values in PSU during low water slack in the mouth of the Haringvliet and around the Maasmond. The gradients west of the Slufter are about 3 PSU/km. T0 scenario, year averaged river outflow and wind and tidal conditions.

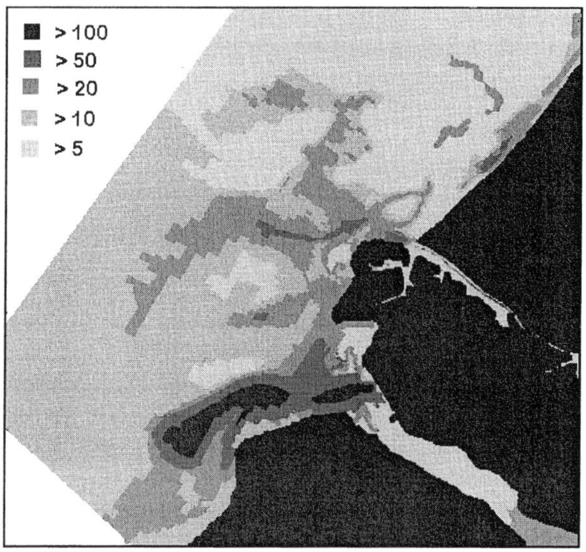

Figure 4 Vertically averaged SPM concentrations in mg/l during low water slack from the T0 computation.

1992). Because there is always more cohesive sediment in the lower half of the water column than in the upper half, this near bed current causes a shoreward net flux of SPM and accumulation of fine sediment in a narrow coastal strip of several km (Visser et al., 1991).

Along the Belgian coast the cross shore salinity gradients have on average values around 0.1 PSU/km, but near the Maasmond and the mouth of the Haringvliet the averages can reach values of more than 1 PSU/km. Instantaneous surface salinity gradients can be as high as 10 PSU/km in the river plume front. As a result the near bed residual currents are very strong in this area and directed towards the outflow points of fresh water.

Strong salinity gradients and estuarine near bed currents occur also in the mouth of the Haringvliet, in the Maasmond and in the Rotterdam Water Way. Residual velocities are as high as 0.4 m/s and are generally eastward directed.

In the direct outflow area of Rhine and Meuse a strong salinity stratification exists most of the time, causing the damping of turbulence and leading to very slow vertical mixing of SPM. This results in SPM concentrations in the lower water layers that are several times higher than in the upper layers. Therefore most of the SPM transport in this area takes place in the lower layers following the near bed residual current pattern.

This leads to a high import of marine cohesive sediment into the Maasmond. The average distance of the flocs to the bed is relatively small. Around the turn of the tide most flocs with fall velocities above $5.\ 10^{-4}$ m/s will merge with the fluid mud layer at the bed within an hour. From there practically no resuspension takes place.

3. RIVER WATER AND 3D DENSITY DISTRIBUTION

The sluices in the Haringvliet are opened (within a time frame around low water) only when the river discharge through the Rotterdam Water Way exceeds 1500 m^3/s. The remainder of the Rhine/Meuse run off is discharged then through the Haringvliet. The long term average of this discharge is less than 700 m^3/s, but the outflow is very pulsed, with high peaks during ebb tides in the winter and spring seasons. Peak values can be as high as 10,000 m^3/s.

River water that is running off through the Maasmond flows along the Dutch coast towards the North-East with an average speed of 10 km a day. River water that is running off through the Haringvliet remains more than a day in the mouth of the Haringvliet before it joins the Dutch coastal current north of the Maasmond.

This means that outflow of river water through the Haringvliet leads to a salinity structure that is different from that, caused by outflow of the same amount of fresh water through the Maasmond.

Outflow through the Haringvliet leads to less saline water in the mouth of the Haringvliet. Outflow through the Maasmond leads to stratification and less saline surface water north of the Maasmond.

A different salinity structure of the coastal water leads also to a different residual current structure and a different transport pattern of cohesive sediment. As we will see, numerical model results indicate, that a small shift of the SPM transport paths to the west can have a large impact on cohesive sediment intrusion into the Maasmond.

For ecological reasons the Dutch government is considering a management scenario for the Haringvliet sluices, in which several sluice gates are permanently opened, also at flood time. This will partly restore the estuarine character of the Haringvliet, bringing back salinity gradients and a modest tidal motion. It will also result in an increase of the year average discharge through the Haringvliet to 1100 m^3/s. As a consequence the year average discharge through the Maasmond will decrease from more than 1500 to 1100 m^3/s.

This will result in a different salinity structure of the coastal water around both river mouths, and in different SPM transport patterns.

To asses the impact of these changes on siltation of harbours and fairways in the Maasmond area a three-dimensional numerical model for hydrodynamics and SPM transport is used to study the supply of cohesive sediment to the Maasmond.

4. MODEL DESCRIPTION

The primitive equations for 3-D hydrostatic incompressible free surface flow with Boussinesq approximation for density gradients (see e.g. Csanady, 1984) can be integrated over non horizontal layers, using Leibniz rule. After minor simplifications this yields :

$$\frac{\partial u_k}{\partial t} + u_k\frac{\partial u_k}{\partial x} + v_k\frac{\partial u_k}{\partial y} + \frac{\omega_{tk}u'_{tk} - \omega_{dk}u'_{dk}}{h_k} - fv_k + \frac{1}{\rho_k}(\frac{\partial P}{\partial x})_k = \nu_h\nabla^2 u_k + \frac{\tau_{x,tk} - \tau_{x,dk}}{\rho_k h_k} \qquad (1),$$

$$\frac{\partial v_k}{\partial t} + v_k\frac{\partial v_k}{\partial y} + u_k\frac{\partial v_k}{\partial x} + \frac{\omega_{tk}v'_{tk} - \omega_{dk}v'_{dk}}{h_k} + fu_k + \frac{1}{\rho_k}(\frac{\partial P}{\partial y})_k = \nu_h\nabla^2 v_k + \frac{\tau_{y,tk} - \tau_{y,dk}}{\rho_k h_k} \qquad (2),$$

$$\frac{\partial h_k}{\partial t} + \omega_{tk} - \omega_{dk} + \frac{\partial h_k u_k}{\partial x} + \frac{\partial h_k v_k}{\partial y} = 0 \qquad (3),$$

The layer integrated transport equation for dissolved and suspended matter reads :

$$\frac{\partial h_k S_k}{\partial t} + \omega_{tk}S_{tk} - \omega_{dk}S_{dk} + \frac{\partial h_k u_k S_k}{\partial x} + \frac{\partial h_k v_k S_k}{\partial y} =$$

$$D_h(\frac{\partial h_k \partial S_k}{\partial x^2} + \frac{\partial h_k \partial S_k}{\partial y^2}) + D_z(\frac{\partial S}{\partial z})_{tk} - D_z(\frac{\partial S}{\partial z})_{dk} + So - Si \qquad (4).$$

$\boldsymbol{u}$ = (u,v,w), velocity vector,
P = pressure,
ν = turbulent viscosity coefficient tensor,
S = concentration of dissolved or suspended matter,
$\boldsymbol{D}$ = turbulence diffusion coefficient tensor,
ρ = water density,
g = acceleration of gravity,
h_k = depth of layer k, k=1,....,b from surface to bottom,
u_k, v_k, ρ_k, S_k : vertically averaged over h_k,

$u'_k = u - u_k$, $v'_k = v - v_k$,

$\omega_{tk} = w_{tk} - u_{tk}\frac{\partial z_k}{\partial t} - u_{tk}\frac{\partial z_k}{\partial x} - v_{tk}\frac{\partial z_k}{\partial y}$ vertical transport velocity through layer interface,

z_k = position of top of layer k,
z_s = z_1 = position of watersurface.
$\cdot_{tk}$: at top of layer k,
$\cdot_{dk}$: at bottom of layer k,
$\cdot_{dk}$ = $\cdot_{tk+1}$
f = Coriolis parameter,
∇^2 = horizontal Laplace operator,
ν_h, D_h = horizontal turbulence viscosity and diffusion coefficients (assumed uniform),
ν_z, D_z = vertical turbulence viscosity and diffusion coefficients,
τ_x $= \rho\nu_z\frac{\partial u}{\partial z}, \tau_y = \rho\nu_z\frac{\partial v}{\partial z}$ shear stress between layers,
$\tau_{x,y,t1}$ = $C_d\rho_a|\mathbf{wi}|wi_{x,y}$: windstress, at the surface,
C_d = wind coefficient,
ρ_a = air density,
$|\mathbf{wi}|$ = magnitude of wind speed,
$wi_{x,y}$ = wind velocity component in x or y direction,
$\tau_{x,db}$ = $\rho_b|\boldsymbol{u}_b|u_b k_b$, $\tau_{y,db} = \rho_b|\boldsymbol{u}_b|v_b k_b$, bed stress, in the bottom layer b,
k_b = bottom friction coefficient,
So = source term, erosion or resuspension
Si = sink term, deposition

The pressure gradient terms contain both barotropic and baroclinic contributions.
Equations (1) and (2) apply to horizontal momentum only.

Density and salinity are coupled via an equation of state with constant temperature.

Equation (4) is solved every time step for salinity and for turbulence kinetic energy (k) and energy dissipation (ε) after which density fields and vertical mixing coefficients are computed.

The numerical model uses orthogonal curvilinear horizontal coordinates on a C-type grid and a σ-layer approach in the vertical. The difference scheme is alternating implicit/explicit in the horizontal, allowing for relatively large time step sizes (Stelling and Leendertse, 1991). The horizontal gradient terms are taken within the σ-plane. This means that σ-planes have to be approximately horizontal.

The SPM transport model is based on the advection-diffusion equation (4), using a fall velocity that is in principle fixed in time and space. The fall velocity w_f is subtracted from ω_{tk} at the layer interfaces.

The mass M of deposited sediment is stored in a separate bed layer for each horizontal grid point. Deposition takes place if the velocity magnitude in the lowest water layer $|\boldsymbol{u}_b| < \boldsymbol{u}_{d,crit}$ The deposition rate is $Si_b = w_f S_b$ with S_b the SPM concentration in the lowest water layer.

Resuspension takes playes if $|u_b| > u_{e,crit}$, and M > 0. The resuspension rate So_b = E, with values between 10^{-3} and 10^{-6} kg/m^{2}s, independent of the flow velocity, but depending on local bed conditions.
$u_{d,crit}$ may be different from $u_{e,crit}$.
These formulations differ from those by Krone and Partheniades, because cohesive sediment layers are never observed on the sea bed in the studied area and settling of cohesive sediment for longer than several hours is only supposed to occur in the Maasmond and in a few former tidal channels in the mouth of the Haringvliet.
Vertical mixing of SPM is modeled via the vertical diffusion term in eqation 4. The diffusion coefficient is obtained from the k-ε model in the hydrodynamic part.
The coupling with the hydrodynamic model goes only one way. Fields of velocities, water levels and vertical diffusion coefficients are transferred to the SPM model. Density effects related to SPM concentration gradients are therefore not represented in the hydrodynamic model.
In the SPM model it is possible to define subareas where a domain dependent set of values of w_f, $u_{d,crit}$, $u_{e,crit}$ and E can be imposed. The definition of the subareas can also be made dependent on local and instantaneous water depths. In this way the model can account for a specific local bed composition, wave conditions or sediment properties.
The numerical scheme of the SPM model is positive, non dispersive and it is able to reproduce very high gradients and higher derivatives without numerical diffusion or strong spurious oscillations (de Kok, 1992). This is especially needed, since SPM concentrations can have differences of an order of magnitude over a few grid cells.

5. SIMULATION OF THE PRESENT SITUATION (T0)

With the model a simulation of the present situation with respect to the river discharges was run. As forcing for the hydrodynamic model year average conditions were chosen. Periodic water level elevations with an average tidal amplitude were imposed on the open sea boundaries (Figure 1). Observed river discharges of 1800 m^3/s in the Rotterdam Water Way and 400 m^3/s in the Haringvliet were imposed on the river boundaries. The Haringvliet sluices were closed as soon as the computed water level at the seaward side became higher than at the landward side, to be opened (partly) again in the reversed situation. A constant uniform long term averaged wind stress (5 m/s) was imposed on the surface. Also year averaged salinities were imposed on the open boundaries.
Critical current velocity values for deposition and resuspension were 0.15 m/s in the sea areas. In the Maasmond and in the harbour basins values of 0.3 m/s and higher were used, to account for the absence of wave activity. The latter parameters were used to calibrate the model to reproduce the observed siltation. The used global particle fall velocity was 5. 10^{-4} m/s.
Eight σ-layers were used, with high resolution near the bed. Horizontal mesh widths ranged from 100 m in the Maasmond until 1000 m at the western open boundary.
The hydrodynamic computation resulted in a realistic reproduction of observed phenomena such as tidal elevations, tidal current profiles and amplitudes, direction and magnitude of

Figure 5 Computed net deposition of cohesive sediment during 1 tidal period. T0-scenario.
Deposition in the mouth of the Haringvliet takes place in tidal channels and amounts to 4000 tonnes dry weight. This is higher than the computed net sedimentation in the Maasmond (3000 tonnes).
During periods with high wave activity a part of the sediment in the mouth of the Haringvliet will be resuspended, which is not the case in the Maasmond.

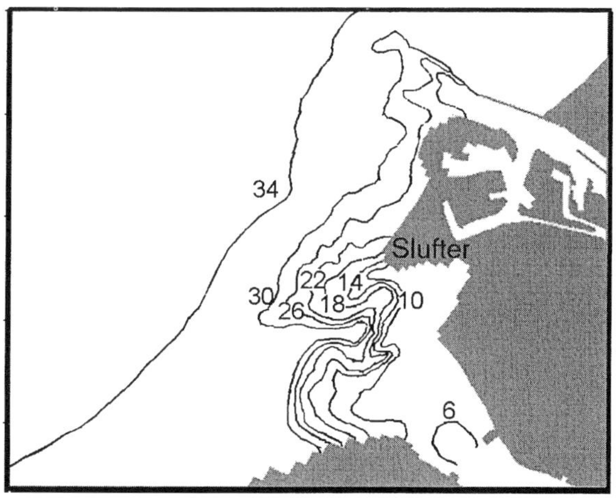

Figure 6 Computed salinity values in PSU during low water slack in the mouth of the Haringvliet and around the Maasmond. The gradients west of the Slufter are about 6 PSU/km. T1 scenario, year averaged total river outflow and wind and tidal conditions.

residual currents, horizontal and vertical salinity distributions, length of the salt wedge and extent of the stratified area.

The direction of the residual near bed currents is schematically shown in Figure 2. The current pattern suggests a transport path from the South-West directly leading towards the Maasmond. The salinity gradients west of the Maasvlakte are negative in the direction of the Maasmond, inducing a positive near bed current in the same direction (Figure 3).

The initial condition for SPM concentrations was a zero field, both for bed and water column. A period of 20 days was simulated with long term averaged observed SPM concentrations on the boundaries. After 16 days the computed SPM concentration fields showed a periodic behaviour.

The computed SPM concentration fields show high gradients both in the vertical and in the horizontal (Figure 4). The computed concentration pattern is very patchy, which is in agreement with field observations. Concentration values above 800 mg/l are computed in the lowest model layer in a residual gyre west of the mouth of the Haringvliet. No deposition occurs here, as a result of the high tidal velocties.

Permanent sedimentation occurs in the former tidal channels in the mouth of the Haringvliet and in the Maasmond (Figure 5), which is in accordance with field observations. After tuning of the local parameters the computed deposited amount in the Maasmond was equal to the observed sedimentation of cohesive sediment in the different sectors of the area.

Almost 75 % of the net import of SPM took place in the two lowest model layers (2500 tonnes dry weight/ tidal period). The main supply route of SPM is near the bottom, west of the Slufter following the encircled current vectors in Figure 2.

6. COMPUTATION WITH "OPEN SLUICE" SCENARIO (T1)

The T0-computation was repeated with different discharge conditions on the river boundaries. The average discharge on the Rotterdam Water Way boundary was 1100 m^3/s, equal to the average discharge on the Haringvliet boundary. Also one third of the sluice gates were opened now permanently. The remaining gates were operated in the same way as in the T0-situation.

This caused a considerable increase of the fresh water content of the mouth of the Haringvliet (Figure 6). West of the Slufter the computed salinity gradients are increasing to more than 6 PSU/km around low water slack. This increase causes a reversal of the residual current west and south-west of the Slufter (Figure 7). The SPM transport path is now directed towards the South-East, to the mouth of the Haringvliet, not to the Maasmond.

This causes a considerable decrease of the net import of SPM into the Maasmond. In the T1-scenario it dropped down to 1/3 of the import in the T0-situation. Scenario computations with stormy conditions and open sluices resulted in a decrease of 30 % of the computed net SPM import.

The increased SPM supply to the mouth of the Haringvliet in the T1-scenario does not lead to increased deposition there, because the tidal current velocities did increase as well, as a result of the occurrence of tidal motion in the Haringvliet. A tidal volume is now going again through the old tidal channels and critical velocities for deposition and resuspension are

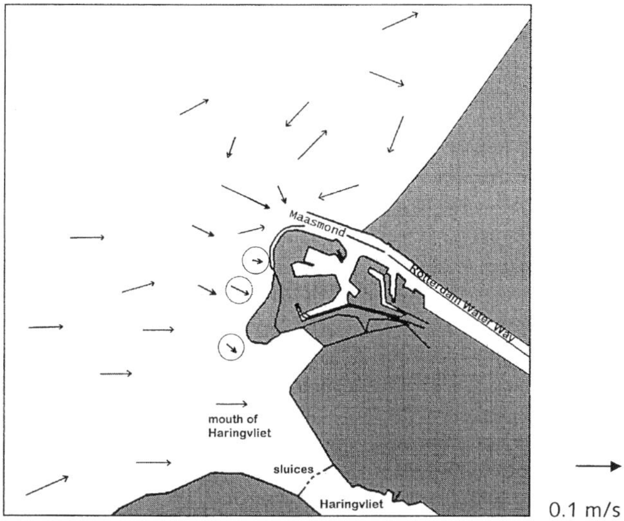

Figure 7 Schematic view of computed residual near bed currents near the Maasmond and the Mouth of the Haringvliet. West and north-west of the Slufter the circulation is now directed towards the mouth of the Haringvliet as a result of the strong salinity gradients in that area. T1 scenario, year averaged total river outflow and wind and tidal conditions.

Figure 8 Computed net deposition of cohesive sediment during 1 tidal period in the T1-scenario. Deposition in the Maasmond is now 1100 tonnes dry weight.

exceeded intermittently on most places in the mouth of the Haringvliet. Net settling occurs only at a few places now (Figure 8).

The mean SPM concentration values at some distance to the coast are higher. As a consequence the main north-eastward going transport of SPM now takes place at a greater distance to the coast and less SPM has the chance to enter the Maasmond.

7. CONCLUSIONS AND DISCUSSION

A three-dimensional numerical model for hydrodynamics and transport of suspended matter was used to assess the impact of a change of the management of the Haringvliet sluices. This modeled change of management led to an increase of the fresh water run off through the Haringvliet by 700 m^3/s and a decrease of the run off through the Maasmond by the same amount.

As a result the salinity gradients in the area between Haringvliet sluices and Maasmond decreased significantly and the residual near bed current near the Slufter turned towards the South-East. The numerical model results indicated a decrease by at least 30 % of the net cohesive sediment import from the sea into the Maasmond as a result of this change.

The model was calibrated to the known year averaged siltation rate. Vertical SPM concentration profiles were roughly tuned to averaged values, known from moorings and ship borne measurements.

Long term averaged surface SPM concentration values were used for boundary conditions. These were assumed to be vertically homogeneous.

The total river run off was constant and fixed at the long term average.

The wind forcing was constant and uniform. However, separate runs were done for different wind speeds and directions. The computed cohesive sediment import into the Maasmond during a SW-storm was comparable to the observed siltation during such periods.

Only the most elementary physical processes were modeled, using the most simple concepts. Wave effects, aggregation, hindered settling, strength development, fluid mud, bed forms and sediment properties were all parameterised in the most simple way by lack of information on sediment behaviour.

This makes the prognostic capabilities of the model questionable.

However, more uncertainty results from the fact that the sediment import in the Maasmond depends strongly on wind, wave conditions and on river run off. The present model does not properly account for the time variability of these conditions.

This model study shows already, that a change in discharge distribution over the two outlets can have large effects on sediment import. Ongoing studies show, that the value of the total river run off has a comparable influence.

Significant wave heights (especially above 2 m, Roskam, 1995) have a very large effect on near bed suspended sediment concentrations (Jago and Jones, 1998) and on resuspension of cohesive sediments, deposited during quiet periods. Not enough information is available, however, to validate the model results with respect to these points.

As the computed siltation in the harbour area strongly depends on these processes, improvement of this specific model can only be obtained by frequent survey of the deposition and resuspension areas in the mouth of the Haringvliet and by including realistic time variation in the modelling of wave effects and river run off.

ACKNOWLEDGEMENTS

The hydrodynamic computations were performed by Delft Hydraulics, using their Delft3D-code.
The SPM transport computations were performed by Aqua Vision, using the SLIB3D-code of RIKZ.
The computations were done in the framework of a broader study to the impact of possible land reclamation projects near the Maasmond, commisioned by the Dutch Ministry of Transport and Public Works and the Port of Rotterdam.

REFERENCES

Csanady, G.T., 1984, *Circulation in the Coastal Ocean*. Reidel, Dordrecht.
de Kok, J.M., 1992, A 3-D Finite Difference Model for The Computation of Near- and Far-field Transport of Suspended Matter near a River Mouth. *Continental Shelf Research*. 12(5/6) 625-642.
de Kok, J.M., 1996, A two-layer model of the Rhine plume. *Journal of Marine Systems*, 8, 269-284.
de Kok, J.M., 1997, Baroclinic eddy formation in a Rhine plume model. *Journal of Marine Systems*, 12, 35-53.
de Kok, J.M., van der Meulen, A, Wang, Z.B. and Schroevers, M., 1999, Effects of possible land reclamation projects on siltation in the Rotterdam harbour area. A model study. In : *Coastal Engineering and Marina Developments,* 131-141, eds. C. Brebbia and P. Anagnostopoulis, WIT Press.
de Ruijter, W.P.M., van der Giessen, A. and Groenendijk, F.C., 1992, Current and density structure in the Netherlands coastal zone. In: *Dynamics and Exchanges in Estuaries and the Coastal Zone* (Coast.Est. Sci.,40), 255-276, ed. D. Prandle, AGU, Washington, DC.
Jago, C.F. and Jones, S.E., 1998, Observation and modelling of the dynamics of benthic fluff resuspended from a sandy bed in the Southern North Sea. *Continental Shelf Research,* 18, 1255-1282.
Macmanus, J.P. and Prandle, D., 1997, Development of a model to reproduce observed suspended sediment distributions in the Southern North sea using Principal Component Analysis and Muliple Linear Regression, *Continental Shelf Research,* 17, 761-778.
Roskam, B., 1995, Wave climate at EUR- and LEG- observation platform for Maasvlakte studies, *RIKZ report RIKZ/OS-95.111x*, The Hague.
Spanhoff, R. and de Kok, J.M., 1991, 3-D model and field studies of silt transport in the Dutch coastal zone of the North Sea with emphasis on dump sites. *Wat.Sci.Tech.*24(10), 39-43.
Stelling, G.S. and Leendertse, J.J., 1991, Approximation of convective processes by cyclic AOI methods. *Proc. 2nd Int. Conf. On Estuarine and Coastal Modelling,* ASCE, New York.

van Alphen, J., 1990, A mud balance for Belgian-Dutch coastal waters between 1969 and 1986. *Netherlands Journal of Sea Research,*.25(1/2),19-30.
Visser, M., de Ruijter, W.P.M. and Postma, L, 1991, The distribution of suspended matter in the Dutch coastal zone. *Neth, J. Sea. Res*., 27, 127-143.
Winterwerp, J.C. , Uittenbogaard, R.E., de Kok, J.M., 2001, Rapid siltation from saturated mud suspensions, Proceedings of the 5th International Conference on Nearshore and Estuarine Cohesive Sediment Transport, INTERCOH'98, *Proceedings in Marine Science No 3, Coastal and Estuarine Fine Sediment Processes*, ed. W.H. McAnally and A.J. Mehta, Elsevier, Amsterdam, 125-146.

Fine Sediment Dynamics in the Marine Environment
J.C. Winterwerp and C. Kranenburg (Editors)

A process-based sand-mud model

M. van Ledden[a]

[a]Faculty of Civil Engineering and Geosciences, Delft University of Technology, Stevinweg 1, P.O. BOX 5048, NL-2600 GA DELFT, The Netherlands.

Predicting the bed composition in estuaries and tidal lagoons is of great practical importance. At present however, horizontal and vertical bed composition variations are often neglected in sediment transport and morphological models. In this paper a process-based sand-mud model is proposed, the model behaviour is analysed and model results are compared to field measurements. In general, it can be concluded that with such a process-based model, governing time scales and dimensionless parameters can be derived which can significantly increase the physical understanding. Furthermore, an expression is derived for the equilibrium mud content at the bed surface when both deposition and erosion occur during the tidal period. In this expression, the settling velocity for mud, the mud concentration and the erosion rate form an important dimensionless parameter. For low parameter values (< 10), sharp transitions are to be expected between areas with a very low mud content and areas in which the mud content can vary between 0 and 100%. The existence of a sharp transition is confirmed by field data from the Westerschelde estuary (The Netherlands). Finally, model results suggest that a local hydrodynamic parameter is not very useful for predicting the mud content in areas exposed to relatively low bed shear stress. Apart from the local hydrodynamics, the local mud concentration, settling velocity, mixing properties of the bed and the sample depth determine the local mud content.

KEY WORDS
sediment distribution; sand; mud; bed composition; numerical analysis; modelling

1. INTRODUCTION

Predicting the distribution of non-cohesive and cohesive sediments in estuaries and tidal lagoons, which leads to zonation of sand, mud and mixed deposits, is of great practical importance. It determines which areas collect nutrients and pollutants, as these tend to adhere to, or be part of, cohesive sediments and which areas form the right habitat for flora and fauna. Based on the rule-of-thumb "the calmer the water, the finer the bed sediment", a local hydrodynamic parameter is often used for predicting the mud content at a certain location (WL | Delft Hydraulics, 1998). However, a useful relationship between the mud content and a local hydrodynamic parameter is not known at present. Probably, a more process-based approach is a successful way for increasing the understanding and predictability of the sand-mud distribution.

In the present process-based sand models (Van Rijn, 1993) and mud models (Teisson, 1997) one representative grain size is used for calculating sediment transport and morphological changes, and the bed composition is often not taken into account. The goal of this paper is an attempt to bridge the gap between sand and mud models by proposing a process-based sand-mud model. First, knowledge of the incorporated processes is discussed (section 2). Thereafter, the process equations are presented (section 3) and the model behaviour is analysed (section 4). Finally, model results are compared to field measurements and the usefulness of a local hydrodynamic parameter for predicting mud content is discussed (section 5).

2. PROCESS KNOWLEDGE

2.1. Incorporated processes

In estuaries and tidal lagoons, bed sediments are continuously picked up, transported and deposited. Due to different sediment properties, the flux and direction of transport of non-cohesive and cohesive sediments are often not equal (Aubrey, 1986; Dronkers, 1984). The result is a strongly varying mud content in the bed in both horizontal and vertical directions. A large number of sediment processes are involved in the distribution of sediments. Erosion and deposition are the most obvious processes, but consolidation, flocculation and physical and biological mixing within the bed may also affect the bed composition. In an ideal process-based model, all these processes should be taken into account for predicting the bed composition in time and space. However, the present knowledge of these processes for mixtures of non-cohesive and cohesive sediments is limited and constructing a process-based sand-mud model with all these processes is far beyond the present state-of-the-art. Simplifications are therefore inevitable and consequently, the validity of the presented sand-mud model herein is limited. Moreover, the limited availability of data is another problem for verification of the model results.

In our modelling approach, the tide is assumed to be the dominant forcing for water motion. Short waves and density currents are neglected. Therefore, a depth-averaged modelling approach is assumed to be sufficient. Furthermore, the mud concentration is assumed to be low (< a few hundred mg/l). Flocculation and consolidation are not taken into account. For the deposition, erosion and mixing processes, process equations are needed to describe the behaviour of non-cohesive and cohesive sediments. Recently, laboratory experiments with mixtures of sand and mud have been reported in the literature. The observed behaviour of the different processes for sand-mud mixtures is discussed separately in the following subparagraphs in order to support the process equations used in the model. These equations are presented in section 3.

2.2. Deposition

Deposition experiments showed that in the case of sediment concentrations below the gel point, non-cohesive sand particles and cohesive mud flocs behave more or less independently during deposition (Torfs et al., 1996). Around the gel point, however, mud flocs form a network in the water column and sand particles are trapped. Typical gel point concentrations are in the range of 30 - 180 g/l (Winterwerp, 1999). It seems reasonable that for lower concentrations the separate deposition formulae for mud and sand are valid and can be applied within our model.

Table 1
Critical mud content and clay content for the experiments

Experiment	Clay type	Critical mud content	Critical clay content
Torfs (1995)	Kaolinite	4%	3%
Torfs (1995)	Montmorillonite	13%	4%
Panagiotopoulos et al. (1997)	Combwich	30%	10.8%

2.3. Erosion

Recently, erosion experiments were carried out with mixtures of non-cohesive and cohesive sediments (Torfs, 1995; Panagiotopoulos et al., 1997). By adding mud to a sand bed, Torfs (1995) observed that the non-cohesive behaviour of the bed was increasingly suppressed. Above a critical mud content (% < 0.063 mm), the bed behaved cohesively. However, the critical mud content was not the same for different sand-mud mixtures. Panagiotopoulos et al. (1997) observed a critical clay content (% < 0.002 mm) of 10.8%. The experimental results are summarised in Table 1. The critical clay content for the experiments of Torfs (1995) was calculated by a re-analysis of these experiments.

Despite the great difference in the critical mud content, the critical clay content is about the same for the experiments of Torfs (1995). Previously, Dyer (1986) and Raudkivi (1990) stressed the role of clay particles in the erosional behaviour of mixed beds. Dyer (1986) gave a transition range for non-cohesive to cohesive behaviour from 5 - 10% clay content by dry weight. This range agrees with the experimental results.

Torfs (1995) also measured the critical shear stress for erosion of sand-mud mixtures with a varying mud content. The critical shear stress is plotted in Figure 1, both as a function of the mud content and the clay content. The clay content was calculated by a re-analysis of the experimental data. The critical shear stress differs significantly for the two sand-mud mixtures at the same mud content, but is about the same at the same clay content. The observations of the transition between non-cohesive and cohesive behaviour as well as the critical erosion shear stress suggest that the erosion process is dominated by the clay content and not the mud content itself.

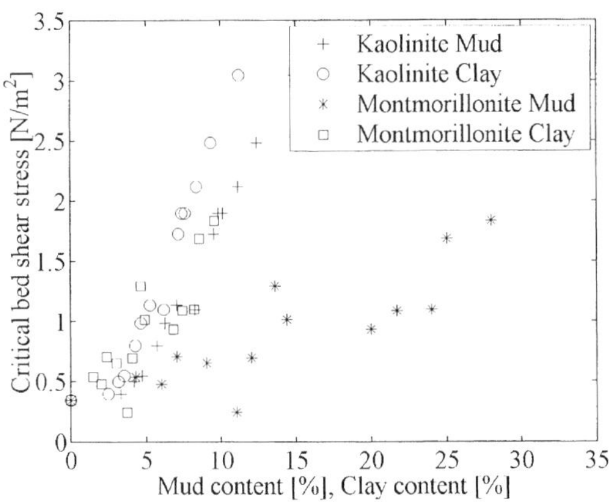

Figure 1. Measured critical bed shear stress vs. mud and clay content.

From a modelling point of view, the traditional erosion formulae for sand and mud alone must be adapted because of this behaviour. Depending on the clay content in the bed, two regimes must be distinguished: a non-cohesive and a cohesive regime. The transition between both regimes is given by a clay content of 5 - 10%. A critical mud content can also be used, because bed composition measurements suggest a strong correlation between the clay content and silt content (% 0.002 - 0.063 mm) within the bed (Van Ledden, 2000). However, it must be kept in mind that a critical mud content seems to be a site specific value, depending on the clay/silt ratio.

2.4 Mixing

Mixing within the bed can have a physical as well as a biological origin. Physical mixing in a sand bed occurs by small-scale bed features and it is generally assumed that the intensity of mixing increases with increasing velocity and decreases further into the bed (Armanini, 1995). Mixing experiments for sediment beds of non-cohesive and cohesive sediments have not been undertaken to the author's knowledge. However, it can be argued that physical mixing must also be dependent on the mud content within the bed. Erosion experiments showed that small-scale bed features are suppressed when the mud content increases (Torfs, 1995). Therefore, physical mixing probably decreases with increasing mud content.

Two types of biological mixing are often distinguished (Boudreau, 1997): local mixing and non-local mixing. Local mixing is caused by organisms which move through the bed and therefore mix the sediment particles. Non-local mixing is caused by organisms that transport a specific sediment fraction from one level to another in the bed. An example is the *Heteromastus filiformis*, which transports only fines from ten centimetres depth to the bed-water interface (Herman, 2000).

Physical mixing and local, biological mixing can be modelled as a diffusion process. For both mixing mechanisms, appropriate mixing coefficients must be defined. Non-local mixing by organisms causes sources and sinks of specific sediment fractions at certain depth within the bed. In our modelling approach, non-local mixing is neglected.

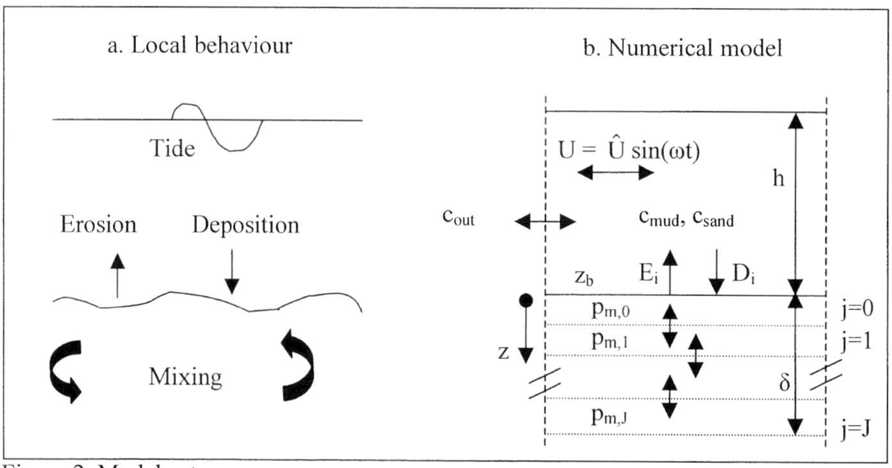

Figure 2. Model set-up.

3. MODEL EQUATIONS

3.1. Local behaviour

In this paper, only the so-called 'local behaviour' is discussed. This situation is sketched in Figure 2a. It is assumed that the behaviour of the sand and mud concentration in the water column, the bed level and the bed composition are dominated by deposition, erosion and mixing within the bed only. Non-local sediment transport processes (advection, diffusion) are assumed to be of minor importance and are neglected in the local model analysis.

Because of this local approach, bed load transport is not taken into account. For sand in suspension, the local approach seems to be valid, because its settling time is generally much smaller than the tidal period. For mud however, the settling time is often of the order of the tidal period and a local approach for mud is not realistic. Therefore, horizontal advection and diffusion are also taken into account for mud in a parameterised way (see section 3.3).

A process-based model basically consists of three parts: a water motion module, a sediment transport module and bed module. For each module, the process equations are discussed separately in the following subsections. Because of the local approach, the derivatives in the horizontal directions are neglected within the governing equations.

3.2. Water motion

The water depth h is taken constant and the varying depth-averaged flow velocity U during the tide is given as a function of time (Figure 2b):

$$U = \hat{U}\, sin(\omega t) \tag{1}$$

where $\hat{U}$ is the amplitude of the velocity, ω the angular frequency of the tidal period and t time. The bed shear stress τ_b is given by:

$$\tau_b = \rho \frac{g}{C^2} U^2 \tag{2}$$

where ρ is the water density, g the gravitational acceleration and C the Chézy-coefficient. The Chézy-coefficient C strongly depends on the bed features and composition. For sandy environments, typical values range from 40 to 60 $m^{1/2}/s$. For muddy environments, the Chézy-coefficient is often much larger, ranging from 60 to 100 $m^{1/2}/s$ (Van Rijn, 1993). Although the roughness strongly depends on the bed composition, the Chézy-coefficient is assumed to be constant at present (C = 60 $m^{1/2}/s$).

3.3. Sediment transport

For sediment transport, two sediment fractions are taken into account: a sand fraction (subscript s) and a mud fraction (subscript m). For the description of the sand and mud transport in suspension, the depth-averaged advection-diffusion equation is used. For a local approach, the equation reduces to:

$$\frac{\partial hc_i}{\partial t} = E_i - D_i \tag{3}$$

where c_i is the depth-averaged volumetric concentration of fraction i, D_i the deposition rate of fraction i and E_i the erosion rate of fraction i. The terms E_i and D_i on the right hand side give the net exchange of sediment fraction i between the bed and the water column: the downward deposition flux D_i and the upward erosion flux E_i (Figure 2b). As was discussed in section 3.1, the above equation can be applied for the sand fraction, but is not realistic for the mud fraction. In a next section, it will be shown that the local model behaviour shows trivial or non-realistic solutions by applying (3) for the mud fraction. Therefore, an extra transport term is introduced to account for horizontal advection and diffusion. The mud concentration equation becomes:

$$\frac{\partial hc_m}{\partial t} = E_m - D_m + k_m(c_{out} - c_m) \tag{4}$$

where c_{out} is the mud concentration outside and k_m is a transport coefficient. The last term on the right hand side in (4) is a transport term by which the mud concentration c_m is affected by the concentration outside the model c_{out} by a transport coefficient k_m (Figure 2b). When the inner concentration is lower than the outside concentration, it is assumed that by advective and diffusive transport mud is imported from outside and vice versa.

Exchange of sediment between the bed and the water column occurs by deposition to and erosion from the so-called 'exchange layer'. The exchange layer is the layer at the bed surface with index j = 0 (Figure 2b). As was concluded from the laboratory experiments, the erosion and deposition flux for sand and mud must be prescribed for two regimes: a non-cohesive and a cohesive regime (see section 2.3). A critical mud content within the bed can be used for the transition between both regimes. In this model, the critical mud content for the transition between both regimes is denoted as $p_{m,crit}$ and the mud content in the exchange layer, denoted as $p_{m,0}$, determines the erosion regime (Figure 2b). The erosion and deposition terms for both regimes are discussed successively below.

- **Non-cohesive regime ($p_{m,0} < p_{m,crit}$)**

In the non-cohesive regime sand and mud are assumed to behave independently during erosion and deposition. The classical sand transport formulae for the sand fraction are more or less valid in this regime (Torfs, 1995). The exchange of sand between the bed and the water column is given by:

$$E_s - D_s = \gamma w_s(c_{e,s} - c_s) \tag{5}$$

where γ is a form coefficient, w_s the settling velocity of the sand fraction and $c_{e,s}$ the depth-averaged equilibrium concentration of the sand fraction. The form coefficient γ takes into account the vertical distribution of the velocity and concentration profile (Galappatti, 1983). The depth-averaged equilibrium concentration ($c_{e,s}$) is given by the Engelund-Hansen transport formula:

$$c_{e,s} = \frac{0.05}{\sqrt{g\Delta^2 C^3 d_s}} \frac{U^4}{h} \tag{6}$$

where Δ is the specific gravity and d_s the sand diameter. In contrast to other sediment transport formulae, no threshold for the initiation of motion is included in the equilibrium concentration of Engelund-Hansen. A sediment transport formula with a threshold for sand can also be applied for the depth-averaged equilibrium concentration, e.g. Van Rijn's transport formula. However, it is not expected that including a threshold will significantly change the observed behaviour, discussed in the next paragraphs. Therefore, the Engelund-Hansen transport formula is considered to be sufficient as a first step. Further research is recommended in order to study the erosion process of sand-mud mixtures in detail and propose more realistic erosion formulations.

For the erosion process of mud, the erosion formula of Partheniades is assumed to be valid. Because the mud content can vary, the mud content in the exchange layer is added to the traditional formula. During erosion experiments in the non-cohesive regime, clay and silt particles are easily washed out from the top layer (Murray, 1977; Torfs, 1995). Thus, the critical erosion shear stress in this regime ($\tau_{e,nc}$) can be very low. For mud deposition the well-known deposition formula of Krone is applied. The exchange of mud is given by:

$$E_m - D_m = p_{m,0} M \left[\frac{\tau_b}{\tau_{e,nc}} - 1\right] H\left[\frac{\tau_b}{\tau_{e,nc}} - 1\right] - w_m c_m \left[1 - \frac{\tau_b}{\tau_d}\right] H\left[1 - \frac{\tau_b}{\tau_d}\right] \tag{7}$$

where $p_{m,0}$ is the mud content in the exchange layer, M the erosion rate, $\tau_{e,nc}$ the critical erosion shear stress for non-cohesive regime, H the Heavyside function, w_m the settling velocity of mud and τ_d the critical deposition shear stress.

- **Cohesive regime** ($p_{m,0} > p_{m,crit}$)

In the cohesive regime, the erosional behaviour of sand and mud is not independent. It is assumed that sand and mud particles are eroded in the same way, but once in the water column behave independently again. The Partheniades formula seems the most obvious formula for describing the erosion of mud and sand particles. This formula is adapted by introducing a parameter for the sand and mud content in the exchange layer. The deposition formulations for sand and mud remain the same as for the non-cohesive regime. The exchange of sand and mud in the cohesive regime are respectively given by:

$$E_s - D_s = p_{s,0} M \left[\frac{\tau_b}{\tau_{e,c}} - 1\right] H\left[\frac{\tau_b}{\tau_{e,c}} - 1\right] - \gamma w_s c_s \tag{8}$$

$$E_m - D_m = p_{m,0} M \left[\frac{\tau_b}{\tau_{e,c}} - 1\right] H\left[\frac{\tau_b}{\tau_{e,c}} - 1\right] - w_m c_m \left[1 - \frac{\tau_b}{\tau_d}\right] H\left[1 - \frac{\tau_b}{\tau_d}\right] \tag{9}$$

where $p_{s,0}$ is the sand content in the exchange layer and $\tau_{e,c}$ is the critical erosion shear stress for the cohesive regime.

3.4. Bed level and composition

The bed level change is determined by the flux of sediment to and from the bed from both sand and mud. The bed density is assumed to be constant in time and space. The bed level change is given by:

$$\frac{\partial z_b}{\partial t} = \frac{\rho_s}{\rho_b}\left[D_m - E_m + D_s - E_s\right] \tag{10}$$

where z_b is the bed level, ρ_s the sediment density and ρ_b the bed density.

Armanini (1995) proposed a continuous formulation for modelling the bed composition. The local change in sand content in a Langrangian co-ordinate system can be given by an advection-diffusion equation:

$$\frac{\partial p_s}{\partial t} + \frac{\partial}{\partial z}\left[p_s\frac{\partial z_b}{\partial t} - \varepsilon_z\frac{\partial p_s}{\partial z}\right] = 0 \tag{11}$$

where p_s is the sand content, z the distance from the bed surface (positive downwards) and ε_z a mixing coefficient within the bed. It is important to note that in (11) the origin of the vertical coordinate z is at the bed surface and z is positive in downward direction (Figure 2b). Conversely, the bed level rate $\partial z_b/\partial t$ is positive if the bed level rises.

The first term in (11) represents the local change in sand content. The net flux of sand by bed level variations is given by the second term. The third term is the net flux by small-scale variations in the bed level, e.g. ripples. Armanini (1995) suggested that the mixing coefficient ε_z in (11) decreases with increasing distance below the bed surface, because the influence of the small-scale variations decreases. The mixing coefficient was assumed to be an exponentially decreasing function:

$$\varepsilon_z = \varepsilon_{z,0}e^{-\alpha_1\frac{z}{d_s}} \tag{12}$$

where $\varepsilon_{z,0}$ is the mixing coefficient at the bed surface, d_s the sand grain size and α_1 a coefficient. The mixing coefficient $\varepsilon_{z,0}$ was assumed to be proportional to the friction velocity u_* and the sand grain size d_s:

$$\varepsilon_{z,0} = \alpha_0 u_* d_s \tag{13}$$

where α_0 is a coefficient.

The advection-diffusion equation (11) needs two boundary conditions. At the bed surface ($z = 0$) the sediment flux is prescribed. This boundary condition reads:

$$\left[p_s \frac{\partial z_b}{\partial t} - \varepsilon_z \left(\frac{\partial p_s}{\partial z} \right) \right]_{z=0} = \frac{\rho_s}{\rho_b} \left[D_s - E_s \right] \tag{14}$$

Deep below the bed surface the diffusive flux is assumed to be zero. The boundary condition is given by:

$$\left[\frac{\partial p_s}{\partial t} + \frac{\partial z_b}{\partial t} \frac{\partial p_s}{\partial z} \right]_{z=\infty} = 0 \tag{15}$$

For the mud content within the bed, the same equation can be solved. In our model however, only two sediment fractions are applied. Therefore, the mud content can be easily expressed in terms of the sand content:

$$p_m = 1 - p_s \tag{16}$$

4. MODEL ANALYSIS

4.1. Scale parameters

The model described in section 3 consists of four dependent variables: two sediment concentrations (c_s) and (c_m) in the water column, the bed level (z_b) and the mud content within the bed (p_m). For analysing the behaviour of this model, it is important to derive the governing external and internal scale parameters, which are elaborated below.

The external forcing is characterised by two parameters: the amplitude of the bed shear stress gives the level of the forcing ($\hat{\tau}_b$) and the time variation of the forcing is characterised by the tidal period (T_{tide}). Within the model, three critical shear stress parameters must be defined: τ_d for mud deposition, $\tau_{e,nc}$ for non-cohesive mud erosion and $\tau_{e,c}$ for erosion of sand and mud in the cohesive regime. These parameters can be made dimensionless by using the maximum bed shear stress. In general, the critical shear stresses are divided by the level of forcing. For this purpose, the reciprocal term is used, because the level of forcing can be zero:

$$\tilde{\tau}_d = \frac{\hat{\tau}_b}{\tau_d} \qquad \tilde{\tau}_{e,nc} = \frac{\hat{\tau}_b}{\tau_{e,nc}} \qquad \tilde{\tau}_{e,c} = \frac{\hat{\tau}_b}{\tau_{e,c}}$$

By using these scale parameters, three situations can be distinguished. These situations are shown in Figure 3. The bed shear stress is plotted during half a tidal period for the three situations. Also the three critical shear stresses are given with horizontal lines. The symbols denote whether or not mud erosion or deposition takes place. Sand erosion and deposition can always occur, because no threshold shear stress is included.

The model results of the three situations are discussed in the next subsections. Besides the external scale parameters, internal time scales also govern the model behaviour. These time scales can be derived from the underlying equations. The time scales are used for explaining the model behaviour and given in the next subsections.

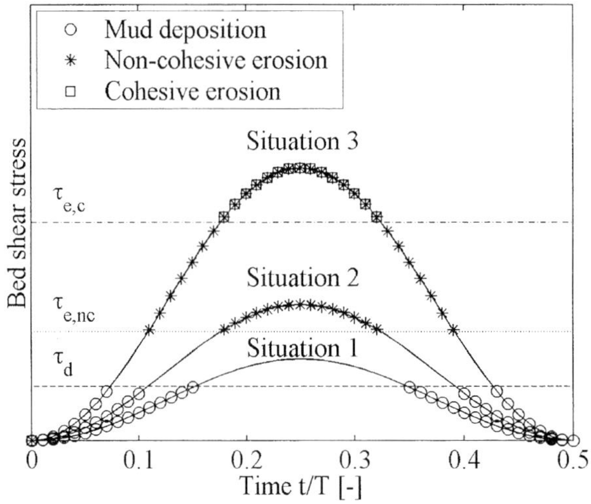

Figure 3. Overview of situations.

4.2. Situation 1

In situation 1 erosion of mud from a non-cohesive bed as well as erosion of sand and mud from a cohesive bed are not possible, because the maximum bed shear stress is always lower than the critical bed shear stress for erosion during the tidal period. However, deposition of mud is possible, because the bed shear stress is lower around slack water than the critical bed shear stress for deposition. Sand deposition is always possible, because the deposition term does not depend on the local hydrodynamics. Erosion of sand from the non-cohesive bed can occur, because the equilibrium concentration formulation does not have a threshold shear stress (Figure 3).

- **Mud concentration**

For the mud concentration, only two time scales are important, the deposition time scale (h/w_m) and the transport time scale (h/k_m), because erosion does not play a role in this situation. For a constant water depth, the behaviour of the mud concentration is governed by the dimensionless ratio k_m/w_m. In Figure 4, results of the mud concentration are given for different values of k_m/w_m. The mud concentration is made dimensionless by using the outside concentration c_{out}.

It can be observed that the (tidally averaged) mud concentration in the water column tends to a finite value for all ratios of k_m/w_m. When the ratio k_m/w_m is higher, the equilibrium value is also higher. The equilibrium mud concentration is always lower than the concentration outside c_{out}.

The (tidally averaged) equilibrium mud concentration can be explained from a balance between the inward horizontal flux due to transport and an outward downward flux due to deposition over one tidal period. The equilibrium concentration can be solved analytically when the adaptation time for settling (h/w_m) is at least in the order of the tidal period.

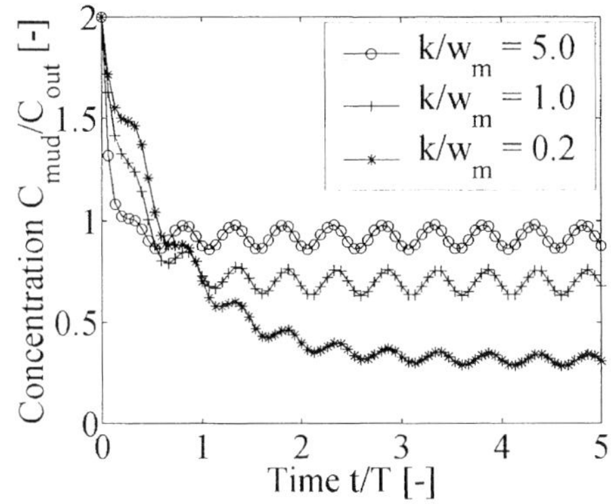

Figure 4. Mud concentration for different values of k_m/w_m.

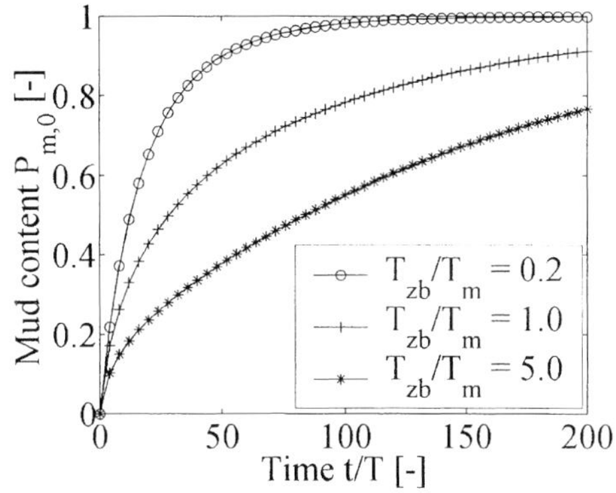

Figure 5. Mud content in exchange layer for different values of T_{zb}/T_m.

In this case, a more or less constant concentration can be assumed during the tidal period and the equilibrium concentration $c_{m,eq}$ is given by:

$$c_{m,eq} = \frac{1}{\frac{w_m}{k_m} a_D + 1} c_{out} \qquad a_D = \left[\left(1 - \frac{1}{2}\tilde{\tau}_d \right) \left(\frac{2}{\pi} \arcsin\left(\frac{1}{\sqrt{\tilde{\tau}_d}} \right) \right) + \frac{1}{\pi}\sqrt{\tilde{\tau}_d - 1} \right] \tag{17}$$

where α_D is a deposition coefficient and $c_{m,eq}$ the equilibrium mud concentration. The coefficient α_D is solved analytically and is a function of the dimensionless shear stress $\tilde{\tau}_d$. By using this expression, it can be seen that for the equilibrium mud concentration in the water column three situations exist:

1. $k_m \ll w_m$
In this case, the supply of mud from outside is very low with respect to the deposition of mud. Therefore, the equilibrium mud concentration is very small, for reaching a balance between the deposition and transport flux during the tide.

2. $k_m \approx w_m$
Because the settling velocity w_m and the transport coefficient k_m have the same order of magnitude, the flux by deposition and transport have the same order of magnitude for an equilibrium mud concentration between 0 and c_{out}.

3. $k_m \gg w_m$
In this case, the supply of mud from outside is very fast with respect to the deposition of mud. Therefore, the equilibrium concentration is about equal to the concentration outside c_{out}.

From the equilibrium mud concentration equation (17), it can easily be seen that when no transport ($k_m = 0$) is included, the model gives a trivial solution. The only flux is caused by deposition and the final mud concentration is zero. The total deposition flux depends on the initial mud concentration.

- **Sand concentration**

The governing time scale for the sand concentration is h/w_s. This time scale is in general much smaller than the tidal period. For the non-cohesive regime, the sand concentration shows a small phase shift with respect to the equilibrium concentration $c_{e,s}$ because of its adaptation time (h/w_s). For the cohesive regime, no erosion of sand can occur. If the initial sand concentration is non-zero, the sand concentration will drop very fast to zero and does not change anymore.

- **Bed level**

Due to the continuous deposition of mud particles and no erosion, the bed level is continuously increasing due to accretion. The net sand exchange during one tidal period is equal to zero. In case of a non-cohesive bed, the speed of the bed level rise is thus determined by the equilibrium mud concentration and the settling velocity. The bed level rise time scale is given by:

$$T_{z_b} = \frac{\rho_b}{\rho_s} \frac{\delta}{\alpha_D w_m c_{m,eq}} = \frac{\rho_b}{\rho_s} \frac{\delta}{c_{out}} \left[\frac{1}{\alpha_D w_m} + \frac{1}{k_m} \right] \tag{18}$$

where T_{zb} is the time scale of bed level rise and δ the total thickness of the bed layer. In (18) the parameter δ is the total thickness of the bed layer, in which the bed composition changes. It is obvious that for an increasing transport coefficient (k_m), settling velocity (w_m) and outside concentration (c_{out}), the time scale of bed level rise (T_{zb}) decreases.

- **Bed composition**

Because of the continuous deposition of mud, the equilibrium composition consists of a mud bed without sand. However, for the development of the bed composition two time scales are important: the time scale of bed level rise (18) and the mixing time scale (T_m). The mixing time scale is given by:

$$T_m = \frac{\delta^2}{\varepsilon_{z,0}} \tag{19}$$

where T_m is mixing time scale and $\varepsilon_{z,0}$ the mixing coefficient at the bed surface.

In Figure 5 the mud content in the exchange layer is given for different ratios of T_{zb}/T_m. In Figure 6 and Figure 7 the bed composition profile is given in a fixed co-ordinate system at different times for T_{zb}/T_m = 0.2 and 5.0 respectively. The depth within the bed is made dimensionless with the constant thickness of the bed layer δ. Due to the fixed co-ordinate system and the continuous bed level rise in this situation, the bed layer in which the bed composition is computed, shifts upward in time (Figure 6 and Figure 7).

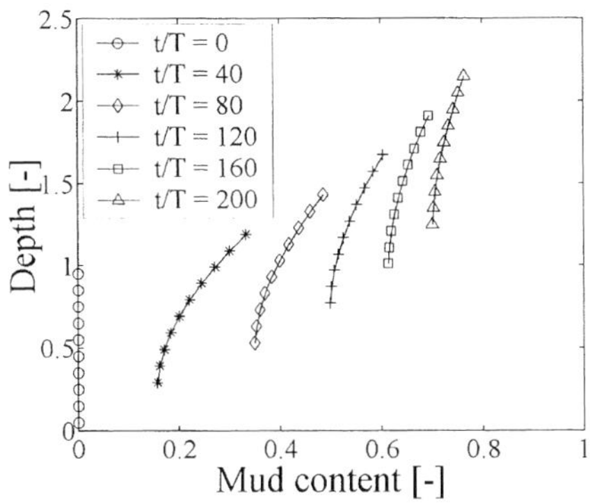

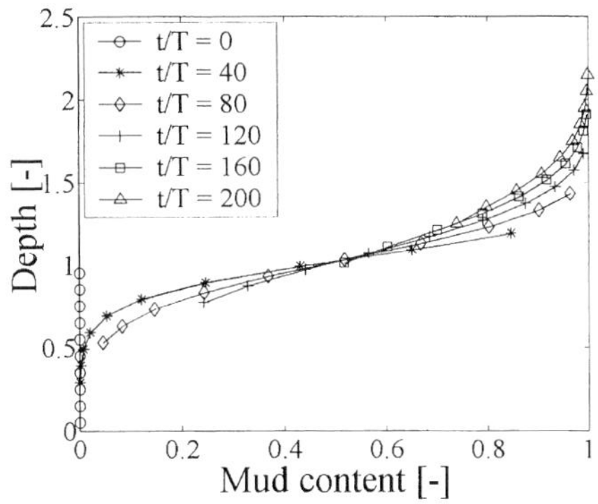

Figure 6. Bed composition for $T_{zb}/T_m = 5.0$. Figure 7. Bed composition for $T_{zb}/T_m = 0.2$.

It can be observed that when the ratio of bed level rise time scale to mixing time scale is relatively large (T_{zb}/T_m = 5.0), the time needed to reach a 100% muddy exchange layer is relatively long (Figure 5). Moreover, the bed composition profile with depth is more or less constant (Figure 6). For the opposite situation (T_{zb}/T_m = 0.2) a 100% muddy top layer is reached very fast (Figure 5) and the bed composition profile shows a stratified profile (Figure 7).

The explanation for this behaviour is that when the net downward flux due to deposition is much larger than the downward transport by mixing, all mud deposited on the exchange layer is conserved in the upper part of the bed. For this situation, the bed composition profile is strongly stratified and a 100% muddy exchange layer is reached very fast.

4.3. Situation 2

In situation 2 erosion of sand and mud from a cohesive bed is not possible. Thus, the situation for the cohesive regime is equal to situation 1. However, mud and sand erosion from a non-cohesive bed is possible (Figure 3). The behaviour of mud concentration, bed level and bed composition in this regime is discussed below.

- **Mud concentration**

In the non-cohesive regime the mud concentration in the water column always tends to the mud concentration outside c_{out}. An equilibrium is reached when the outward deposition flux is equal to the inward erosion and transport flux during a tidal period. When it is assumed that the concentration in the water column and the mud content in the top layer are more or less constant during the tidal period in the equilibrium situation, the equilibrium mud concentration is given by:

$$c_{m,eq} = \frac{p_{m,eq,0} M \alpha_{E,NC} + k_m c_{out}}{w_m \alpha_D + k_m} \qquad \alpha_{E,NC} = \left(\frac{1}{2}\tilde{\tau}_{e,nc} - 1\right)\left(1 - \frac{2}{\pi} arcsin\left(\frac{1}{\sqrt{\tilde{\tau}_{e,nc}}}\right)\right) + \frac{1}{\pi}\sqrt{\tilde{\tau}_{e,nc} - 1} \quad (20)$$

where $\alpha_{E,NC}$ is the erosion coefficient for non-cohesive regime and $p_{m,eq,0}$ the equilibrium mud content in exchange layer. The coefficient $\alpha_{E,NC}$ is solved analytically and is a function of the dimensionless critical shear stress for erosion in the non-cohesive regime ($\tilde{\tau}_{e,nc}$). Thus, compared to the equilibrium mud concentration in situation 1 (17), an extra term is introduced due to erosion in (20). However, the equilibrium mud content ($p_{m,eq,0}$) is not known yet. In the discussion of the bed level and composition below, it will be shown that the equilibrium mud content is equal to:

$$p_{m,eq,0} = \frac{w_m c_{m,eq}}{M} \frac{\alpha_D}{\alpha_{E,NC}} \tag{21}$$

By using expression (21) in equation (20), the final equilibrium mud concentration in the water column is:

$$c_{m,eq} = \frac{\frac{w_m c_{m,eq}}{M} \frac{\alpha_D}{\alpha_{E,NC}} M\alpha_{E,NC} + kc_{out}}{w_m \alpha_D + k} = c_{out} \tag{22}$$

Thus, the equilibrium concentration in this situation is always equal to the concentration outside. However, this is only valid for the non-cohesive regime. When the cohesive regime is reached, only transport and deposition (see situation 1) determine the equilibrium. In this case, the equilibrium concentration of situation 1 is reached.

- **Bed level and composition**

For the non-cohesive regime, equilibrium is reached when the total flux of sand and mud to and from the bed are equal during one tidal period. When this situation is reached, the bed level remains constant. At this equilibrium, the mud content in the exchange layer for the non-cohesive regime has also an equilibrium value. An equilibrium exists because of a balance between the erosion and deposition flux for mud during one tidal period. Assuming a more or less constant mud concentration and mud content during one tidal period, the equilibrium mud content can be expressed by:

$$p_{m,eq,0} = \frac{w_m c_{m,eq} \frac{1}{T}\int_0^T [1-\tilde{\tau}_d] H[1-\tilde{\tau}_d] dt}{M \frac{1}{T}\int_0^T [\tilde{\tau}_{e,nc} - 1] H[\tilde{\tau}_{e,nc} - 1] dt} = \frac{w_m c_{m,eq}}{M} \frac{\alpha_D}{\alpha_{E,NC}} \tag{23}$$

As already mentioned, the equilibrium mud concentration is about equal to the concentration outside. Thus, the equilibrium mud content in the exchange layer is:

$$p_{m,eq,0} = \frac{w_m c_{out}}{M} \frac{\alpha_D}{\alpha_{E,NC}} \tag{24}$$

In the final expression for the equilibrium mud content (24), the deposition flux during one tide is recognised in the numerator, the erosion flux for a 100% mud bed ($p_{mud,0} = 1.0$) during one tide in the denominator.

It is important to note that when the equilibrium mud content is larger than the critical mud content for the transition between non-cohesive and cohesive behaviour ($p_{m,eq,0} > p_{mud,crit}$), the non-cohesive equilibrium mud content will never be reached. When starting with a non-cohesive bed, the mud content in the top layer tends to the equilibrium in the cohesive regime. When reaching the critical mud content the behaviour changes into a cohesive behaviour. When no erosion of the cohesive bed is possible, the final situation is a total mud bed. In fact, the same situation as in situation 1 is reached with only deposition and no erosion.

4.4. Situation 3

In situation 3, the model behaviour is strongly comparable to situation 2. The only extra possibility is erosion in the cohesive regime (Figure 3). Therefore, two equilibria exist for the mud content in the exchange layer (24), indicated with indices *nc* and *c*: a non-cohesive equilibrium $p_{m,0,eq,nc}$ and a cohesive equilibrium $p_{m,0,eq,c}$. In analogy to the non-cohesive equilibrium, the cohesive equilibrium is also defined by (24), because the only difference in the erosion and deposition formulae is the critical erosion shear stress. For the cohesive equilibrium ($p_{m,0,eq,c}$) the dimensionless critical shear stress for erosion in the cohesive regime ($\tilde{\tau}_{e,c}$) must be used. In situation 3, the final equilibrium depends on the values of the non-cohesive and cohesive equilibrium compared to the value of the critical mud content ($p_{mud,crit}$). Three different situations can be distinghuished and are summarised in Table 2.

The final equilibrium for these situations can be explained as follows:

- Situation 3.1: the non-cohesive equilibrium is never reached, because the non-cohesive equilibrium mud content is higher than the critical mud content. When the mud content is lower than the critical mud content, the mud content increases to reach the non-cohesive equilibrium value. When the critical mud content is reached, the equilibrium changes to the cohesive equilibrium and this is also the final equilibrium.
- Situation 3.2: this situation is comparable to situation 3.1, but in this case the cohesive equilibrium is never reached because its value is in the non-cohesive range. The final equilibrium is the non-cohesive equilibrium.
- Situation 3.3: both equilibria can exist and the final situation depends on the initial mud content in the exchange layer. If the initial mud content is higher than the critical mud content, the cohesive equilibrium is reached. If the initial mud content is lower than the critical mud content, the non-cohesive equilibrium is reached.

Table 2
Overview of equilibria in situation 3

Situation	Cohesive equilibrium ($p_{m,0,eq,c}$)	Non-cohesive equilibrium ($p_{m,0,eq,nc}$)	Final equilibrium
3.1	$> p_{mud,crit}$	$> p_{mud,crit}$	Cohesive
3.2	$< p_{mud,crit}$	$< p_{mud,crit}$	Non-cohesive
3.3	$> p_{mud,crit}$	$< p_{mud,crit}$	Depends on initial conditions

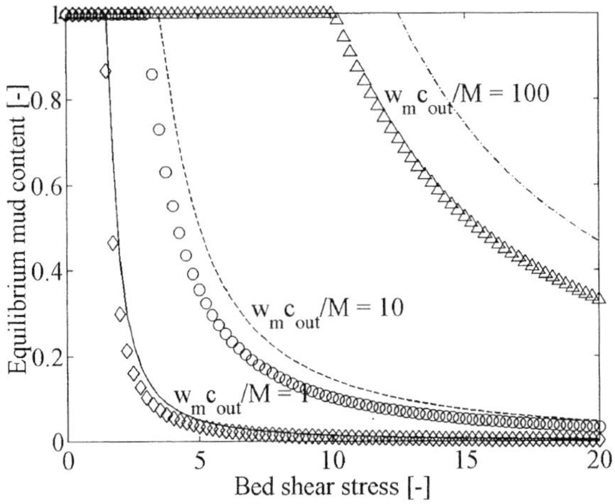

Figure 8. Equilibrium mud content.

5. COMPARISON TO FIELD MEASUREMENTS

Correlations between the maximum bed shear stress and the mud content at a certain location of the Westerschelde estuary (The Netherlands) showed the following pattern (WL | Delft Hydraulics, 1998): for high bed shear stress, the mud content was always low (< 10%). For low bed shear stress, the mud content varied between 0 and 100% and no useful relationship could be determined. Often, a sharp transition between these regimes was observed and a critical transition value could be defined. However, physical explanations for this typical pattern are not available at present.

In the previous chapter an equilibrium mud content in the exchange layer was defined (24). Similarly to the above described correlation, this equilibrium mud content can be given as a function of the maximum bed shear stress during the tide (Figure 8). The equilibrium mud content is given for different values of $w_m c_{out}/M$. The maximum bed shear stress is made dimensionless by the critical erosion shear stress. It is assumed that the critical erosion shear stress for the non-cohesive regime is equal to the critical erosion shear stress for the cohesive regime. The critical shear stress for deposition must also be known to calculate the coefficient α_D. The equilibrium mud content is given for two ratios of $\tau_e/\tau_d = 5$ (lines) and $\tau_e/\tau_d = 10$ (open symbols). For small values (order 1) of $w_m c_{out}/M$ a sharp transition exists between a full mud bed and a full sand bed (

Figure 8). Beyond a dimensionless bed shear stress of about 3, the mud content is less than 10%, while below a dimensionless bed shear stress of about 2, the equilibrium value is about 100%. With increasing values of $w_m c_{out}/M$ the transition becomes less sharp.

For the Westerchelde area, typical values for the parameters are $w_m = 5*10^{-4}$ m/s, $M = 1*10^{-8}$ m/s and $c_{out} = 4*10^{-5}$ ($\approx$ 100 mg/l). The parameter-value $w_m c_{out}/M$ has a value of about 2. Thus, a sharp transition has to be expected in this area. This is confirmed by the observed correlation between the bed shear stress and the mud content (WL | Delft Hydraulics, 1998).

It is important to note that when the critical shear stress for erosion in the non-cohesive and cohesive regime are not equal, the non-cohesive and cohesive equilibrium mud content are not equal (see section 4.3). In this case, the equilibrium mud content as a function of the maximum bed shear stress is discontinuous. The discontinuity arises at the critical mud content between the non-cohesive and cohesive regime.

The presented sand-mud model also implies some explanation for the observed large scatter in areas with a low bed shear stress. First, the relatively long time scale for reaching the equilibrium mud content suggests that the measured mud content at the sample points is probably not at its equilibrium value, but evolves towards an equilibrium value. The time scale for reaching the equilibrium mud content can be estimated by using (19). The parameters for the time scale T_{zb} are assumed to be as follows: $\delta = 0.10$ m, $\rho_b = 1200$ kg/m^3, $\rho_s = 2650$ kg/m^3, $w_m = 5*10^{-4}$ m/s, $c_{out} = 4*10^{-5}$ (≈ 100 mg/l). Thus, the time scale for reaching a total mud bed is about 50 days, about two months. When mixing within the bed is included, the time scale becomes even much larger. Thus, the measured mud content at a single site not only depends on the actual hydrodynamic conditions, but also on the hydrodynamic conditions in last months. The equilibrium mud content can be seen as the upper limit for the mud content in a certain area. Second, scatter can be caused by differences between the real velocity profile and water depth during the tide and the assumed sinusoidal velocity profile and constant water depth (1). Especially for the intertidal areas, the velocity profile and water depth can be quite different from the assumed hydrodynamical situation due to flooding and drying.

6. CONCLUSIONS

In this paper, a process-based sand mud model is proposed and analysed. In general, it can be concluded that with such a process-based model, governing time scales and dimensionless parameters can be derived which can increase the physical understanding of the bed composition significantly. Furthermore, an equilibrium mud content within the exchange layer was found when both deposition and erosion occur during the tidal period (24). In this equation, the settling velocity for mud (w_m), the mud concentration (c_{out}) and the erosion rate (M) form an important dimensionless parameter ($w_m c_{out}/M$). This parameter expresses the ratio between the deposition and erosion flux capacity.

In earlier studies, correlations between the maximum or mean shear stress and the mud content often showed the following characteristic picture. A critical shear stress seems to exist, below which the mud content can vary between 0 and 100%. Above this value, the mud content is always low. A sharp transition is sometimes observed between both regimes. This pattern can be explained with the presented sand-mud model. Model results suggest that the sharp transition between areas with a very low mud content and other areas depends on the aforementioned dimensionless parameter. For low values (< 10) the transition is sharp, while for higher values the transition becomes more and more gradual. The observed sharp transition in field data in earlier studies follows from the low value of the dimensionless parameter for these areas.

Two explanations are given for the variation in mud content between 0 and 100% at sample sites with low bed shear stress. First, the actual mud content at the sample site is probably not in equilibrium due to relatively large adapting time scales. Second, the scatter is probably caused by the difference between the actual hydrodynamic situation and the assumed sinusoidal velocity profile and the constant water depth.

Finally, the model results also suggest that a local hydrodynamic parameter (e.g. maximum bed shear stress) for predicting the mud content at a certain location is not very useful for areas exposed to a relatively low bed shear stress. Apart from the local hydrodynamics, the local mud concentration, the settling velocity, the mixing properties within the bed and the sample depth are parameters which determine the local mud content.

ACKNOWLEDGEMENTS

The author is grateful to dr Z.B. Wang for the extensive discussions and valuable comments. Also the comments from dr J.C. Winterwerp and prof dr H.J. de Vriend are highly appreciated. This research was supported by the Technology Foundation STW, applied science division of NWO and the technology programme of the Ministry of Economic Affairs.

REFERENCES

Armanini, A., 1995, Non-uniform sediment transport: dynamics of the active layer, *Journal of Hydraulic Research*, 33(5), 611-622.

Aubrey, D.G., 1984, Hydrodynamic controls on sediment transport in well-mixed bays and estuaries, In: *Physics of Shallow Estuaries and Bays*, ed. J. v.d. Kreeke, 245-258.

Boudreau, B.P., 1997, Diagenetic Models and Their Implementation - Modelling Transport and Reactions in Aquatic Sediments, Springer Verlag.

Dronkers, J., 1984, Tide-induced Transport of Fine Sediment, In: *Physics of Shallow Estuaries and Bays*, ed. by J. van de Kreeke, Springer Verlag, 228-244.

Dyer, K.R, 1994, Estuarine sediment transport and deposition, In: *Sediment Transport and Depositional Processes*, ed. by K. Pye, Blackwell Scientific Publications, 193-218.

Herman, P., 2000, personal communication.

Murray, W.A., 1977, Erosion of coarse sand-clayey silt mixtures, *Journal of Hydraulic Division*, 1222-1227.

Panagiotopoulos, I., Voulgaris, G., Collins, M.B., 1997, The influence of clay on the threshold of movement on fine sandy beds, *Coastal Engineering*, 32, 19-43.

Rijn, L.C. van, 1993, Principles of sediment transport in rivers, estuaries and coastal seas, Aqua Publications, Amsterdam.

Teisson, C., 1997, A review of cohesive sediment transport models, In: *Proceedings of 4th Nearshore and Estuarine Cohesive Sediment Transport Conference*, Wallingford, England.

Torfs, H., 1995, Erosion of mud/sand mixtures. Ph.D. thesis, Katholieke Universiteit Leuven, faculteit der Toegepaste Wetenschappen, Departement Burgelijke Bouwkunde, Laboratorium voor Hydraulica.

Torfs, H., Mitchener, H., Huysentruyt, H., Toorman, E., 1996, Settling and consolidation of mud/sand mixtures, *Coastal Engineering*, 29, 27-45.

Van Ledden, M., 2000, Sediment segregation in estuaries and tidal lagoons, a literature survey, Delft University of Technology, The Netherlands.

Winterwerp, J.C., 1999, On the dynamics of high-concentrated mud suspensions, Ph.D. thesis, Delft University of Technology, Delft.

WL | Delft Hydraulics, 1998, A tool for mud flat classification (Z2037.50), Prepared for: European Commision, MAST3 Programme.

Fine Sediment Dynamics in the Marine Environment
J.C. Winterwerp and C. Kranenburg (Editors)

3-D numerical modelling of mud and radionuclide transport in the Chernobyl Cooling Pond and Dnieper - Boog Estuary

N. Margvelashvili, V. Maderich, S.Yuschenko and M. Zheleznyak

Institute of Mathematical Machine and System Problems,
Glushkova av. 42, Kiev, 03187, The Ukraine.

The 3-D model THREETOX, that includes modules of hydrodynamics, sediment and pollutant transport, was developed to simulate the radionuclide fate in a deep stratified water body. This paper describes the methodology and results of simulation of the radionuclide transport and fate in the cooling pond of the Chernobyl Nuclear Power Plant and in the Dnieper – Boog Estuary. The analysis of the efficiency of the chosen sediment transport model is based on the use of radionuclides from the Chernobyl accident as tracer.

Modelling; sediment transport; mud; radionuclide transport; Chernobyl accident

1. INTRODUCTION

In natural streams affected by human pollution – heavy metals, radionuclides, PCB, nutrients and others - suspended sediments play a role as the carriers of contaminants over long distances from the release areas. The processes of sedimentation and erosion of contaminated sediments drive the re-distribution of the pollutant in the bed sediment. The magnitude of the partition coefficient that describes the distribution of contaminant between the liquid and solid phases increases usually with decreasing sediment grain size. Therefore, models describing the transport and fate of suspended sediments are the important parts of the modelling system for the simulation of pollutant transport in water bodies (Onishi et al., 1981; Santschi et al., 1989; Perianez, 2000). However, immediate data on cohesive sediment transport are rarely available for many natural water bodies. In this respect radioactive isotopes, treated as tracers, could provide an efficient tool for studying the mud behaviour in surface waters.

After the Chernobyl accident, a set of models was developed to simulate the fate and behavior of the Chernobyl radionuclides in water bodies in the vicinity of the Chernobyl Nuclear Power Plant (NPP) and in the Dnieper river/reservoir system (Zheleznyak et al., 1992;1997). The recently developed 3-D model THREETOX, that includes modules for hydrodynamics, sediment transport and pollutant transport, has been applied to deep stratified water bodies contaminated after the Chernobyl accident (Margvelashvili et al., 1997; Koziy et al., 1998; Margvelashvili et al., 1999).

The aim of this paper is the extension of the THREETOX model for the simultaneous description of different sediment fractions, including cohesive sediments. The model is applied to study the fate and behavior of radionuclides in the highly contaminated Cooling Pond of the NPP and in the Dnieper-Boog Estuary. The analysis of the efficiency of the chosen sediment transport model is based on the use of the radionuclides released during the Chernobyl accident as tracer.

2. MODEL

2.1. Hydrodynamics

The hydrodynamics of THREETOX model are based on the three-dimensional, time-dependent, free surface model of Blumberg and Mellor (1987). The prognostic variables of the hydrodynamics code are the three components of velocity, the temperature, salinity and surface elevation fields. The governing equations are:

$$\vec{\nabla} \cdot \vec{U} = 0, \tag{1}$$

$$\frac{\partial U}{\partial t} + \vec{U} \cdot \vec{\nabla} U - fV = -\frac{1}{\rho_0}\frac{\partial P}{\partial x} + \frac{\partial}{\partial z}\left(\nu \frac{\partial U}{\partial z}\right) + A\Delta U, \tag{2}$$

$$\frac{\partial V}{\partial t} + \vec{U} \cdot \vec{\nabla} V + fU = -\frac{1}{\rho_0}\frac{\partial P}{\partial y} + \frac{\partial}{\partial z}\left(\nu \frac{\partial V}{\partial z}\right) + A\Delta V, \tag{3}$$

$$\frac{\partial (T,S)}{\partial t} + \vec{U} \cdot \vec{\nabla}(T,S) = \frac{\partial}{\partial z}\left(\nu_T \frac{\partial (T,S)}{\partial z}\right) + A_T \Delta (T,S), \tag{4}$$

$$\rho = \rho(T,S), \tag{5}$$

$$P(x,y,z,t) = P_0 + g\rho_0(\eta - z) + g\int_z^{\eta} \rho(x,y,z',t)dz' \tag{7}$$

The concept of eddy viscosity (ν), diffusivity (ν_T) and the Prandtl hypothesis, with variable turbulence length scale, are used to define the turbulence stresses and fluxes. The vertical turbulent exchange coefficients are:

$$\begin{cases} \nu_T = k_1^2 z^2 \left(1 - \frac{z}{H}\right)^2 \left|\frac{\partial \vec{U}}{\partial z}\right| \left(1 - \frac{Ri}{Ri_c}\right)^{1/2}, & 0 < Ri < Ri_c \\ \nu_T = 0.0, & Ri \geq Ri_c \\ \nu = \nu_T (1 + Ri) & \end{cases} \tag{8}$$

Here $\vec{U} = (U,V,W)$ is the velocity vector, T is the temperature, S is the salinity, η is the free surface elevation, f is the Coriolis parameter, P is the pressure, ρ and ρ_o are the calculated and the standard density of water, g is the acceleration due to gravity, A and A_T are the coefficients of turbulent horizontal viscosity and the temperature diffusivity, respectively, and $Ri = -\frac{g(\partial \rho / \partial z)}{\rho_0 (\partial \vec{U} / \partial z)^2}$ is the Richardson number, $k_1 = 0.1$, $Ri_c = 10$.

At the free surface all fluxes (momentum, heat, etc.) are prescribed. At the bottom and at the land boundaries, the conditions of zero diffusive fluxes of all properties are used. The open lateral boundary conditions are modified radiation conditions.

2.2. Sediment transport

All particulate matter is represented by three different grain size fractions. The effect of sediment on the hydrodynamics is neglected. Suspended sediment transport is described by the advection- diffusion equation, taking into account the settling velocity of the sediment:

$$\frac{\partial S_{di}^{W}}{\partial t}+U\frac{\partial S_{di}^{W}}{\partial x}+V\frac{\partial S_{di}^{W}}{\partial y}+(W-W_{gi})\frac{\partial S_{di}^{W}}{\partial z}=\frac{\partial}{\partial z}\left(\nu_T\frac{\partial S_{di}^{W}}{\partial z}\right)+A(\Delta S_{di}^{W}), \tag{9}$$

where S_{di}^{w} is the concentration of i-th fraction of sediment in the water column, and W_{gi} is settling velocity of solid particles. At the free surface, zero vertical sediment flux is assumed, i.e:

$$(W-W_{gi})S_{di}^{W}=\nu_T\frac{\partial S_{di}^{W}}{\partial z},\quad z=\eta. \tag{10}$$

The vertical flux of suspended sediments at the bottom is equal to the difference between the resuspension and sedimentation rate:

$$\nu_T\frac{\partial S_{di}^{W}}{\partial z}+W_{gi}S_{di}^{W}=q_i^{W}-q_i^{b},\qquad z=-h+z_0, \tag{11}$$

where h is the depth, z_0 is the bottom roughness, q_i^{w}, q_i^{b} are the sedimentation and resuspension rates, respectively. The concentration of sediments in the upper bottom layer is described by the following equation:

$$\frac{\partial(m_{di}^{b})}{\partial t}=q_i^{W}-q_i^{b}, \tag{12}$$

where $m_{di}^{b}=S_{di}^{b}Z_*$ is the mass of i-th fraction of bottom sediment per unit area, and Z_* is the depth within the bottom sediment top layer. The thickness of the bottom layer and bed porosity are defined from the relationships:

$$\tilde{Z}=\frac{1}{1-\varepsilon}\sum_{i\ solid}\frac{m_{di}^{b}}{\rho_{Si}},\qquad \tilde{\varepsilon}=\frac{V_{tot}-\sum_{i\ solid}V_i}{V_{tot}}=\frac{Z_{\min}-\sum_{i\ solid}\frac{m_{di}^{b}}{\rho_{Si}}}{Z_{\min}}, \tag{13}$$

$$Z_*=\begin{cases}\tilde{Z},\ \varepsilon=\varepsilon_0=const,\ \tilde{Z}>Z_{\min}\\[2ex] Z_{\min},\ \varepsilon=\tilde{\varepsilon},\ \tilde{Z}\le Z_{\min}\end{cases} \tag{14}$$

where ε is the porosity of the bed layer, V_i is the volume of i-th fraction of the sediments, ρ_{Si} is the density of the solid particles. From eqns. (12) and (14) the concentration of the i-th fraction of sediment in the bottom layer is defined as:

$$S_{di}^{b}=\frac{m_{di}^{b}}{Z_*}. \tag{15}$$

For cohesive sediment the erosion and deposition rates are modelled by using the formulae of Ariathurai and Krone (1976). For non-cohesive sediments the bottom boundary condition describes the resuspension or settling of sediments, which depends on the ratio between the actual and the near-bed equilibrium concentration of the sediments. The latter one is estimated according to Van Rijn (1984). The effective bottom shear stresses induced by currents and wind waves are summed. The formula derived on the basis of numerical solution of the equations for oscillating bottom boundary layer (Zheleznyak, 1988) was used to parameterise the wave induced stresses.

2.3. Radionuclide transport

The sub-model of radionuclide transport describes the specific water-sediment sorption processes. It includes the advection-diffusion equations for dissolved ($C_s^{\,w}$) and adsorbed by suspended sediment ($C_p^{\,w}$) radioactivity in the water column, and the equations for concentration of the dissolved ($C_s^{\,b}$) and adsorbed ($C_p^{\,b}$) radioactivity in the bottom deposits:

$$\frac{\partial C_s^{\,w}}{\partial t}+\frac{\partial UC_s^{\,w}}{\partial x}+\frac{\partial VC_s^{\,w}}{\partial y}+\frac{\partial WC_s^{\,w}}{\partial z}= \frac{\partial}{\partial z}\left(\nu_T\frac{\partial C_s^{\,w}}{\partial z}\right)+A(\Delta C_s^{\,w})-\lambda C_s^{\,w}-a_{1,2}^{\;w}(S_d^{\,w}K_d^{\,w}C_s^{\,w}-C_p^{\,w}), \tag{16}$$

$$\frac{\partial C_p^{\,w}}{\partial t}+\frac{\partial UC_p^{\,w}}{\partial x}+\frac{\partial VC_p^{\,w}}{\partial y}+\frac{\partial (W-W_g)C_p^{\,w}}{\partial z}= \frac{\partial}{\partial z}\left(\nu_T\frac{\partial C_p^{\,w}}{\partial z}\right)+A(\Delta C_p^{\,w})-\lambda C_p^{\,w}+a_{1,2}^{\;w}(S_d^{\,w}K_d^{\,w}C_s^{\,w}-C_p^{\,w}), \tag{17}$$

$$\frac{\partial m_s^{\,b}}{\partial t}=\frac{F_{dif}}{Z_*}(C_s^{\,w}-\frac{C_s^{\,b}}{\varepsilon})-a_{1,2}^{\;b}(\frac{S_d^{\,b}}{\varepsilon}K_d^{\,b}C_s^{\,b}-C_p^{\,b})\cdot Z_*-\lambda m_s^{\,b}, \tag{18}$$

$$\frac{\partial m_p^{\,b}}{\partial t}=(\frac{C_p^{\,w}\sum_{i\,solid}q_i^{\,w}}{\sum_{i\,solid}S_{di}^{\,w}}-\frac{C_p^{\,b}\sum_{i\,solid}q_i^{\,b}}{\sum_{i\,solid}S_{di}^{\,b}})+a_{1,2}^{\;b}(\frac{S_d^{\,b}}{\varepsilon}K_d^{\,b}C_s^{\,b}-C_p^{\,b})\cdot Z_*-\lambda m_p^{\,b}, \tag{19}$$

where $m_s^{\,b}=C_s^{\,b}Z_*$, $m_p^{\,b}=C_p^{\,b}Z_*$, $S_d^{\,w}=\sum_{i\,solid}S_{di}^{\,w}$, $C_p^{\,w}=\sum_{i\,solid}C_{pi}^{\,w}$, $W_g=\sum_{i\,solid}W_{gi}S_{di}^{\,w}/S_d^{\,w}$, $S_d^{\,b}=\sum_{i\,solid}S_{di}^{\,b}$, $C_p^{\,b}=\sum_{i\,solid}C_{pi}^{\,b}$, λ is the radionuclide decay constant, and F_{dif} is the diffusion coefficient. The governing equations of the model are similar to those used in the FLESQOT model (Onishi et al., 1989).

The exchanges between the different phases are described by diffusion, sorption, and sedimentation-resuspension processes. Adsorption and desorption of radionuclides between liquid and solid phases are described by the radionuclide exchange rates $a_{1,2}^{\;w}$ $a_{1,2}^{\;b}$, and by the distribution coefficients $K_d^{\,w}, K_d^{\,b}$ (Onishi et al., 1981; Santschi and Honeyman, 1989;

IAEA, 1985; Carrol and Harms 1999), which are defined, under steady hydraulic conditions, as

$$S_d^w K_d^w = \lim_{t\to\infty}\left(\frac{C_p^w}{C_s^w}\right), \quad \frac{S_d^b}{\varepsilon} K_d^b = \lim_{t\to\infty}\left(\frac{C_p^b}{C_s^b}\right). \tag{19}$$

At the free surface $z = \eta$ the boundary conditions are:

$$\nu_T \frac{\partial C_s^w}{\partial z} = WC_s^w, \quad (W - W_g)C_p^w - \nu_T \frac{\partial C_p^w}{\partial z} = 0, \tag{20}$$

The fluxes into the bottom $z = -h + z_0$ are:

$$\nu_T \frac{\partial C_s^w}{\partial z} = \frac{F_{dif}}{Z_*}(C_s^w - \frac{C_s^b}{\varepsilon}), \tag{21}$$

$$\nu_T \frac{\partial C_p^w}{\partial z} + (W - W_g)C_p^w = \frac{C_p^w \sum_{i\ solid} q_i^w}{\sum_{i\ solid} S_{di}^w} - \frac{C_p^b \sum_{i\ solid} q_i^b}{\sum_{i\ solid} S_{di}^b}, \tag{22}$$

In THREETOX, σ-coordinate transformation is applied to avoid difficulties in the numerical solution for realistic bottom topography. Splitting of the barotropic and baroclinic modes are imposed in the code. The governing equations together with the boundary conditions are solved by finite difference techniques. A horizontally and vertically staggered mesh is used for the computations.

3. CASE STUDIES

3.1. Chernobyl NPP cooling pond

The Chernobyl Cooling Pond (ChCP) was constructed in 1972 in the floodplain of the River Pripyat. The length is 11.5 km, the maximum width is 2.2 km, the maximum depth is 18 m, and the volume is 0.16 km^3 (Figure 1). The water level in the ChCP, 6 m above the River Pripyat level, is kept constant by pumping of the river water into the cooling pond to compensate the losses due to seepage and evaporation. Currents in the ChCP were driven by the discharge and intake of cooling water, and by wind and baroclinic effects. During the operation of the NPP, the water mass of the ChCP had significant temperature stratification with temperature gradients up to 10°-12°C. The upper layer of the bed consists mainly of mud. The ChCP was heavily contaminated by the Chernobyl accident in April-May, 1986. Till today, the concentrations of ^{137}Cs and ^{90}Sr in the bottom sediments are rather high. Field measurements revealed redistribution and accumulation of the radionuclide in deepest parts of the cooling pond. At present, the model studies on the fate and behaviour of radionuclide in the ChCP are stimulated by the needs to assess the radionuclide re-distribution after the closure of the Chernobyl NPP in December, 2000.

The dynamics of ^{137}Cs in the ChCP over one year after the accident has been simulated. The regime of the water discharge used in the simulations as boundary condition, is shown in Table 1. The temperature of cooling water entering the Cooling Pond was set 10° C above the temperature in the water intake channel. Ten days averages of the temperature of the atmosphere, and of wind characteristics were used to specify the boundary conditions at the surface.

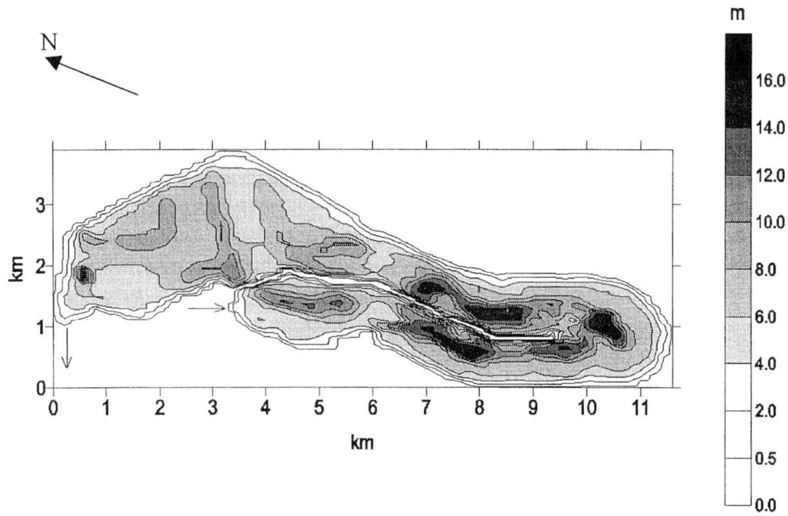

Figure 1. Depth distribution in the ChCP. The arrows show the cooling water intake and outfall.

The atmospheric fallout of the radionuclide at the surface of the pond was modelled by instantaneous, non-homogeneous deposition (Figure 2) based on the interpolation of the density of the ^{137}Cs deposition on the soil around the ChCP (Shiklomanov, 1992). Sediment and radionuclide fluxes in the water discharge channel were set equal to those in the water intake channel. The parameter values were $K_d^{\,b}$=3 m^3/kg, $a_{1,2}^{\,w} = 1\,d^{-1}$ and $a_{1,2}^{\,b} = 1\,y^{-1}$. To estimate the sensitivity of the computational results to variations in the distribution coefficient, the calculations were carried out for $K_d^{\,w}$ = 3 m^3/kg and for $K_d^{\,w}$=15 m^3/kg as well.

Simulations have been carried out for three suspended sediment diameters: 5, 15 and 30 μm, and an aerosol sediment flux at the water surface $5\cdot10^{-6}$ kg $m^{-2}s^{-1}$. The resolution of the numerical grid in the horizontal directions was 500 × 150 m. Vertical coordinate was resolved by 11 equally-spaced levels. The spatial distribution of the different fractions of the bottom sediments of the pond measured in 1983 (Shiklomanov, 1992) shows an intensive accumulation of the clay sediments in the deepest parts of the pond (Figure 3).

Table 1.
Water flow through the discharge channel in 1986.

Date	1-26 April	27 Apr.-30 Sept.	1 -31Oct	1 –30 Nov	1 - 31 Dec.
Discharge (м^3/s)	200	15	50	100	150

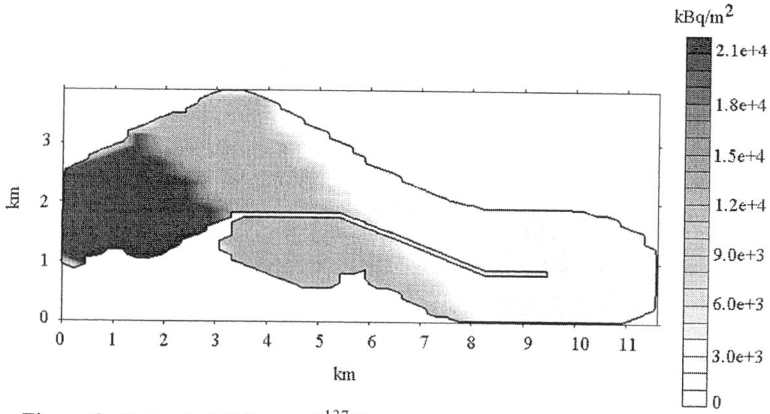

Figure 2. Estimated fallout of ^{137}Cs at the surface of ChCP in April 1986.

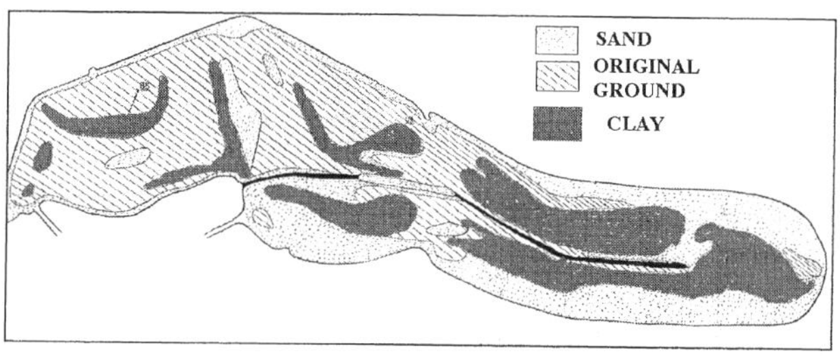

Figure 3. Observed distribution of bottom sediments in ChCP in 1983 (Shiklomanov, 1992).

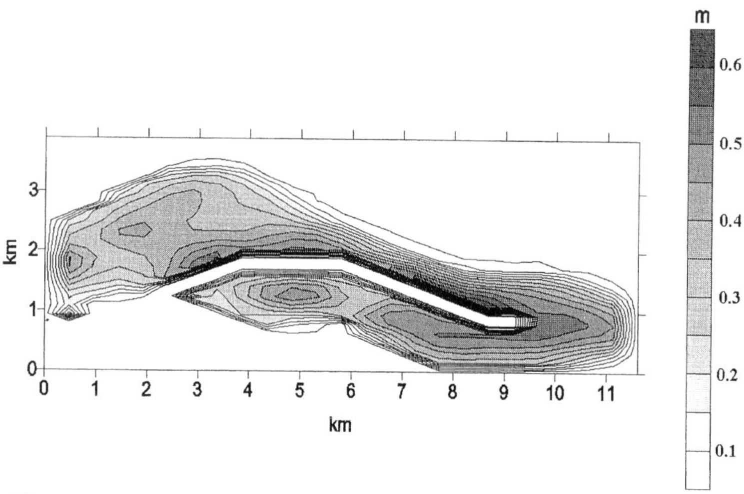

Figure 4. The simulated deposition rate integrated over 10 years period.

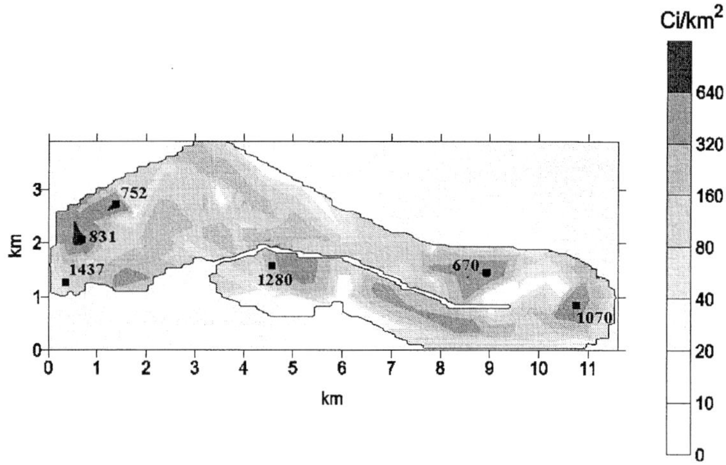

(a)

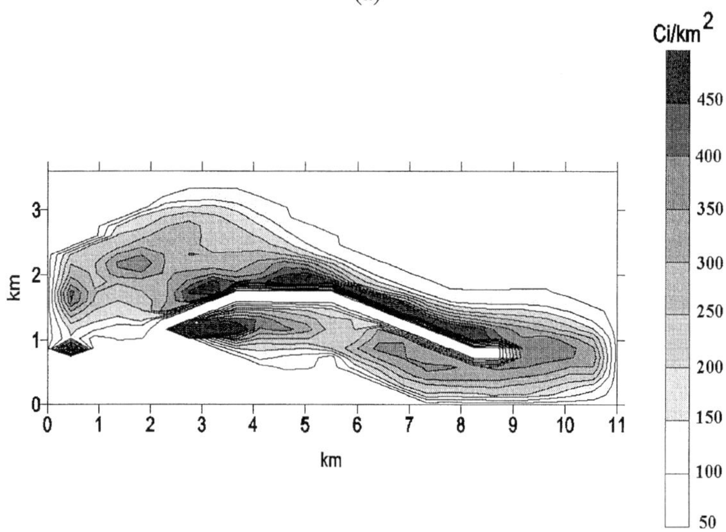

(b)

Figure 5. Measured (Shiklomanov, 1992) (a) and computed (b) distributions of ^{137}Cs in the bottom sediments of ChCP.

The deposition rate was simulated by THREETOX for one year and then linearly extrapolated for a period of 10 year. The computed spatial distribution deposition in the pond (Figure 4) is similar to the measurements (Figure 3). The distributions of ^{137}Cs in the bottom sediments as measured in 1989 (Shiklomanov, 1992) and simulated over one year (from 1986 to 1987) are shown in Figure 5. Both computed and measured distributions of the bottom contamination correlate with the pattern of the sedimentation rate in the ChCP as shown in Figure 3.

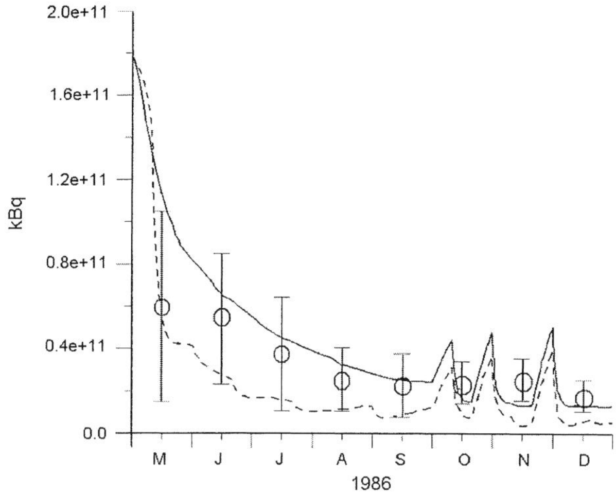

Figure 6. Computed and measured contents of the ^{137}Cs in the water column of CPP. The solid line corresponds to K_d^w = 3 m^3/kg, the dashed line corresponds to K_d^w=15 m^3/kg.

Measured (Shiklomanov, 1992) and computed ^{137}Cs contents in the water column of the pond are depicted in Figure 6. Peaks on the curves in the autumn and winter are explained by the storm events resulting in resuspension of contaminated bottom sediments. According to the simulations, more than 95 % of total ^{137}Cs in the cooling pond had been deposited into the bottom sediments before the end of 1986. This is in agreement with the estimate BIOMOVS (1996).

3.2. Dnieper-Boog estuary

The Dnieper-Boog Estuary (DBE), located on the north-west coast of the Black Sea, is the largest estuary of this sea, with a surface area of 1006.3 km^2 and a volume of 4.24 km^3 (Figure 7). It is connected with the Black Sea through the Kinbourn Strait, located at the left-hand side of the figure. The regime of this drowned-river estuary varies from stratified to partially mixed. The sources of freshwater discharge are the River Dnieper and the River Southern Boog. DBE is at the end of Chernobyl's riverine radionuclide transport from the Chernobyl accident area to the Black Sea. The bottom sediments in the DBE are sandy at the river mouths. The bottom sediments in the other parts of the DBE mainly consist of cohesive sediments.

The simulation of the dispersion of radionuclides that entered the DBE after the Chernobyl accident was carried out for the period May 1986 - April 1988. To diminish the effect of the uncertainty sea level variations at the open boundary, the calculations were carried out in two nested areas. Temperature, salinity, velocity and sea elevation fields of the North -Western Black Sea resulting from a large area model calculation, were used as open boundary conditions for a nested model of higher resolution. The monthly-averaged wind, with a stochastic component, and the monthly-averaged air temperature were specified according to Simonov and Altman (1991).

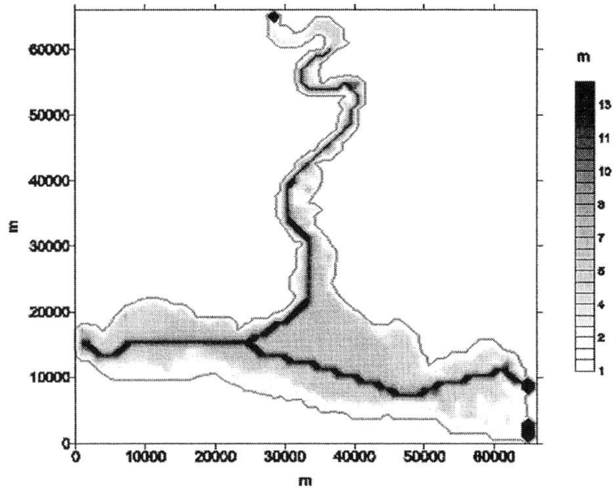

Figure 7. The bathymetry of the DBE.

The monthly-averaged concentrations of suspended sediment at the mouth of the River Dnieper and at the mouth of the River Southern Boog were specified according to data from the State Water Cadastre. The initial thickness of the upper bottom sediment layer was set at 2 cm. Monthly-averaged concentrations of dissolved and adsorbed radionuclides in the river water at the mouths of the Dnieper and the Southern Boog were prescribed by using data from Batrakov et al. (1994), Polikarpov et al. (1988) and Polikarpov et al. (1992).

The computed salinity, presented in Figure 8, is compared with survey data that are represented in the figure by the values with decimals. The salt-water intrusion into the DBE takes place mainly in the summer season, when the water discharge from the River Dnieper is low. Along the navigational channel of the DBE, the density-induced deep undercurrent results in a wedge of salty Black Sea water.

In Figure 9, the computed vertical distributions of dissolved ^{137}Cs and ^{90}Sr along the DBE and adjacent shelf are shown for July 1987. The patterns of isolines for the ^{137}Cs and ^{90}Sr concentrations are similar, but the concentration gradients are opposite for the two radionuclides. The concentrations of ^{137}Cs increase seawards and towards the bottom, while those of ^{90}Sr decrease. Differences between distribution of dissolved ^{137}Cs and ^{90}Sr concentrations in the mouth of the River Dnieper are explained by the differences in atmospheric fallout and by the behaviour of these radionuclides throughout the River Dnieper basin. As noted by Voitsekhovich (1997) almost 100% of ^{137}Cs, but 70% of ^{90}Sr having reached the River Dnieper, was deposited in the chain of reservoirs of the river.

The correlation of the dissolved radionuclide concentration and the salinity describes the mixing processes in an estuary. Deviations of the salinity *(S)* and the radionuclide concentration *(C)* from the area-averaged values ($\overline{S}$ and $\overline{C}$, respectively) were normalised to the maximum difference in salinity, ΔS, and to the radionuclide concentration ΔC in the area, respectively, to produce a non-dimensional representation (Figure 10). The computed data points represent the spring, summer, autumn and winter seasons of 1987. Both ^{137}Cs and ^{90}Sr data points converge into characteristic curves. To clarify the reason for the non-linear character of the dependence between ^{137}Cs and salinity, an additional simulation was carried out without exchange between the dissolved and attached phases. The calculated points showed a perfect linear correlation, represented by the straight line (2) in Figure 10b. Consequently, the non-linear character of the correlation between the salinity and the dissolved ^{137}Cs concentration in water was due to the exchange of ^{137}Cs with bottom and suspended sediments.

Field data of Polikarpov et al. (1988), collected over the north-western shelf in 1989, are represented by crosses in Figure 10b. They correlate quite well with the computational results. Due to a relatively weak exchange of dissolved ^{90}Sr with sediments, the correlation between salinity and dissolved radionuclide concentration was nearly linear (Figure 10b). The relatively weak quadratic deviations from linearity are due to the non-equilibrium mixing processes in the spring and in the autumn of 1987.

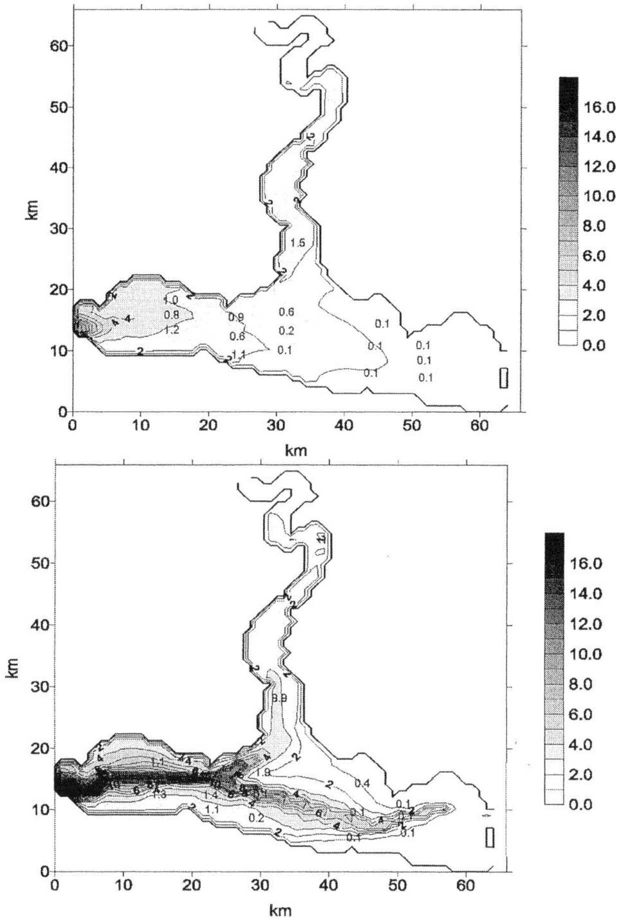

Figure 8. Calculated and observed salinity at the surface and on the bottom of the DBE in June 24, 1987.

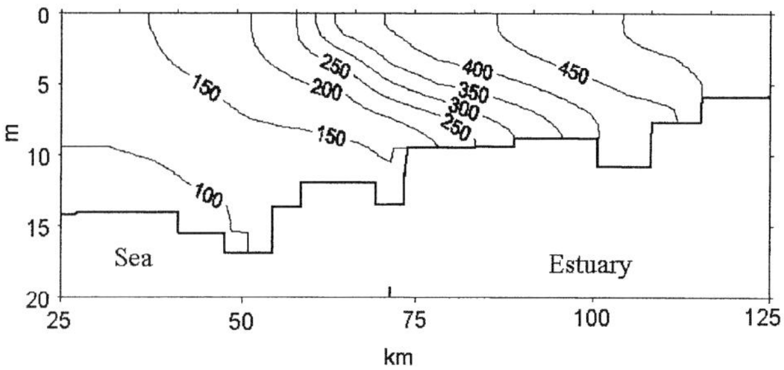

(a)

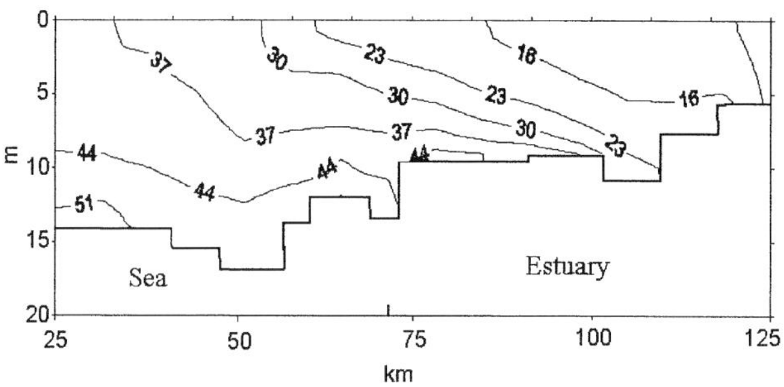

(b)

Figure 9. Computed vertical distribution of dissolved ^{137}Cs (a) and ^{90}Sr (b) along the DBE and north-western shelf in July 1987.

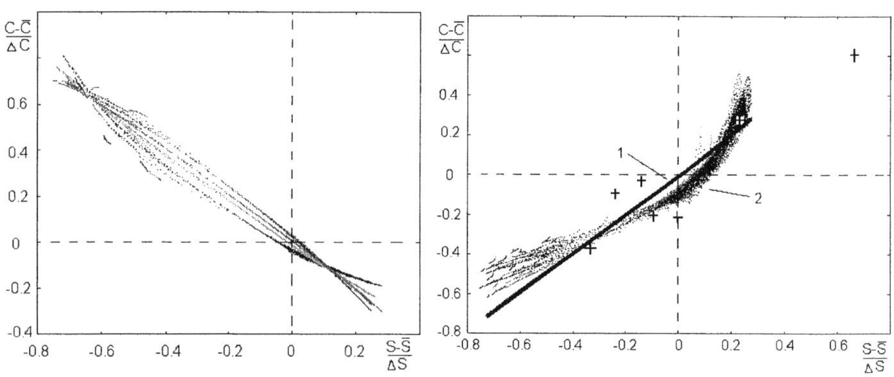

(a) (b)

Figure 10. Correlation between the dissolved ^{90}Sr concentration and salinity (a); Correlation between the dissolved ^{137}Cs concentration and salinity (b).

Calculations showed, that in the spring of 1988 more than 90% of the total ^{137}Cs inventory (0.72 TBq) in the DBE was deposited in the bottom sediments, while 76% of the total ^{90}Sr inventory (1.3 TBq) was in the dissolved phase. The ^{137}Cs flux from the DBE to the Black Sea, during May 1986 till April 1988, was found to be equal to 0.67 TBq, 48 % from the total flow into the DBE (1.39 TBq), while the ^{90}Sr outflow from the DBE to the Black Sea was equal to 15.5 TBq, 92 % from the total inflow (16.8 TBq).

4. DISCUSSION AND CONCLUSIONS

The 3-D simulation of the fate of radionuclides in the Cooling Pond of the Chernobyl NPP in the period 1986-1992 was performed on the basis of initial atmospheric fallout data. There is reasonable agreement between measured and computational data for the radionuclide concentration in the water and in the bottom sediments. Both cohesive sediments and radionuclides were more intensively deposited in deepest parts of the cooling pond. The model reproduced the role of mud sedimentation as a main factor determining the location of the most contaminated spots of bottom sediments of this deep water body.

The Chernobyl NPP has been closed in December 2000 as an installation for nuclear energy production. However, because of technological reasons the cooling pond will be maintained in the current condition for the next several years. After this period, the pumping of river water into the cooling pond will be terminated and its water level will drop to 6 m, leaving 50% of the sediments exposed to the atmosphere. This is considered as a potential source of radiological risk for further wind resuspension. A modelling tool should be developed to predict the ^{137}Cs redistribution in the bed during this period and to support the remediation strategies. The results of the present study show that the proposed approach based on the simulation of dynamics of 3-D fields of suspended sediments and radionuclides in the pond could be used as background for such a tool. The dominant role of mud sedimentation in the redistribution of ^{137}Cs in bottom deposition of the Chernobyl Cooling Pond was confirmed by this study. Therefore, 3-D models that could perform an accurate simulation of mud dynamics should be used to simulate the dynamics of the bed contamination according to the scenarios of the diminishing of the pond's water level and to support the remediation activities.

At the edge of the salt intrusion into an estuary, the radionuclide deposition rate into the bottom increases under the influence of two processes. The first process is flocculation in these areas that intensifies the sedimentation rate. The second process is the increase of the sediment contamination due to higher values of the distribution (partition) coefficient in salt water in comparison with the typical values for fresh water. While some studies were undertaken to show that the distribution coefficients are dependent upon salinity, it is still difficult to define a certain parameterization for this complicated, sorption driven, mechanism (Carrol and Harms, 1999). Therefore during these model runs, a single distribution coefficient value was used for the whole estuary. The ranges of the distribution coefficient values, as used within this study, cover the overlapping ranges of estimates of the fresh and salt water distribution coefficients (Onishi et al. 1981; IAEA, 1985). The undertaken 3-D modeling study shows that differences in the total fluxes and distribution of radionuclide having mid-magnitude (^{137}Cs) and low-magnitude (^{90}Sr) distribution coefficient values could be quantified on the basis of the chosen schematization of the sediment transport and of the radionuclide – sediment exchange processes.

The model should be improved taking into account the influence of flocculation on 3-D mud transport and using distribution coefficients depending on salinity to simulate the increase of radionuclide scavenging in the mixing zones of fresh and salt water. The collection and processing of data on radionuclide concentration in bottom sediments in estuaries together with hydrological data could lead to the basis for such model refinement.

ACKNOWLEDGEMENTS

We would like to thank Rudie Heling (NRG, The Netherlands) for valuable comments on the manuscript. This article benefited from the comments and suggestions of two anonymous reviewers. The work was partially supported by the EU Contract RODOS, INTAS 97-31278 and by contract of the Ukrainian Ministry of Emergencies and Population Protection Affairs from Consequences of the Chernobyl Catastrophe.

REFERENCES

Ariathurai, R. and Krone, R. B., 1976, Finite element model for cohesive sediment transport. *Journal of Hydraulic Division ASCE*, (104) 2, 323-328.

Batrakov, G.F., Eremeev, V.N., Chudinovskikh, T.V. and Zemlyanoy, A. D., 1994, Radioactivity of the Black Sea, Ecosi-Hydrophysics, Sevastopol.

BIOMOVS II, 1996, Technical Report No. 10, Assessment of the Consequences of the radioactive Contamination of Aquatic Media and Biota. Model Testing Using Chernobyl Data, Swedish Radiation Protection Institute, Stockholm.

Blumberg, A.F. and Mellor, G.L., 1987. A description of a three dimensional coastal ocean circulation model, In: Three-Dimensional Coastal Ocean Models, N. Heaps (ed), Am. Geoph. Union, Washington, D.C. 1-16.

Carroll, J. and Harms, I.H., 1999, Uncertainty analysis of partition coefficients in a radionuclide transport model, *Water Research*, (33) 11, 2617-2626.

IAEA. Sediment Kd and concentration factors for radionuclides in the marine environment. IAEA Technical Report No.247, International Atomic Energy Agency, Vienna, 1985.

Koziy, L., Maderich, V., Margvelashvili, N. and Zheleznyak, M., 1998, Three-Dimensional model of the radionuclide dispersion in the estuaries and shelf seas. *Journal of Environmental Modeling and Software*, (13) 5-6, 413-421.

Margvelashvili, N., Maderich, V., and Zheleznyak, M., 1997, THREETOX - a computer code to simulate three-dimensional dispersion of radionuclides in stratified water bodies, *Radiation Protection Dosimetry*, (73) 1-4, 177-180.

Margvelashvili, N., Maderich, V.and Zheleznyak, M., 1999, Simulation of radionuclide flux from Dnieper-Bug Estuary into the Black sea, *Journal of Environmental Radioactivity*, (43) 2, 157-171.

Rijn van, L.C., 1984, Sediment transport. Part II: Suspended load transport, *Journal of Hydraulic Engineering*, (110) 11, 1613-1641.

Onishi, Y., Dummuller, D.C. and Trent, D.S., 1989, Preliminary Testing of Turbulence and Radionuclide Transport Modeling in Deep Ocean Environment, Report PNL-6853, Pacific Northwest Laboratory, Richland, Washington.

Perianez, R., 2000, Modelling the tidal dispersion of ^{137}Cs and ^{239,240}Pu in the English Channel, *Journal of Environmental Radioactivity*, (49) 3, 259-277.

Polikarpov, G.G., Timoschuk, V.I. and Kulebakina, L.G., 1988, Concentration of ^{90}Sr in the aquatic environment of Lower Dnieper toward the Black Sea, *Dopovidi (Proceedings) of National Academy of Sciences of Ukraine*, ser. B, 3, 75-76 .

Polikarpov, G.G., Livingston, H.D., Kulebakina, L.G., Buesseler, K.O., Stokozov, N.A. and Casso, S.A., 1992, Inflow of Chernobyl ^{90}Sr to the Black Sea from the Dniepr river, *Journal of Estuarine, Coastal and Shelf Science*, (34) 2, 315-320.

Santschi, P.H., and Honeyman, B.D., 1989, Radionuclides in aquatic environments, *Radiation Physics and Chemistry*, (34) 2, 213-240.

Shiklomanov, I.A. (ed.), 1992, Hydrological, thermal, chemical and radiological regime of the Cooling Pond of Chernobyl NPP, Tech. Report, State Hydrological Institute, Leningrad.

Simonov, A.I. and Altman, E.N. (Eds.), 1991, Hydrometeorology and hydrochemistry of seas of USSR. v.IV, Black sea, 1, Hydrometeorological conditions, Hydrometeorological Publ., S.-Petersburg.

Voitsekhovich, O.V. (ed.), 1997, Radioecology of water objects of the Chernobyl NPP accident impact area, Chernobylinterinform, Kiev.

Zheleznyak, M., 1988. Structure of the bottom turbulent boundary layer under the waves, *Hydromechanics*, (58), 1-8.

Zheleznyak, M. , Demchenko, R., Khursin, S., Kuzmenko, Yu., Tkalich, P. and Vitjuk, N., 1992, Mathematical modeling of radionuclide dispersion in the Pripyat-Dnieper aquatic system after the Chernobyl accident, *The Science of the Total Environment* (112), 1, 89-114.

Zheleznyak, M., Shepeleva, T., Sizonenko, V. and Mezhueva, I., 1997, Simulation of countermeasures to diminish radionuclide fluxes from Chernobyl zone via aquatic pathways, *Radiation Protection Dosimetry*, (73) 1-4, 181-186.

Episodic transport of organic-rich sediments in a microtidal estuarine system

F. G. Marván[a], S. G. Wallis[a] and A. J. Mehta[b]

[a]Department of Civil & Offshore Engineering, Heriot-Watt University, Riccarton, Edinburgh, EH14 4AS, UK*

[b]Department of Civil & Coastal Engineering, University of Florida, 345 Weil Hall, Gainesville, Florida 32611, USA

Episodic transport of organic-rich sediment was examined in the Ortega/Cedar estuary system in Florida using a newly developed 2D horizontal model for fine cohesive sediment transport. Bed sediment samples were analyzed to provide information on organic content, settling velocity, erodibility and consolidation for inclusion in the model. Only a rudimentary calibration of the model was possible due to lack of data. Nevertheless the model reproduced the main tidal and sediment transport features of the estuary system, including the predominantly depositional nature of the system. In a more sophisticated version of the model, sediment erosion and deposition were manipulated by implementing organic content dependent functions, derived from the analysis of data from several sites. This allowed the model to be used to investigate the sensitivity of the sediment transport to the organic content. The changes in deposition rates caused by varying the organic content were found to be significant during large river discharge events, but during normal discharges they were insignificant. At approximately 40% organic content, there is a tendency for sediment accumulation rates to decrease and, in places, erosion becomes the dominant sediment transport process.

Keywords

Modeling, Ortega/Cedar Estuary, erosion, deposition, organic content

1. INTRODUCTION

In Florida's highly biologically active estuarine and lacustrine environments, the fraction of fine-grained sediment that is organic is often of the order of 20-60% by weight and sometimes as high as 90-95%. There are three main sources of this organic matter. Terrestrial systems tend to be abundant in carbon (C), and the biomass produced by woodland and grassland is of the order of 50g C/m^2 (Mehta et al., 1997). Much of this material is degraded within the soil but some of it is washed away and introduced into fresh water and marine

*contact email: stevew@civ.hw.ac.uk

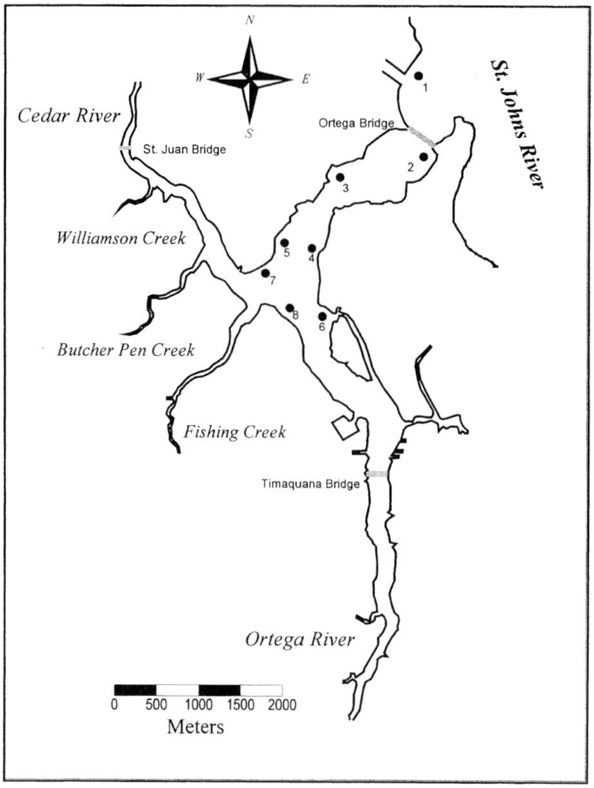

Figure 1. The study area showing the Ortega and Cedar rivers adjacent to the St Johns River: numbers indicate location of core sites (after Cooper & Donoghue, 1999).

environments. The composition of this material is mainly cellulose that degrades very slowly. Aquatic plants, although breaking down more easily, also contribute to the input of organic matter. The third source of organic matter is provided by phytoplankton, which typically has a biomass of 1.5g C/m^2 with 5-6 crops/year for the Florida region. Trefry et al. (1992) state that the coastal waterways in Florida are stressed by inputs of fine-grained organic-rich sediments from riverine systems. Besides the alterations to the benthic community that this input causes, there are indirect problems associated with organic sediment such as sorption of contaminants like Cd, Cu, Hg, Pb, Zn and PCB's. For example, within the study area of this paper (see Figure1), PCB's in sediments of the Cedar river have been documented at up to 0.023 ppm (Campbell et al., 1993), and detectable amounts (up to 0.055 ppm) are also found in every species of fish collected from the area.

Most of Florida's estuaries are microtidal, hence an important hydrodynamic feature of the region is the occurrence of episodic events such as heavy rainfall and storms which act as natural dredging mechanisms due to the strong currents they generate.

The primary aim of the study reported here was to investigate the role of organic matter on the transport of the suspended sediment in the Ortega/Cedar system. This was achieved by (a) developing a numerical model of the estuary; (b) identifying the dependence of sediment settling velocity and erodibility on organic content; and (c) studying the sensitivity of the model output to the organic content. Two river discharge conditions were considered, representing typical (dry weather) and extreme (storm) events. The role of bed sediment consolidation was omitted in the results presented here.

The average annual rainfall in the Ortega/Cedar basin is approximately 132 cm and the major portion of it falls between June and September (Campbell et al. 1993). Water depth in the study area ranges between 1m and 7m. At the mouth of the Ortega River (where it joins the St. Johns River), the semidiurnal tidal amplitude varies from 0.14 m (neap tide) to 0.28 m (spring tide) with a mean of 0.18m. The bottom and suspended sediment is mostly a mixture of clay, silt and organic matter. Typical suspended sediment concentration is approximately 15mg/l; however, during storm runoff events it may be as much as 105 mg/l.

2. MODEL DESCRIPTION

Both the velocity field and the sediment transport processes in the model of the Ortega/Cedar river system were modeled using a 2D xy (i.e. depth-averaged) MATLAB based code. The advantages of this coding language are that the conventional use of nested loops for determining the solution are avoided by the use of matrix notation, and when dealing with implicit schemes the solution is easily obtained. In addition, the output is easily manipulated to obtain graphical representation. A disadvantage encountered was the speed at which calculations where done and some limitations associated with large matrices.

The hydrodynamic equations in the model were solved with the finite difference semi-implicit algorithm developed by Casulli (1990), in which the water surface elevation is obtained implicitly and the velocities are determined in an explicit fashion. Advective-diffusive transport was calculated with (1) below using a finite-volume explicit method based on the quadratic upstream interpolation (QUICKEST) method of Leonard (1979):

$$\frac{\partial(hC)}{\partial t}+\frac{\partial(huC)}{\partial x}+\frac{\partial(hvC)}{\partial y}-\frac{\partial}{\partial x}\left(D_{xx}\frac{\partial(hC)}{\partial x}+D_{xy}\frac{\partial(hC)}{\partial y}\right)-\frac{\partial}{\partial y}\left(D_{yx}\frac{\partial(hC)}{\partial x}+D_{yy}\frac{\partial(hC)}{\partial y}\right)=S \quad (1)$$

where C is the depth-averaged suspended sediment concentration, u and v are the longitudinal and transverse depth-averaged velocities, h is the water depth, S is a source-sink term, x and y are the longitudinal and transverse co-ordinate directions and t is time. The dispersion coefficients D_{xx}, D_{xy} and D_{yy} are treated as follows (Preston, 1985):

$$D_{xx} = \frac{K_l u^2 + K_t v^2}{C_z\sqrt{u^2+v^2}} h\sqrt{g} \quad (2)$$

$$D_{yy} = \frac{K_l v^2 + K_t u^2}{C_z\sqrt{u^2+v^2}} h\sqrt{g} \quad (3)$$

$$D_{xy} = D_{yx} = \frac{(K_l - K_t)\,uv}{C_z\sqrt{u^2+v^2}} h\sqrt{g} \quad (4)$$

where K_l and K_t are the dispersion coefficients in the longitudinal and transverse directions taken to be 13 and 1.2 respectively (Lin and Falconer, 1995), C_z is the Chézy coefficient and g is the acceleration due to gravity. The source-sink term in (1) accounts for erosion and deposition in the following way (Teisson and Latteux, 1986):

$$S = Q_e + Q_d \tag{5}$$

in which Q_e is the erosion flux and Q_d is the corresponding deposition flux, expressed according to Krone (1962) as:

$$Q_d = -W_s C p \tag{6}$$

where W_s is the sediment settling velocity and p is the probability for deposition defined as $(1-\tau_b/\tau_d)$ in which τ_b is the bed shear stress and τ_d is a critical shear stress for deposition. Following common practice, the parameter τ_d was set to a value above the highest shear stress found in the modeled system, allowing deposition to occur at all times. In contrast, Q_e can not be treated in this way because erosion depends on the shear strength of the bed sediment and is therefore considered in the following form:

$$Q_e = \in_N (\tau_b - \tau_s) \tag{7}$$

where $\in_N$ is the erosion rate constant and τ_s is the bed shear strength.

When sediment deposits it begins to consolidate, thus changing the bed sediment density and consequentially the shear strength of the material. Consolidation can be introduced into the model by calculating the settling rate of the deposited material. Experimental results of Toorman and Berlamont (1993) show that this rate (W_{sc}) can be divided in two identifiable modes, namely loose soil and compacted soil consolidation. When combined and expressed in terms of concentration (instead of excess density), these two modes give the following expression (Jiang, 1999):

$$W_{sc} = W_{sc1} \exp\left(-\frac{C}{C_{s1}}\right) F_t + W_{sc2} \left(1 - \frac{C}{C_{s2}}\right)^{mt} (1 - F_t) \tag{8}$$

where F_t is a characteristic mode transition (loose/compact soil) function.

$$F_t = \exp\left[-\left(\frac{C}{C_t}\right)^{nt}\right] \tag{9}$$

In (8) and (9) W_{sc1} and W_{sc2} are the settling rates of the first and second consolidation modes respectively. C_t is the transition concentration, C_{s1} corresponds to the maximum settling flux concentration, C_{s2} is the saturation concentration (maximum compaction concentration) and mt and nt are sediment dependant constants.

For practical reasons, only the settling and erosion were considered in this initial study reported here. Although consolidation was implemented in the model, it was deactivated since it made the calculations much slower.

3. DATA ANALYSIS

Bed sediment samples were taken from the study area and were analyzed for organic content, settling velocity, erodibility and consolidation. The mean organic content (Oc) was found to be 28%. Previous samples reported in Mehta et al. (2000) show similar results for the sampled area, having values between 22 and 36%. Measurements were also obtained from the St. Johns River Water Management District (SJRWMD), which showed less organic content within the study area (varying between 8 and 22%).

The settling velocity was determined by using a 2m settling column with 9 withdrawal ports. Five different initial concentrations (2, 5, 10, 14.5 and 20 kg/m^3) were used to cover the settling ranges that characterize fine sediment. Samples from the ports were taken over a 3 hour period, and concentrations evaluated using gravimetric analysis. By plotting concentration against settling velocity the curve given by (10) (Hwang, 1989) was fitted to the data points (Figure 2):

$$W_s = \frac{aC^n}{\left(C^2 + b^2\right)^m} \tag{10}$$

where a, b, n and m are empirical coefficients (see Figure 2) and C is expressed in kg/m^3 and W_s in m/s. For the free settling range, a constant settling velocity, Wsf, was provided at $C_1 \leq$ 0.25 kg/m^3.

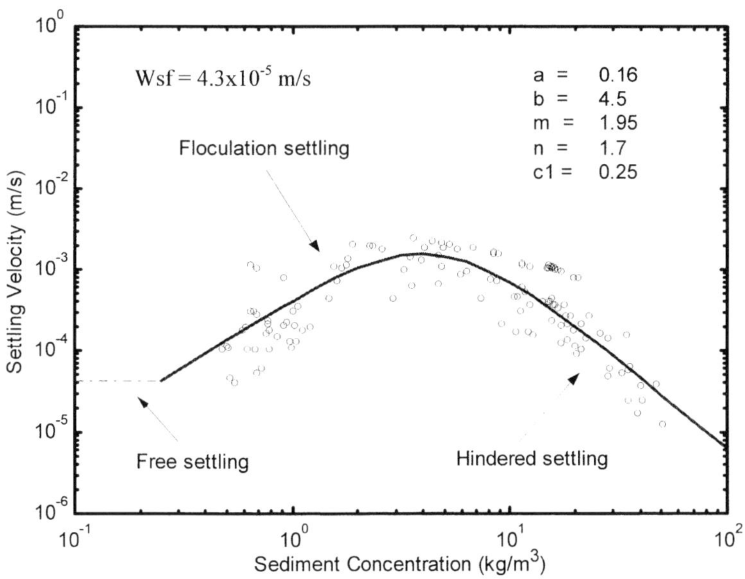

Figure 2. Variation of settling velocity with sediment concentration: symbols are experimental data and the line is equation (10) using empirical coefficient values shown.

Erodibility of the sediment was determined using a Particle Erosion Simulator described by Tsai and Lick (1986). By increasing the oscillation frequency of the device, at 45 minutes intervals, increasing shear stresses were applied to a deposited bed. Within each interval, suspended sediment concentration samples were taken at 5, 10, 15, 25, 35, and 45 minutes. Two tests were carried out using 24h and 75h pre-erosion settling periods, and the following observations were made.

After 45 minutes, the suspended concentration of eroded material for the 24h case was almost twice the value for the 75h case, indicating that the 75h test needed a higher shear stress to achieve the same degree of erosion. This is consistent with the 75h case bed being more consolidated. A stratified erosional behavior was observed for this sample in sympathy with consolidation starting at the bottom of the sediment bed and migrating upwards.

By plotting the erosion rate against bed shear stress, the material's shear strength τ_s was found to be 0.114 Pa, and the erosion rate constant $\in_N$ was 1.024×10^{-3} kg/N s. In the model, τ_s was determined with the following expression (Mehta, 1991), when necessary:

$$\tau_s = \alpha(\phi - \phi_l)^{\beta} \tag{11}$$

where ϕ is the solids volume fraction $=(\rho_D/\rho_s)$, ρ_D is the dry density, ρ_s is the grain density, ϕ_l is a limiting value of ϕ at which $\tau_s = 0$ and α and β are sediment-specific coefficients.

Boundary data for the model was supplied as follows. A mean tide curve was synthesized at the mouth of the Ortega river using published tidal constituents. River discharges for the Ortega river (mean of $1.4 m^3/s$; maximum of $112 m^3/s$) were available from a gauging station and discharges in the other main river inputs (Fishing Creek, Butcher Pen Creek, Williamson Creek, Cedar River) were estimated from their watersheds. A frequency analysis of the Ortega discharge data enabled two flow conditions to be specified: "normal discharge" – $0.85 m^3/s$ occurring 93.8% of the time and "storm discharge" – $78 m^3/s$ occurring 0.16% of the time. The corresponding flows in all the rivers are shown in Table 1.

The landward suspended sediment concentration boundary conditions were furnished from a rating curve, developed by relating values of measured concentration of total suspended solids C to the corresponding values of river discharge Q in the Ortega and Cedar rivers, obtaining the following relation.

$$C = 2 \times 10^{-5} Q^{2.23} \tag{12}$$

At the mouth, this type of boundary condition was also applied but only when the flow was entering the system from the St. Johns River. Since no concentration data were available from

Table 1
Tributary discharges in m^3/s.

Tributary	Normal conditions	Storm runoff event
Ortega River	8.50×10^{-1}	7.80×10^{1}
Fishing Creek	1.60×10^{-1}	1.46×10^{1}
Butcher Pen Creek	4.00×10^{-2}	3.75×10^{0}
Williamson Creek	3.90×10^{-2}	3.64×10^{0}
Cedar River	6.50×10^{-1}	5.52×10^{1}

this site, the following rating curve was developed using measured sediment characteristics and a zero-dimensional re-suspension model (Mehta and Li, 1999):

$$C = 1.65 \times 10^{-2} Q^{0.52} \tag{13}$$

Deposition rate data were obtained from SJRWMD. Bed sediment cores from eight sites (Figure 1) were analyzed by the lead-210 and cesium-137 methods (Cooper and Donoghue, 1999).

The estuary was modeled on a rectangular square grid of size 60m, with bed elevations in each grid square being evaluated from nautical charts. Simulations were undertaken using a time step of 90s, which was small enough to resolve the temporal features of the tidal flow.

4. INITIAL MODEL CALIBRATION FOR THE ORTEGA/CEDAR SYSTEM

Two main scenarios were used for modeling the Ortega/Cedar system: (a) the mean tide curve at the mouth with the "normal discharges" and (b) the mean tide curve at the mouth with the "storm" discharges. In both cases the settling was modeled using (10) with the constants shown in Figure 2. The bed sediment shear strength was set at the experimentally determined value (0.114 Pa).

Several runs were carried out for these two scenarios in which the bed shear stress was adjusted by manipulating the (domain constant) Manning's friction coefficient *n* so that different annual deposition rates could be simulated at the core sites shown in Figure 1. Note that only (flow induced) bed shear stresses were required because all other settling/erosion/deposition parameters were based on the laboratory studies of the estuary sediment. It may appear unconventional to calibrate a sediment transport model by varying a roughness coefficient, but it is entirely logical since the coefficient describes the bed roughness, which is influenced by the sedimentary material, and it also controls the flow velocities generated by the hydrodynamic model, which are fed into the sediment transport model both directly into the advective transport terms and indirectly (via the shear stress) into the erosion and deposition fluxes. The accumulation rate obtained for each scenario was multiplied by the corresponding frequency of discharge and the sum of both scenarios was compared with the core data. In order to determine which value of Manning's coefficient gave closest agreement to the measured deposition rates, the following relation for the root mean square error (e_{RMS}) was applied:

$$e_{RMS} = \left[\Sigma \frac{(h_m - h_p)^2}{N-1} \right]^{1/2} \tag{14}$$

where h_m is the measured deposition thickness h_p is the predicted thickness and N is the number of core sites.

The most accurate results were obtained when the Manning's coefficient was set to 0.027, as shown in the last row in Table 2. This is an appropriate value in view of the fine-grained composition of the bed and the size and shape of the channels (Chow, 1959). The model also reproduced the main tidal features of the estuary, but a detailed hydrodynamic calibration was not possible due to lack of suitable data.

Table 2
Sediment accumulation rate at core sites in mm/year for different Manning coefficients.

Site	Manning's n					Data
	0.024	0.026	0.027	0.028	0.030	
1	10	11.6	12.4	12.8	15	7.56
2	11	12	12.5	13	15.6	4.34
3	10.5	16.6	20.7	23.7	22.6	19.60
4	4.2	6	4.4	2.4	17.8	4.52
5	18.9	8.8	15	15.1	23.8	18.80
6	1.4	1	0.5	1.3	3.9	1.32
7	9	12.2	15	18	19.5	19.35
8	8.4	7.9	9.9	11	13.3	4.90
e_{RMS}	6.00	5.93	4.63	5.02	8.19	

5. IMPROVED REPRESENTATION OF SEDIMENT TRANSPORT PROCESSES

An enhanced representation of the sediment transport processes was sought by relating the settling velocity and the bed sediment shear strength to the content of organic matter in the sediment. Settling velocity data for three similar fine-sediment systems (Hwang, 1989; Burt and Stevenson, 1983; Marván, 2001) were analyzed, see Table3.

A correlation was found between organic content and coefficient a in (10). The manner in which a influences the settling velocity is by a vertical displacement of the settling velocity curve (Figure 2), so that as the organic content increases the settling velocity decreases. For organic content in the range 20%-50% a parabolic equation was fitted through the data points as given by (15) and a constant value of a (0.171) was used for organic content below 20%. The relationship is shown in Figure 3.

$$a = -0.00032Oc^2 + 0.0144Oc + 0.01 \qquad (15)$$

Organic rich sediments are generally fine with agglomerate sizes below 63μm (Mehta et al., 1997). Their density is low compared with silt and clay, and this decreases with increasing organic content. By analyzing data from two sites in Florida (Mehta et al., 1994; Rodriguez et

Table 3
Variation of coefficient a in (10) with organic content (Oc).

Investigator(s)	Oc (%)	a	Aquatic body
Burt and Stevenson (1983)	13	0.17	Thames River, UK
Hwang (1989)	38	0.09	Lake Okeechobee (FL), USA
	40	0.08	
	43	0.027	
Marván (2001)	28	0.16	Ortega River (FL), USA

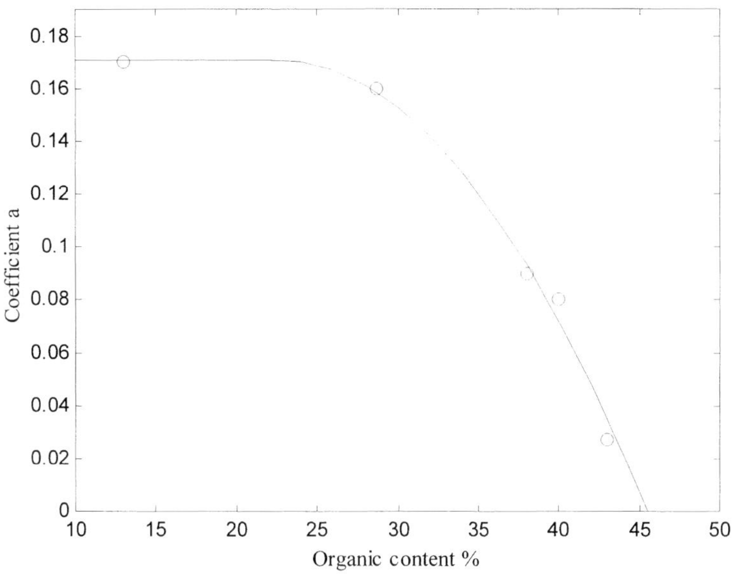

Figure 3. Variation of coefficient a in equation (10) with increasing organic content: symbols are experimental data and the line is equation (15).

al., 1997), correlations were observed between (a) grain density and (b) bulk density versus organic content, as shown by:

$$\rho_s = -16.5\,Oc + 2650 \tag{16}$$

$$\rho_b = 1568\,e^{-0.35\,Oc} - \frac{Oc}{0.9} + 1114 \tag{17}$$

By applying the mass balance equation $[\rho_D=((\rho_b-\rho_w)/(\rho_s-\rho_w))\rho_s]$, where ρ_w is the water density, the dry density ρ_D was obtained, for use in (11). As observed in Figure 4, these equations match the data in a reasonable way.

Shear strength and organic content data from Mehta and Parchure (2000) for different sites (Rodman Reservoir, Kissimmee River and the Ortega/Cedar river system) were assembled with the Ortega/Cedar data. By using (11) with $\alpha = 1.15$ and $\beta = 0.83$, allowing ϕ to vary with organic content (as in (16) & (17)) and setting ϕ_l to be zero, a good fit with this data was found, see Figure 5. A relatively small variation in τ_s can be observed for organic contents above 20%, ranging from 0.112 to 0.107 Pa. Perhaps a better fit could have been obtained but it was important for the curve to pass through 0.114 Pa when the organic content is 28%, because these were the experimentally derived values for the Ortega/Cedar sediment.

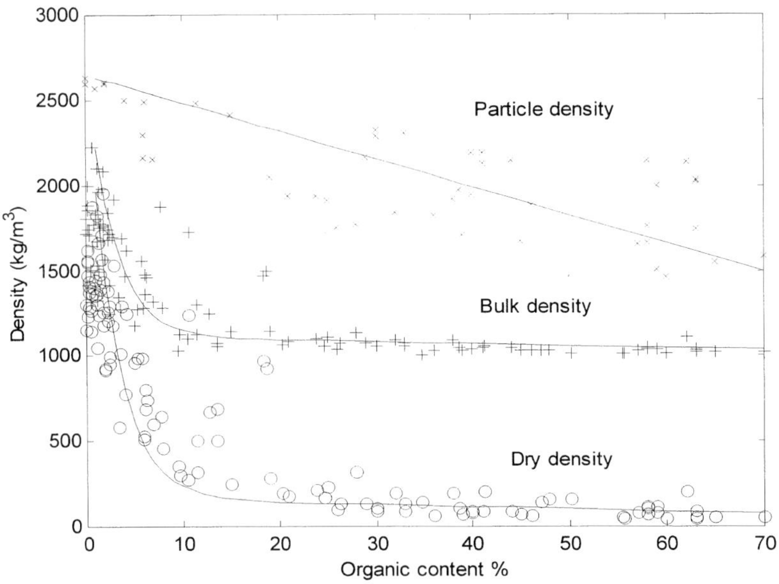

Figure 4. Variation of sediment density (particle, bulk and dry) with increasing organic content: symbols are experimental data and the lines are fitted equations (see text).

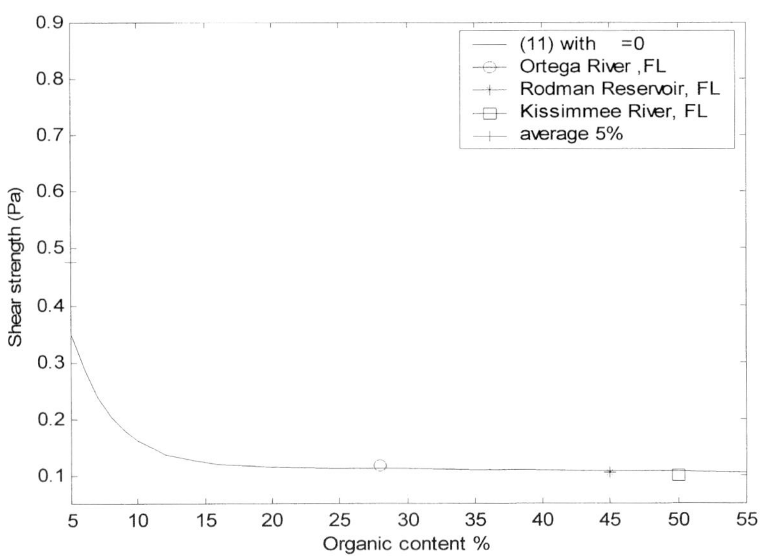

Figure 5. Variation of shear strength with increasing organic content: symbols are experimental data and the line is equation (11) using the empirical coefficients given in the text. (Average 5% refers to lowest organic content data reported in Mehta & Parchure, 2000).

The trend shown in Figure 5 indicates that the shear strength of the material decreases with increasing organic content, suggesting that even if inter-bonding of the agglomerates by biogenic agents exists, it does not provide as much strength as the electrochemical bonds normally associated with highly cohesive sediments.

The model was now run for the two discharge scenarios with a Manning's n of 0.027 and allowing the density, settling velocity and shear strength to vary according to (15) – (17). After running the model for various organic content scenarios (10-45%Oc), significant differences in the accumulation rate at the core sites shown in Figure 1 were found for the "storm" discharge case but virtually no differences were found for the "normal" discharge case. Table 4 shows the sediment parameter values and Table 5 shows yearly accumulation rates derived from the two discharge cases as before. A measure of the sensitivity of the (core site) average sediment accumulation rate to organic content is given in the last row of the table using a mean square error criterion evaluated as before.

Looking at the whole estuarine system in terms of the e_{RMS} values, its sedimentary behaviour appears to be rather insensitive to organic content except when this is greater than 35%. However, examination of individual sites reveals a more complex situation, as discussed below.

Figure 6 shows results for the "storm" discharge case and with 28% organic content. The map shows typical suspended sediment concentrations, and the individual rectangular plots indicate the time history of sediment accumulation at a number of locations over two tidal

Table 4
Variation of sediment parameters with organic content.

	Oc (%)							
	10	15	20	25	28	35	40	45
ρ_D (kg/m^3)	233.4	161.8	144.2	136.2	132.2	123.5	117.3	111.0
ϕ	0.094	0.067	0.062	0.061	0.060	0.060	0.059	0.058
τ_s (Pa)	0.162	0.123	0.115	0.113	0.112	0.111	0.110	0.109
a	0.171	0.171	0.171	0.170	0.162	0.122	0.074	0.010

Table 5
Sediment accumulation rate at core sites in mm/year for different organic content.

Site	Oc (%)							
	10	15	20	25	28	35	40	45
1	8.8	11	12	12.3	12.6	13.7	14.7	11.8
2	10.2	12	12.5	12.6	12.8	13.2	13.8	11.9
3	15.1	27	30.6	29	27.9	18.2	6.3	-23.4
4	11.2	7.6	5.3	4.1	3.5	1.7	-0.2	-6.8
5	13.1	22	23.5	22.3	20.7	13.4	2.5	-21.1
6	6.9	2.8	0.4	-0.7	-1.1	-2	-2.7	-6.3
7	12.1	21.2	22.9	22	20.5	13.4	3.1	-16.1
8	8.5	9.7	10.5	10.8	10.9	10.8	10.3	7
e_{RMS}	5.73	4.98	6.27	5.82	5.53	5.81	11.43	26.63

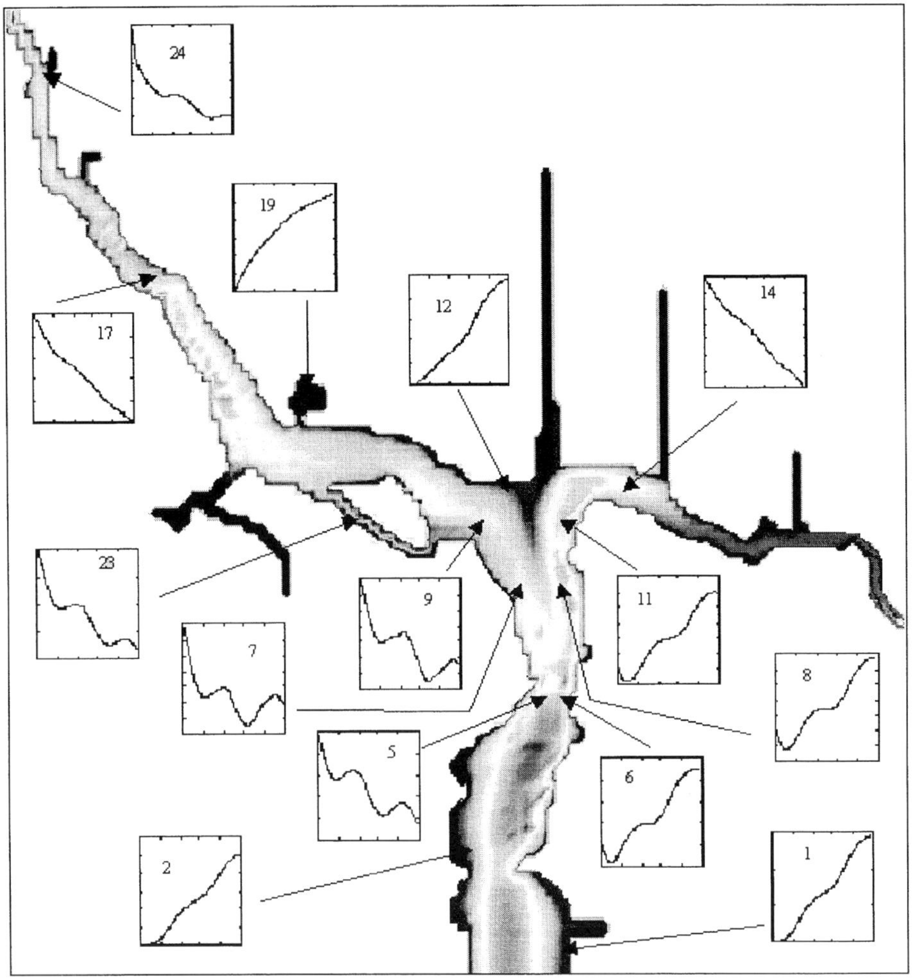

Figure 6. Sediment accumulation in Ortega/Cedar system for "storm" discharge case predicted with an organic content of 28%. Dark shading is low concentration; light shading is high concentration. Boxes show deposition (rising trend) and erosion (falling trend) on a qualitative basis.

cycles. Deposition can be observed near the north bank of the Ortega (sites 1, 6, 8 and 11), while on the south side erosion can be observed (sites 5, 7 and 9). At the lower end of the Cedar river (site 14), a highly erosive zone is observed. Another erosional area is in the upper reach of the Ortega River (site 17), where the water depth is shallower than the adjacent areas. Site 24 also shows erosion but is not as prominent. Site 12, which corresponds to core site 8, is located in a shallow area (approximately 1m deep) where there is also a relatively low flow (0.09 m/s); hence, the area is characterised by deposition but only at small rates.

The tidal influence is strongest near the confluence of the Ortega and Cedar rivers. Note the periods of erosion and deposition in the time history plots at sites 5,6,7,8,9, & 11. At site 19 (within a small marina), however, a nearly monotonic accumulation rate can be observed, which suggests that there is no significant tidal influence and that the sediment transport may be due to turbulent diffusion. Site 6, which corresponds to core site 3, shows the highest accumulation, which is consistent with the core site measurements. This behavior can be attributed to the fact that there is an erosional area upstream of site 6 where the estuary is narrow. As the estuary widens and deepens, so the flow becomes slower and the material tends to deposit.

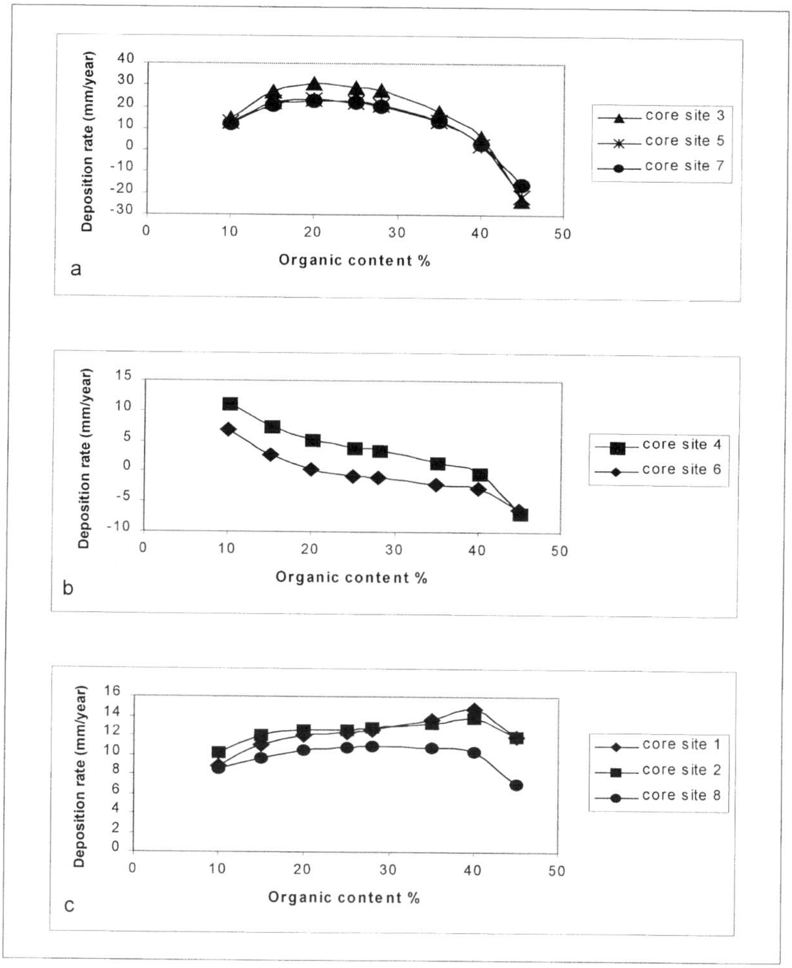

Figure 7. Predicted sediment deposition rates at the core sites with increasing organic content.

Considering the above factors, an explanation can be given as to the behavior observed when the organic content is manipulated. This explanation can be considered in terms of the three plots shown in Figure 7. For sites 3, 5 and 7 a parabolic trend can be observed (Figure 7a). As the organic content increases, in the interval between 10 and 20%, the deposition rate also increases. This behavior can be related to the fact that a highly erosional area is located upstream from where sediment is transported and deposited in these areas (sites 3, 5 and 7). As organic content continues to increase the soil becomes softer so these areas experience erosion as well and deposition starts to decrease, so that these sites become erosional when the organic content rises above 40%. Sites 9 and 7, which correspond to core sites 4 and 6 respectively, behave as erosive areas when storm events occur. As the organic content increases, the soil becomes looser and more erosion occurs resulting in less deposition when the two scenarios (storm and normal conditions) are combined together (Figure 7b).

Sites 1, 2 and 12 (cores 1, 2 and 8) are situated in low flow zones (velocity below 0.1m/s) so most of the sediment transported to these areas is by turbulent diffusion (Figure 7c). When the organic content increases more sediment is suspended in the areas adjacent to these sites and more deposition occurs. The decrease in deposition rate found when the organic content is above 40% can be attributed to erosion taking place at these sites as well as to the decrease in the settling velocity associated with increasing organic content.

6. CONCLUSIONS

A numerical model for cohesive sediment transport was applied to the Ortega/Cedar estuary system in Florida, USA. Using laboratory experiments on samples of the estuary sediment (together with similarly derived data from other estuaries), the effect of the organic content of the sediment on sediment density, settling velocity and shear strength was included in the model. The sensitivity of sediment accumulation rates to the organic content of the sediment was explored through model simulations and some important local variations were found.

In simple terms, increasing the organic content of the sediment reduced the settling velocity, the density and the shear strength. The changes in deposition rates caused by varying the organic content were found to be significant during large river discharge events, but during normal discharges they were insignificant. The variation of deposition rate with organic content is consistent with the flow regime and the morphology of the system, and three different main types of behavior were found. Deposition dominates for organic content less that 35%, but once it exceeds 40% the model indicates a tendency for sediment accumulation rates to decrease and, in places, erosion becomes the dominant sediment transport process.

Further modeling work should include activating the consolidation routine. Another enhancement would be to implement a particle tracking method, so that the movement of sediments with different organic contents can be included. By implementing such ideas, more realistic simulations can be expected to be obtained.

ACKNOWLEDGEMENT

The authors would like to thank Dr. Chandy John and Mr. John Higman of St. Johns River Water Management District, Palaka, Florida for providing relevant data from the study area.

REFERENCES

Burt, T. N., and Stevenson, W.R., 1983. Field settling velocity of Thames mud. *Report IT 251*, Hydraulics Research, Wallingford, UK, 9p.

Campbell, D., Bergman, M., Brody, R., Keller, A., Livingston-Way, P., Morris, F., and Watkins, B., 1993 Lower St. Johns river basin SWIM plan. St. Johns River Water Management District.

Casulli, V., 1990. Semi-implicit finite difference methods for the two-dimensional shallow water equations. *Journal of Computational Physics*, 86, 56-74.

Chow, V.T., 1959. Open-Channel Hydraulics. McGraw-Hill, New York, 680p.

Cooper, W. T., and Donoghue, J. F., 1999. Investigation of historic sedimentation rates in the lower St. Johns River. *Draft report to the St. Johns River Water Management District*, Florida State University, Talahassee, Florida 40p

Hayter, E. J., and Mehta, A.J., 1986. Modelling cohesive sediment transport in estuarial waters. *Applied Mathematical Modelling*, 10, 294-303.

Hwang, K., 1989. Erodibility of fine sediment in wave-dominated environments. *Ph.D. Thesis*, Coastal and Oceanographic Engineering Department, University of Florida, Gainesville, USA.

Jiang, J., and Mehta, A.J., 1999. Consolidation modelling for cohesive sediment transport. *Report UFL/COEL-99/006*, Coastal and oceanographic Engineering Department, University of Florida, Gainesville USA, 25p

Krone, R.B., 1962. Flume studies of the transport of sediment in estuarial shoaling processes. *Final Report*, Hydraulics Engineering Laboratory and Sanitary Engineering research laboratory, University of California, Berkeley, USA .

Leonard, B.P., 1979. A stable and accurate convective modelling procedure based on quadratic upstream interpolation. *Computer Methods in Applied Mechanics and Engineering*, 19, 59-98.

Lin, B. and Falconer, R.A. 1995. Modelling sediment fluxes in estuarine waters using a curvilinear co-ordinate system. Estuarine, Coastal and Shelf Science, 41, 413-428.

Marván, F.G., 2001. A two-dimensional numerical transport model for organic-rich cohesive sediments in estuarine waters. *Ph.D. Thesis*, Heriot-Watt University, Edinburgh, UK.

Mehta, A.J., 1991. Characterization of cohesive soil bed surface erosion, with special refference to the relationship between erosion, shear strength and bed density. Report UFL/COEL/MP-91/4, Coastal and Oceanographic Engineering Department, University of Florida, Gainesville, 83p.

Mehta, A.J., Lee, S.C., Li, Y., Vinzon, S.B., and Aberu, M.G., 1994. Analysis of some sedimentary properties and erodibility characteristics of bottom sediment for the Rodman Reservoir, Florida. *Report No. UFL/COEL-90/008,* Coastal and Oceanographic Engineering Department, University of Florida, Gainesville, USA.

Mehta, A.J., Kirby, R., and Hayter, E.J., 2000. Ortega/Cedar River basin, Florida, restoration: Work plan to assess sediment-contaminant dynamics. *Report No. UFL/COEL-99/019,* Coastal and Oceanographic Engineering Department, University of Florida, Gainesville, USA, 30p.
Mehta, A.J., Kirby, R., Stuck, J.D., Jiang, J., and Parchure, T.M. 1997. Erodibility of organic-rich sediments: A Florida perspective. *Report UFL/COEL/MP-97/01,* Coastal and Oceanographic Engineering Department, University of Florida, Gainesville, USA, 60p .
Mehta, A.J., and Parchure, T. M., 2000. Surface erosion of fine-grained sediment revisited. In: Muddy coast Dynamics and Resource Management, B.W. Flemming, M.T. Delafontaine and G. Liebezeit, eds., Elsevier, Oxford, UK.
Preston, R.W., 1985. The representation of Dispersion in two-dimensional shallow-water flow. *Report TPRD/L/2783/N84,* Technology Planning and Research Division, Central Electricity Research Laboratories, 13p.
Rodriguez, H.N., Jiang, J., and Mehta, A.J., 1997. Determination of selected sediment properties and erodibility of bottom sediments from the lower Kissimmee River and Taylor Creek-Nubbin Slough basins, Florida. *Report UFL/COEL-97/09,* Coastal and Oceanographic Engineering Department, University of Florida, Gainesville, USA,
Toorman, E.A., and. Berlamont, J.E, 1993. Mathematical modelling of cohesive sediment settling and consolidation. In: *Nearshore estuarine cohesive sediment transport*, A.J. Mehta, ed., American Geophysical Union, Washington DC, 167-184.
Trefry, J.H., Chen, N.C., Trocine, R.P., and Metz, S., 1992. Impingement of organic-rich, contaminated sediments on Manatee Pocket, Florida. *Florida Scientist*, 55(3), 160-171.
Tsai, C., and Lick, W., 1986. A portable device for measuring sediment resuspension. *Journal of Great Lakes research*, 12(4), 314-321.

Fine Sediment Dynamics in the Marine Environment
J.C. Winterwerp and C. Kranenburg (Editors)

An Adaptive Finite Element Solution for Cohesive Sediment Transport

David A. Mayne, Asif S. Usmani and Martin Crapper

School of Civil and Environmental Engineering, University of Edinburgh

A 2DV h-adaptive finite element code for solving coupled Navier-Stokes and scalar-transport equations for application to estuarine cohesive sediment transport is described. The program is tested against a well known benchmark problem, the thermally driven cavity problem, and provides results that compare well with existing solutions, generating confidence in the coupling of the governing equations and in the h-adaptive re-meshing routines. Numerical models for varying viscosity non-Newtonian flow, flocculation and settling are described. The program is also used to solve settling in slack estuarial water. The settling speed, general form and development of the hindered settling layer is found to be close to that observed in experimental and field data. h-adaptivity also enables effective capture of the lutocline as it settles.

Key words: Finite element method, h-adaptive re-meshing, modelling, flocculation.

1. INTRODUCTION

The code was developed with the following objectives in mind: (i) to use adaptive grid techniques to refine the simulation of hindered settling lutoclines; (ii) to allow floc formation and break-up to be directly simulated; (iii) to allow the variation in constitutive behaviour with increasing concentration suspensions to be modelled; and (iv) to test simulations of laminarisation of flow due to increasing concentration suspension.

1.1 Governing equations

The governing equations have been solved for a constant density, incompressible Newtonian fluid using the Boussinesq approximation to model buoyancy. This involves solution of the coupled Navier-Stokes and scalar-transport equations, allowing for variable viscosity, and non-Newtonian effects.

Continuity

$$\nabla \cdot \mathbf{v} = 0 \tag{1}$$

where $\mathbf{v}$ represents the velocity.

Navier-Stokes

$$\rho\left(\frac{\partial \mathbf{v}}{\partial t} + \mathbf{v}\cdot\nabla\mathbf{v}\right) + \nabla P = \nabla\cdot\mu\left[\nabla\mathbf{v} + (\nabla\mathbf{v})^T\right] - \rho\mathbf{g}\beta\left(C - C_r\right) \tag{2}$$

subject to boundary conditions:

$$\mathbf{F} = P\mathbf{n} - \mu \left[\nabla\mathbf{v} + (\nabla\mathbf{v})^T\right] \cdot \mathbf{n} \tag{3}$$

$$\mathbf{v} = \bar{\mathbf{v}}\,(x, y, t) \tag{4}$$

and initial conditions:

$$\mathbf{v}\,(t = 0) = \mathbf{v_o} \qquad \text{with} \qquad \nabla \cdot \mathbf{v_o} = 0 \tag{5}$$

μ is the dynamic viscosity, $\mathbf{g}$ is the acceleration due to gravity, β is the volumetric coefficient of expansion, C is the mass concentration, C_r is a reference mass concentration, $\mathbf{F}$ represents the applied tractions on the boundary, ρ is the fluid density and n is the unit normal vector.

Scalar-Transport

$$\frac{\partial C}{\partial t} + \mathbf{v} \cdot \nabla C = \nabla \cdot \alpha \nabla C \tag{6}$$

subject to boundary conditions:

$$\mathbf{n} \cdot (\alpha \nabla C) = q \tag{7}$$

$$C = \bar{C}\,(x, y, t) \tag{8}$$

and initial conditions:

$$C\,(t = 0) = C_o \tag{9}$$

where q is a normal mass flux and α is the diffusivity.

Discretisation in the time domain is achieved by applying the generalised midpoint rule (T.J.R.Hughes, 1983; T.J.R.Hughes, 1987). The choice of unconditionally stable implicit methods is enforced by the use of h-adaptivity as the smallest elements determine the stability of conditionally stable explicit methods, which makes them impractical for use in this context. The formulations described above were implemented in the implicit transient 2DV FE code CADTRAS (Coupled Advective Diffusive TRAnSport model). The code incorporates an unstructured Delaunay triangulation based mesh generator (H.C.Huang and A.S.Usmani, 1994), which allows automatic adaptive re-meshing to take place at each time step if necessitated by the *a-posteriori* error estimation algorithm. Quadratic six-node triangular elements are used for all the meshes resulting in second order accuracy.

1.2 Adaptivity

The finite element method can be optimised by using mesh adaptivity. h-Adaptivity involves altering the size and placement of elements in the domain, the program determining which region needs refining and automatically adapting the mesh to suit the problem. Adaptivity removes the need for trial and error mesh design, automatically designing the optimal mesh for the problem and increasing the accuracy and reliability of finite element analysis. The h-adaptive finite element method is ideally suited to modelling cohesive sediment transport, effectively capturing important flow features that characterise its behaviour.

Adaptivity automatically produces an optimal mesh based on a user specified discretisation error thus saving computational time and focusing effort *intelligently* over successive time steps on areas of high scalar gradients. There are five distinct steps to the iterative adaptive process used here:

1. Solution of the coupled system.
2. Recovery of smoothed scalar gradients using the super-convergent patch recovery (SPR) method (O.C.Zienkiewicz and J.Z.Zhu, 1991).
3. Error Estimation using the *a-posteriori* error calculated at all nodes in the mesh for the scalar field.
4. Re-meshing based on the mesh sizes produced from the previous step.
5. Transfer of all data to the new mesh.

Recovery - In order to calculate an error for the mesh, the finite element approximate scalar field has to be compared to an 'exact' solution, this is not readily available but a more accurate solution can be calculated. Hinton and Campbell (E.Hinton and J.S.Campbell, 1974) show that finite elements produce superior values of scalar gradient at node points after application of a *smoothing* procedure. Their method was based on a global smoothing scheme requiring the solution of a large system of equations. Zienkiewicz and Zhu (O.C.Zienkiewicz and J.Z.Zhu, 1987) state that a globally smoothed value can be used as an higher order approximation of the scalar field. The scalar field generated by the finite element method is most accurate at nodal points whereas the scalar gradients are most accurate at Gaussian integration points, known as the super-convergence phenomenon, see Zlamal (M.Zlamal, 1978). A more efficient and effective procedure was introduced by Zienkiewicz and Zhu (O.C.Zienkiewicz and J.Z.Zhu, 1991), called super-convergent patch recovery (SPR). The smoothed nodal gradients are calculated from the Gauss points on a patch of elements surrounding a node, using a least squares interpolation, for each node in the mesh.

Error Estimation - Once the 'exact' solution has been calculated it can be compared against the unsmoothed solution and prediction of the level of refinement needed to satisfy the error limit can be made. This is achieved by calculating the error in discretisation over the whole domain and specifying new element sizes based on the magnitude of the local errors. Areas of high scalar-gradient tend to need the most refinement as the greatest discretisation errors occur there. The *a-posteriori* error is based upon an *energy norm* (see (R.W.Lewis et al., 1991)).

$$||e||^2 = \int_\Omega (\nabla C)^T \alpha \nabla C d\Omega - \int_\Omega (\nabla \hat{C})^T \alpha \nabla \hat{C} d\Omega \tag{10}$$

if we define,

$$||Q||^2 = \int_\Omega (\nabla C)^T \alpha \nabla C d\Omega \tag{11}$$

$$||\hat{Q}||^2 = \int_\Omega (\nabla \hat{C})^T \alpha \nabla \hat{C} d\Omega \tag{12}$$

then Equation (10) can be rewritten as

$$||e||^2 = ||Q||^2 - ||\hat{Q}||^2 \tag{13}$$

Such a definition allows a practical representation of the error norm in terms of a percentage error η,

$$\eta = \frac{||e||}{||Q||} \text{x} 100\% \tag{14}$$

Re-meshing and mesh generation - Specification of a permissible discretisation error determines the level of refinement throughout the mesh, leading to a predicted reduction or increase in the element sizes so that the new mesh may possess an approximately equal distribution of error. The maximum permissible error for each element is calculated as,

$$||\hat{e}||_e = \bar{\eta} \left(\frac{||Q||^2}{m} \right)^{\frac{1}{2}} \tag{15}$$

where m is the number of elements, $\bar{\eta}$ is the specified maximum percentage error. Dividing $||\hat{e}||_e$ by the calculated error in an element yields a parameter ξ_e as follows,

$$\xi_e = \frac{||e||_e}{||\hat{e}||_e} \tag{16}$$

i.e. if $\xi_e > 1$ the mesh must be refined in the vicinity of element e, conversely, if $\xi_e < 1$ the mesh may be coarsened. The new element size is calculated using,

$$\bar{h_e} = \frac{h_e}{\xi_e^{\frac{1}{p}}} \tag{17}$$

where h_e is the original element size and p is the order of the element shape functions.

Mesh data transfer - Ensuring proper transfer of variables between meshes is crucial for conservation of quantities such as energy and momentum. A transfer strategy using local coordinates of nodal points and elemental shape functions has been used that maps the mesh data accurately. The local coordinates $(\xi - \eta)$ of each node in the adapted mesh are determined with respect to the elements of the previous mesh. Element shape functions are then used to interpolate the data onto the new mesh nodes. This ensures global conservation of mass, momentum and concentration over successive time steps.

1.3 The thermally driven cavity benchmark problem

Thorough testing and benchmarking of the coupled Navier-Stokes and scalar transport equations is crucial in generating confidence in the program's predictive ability. Successful solution of the benchmark over a range of Rayleigh numbers generates confidence in the coupling of the governing equations and in the h-adaptive process. The problem involves modelling fluid flow in a two dimensional square cavity of typical dimension L with the two vertical walls being maintained at a temperature difference of $\Delta\phi$ (see Figure 1). The top and bottom walls are insulated and the velocities at all boundaries set to zero. The fluid inside the cavity is initially at rest and at a temperature which is the mean of the temperatures on the vertical walls. Figure 2 shows the top half of the cavity for four

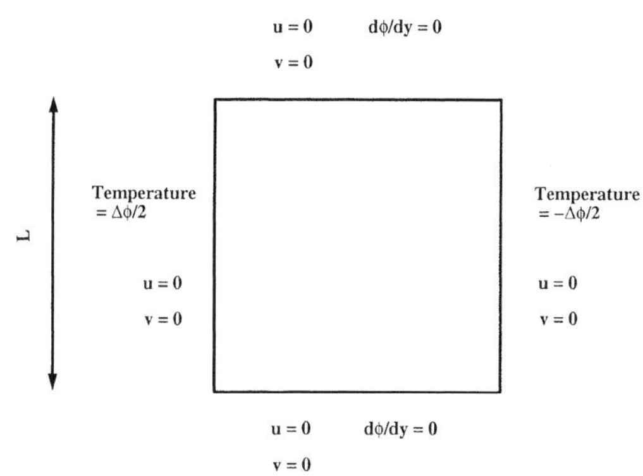

Figure 1: Boundary conditions for thermal cavity benchmark problem.

different dimensionless times. It can be seen that the mesh adapts to *follow* the high temperature gradient front as it passes the departing corner. The refinement of the mesh around the side walls at $t = 0.0001$ (Figure 2(a)) is due to a pre-adaptive loop where the mesh is refined based on the the initial conditions. Figure 2(a) shows mesh refinement around the tightly bunched isotherms at the side walls. Effectively capturing the thin boundary layers that characterise this benchmark is crucial in its successful solution.

2. NUMERICAL MODELS FOR COHESIVE SEDIMENT

This section details the numerical algorithms used to model three complex physical phenomena associated with cohesive sediment: non-Newtonian flow, flocculation and settling.

2.1 Non-Newtonian flow

The material viscosity μ_m varies throughout the flow field. It is an apparent viscosity of the mud/water mixture treated as a continuum and has be calculated using an empirical equation derived by Crapper (M.Crapper, 1995).

$$\mu_m = \rho e^{(0.1096\rho_m - 123.487)} \tag{18}$$

where ρ_m is the bulk density. The constitutive equation is expressed using the four parameter Moore model as set out by Toorman (E.A.Toorman, 1994). This allows the constitutive model to be set by changing four parameters, figure 3 shows the model. Thus the constitutive relationship becomes.

$$\tau = \tau_{ty} + \mu_\infty \dot{\gamma} + \frac{\Delta\tau}{1 + \frac{Y_o}{\dot{\gamma}}} \tag{19}$$

Where $\dot{\gamma}$ is the shear rate; τ is the shear stress; $\Delta\tau$ is the Bingham shear stress τ_B minus the true yield stress τ_{ty}; Y_o is a shear rate given by $\Delta\tau$ / $\Delta\mu$; where $\Delta\mu$ is μ_∞ the Bingham viscosity, minus μ_o, the initial differential viscosity. This allows a smooth transition to non-Newtonian flow based on local aggregate concentration of all size classes present at a given node in the solution domain.

2.2 Flocculation

The numerical flocculation model includes the major contributors to flocculation and floc behaviour: particle geometry, particle numbers, collision mechanisms and inter-particle adhesion. Collision mechanisms in terms of particle size and flow characteristics are examined and the adhesion coefficient is used model adhesion between colliding particles. The effects of temperature, salinity and pH on the flocculation process are not modelled explicitly, they are assumed to be constant across the flow domain. Particle binding organics are taken to reduce the overall density of the primary sediment particles.

The concept of fractal dimension is used to model the ability of fine particles to fill the space as a function of the overall size of the floc created, demonstrated by Kranenburg (C.Kranenburg, 1994), Huang (H.Huang, 1994) and Winterwerp (J.C.Winterwerp, 1999).

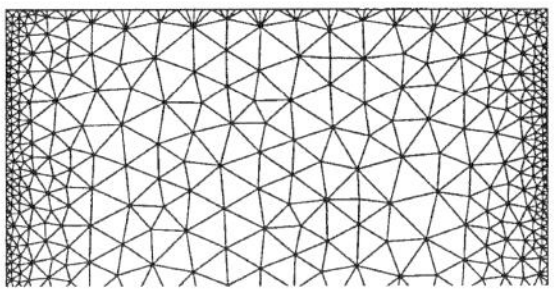

(a) $t^* = 0.0001$, mesh

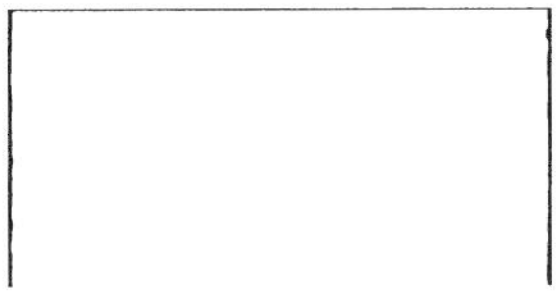

(b) $t^* = 0.0001$, isotherms

(c) $t^* = 0.0001$, velocity vectors

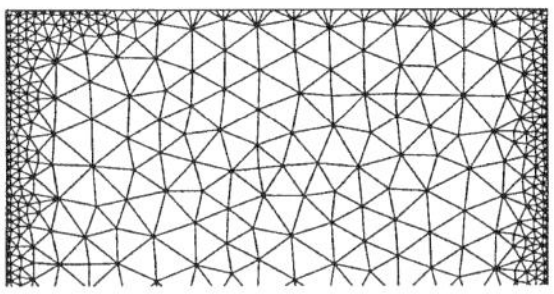

(d) $t^* = 0.0005$, mesh

(e) $t^* = 0.0005$, isotherms

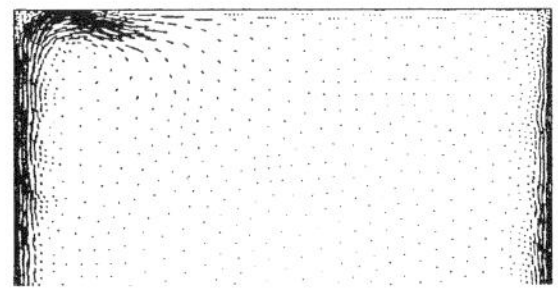

(f) $t^* = 0.0005$, velocity vectors

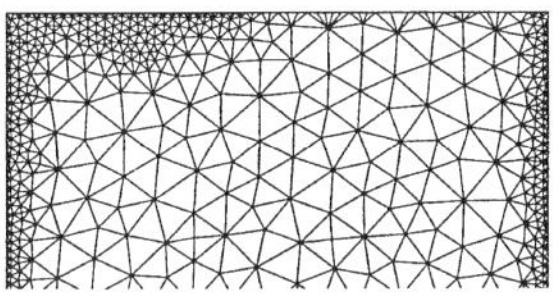

(g) $t^* = 0.0008$, mesh

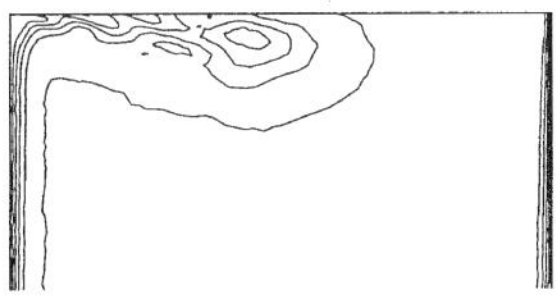

(h) $t^* = 0.0008$, isotherms

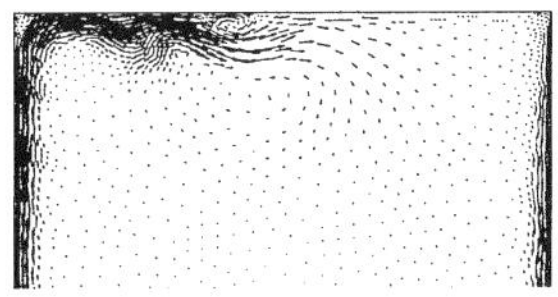

(i) $t^* = 0.0008$, velocity vectors

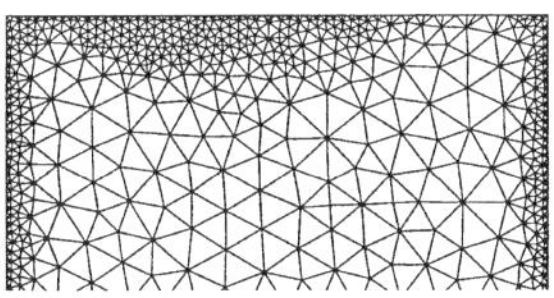

(j) $t^* = 0.001$, mesh

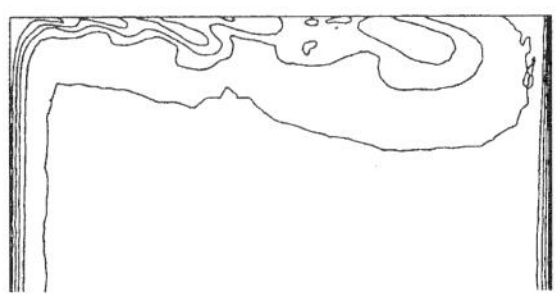

(k) $t^* = 0.001$, isotherms

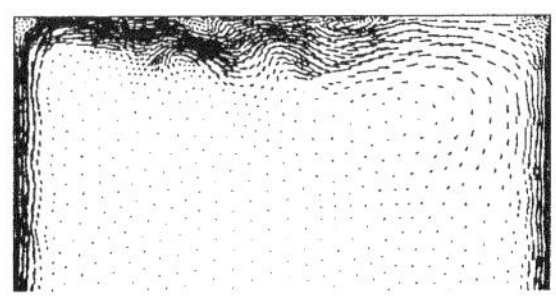

(l) $t^* = 0.001$, velocity vectors

Figure 2: Mesh development for $\mathrm{Ra} = 10^8$.

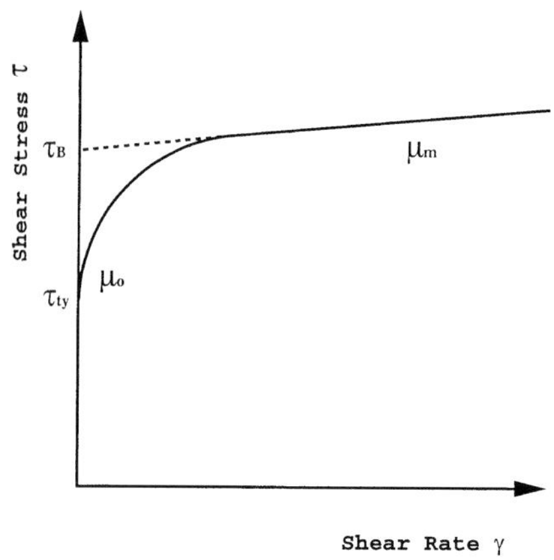

Figure 3: Non-Newtonian four parameter Moore model.

size class	number of particles	floc diameter	floc volume	floc density
1	1	d	V_1	2300.00
2	2	1.414d	$2.83V_1$	1919.23
3	4	2.0d	$8.0V_1$	1650.00
4	8	2.828d	$22.61V_1$	1459.62
5	16	4.0d	$64.0V_1$	1325.00
6	32	5.656d	$180.94V_1$	1229.81
ns	$2\,(ns-1)$	$(I_{ns})^{\frac{1}{nf}}\,d$	$\frac{\pi}{6}d_{ns}^3$	$\left((\rho_s-\rho)\;I_{ns}^{1-\frac{3}{nf}}\right)+\rho$

Table 1: Floc size relationship.

Following Krishnappan (B.G.Krishnappan, 1990; B.G.Krishnappan, 1991) the continuous range of floc sizes contained in the suspension is discretised into a finite number of bins. Table 1 shows the theoretical relationship between each size class in terms of number of particles, floc diameter, floc volume, floc density and the range of each size class where I_{ns} is the number of particles in size class ns. Each size class is treated as a separate set of scalar values to be transported with a scalar-transport equation being solved at each node for each size class. In total there are five parameters that control the number of new flocs created at any node in the numerical model.

$$PN = N_1\; N_2\; \alpha_c\; \Delta T\; K \tag{20}$$

where PN is the number of new flocs created, N_1 and N_2 are the number of particles in the two given size classes interacting, α_c is the adhesion coefficient, ΔT is the time considered for collisions to take place (time step size) and K is the total collision frequency function. K can be written as (B.G.Krishnappan, 1990):

$$K \;=\; K_b + \left[\left(K_{sh}^2\right) + \left(K_i^2\right) + \left(K_d^2\right)\right]^{0.5} \tag{21}$$

Where K_b, K_{sh}, K_i and K_d are the probabilities of inter-particle collision based on the collision mechanisms of Brownian motion, fluid shear, inertial encounters and differential settling respectively. Each collision mechanism has an associated equation *i.e.* K_d can be expressed as:

$$K_d = \frac{2}{9}\frac{\pi g}{\nu}\left(\frac{\rho_s - \rho}{\rho}\right)(r_i + r_j)^2 \; |r_i^2 - r_j^2| \tag{22}$$

where ρ_s is the density of the sediment, ν is the kinematic viscosity and r_i and r_j are the floc diameters in size classes i and j. Further details on the collision mechanism equations can be found in Dyer (K.R.Dyer, 1986), Krishnappan (B.G.Krishnappan, 1991) and van Leussen (Leussen, 1986). The adhesion coefficient α_c varies from 0 to 1 and sets the probability of a floc being created from the collision of two particles.

2.3 Settling velocity

The settling velocity along with hindered settling effects are calculated for each size class at each node in the problem domain using an equation presented by Winterwerp (J.C.Winterwerp, 1999).

3. SLACK WATER HINDERED SETTLING

Settling, flocculation and formation of a hindered settling layer can be studied by solving an estuarial slack water settling problem. The problem involves solution of the uncoupled scalar-transport equations with a uniform concentration of sediment allowed to settle in a 'tank' with a sloped base, see figure 4. This effectively represents a 2-D vertical slice of estuary from water surface to bed. The initial concentration field is set to 20 kg m^{-3} following the floc distribution set out in Roberts and Feates (W.Roberts and N.G.Feates, 1999). The simulation contains seven size classes and assumes flocs to have a fractal dimension of 1.8. Figure 5 shows the evolution of the concentration profile over time at x = 0. Three distinct zones develop: a cleared layer containing mainly unflocculated particles, a hindered settling layer and a settled bed. Moving fluid mud is not present in this test as there is no coupling with the Navier-Stokes equations that provide the negatively buoyant forces and therefore no mechanism for its creation. The collision mechanisms K_{sh} and K_i are effectively zero as they depend on fluid shear, inter-particle collisions being entirely due to differential settling and Brownian motion. The top of the hindered settling layer is clear from the first graph (figure 5(a)) and remains clear throughout the simulation. The layer settles at approximately 0.1 mm/s, this is consistent with experimental data obtained by Ali and Crapper (K.H.M.Ali and M.Crapper, 1993). The shape and development of the concentration profile compares well with experimental data and field measurements generally reported in the literature, see Odd and Rodger (N.V.M.Odd and J.G.Rodger, 1986).

3.1 Adaptivity

Figures 6 and 7 show a sequences of adaptive meshes with the corresponding isolute

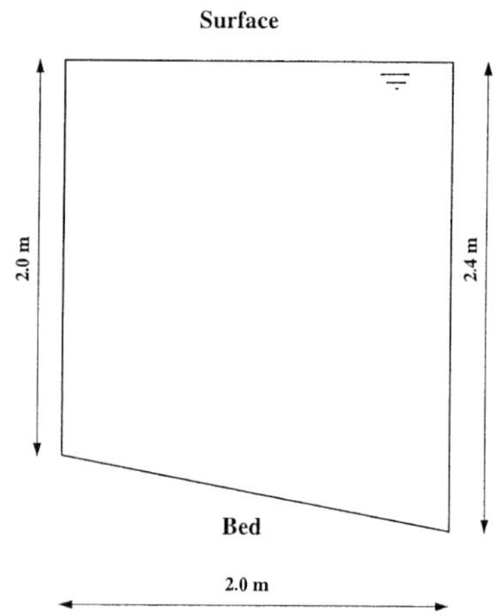

Figure 4: Settling domain geometry.

contour plot for that particular time step. A total of ten adaptive meshes are produced over the 200 time steps that make up the simulation, six are shown. It can be seen from the isolute plots, figures 6(b) to 7(f), that the hindered settling layer forms at the start and persists throughout the run. The sharp nature of the concentration gradients is revealed in the tight bunching of the contours around the lutocline. It can also be seen that the mesh adapts to follow the top of the hindered settling layer as it settles, figures 6(a) to 7(e), effectively capturing the steep concentration gradients. It can also be seen that the bed layer increases in thickness over time. The mesh initially refines to capture sharp gradients at the bottom boundary, the level of discretisation decreasing as the concentration gradient decreases, figures 6(b) to 7(f).

4. CONCLUSIONS

A robust and efficient adaptive finite element model incorporating flocculation modelling and non-Newtonian flow and the damping of turbulence at high concentrations has been developed for application to cohesive sediment transport problems. The program has been successfully benchmarked against the thermally driven cavity problem, generating confidence in the coupling of the governing equations and in the h-adaptive process. Settling of cohesive sediment in quiescent waters was modelled using an initial uniform concentration of 20 kg m^{-3} and results compared qualitatively with experimental data. The settling speed, general form and development of the hindered settling layer was found to be close to that observed in experimental and field data. h-adaptivity also enabled effective capture of the lutocline as it settled.

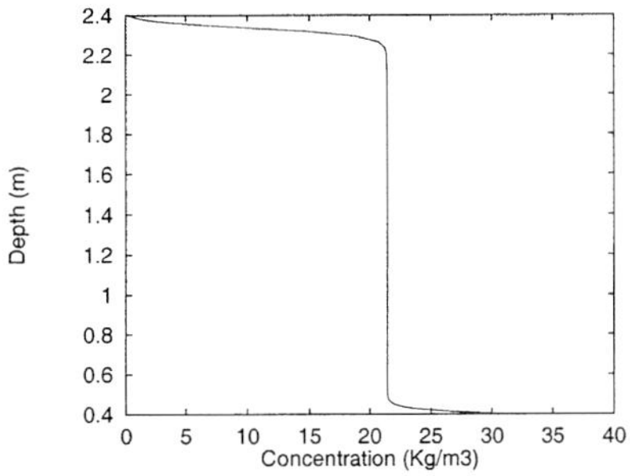

(a) Time = 576 seconds

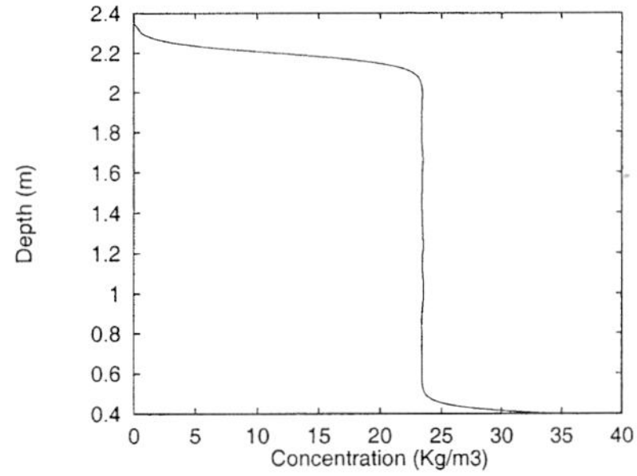

(b) Time = 1680 seconds

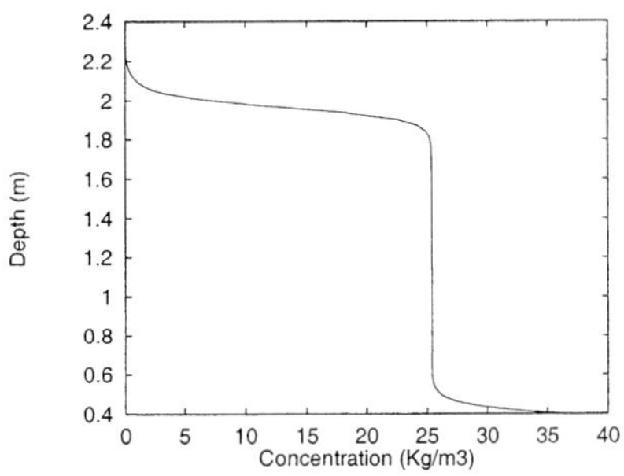

(c) Time = 3600 seconds

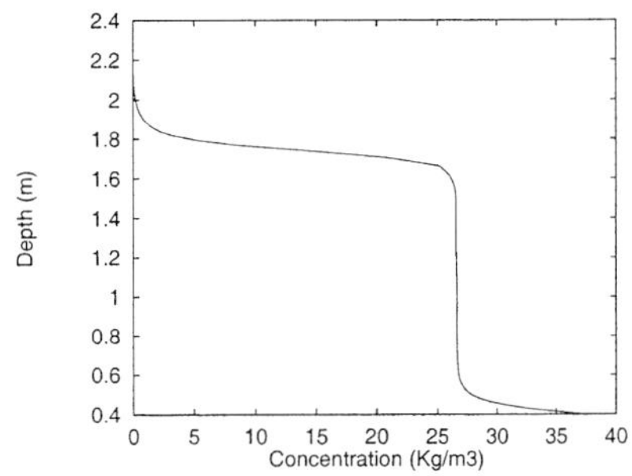

(d) Time = 5520 seconds

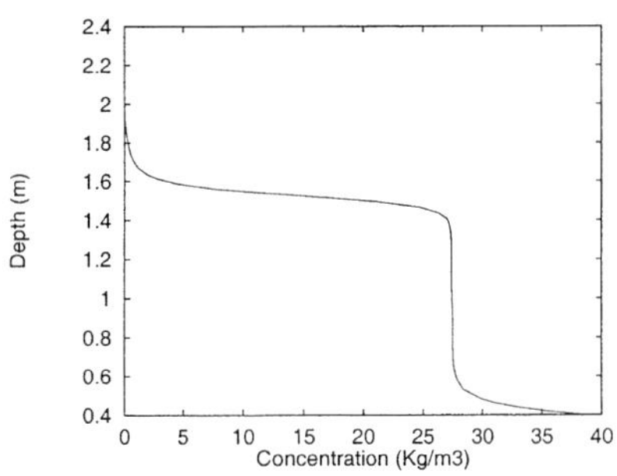

(e) Time = 7392 seconds

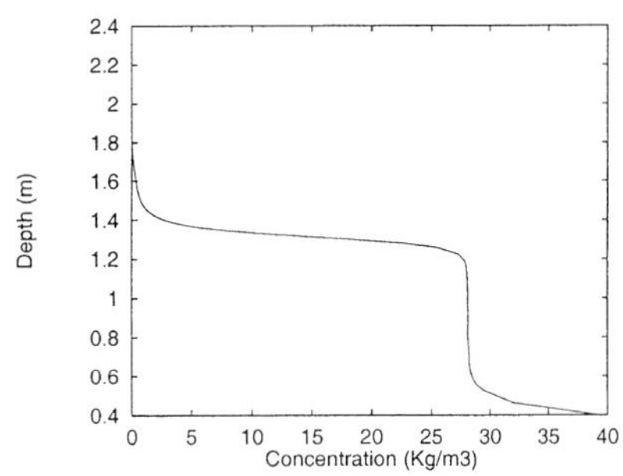

(f) Time = 9264 seconds

Figure 5: Variation in concentration profile over time.

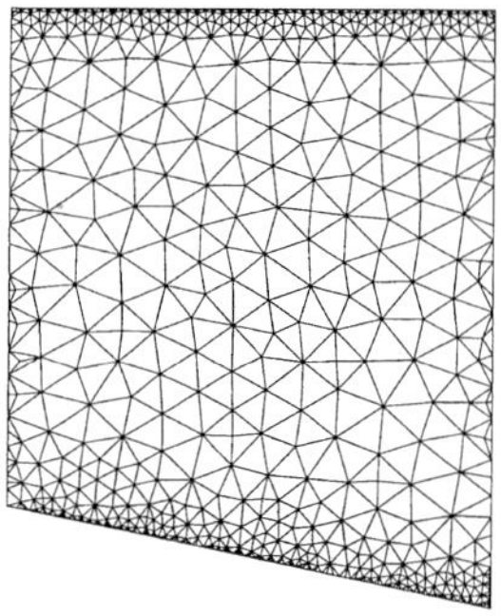

(a) Time = 24 seconds

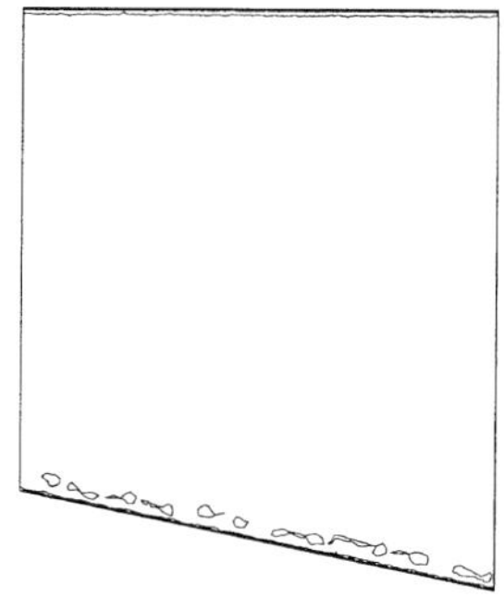

(b) Time = 24 seconds

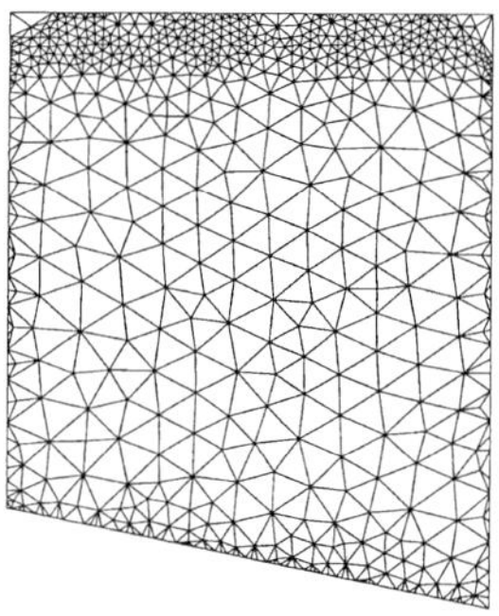

(c) Time = 1368 seconds

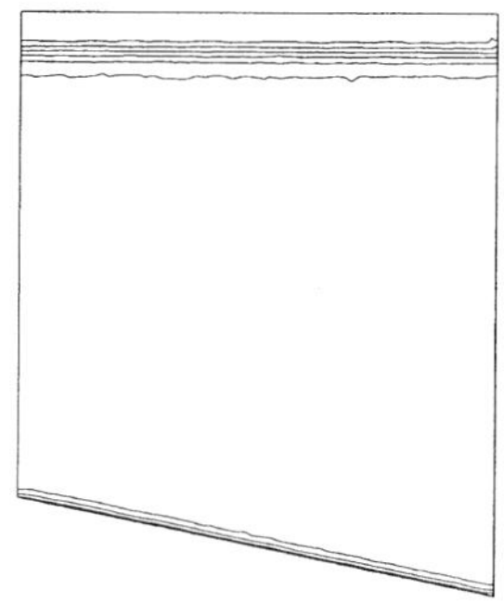

(d) Time = 1368 seconds

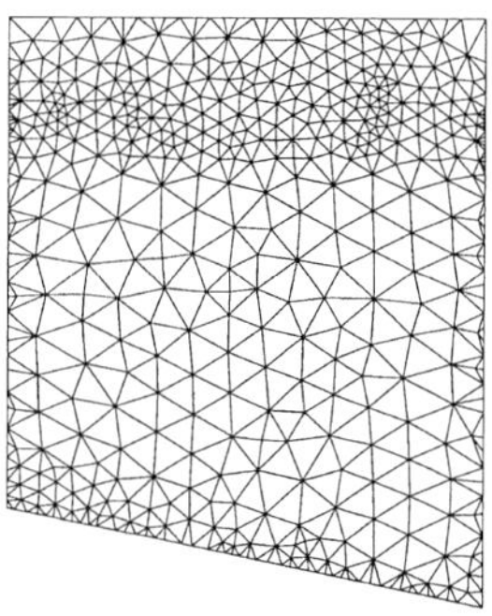

(e) Time = 3648 seconds

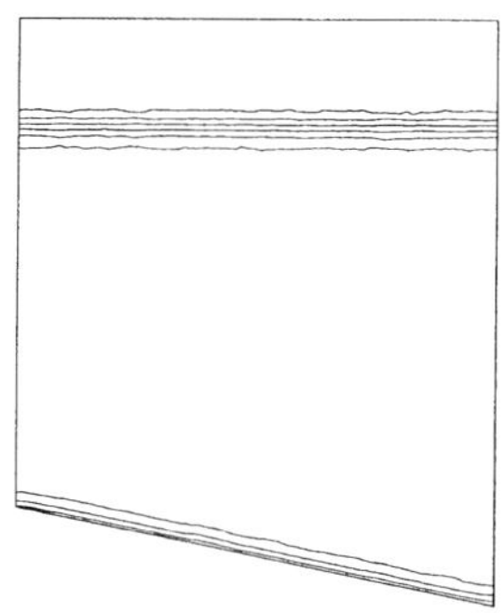

(f) Time = 3648 seconds

Figure 6: Evolution of mesh with changes in concentration field, meshes 3, 6 and 7.

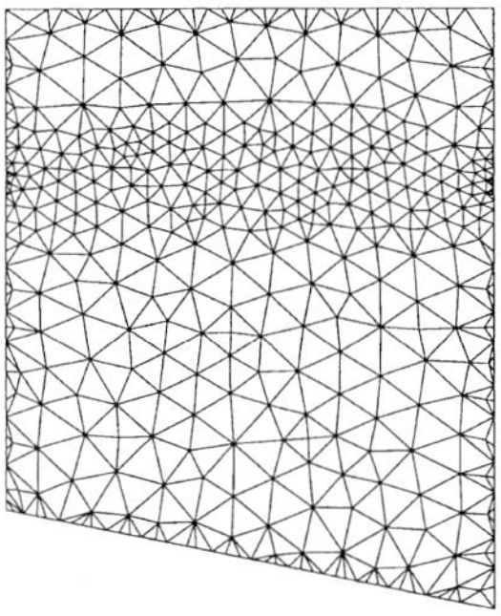

(a) Time = 5676 seconds

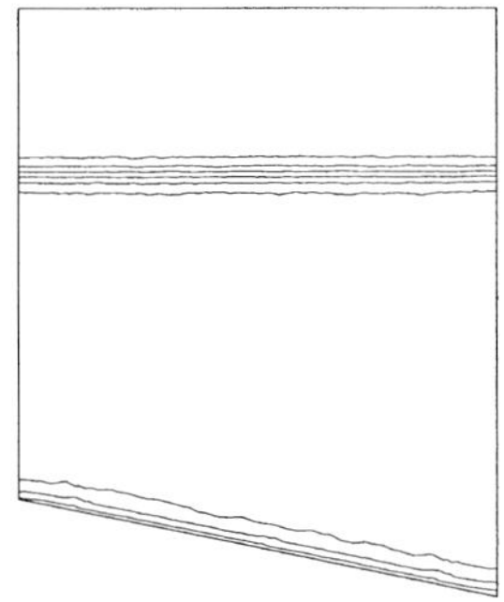

(b) Time = 5676 seconds

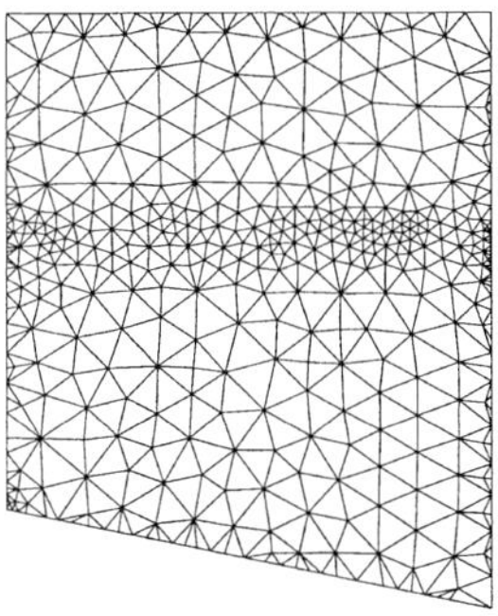

(c) Time = 7668 seconds

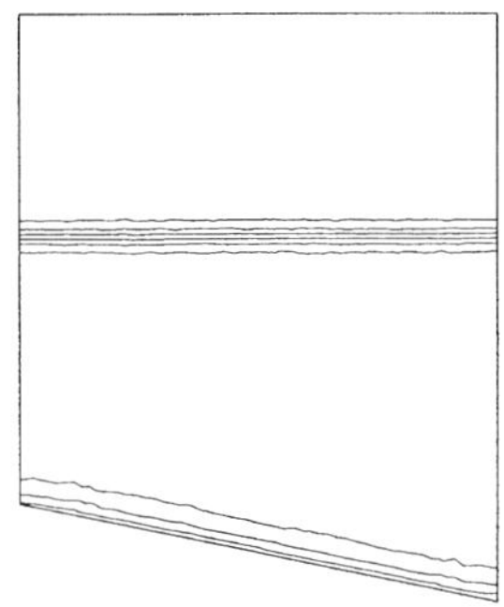

(d) Time = 7668 seconds

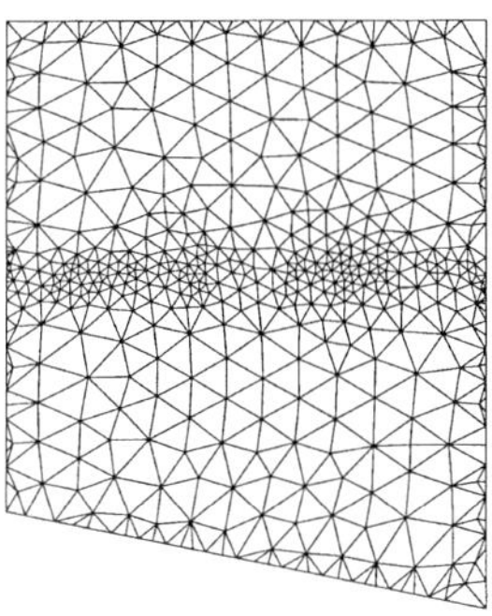

(e) Time = 8928 seconds

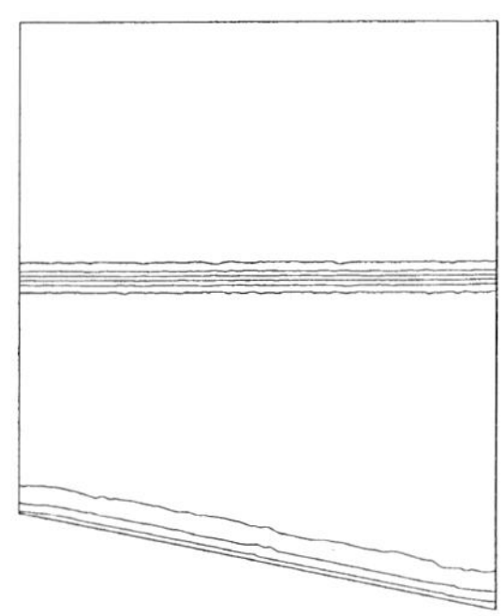

(f) Time = 8928 seconds

Figure 7: Evolution of mesh with changes in concentration field, meshes 8, 9 and 10.

REFERENCES

B.G.Krishnappan (1990). Modelling of settling and flocculation of fine sediments in still water. *Canadian Journal of Civil Engineering*, 17:763–770.

B.G.Krishnappan (1991). Modelling of cohesive sediment transport. In *International Symposium on the Transport of Suspended sediments and its Mathematical Modelling*, pages 433–448, florence, Italy.

C.Kranenburg (1994). The fractal structure of cohesive sediment aggregates. *Estuarine, Coastal and Shelf Science*, 39:415–460.

E.A.Toorman (1994). A review of the use of the concentric cylinder viscometer for cohesive sediment suspensions. In *Cohesive Sediments - 4th Nearshore and Estuarine Cohesive Sediment Transport Conference INTERCOH '94*, Wallingford.

E.Hinton and J.S.Campbell (1974). Local and global smoothing of discontinuous finite element functions using a least squares method. *International Journal for Numerical Methods in Engineering*, 8:461–480.

H.C.Huang and A.S.Usmani (1994). *The Finite Element Analysis for Heat Transfer*. Springer-Verlag.

H.Huang (1994). Fractal properties of flocs formed by fluid shear and differential settling. *Physics of Fluids*, 6:3229–3234.

J.C.Winterwerp (1999). Flocculation and settling velocity. Technical report, Delft Hydraulics.

K.H.M.Ali and M.Crapper (1993). Measuring techniques including the application of medical ultrasound technology to the laboratory study of fluid mud. In *Proceedings of the 25th IAHR Congress*, pages 166–173, Tokyo, Japan.

K.R.Dyer (1986). *Coastal and Estuarine Sediment Dynamics*, chapter 8, pages 203–230. John Wiley and Sons.

Leussen, W. (1986). Aggregation of particles, settling velocity of mud flocs. In *Physical processes in Estuaries*, Netherlands.

M.Crapper (1995). *Fluid Mud Modelling*. PhD thesis, University of Liverpool.

M.Zlamal (1978). Superconvergence and reduced integration in the finite element method. *Mathematics of Computation*, 32:663–685.

N.V.M.Odd and J.G.Rodger (1986). An analysis of the behaviour of fluid mud in estuaries. Technical report, Hydraulics Research Wallingford Limited.

O.C.Zienkiewicz and J.Z.Zhu (1987). A simple error estimator and adaptive procedure for practical engineering analysis. *International Journal for Numerical Methods in Engineering*, 24:337–357.

O.C.Zienkiewicz and J.Z.Zhu (1991). Adaptivity and mesh generation. *International Journal for Numerical Methods in Engineering*, 32:783–810.

R.W.Lewis, H.C.Huang, A.S.Usmani, and J.T.Cross (1991). Finite element analysis of heat transfer and flow problems using adaptive remeshing including application to solidification problems. *International Journal for Numerical Methods in Engineering*,

32:767–781.
T.J.R.Hughes (1983). Analysis of transient algorithms with particular reference to stability behaviour. In *Computational Methods for Transient Analysis*. Elsevier Science Publishers.
T.J.R.Hughes (1987). *The Finite Element Method - Linear Static and Dynamic Finite Element Analysis*. Prentice-Hall International, Inc., New Jersey.
W.Roberts and N.G.Feates (1999). Flocculation field experiment. In *Cosinus, 2nd Annual General Meeting - Book of abstracts*, pages 18–21. Katholieke Universiteit Leuven.

Fine Sediment Dynamics in the Marine Environment
J.C. Winterwerp and C. Kranenburg (Editors)

Numerical modelling of Mud Transport Processes in the Tamar Estuary

Ole Petersen[a], H. Jacob Vested[a], Andy Manning[b], Malcolm Christie[b] and Keith Dyer[b]

[a] DHI Water & Environment, Agern Allé 11, DK-2970 Hørsholm, Denmark, osp@dhi.dk
[b] Institute of Marine Studies, University of Plymouth, Plymouth PL4 8AA, United Kingdom

Transport processes of fine-grained sediments in the Tamar Estuary, UK, are studied using a combination of two- and three-dimensional numerical models and a comprehensive observational data set, collected during a COSINUS field campaign in 1999. The three-dimensional model is based on a hydrostatic version of MIKE 3, combining models for flow, stratification, turbulence and mud transport. Using a two-dimensional flow model of the whole estuary to provide boundary information, a high-resolution three-dimensional model is set up for a section of the upper estuary, containing a pronounced turbidity maximum. The model is calibrated using the observations. A sensitivity analysis is carried out, where various formulations of flocculation effects and of buoyancy effects on the turbulence are investigated. The conclusions are that the models can provide a realistic picture of the mud transport processes, but are sensitive to the specific parameterisation of flocculation.

KEY WORDS
cohesive sediments, fluid mud, numerical modelling, turbulence, experiments

1. INTRODUCTION

Transport of fine-grained sediments as mud and silts is a prominent feature in estuarine coastal regions, where it is responsible for e.g. establishment of intertidal mud flats and may have adverse effects on manmade installations as siltation in harbours. To improve the fundamental knowledge of cohesive sediment transport and increase the technical ability to deal with effects of cohesive sediments, the EC MAST III research project COSINUS has been established. DHI has, as part of the project, applied a three-dimensional cohesive sediment transport model. The present paper describes the application of this model to the Tamar Estuary, UK, where an extensive field measurement program has been undertaken as part of the COSINUS project. The objective of the work has been to provide a realistic comparison of the model with observations in general, and, in particular, of the descriptions of cohesive sediment processes that have been developed as part of COSINUS. The model setup has consequently been focused on the proper resolution of the sediment processes, i.e. the vertical exchange and horizontal advection of sediment, rather than modelling of the estuary as a whole.

The observational basis for the study has been a unique set of comprehensive field measurements of hydrodynamic and sedimentological parameters, collected during a spring and a neap tidal cycle at two stations.

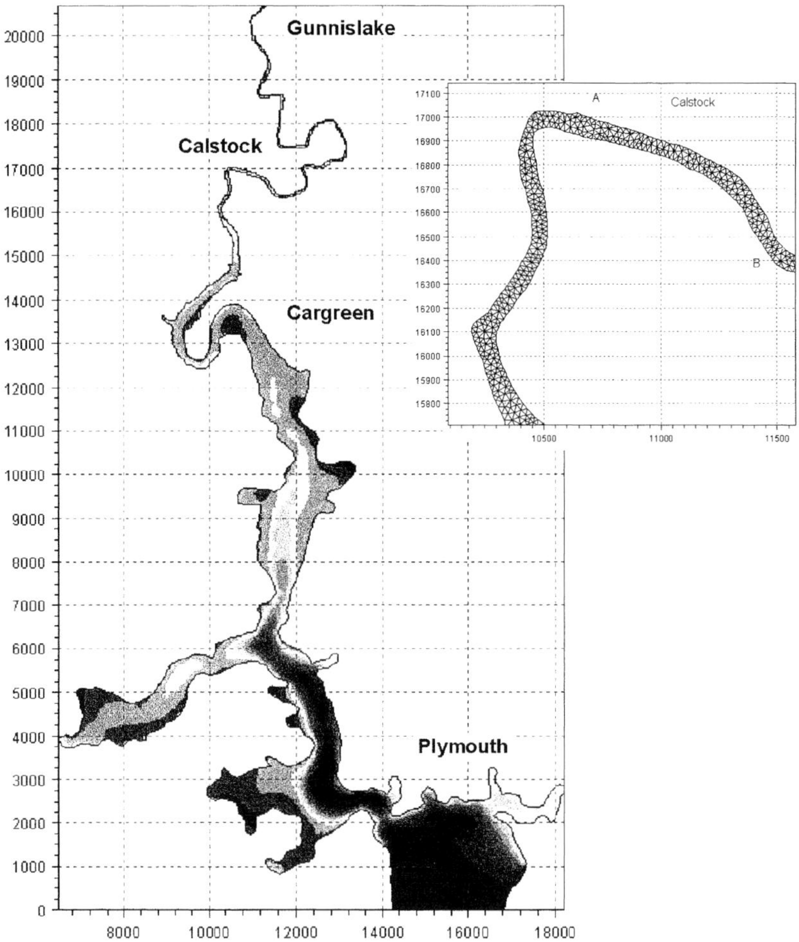

Figure 1. The Tamar Estuary, UK. The inset shows the unstructured computational mesh around Calstock with the two observation stations, Station A and Station B, included.

2. THE TAMAR ESTUARY

The Tamar Estuary is an approximately 30 km long estuary on the southern coast of England, with a relatively wide entrance on the southern coast of England, which narrows down to a 100 m wide tidal channel in the upper 20 km. Tidal effects are prominent all over the estuary, and it contains a considerable amount of fine-grained sediments, with a turbidity maximum extending approximately 5 km around Calstock. The estuary has been the subject of numerous scientific studies on the general hydrodynamics and morphological characteristics, which are described in a number of scientific papers (Uncles and Stephens, 1989; Uncles and Stephens, 1993).

The present detailed field measurements have been made on September 16 and 22, 1999 in two sections, A and B, marked on the map in Figure 1, located approximately 7 and 8 km downstream from the head of the estuary at the Gunnislake weir, where it forms a narrow tidal channel. The measurements cover a spring and a neap tidal period. The location is chosen within the reach where the turbidity maximum forms. The upper reach of the estuary is relatively narrow, with widths from 40 to 100 m and depths in the order of 1-5 m measured from MSWL. The tides in the estuary display some spring-neap variation with a tidal range from 2 to 3 m in the upstream part. Salinity intrusions extend during flood up past Calstock, thus some buoyancy influence on the flow is present.

The field measurements consist of hydrographic observations as water levels and currents during a neap and a spring tidal cycle in the two stations. Further, high-resolution profiles of salinity, suspended sediment and specialised measurements of settling velocity, in-situ floc sizes and other sediment properties are made. The focus of the measurement program has thus been on the local description of sediment processes affecting the vertical exchange rather than the description of the estuary in general. The observations are described in detail elsewhere (Roberts and Feates, 2000; Dyer *et al.*, 2000).

3. THE COHESIVE SEDIMENT MODEL

It has been decided that in order to get the best utilisation of the detailed observations, the flow and sediment processes in the estuary are described using a two-dimensional vertical hydrodynamic model, covering the upper 16 km of the estuary, as this enables a high resolution of the vertical processes as well as an inclusion of advective processes at the same time. The drawback of this approach is that cross sectional variations are not included and that proper seaward boundary conditions are more complicated to establish.

The model is set up using DHI's general three-dimensional model system, MIKE 3. The hydrodynamic model is based on a three-dimensional hydrostatic solution on a vertical sigma co-ordinate system, which is able to adapt to the very large variations in water depth encountered. The horizontal grid is an equidistant Cartesian grid. The model includes transport of salt and temperature, and vertical mixing is based on a $k\varepsilon$-model with sediment induced buoyancy terms included. An implicit solution of the hydrodynamic equations are used and the explicit QUICKEST scheme for transport-diffusion equations.

The transport of suspended sediment is described using

$$D_t c = \partial_x(\Gamma_x \partial_x c) + \partial_y(\Gamma_y \partial_y c) + \partial_z(\Gamma_z \partial_z c) + \partial_z(w_s c) + S_c \quad (1)$$

where D_t is the material derivative, c is the suspended sediment concentration by mass, x, y and z are Cartesian co-ordinates, Γ is sediment diffusivity, w_s is the settling velocity and S is sediment source.

The sediment model describes the vertical distribution of the settling suspended sediment, where the settling rate may depend on concentration and turbulence through flocculation. The sediment exchange with the bed is formulated in terms of deposition and erosion.

The settling is described as by Petersen and Vested (2000), where the implementation is made such that formation of a concentrated benthic suspension layer (CBS) is possible. In order to accomplish this, an upwind implementation of the discrete settling flux F_{SED} between element number i and $i+1$ (i positive upwards) is used such that

$$F_{SED} = w(c_i) \cdot c_{i+1} \tag{2}$$

where $w(c_i)$ is the settling velocity from element $i+1$ to i. This implementation makes it possible to model the formation of CBS layers using a reasonable vertical resolution to resolve the steep density gradients across the interface.

The fixed bed consists of a layered bed with a mass balance for each layer, such that the layer thickness h grows and shrinks according to the net deposition from the suspension.

The upper bed layer is updated according to

$$d_t(h\rho_B) = D - E \tag{3}$$

where ρ_B is density of the material in the bed layer.

The deposition D is calculated as

$$D = w_s c_b p_D \tag{4}$$

where c_b is the concentration just above the bed and p_D is the probability of deposition, depending on the bed stress and floc properties, defined as $p_D = 1 - \tau_b / \tau_D$, τ_D is the critical stress for deposition.

The erosion from the settled bed is described as

$$E = E_o(\tau_b / \tau_E - 1)^m \tag{5}$$

where τ_b is the bed shear stress, τ_E is the critical stress for erosion and E_o and m are calibration constants.

4. MODEL SET UP

The model is set up as a two-dimensional vertical model, with a 100 m horizontal spacing and 15 to 50 layers in the vertical. The model is made such that the surface area upstream Calstock corresponds to the actual area. Boundary conditions for the model are the fresh water discharge at the head and the tidal water level variations and salinity at the seaward boundary. Due to the complex bathymetry, the tide becomes strongly asymmetric, with a rapid, intense flooding phase and a longer ebb phase with slower and more constant currents. In order to get a reasonable estimate of the very dynamic tidal variation at the boundary, simulations covering the whole estuary have been made using a simpler vertically averaged hydrodynamic model, which is based on an unstructured mesh, capable of resolving the narrow upper part of the estuary.

Figure 1 shows the model domain and the mesh around Calstock. The model is forced at Plymouth using idealised tidal waves, and it does not contain meteorological forcing.

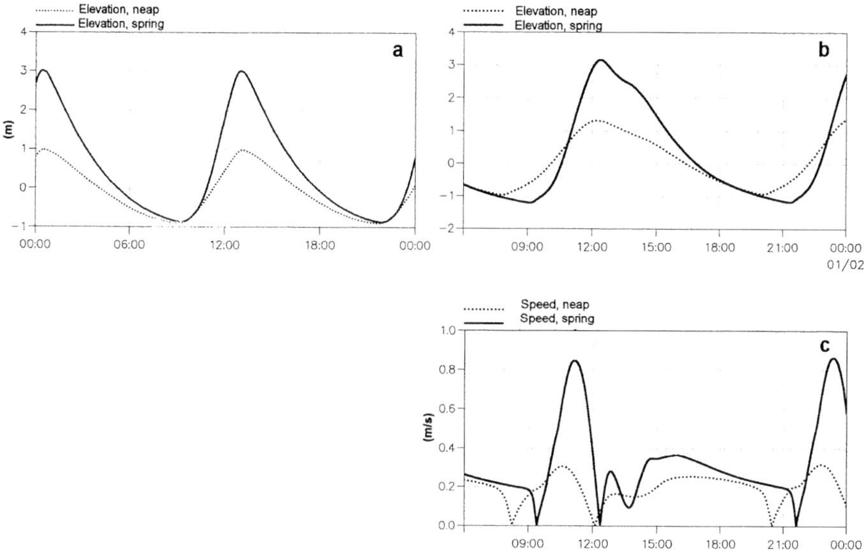

Figure 2. Modelled water level and mean current speed at Cargreen (a) and at Station A during neap (b) and spring tide (c).

From simulations of a spring and a neap tidal cycle, water level variations have been extracted at the boundary location for the three-dimensional model at Cargreen, as shown in Figure 2 (a). Corresponding currents and water level from Calstock are also shown in Figure 2 (b) and 2 (c).

Salinities at the seaward boundary are set at a constant value, such that the salinity variations at Calstock correspond to the observations (not shown). For sediment an SPM concentration of 0 on the seaward boundary is assumed.

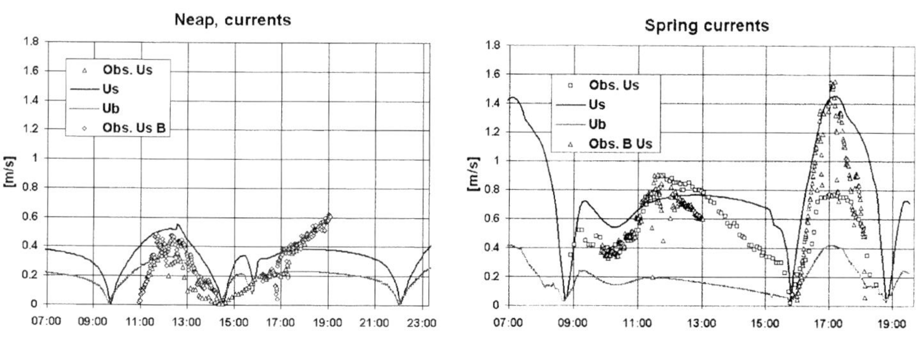

Figure 3. Modelled and observed current speed during spring and neap tide.

5. CALIBRATION OF THE HYDRODYNAMIC MODEL

A calibration of the hydrodynamic parameters, i.e. the bed friction in the form of an effective roughness height, is made using measured current speeds near the surface and the bed in Station A during ebb and flood tide. Examples of modelled and measured time series of velocities are shown in Figure 3 for a spring and a neap period. The calibration shows that the model does reproduce the general pattern of the observed currents. As the model is forced using an idealised tidal wave, some of the discrepancy, especially during neap tide, may arise from local conditions as wind setup or variations in fresh water runoff, which are not included in this idealised model setup.

The simulations cover 3-5 tidal periods, such that the hydrodynamic fields used for the sediment simulations are established after the first period.

The modelled salinity variation presented in Figure 5 shows that during spring tide a vertically mixed saline front advances upstream Station A, producing a very weak stratification at Calstock. At neap tide, the front advances upstream in a similar way as during flood tide, but at high tide a distinct stratification and a corresponding baroclinic circulation develop. The duration of this is approximately 3 hours, corresponding to the observations. Figure 4 shows velocity profiles during the spring period, indicating the existence of a baroclinic current of approximately 0.25 m/s.

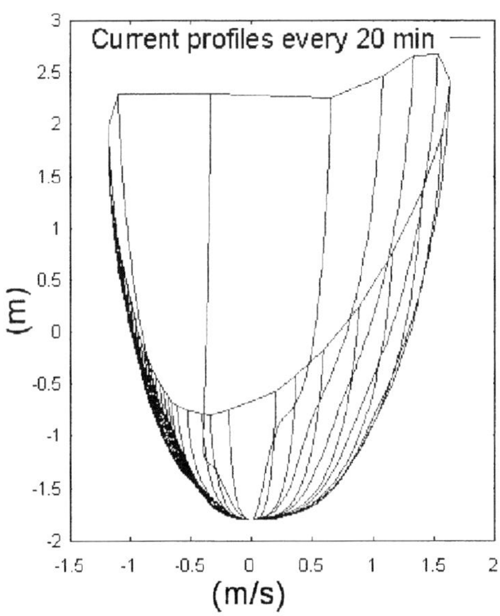

Figure 4. Velocity profiles during a spring tidal cycle at Station A.

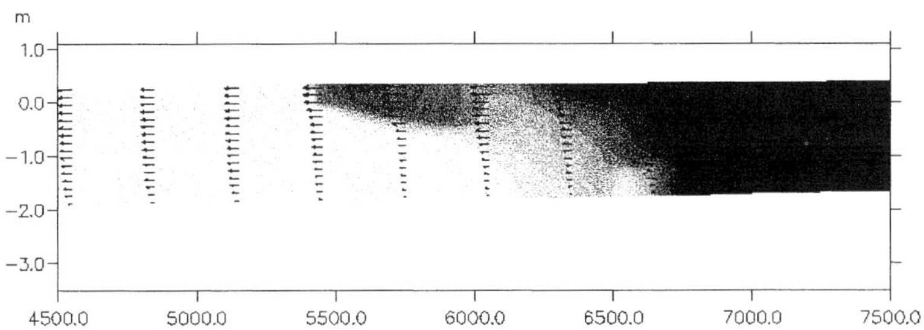

Figure 5. Head of saltwater front at slack water during ebb tide. Light shading indicates high salinity and dark shading indicates low salinity.

The height of the velocity profiles corresponds roughly to the water depth, thus the very large variations in depth are also seen from the figure.

6. COHESIVE SEDIMENT DISTRIBUTION

Initially, a uniform 120 kg/m^2 sediment layer extending 4 km is placed around Calstock. This distribution is estimated from the observations and cannot be expected to represent an equilibrium, but as the main focus here is on the local processes, this may be a reasonable approximation. The parameters for the sediment processes are further chosen with reference to the observations, but adjusted such that the observed concentration levels are reproduced. The resulting set of parameters is shown in Table 1.

Table 1.
Sediment model parameters

	Spring	Neap
Density, ρ_s [kg m^{-3}]	2650	2650
Fall velocity, w_s [mm s^{-1}]	5	5
Critical erosion stress, τ_E [N m^{-2}]	0.5	0.2
Erosion rate, E_o [kg m^{-2} s^{-1}]	1 10^{-4}	5 10^{-5}
Critical deposition stress, τ_D [N m^{-2}]	0.5	0.5
C_H [g l^{-1}]	5	5
Density of bed material, ρ_B [kg m^{-3}]	1200	1200
Equivalent bed roughness, k_s [mm]	5	5

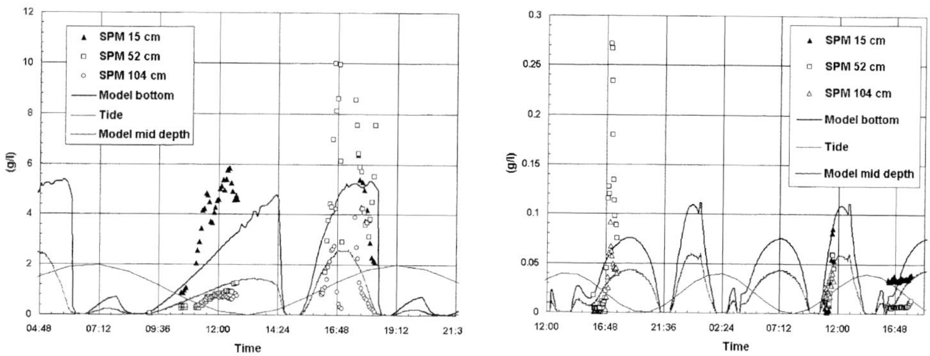

Figure 6. Modelled and observed SPM concentrations at different levels above the bed during spring (left) and neap (right) tide. Tide is tidal elevation shown on an arbitrary scale.

Figure 6 shows modelled and observed development of the SPM concentration in Station A at neap tide and at spring flood. Generally, the model gives concentrations in range with the observations, with some differences in the duration of the peaks.

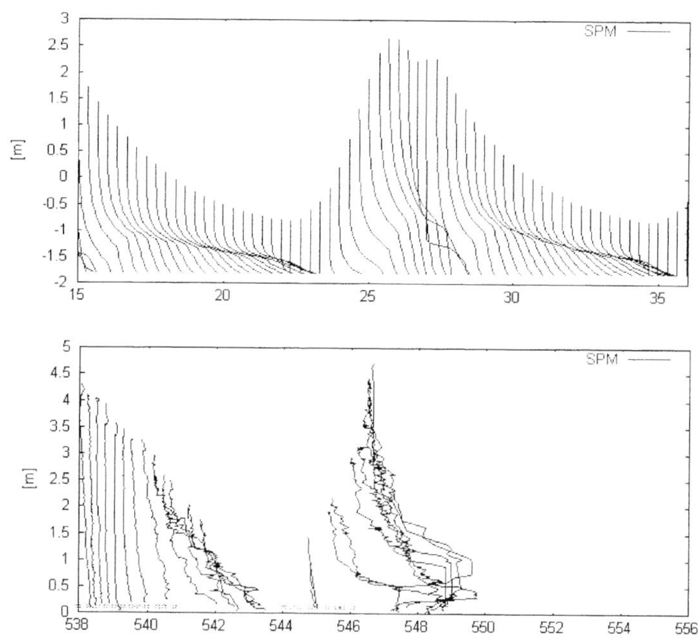

Figure 7. Modelled (upper) and observed (lower) vertical distribution of SPM at Station B during spring tide.

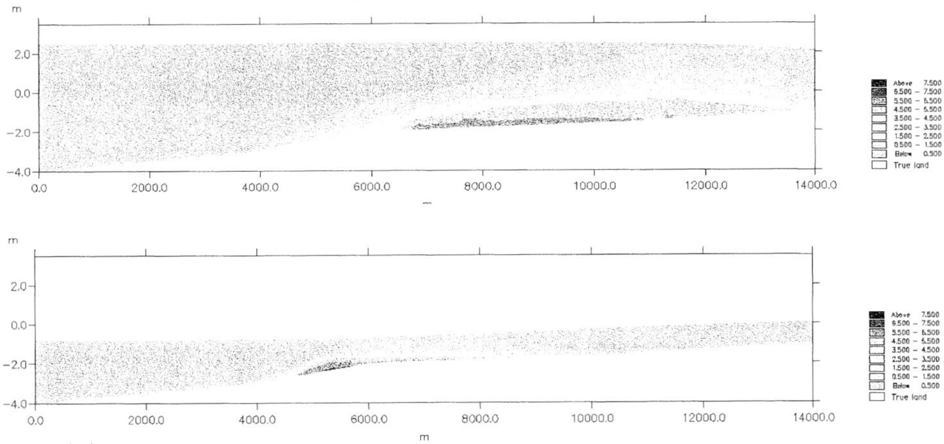

Figure 8. Modelled distribution of sediment in the estuary during spring tide.

The vertical distribution of sediments is shown in Figure 7, where the development of modelled and observed sediment profiles in Station B are shown during spring ebb and flood. The height of the profiles corresponds roughly to the water depth. A deposition phase around low slack water is apparent, although it appears to be stronger in the model than in the observations, and again around high slack water a shorter deposition phase is seen, where a relatively thick (approximately 1.5 m) CBS layer forms.

Figure 8 shows computed examples of the distribution of suspended sediments in the turbidity maximum at ebb and flood tide. The simulations show that the turbidity maximum extends approximately 4 km and contains up to 80 ton of sediment at spring tide; figures that are in the range with those estimated from observations.

7. EFFECTS OF FLOCCULATION AND STRATIFICATION

As part of the COSINUS project, parameterisations for key cohesive sediment processes have been developed based on theoretical considerations, field data and laboratory experiments, reported in several papers (see these proceedings).

A series of simulations are made here, where the sensitivity of the result to different formulations of sediment processes is investigated.

The settling velocity is given by

$$w_s = w_{s0}(1-\Phi)^n, \Phi < \Phi_c$$
$$w_s = w_{s0} \cdot \xi, \Phi \geq \Phi_c \tag{6}$$

where Φ is the volume concentration and ξ, Φ_c are calibration constants.

In-situ settling

The net effect of flocculation and particle break-up is described using an empirical relation for settling velocity, derived from in-situ measurements in Tamar by Dyer *et al.* (2000), relating settling velocity w_s [m/s] to local shear and sediment concentration, as

$$w_s = w_{so} + a_1 C + a_2 \tau + a_3 \tau^2 \tag{7}$$

where C [g/l] is sediment concentration by mass, τ [Pa] is the local shear stress and the a's are empirical constants. The in-situ calibration gives w_{s0} = 0.7 [mm s^{-1}], a_1 = 0.5 [mm • m^3 s^{-1} kg^{-1}], a_2= 7.9 [mm s^{-1} Pa^{-1}] and a_3 = 1.4 [mm s^{-1} Pa^{-2}]. This relation follows the conceptual form suggested in Dyer (1989).

Flocculation I

Alternatively is applied a relation derived by Winterwerp (2000), based on an equilibrium floc size, giving

$$w_s = w_{s0} + a_1 \frac{k_A}{k_B} \frac{\Delta g}{\nu \sqrt{G}} D_p c \tag{8}$$

where a_1 = 0.8175 [s^{-1}] k_A = 14.6 [m^2 kg^{-1}], k_B = 14.0 10^3 [s$^{0.5}$ m^{-2}] are empirical constants, Δg is reduced gravity based on sediment density, ν is kinematic viscosity, G is turbulent shear defined as $G = \sqrt{\varepsilon / \nu}$, D_P is particle size of the primary particles and c mass SPM concentration.

Flocculation II

A heuristic relation linking effective settling velocity to shear and SPM concentrations has been suggested by Malcharek (1995) based on data from the Weser Estuary. He suggested that

$$w_s = w_{s0}(1 + a_1 G)/(1 + a_2 G^2) \tag{9}$$

where a_1 = 0.3 and a_2 = 0.09.

Table 2.
Sensitivity test of fall velocities

	Case	Settling velocity
1	Reference	w_s =5 mm/s and Eq. 2
2	In-situ	Eq. 7 and Eq. 2
3	Flocculation I	Eq. 8 and Eq. 2
4	Flocculation II	Eq. 9 and Eq. 2
5	Constant	w_s =5 mm/s and centred differences for F_{SED}
6	No buoyancy	as 1
7	Mixing length	as 1

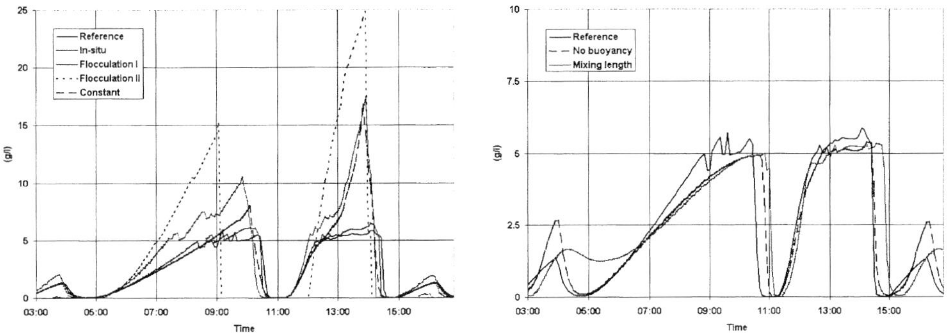

Figure 9. Sensitivity of modelled SPM concentrations to different parameterisations of the settling velocity (left) and turbulence descriptions (right).

The results are shown in Figure 9 (left) as time series of bottom SPM concentration during a spring tide using the 5 different models for the settling velocity. The basic setup corresponds to the one shown in Table 1, and the 5 parameterisations are shown in Table 2.

The comparison shows that the two flocculation formulations and the one without limitations in the near bottom settling (3, 4 and 5), gives relatively high concentrations near the bed during flood tide. For the latter case, the high concentrations arise because eventually all the suspended sediment becomes contained in the lowest grid cell, and for the two former cases, the high concentrations arise due to the vertical variations in settling velocity. For cases 1 and 2, the concentration levels are very similar.

In Figure 9 (right) are shown time series of bottom concentrations for cases 6 and 7, where the settling velocity is unchanged, but the turbulence description altered. In case 6, the sediment induced buoyancy is removed, affecting mainly the turbulence damping above the CBS layers. It is seen that compared to the reference, a smoother solution is obtained, while the concentration level is nearly unchanged.

For case 7, the $k\varepsilon$-model has been replaced by a simplified Prandtl mixing length model that includes damping terms as suggested by Toorman (2000). The mixing length l is related to water depth H as $l = \kappa\, z\, (1-z/H)^{0.5}$ and the effects of buoyancy are parameterised by a damping function on the viscosity, where $\nu_T = \nu_O(1+100R_i)^{-0.5}$, and on the Prandtl number $\sigma_T = \sigma_O(1+21R_i)^{0.33}$

The solution is seen to be very similar to case 6, except at the beginning of the ebb tide, where the vertical mixing apparently is stronger with the mixing length model.

8. DISCUSSION AND CONCLUSIONS

The application of a three-dimensional numerical model to a reach of the Tamar Estuary around the location of the turbidity maximum has shown that it is possible to provide a realistic description of the cohesive sediment dynamics in a macrotidal estuary, in the sense that the general variations of the flow and the suspended sediments are reproduced, but with significant differences, especially in the peak concentrations. This may be due to the very steep vertical gradients, which develop during slack water, when the material settles toward the bed. Investigations of the effect of parameterisations concerning the vertical exchange of sediments, i.e. flocculation, the occurrence of fluid mud layers and effects of sediment induced buoyancy indicate that it is important to include a description of the formation of CBS layers, but firm conclusions must await higher resolved near-bed measurements than available here.

The different formulations of the settling velocity do give somewhat different results, especially regarding peak near-bed concentrations, which are sensitive to the relationship between concentration and settling speed. Apparently, the relations without a realistic parameterisation of the hindered settling tend to predict too high bed concentrations during slack water. However, all the used formulations do preserve the general tidal variation and predict similar levels of concentrations (within at least a factor of 4).

One should, however, be careful to draw quantitative conclusions, as the observational basis is limited.

ACKNOWLEDGEMENTS

This work is co-financed by the European Commission, Directorate XII for Science, Research & Development, through the COSINUS project within the framework of the MAST 3 programme, contract MASC3-CT97-0082.

REFERENCES

Roberts and Feates (2000), Measurement of floc size and settling velocity at Calstock on the Tamar Estuary, Proceedings *INTERCOOH-2002, Elsevier, Coastal and Estuarine Fine Sediment Processes*, ed. J. C. Winterwerp and C. Kranenburg, this volume.

Dyer, K. R., Bale, A. J., Christie, M. C., Feates, N., Jones, S. and Manning, A. (2000), The dynamics of suspended sediment in an estuarine turbidity , Proceedings *INTERCOOH-2002, Elsevier, Coastal and Estuarine Fine Sediment Processes*, ed. J. C. Winterwerp and C. Kranenburg, this volume.

Uncles, R. J. and Stephens, J. A. (1993), The nature of the turbidity maximum in the Tamar Estuary, UK, *Estuarine Coastal Shelf Science*, 36, 413-431.

Uncles, R. J. and Stephens, J. A. (1989), Distributions of suspended sediments at high water in a macrotidal estuary, *Journal of Geophysical Research*, (94) C10, 14395-14406.

Winterwerp, J. C. (2000), *On the dynamics of high-concentrated mud suspensions.* Thesis, Judels Brinkman & Ammerlaan, Delft.

Dyer, K. R. (1989), Sediment processes in estuaries: future research requirements, *Journal of Geophysical Research*, (94) C10, 14327-14339.

Malcharek, A. (1995), Matematische modellierung von strömungen und stofftransport in ästuaren, *Bericht 44, Inst. für Strömungsmechanik und elektron. rechnen im bauwesen der Universität Hannover*, Hannover, pp 200.

Toorman, E. (2000), Personal communication.

Fine Sediment Dynamics in the Marine Environment
J.C. Winterwerp and C. Kranenburg (Editors)

Dynamics of the turbidity maximum in the Changjiang Estuary, China

Z. Shi

Department of Harbour & Coastal Engineering, Shanghai Jiao Tong University, 1954 Hua Shan Road, Shanghai 200030, People's Republic of China

Observations have shown that the Changjiang Estuary has a turbidity maximum zone. Vertical profiles of horizontal current speed/direction, salinity and cohesive suspended sediment concentration were measured in the Changjiang Estuary. Cohesive suspended sediment concentrations were also acoustically monitored. A two-dimensional depth-integrated horizontal (2DH) numerical model was developed to predict tidal currents and cohesive sediment transport processes within the turbidity maximum at the South Channel of the Changjiang Estuary. The Alternating Direction Implicit (A.D.I.) scheme was used to solve the governing equations. Those observational and modeled data were analyzed for the mechanisms for the formation of the turbidity maximum and intratidal variations in cohesive sediment transport processes, with special emphasis on near-bed processes in the Changjiang Estuary.

Four dominant mechanisms responsible for the formation of the turbidity maximum are 1) tidal asymmetry and 2) gravitational circulation at the seaward end of the North Passage, 3) near-bed periodic tidal resuspension at the landward end of the North Passage, and 4) turbulence suppression by cohesive suspension/salinity stratification at the North Passage in the Changjiang Estuary. In addition, three dominant cohesive sediment transport processes were acoustically identified within the near-bed high concentrated mud suspensions: 1) long-period resuspension events superimposed on 2) short period bursts close to the cohesive mud bed; and 3) re-entrainment of the near-bed high concentrated mud suspensions by tidal shear flow.

Key words: cohesive sediment transport, turbidity maximum, the Changjiang Estuary

1. INTRODUCTION

A turbidity maximum zone has been found and studied in many turbid estuarine environments. Those studies include both field measurements (Grabemann & Krause, 1989; Uncles *et al.*, 1993) and numerical modeling (Li *et al.*, 1994; Pickens *et al.*, 1994; Brenon & Le Hir, 1999). A number of mechanisms governing the cohesive suspended sediment transport in the turbidity maximum have been proposed: 1) wind (Weir & McManus, 1987), 2) flood/ebb tidal asymmetry (Allen *et al.*, 1980; Jay & Smith, 1990), 3) estuarine circulation (Schubel, 1968; Jay & Smith, 1990), 4) tidal pumping, i.e., the difference between flood and ebb suspended sediment transport (Uncles *et al.*, 1985), 5) flocculation (Wolanski & Gibbs, 1995), and 6) turbulence suppressions of cohesive suspended sediment/salinity stratifications (Kirby & Parker, 1977; Wolanski *et al.*, 1988; Hamblin, 1989; Uncles & Stephens, 1989, 1993; Geyer, 1993).

The Changjiang Estuary (Figure 1) is a highly turbid mesotidal estuary with mean tidal range of 2.8 m. The annual mean suspended sediment load from the river reaches 4.9×10^8 tons. 40 percentage of the sediment load is deposited in the estuary (Milliman *et al.*, 1985). Studies of cohesive suspended sediment transport have been carried out in the Changjiang Estuary since the 1980s (Yun & Wan, 1982; Su & Wang, 1986; Shi *et al.*, 1996, 1997, 1999, 2001). Turbidity maximum has also been studied in the Changjiang Estuary (Zhou & Wu, 1994, 1996; Li & Zhang, 1998).

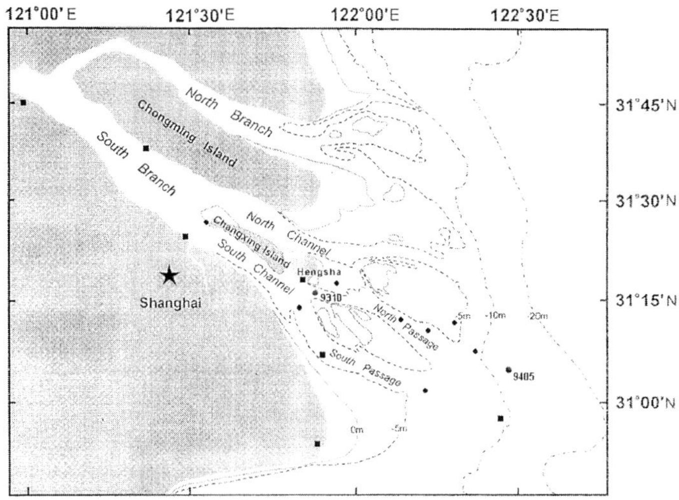

Figure 1. The Changjiang Estuary and monitoring stations (black dot).
Note that squares are water level stations.

Nevertheless, the mechanisms responsible for the formation of the turbidity maximum and cohesive sediment transport processes within the turbidity maximum in the Changjiang Estuary are complex and still poorly understood. In this study, both field observations and 2DH numerical modeling were undertaken 1) to investigate the dominant mechanisms for the formation and maintenance of the turbidity maximum, 2) to identify near-bed cohesive suspended sediment transport processes within the turbidity maximum in the Changjiang Estuary.

2. FIELD OBSERVATIONS

2.1 Methods

Annual freshwater discharge into the Changjiang Estuary is $9.2\times10^{11}m^3$. In the summer, freshwater discharge each month could reach 12% of annual discharge. In the winter, freshwater discharge each month could be 4% of annual discharge. All monitoring stations used in the present study are shown in Figure 1. The present study area was within the South Channel-North Passage-South Passage (Figure 1). Hydrographic and suspended sediment concentration data were obtained at three spring tides (30-31 October 1993; 28 May-3 June 1994; 13-14 September 1996). A spring tide was monitored for 23 hrs on 30-31 October 1993. Spring tide was monitored for 28 hrs on 28-29 May 1994, and 28 hrs on 13-14 September 1996, respectively. Field observations and data analysis were particularly confined to the North Passage (Figure 2), because of the ongoing deepwater navigational waterway engineering project there.

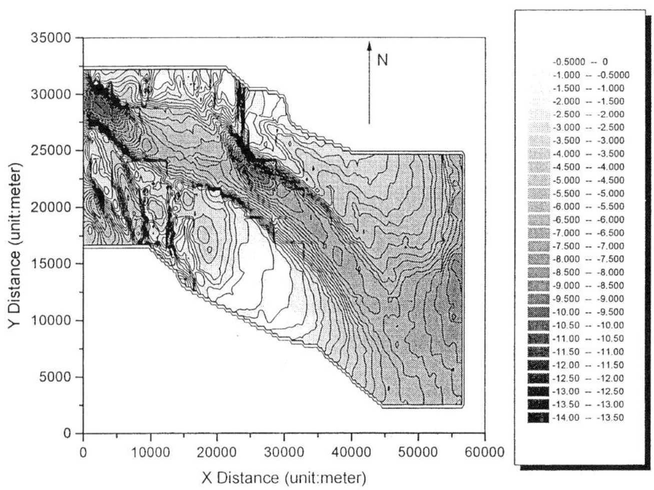

Figure 2 The bathymetry of the North Passage of the Changjiang Estuary.

Field experiments were completed aboard. Horizontal current speed and direction were hourly measured at $0.0H$, $0.2H$, $0.4H$, $0.6H$, $0.8H$ and $1.0H$ (H: the total water depth) using an electromagnetic current meter. Water sampling was also made at the same levels each hour. The water sampler consists of a horizontal steel tube, which collects near-instantaneous point samples of water and suspended sediment. The water samples were then passed through 63 μm sieves to remove the coarse suspended fraction in the laboratory. These samples were filtered to retain the suspended material. The samples were then rinsed with distilled water, and oven-dried at $40^{\circ}C$. The filters were dried to $70^{\circ}C$ to remove additional moisture and weighed. Cohesive suspended sediment concentration was calculated from final sediment weights and total water volumes filtered. Particle sizes of the cohesive suspended sediments were analyzed by a Coulter Counter analyzer. Salinity was measured from the water samples.

Cohesive suspended sediment concentration, at high temporal and spatial resolutions, was also measured by the acoustic suspended sediment monitor at station 9405 (122° 27′ 04″ E, 31° 05′ 15″ N, 28 May-3 June 1994). The details of the present acoustic instrumentation can be found in Shi *et al.* (1996, 1997, 1999). It is composed of a 0.5 or 1.5 MHz acoustic transducer/receiver. The entire system is under PC-computer control for the synchronization of sampling and preliminary data reduction and storage. It has a pulse length of about 40 μs. It measured the vertical profile of sound scattered from cohesive suspended sediments in range bins with a vertical resolution of 10 cm and temporal resolution of 1.5 s. The data were sampled at a rate of approximately 75 KHz for a 12 min burst each hour. Each burst consisted of 480 profiles of backscattered acoustic energy from the suspended sediments between the seabed and the acoustic sensors.

2.2 Results

Two stations (9310 and 9405) at the two ends of the North Passage were selected for analyzing intratidal variations in time series of measured current speed, salinity and suspended sediment concentration (Figure 1). Measured data from other stations are not presented here. Semidiurnal tides prevail. Current speed varied from 0.4 to 2.0 $m \cdot s^{-1}$. Current speeds were larger between 2400 hrs and 0400 hrs, 1400 hrs and 2100 hrs (Figure 3a). Flood/ebb tidal asymmetry was evident. Flood tide duration was shorter than ebb tide. Current speed at ebb tide was larger than that at flood tide (Figure 3a). Low salinity varied from 0.1 to 2.0 psu during the first flood/ebb tidal cycle (2000 hrs-0500 hrs), and from 0.1 to 4.0 psu during the second flood/ebb tidal cycle (0800 hrs-1800 hrs) (Figure 3b). Saltwedge was present at maximum ebb tide, especially apparent at the second maximum ebb tide (Figure 3b). Cohesive suspended sediment concentration varied from 0.1 to 1.5 g L^{-1} during the first flood/ebb tidal cycle (2000 hrs-0500 hrs), and from 0.3 to 4.0 g L^{-1} during the second flood/ebb tidal cycle (0800 hrs-1800 hrs) (Figure 3c). There were at least two tidal resuspension events at 0000 hrs-0600 hrs and 1100 hrs-2000 hrs, respectively (Figure 3c). Cohesive suspensions were highly stratified (Figure 3c). There was also tidal variation of stratification. Stratification at ebb tide seemed to be stronger than that at flood tide.

As shown in Figure 4a, two flood/ebb tidal cycles were monitored at station 9405 on 28-29

May 1994. The observed horizontal current speeds decreased generally from surface to bottom through the water column at each sample interval (Figure 4a). The maximum current speed zone was below the water surface at flood tide (0900 hrs-1400 hrs, 28 May 1994; 2000 hrs-0100 hrs, 28-29 May 1994), while the maximum current speeds reached the water surface at ebb tides (1500 hrs-1900 hrs, 28 May 1994; 0200 hrs-0800 hrs, 29 May 1994) (Figure 4a). Salinity varied from 15 to 27 psu, but with a differing vertical gradient and a high degree of stratification. Salinity increased from surface to bottom, indicating a near-bed saltwedge (Figure 4b). At the first flood tide (0900 hrs-1400 hrs, 28 May 1994), the vertical gradient of salinity decreased through time. At the following ebb tide (1500 hrs-1900 hrs, 28 May 1994), the vertical gradient of salinity increased through time. At the second flood tide (2000 hrs-0100 hrs, 28-29 May 1994), the thickness of the near-bed saltwedge increased significantly (Figure 4b). Near-bed high concentration suspensions were present for prolonged periods on 28-29 May 1994 (Figure 4c). There were two tidal resuspension processes within a flood/ebb tidal cycle (Figure 4c). They were mainly present at maximum flood tide (0800hr-1100hr, 28 May 1994; 2000 hrs-0200 hrs, 29 May 1994) and maximum ebb tide (1200 hrs-1800 hrs, 0400 hrs-0900 hrs, 29 May 1994). Tidal resuspension process

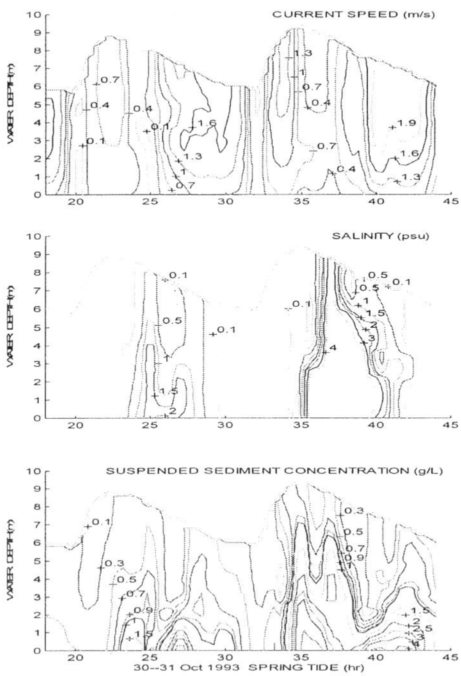

Figure 3 Time series of measured current speed, salinity and suspended sediment concentration at station 9310 (30-31 October 1993, Spring tide).

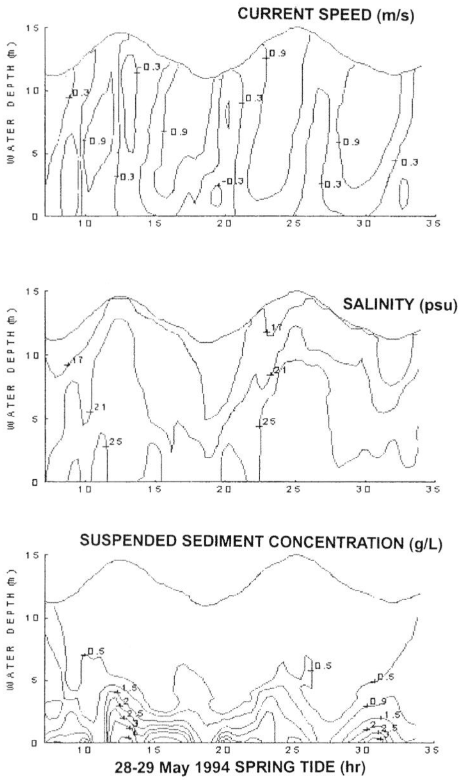

Figure 4 Time series of measured current speed, salinity and suspended sediment concentration at station 9405 (28-29 May 1994, Spring tide).

lasted longer, associated with high suspended sediment concentration (Figure 4c). Bottom concentration maximum was coincident with both the maximum flood and ebb tides. Tidal resuspension processes contributed to the concentration maximum at ebb tide (Figure 4c).

Several selected acoustic graphic images reveal the settling, diffusion, re-entrainment (erosion of near-bed high concentration suspension), and resuspension (erosion of muddy bed) processes although the dynamic nature of cohesive sediment transport processes are complex (Figures 5a, 5b and 5c). Cohesive sediment transport was confined to the near-bed zone. The near-bed high-concentrated mud suspensions were moved as high turbidity clouds by tide- and wave-induced forcing. There was measurable vertical stratification of cohesive suspension concentration through the time (Figures 5 and 6). A thick fluid mud layer was present close to the bottom at 0100 hrs (29 May 1994) (Figure 5a). At 0800 hrs (29 May 1994), the cohesive suspended sediment transport processes were dominated by resuspension and re-entrainment (Figure 5b).

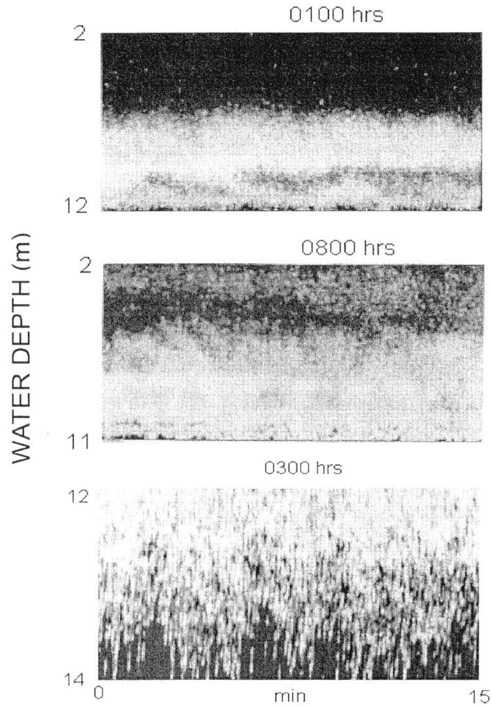

Figure 5 Acoustic imaging of cohesive sediment transport processes at station 9405.

Two dominant cohesive sediment transport processes were shown at the bottom (Figure 5c, 0300 hrs, 29 May 1994). These are cautiously interpreted as the long-period (a few minutes) erosion events superimposed on the short-period (a few seconds) burst events (high frequency interfacial waves) within the near-bed high-concentrated mud suspensions, as found in the Hangzhou Bay (Shi, 1998). It is unclear that under what conditions does one process dominate over the other one. It is also difficult to quantify the amount of cohesive sediment re-entrained by these two processes.

Figure 5c also shows periods of erosion (dark) and periods with no erosion (light colour) near the bottom. The frequency of such events is up to 1 Hz. There is ca 50% (intermittency factor 0.5) in 15 min sampling. This intermittency factor is thought to vary from one sampling period to another sampling period. It is probably higher at accelerating tide than at decelerating one, both flooding accelerating or ebbing accelerating tide. Intermittency of the re-entrainment process could be associated with interfacial instabilities of the kind involved in the re-entrainment process.

III NUMERICAL MODELLING

3.1 Governing equations

Two-dimensional horizontal model has been developed to calculate tidal elevations and currents. The Bossinesq approximation and the hydrostatic condition are introduced in this model. The governing equations were studied using the finite difference scheme on a regular square mesh in the horizontal plane. A staggered grid was used for the grid in the Alternating Direction Implicit scheme (Leendertse & Gritton, 1971). The calibration of the hydrodynamics has been performed with measured tidal elevations at different locations, so as to model and verify 48 hrs tidal elevations within a complete tidal cycle. Sensitivity analyses of the results and modeled phenomenae to the spatial resolution are required for future studies.

The shear stresses can be described as follows:

$$\tau_{wx} = \rho_a c_w \left| W - \sqrt{U^2 + V^2} \right| (w_x - U) \quad (1)$$

$$\tau_{wy} = \rho_a c_w \left| W - \sqrt{U^2 + V^2} \right| (w_y - V) \quad (2)$$

$$\tau_{bx} = \frac{\rho g U \sqrt{U^2 + V^2}}{C^2} \quad (3)$$

$$\tau_{by} = \frac{\rho g V \sqrt{U^2 + V^2}}{C^2} \quad (4)$$

where ρ_a is the density of air ($kg \cdot m^{-3}$); c_w the drag coefficient between the water and air (2.6×10^{-3}); $W = \sqrt{w_x^2 + w_y^2}$ the wind speed, w_x and w_y the components of W in the x and y directions; τ_{bx} and τ_{by} the components of bottom shear stress in the x and y directions.

Depending on the variations in τ_b, the following bottom deposition/erosion processes can be defined (Krone, 1962; Partheniades, 1965):

$$D = \alpha' w_s C \left(1 - \frac{\tau_b}{\tau_d}\right) \qquad \tau_b \le \tau_d \quad (5)$$

$$E = M\left(\frac{\tau_b}{\tau_e} - 1\right) \qquad \tau_e \le \tau_b \qquad (6)$$

where α' is the probability of flocs settling (0.67-0.84); M a coefficient for sediment erosion ($6.4 \times 10^{-3}\ kg \cdot m^{-2} \cdot s^{-1}$).

Discharges per unit width can be calculated at the grid boundaries. Tidal elevation and suspended sediment concentration can be calculated in the center of the grid. Hydrodynamic boundary conditions are time series of tidal elevation or depth-averaged current velocity in the x and y directions. Suspended sediment boundary conditions are time series of the measured suspended sediment concentrations. Because the modeling time is very short, the suspended sediment concentrations at the upper and lower boundaries are assumed to remain unchanged. The suspended sediment concentration at the upper boundary is set at 1.0 gL^{-1} and 0.1 gL^{-1} at the lower boundary in the modeling.

3.2 Results

The computed area is within the South Channel-the North Passage-the South Passage of the Changjiang Estuary (Figure 1). The direction of x axis is toward 130°. The open boundary of the upper reach was assumed to be at Xuliujing (Figure 1). The open boundary of the lower reach was through the y axis at Luhuashan. The South and North boundaries are the south ends of the Chongming Island and Luchaogang, respectively. A square mesh was used with length of $1000\,m$ and time step of $90\,s$. Generally, the predicted velocity fields show the complex hydrodynamic processes at 0100 hrs, 0400 hrs, and 1300 hrs.

Predicted tidal velocity field is shown in Figure 6 at 0100 hrs during flood tide. Tidal currents outside the estuary were predominantly westward. Flowing into the estuary, tidal currents were gradually decreasing due to the river runoff and topography within the South Channel-the North and South Passages.

Tidal velocity field prediction is shown in Figure 7 at 0400 hrs during maximum flood tide. The tidal currents outside the estuary were predominantly westward. Tidal flow was still westward within the North Passage, and northwestward within the South Passage. Tidal currents were streamline within both the North and South Passages (Figure 7).

Figure 8 shows the predicted tidal velocity field at 1300 hrs during ebb tide. Tidal currents were generally low. The tidal currents near the estuary were predominantly eastward.

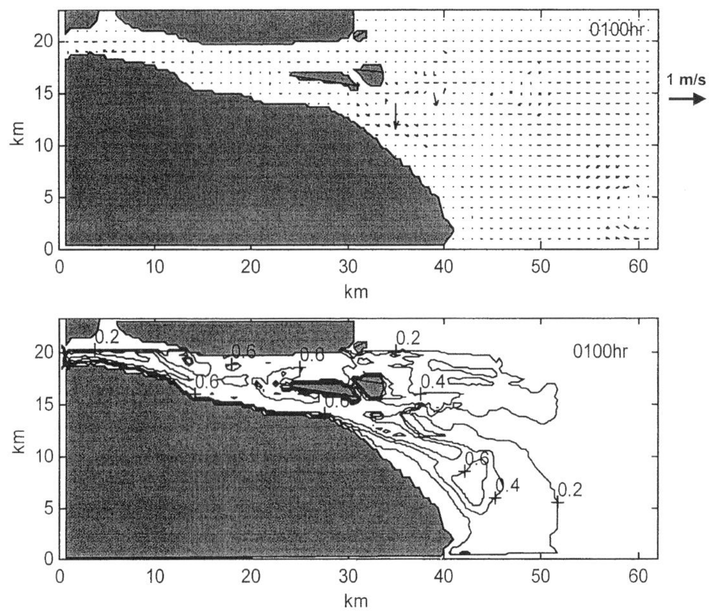

Figure 6 Modeled tidal current velocity vector (top) and suspended sediment concentration field (bottom) (0100 hrs). Note: the distance ($\times 3$) in the x direction.

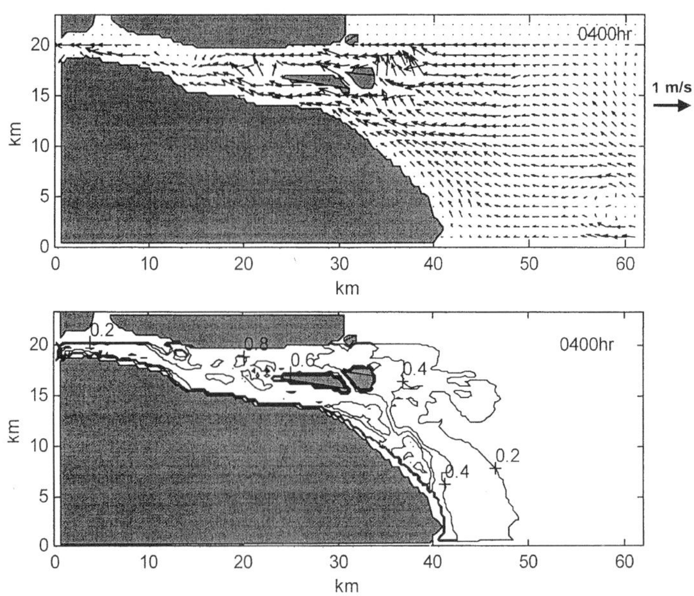

Figure 7 Modeled tidal current velocity vector (top) and suspended sediment concentration field (bottom) (0400 hrs). Note: the distance ($\times 3$) in the x direction.

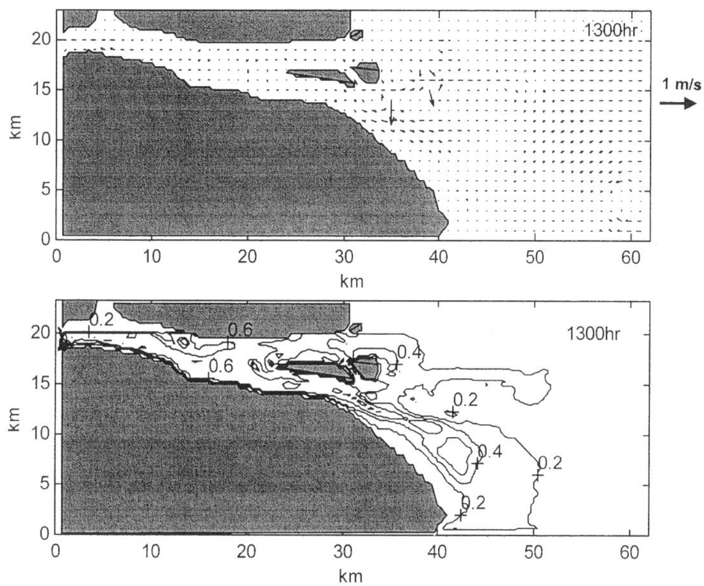

Figure 8 Modeled tidal current velocity vector (top) and suspended sediment concentration field (bottom) (1300 hrs). Note: the distance ($\times 3$) in the x direction.

There was a general eastward boundary at the mouth of the estuary (Figures 6, 7 and 8).

At the mouth of the South Passage, the turbidity maximum moved both eastward and southward at 0100 hrs (Figure 6). The suspended sediment concentration was up to 0.6 g L^{-1}. The turbidity maximum moved eastward near the mouth of the North Passage (Figure 6). The suspended sediment concentration was up to 0.6 g L^{-1}.

Due to the tidal propagation, the turbidity maximum moved slightly westward at 0400 hrs during maximum flood tide (Figure 7). At the mouth (seaward end) of the South Passage, the turbidity maximum moved both eastward and southward at 0400 hrs (Figure 7). The concentration was up to 0.8 gL^{-1}. The turbidity maximum moved southeastward near the mouth of the North Passage (Figure 7). The suspended sediment concentration was up to 0.8 gL^{-1}.

At the mouth of the South Passage, the turbidity maximum moved both eastward and southward at 1300 hrs (Figure 8). The concentration was up to 0.6 g L^{-1}. A suspension jet was present at the mouth of the North Passage (Figure 8). The suspended sediment concentration was up to 0.4 g L^{-1}.

4 DISCUSSION

Short duration and strong flood tidal current speeds and longer duration but weaker ebb tide current speeds were evident at the landward end of the North Passage (Figure 3, station 9310). They were attributed to the tidal asymmetry caused by the interaction between the incoming tidal waves and the estuarine topography. Tidal asymmetry does govern the longitudinal transport of estuarine cohesive suspended sediment at the landward end. However, tidal asymmetry was not apparent at the seaward end of the North Passage (Figure 4, station 9405). The mixing of river runoff and seawater created longitudinal and vertical density gradients. In addition, the bottom saltwedge was also present at the seaward end (Figure 4). As a result, the estuarine gravitational circulation was longitudinally present at the landward end of the North Passage.

One effect of saltwedge on the cohesive suspended sediment transport is via flocculation. As shown in Figures 3 and 4, the salinity distributions were highly stratified. Salinity stratification suppresses the turbulence and enhances the trapping of bottom fine suspended sediment, contributing to the formation of turbidity maximum at the North Passage of the Changjiang Estuary. Moreover, the suppression of turbulence by the highly stratified suspensions could also enhance the formation of near-bed high concentrated mud suspension (Figures 3 and 4). Stronger stratification seemed to occur during ebb flow and weaker stratification occurs during flood tide (Figures 3 and 4). An indication of the stratification effect is an abrupt decrease in the fine suspended sediment concentration in the upper layer as it overrides the saltwedge (Figures 3 and 4).

Turbidity maximum was consistently observed during spring tide at the North Passage of the Changjiang Estuary (Figures 3 and 4). The magnitude and location of the turbidity maximum varied with the tidal phase and the freshwater-saltwater interface (Uncles & Stephens, 1989, 1993). Within a flood/ebb tidal cycle, there were several peak concentrations (Figures 3 and 4). It suggested that the main factor controlling the suspended sediment concentration was tidal resuspension process. These tidal resuspension processes (concentration maxima) were mainly related to high tidal current speed, i.e., bottom boundary-induced turbulence (Figures 3 and 4) and high salinity (Figures 3 and 4), especially at the seaward end of the North Passage, as observed in other estuaries (Schubel, 1968; Wellershaus, 1981). These processes probably refer to 'tidal pumping' (Uncles *et al.*, 1985). Near-bed tidal resuspension processes provide the necessary fine suspended sediment sources for the formation of the turbidity maximum. Kelvin-Helmholtz instability and random internal wave interactions are also important in vertical exchange in stratified suspensions of the Changjiang Estuary.

Figures 5a, 5b and 5c have also demonstrated two different processes: 1) the resuspension of the newly-deposited cohesive mud bed, and 2) the re-entrainment of near-bed high concentrated mud suspensions. The former process probably results from strong bottom shear velocity (stress), while the latter process is due to turbulent shear flow-induced interfacial instabilities. Furthermore, resuspension of the cohesive mud bed is highly episodic (Figure

5c). Such resuspension events lasted only a few seconds. When the resuspension flux exceeded lower layer (high concentrated mud suspension) re-entrainment into the upper water column, a near-bed high concentrated mud suspension (mobile fluid mud layer) can form (Figure 5a).

Turbidity maximum was simulated with a depth-averaged sediment transport model at the South Channel (Figures 6, 7 and 8). Modeled results suggested that the turbidity maximum was strongly influenced by tidal hydrodynamics (Figures 6, 7 and 8), similar to other studies in the literature (Allen *et al.*, 1980; Dyer & Evans, 1989; Brenon & Le Hir, 1999). Two processes mainly related to the vertical mixing can not be studied using a depth-averaged model. In other words, both field observations and numerical studies have not been well integrated. 2DV numerical model should be developed in the future. Both wave-induced resuspension and the density gradients have not been taken into account in the numerical model, a three-dimensional model of hydrodynamics and cohesive sediment transport is needed in the future study. It should be pointed out that the predicted current velocity vectors were quantitatively consistent with the field measurements, but the predicted suspended sediment concentration distribution was qualitatively comparable with the field observations.

5 CONCLUSION

Field observations and modeling of cohesive sediment transport have been carried out to investigate the mechanisms for the formation of the turbidity maximum, with special emphasis on near-bed processes, in the Changjiang Estuary. In summary, four dominant mechanisms responsible for the formation of the turbidity maximum are 1) tidal asymmetry and 2) gravitational circulation at the seaward end of the North Passage, 3) near-bed periodic tidal resuspension at the landward end of the North Passage, and 4) turbulence suppression by suspension/salinity stratification at the North Passage. In addition, three dominant cohesive sediment transport processes were acoustically identified within the near-bed high concentrated mud suspensions: 1) long-period resuspension events superimposed on 2) short period bursts close to the cohesive mud bed; and 3) re-entrainment of the near-bed high concentrated mud suspensions by tidal shear flow.

ACKNOWLEDGMENTS

This research was supported by the Marine Sciences (Estuarine and Coastal) Program, the National Natural Science Foundation of China (Grant No. 49806005) and the State Education Ministry of China. Mr.H.Q. Zhou and Dr.Y.G. Zhang are thanked for their help with the numerical modeling. Two anonymous reviewers are thanked for their constructive comments.

REFERENCES

Allen, G.P., Salomon, J.C., Bassoulet, P., Du Penhoat, Y. and DeGrandpre, C. 1980 Effect of tides on mixing and suspended sediment transport in macrotidal estuaries. *Sedimentary*

Geology 26: 69-90.

Brenon, I. and Le Hir, P. 1999 Modelling the turbidity maximum in the Seine Estuary (France): identification of formation processes. *Estuarine, Coastal and Shelf Science* 49: 525-544.

Dyer, K.R. and Evans, E.M. 1989 Dynamics of turbidity maximum in a homogeneous tidal channel. *Journal of Coastal Research* SI 5: 23-30.

Geyer, W.R. 1993 The importance of stratification on the formation of the turbidity maximum. *Estuaries* 16: 113-125.

Grabemann, I. and Krause, G. 1989 Transport processes of suspended matter derived from time series in a tidal estuary. *Journal of Geophysical Research* 94 (C10): 14,373-14,380.

Hamblin, P.F. 1989 Observations and model of sediment transport near the turbidity maximum of the upper Saint Lawrence estuary. *Journal of Geophysical Research* 94 (C10): 14,419-14,428.

Jay, D.A. and Smith, J.D. 1990 Residual circulation in shallow estuaries: 1. Highly stratified, narrow estuaries. *Journal of Geophysical Research* 95 (C1): 711-731.

Kirby, R. and Parker, W.R. 1977 The physical characteristics and environmental significance of fine-sediment suspensions in estuaries. In: C.B. Officer (Ed.), *Estuaries, Geophysics and the Environment.* National Academy Press, Washington, D.C., 110-120.

Krone, R.B. 1962 Flume studies of the transport of sediment in estuarial shoaling processes. *Final Report, Hydrologic Engineering Laboratory and Sanitary Engineering Research Laboratory*, Berkeley: University of California, 110.

Leendertse, J.J. and Gritton, E.C. 1971 *A water quality simulation model for well mixed estuaries and coastal seas.* Vol. II Computation Procedures. Rand Corporation, 53.

Li, J.F. and Zhang, C. 1998 Sediment resuspension and implications for turbidity maximum in the Changjiang Estuary. *Marine Geology* 148: 117-124.

Li, Z.H., Nguyen, K.D., Brun-Cottan, J.-C. and Martin, J.-M. 1994 Numerical simulation of the turbidity maximum transport in the Gironde estuary (France). *Oceanologica Acta* 17 (5): 479-500.

Milliman, J.D., Shen, H.T., Yang, Z.S. and Meade, R.H. 1985 Transport and deposition of river sediment in the Changjiang estuary and adjacent continental shelf. *Continental Shelf Research* 4: 37-45.

Partheniades, E. 1965 Erosion and deposition of cohesive soils. *Journal of the Hydraulics Division*, ASCE, 91 (HY1): 105-139.

Pickens, K., Chroneer, Z., Patel, P. and Lick, W. 1994 The formation of a turbidity maximum in an estuary. In: M.L. Spaulding, K. Bedford, A. Blumberg, R.T. Cheng and C. Swanson (Eds.), *Estuarine and Coastal Modeling*, ASCE, New York, 187-201.

Schubel, J.R. 1968 Turbidity maximum of the Northern Chesapeake Bay. *Science* 161: 1013-1015.

Shi, Z. 1998 Acoustic observations of fluid mud and interfacial waves in the Hangzhou Bay, China. *Journal of Coastal Research* 14 (4): 1348-1353.

Shi, Z., Ren, L.F. and Lin, H.L. 1996 Vertical suspension profile in the Changjiang Estuary. *Marine Geology* 130: 29-37.

Shi, Z., Ren, L.F. and Hamilton, L.J. 1999 Acoustic profiling of fine suspension concentration in the Changjiang Estuary. *Estuaries* 22 (3A): 648-656.

Shi, Z., Hamilton, L.J. & Ren, L.F. 2001 Reply to discussion of "Acoustic profiling of fine suspension concentration in the Changjiang Estuary." *Estuaries* 24 (1): 136-137.

Shi, Z., Ren, L.F., Zhang, S.Y. and Chen, J.Y. 1997 Acoustic imaging of cohesive sediment re-entrainment and resuspension in the Changjiang Estuary, East China Sea. *Geo-Marine Letters* 17: 162-168.

Su, J.L and Wang, K.S. 1986 The suspended sediment balance in Changjiang Estuary. *Estuarine, Coastal and Shelf Science* 23: 81-98.

Uncles, R.J., Elliot, R.C.A. and Weston, S.A. 1985 Observed fluxes of water, salt and suspended sediments in a muddy, macrotidal estuary. *Estuarine, Coastal and Shelf Science* 20: 147-167.

Uncles, R.J. and Stephens, J.A. 1989 Distributions of suspended sediment at high water in a macrotidal estuary. *Journal of Geophysical Research* 94 (C10): 14395-14405.

Uncles, R.J. and Stephens, J.A. 1993 Nature of the turbidity maximum in the Tamar estuary, U.K. *Estuarine, Coastal and Shelf Science* 36: 413-431.

Weir, D.J. and McManus, J. 1987 The role of wind in generating turbidity maximum in the Tay Estuary. *Continental Shelf Research* 7 (11/12): 1315-1318.

Wellershaus, S. 1981 Turbidity maximum and mud shoaling in the Weser estuary. *Archiva Hydrobiologica* 92: 161-198.

Wolanski, E. and Gibbs, R.J. 1995 Flocculation of suspended sediment in the Fly river estuary, Papua New Guinea. *Journal of Coastal Research* 11 (3): 754-762.

Wolanski, E., King, B. and Galloway, D. 1995 Dynamics of the turbidity maximum in the Fly River estuary, Papua New Guinea. *Estuarine, Coastal and Shelf Science* 40: 321-337.

Wolanski, E., Chappell, J., Ridd, P. and Vertessy, R. 1988 Fluidization of mud in estuaries. *Journal of Geophysical Research* 93 (C3): 2351-2361.

Yun, C.X. and Wan, J.R. 1982 A study of diffusion of upper layer suspended sediments in discharges from the Changjiang estuary into the sea, based on satellite imagery. In: V.S. Kennedy (Ed.), *Estuarine Comparisons*, Academic Press, New York, 693-704.

Zhou, H.J. and Wu, S.R. 1994 Numerical modeling of suspended sediment in the Changjiang Estuary. *Proceedings of International Symposium on Waves-Physical and Numerical Modeling*, Vancouver, Canada, 1636-1644.

Zhou, H.J. and Wu, S.R. 1996 Mechanisms of formation and maintenance of the turbidity maximum in the Changjiang Estuary. *Proceedings of the Fourth Pacific/Asia Offshore Mechanics Symposium*, Pusan, Korea, ISOPE, 99-104.

Fine Sediment Dynamics in the Marine Environment
J.C. Winterwerp and C. Kranenburg (Editors)

Numerical Assessment of Source and Sink Terms for Cohesive Sediments

C. Schweim[a], J.V. Prochnow[a] and J. Köngeter[a]

[a] Aachen University of Technology, Institute of Hydraulic Engineering and Water Resources Management (IWW), Kreuzherrenstrasse, D-52056 Aachen

Commonly used formulae for the description of erosion and deposition of cohesive sediments are assessed concerning their applicability within a numerical simulation. The experimental database for the evaluation of the formulations in question (Partheniades and Parchure and Mehta for erosion, Krone for deposition with approaches of Krone, van Leussen and Richardson and Zaki for the settling velocity) results from erosion and deposition experiments with reservoir sediments in an annular flume. The detailed knowledge of the flow field in the annular flume, resulting from a 3D-LES simulation, is a prerequisite for the assessment of the terms as the 2DV sediment transport model represents a radial section of the annular flume. Comparison of the simulation of erosion and deposition experiments with the measurements show the need for further research as well as the suitability of the presented strategy.

KEYWORDS: cohesive sediment, erosion, deposition, numerical simulation

1. INTRODUCTION

Due to their ability to bind pollutants like heavy metals, polycyclic aromatic hydrocarbons or chlorinated hydrocarbons, cohesive sediment deposits can constitute a considerable danger for the aquatic environment. Erosion of such contaminated sediment deposits during extreme hydraulic events leads to high concentrations of pollutants in the water, with disastrous effects on the ecosystem.

With increasing computer capabilities and advances in numerical methods, tools to numerically simulate hydraulic processes are of growing relevance in hydraulic engineering. Numerical models including cohesive sediment transport are more and more applied to forecast the formation of cohesive sediment deposits and their resuspension, thus estimating the danger for the environment and the benefit of constructive or operational measures. For the development of terms describing erosion and deposition of cohesive sediments, experimental research on sediment behaviour is still necessary due to the limited understanding of physical, biological and chemical processes involved.

Annular flumes are especially well-suited for experimental investigation of cohesive sediments whose ability to form aggregates can be taken into account in the endless flow which is generated in these flumes without using floc-destroying pumps. Experiments on erosion and deposition of cohesive sediments were carried out in the annular flume of the IWW at Aachen University of Technology with different cohesive sediments containing clay, silt, fine sand

and a high content of organic matter. Results from these experiments are used here to test the applicability of commonly used formulae for erosion and deposition of cohesive sediments within a numerical simulation.

The formulations used are those of Partheniades respectively Parchure and Mehta describing erosion and those of Krone, van Leussen and Richardson and Zaki for deposition. Flow properties of the annular flume flow, procedure of erosion and deposition experiments and the numerical model used for simulation of the experiments are subsequently presented. Afterwards, the evaluation of empirical parameters needed in the formulae is shown and the results are compared to values found in the literature. Finally, simulation results are given and discussed.

2. EXPERIMENTAL SET-UP

2.1. Properties of the annular flume flow

The annular flume of the IWW is a ring-like channel with a width of 0.25 m and a mean diameter of 3.25 m. A lid with the same dimensions is adjusted to touch the water surface, resulting in a flow depth of 0.175 m. Channel and lid rotate in opposite directions, generating an endless flow in tangential direction by the relative velocity between both parts.

Centrifugal forces occurring in the circular movement lead to a distinct secondary flow and with that to a complex three-dimensional flow field. Tangential flow velocities are mainly dependent on the relative velocity between channel and lid. Properties of the secondary flow can be influenced by a variation of the ratio of rotational speeds of channel (ω_f) and lid (ω_l). This fact allows to calibrate the flow to get optimal conditions for either deposition or erosion experiments. Calibration was performed with help of extensive LDV-measurements of the three-dimensional flow field (Spork et al. 1998). Optimal ratios of rotational speeds were determined for deposition ($\omega_l/\omega_f = -1.8$, minimised secondary flow velocities) and for erosion ($\omega_l/\omega_f = -1.6$, uniform distribution of bottom shear stress).

2.2. Experiments in the annular flume

Experiments on deposition and erosion behaviour of fine cohesive sediments in the annular flume are carried out as follows:

Deposition is started by mixing water and sediment in the flume to form a suspension at a predefined, homogeneous concentration. Rotational speeds of flume and lid (hence bottom shear stress) are then reduced stepwise in order to allow sediments to deposit.

Before starting the erosion experiment, a sediment bed has to be prepared. This is accomplished by deposition of sediment from a homogeneous suspension under no-flow conditions. After 72 hours of deposition and consolidation, the bed is eroded by a stepwise increase of bottom shear stress.

During the experiments, sediment-water mixture (taken at mean flow depth at the inner wall of the channel) is pumped continuously through a turbidity meter, thus recording the time-dependent development of sediment concentration in the water column. From the experimental results, sediment characterising parameters like critical bed shear stresses, deposition or erosion rates and settling velocities can be derived. Also, the vertical density profile of a sediment bed, prepared at the same time in a tank under the same conditions as in the annular flume, is measured to correlate the resulting sediment parameters with the state of consolidation of the bed. This is done with an ultrasonic high concentration meter (UHCM) made by Delft Hydraulics, The Netherlands.

The sediments used in the present investigation are taken from the reservoirs Heimbach (deposition test) and Obermaubach (erosion test). These are located in the Eifel, a low mountain range near Aachen. Some parameters characterising these sediments are given in Table 1. In order to keep the environmental conditions as natural as possible, original water from the reservoirs is used in the experiments.

Table 1
Properties of investigated sediments

	d_{50} [mm]	d_{60}/d_{10} [-]	d > 63 μm [%]	clay [%]	organic matter [%]
Heimbach	0.0373	12.2	27.5	6.37	7
Obermaubach	0.0393	4.1	38.8	0.72	12

3. FLOW SIMULATION MODEL

Basis for the evaluation of sediment transport terms is an accurate knowledge of the flow field: All flow parameters governing the sediment transport (velocity, bottom shear stress and turbulence) should be reproduced as precise as possible. As measurements of the flow field could not be carried out for all flow situations occurring during the experiments, a reliable flow simulation model is needed.

Simulation of the flow in the annular flume is performed with the three-dimensional finite element code Pastis-3D (Projection Algorithm Solver for Transient Incompressible Flow Simulations in 3 Dimensions, Daniels 1992). This code uses a large eddy approach with a Smagorinsky model for subgrid scales of turbulence.

In Figure 1, simulated and measured secondary flow fields (temporally averaged) in a section of the annular flume are compared, showing a very good agreement.

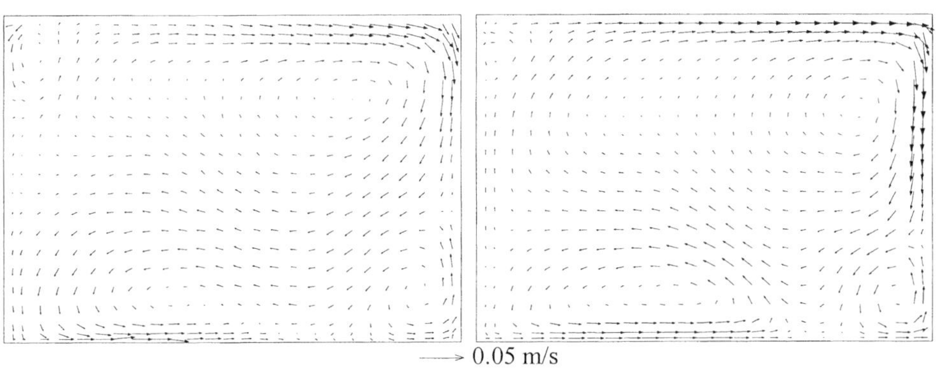

Figure 1. Comparison of measured (left) and simulated (right, interpolated on measuring grid) secondary flow field at ω_f = -2.0 rpm and ω_l = 3.6 rpm, left side represents inner wall.

Also, the simulated mean bottom shear stress (averaged in space and time) is very close to the measurement, as shown in Figure 2. The bottom shear stress is not measured directly but derived from the flow velocity using the logarithmic law of the wall.

The accurate reproduction of flow in the simulation allows the conclusion that deviations between measured and simulated sediment concentrations are actually due to the used source and sink terms and not due to an inaccurate input of flow parameters. Detailed information on the annular flume flow simulation is given in Schweim et al. (2000).

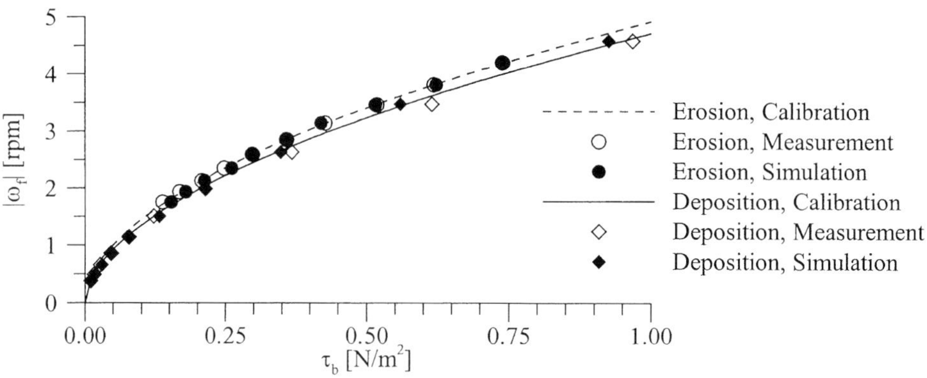

Figure 2. Comparison of measured and simulated mean bottom shear stress. The calibration functions are fitted to the measurements (Spork 1997) and used to apply specific bottom shear stresses in the experiments by controlling the speeds of rotation of channel and lid.

4. TRANSPORT SIMULATION MODEL

Within the transport simulation, the sediment transport equation is solved independently from the flow simulation. The simulated flow velocities (temporally averaged for each bottom shear stress step) serve as input for the solution of the advection-diffusion-equation for transport of suspended sediment that is classically written as:

$$\frac{\partial C}{\partial t} + v_i \frac{\partial C}{\partial x_i} + \frac{\partial w_s C}{\partial z} - \frac{\partial}{\partial x_i}\left(D \frac{\partial C}{\partial x_i} \right) = S \qquad (1)$$

In Equation 1, C [g/l] is the sediment concentration, v [m/s] the flow velocity and S [kg/m^3/s] a source or sink of sediment within the domain. At the present state, the settling velocity w_s [m/s] is assumed to be zero during the simulation of erosion experiments. For the simulation of deposition, different formulations to calculate the settling velocity are tested (chapter 6). D [m^2/s] is the diffusion coefficient that can be calculated with the molecular diffusion coefficient D_m and the turbulent diffusion coefficient D_t (Forkel 1995):

$$D = D_m + 2D_t \qquad (2)$$

The molecular diffusion coefficient is calculated with the Boltzmann constant k [kg·m^2/s^2/K], the absolute water temperature T [K], the dynamic viscosity μ [kg/m/s] and the particle size d [m] (Winterwerp 1999):

$$D_m = \frac{kT}{6\pi\mu d} \tag{3}$$

The turbulent diffusion coefficient is calculated with (Bergen 1999):

$$D_t = c_t^2 \Delta^2 \left(S_{ij} S_{ij}\right)^{1/2} \tag{4}$$

In equation 4, c_t [-] is an empirical parameter that is set to $c_t = 10\ c_s$ (Forkel 1995) with c_s being an empirical parameter from the subgrid scale turbulence model calibrated with the flow simulation. Δ [m] is the filter width of the flow simulation and S_{ij} [s^{-1}] the deformation tensor:

$$S_{ij} = \frac{1}{2}\left(\frac{\partial v_i}{\partial x_j} + \frac{\partial v_j}{\partial x_i}\right) \tag{5}$$

A consideration of the particularity of the annular flume flow leads to a remarkable simplification of the transport simulation model. Regarding the axial symmetry of the endless flow, it is obvious that both the mean velocity gradient and the mean concentration gradient in tangential direction are zero. The three-dimensional domain of the flow simulation can hence be reduced to a two-dimensional vertical domain in the transport simulation.

Consequently, Equation 1 is solved at a finite element grid corresponding to a section of the three-dimensional grid of the flow simulation. Accordingly, the spatial variables are the radial (x-) and the vertical (z-) direction. The tangential velocity gradients dv/dx and dv/dz must nevertheless be taken into account in the calculation of the deformation tensor and hence the turbulent diffusion coefficient (Eqs. 5 and 4). The tangential velocity is as well crucial in the calculation of bottom shear stress with the logarithmic law of the wall.

The boundary conditions guarantee that no sediment is passing through the free surface:

$$w_s C + D\frac{\partial C}{\partial z} = 0 \tag{6}$$

and through the vertical walls:

$$D\frac{\partial C}{\partial x} = 0 \tag{7}$$

At the bottom, the flux of sediment Φ [kg/m^2/s] corresponds to flux due to deposition (Φ = σ) respectively erosion (Φ = ε):

$$w_s C + D\frac{\partial C}{\partial z} = \phi \tag{8}$$

The illustration of spatially and temporally averaged bottom shear stress in chapter 3 was chosen for presentation. In the calculation of sediment flux Φ in the transport simulation, the bottom shear stress is derived from simulated near wall velocities, slightly varying in radial direction.

A change in the flow section of the flume due to deposition or erosion is not implemented in the model at the present state. To account for this change, a new simulation of the flow field would be necessary if changes in the height of the sediment bed and with that in the flow depth lead to remarkably changed flow conditions. This proceeding would be extremely time consuming due to the considerable computational resources needed for the three-dimensional large eddy simulation of the flow. It is hence assumed that small changes in the flow depth occurring during the experiments (< 20 mm) have no significant influence on the flow field.

The influence of the secondary current on the concentration field is such that no noticeable concentration gradient in the water column develops (not shown here). This finding is in accordance with recent measurements of the concentration gradient in the annular flume of the IWW. Hence, the mean concentration in the annular flume shown in the following is representative for erosion respectively deposition occurring during both experiments and simulations.

5. SIMULATION OF COHESIVE SEDIMENT EROSION

5.1. Formulation of Partheniades

Erosion of homogeneous beds is commonly described following Partheniades (1984):

$$\varepsilon = M\frac{\tau_b - \tau_{ce}}{\tau_{ce}} \quad \text{for } \tau_b \geq \tau_{ce} \tag{9}$$

where ε [kg/m^2/s] is the erosion rate, M [kg/m^2/s] is an erosion parameter, τ_b [N/m^2] is the bottom shear stress and τ_{ce} [N/m^2] the critical bottom shear stress for erosion. Equation 9 implies that the erosion rate is constant for a constant bottom shear stress. The suspended sediment concentration in the water increases hence linearly with time which can only be true for homogeneous sediment beds with a constant shear strength with depth, occuring rarely in nature.

However, as this formulation is commonly used for calculation of erosion rates of non-homogeneous beds, its applicability is tested here by comparing results of the simulation with results from erosion experiments with sediments from Obermaubach. The parameters τ_{ce} and M were evaluated from the concentration-time-curve measured during the experiment and are compared to values from the literature in Table 2.

A comparison of the measured and simulated concentration development in the water column with time is given in Figure 3.

Table 2
Critical bottom shear stress and erosion constant, this investigation and values from literature

	τ_{ce} [N/m^2]	M [kg/m^{2}s]
this investigation, lake mud	0.2	$3.48 \cdot 10^{-5}$
Partheniades (1962), bay mud	0.5	$3.00 \cdot 10^{-5}$
Mulder and Udink (1990), estuarial mud	0.4	$71.0 \cdot 10^{-5}$

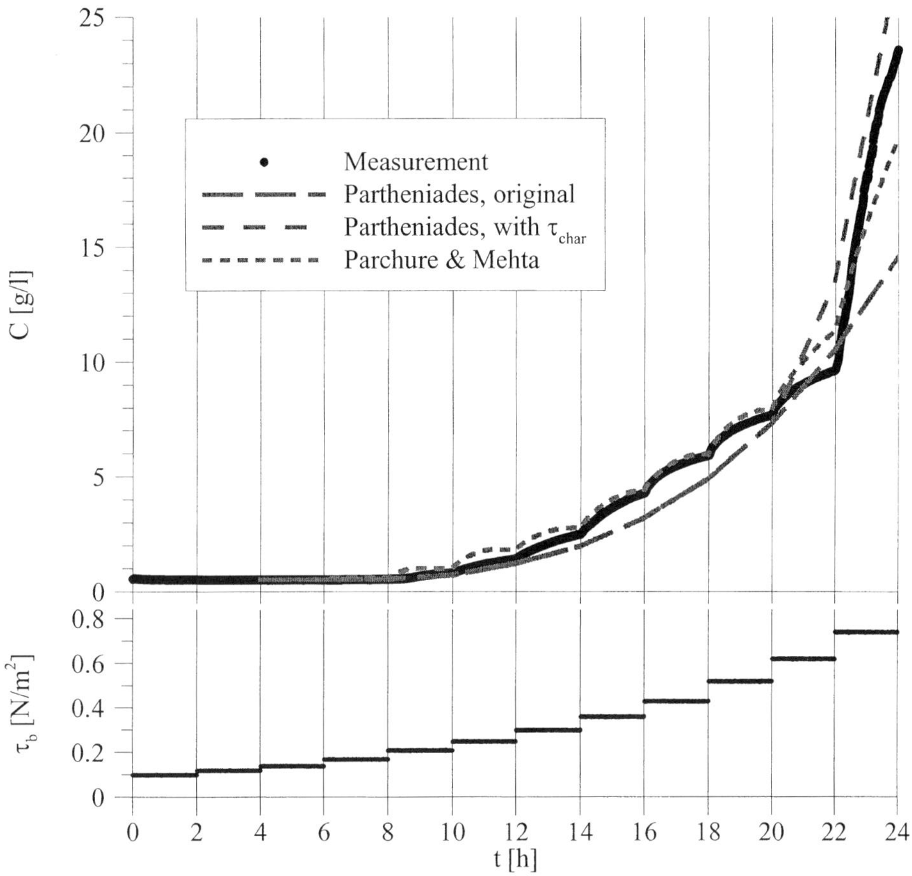

Figure 3. Comparison of measured and simulated concentrations in the water column during an erosion experiment with sediments from the reservoir Obermaubach (upper diagram) and mean bottom shear stress history during the experiment (lower diagram)

The concentration simulated with the original formulation of Partheniades (Eq. 9) is in the same order of magnitude as in the measurement. However, this formulation is ignoring the typical erosion characteristics of stratified sediment beds, where mass-erosion is caused by a sudden increase of τ_b (at the beginning of a bottom shear stress step) and floc-erosion occurs when the bottom shear stress stays constant for a longer time (to the end of a step). This fact would lead to a considerable over-estimation of concentrations in the case that erosion during longer periods with constant τ_b (and hence smaller mean erosion rates) would be simulated with the same erosion parameter M (and hence the same erosion rates).

Additionally, the erosion rates in the last bottom shear stress step are considerably smaller in the simulation than in the measurement. This is probably due to the fact, that τ_b exceeds the maximal shear strength of the bottom. In order to account for this, a characteristic bottom shear stress τ_{char} may be introduced (Spork 1997). As an exact value of the maximal shear strength of the bottom can not be evaluated from the measurements, τ_{char} is assumed to be equal to the mean value of the last two steps ($\tau_{char} = 0.678$). In the case that τ_b exceeds τ_{char}, a changed value of M ($1.29 \cdot 10^{-4}$ kg/m^2/s) is used in the calculation of ε with Equation 9 to account for the higher effective shear stress. Simulated concentrations with this method show a better agreement than the original formulation (Figure 3). However, the validity of this approach can not be proven with the given experimental data as τ_{char} is only exceeded for a short time.

5.2. Formulation of Parchure and Mehta

Beds of cohesive sediments that developed from deposition of flocculated sediment are non-homogeneous and show an increasing shear strength with depth. This leads to decreasing erosion rates with ongoing erosion at constant bottom shear stresses and the resulting increase in shear strength at the bed surface. Erosion rates of such beds may be described with the formula by Parchure and Mehta (1985):

$$\varepsilon = \varepsilon_0 \cdot \exp\left(\alpha\left(\tau_b - \tau_s\right)^{1/2}\right) \quad \text{for } \tau_b \geq \tau_s \tag{10}$$

ε_0 is the floc-erosion rate, α is an erosion coefficient [m/N$^{0.5}$] and τ_s [N/m^2] is the bed shear strength varying with depth. The floc-erosion rate describes the erosion that is due to the stochastic nature of τ_b and τ_s and that occurs when $\tau_b = \tau_s$.

To determine the parameters ε_0 and α and the shear strength $\tau_s(z)$ of the eroded sediment bed, the following procedure was applied (compare Parchure and Mehta 1985 and Spork 1997): ε_0 was calculated as the mean of the measured erosion rates during the last 10 minutes of every bottom shear stress step with $8\ h < t < 22\ h$, assuming that floc-erosion occurs at the end of each of these steps. This assumption is also required to evaluate the bed shear strength as it implicitly contains the fact that the shear strength of the bed surface at the end of the step i $\tau_s(z_i)$ is equal to the bottom shear stress $\tau_{b,i}$. With the known dry density $\rho(z)$ (measured at the end of consolidation time and before eroding the bed) and the known change of concentration ΔC_i, the erosion depth Δz_i can be calculated with Equation 11 where h is the flow depth:

$$\Delta z_i = h \cdot \frac{\Delta C_i}{\rho(z)} \tag{11}$$

As the dry density of the sediment bed is variable with depth, the calculation of Δz_i is performed iteratively. Density is interpolated linearly between the measured values.

The erosion coefficient α is calculated considering that at the beginning of a bottom shear stress step i (with $\tau_{b,i}$), τ_s is equal $\tau_{b,i-1}$, ε_0 is equal $\varepsilon_{f,i-1}$ and ε is equal $\varepsilon_{m,i}$ (mass-erosion rate of step i, calculated as mean erosion rate in the first 10 minutes of the step). Equation 10 can hence be converted to calculate an erosion coefficient α_i for each step. The erosion coefficient α used in the simulation is determined as the mean value of all α_i of the steps with $10\ h < t < 24\ h$.

The parameters ε_0 and α used in this investigation are given and compared to values from the literature in Table 3. Dry density and bottom shear strength are given in Figure 4.

Table 3
Floc erosion rate ε_0 and erosion coefficient α used in this investigation and from the literature

	ε_0 [kg/m^2/s]	α [m/N$^{0.5}$]
this investigation, lake mud	$15.0 \cdot 10^{-6}$	7.27
Parchure and Mehta (1985), lake mud	$5.3 \cdot 10^{-6}$	13.6
Partheniades (1965), bay mud	$0.07 \cdot 10^{-6}$	8.3

The concentration resulting from erosion simulated with the formula of Parchure and Mehta is given in Figure 3. It is evident that the processes of mass- and floc-erosion are quite well simulated using this formula. The considerable deviation from the measurement in the second last step where no floc erosion is simulated is due to the fact that mean bottom shear stresses were used for the evaluation of parameters. In the simulation of this step, the mean bottom shear stress used in the calculation of the shear strength profile is locally exceeded, leading to considerably higher erosion rates than that forming the basis of the parameter evaluation. However, the used formula is suited to deliver a good reproduction of increasing concentrations due to erosion of stratified beds with the limitation of three empirical parameters (ε_0, α and $\tau_s(z)$) to be determined.

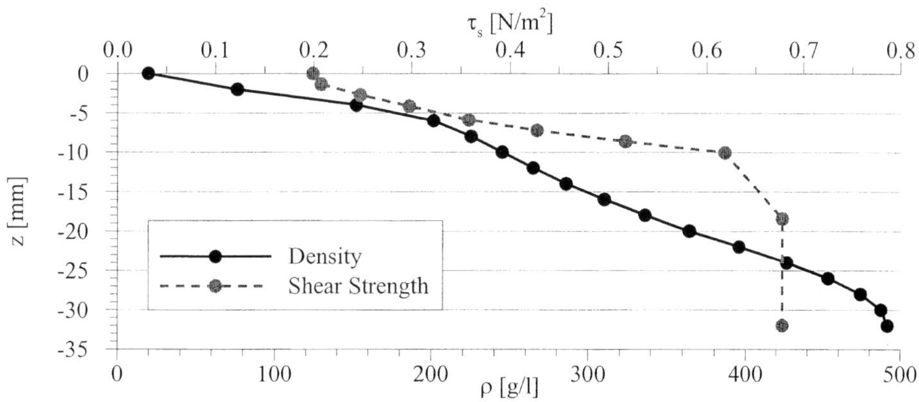

Figure 4. Measured density profile and calculated bottom shear strength of sediment bed eroded in the experiment with sediments from Obermaubach

6. SIMULATION OF COHESIVE SEDIMENT DEPOSITION

6.1. Deposition rate

The deposition rate σ [$kg/m^2/s$] as the flux of sediment from the water column to the sediment bed is generally formulated as (Krone 1962):

$$\sigma = w_s \cdot C \cdot \left(1 - \frac{\tau_b}{\tau_{cd}}\right) \quad \text{for } \tau_b \leq \tau_{cd} \tag{12}$$

where w_s [m/s] is the settling velocity, C [g/l] is the sediment concentration and the term in brackets with τ_{cd} [N/m^2] as the critical bottom shear stress for deposition is the probability that a floc reaching the bed will stick to it. From the experimental data, $\tau_{cd} = 0.615\ N/m^2$ as the bottom shear stress of the step in which the decrease in concentration starts was estimated.

Equation 12 is used to calculate σ in all simulations shown below. Concerning the settling velocity, different formulations are fitted to the experimental data and applied in a simulation.

6.2. Settling velocity

To estimate the settling velocities occurring during the experiment, Equation 13 (Lau 1994) was used to calculate average settling velocities for each bottom shear stress step:

$$w_{s,av} = \frac{h}{\Delta t} \cdot \ln\left(\frac{C(t_0)}{C(t_0 + \Delta t)}\right) \tag{13}$$

h [m] is the flow depth, Δt [s] is the length of the step, C [g/l] is the sediment concentration and t_0 is the time at the beginning of the step. The average settling velocities calculated with Equation 13 are given in Figure 5.

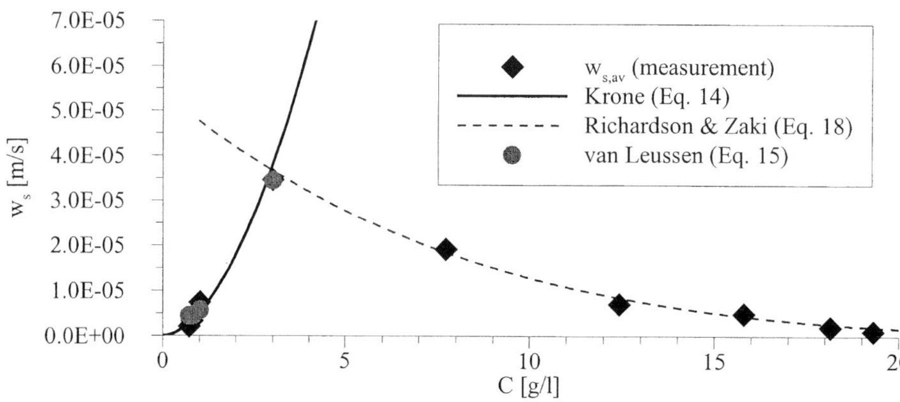

Figure 5. Settling velocities vs. suspended sediment concentration, measurement and fitted functions used in the simulations

In a first simulation (S1), a constant settling velocity calculated as the mean of $w_{s,av}$ of each step is used (w_s = const = $9.06 \cdot 10^{-6}$ m/s).

To calculate the settling velocity during a second simulation (S2), Equation 14 (Krone 1962) which describes w_s as a function of the sediment concentration C is used:

$$w_s = K \cdot C^m \tag{14}$$

with K and m as empirical constants. In using equation 14 it has to be considered that effects of hindered settling, commonly assumed for C > 3-10 g/l, are not taken into account. In the given experiment, hindered settling occurs in the steps 2-6 with a concentration > 5 g/l (compare Figures 5 and 6). Consequently, fitting of parameters K and m is performed to the experimental data of steps with t > 12 h (last 4 steps). The simulation S2 hence started at t = 12 h with an initial concentration C corresponding to the experiment.

With w_s being a function of the concentration, it is obvious that equation 14 is an estimation of the settling velocity where effects of turbulence or differential settling are not taken into account. To take the effect of turbulence into consideration, van Leussen (1994) proposed an extension of equation 14:

$$w_s = K \cdot C^m \frac{1 + aG}{1 + bG^2} \tag{15}$$

a and b are empirical coefficients and G is a dissipation parameter defined by the turbulent dissipation rate per unit mass ε, the kinematic viscosity ν and the Kolmogorov microscale of turbulence η:

$$G = \sqrt{\frac{\varepsilon}{\nu}} = \frac{\nu}{\eta^2} \tag{16}$$

Within the simulation using Equation 15 (S3), the dissipation parameter G is calculated as the "absolute velocity gradient" from the flow field following Camp and Stein (1943):

$$G^2 = \left(\frac{\partial u}{\partial y} + \frac{\partial v}{\partial x}\right)^2 + \left(\frac{\partial u}{\partial z} + \frac{\partial w}{\partial x}\right)^2 + \left(\frac{\partial v}{\partial z} + \frac{\partial w}{\partial y}\right)^2 \tag{17}$$

The coefficients a and b were evaluated by fitting the right hand side of Equation 15 to the measured settling velocities $w_{s,av}$. Therefore, the parameters K and m from simulation 2 are used together with mean dissipation parameters G of every bottom shear stress step. Results of this fit are shown in Figure 5. Analogous to the simulation with the equation of Krone (S2), the parameter fit and the simulation of S3 is carried out for the last 4 steps of the experiment, ignoring the part where hindered settling occurs.

A further simulation (S4) is carried out to account for the effects of hindered settling. Therefore, the expression of Richardson and Zaki (1954) is fitted to the average settling velocities of steps with t < 12 h:

$$w_s = w_0 (1 - k \cdot C)^{\beta} \qquad \text{for } C > C_h \tag{18}$$

The parameters in Eq. 18 are: $w_0 = 5.4 \cdot 10^{-5}$ m/s, k = 0.025 l/g and $\beta = 5$ (see Fig. 5). C_h is determined as the intersection of Equations 14 and 18 ($C_h = 2.95$ g/l). For $C < C_h$, Eq. 14 is used to calculate settling velocities in S4. The parameters K, m, a and b used in this investigation are given and compared to values from the literature in Table 4.

Table 4
Parameters for calculation of settling velocity used in this investigation and from the literature

	K	m	a	b
this investigation, S1, lake mud	$9.06 \cdot 10^{-6}$	0	0	0
this investigation, S2, lake mud	$4.98 \cdot 10^{-6}$	1.85	0	0
this investigation, S3, lake mud	$4.98 \cdot 10^{-6}$	1.85	3.24	0.50
Krone (1962), bay mud	1	1.33	0	0
Malcherek (1995), estuarial mud	3.5	1	0.3	0.09
Mulder and Udink (1990), estuarial mud	$1.0 \cdot 10^{-3}$	0	0	0

6.3. Simulation results

The concentration-time curves resulting from simulations 1, 2, 3 and 4 are given and compared to the experimental data in Figure 6.

Analysing the given experimental and numerical results, it can be stated, that the effect of differential settling (settling and deposition of different grain size fractions with different settling velocities and critical bottom shear stresses, especially occurring with sediment mixtures containing non-cohesive fractions) that leads to the characteristic concentration-time-curve in the experiment with small concentration gradients respectively an equilibrium concentration at the end of the steps, cannot be reproduced in the simulations. Accordingly, the given experiment is not suited to estimate settling velocities as shown before. The calculation of $w_{s,av}$ with Equation 13 is decisively influenced by the length of the bottom shear stress steps, in a way that considerable time periods without substantial deposition are included in the averaging.

The simulation with constant settling velocities (S1) roughly gives a concentration in the same order of magnitude as the experiment. However, simulation of an experiment with longer time-steps and hence longer periods with small concentration gradients would result in a considerable under-estimation of the concentration as the deposition rates within the steps are nearly constant in the simulation so that no equilibrium concentration will be reached.

The resulting concentration development of simulation 2 is in good agreement with the measurement. That's most likely due to the fact, that effects of differential settling are considerably smaller to the end of the experiments where particles with bigger sizes already deposited.

In simulation 3 as well as in simulation 4, mean settling velocities are partly under-estimated compared to the measurement, leading to a considerable over-estimation of the concentration in the water column.

However, with respect to the limitations in parameter estimation as described before, the given results do not allow final statements about the applicability of the tested formulations for the settling velocity.

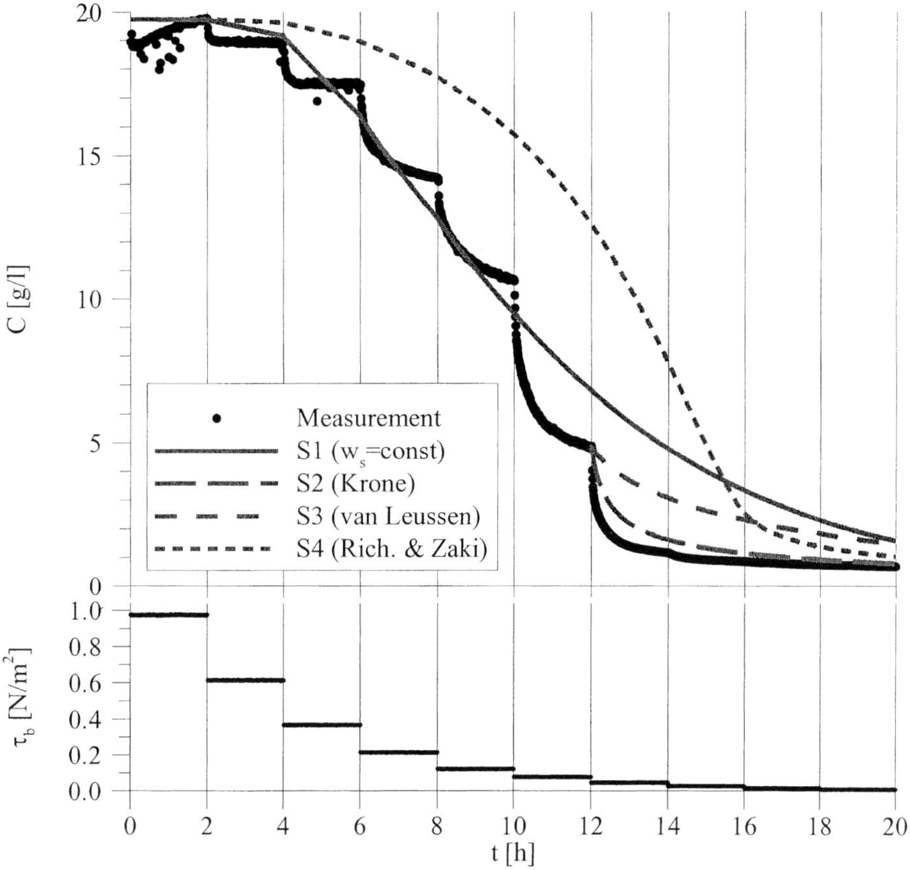

Figure 6. Comparison of measured and simulated concentrations in the water column during a deposition experiment with sediments from the reservoir Heimbach (upper diagram) and mean bottom shear stress history during the experiment (lower diagram).

7. CONCLUSIONS

Comparison of simulations with experimental results show the limits of the applied formulae very clearly. In some simulations, a satisfactory agreement concerning the order of magnitude of concentrations between experiments and simulations could be reached. However, this agreement is rather due to the use of parameters fitted directly to the experimental database than to the correct interpretation of physical phenomena involved. The considerable deviation detected in the comparison of parameters used in this investigation and in the literature underlines this hypothesis.

The formula of Parchure and Mehta seems to be an exception from the above cited hypothesis as the phenomena of mass- and floc-erosion are well reproduced in the simulation. This shows that for a realistic simulation of erosion of cohesive sediments, it is indispensable to know the state of consolidation of the bed being the determining factor for the shear strength. It can consequently be postulated that consolidation models should aim at predicting the density profile more than the temporal development of height of the bed surface (Teisson 1997).

Use of Partheniades' formulation can lead to considerable errors if it is used for simulation of erosion of the top layers of cohesive sediment beds with increasing shear strength. For erosion of lower layers with more or less constant shear strength at high bottom shear stresses, it is appropriate if M and τ_{ce} are carefully adapted to the sediment properties.

Concerning the simulation of deposition of cohesive sediments, it is obvious that the formula of Krone which connects the settling velocity with the sediment concentration and that hereby neglects important phenomena like turbulence effects on aggregation, hindered settling or differential settling can only deliver an estimation of the concentration development. To introduce such phenomena, flocculation models may be used (e.g. Winterwerp 1999, giving a formulation for the settling velocity as a function of floc size and residence time of flocs until break-up).

Continuing the presented combined experimental and numerical investigation, the experimental database will be widened in the future in order to support the statements made in this investigation. In this way, the presented strategy of direct adaptation of results from comprehensive experiments in mathematical and numerical formulations may help to improve the numerical prediction of cohesive sediment transport.

REFERENCES

Bergen, O., 1999, Die Large-Eddy Simulation von Strömungen in natürlichen Seen und Talsperrenspeichern mit der Finite Elemente Methode. In J. Köngeter (ed.), *Mitteilungen des Lehrstuhls und Instituts für Wasserbau und Wasserwirtschaft der RWTH Aachen*: 119. Aachen: Verlag Mainz. (in German)

Camp, T.R. and Stein, P.C., 1943, Velocity Gradients and Internal Work in Fluid Motion. *J. of the Society of Civil Engineers* 30(10): 219-237.

Daniels, H., 1992, Pastis-3D, Implementation Aspects and Users Manual. UCRL-MA-111833, *Lawrence Livermore National Laboratory*, USA.

Forkel, C., 1995, Die Grobstruktursimulation turbulenter Strömungs- und Stoffausbreitungsprozesse in komplexen Geometrien. In J. Köngeter (ed.), *Mitteilungen des Lehrstuhls und Instituts für Wasserbau und Wasserwirtschaft der RWTH Aachen*: 102. Aachen: Verlag Mainz. (in German)

Krone, R.B., 1962, Flume Studies of the Transport of Sediment in Estuarial Shoaling Processes. *Hydraulic Engineering Laboratory and Sanitary Engineering Research Laboratory*: Report. University of California.

Lau, Y.L., 1994, Temperature Effect on Settling Velocity and Deposition of Cohesive Sediments. *J.Hydraulic Research* 32(1): 41-51

Leussen, W. van, 1994, Estuarine Macroflocs and their Role in Fine-Grained Sediment Transport. *PhD-thesis*. University of Utrecht.

Malcherek, A., 1995, Mathematische Modellierung von Strömungen und Stofftransportprozessen in Ästuaren. *Institut für Strömungsmechanik und Elektronisches Rechnen im Bauwesen der Universität Hannover*: Bericht Nr. 44/1995. Universität Hannover. (in German)

Mulder, H.P.J. and Udink, C., 1990, Modelling of Cohesive Sediment Transport. A Case Study: the Western Scheldt Estuary. In B.L. Edge (ed.), *22nd Coastal Engineering Conf.*, Delft, 2-6 July 1990. New York: ASCE

Parchure, T.M. and Mehta, A.J., 1985, Erosion of Soft Cohesive Sediment Deposits. *J. Hydraulic Engineering* 111(10): 1308-1326.

Partheniades, E., 1962, A Study of Erosion and Deposition of Cohesive Soils in Salt Water, Dissertation: University of California, Berkeley.

Partheniades, E., 1965, Erosion and Deposition of Cohesive Soils. *ASCE, J. of the Hydraulics Division* 91(1): 105-139

Partheniades, E., 1984, A Fundamental Framework for Cohesive Sediment Dynamics. In A.J. Mehta (ed.), *Lecture Notes on Coastal and Estuarine Studies 14, Estuarine Cohesive Sediment Dynamics*: 219-250. Berlin, Heidelberg, New York, Tokyo: Springer.

Richardson, J.F. and Zaki, W.N., 1954, The Sedimentation of a Suspension of Uniform Spheres under Conditions of Viscous Flow. Chem. Eng. Sci. 3, 65-73.

Schweim, C., Zhou J., Spork, V., Prochnow, J.V. and Köngeter, J., 2000, Large Eddy Simulation of a Lid-Driven Annular Flume Flow. In J.A. Odgaard (ed.), *Hydroinformatics 2000 - 4th Int. Conf. Hydroinformatics, Iowa, 23-27 July 2000*. Preprint CD-ROM

Spork, V., 1997, Erosionsverhalten feiner Sedimente und ihre biogene Stabilisierung. In J. Köngeter (ed.), *Mitteilungen des Lehrstuhls und Instituts für Wasserbau und Wasserwirtschaft der RWTH Aachen*: 114. Aachen: Verlag Mainz. (in German)

Spork, V., Eisler, R. and Köngeter, J., 1998. Optimisation of Experimental Conditions for Annular Flumes by LDV Measurements. In A.W. Jayawardena, J.H.W. Lee & Z.Y. Wang (eds), *River Sedimentation - Proc. 7th Int. Conf. River Sedimentation, Hong Kong, 16-18 December 1998*: 329-335. Rotterdam: Balkema.

Teisson, C., 1997, A Review of Cohesive Sediment Transport Models. In N. Burt, R. Parker & J. Watts (eds), *Cohesive Sediments - Proc. 4th Nearshore and Estuarine Cohesive Sediment Transport Conf., Wallingford, 11-15 July 1994*: 367-381. Chichester, New York, Weinheim, Brisbane, Singapore, Toronto: John Wiley & Sons.

Winterwerp, H., 1999, On the Dynamics of High-Concentrated Mud Suspensions. *Communications on Hydraulic and Geotechnical Engineering*: Report No. 99.3. Delft University of Technology.

Fine Sediment Dynamics in the Marine Environment
J.C. Winterwerp and C. Kranenburg (Editors)

Modeling the sediment concentration profiles at the Amazon Shelf

Susana B. Vinzon[(a)] and Afonso M. Paiva[(b)]

[(a)] DRHIMA / Escola de Engenharia / Universidade Federal do Rio de Janeiro, 21945-970, Rio de Janeiro, RJ, Brazil

[(b)] Programa de Engenharia Oceânica / COPPE / Universidade Federal do Rio de Janeiro, 21945-970, Rio de Janeiro, RJ, Brazil

Experimental data on flow and sediment dynamics on the Amazon Shelf, obtained under AMASSEDS (A Multidisciplinary Amazon Shelf Sediment Study), have been interpreted with the help of a modeling approach in order to examine the vertical structure of flow-sediment interaction. Long-term accumulation mechanisms as well as short-term processes related to the tides are examined in the context of the vertical transport processes. Tidal signatures indicate that sediment dynamics over short time scale is strongly influenced by resuspension events governed by tidal forcing.

A one-dimensional, vertical sediment transport model was developed, assuming a local mass balance in the water column over the tidal time-scale. The model solves the sediment transport equation following the particle tracking method. The sediment settling velocity is considered to be dependent upon the sediment concentration, and the erosion flux function is fitted using in situ near-bed measurements of velocity and sediment concentration. Salt stratification damping of turbulence is also included, and the relationship between shear strength and bed density, required for the calculation of the erosion flux, is derived from laboratory analysis of Amazon sediment samples. The simulated vertical concentration profiles and the near-bed horizontal sediment transported compares well with the observations.

Keywords: Cohesive sediment dynamics, Amazon shelf

1. INTRODUCTION

Amazon River, by far the largest river discharge on the planet, meets the seawater on the shallow continental shelf 100-200 km seaward of the mouth (Geyer, 1995). In addition to the large input of fresh water, strong tidal currents, associated with tidal waves of about 5 m amplitude (Gibbs, 1970), and wind-driven currents contribute to one of the most energetic continental shelf environments in the world. Spring-neap variations in the vertical density structure on the Amazon shelf have been observed (Geyer, 1995). During neap tides the water column is strongly stratified. During spring tides, however, stratification is greatly reduced (Kineke, 1993).

A comprehensive research program of sediment transport on the Amazon shelf occurred from 1989 to 1991 as part of the AMASSEDS project (A Multidisciplinary Amazon Shelf Sediment Study, Amasseds, 1990). Synchronous profiles of current, sediment concentration, salinity and temperature over the tidal cycle obtained during AMASSEDS (Figure 1) were available for the present study, offering the opportunity to take a close look at the mechanisms governing sediment transport in high concentration environments.

Figure 2 shows instantaneous vertical profiles of advective sediment fluxes in one of the observational sites. The sediment flux within the lower 2 meters is 1-3 orders of magnitude stronger than in the remaining part of the water column, demonstrating the importance of measuring velocity and sediment concentration close to the bottom and understanding its dynamics. In this study, in order to describe the vertical sediment profile, a one-dimensional fine-sediment transport model is formulated. The model assumes a local mass balance in the water column over a tidal cycle, and accounts for the main processes occurring in the water column (erosion/deposition, salt stratification, sediment flocculation, interparticle interaction, turbulence damping, non newtonian behavior of the water-sediment mixture), including those in the near-bed layers responsible for the larger part of the horizontal sediment transport. Due to the significant turbulence damping of high-density fluid mud-like suspension, which dominates the lower water column, the flow in the bottom ~1m is described in terms of an oscillatory viscous flow model. In contrast, in the remainder of the water column, a salt stratified turbulent flow condition is assumed. All the model parameters were determined in laboratory experiments based on sediment samples and field data from the Amazon Shelf.

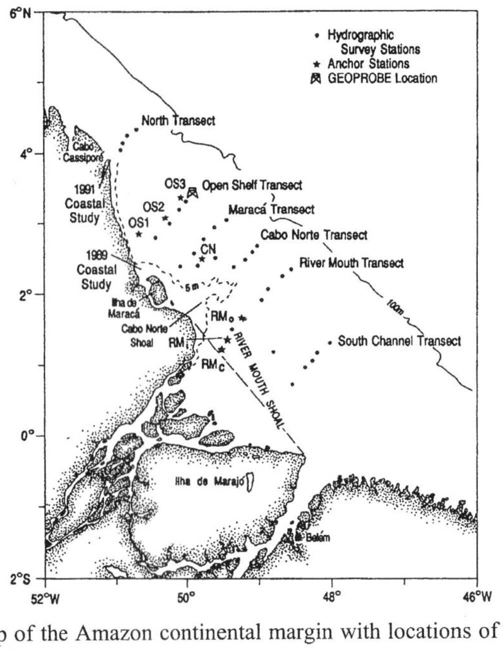

Figure 1. Topographic map of the Amazon continental margin with locations of AMASSEDS anchor stations (from Kineke, 1993).

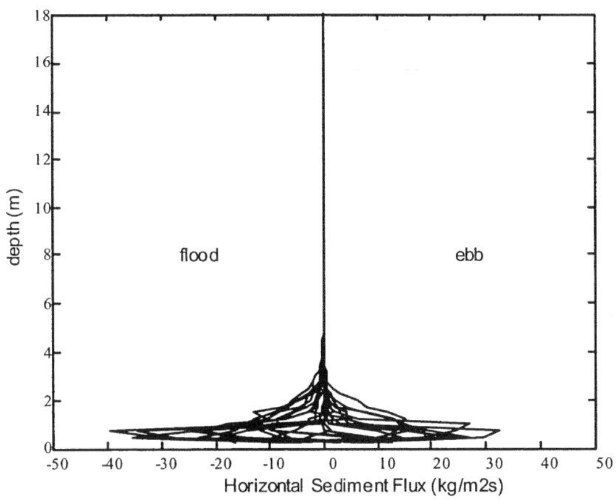

Figure 2. Sediment flux profiles over two tidal cycles in the Amazon Shelf at station CN.

This paper is organized as follows. In Section 2 the relevant mechanisms occurring in high-concentration fine-sediment environments are examined in the context of the Amazon Shelf. In Section 3 the mathematical and numerical models are presented, and the model parameters are calculated. The results are compared to observations and the principal conclusions from this work are then presented in Section 4.

2. MUD DYNAMICS IN AMAZON ESTUARY

Fine sediments tend to be "trapped" in the region of the turbidity maximum, which is a distinctive feature of estuarine regions. Suspended sediment concentrations in this region are typically higher than those upstream as well as those downstream (Mehta, 1989). Because of the varying forcing functions, turbidity maxima exhibit highly dynamic behavior, with position and strength varying during the tidal period, and also over longer time scales due to, for example, variations in the river discharge. Associated with the turbidity maximum there can be extremely high concentration layers at the bottom, especially in meso-, macro-, and hyper- tidal estuaries where fine sediments are abundant, as in the Amazon estuary. During the AMASSEDS surveys, dense (>~10g/l) near-bed suspensions (thickness of the order of a few meters) were found on the inner and middle shelf covering an area ranging from 5,700 to 10,000 km^2 (Kineke, 1993). Typical Amazon River solids concentrations at the river's mouth are of the order of 2 to 20 mg/l.

A raw sediment mass balance for the long term sediment transport can be carried out by taking the average Amazon sediment discharge [Q_s = 11x10^8 tons/year (Meade et al., 1985)], the mean mass in suspension calculated at the Amazon Shelf using data collected during Amasseds (m= 96 kg/m^2), and the deposition area (Ω ranging from 5,700 to 10,000 km^2). Equating the mass delivered by the river and that found over the shelf, the time interval of sediment supply necessary to obtain the observed quantity of sediment over the shelf is given by the equation $t = m\Omega / Q_s$, and with a range of 6 to 10 months. In turn, mechanisms responsible for retaining these sediments in the Amazon Shelf are necessary. Data are insufficient to discern the genesis of turbidity maxima over the Amazon Shelf. Since the residual circulation, related to salt stratification and tidal asymmetry changes along the estuary, and since the degree of stratification changes over the spring-neap cycle, different processes may be active at different locations and at different times. In the Amazon estuary residual circulation may be important at neaps with high river flow, but may not be significant at spring tides. Other mechanisms related to rheologic behavior of Amazon mud changing from shear thinning at relatively low concentrations to shear-thickening at concentrations of about 300 g/l (Faas 1985) can also be important controlling factors making tidal trapping an effective mechanism for mud accumulation.

Variations of sediment concentrations over time-scales of the tidal cycle have been observed at the Amazon Shelf. Associated with the sediment settling velocity, a time scale can be defined related to water column processes. The flocs can have sizes ranging from that of an individual particle, around 4μm to as much as 500μm or more, with associated density variations. Thus, a wide range of settling velocities characterizes the fine sediment transport. Taking a settling velocity range from 0.02 to 2 mm/s, characteristic of the Amazon Shelf (Gibbs, 1985; Gibbs and Konwar, 1986; Kineke, 1993), and a reference water depth of 10 m, the time-scale associated with settling would range from 6 days to 1 hour, respectively. Accordingly, the water column concentration profile variations can be expected to reflect tidal action as well as longer period meteorological effects. Short period gravity waves, with a characteristic time-scale on the order of 10 s, can be expected to have only a residual, wave-mean, effect.

Figure 3 shows flow velocity and sediment flux from the bottom for two anchor stations at the Amazon Shelf. The good correlation between bottom sediment fluxes and tide forcing observed in this picture, as well as in other anchor stations, highlights the importance of tidal erosion/deposition mechanisms. However, the phase difference between the sediment flux and the velocity is not always as expected [at some instances, the sediment flux increases while the velocity still decreases at CN (2428)], which is possibly explained by the low sampling time resolution and by the use of a non-fixed measuring device (Vinzon, 1998). Other evidence of the tidal forcing effect on the sediment concentration profiles is presented in Figure 4, which shows the maximum elevation of the 0.1 and 1.0 g/l concentration isolines reached during the tidal cycle as a function of the tidal range, for all the measurement sites.

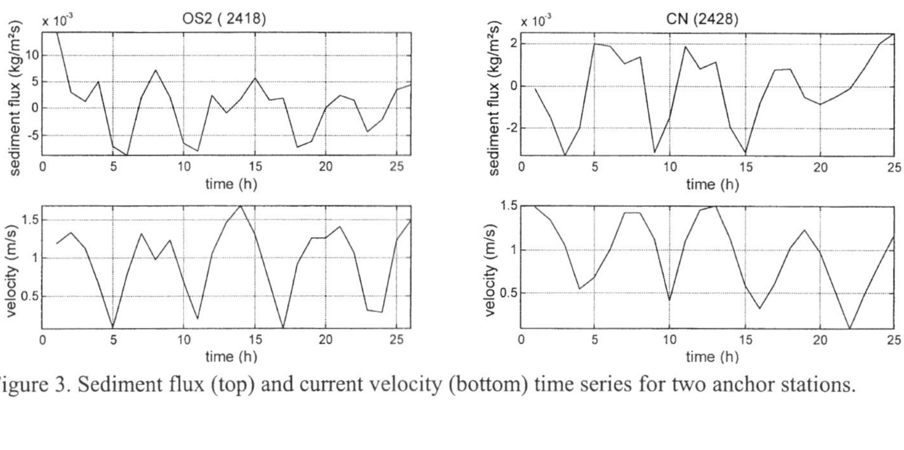

Figure 3. Sediment flux (top) and current velocity (bottom) time series for two anchor stations.

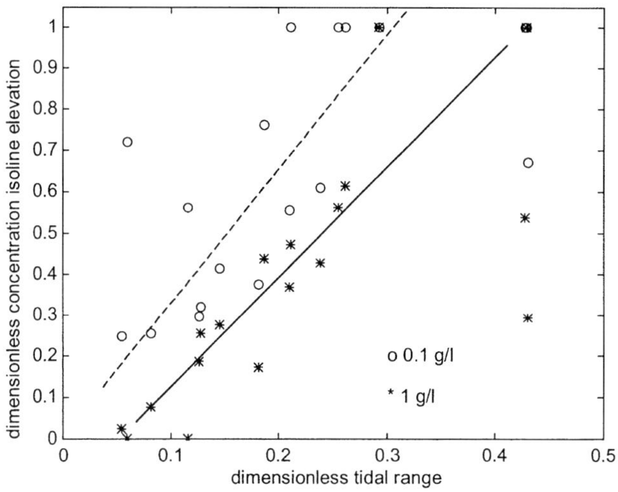

Figure 4. Maximum elevation of the 0.1 and 1.0 g/l concentration isolines reached during the tidal cycle as a function of the tidal range (both made dimensionless by dividing by the local depth) for all the measurement sites. The lines in the figure indicate the general trends in the data.

3. 1-D VERTICAL MODULE FOR FINE SEDIMENT TRANSPORT MODELING

The governing equation for the vertical transport of suspended sediment is the conservation of mass,

$$\frac{\partial C}{\partial t} = \frac{\partial}{\partial z}\left[C(W_s - w) + \varepsilon_s \frac{\partial C}{\partial z} \right] \tag{1}$$

in which C is the suspended sediment concentration, W_s is the settling velocity, ε_s is the mass diffusion coefficient, and w is the vertical component of the flow velocity.

Equation (1) requires two boundary conditions, one at the free surface and another at the bottom level. At the free surface, the boundary condition corresponds to no net sediment flux. For the bottom condition, the model assumes that bed-suspension sediment exchange occurs only in one direction, i.e. erosion or deposition (erosion and deposition are not considered to occur simultaneously). For erosion, i.e., when the flow bed shear stress, τ_o, is greater than the shear strength of the overlying bed layer, τ_s, a linear rate of erosion is prescribed according to

$$z = 0, \quad \tau_o > \tau_s \qquad E = M\left(\frac{\tau_o}{\tau_s} - 1 \right), \tag{2}$$

in which M is an empirical erosion rate constant. When no erosion is occurring, a deposition rate is prescribed according to

$z = 0,\ \tau_o \leq \tau_s \qquad D = C(W_s - w) \qquad (3)$

To solve Equation (1) with the corresponding boundary conditions, it is necessary to prescribe the settling velocity (W_s), the bottom shear stress (τ_o), the bottom shear strength (τ_s), the erosion rate constant (M), and the mass diffusivity coefficient (ε_s), and to evaluate the vertical velocity (w). These parameters depend on the sediment and flow characteristics.

3.1. Settling Velocity (W_s)

Experiments using the multi-depth method were carried out in order to determine the relationship between the settling velocity and the sediment concentrations. During these experiments, the mud particles were aggregated, and therefore an interface was formed between the upper water layer and the top of the suspension. The settling velocity was estimated from the sediment mass conservation equation for a quiescent medium, and from the fall velocity of the water-mud interface, which is constant during the settling stage (Imai, 1981). The results are shown in Figure 5. With increasing concentration, the data points indicate the existence of an increasing velocity region, which is related to flocculation effects, and a decreasing velocity region, associated to hindered settling. From the above experiments, the following empirical relationships between settling velocity and concentration were obtained:

$$W_s = 0.05\,(1.35 - 0.01\,C)^{5.6} \text{ mm/s} \qquad \text{for } C > 1.7 \text{ g/l} \qquad (4)$$

$$W_s = 0.11 C^{1.6} \text{ mm/s} \qquad \text{for } C < 1.7 \text{ g/l} \qquad (5)$$

These relationships are also shown in Figure 5 (solid lines). The equations are similar to those proposed by Ross (1988); however, the coefficients calculated in the present study led to a better agreement between the results obtained with equation (1) and the settling experiments.

Floc sizes measured by Gibbs and Konwar (1986) in the shelf region seaward of the Amazon River mouth had mean values of 50-100μm, and a maximum of 200μm. Photographs of the in situ suspended materials showed modal floc size in the range of 200-500μm (Kineke, 1993). Considering these size ranges, with corresponding densities of 1.3 and 1.01 g/cm^3 (Gibbs, 1985, Kineke, 1993, Fennesy et al., 1994), an estimated floc settling velocity for the Amazon Shelf, according to Stokes' equation, would range between 0.02 to 2.8 mm/s (represented by the shaded area in Figure 5).

The settling velocity of cohesive sediments is influenced by flocculation, which in turn depends on sediment concentration, on turbulence, and on organic contents and physico-chemical properties of the sediments (Dyer, 1989, Winterwerp, 1998). Turbulence can enhance the coagulation process, bringing particles together, or can breakup the already formed flocs, due to the shear stresses. In general, the settling velocities determined in laboratory conditions underestimated those determined from field observations, and this could be explained by the absence of turbulence in the laboratory experiments. Therefore, a fitting factor λ_1, was considered in the modeling process in order to account for the field and lab discrepancies, preserving however the functional relationship between the settling velocity and the sediment concentration. The effect of introducing this fitting factor is also shown in Figure 5.

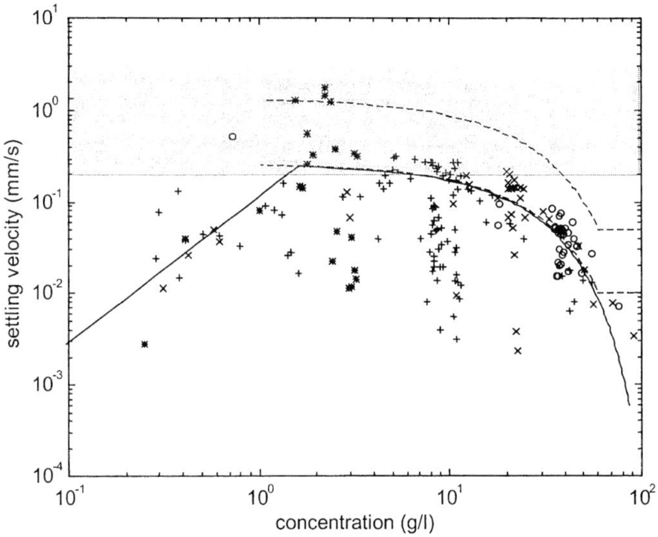

Figure 5. Settling velocity versus sediment concentration obtained from settling experiments. Shadow zone shows settling velocity estimated from field measurements, and dotted line shows the upper limit for the settling velocity allowed in the modeling process (for λ_1 equals 5).

3.2. Bottom shear stress (τ_o)

The high concentration of fine sediments near the bottom found in the Amazon Shelf, and the associated enhanced fluid viscosity, inhibit turbulence development in the near-bed layer. Thus, bottom shear stress is calculated considering the analytical solution for a viscous boundary layer for the oscillatory tidal flow (Nielsen, 1992; Vinzon, 1998; Vinzon and Mehta, 2001):

$$\tau(0,t) = A\sigma e^{i\sigma t}(1+i)\frac{\nu\rho}{\sqrt{2\nu_f/\sigma}} \tag{6}$$

in which σ is the semi-diurnal tidal frequency, and $A\sigma = u_\infty$, is the velocity at the outer edge of the boundary layer. The top of the boundary layer is considered to be at z = 4 m (Geyer, 1995, suggested that the boundary layer in the Amazon Shelf region is confined between 3 to 5 m for neap and spring tides respectively). The kinematic viscosity, ν_f, was obtained from the analysis of dense suspensions of Amazon sediment, taken from the tops of box cores (Faas, 1985, Vinzon and Mehta, 2001).

3.3. Bottom shear strength (τ_s)

Following Migniot (1968), Otsubo and Muraoka (1988) and Dade (1992), among others, it is considered in the present study that the yield stress represents a measure of the interparticle bond strength per unit area. Thus, the bed shear strength, τ_s, can be determined from its correlation with the yield stress measured from Amazon sediment samples. The upper Bingham yield stress, τ_y, is defined from the stress-versus-shear rate flow curve by extrapolation from the low values of shear rate.

Otsubo and Muraoka (1988) performed extensive experiments to relate shear strength to the yield stress for sediments of different mineral compositions and water contents. The functional relationship between shear strength τ_s and yield stress τ_y, obtained for natural clay mixtures similar to the characteristics of the Amazon sediment samples, is given by

$$\tau_s = 0.27\tau_y^{0.6} \text{ Pa} \tag{7}$$

Yield stress values obtained by Faas (1985) and Dade (1992) based on the laboratory analysis of superficial sediments of the Amazon Shelf bottom are presented in Figure 6, as a function of sediment concentration. The best fit line for the combined data is given by:

$$\tau_y = 2.02x10^{-6}C^{2.62} \text{ Pa} \tag{8}$$

where the sediment concentration is given in grams per liter.

Combining equations (7) and (8) a relationship between shear strength and sediment concentration is then obtained:

$$\tau_s = 1.03x10^{-4}C^{1.57} \text{Pa} \tag{9}$$

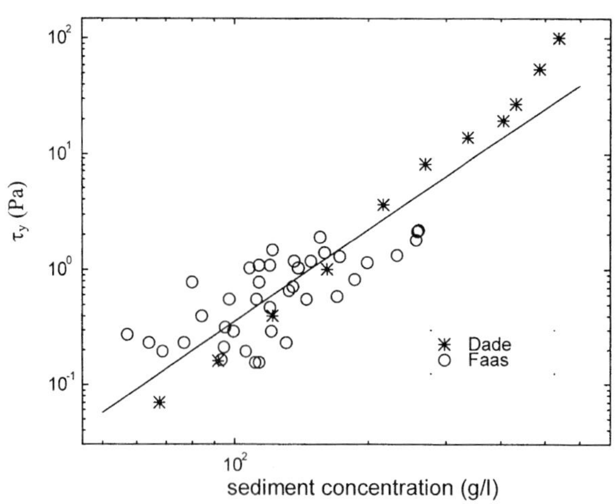

Figure 6. Yield shear stress versus sediment concentration for Amazon Shelf mud (from Dade, 1992 and Faas, 1985).

According to equation (9), for the maximum concentration of 321 g/l observed in the Amazon set data (anchor station OS1), one would obtain a value of shear strength of 0.91Pa. However, a mud layer with a horizontal velocity of 7 mm/s was observed at this site, although the estimated current-induced bottom shear stress reached a maximum value of only 0.42 Pa, lower than the sediment shear strength. Beside the assumptions necessary to obtain both the shear strength and the shear stress, other physical environmental factors may also change the properties determined in the laboratory. Therefore, it is highly likely that in the prototype environment wave action lowers bed shear strength, as well as enhances current-induced bottom shear stress. With the purpose of incorporating these effects, a coefficient that multiplies the bed shear strength ($\theta \leq 1$) was included and evaluated from the measurements as follows.

3.4. Erosion rate (*M*)

The coefficient θ and the erosion rate constant M were determined adjusting equation (2) for data from the anchor station CN (2428). The erosion rate is defined with respect to the positive sediment flux. Bottom shear stress was calculated according to equation (6) and bed shear strength was obtained from equation (9). The best fit between the excess shear stress ($\tau_o - \tau_s$) and the erosion rate for this anchor station gives a coefficient θ=0.4 and an erosion rate constant M=0.02 kg $m^{-2}s^{-1}$.

3.5. Mass diffusivity (ε_S)

Lower water column:

The molecular mass diffusivity is negligible in the lower water column, due to the high viscosity of the sediment-water mixture. Wave action and the increased effective stress developed as a consequence of contact between particles, are the main mechanisms helping to prevent complete deposition of the primary lutocline.

When the sediment-water mixture is truly a suspension, i.e., at relatively low concentrations, effective stresses are absent. At intermediate concentrations the sediment-water mixture does not present a well-defined soil structure, but the drag forces are comparable with effective normal stresses (Pane and Schiffman, 1985). Following Toorman and Berlamont (1991) the presence of effective stresses can be modeled as a diffusive process according to

$$\frac{\partial C}{\partial t} + \frac{\partial}{\partial z}(CW_s) + \frac{\partial}{\partial z}\left(\varepsilon_c \frac{\partial C}{\partial z}\right) = \frac{\partial}{\partial z}\left(\varepsilon_s \frac{\partial C}{\partial z}\right) \quad (10)$$

where ε_S is the mass diffusion coefficient and

$$\varepsilon_c = \left(\frac{W_s}{g}\frac{\partial \sigma'}{\partial C}\frac{\rho_s}{\rho_s - \rho_w}\right) \quad (11)$$

accounts for the effective stress as a diffusive coefficient that prevents rapid deposition, thus allowing for a slow time-rate of dewatering.

There are no measurements of consolidation parameters, i.e., permeability and effective normal stress, for the Amazon Shelf sediments. Anyway, it is difficult to distinguish to which extend, wave action or interparticle support, would be more important in preventing the collapse of the primary lutocline. In order to adjust the diffusion coefficient in the lower layer, the settling experiments described in section 3.1 were simulated numerically. This simulation gave as a result that the coefficient ε_c varies from 0.2×10^{-5} m^2/s to 0.3×10^{-7} m^2/s for sediment concentrations ranging from 25 g/l to 150 g/l, respectively. Taken those values as indicative, the diffusion coefficient in the lower suspension, ε_c, was introduced in the model as a fitting parameter, together with a limiting concentration, C_{lim}, representing the minimum concentration in which the particle interaction occurs.

Upper suspension:

In contrast to the lower zone, in the upper suspension the mass diffusivity is directly related to the momentum diffusivity, which can be large. The vertical mass diffusivity in turbulent flows is a result of the balance between the turbulent kinetic energy production due to shear, and dissipation or damping due to viscosity and buoyancy work. In stratified estuaries, buoyancy effects result in turbulence damping and mixing reduction across the stratified layers. Along these lines, the generation of salinity fine-structure by vertical diffusion has been demonstrated by, among others, Posmentier (1976). Discontinuities in diffusive flux can contribute to the formation of secondary lutoclines (Scarlatos and Mehta, 1990).

Salinity stratification appears to be related to secondary lutocline formation in the upper water column in the Amazon Shelf area. This is illustrated in Figure 7, which shows suspended sediment concentration profiles and water density gradients over the tidal cycle for two observational sites. In the anchor station OS2, on the left, a secondary lutocline was formed in the region of strong salinity stratification, while in the anchor station OS1, on the right, salt stratification was weak and no secondary lutocline is observed.

Above the viscous layer a parabolic eddy viscosity model was considered:

$$\varepsilon_s = \frac{\nu}{S_{ch}} \phi(z) \tag{12}$$

$$\nu(z) = \kappa u_* z \left(1 - \frac{z}{h}\right) \tag{13}$$

and salinity stratification was included through a turbulence damping function (Munk and Anderson, 1948):

$$\phi = (1 + aRi)^b \tag{14}$$

in which a=3.33 and b=-1.5, and Ri is the gradient Richardson number.

Although Ri varies throughout the tidal cycle, a depth dependent tidally averaged turbulence damping function was used in the model simulation, a simplification that was seem to be appropriated in the present case study. The Schmidt number, S_{ch}, which relates momentum and mass diffusivities, was adjusted in the modeling process. Values presented in the literature range from 0.7 to 2.4 (Costa and Mehta, 1990, Yamada, 1975)

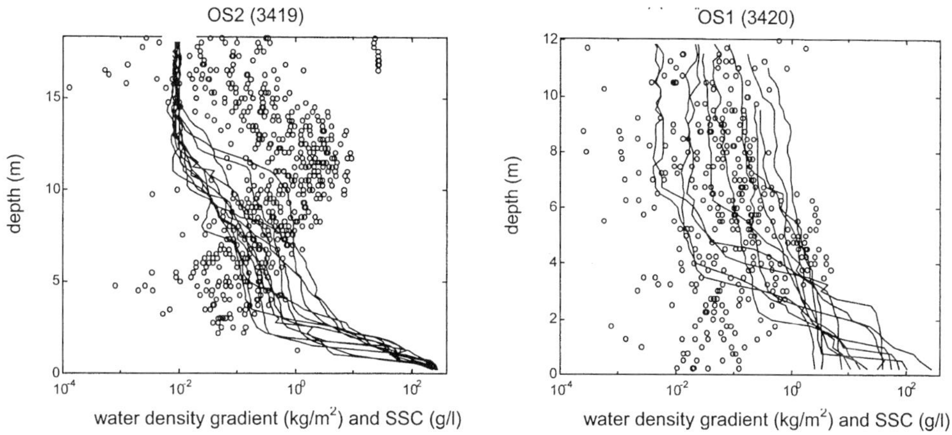

Figure 7. Suspended sediment concentration profiles (solid line), and water density gradients including salinity and temperature effects (circles), for anchor stations in the Amazon Shelf for cases of strong (left) and weak (right) salinity stratification.

3.6. Vertical velocity (*w*)

The vertical component of the flow velocity is usually neglected in shallow water models, but it may be important where noticeable bathymetry variations occur, or where the tides have high amplitudes, which is the case in the Amazon Shelf. Vertical velocity values are low, but they can be of the order of the settling velocity of fine sediments. The vertical velocity was estimated according to:

$$w = \lambda_2 \frac{z}{h}\frac{\partial h}{\partial t}, \tag{15}$$

in which h is the water column depth, and λ_2 is a fitting parameter.

3.7. Numerical model

Fine sediment transport models are very demanding on numerical methods. The quantities involved, like settling velocity or diffusion, vary by several orders of magnitude over the physical domain. In the vertical dimension, strong sediment concentration gradients are observed, and these raise severe problems in solving the transport equation. The particle tracking technique (Jin, 1993) is used in the present work, which has the advantage over traditional Eulerian methods of overcoming artificial diffusion, and wiggles formation. In the particle tracking method the substance is divided into set of small discrete particles, which are advected by the ambient fluid velocities, and dispersed by means of a random walk algorithm that takes into account the mass diffusion coefficient ε_s. This approach is being increasingly used in the modeling of contaminant transport, and its introduction into fine sediment transport modeling has been proved to be very useful (Vinzon, 1998).

4. RESULTS AND DISCUSSION

The sediment fluxes calculated from equations (2) and (3) for the anchor station CN(2428) are compared to observations at the bottom level (z=0.25m) in Figure 8. Deposition, which happens when the bed shear stress is lower than the bed shear strength, is represented by negative flux values. The calculated values agree in order of magnitude with the observed ones, and the time evolution of the observed fluxes is well represented during most of the experiment, with the exception of the period around hour 21. As far as the model results compare well, over all the simulated period, with the observed concentration profiles and horizontal fluxes, measured fluxes could be affected by the imprecision in profiler positioning.

Time evolution of sediment concentration in the water column was simulated for two tidal cycles, with model parameters computed from Amazon Shelf observations. The initial concentration was based on data from anchor station CN(2428), and the simulation started during the flood period. The water depth and the current velocity at the top of the boundary layer were taken from the observations, and varied with time. The settling velocity, current shear stress, bed shear strength, erosion rate, damping stratification function, and momentum diffusivity distribution were computed using the functions described in Section 3. The five fitting parameters, discussed in section 3, were adjusted in order to obtain the better agreement between the computed and observed concentrations. The resulting values were: λ_1=3.26 (settling velocity coefficient), S_{ch}=1.82 (Schmidt number), λ_2=1.15 (correction for the vertical velocity), ε_c=4.85x10^{-5} m^2/s (mass diffusivity in the viscous layer) and C_{lim}=40g/l (limiting concentration for the viscous layer).

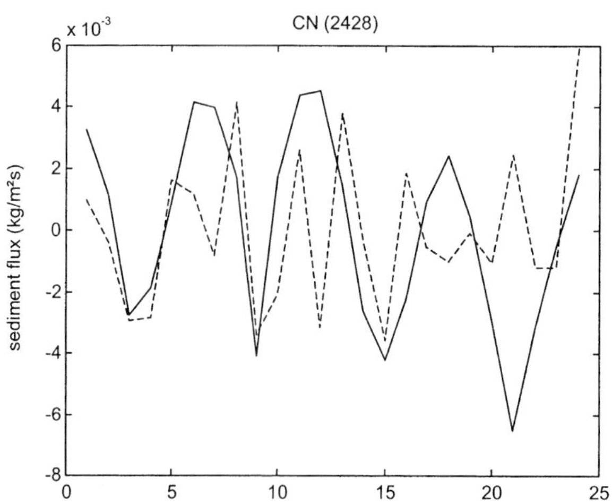

Figure 8. Observed (dotted line) and calculated (solid line) bottom sediment fluxes for anchor station CN (2428) at z=0.25m.

The vertical concentration profiles simulated with the above set of values are shown in Figure 9 and compared with the observed profiles at the different tidal stages. The model results capture the strong variations of sediment concentrations, which are observed to occur along the water column (concentration changes four orders of magnitude from the bottom to about ten-meter depth). The primary lutocline is well represented in most of the vertical profiles. The secondary lutoclines is reasonably well represented during high waters, and to a less extent during the rest of the tidal cycle. Note that the concentrations in the secondary lutocline are of the order of the model resolution, which is 0.01 g/l for this simulation, what may have affected the model capability of representing this feature.

The total horizontal sediment fluxes, computed with observed velocity and simulated concentration over the two tidal cycles, is compared to the fluxes based on observed concentrations in Figure 10. The mean relative error between the two estimates is 30%. The agreement is quite good, considering that some of the model variables vary by more than one order of magnitude within the domain. The results presented in Figures 9 and 10 indicate that the model captures the most significant mechanisms related to the Amazon Shelf sediment dynamics over the tidal time-scale.

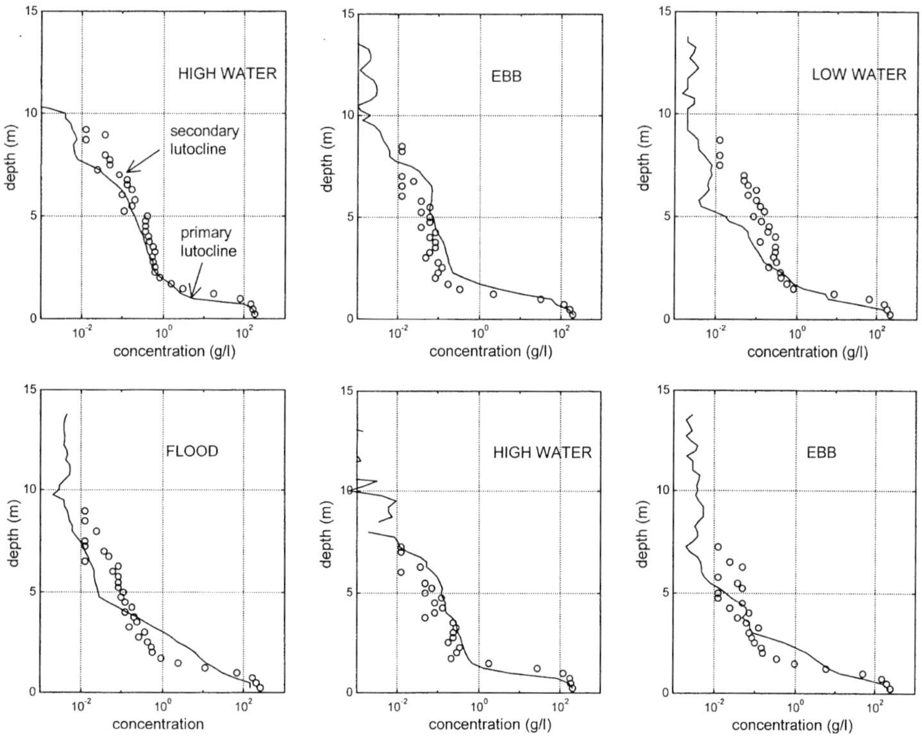

Figure 9. Sediment concentration profiles at anchor station CN(2428), measured (solid line) and simulated (circles).

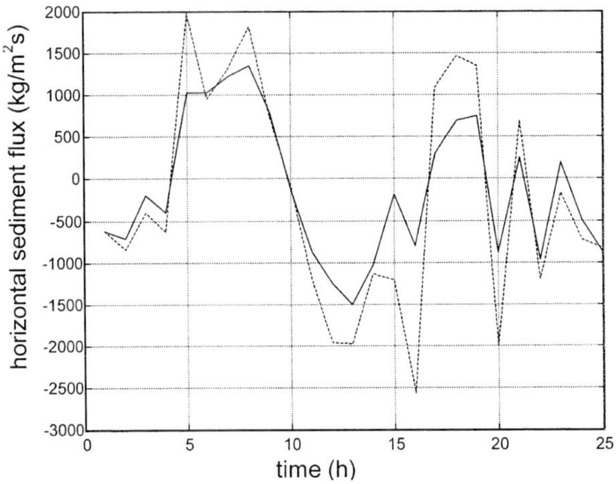

Figure 10. Measured (solid line) and simulated (dotted line) horizontal sediment fluxes over two tidal cycles at anchor station CN (2428).

There are some questions that require further research, especially with respect to the turbulence damping and mixing within the viscous layer. Horizontal sediment fluxes prediction is critically dependent on the dynamics of the viscous high sediment concentration layer. A more refined representation of the transition between turbulent and viscous flow behavior, which occurs when high sediment concentrations are present, could reduce the constrains imposed by the necessity of tuning C_{lim} (the limiting concentration for the viscous layer).

The shallow coastal zone along the inner continental shelf is a very energetic region where water waves and currents interact with the bottom sediments. Over the Amazon Shelf, even comparatively short waves, with periods of 7-8 s, may have an influence upon 40 m of water column, corresponding to the maximum depth of the AMASSEDS profiling survey. Because of the short time scale of wave motion, the associated boundary layer is confined to a relatively thin zone close to the seabed. Under short period waves, high time rates of change in pore-water pressure can promote the pore fluid flow relative to the soil skeleton, which may eventually lead to the rupture of the interparticle bonds (Mehta, 1996). Soil liquefaction, or loss of effective stress, occurs as a result of the build-up of excess pore pressure due to dynamic loading. The lack of synoptic wave data and longer time series of measurements makes it difficult to infer the effects of the wave action on sediment resuspension at the Amazon Shelf. Over a weekly scale, measurements at other sites indicate that wave activity superimposed on tidal action correlates well with suspended sediment concentration variation. The wave action was not explicitly included in the model, but its effect upon the mixing is possibly reflected in the values of the diffusion coefficient near the bottom.

Kineke (1993) suggested that a combination of density-driven estuarine circulation, salt induced stratification, and flocculation enhances the trapping of sediments in fluid muds of the Amazon Shelf. However, as discussed by Vinzon (1998) and shown by the scale analysis of Section 2, those processes are important primarily over long time scales (6 to 10 months). The results of the present study support the view that, over the tidal time scale, the sediment dynamics in the Amazon Shelf is strongly influenced by resuspension events governed by tidal forcing. Understanding and modeling the forcing and the physical processes responsible for the dynamics of the near-bottom high sediment concentration layers is very important, considering that the spreading of fine sediments over the Amazon continental shelf is dominated by the advective transport occurring within these layers.

5. ACKNOWLEDGMENT

The authors deeply appreciate the contributions of Dr. Ashish Mehta of Universtiy of Florida, USA, for his valuable discussions. Special thanks are due to Dr. Gail Kineke of Bostom College, Chestnut Hill, MA, USA, for providing the database. This work was made possible by financial support from CAPES, Brazil.

REFERENCES

AMASSEDS, 1990. A Multidisciplinary Amazon Shelf Sediment Study, EOS, *Transactions of The American Geophysical Union*, 71(45), 771-1777.

Costa, R.G., and Mehta, A.J., 1990. Flow-fine sediment hysteresis in sediment-stratified coastal waters, *Proceedings of the 20th Coastal Engineering Conference*, ASCE, New York, 2047-2060.

Dade W. B., 1992. Studies on Boundary Conditions for Fine-Sediment Transport, Ph.D. Dissertation, University of Washington, Seattle, 260p.

Dyer, K.R., 1989. Sediment processes in estuaries: future research requirements, *Journal of Geophysical Research*, 94(C10), 14327-14339.

Faas, R.W., 1985. Time and density-dependent properties of fluid mud suspensions, NE Brazilian Continental Shelf, *Geo-Marine Letters*, 4, 147-152.

Fennessy, M.J., Dyer, K.R., and Huntley, D.A., 1994. INSSEV: An instrument to measure the size and settling velocity of flocs in situ, *Marine Geology*, 117, 107-117.

Geyer, W., 1995. Tide-induced mixing in the Amazon frontal zone, *Journal of Geophysical Research*, 100(C2), 2341-2353.

Gibbs, R. J., 1970. Circulation in the Amazon river estuary and adjacent Atlantic Ocean, *Journal of Marine Research*, 28, 113-123.

Gibbs, R.J., 1985. Estuarine flocs: their size, settling velocity and density, *Journal of Geophysical Research*, 90(C2), 3249-3251.

Gibbs, R.J. and Konwar, L., 1986. Coagulation and settling of Amazon River suspended sediment, *Continental Shelf Research*, 6(1/2), 127-149.

Imai, G., 1981. Experimental studies on sedimentation mechanism and sediment formation of clay minerals, *Soils and Foundations*, 21(1), 7-20.

Jin, X.-Y., 1993. Quasi-three dimensional numerical modelling of flow and dispersion in shallow water, *Communication on Hydraulic and Geotechnical Engineering*, Department of Civil Engineering, Delft University of Technology, Report 93-3, 174p.

Kineke, G.C., 1993. Fluid muds on the Amazon continental shelf, Ph.D. dissertation, University of Washington, Seattle, 259p.

Meade, R.H., Dunne, T., Richey, J.E., Santos, U.D., and Salati, E., 1985. Storage and remobilization of suspended sediment in the lower Amazon river of Brazil, *Science*, 228, 488-490.

Mehta, A.J., 1989. On estuarine cohesive sediment suspension behavior, *Journal of Geophysical Research*, 94(C10), 14303-14314.

Mehta, A.J., 1996. Interaction between fluid mud and water waves, *Environmental Hydraulics*, V. P. Singh & W.H. Hager Eds., Kluwer Academic Publishers, Dosdrecht, The Netherlands, 153-187.

Mignot, C., 1968. A study of the physical properties of various forms of fine sediment and their behavior under hydrodynamic action, *La Houille Blanche*, 7, 591-620.

Munk, W.H. and Anderson, E.R., 1948. Notes on a theory of the thermocline, *Journal of Marine Research*, 1, 176-295.

Nielsen, P., 1992, Coastal bottom boundary layers and sediment transport, In: *Advanced series in sediment transport*, vol. 4, World Scientific, P.L-F. Liu Ed., JBW Printers & Binders Pte. Ltd., Singapore, 334p.

Otsubo, K. and Muraoka, K., 1988. Critical shear stress of cohesive bottom sediments, *Journal of Hydraulic Engineering*, ASCE, vol. 114, n. 10, pp.1241-1256.

Pane, V. and Schiffman, R.L., 1985. A note on sedimentation and consolidation, *Geotechnique*, 35, 69-72.

Posmentier, E.S., 1976. The generation of salinity fine structure by vertical diffusion, *Journal of Physical Oceanography*, 7, 198-300.

Ross, M.A., 1988, Vertical structure of estuarine fine sediment suspensions, Ph.D. dissertation, University of Florida, Gainesville, 206 p.

Scarlatos P.D. and Mehta, A.J., 1990. Some observations on erosion and entrainment of estuarine fluid muds, *Residual Currents and Long Term Transport*, R.T. Cheng ed., Springer-Verlag, New York, 544 p.

Toorman, E.A., and Berlamont, J.E., 1991. A hindered settling model for the prediction of settling and consolidation of cohesive sediment, *Geo-Marine Letters*, 11, 179-183.

Vinzon, S.B., 1998. A preliminary examination of Amazon Shelf sediment dynamics, Engineer Degree Thesis, University of Florida, Gainesville, 168 p.

Vinzon, S.B. and Mehta, A.J., 2001. Boundary layer effects due to suspended sediment in the Amazon River estuary, in Coastal and Estuarine Fine Sediment Processes, W.H. McAnally and A.J. Mehta Eds., Elsevier, Amsterdam, 507p.

Winterwerp, J.C., 1998. A simple model for turbulence induced flocculation of cohesive sediment, *Journal of Hydraulic Research*, 36(3), 309-326.

Yamada, T., 1975. The critical Richardson number and the ratio of the eddy transport coefficients obtained from a turbulence closure model, *Journal of the Atmospheric Sciences*, 32, 926-933

Contributing authors

T.J. Andersen
Institute of Geography
University of Copenhagen
Øster Voldgade 10
DK-1350 København
K, Denmark

P.G.E.F. Augustinus
Institute for Marine and
Atmospheric Research Utrecht (IMAU)
Dept. of Physical Geography
Utrecht University
P.O. Box 80.115
3508 TC Utrecht
The Netherlands

A.J. Bale
Plymouth Marine Laboratory
Prospect Place
Plymouth PL1 3DH
United Kingdom

C.D. Barrett
Plymouth Marine Laboratory
Prospect Place
West Hoe
Plymouth, PL1 3DH
United Kingdom

J. Berlamont
Hydraulics Laboratory
Civil Engineering Department
Katholieke Universiteit Leuven
Kasteelpark Arenberg 40
B-3001 Leuven
Belgium

S. Bourban
HR Wallingford
Howbery Park
Wallingford
Oxon OX10 8BA
United Kingdom

M.L. Brennan
Environmental Fluid Mechanics Laboratory
Dept. Civil & Environmental Engineering
Stanford University
Stanford
CA 94305-4020
USA

M.B. Brinsley
Plymouth Marine Laboratory
Prospect Place
West Hoe
Plymouth, PL1 3DH
United Kingdom

A.W. Bruens
Fluid Mechanics Section
Civil Engineering Department
Delft University of Technology
PO Box 5048
2600 GA Delft
The Netherlands

J.R. Burau
U.S. Geological Survey
Placer Hall 6000 J St.
Sacramento
CA 95819
USA

F. Cayocca
Institut Français de Recherche
pour l'Exploitation de la Mer (IFREMER)
Centre IFREMER de Brest,
BP 70 29280
Plouzané
France

C. Cheviet
Laboratoire National d'Hydraulique et
Environnement (EDF)
6 quai Watier
78400 Chatou
France

M.C. Christie
Institute of Marine Studies
University of Plymouth
Plymouth PL4 8AA
United Kingdom

M. Crapper
School of Civil and Environmental
Engineering
University of Edinburgh
United Kingdom

M.P. Dearnaley
HR Wallingford
Howbery Park, Wallingford
Oxfordshire OX10 8BA
United Kingdom

E.M.G.T. de Deckere
Netherlands Institute of Ecology
Centre for Estuarine and Coastal Ecology
P.O. Box 140
4400 AC Yerseke
The Netherlands

K.R. Dyer
Institute of Marine Studies
Univ. of Plymouth
Plymouth PL4 1HP
United Kingdom

N. Feates
HR Wallingford
Howberry Park
Wallingford
OX10 8BA
United Kingdom

N. Gratiot
Laboratoire des Ecoulements
Géophysiques et Industriels
BP 53
38041 Grenoble
Cédex 9
France

M. Guismia
Laboratoure National d'Hydraulique
et Environnement (EDF)
6 quai Watier
78400 Chatou
France

G. Gust
Technical University of Hamburg-Harburg
Ocean Engineering 1
Lämmersieth 72DE-22305
Hamburg

C.H.R. Heip
Netherlands Institute of Ecology
Centre for Estuarine and Coastal Ecology
P.O. Box 140
4400 AC Yerseke
The Netherlands

P. Le Hir
Institut Français de Recherche
pour l'Exploitation de la Mer (IFREMER)
Centre IFREMER de Brest,
BP 70 29280
Plouzané
France

E.J. Houwing
Institute for Inland Water Management and
Waste Water Treatment (RIZA)
Rijkswaterstaat
van Leeuwenhoekweg 20
3316 AV Dordrecht
The Netherlands

B.M. Janssen-Stelder
Institute for Marine and
Atmospheric Research Utrecht (IMAU)
Dept. of Physical Geography
Utrecht University
P.O. Box 80.115
3508 TC Utrecht
The Netherlands

J. Jiang
ASL Environmental Sciences Inc.
1986 Mills Road
Sidney
BC V8L 5Y3
Canada

S. Jones
School of Ocean Sciences
Univ. of Wales
Menai Bridge Bangor
Gwynedd LL50 5EY
United Kingdom

T. van Kessel
Marine and Coastal Management Section
WL | Delft Hydraulics
P.O. Box 177
2600 MH Delft
The Netherlands

W. van Kesteren
Marine and Coastal Management Section
WL | Delft Hydraulics
P.O. Box 177
2600 MH Delft
The Netherlands

H.-S. Kim
Department of Civil Engineering
Sun Moon University
100 Kalsan Ri
Tangjeong Myeon
Asan Si
Chungnam 336-840
Korea

J.M. de Kok
National Institute for Coastal and
Marine Management / RIKZ
P.O. Box 20907
2500 EX The Hague
The Netherlands

J. Köngeter
Aachen University of Technology
Institute of Hydraulic Engineering and Water
Resources Management (IWW)
Kreuzherrenstrasse
D-52056
Aachen
Germany

B.A. Kornman
Institute for Marine and
Atmospheric Research Utrecht (IMAU)
Utrecht University
Department of Physical Geography
P.O. Box 80115
3508 TC Utrecht
The Netherlands

C. Kranenburg
Fluid Mechanics Section
Civil Engineering Department
Delft University of Technology
PO Box 5048
2600 GA Delft
The Netherlands

A. Kroon
Institute for Marine and
Atmospheric Research Utrecht (IMAU)
Utrecht University
P.O. Box 80.115
3508 TC Utrecht
The Netherlands

T. Kusada
Department of Urban and
Environmental Engineering
Kyushu University
Hakozaki 6-10-1 (SUIKO)
Higashi-ku
Fukuoka 812-8581
Japan

D.-Y. Lee
Principal Research Scientist
Korea Ocean Research and
Development Institute
P.O. Box 29
Ansan 425-600
Korea

M. van Ledden
Faculty of Civil Eng. and Geosciences
Delft University of Technology
Stevinweg 1
P.O. Box 5048
2600 GA Delft
The Netherlands

K.C. Leurer
Hydraulics Laboratory
Civil Engineering Department
Katholieke Universiteit Leuven
Kasteelpark Arenberg 40
B-3001 Leuven
Belgium

D.G. Lintern
Dept. Engineering Sciences
Univ. of Oxford
Oxford OX1 3PJ
United Kingdom

J.P.-Y. Maa
Assoc. Prof., School of Marine Science
Virginia Institute of marine Science
College of William and Mary
Gloucester Point
VA, 23062
USA

V. Maderich
Institute of Mathematical Machine
and System Problems
Glushkova Av. 42
Kiev 03187
The Ukraine
Russia

A.J. Manning
Institute of Marine Studies
University of Plymouth
Drake Circus
Plymouth
Devon PL4 1HP
United Kingdom

N. Margvelashvili
Institute of Mathematical Machine
and System Problems
Glushkova Av. 42
Kiev 03187
The Ukraine
Russia

M. Markofsky
Institut für Strömungsmechanik
Applestrasse 9A
D-30167 Hannover
Germany

F.G. Marván
Department of Civil & Offshore Engineering
Heriot-Watt University
Riccarton Edinburgh
EH14 4AS
United Kingdom

D.A. Mayne
School of Civil and Environmental
Engineering
University of Edinburgh
United Kingdom

A.J. Mehta
Department of Civil and
Engineering Coastal Engineering
University of Florida
345 Weil Hall
Gainesville
FL 32611-6580
USA

L.M. Merckelbach
Delft University of Technology
Faculty of Civil Engineering and
Geosciences
P.O. Box 5048
2600 GA Delft
The Netherlands

H. Michallet
Laboratoire des Ecoulements
Géophysiques et Industriels
BP 53
38041 Grenoble
Cédex 9
France

S.B. Mitchell
School of the Environment
University of Brighton
Lewes Road
Brighton BN2 4GJ
United Kingdom

S.G. Monismith
Environmental Fluid Mechanics Laboratory
Dept. Civil & Environmental Engineering
Stanford University
Stanford
CA 94305-4020
USA

M. Mory
Ecole Nationale Supérieure en Génie des Technologies Industrielles
rue Jules Ferry
64000 Pau
France

B.W. Nelson
Department of Environmental Sciences
University of Virginia
36 University Circle
Charlottesville
VA 22903
USA

A.M. Paiva
Programa de Engenharia Oceânica / COPPE
Universidade Federal do Rio de Janeiro
21945-970
Rio de Janeiro RJ
Brazil

D.M. Paterson
Gatty Marine Laboratory
The University of St. Andrews
Fife, UK KY16 8LB
United Kingdom

M. Pejrup
Institute of Geography
University of Copenhagen
Øster Voldgade 10
DK-1350 København
Denmark

O. Petersen
DHI Water & Environment
Agern Allé 11
DK-2970 Hørsholm
Denmark

Joern Prochnow
Institute of Hydraulic Engineering and Water Resources Management
Aachen University of Technology
Kreuzherrenstrasse
52065 Aachen
Germany

J.R. Roberti
Ministry of Transport
Public Works and Water Management
Institute for Marine and Coastal Management (RIKZ)
The Netherlands

W. Roberts
HR Wallingford
Howbery Park
Wallingford OX10 8BA
United Kingdom

W.A.C. van Santen
Institute for Marine and Atmospheric Research Utrecht (IMAU)
Dept. of Physical Geography
Utrecht University
P.O. Box 80.115
3508 TC Utrecht
The Netherlands

P.D. Scarlatos
Department of Civil Engineering
Florida Atlantic University
777 Glades Road
Bldg. 36, Boca Raton
Florida 33431
USA

C.A.F. Schettini
Centro de Ciências Technológicas da Terra e do Mar CTTMar-UNIVALI
C.P. 360
Itajai SC
88.302
Brazil

D.H. Schoellhamer
U.S. Geological Survey
Placer Hall
6000 J St.
Sacramento
CA 95819
USA

C. Schweim
Aachen University of Technology
Institute of Hydraulic Engineering and Water Resources Management (IWW)
Kreuzherrenstrasse
D-52056
Aachen
Germany

Z. Shi
Department of Harbour & Coastal Engineering
Shanghai Jiao Tong University
1954 Hua Shan Road
Shanghai 200030
People's Republic of China

G.C. Sills
Department of Engineering Science
Oxford University
Parks Road
Oxford, OX1 3PJ
United Kingdom

J. Spearman
HR Wallingford
Howbery Park
Wallingford OX10 8BA
United Kingdom

N. Staats
Laboratory for Microbiology ARISE,
University of Amsterdam
Nieuwe Achtergracht 127
1018 WS Amsterdam
The Netherlands

L.J. Stal
Netherlands Institute of Ecology
Centre for Estuarine and Coastal Ecology
P.O. Box 140
4400 AC Yerseke
The Netherlands

I.C. Tánczos
WL | Delft Hydraulics
P.O. Box 177
2600 MH Delft
The Netherlands

G.R. Termaat
Institute for Marine and Atmospheric Research Utrecht
Utrecht University
Department of Physical Geography
P.O. Box 80115
3508 TC Utrecht
The Netherlands

E.A. Toorman
Hydraulics Laboratory
Civil Engineering Department
Katholieke Universiteit Leuven
Kasteelpark Arenberg 40
B-3001 Leuven
Belgium

R.J. Uncles
Plymouth Marine Laboratory
Prospect Place
West Hoe
Plymouth, PL1 3DH
United Kingdom

A.S. Usmani
School of Civil and
Environmental Engineering
University of Edingburgh
United Kingdom

H.J. Vested
DHI Water & Environment
Agern Allé 11
DK-2970 Hørsholm
Denmark

S.B. Vinzon
DRHIMA Escola de Engenharia
Universida Federal do Rioa de Janeiro
21945-970
Rio de Janeiro RJ
Brazil

D. Violeau
Laboratoire National d'Hydraulique et
Environnement (EDF)
6 Quai Watier
78400 Chatou
France

M.B. de Vries
WL | Delft Hydraulics
P.O. Box 177
2600 MH Delft
The Netherlands

S.G. Wallis
Department of Civil & Offshore Engineering
Heriot-Watt University
Riccarton Edinburgh
EH14 4AS
United Kingdom

R. Watanabe
Department of Civil Engineering
Fukuoka University
Nanakuma 8-19-1
Jonan-ku
Fukuoka 814-0180
Japan

H. Weilbeer
Institut für Strömungsmechanik
Applestrasse 9A
D-30167 Hannover
Germany

J.R. West
School of Civil Engineering
University of Birmingham
Edgbaston
Birmingham B15 2TT
United Kingdom

J. Widdows
Plymouth Marine Laboratory
Prospect Place
West Hoe
Plymouth, PL1 3DH
United Kingdom

M. van Wijngaarden
Ministry of Transport
Public Works and Water Management
Institute for Inland Water Management
and Waste Water Treatment (RIZA)
PO Box 52
3300 AB Dordrecht
The Netherlands

B. de Winder
Laboratory for Microbiology
ARISE,
University of Amsterdam
Nieuwe Achtergracht 127
1018 WS Amsterdam
The Netherlands

J.C. Winterwerp
Fluid Mechanics Section
Civil Engineering Department
Delft University of Technology
PO Box 5048
2600 GA Delft
The Netherlands
also:
WL | Delft Hydraulics
PO Box 177
2600 MH Delft
The Netherlands

H. Yamanishi
Institute of Lowland Technology
Saga University
Honjyou 1
Saga 840-8502
Japan

K. Yamasaki
Department of Urban and
Environmental Engineering
Kyushu University
Hakozaki 6-10-1
(SUIKO)
Higashi-ku
Fukuoka 812-8581
Japan

S. Yuschenko
Institute of Mathematical Machine
and System Problems
Glushkova Av. 42
Kiev 03187
The Ukraine
Russia

M. Zheleznyak
Institute of Mathematical Machine
and System Problems
Glushkova Av. 42
Kiev 03187
The Ukraine
Russia